Errata and Corrigenda number 3: January 1975 have been entered.
[I think I have entered no 4, which was previously lost a slip stuck in.]
number 5: January 1977
number 6: January 1978
number 7: January 1979
number 8: January 1980
number 9: January 1981

THERMAL CONDUCTIVITY
Nonmetallic Liquids and Gases

THERMOPHYSICAL PROPERTIES OF MATTER
The TPRC Data Series

A Comprehensive Compilation of Data by the
Thermophysical Properties Research Center (TPRC), Purdue University

Y. S. Touloukian, Series Editor
C. Y. Ho, Series Technical Editor

Volume 1. Thermal Conductivity–Metallic Elements and Alloys
Volume 2. Thermal Conductivity–Nonmetallic Solids
Volume 3. Thermal Conductivity–Nonmetallic Liquids and Gases
Volume 4. Specific Heat–Metallic Elements and Alloys
Volume 5. Specific Heat–Nonmetallic Solids
Volume 6. Specific Heat–Nonmetallic Liquids and Gases
Volume 7. Thermal Radiative Properties–Metallic Elements and Alloys
Volume 8. Thermal Radiative Properties–Nonmetallic Solids
Volume 9. Thermal Radiative Properties–Coatings
Volume 10. Thermal Diffusivity
Volume 11. Viscosity
Volume 12. Thermal Expansion–Metallic Elements and Alloys
Volume 13. Thermal Expansion–Nonmetallic Solids

New data on thermophysical properties are being constantly accumulated at TPRC. Contact TPRC and use its interim updating services for the most current information.

THERMOPHYSICAL PROPERTIES OF MATTER
VOLUME 3

THERMAL CONDUCTIVITY
Nonmetallic Liquids and Gases

Y. S. Touloukian
Director
Thermophysical Properties Research Center
and
Distinguished Atkins Professor of Engineering
School of Mechanical Engineering
Purdue University
and
Visiting Professor of Mechanical Engineering
Auburn University

P. E. Liley
Associate Senior Researcher
Thermophysical Properties Research Center
and
Associate Professor of Mechanical Engineering
Purdue University

S. C. Saxena
Professor of Energy Engineering
University of Illinois
Chicago Campus
Formerly
Associate Senior Researcher
Thermophysical Properties Research Center
Purdue University

IFI/PLENUM • NEW YORK-WASHINGTON • 1970

Library of Congress Catalog Card Number 73-129616

SBN (13-Volume Set) 306-67020-8

SBN (Volume 3) 306-67023-2

Copyright © 1970, Purdue Research Foundation

IFI/Plenum Data Corporation is a subsidiary of
Plenum Publishing Corporation
227 West 17th Street, New York, N.Y. 10011

Distributed in Europe by Heyden & Son, Ltd.
Spectrum House, Alderton Crescent
London N.W. 4, England

Printed in the United States of America

"In this work, when it shall be found that much is omitted, let it not be forgotten that much likewise is performed..."

SAMUEL JOHNSON, A.M.
From last paragraph of Preface to his two-volume *Dictionary of the English Language*, Vol. I, page 5, 1755, London, Printed by Strahan.

Foreword

In 1957, the Thermophysical Properties Research Center (TPRC) of Purdue University, under the leadership of its founder, Professor Y. S. Touloukian, began to develop a coordinated experimental, theoretical, and literature review program covering a set of properties of great importance to science and technology. Over the years, this program has grown steadily, producing bibliographies, data compilations and recommendations, experimental measurements, and other output. The series of volumes for which these remarks constitute a foreword is one of these many important products. These volumes are a monumental accomplishment in themselves, requiring for their production the combined knowledge and skills of dozens of dedicated specialists. The Thermophysical Properties Research Center deserves the gratitude of every scientist and engineer who uses these compiled data.

The individual nontechnical citizen of the United States has a stake in this work also, for much of the science and technology that contributes to his well-being relies on the use of these data. Indeed, recognition of this importance is indicated by a mere reading of the list of the financial sponsors of the Thermophysical Properties Research Center; leaders of the technical industry of the United States and agencies of the Federal Government are well represented.

Experimental measurements made in a laboratory have many potential applications. They might be used, for example, to check a theory, or to help design a chemical manufacturing plant, or to compute the characteristics of a heat exchanger in a nuclear power plant. The progress of science and technology demands that results be published in the open literature so that others may use them. Fortunately for progress, the useful data in any single field are not scattered throughout the tens of thousands of technical journals published throughout the world. In most fields, fifty percent of the useful work appears in no more than thirty or forty journals. However, in the case of TPRC, its field is so broad that about 100 journals are required to yield fifty percent. But that other fifty percent! It is scattered through more than 3500 journals and other documents, often items not readily identifiable or obtainable. Nearly 50,000 references are now in the files.

Thus, the man who wants to use existing data, rather than make new measurements himself, faces a long and costly task if he wants to assure himself that he has found all the relevant results. More often than not, a search for data stops after one or two results are found—or after the searcher decides he has spent enough time looking. Now with the appearance of these volumes, the scientist or engineer who needs these kinds of data can consider himself very fortunate. He has a single source to turn to; thousands of hours of search time will be saved, innumerable repetitions of measurements will be avoided, and several billions of dollars of investment in research work will have been preserved.

However, the task is not ended with the generation of these volumes. A critical evaluation of much of the data is still needed. Why are discrepant results obtained by different experimentalists? What undetected sources of systematic error may affect some or even all measurements? What value can be derived as a "recommended" figure from the various conflicting values that may be reported? These questions are difficult to answer, requiring the most sophisticated judgment of a specialist in the field. While a number of the volumes in this Series do contain critically evaluated and recommended data, these are still in the minority. The data are now being more intensively evaluated by the staff of TPRC as an integral part of the effort of the National Standard Reference Data System (NSRDS). The task of the National Standard Reference Data System is to organize and operate a comprehensive program to prepare compilations of critically evaluated data on the properties of substances. The NSRDS is administered by the National Bureau of Standards under a directive from the Federal Council for Science and Technology, augmented by special legislation of the Congress of the United States. TPRC is one of the national resources participating in the National Standard Reference Data System in a united effort to satisfy the needs of the technical community for readily accessible, critically evaluated data.

As a representative of the NBS Office of Standard Reference Data, I want to congratulate Professor Touloukian and his colleagues on the accomplishments represented by this Series of reference data books. Scientists and engineers the world over are indebted to them. The task ahead is still an awesome one and I urge the nation's private industries and all concerned Federal agencies to participate in fulfilling this national need of assuring the availability of standard numerical reference data for science and technology.

EDWARD L. BRADY
Associate Director for Information Programs
National Bureau of Standards

Preface

Thermophysical Properties of Matter, the TPRC Data Series, is the culmination of twelve years of pioneering effort in the generation of tables of numerical data for science and technology. It constitutes the restructuring, accompanied by extensive revision and expansion of coverage, of the original *TPRC Data Book*, first released in 1960 in loose-leaf format, $11'' \times 17''$ in size, and issued in June and December annually in the form of supplements. The original loose-leaf *Data Book* was organized in three volumes: (1) metallic elements and alloys, (2) nonmetallic elements, compounds, and mixtures which are solid at N.T.P., and (3) nonmetallic elements, compounds, and mixtures which are liquid or gaseous at N.T.P. Within each volume, each property constituted a chapter.

Because of the vast proportions the *Data Book* began to assume over the years of its growth and the greatly increased effort necessary in its maintenance by the user, it was decided in 1967 to change from the loose-leaf format to a conventional publication. Thus, the December 1966 supplement of the original *Data Book* was the last supplement disseminated by TPRC.

While the manifold physical, logistic, and economic advantages of the bound volume over the loose-leaf oversize format are obvious and welcome to all who have used the unwieldy original volumes, the assumption that this work will no longer be kept on a current basis because of its bound format would not be correct. Fully recognizing the need of many important research and development programs which require the latest available information, TPRC has instituted a *Data Update Plan* enabling the subscriber to inquire, by telephone if necessary, for specific information and receive, in many instances, same-day response on any new data processed or revision of published data since the latest edition. In this context, the TPRC Data Series departs drastically from the conventional handbook and giant multivolume classical works, which are no longer adequate media for the dissemination of numerical data of science and technology without a continuing activity on contemporary coverage. The loose-leaf arrangements of many works fully recognize this fact and attempt to develop a combination of bound volumes and loose-leaf supplement arrangements as the work becomes increasingly large. TPRC's *Data Update Plan* is indeed unique in this sense since it maintains the contents of the TPRC Data Series current and live on a day-to-day basis between editions. In this spirit, I strongly urge all purchasers of these volumes to complete in detail and return the *Volume Registration Certificate* which accompanies each volume in order to assure themselves of the continuous receipt of annual listing of corrigenda during the life of the edition.

The TPRC Data Series consists initially of 13 independent volumes. The initial ten volumes will be published in 1970, and the remaining three by 1972. It is also contemplated that subsequent to the first edition, each volume will be revised, updated, and reissued in a new edition approximately every fifth year. The organization of the TPRC Data Series makes each volume a self-contained entity available individually without the need to purchase the entire Series.

The coverage of the specific thermophysical properties represented by this Series constitutes the most comprehensive and authoritative collection of numerical data of its kind for science and technology.

Whenever possible, a uniform format has been used in all volumes, except when variations in presentation were necessitated by the nature of the property or the physical state concerned. In spite of the wealth of data reported in these volumes, it should be recognized that all volumes are not of the same degree of completeness. However, as additional data are processed at TPRC on a continuing basis, subsequent editions will become increasingly more complete and up to date. Each volume in the Series basically comprises three sections, consisting of a text, the body of numerical data with source references, and a material index.

The aim of the textual material is to provide a complementary or supporting role to the body of numerical data rather than to present a treatise on the subject of the property. The user will find a basic theoretical treatment, a comprehensive presentation of selected works which constitute reviews, or compendia of empirical relations useful in estimation of the property when there exists a paucity of data or when data are completely lacking. Established major experimental techniques are also briefly reviewed.

The body of data is the core of each volume and is presented in both graphical and tabular format for convenience of the user. Every single point of numerical data is fully referenced as to its original source and no secondary sources of information are used in data extraction. In general, it has not been possible to critically scrutinize all the original data presented in these volumes, except to eliminate perpetuation of gross errors. However, in a significant number of cases, such as for the properties of liquids and gases and the thermal conductivity of all the elements, the task of full evaluation, synthesis, and correlation has been completed. It is hoped that in subsequent editions of this continuing work, not only new information will be reported but the critical evaluation will be extended to increasingly broader classes of materials and properties.

The third and final major section of each volume is the material index. This is the key to the volume, enabling the user to exercise full freedom of access to its contents by any choice of substance name or detailed alloy and mixture composition, trade name, synonym, etc. Of particular interest here is the fact that in the case of those properties which are reported in separate companion volumes, the material index in each of the volumes also reports the contents of the other companion volumes.* The sets of companion volumes are as follows:

Thermal conductivity:	Volumes 1, 2, 3
Specific heat:	Volumes 4, 5, 6
Radiative properties:	Volumes 7, 8, 9
Thermal expansion:	Volumes 12, 13

The ultimate aims and functions of TPRC's Data Tables Division are to extract, evaluate, reconcile, correlate, and synthesize all available data for the thermophysical properties of materials with the result of obtaining internally consistent sets of property values, termed the "recommended reference values." In such work, gaps in the data often occur, for ranges of temperature, composition, etc. Whenever feasible, various techniques are used to fill in such missing information, ranging from empirical procedures to detailed theoretical calculations. Such studies are resulting in valuable new estimation methods being developed which have made it possible to estimate values for substances and/or physical conditions presently unmeasured or not amenable to laboratory investigation. Depending on the available information for a particular property and substance, the end product may vary from simple tabulations of isolated values to detailed tabulations with generating equations, plots showing the concordance of the different values, and, in some cases, over a range of parameters presently unexplored in the laboratory.

The TPRC Data Series constitutes a permanent and valuable contribution to science and technology. These constantly growing volumes are invaluable sources of data to engineers and scientists, sources in which a wealth of information heretofore unknown or not readily available has been made accessible. We look forward to continued improvement of both format and contents so that TPRC may serve the scientific and technological community with ever-increasing excellence in the years to come. In this connection, the staff of TPRC is most anxious to receive comments, suggestions, and criticisms from all users of these volumes. An increasing number of colleagues are making available at the earliest possible moment reprints of their papers and reports as well as pertinent information on the more obscure publications. I wish to renew my earnest request that this procedure become a universal practice since it will prove to be most helpful in making TPRC's continuing effort more complete and up to date.

It is indeed a pleasure to acknowledge with gratitude the multisource financial assistance received from over fifty of TPRC's sponsors which has made the continued generation of these tables possible. In particular, I wish to single out the sustained major support being received from the Air Force Materials Laboratory–Air Force Systems Command, the Office of Standard Reference Data–National Bureau of Standards, and the Office of Advanced Research and Technology–National Aeronautics and Space Administration. TPRC is indeed proud to have been designated as a National Information Analysis Center for the Department of Defense as well as a component of the National

---

*For the first edition of the Series, this arrangement was not feasible for Volume 7 due to the sequence and the schedule of its publication. This situation will be resolved in subsequent editions.

Standard Reference Data System under the cognizance of the National Bureau of Standards.

While the preparation and continued maintenance of this work is the responsibility of TPRC's Data Tables Division, it would not have been possible without the direct input of TPRC's Scientific Documentation Division and, to a lesser degree, the Theoretical and Experimental Research Divisions. The authors of the various volumes are the senior staff members in responsible charge of the work. It should be clearly understood, however, that many have contributed over the years and their contributions are specifically acknowledged in each volume. I wish to take this opportunity to personally thank those members of the staff, research assistants, graduate research assistants, and supporting graphics and technical typing personnel without whose diligent and painstaking efforts this work could not have materialized.

Y. S. TOULOUKIAN

Director
Thermophysical Properties Research Center
Distinguished Atkins Professor of Engineering

Purdue University
Lafayette, Indiana
July 1969

Introduction to Volume 3

This volume of *Thermophysical Properties of Matter*, the TPRC Data Series, covering the thermal conductivity of fluids, presents the data on nonmetallic materials which are in the fluid state at normal temperature and pressure. It is not as comprehensive as its two companion volumes, Volumes 1 and 2, primarily because we have not been able to cover an extensive number of fluids due to lack of technological interest.

The volume comprises three major sections: the front text material together with its bibliography, the main body of numerical data with its references, and the material index.

The text material is intended to assume a role complementary to the main body of numerical data, the presentation of which is the primary purpose of this volume. It is felt that a moderately detailed discussion of the theoretical nature of the property under consideration together with a review of predictive procedures and recognized experimental techniques will be appropriate in a major reference work of this kind. The extensive reference citations given in the text should lead the interested reader to a highly comprehensive literature for a detailed study. It is hoped, however, that enough detail is presented for this volume to be self-contained for the practical user.

The main body of the volume consists of the presentation of numerical data compiled over the years in a most meticulous manner. The coverage includes a selected number of pure substances, identical to those covered in Volume 6 of this Series, and a number of mixtures which are felt to be of greatest engineering importance. The extraction of all data directly from their original sources ensures freedom from errors of transcription. Furthermore, a number of gross errors appearing in the original source documents have been corrected. The organization and presentation of the data, together with other pertinent information on the use of the tables and figures, are discussed in detail in the section entitled *Numerical Data*.

The data covering pure substances have been critically reviewed and analyzed, and "recommended reference values" are presented. It is hoped that in future editions of this volume the data on mixtures will also receive similar critical scrutiny and "recommended reference values" will be incorporated.

As stated earlier, all data have been obtained from their original sources and each data set is so referenced. TPRC has in its files all documents cited in this volume. Those that cannot readily be obtained elsewhere are available from TPRC in microfiche form.

The material index at the end of this volume covers the contents of all three companion volumes (Volumes 1, 2, and 3) on thermal conductivity. It is hoped that the user will find these comprehensive indices helpful.

This volume has grown out of the activities made possible initially by TPRC's Founder Sponsors, and, since 1960, principally through the support of the Air Force Materials Laboratory–Air Force Systems Command, under the monitorship of Mr. John H. Charlesworth. The effort on the critical analysis of the data on the elements was made possible through the Office of Standard Reference Data–National Bureau of Standards, under the monitorship of Dr. Howard J. White, Jr. The authors wish to acknowledge with pleasure the cooperation of their colleague, Dr. R. W. Powell, who has contributed the portion of the text covering the liquid state. Extensive personal inquiries have been made to the authors of research papers and reports requesting clarifications and/or original data. Their enthusiastic response to these inquiries is gratefully acknowledged.

Inherent to the character of this work is the fact that in the preparation of this volume we have drawn most heavily upon the scientific literature and feel a debt of gratitude to the authors of the referenced articles. While their often discordant results have caused us much difficulty in reconciling their findings, we consider this to be our challenge and our contribution to negative entropy of information, as an effort is made to create from the randomly distributed data a condensed, more orderly state.

While this volume is primarily intended as a reference work for the designer, researcher, experimentalist, and theoretician, the teacher at the graduate level may also use it as a teaching tool to point out to his students the topography of the state of knowledge on the thermal conductivity of fluids. We believe there is also much food for reflection by the specialist and the academician concerning the meaning of "original" investigation and its "information content."

The authors are keenly aware of the possibility of omissions or errors in a work of this scope. We hope that we will not be judged too harshly and that we will receive the benefit of suggestions regarding references omitted, additional material groups needing more detailed treatment, improvements in presentation, and, most important, any inadvertent errors. If the *Volume Registration Certificate* accompanying this volume is returned, the reader will assure himself of receiving annually a list of corrigenda as possible errors come to our attention.

Lafayette, Indiana
July 1969

Y. S. TOULOUKIAN
P. E. LILEY
S. C. SAXENA

Contents

Foreword vii
Preface ix
Introduction to Volume 3 xiii
Grouping of Materials and List of Figures and Tables xviii

Theory, Estimation, and Measurement

Notation 1a
Thermal Conductivity of Liquids 3a
  1. Introduction 3a
  2. Noteworthy Supplementary Publications 3a
    A. Data Compilations 3a
    B. Special Groups of Fluids 4a
      a. Organic Liquids 5a
      b. Refrigerants 6a
      c. Oils 6a
      d. Biological Fluids 6a
      e. Fruit Juices and Sugar Solutions 6a
      f. Aqueous Solutions of Organic Compounds: Mixtures of Organic Liquids . . . 6a
      g. Aqueous Solutions of Inorganic Salts 7a
      h. Metal–Ammonia Solutions 7a
      i. Some Polyphenyls and Mixtures Thereof 7a
      j. Molten Salts 7a
    C. Influence of Environmental and Structural Factors 8a
      a. Conduction 8a
      b. Effect of Convection 8a
      c. Effect of Radiation 8a
      d. Possible Effect of an Applied Electrostatic Potential 10a
      e. The Soret Effect 11a
      f. Possible Effect of Molecular Orientation 11a
  3. Theory and Empirical Equations 12a
    A. Theories for Electrically Nonconducting Liquids 12a
    B. Empirical Equations for Electrically Nonconducting Liquids 13a
    C. Electrically Conducting Liquids: Molten Metals 18a
  4. Experimental Methods 19a
    A. Electrically Nonconducting Liquids 19a
      a. Horizontal Plane Layer 19a
      b. Concentric Cylinder 20a
      c. Concentric Sphere 20a
      d. Hot-Wire Methods 21a
      e. Thermal-Comparator Method 21a
    B. Methods Applicable to Molten Metals 21a

   a. Longitudinal Heat Flow (Comparative) 21a
   b. Direct Electrical Heating Methods 22a
Thermal Conductivity of Gases and Gas Mixtures 23a
 1. Introduction 23a
 2. Experimental Methods 23a
  A. The Cooling Method 24a
  B. The Hot-Wire Method 24a
   a. Effect of Convection 26a
   b. Effect of Radiation 26a
   c. Effect of Temperature Jump 26a
   d. Effect of Finite Thickness of the Cell Tube 27a
   e. Effect of Nonaxiality of the Wire 27a
  C. The Parallel-Plate Method 28a
  D. The Concentric-Cylinder Method 29a
  E. The Concentric-Sphere Method 29a
  F. The Concentric Sphero-Cylinder Method 30a
  G. The Line-Source Flow Method 30a
  H. The Hot-Wire Thermal Diffusion Column Method 31a
  I. The Shock-Tube Method 31a
  J. The Arc Method 32a
  K. The Ultrasonic Method 32a
  L. The Prandtl Number Method 32a
  M. The Dilatometric Method 33a
  N. The Thermal-Comparator Method 33a
  O. The Frequency-Response Technique 33a
  P. The Transient Hot-Wire Method 34a
  Q. The Unsteady-State Method of Lindsay and Bromley 34a
  R. The Flow Methods 34a
 3. Theoretical Methods 34a
  A. Introduction 34a
  B. The Mean-Free-Path Theories 35a
  C. The Rigorous (Chapman–Enskog) Theories 35a
   a. Pure Monatomic Gases 35a
   b. Pure Polyatomic and Polar Gases 36a
   c. Multicomponent Systems of Monatomic Gases 37a
   d. Multicomponent Systems of Polyatomic (and Polar) Gases 38a
   e. Multicomponent Systems of Dissociating Gases 39a
   f. Multicomponent Systems of Partially or Fully Ionized Gases 40a
   g. Conclusions 41a
   h. Effect of Density (Pressure) 41a
 4. Estimation Methods 43a
  A. Introduction 43a
  B. Pure Gases 43a
  C. Multicomponent Gas Mixtures 44a
   a. Methods Based on the Linear and Reciprocal Mixing Rules . . . 44a
   b. Methods Based on Empirical Functions for Binary Systems 45a
   c. Methods due to Lindsay and Bromley 46a
   d. Method due to Saxena and Tondon 47a
   e. Methods due to Hirschfelder, Curtiss, and Bird 47a
   f. Method due to Ulybin, Bugrov, and Il'in 47a
   g. Methods due to Cheung, Bromley, and Wilke 48a
   h. Method due to Mason and Saxena 48a

	i.	Method due to Srivastava and Saxena	49a
	j.	Method due to Mason and von Ubisch	49a
	k.	Method due to Wright and Gray	49a
	l.	Method due to Saxena and Gambhir	50a
	m.	Method due to Gambhir and Saxena	50a
	n.	Method Based on Kinetic Theory and other Data	50a
	D. Summarizing Remarks	50a
5.	Sutherland Coefficients	51a
6.	Sources of Further Information	81a
References to Text	83a

Numerical Data

Data Presentation and Related General Information 103a
1. Scope of Coverage 103a
2. Presentation of Data 104a
3. Symbols and Abbreviations Used in the Figures and Tables 105a
4. Convention for Bibliographic Citation 105a
5. Name, Formula, Molecular Weight, Transition Temperatures, and Physical Constants of Elements and Compounds. 105a
6. Conversion factors 108a

Numerical Data on Thermal Conductivity of Nonmetallic Liquids and Gases (see pp. xviii to xxii for detailed listing of entries for each of the following groups of materials). 1
1. Elements 1
2. Inorganic Compounds 95
3. Organic Compounds 129
4. Binary Systems 247
   A. Monatomic–Monatomic Systems 251
   B. Monatomic–Nonpolar Polyatomic Systems 295
   C. Nonpolar Polyatomic–Nonpolar Polyatomic Systems. 381
   D. Polar–Nonpolar Polyatomic Systems 440
   E. Polar–Polar Systems 470
5. Ternary Systems. 477
   A. Monatomic Systems 478
   B. Monatomic and Nonpolar Polyatomic Systems 484
   C. Nonpolar Polyatomic Systems 497
   D. Nonpolar and Polar Systems 499
6. Quaternary Systems 503
   A. Monatomic Systems 504
   B. Monatomic and Nonpolar Polyatomic Systems 505
7. Multicomponent Systems 511

References to Data Sources 515

Material Index

Material Index to Thermal Conductivity Companion Volumes 1, 2, and 3 . . . . . A1

GROUPING OF MATERIALS AND LIST OF FIGURES AND TABLES

1. ELEMENTS

Figure and/or Table No.	Name	Symbol	Physical State*	Page No.
1	Argon	Ar	S, L, V, G	1
2	Bromine	Br_2	-, L, V, G	13
3	Chlorine	Cl_2	-, L, V, G	17
4	Deuterium	D_2	-, L, V, G	21
5	Fluorine	F_2	-, L, V, G	26
6	Helium	He	S, L, -, G	29
7	Hydrogen	H_2	S, L, V, G	41
8	Krypton	Kr	S, L, V, G	50
9	Neon	Ne	S, L, V, G	56
10	Nitrogen	N_2	S, L, V, G	64
11	Oxygen	O_2	-, L, V, G	76
12	Radon	Rn	-, L, V, G	84
13	Tritium	T_2	-, L, -, -	87
14	Xenon	Xe	S, L, V, G	88

2. INORGANIC COMPOUNDS

15	Ammonia	NH_3	-, L, G, -	95
16	Boron Trifluoride	BF_3	-, -, G, -	99
17	Hydrogen Chloride	HCl	-, -, G, -	101
18	Hydrogen Iodide	HI	-, -, G, -	103
19	Hydrogen Sulfide	H_2S	-, -, G, -	104
20	Nitric Oxide	NO	-, -, G, -	106
21	Nitrogen Peroxide	NO_2	-, L, G, -	108
22	Nitrous Oxide	N_2O	-, -, G, -	114
23	Sulfur Dioxide	SO_2	-, L, G, -	116
24	Water	H_2O	-, L, G, -	120

3. ORGANIC COMPOUNDS

25	Acetone	$(CH_3)_2CO$	-, L, G, -	129
26	Acetylene	CHCH	-, -, G, -	133
27	Benzene	C_6H_6	-, L, G, -	135
28	i-Butane	$i\text{-}C_4H_{10}$	-, -, G, -	139
29	n-Butane	$n\text{-}C_4H_{10}$	-, L, G, -	141
30	Carbon Dioxide	CO_2	-, L, G, -	145

*S = solid, L = saturated liquid, V = saturated vapor, G = gas.

3. ORGANIC COMPOUNDS (continued)

Figure and/or Table No.	Name	Formula	Physical State*	Page No.
31	Carbon Monoxide	CO	L, G	151
32	Carbon Tetrachloride	CCl_4	L, G	156
33	Chloroform	$CHCl_3$	L, G	161
34	n-Decane	$C_{10}H_{22}$	L, G	164
35	Ethane	C_2H_6	-, G	167
36	Ethyl Alcohol	C_2H_5OH	L, G	169
37	Ethylene	CH_2CH_2	L, G	173
38	Ethylene Glycol	CH_2OHCH_2OH	L, -	177
39	Ethyl Ether	$(C_2H_5)_2O$	L, G	179
40	Freon 11	Cl_3CF	L, G	183
41	Freon 12	Cl_2CF_2	L, G	187
42	Freon 13	$ClCF_3$	-, G	191
43	Freon 21	Cl_2CHF	L, G	193
44	Freon 22	$ClCHF_2$	L, G	197
45	Freon 113	CCl_2FCClF_2	L, G	201
46	Freon 114	$CClF_2CClF_2$	L, G	205
47	Glycerol	$CH_2OHCHOHCH_2OH$	L, -	209
48	n-Heptane	C_7H_{16}	L, G	211
49	n-Hexane	C_6H_{14}	L, G	214
50	Methane	CH_4	L, G	218
51	Methyl Alcohol	CH_3OH	L, G	223
52	Methyl Chloride	CH_3Cl	L, G	227
53	n-Nonane	C_9H_{20}	L, G	230
54	n-Octane	C_8H_{18}	L, G	233
55	n-Pentane	C_5H_{12}	L, G	236
56	Propane	C_3H_8	-, G	240
57	Toluene	$C_6H_5CH_3$	L, G	242

4. BINARY SYSTEMS

A. Monatomic - Monatomic Systems

58	Argon and Helium	Ar and He	-, G	251
59	Argon and Neon	Ar and Ne	-, G	258
60	Argon and Krypton	Ar and Kr	-, G	263
61	Argon and Xenon	Ar and Xe	-, G	267
62	Helium and Neon	He and Ne	-, G	271
63	Helium and Krypton	He and Kr	-, G	276
64	Helium and Xenon	He and Xe	-, G	280
65	Krypton and Neon	Kr and Ne	-, G	284
66	Krypton and Xenon	Kr and Xe	-, G	288
67	Neon and Xenon	Ne and Xe	-, G	291

*L = saturated liquid, G = gas.

4. BINARY SYSTEMS (continued)

   B. Monatomic - Nonpolar Polyatomic Systems

Figure and/or Table No.	Name	Formula	Physical State*	Page No.
68	Argon and Benzene	Ar and C_6H_6	G	295
69	Argon and Carbon Dioxide	Ar and CO_2	G	297
70	Argon and Deuterium	Ar and D_2	G	299
71	Argon and Hydrogen	Ar and H_2	G	301
72	Argon and Methane	Ar and CH_4	G	304
73	Argon and Nitrogen	Ar and N_2	G	306
74	Argon and Oxygen	Ar and O_2	G	311
75	Argon and Propane	Ar and C_3H_8	G	316
76	Helium and Air	He and Air	G	318
77	Helium and n-Butane	He and C_4H_{10}	G	320
78	Helium and Carbon Dioxide	He and CO_2	G	322
79	Helium and Cyclopropane	He and C_3H_6	G	325
80	Helium and Deuterium	He and D_2	G	327
81	Helium and Ethane	He and C_2H_6	G	329
82	Helium and Ethylene	He and C_2H_4	G	331
83	Helium and Hydrogen	He and H_2	G	333
84	Helium and Methane	He and CH_4	G	338
85	Helium and Nitrogen	He and N_2	G	340
86	Helium and Oxygen	He and O_2	G	343
87	Helium and Propane	He and C_3H_8	G	345
88	Helium and Propylene	He and C_3H_6	G	347
89	Krypton and Deuterium	Kr and D_2	G	349
90	Krypton and Hydrogen	Kr and H_2	G	351
91	Krypton and Nitrogen	Kr and N_2	G	354
92	Krypton and Oxygen	Kr and O_2	G	356
93	Neon and Carbon Dioxide	Ne and CO_2	G	358
94	Neon and Deuterium	Ne and D_2	G	360
95	Neon and Hydrogen	Ne and H_2	G	362
96	Neon and Nitrogen	Ne and N_2	G	365
97	Neon and Oxygen	Ne and O_2	G	368
98	Xenon and Deuterium	Xe and D_2	G	371
99	Xenon and Hydrogen	Xe and H_2	G	374
100	Xenon and Nitrogen	Xe and N_2	G	377
101	Xenon and Oxygen	Xe and O_2	G	379

   C. Nonpolar Polyatomic - Nonpolar Polyatomic Systems

102	Acetylene and Air	C_2H_2 and Air	G	381
103	Air and Carbon Monoxide	Air and CO	G	383
104	Air and Methane	Air and CH_4	G	385
105	Benzene and Hexane	C_6H_6 and C_6H_{14}	G	387
106	Carbon Dioxide and Ethylene	CO_2 and C_2H_4	G	389
107	Carbon Dioxide and Hydrogen	CO_2 and H_2	G	391
108	Carbon Dioxide and Nitrogen	CO_2 and N_2	G	396
109	Carbon Dioxide and Oxygen	CO_2 and O_2	G	401
110	Carbon Dioxide and Propane	CO_2 and C_3H_8	G	403

*G = gas.

4. BINARY SYSTEMS (continued)

C. Nonpolar Polyatomic - Nonpolar Polyatomic Systems (continued)

Figure and/or Table No.	Name	Formula	Physical State*	Page No.
111	Carbon Monoxide and Hydrogen	CO and H_2	G	405
112	Deuterium and Hydrogen	D_2 and H_2	G	407
113	Deuterium and Nitrogen	D_2 and N_2	G	410
114	Ethylene and Hydrogen	C_2H_4 and H_2	G	413
115	Ethylene and Methane	C_2H_4 and CH_4	G	415
116	Ethylene and Nitrogen	C_2H_4 and N_2	G	417
117	Hydrogen and Nitrogen	H_2 and N_2	G	419
118	Hydrogen and Nitrous Oxide	H_2 and N_2O	G	427
119	Hydrogen and Oxygen	H_2 and O_2	G	429
120	Methane and Propane	CH_4 and C_3H_8	G	432
121	Nitrogen and Oxygen	N_2 and O_2	G	434
122	Nitrogen and Propane	N_2 and C_3H_8	G	438

D. Polar - Nonpolar Polyatomic Systems

Figure and/or Table No.	Name	Formula	Physical State*	Page No.
123	Acetone and Benzene	C_3H_6O and C_6H_6	G	440
124	Ammonia and Air	NH_3 and Air	G	442
125	Ammonia and Carbon Monoxide	NH_3 and CO	G	444
126	Ammonia and Ethylene	NH_3 and C_2H_4	G	446
127	Ammonia and Hydrogen	NH_3 and H_2	G	448
128	Ammonia and Nitrogen	NH_3 and N_2	G	451
129	Ethanol and Argon [Dimethyl ether]	C_2H_6O and Ar [$(CH_3)_2O$]	G	454
130	Ethanol and Propane [Dimethyl ether]	C_2H_6O and C_3H_8 [$(CH_3)_2O$]	G	456
131	Methanol and Argon	CH_4O and Ar	G	458
132	Methanol and Hexane	CH_4O and C_6H_{14}	G	460
133	Methyl Formate and Propane	$C_2H_4O_2$ and C_3H_8	G	462
134	Steam and Air	H_2O and Air	G	464
135	Steam and Carbon Dioxide	H_2O and CO_2	G	466
136	Steam and Nitrogen	H_2O and N_2	G	468

E. Polar - Polar Systems

Figure and/or Table No.	Name	Formula	Physical State*	Page No.
137	Chloroform and Ethyl Ether	$CHCl_3$ and $C_4H_{10}O$	G	470
138	Diethylamine and Ethyl Ether	$C_4H_{10}NH$ and $C_4H_{10}O$	G	472
139	Ethanol and Methyl Formate [Dimethyl ether]	C_2H_6O and $C_2H_4O_2$ [$(CH_3)_2O$]	G	474

5. TERNARY SYSTEMS

A. Monatomic Systems

Figure and/or Table No.	Name	Formula	Physical State*	Page No.
140	Neon-Argon-Krypton	Ne-Ar-Kr	G	478
141	Helium-Argon-Xenon	He-Ar-Xe	G	479
142	Helium-Krypton-Xenon	He-Kr-Xe	G	480
143	Helium-Argon-Krypton	He-Ar-Kr	G	481
144	Helium-Neon-Xenon	He-Ne-Xe	G	482
145	Argon-Krypton-Xenon	Ar-Kr-Xe	G	483

* G = gas.

xxii *Grouping of Materials and List of Figures and Tables*

5. TERNARY SYSTEMS (continued)

   B. Monatomic and Nonpolar Polyatomic Systems

Figure and/or Table No.	Name	Formula	Physical State*	Page No.
146	Helium–Oxygen–Methane	He–O_2–CH_4	G	484
147	Argon–Oxygen–Methane	Ar–O_2–CH_4	G	485
148	Helium–Argon–Nitrogen	He–Ar–N_2	G	486
149	Helium–Nitrogen–Methane	He–N_2–CH_4	G	487
150	Argon–Krypton–Deuterium	Ar–Kr–D_2	G	488
151	Helium–Neon–Deuterium	He–Ne–D_2	G	489
152	Neon–Argon–Deuterium	Ne–Ar–D_2	G	490
153	Neon–Krypton–Deuterium	Ne–Kr–D_2	G	491
154	Neon–Hydrogen–Oxygen	Ne–H_2–O_2	G	492
155	Argon–Hydrogen–Nitrogen	Ar–H_2–N_2	G	493
156	Neon–Hydrogen–Nitrogen	Ne–H_2–N_2	G	494
157	Neon–Nitrogen–Oxygen	Ne–N_2–O_2	G	495
158	Argon–Krypton–Hydrogen	Ar–Kr–H_2	G	496

   C. Nonpolar Polyatomic Systems

159	Nitrogen–Oxygen–Carbon Dioxide	N_2–O_2–CO_2	G	497
160	Hydrogen–Nitrogen–Oxygen	H_2–N_2–O_2	G	498

   D. Nonpolar and Polar Systems

161	Argon–Propane–~~Ethanol~~ [Dimethyl ether]	Ar–C_3H_8–~~C_2H_6O~~ [$(CH_3)_2O$]	G	499
162	Hydrogen–Nitrogen–Ammonia	H_2–N_2–NH_3	G	500

6. QUATERNARY SYSTEMS

   A. Monatomic Systems

163	Neon–Argon–Krypton–Xenon	Ne–Ar–Kr–Xe	G	504

   B. Monatomic and Nonpolar Polyatomic Systems

164	Argon–Krypton–Xenon–Hydrogen	Ar–Kr–Xe–H_2	G	505
165	Argon–Krypton–Xenon–Deuterium	Ar–Kr–Xe–D_2	G	506
166	Argon–Hydrogen–Deuterium–Nitrogen	Ar–H_2–D_2–N_2	G	507
167	Argon–Hydrogen–Nitrogen–Oxygen	Ar–H_2–N_2–O_2	G	508
168	Neon–Argon–Hydrogen–Nitrogen	Ne–Ar–H_2–N_2	G	509
169	Argon–Xenon–Hydrogen–Deuterium	Ar–Xe–H_2–D_2	G	510

7. MULTICOMPONENT SYSTEMS

170	Air		G	512

---

* G = gas.

Theory, Estimation, and Measurement

Notation

a	Numerical constant; Accommodation coefficient
A	Numerical quantity dependent on temperature in equation (19L)*; Constant; Cross-sectional area
A_{ij}	Collision integral function
b	Numerical constant
B	Constant term $(A_0 C_{p0} M^{-1/3})$ of equation (21L)
B_{ij}	Collision integral function
c	Numerical constant
C	Specific heat
C_p	Specific heat at constant pressure
C_v	Specific heat at constant volume
$C_{v\,\text{int}}$	Contribution of internal degrees to C_v
$\bar{C}$	Molar heat capacity (subscripts p or v indicate constant pressure or volume)
d	Density; Distance
d_0	Density at 0 C, equation (12L)
$\bar{d}$	Molar density
D	Trouton's constant; Coefficient of self-diffusion
D_{ij}	Diffusion coefficient
e	Emissivity
Δe	Energy interchange, equation (3L)
E	Constant, equation (82G)*
E_i	Gas mixture function, equation (73G)
f	Independent function; Eucken factor
f_k^i	Correction factor to k for higher approximations and weakly dependent upon reduced temperature
f_η	Correction factor to η for higher approximations
f_E	Eucken factor
f_H	Eucken-type correction factor due to Hirschfelder
f_M	Eucken-type correction factor due to Mason and Monchick
f_S	Eucken-type correction factor due to Saxena, Saksena, and Gambhir

*Equations in the sections on liquids and gases are here, and here only, differentiated by letters L and G respectively.

F	Numerical constant
g	Gravitational acceleration
g_{ij}	Multicomponent gas mixture thermal conductivity function, equation (62G)
G	Parameter of equation (26L)
h, h_r, h_k	Heat loss per unit area of the wire for unit difference in temperature of the wire and surroundings, such a loss by radiation only, such a loss by conduction
H	Structural hindrance factor of equations (26L) and (27L); Enthalpy
ΔH	Heat of reaction
ΔH_b	Latent heat of vaporization at boiling point
ΔH_I	Heat of ionization
ΔH_v	Latent heat of vaporization
I	Electrical current
k	Thermal conductivity
k^0	Frozen thermal conductivity
k_L	Liquid thermal conductivity
k_P	Thermal conductivity at a pressure P
k_w	Thermal conductivity of the wall material of the cell tube
k_r	Thermal conductivity of a reacting gas system
k_S	Solid thermal conductivity
k_mix	Thermal conductivity of a mixture
k_mix^0	Thermal conductivity of a mixture with frozen internal degrees of freedom
k_0	Thermal conductivity at 0 C, equation (12L); Limiting value of k as density tends to zero
k_1	Wire conductivity
k_2	Gas conductivity
$\bar{k}$	Boltzmann constant
K	Kelvin
l	Half length of wire
L	Intermolecular parameter
L_{ij}	Component of matrix
m	Number of carbon atoms
M	Molecular weight
n	Number of atoms in a molecule; Concentration; Number of components in a mixture

Notation

n_{ik}	Stoichiometric coefficient for species k in reaction i	x	Molecular fraction of constituent; Distance
N	Avogadro's number	x_a	Mole fraction of an atom
p	Exponent ($p = 3.77 - 2.94T_r$)	x_i	Mole fraction of an ion
P	Pressure	Y	Numerical constant, equation (12L)
Pr	Prandtl number	z	Distance
Q	Heat quantity; Heat quantity conducted in unit time	Z	Compressibility coefficient; Number of collisions during τ
r	Molecular radius	α	Coefficient of volume expansion; Coefficient of resistance; Small correction factor; Thermal diffusivity
r_h	Half width at half maximum		
r_i	Inner radius	β	Coefficient indicating degree of association
r_o	Outer radius	γ	Grüneisen constant; Ratio of two specific heats of a gas
r_1	Wire radius		
r_2	Tube radius (inner)	δ	Mean distance between centers of molecules; Small correction factor
r_3	Outer radius of the tube		
R	Gas constant; Rayleigh number (product of Grasshof and Prandtl numbers); Refrigerant number	Δ	Increment
		ϵ	Thermal conductivity factor, equivalent to ω^p; Intermolecular force parameter
s	Exponent; Molecular diameter	η	Viscosity
S	A characteristic dimension, such as liquid film thickness; Sutherland constant	θ	Characteristic temperature; Excess temperature
ΔS_v	Entropy of vaporization at the normal boiling point	θ_1, θ_2	Constants of equation (9L) having values depending on the structure of the liquid and of the homologous series to which it belongs
ΔS^*	Modified Everett entropy of vaporization $[\Delta S^* = \Delta H_v T_b^{-1} + R \ln 273 T_b^{-1}]$		
t	Time; Temperature, C, in equations (1L) and (4L) etc.	θ_w	Temperature difference between the inner and outer walls of the cell tube
T	Temperature, K	λ	Mean free path
T^*	Reduced temperature	μ	Number of chemical species
T_b	Boiling point	π	Equal to 3.14159...
T_c	Critical temperature	ν	Frequency of molecular vibration; Number of independent chemical reactions; Numerical constant
T_i	Incident gas stream temperature; Temperature of the inner surface		
T_m	Melting point	ρ	Resistance per unit length; Density
T_r	Reflected gas stream temperature; Reduced temperature ($T_r = T/T_c$)	σ	Molecular diameter; Lennard-Jones potential constant; Electrical conductivity; Stefan–Boltzmann constant
T_{rb}	Reduced boiling point ($T_{rb} = T_b/T_c$)		
T_R	$(T - T_m)/(T_c - T_m)$	σ_R	Reduced electrical conductivity, equation (39L)
T_w	Temperature of the wall		
T_o	Temperature of the outer surface	τ	Relaxation time characterizing translational energy exchange
T_1	Hot-wall temperature		
T_2	Cold-wall temperature	τ_c	Time between two successive collisions
ΔT	Temperature difference	ϕ	Exponent; Coefficient
u	Concentration; Gas velocity	ϕ_{ij}	Wassiljewa coefficients
U	Velocity of sound in a liquid	ω	Expansion factor, equation (34L); Acentric factor
$\bar{v}$	Mean speed		
V_m	Molecular volume of liquid at melting point	$\Omega_{ij}^{(l,n)*}$	Reduced collision integral

Thermal Conductivity of Liquids

1. INTRODUCTION

The substances considered in Volume 3, *Thermal Conductivity of Nonmetallic Liquids and Gases*, are those that occur in the liquid or gaseous state at normal temperature and pressure. Mercury is not included here, since it is a metallic element. The thermal conductivity of liquid mercury, and of other molten metals, may be found in Volume 1 which is devoted to the metallic elements and alloys. For a few of the substances included in Volume 3 values for the solid state are also given. This has been done to make the data section as complete as possible but the present text will not deal with the experimental methods used for the solid phase.

Volume 3 has a somewhat different format to that which has been adopted in Volumes 1 and 2. For each of the fluids treated, there is a short discussion of the available data. These data have often been fitted by a curve for which a formula may be given. Recommended values evaluated from this equation, or graphically, usually at 10 K intervals, are tabulated and a graphical plot is included for each liquid which shows the percent departures of the available experimental data from those given by the equation.

Since detailed analyses of this kind have not yet been completed for all liquids, it is felt desirable to include in this text such supplementary information as may assist the reader in obtaining thermal conductivity values for other liquids. For liquids which may not feature in the data section that follows, the reader should consult another TPRC work, the *Thermophysical Properties Research Literature Retrieval Guide* [598],* in the hope that one or more papers containing the desired thermal conductivity will be located among the 33,700 referenced papers that have been published prior to July 1964. The remaining sections of this introductory text on liquid thermal conduction are also designed to be of assistance by serving to supplement the data section that follows.

Section 2 contains references to compilations of thermal conductivity data, together with a brief outline of their scope, and to a few selected papers thought likely to be of value, either because they contain values for particular groups of liquids or because they relate to other factors likely to influence thermal conductivity determinations, and which should be taken into consideration when heat transfer estimations are being made.

Section 3 deals very briefly with the theoretical position and more extensively with many empirical equations that have been proposed. On the basis of this information it should be possible to make thermal conductivity estimations that should prove helpful for many liquids for which no experimental thermal conductivity values can be located.

Section 4 relates to the experimental methods and is intended for those who wish to have some insight into this phase of the subject, to contribute to further knowledge in this field, or to become equipped to make measurements on liquids for which the required information does not appear to be available.

2. NOTEWORTHY SUPPLEMENTARY PUBLICATIONS

A. Data Compilations

This section lists a number of collections of liquid thermal conductivity data together with a brief outline of their scope. Each account is headed by the name(s) of the author(s) and chronological order has been adopted.

Sakiadis and Coates [454, 456] published two survey papers containing outlines of the experimental methods used by earlier workers for the determination of liquid thermal conductivities. An attempt is made to assess the reliability of the results but some of their ratings need modification in the

---

*References appear under the heading *References to Text*.

light of later publications. Tables are included of the available data for the thermal conductivity of pure organic and inorganic liquids and for their aqueous solutions. A miscellaneous section includes several oils, but helium and molten metals and alloys are not treated. These two papers serve to give a nearly complete presentation of the data available up to 1954.

Filippov [151] published his own results obtained from new determinations of the thermal conductivity for 41 organic liquids at 30 C. He also lists most probable values at this temperature for 150 organic liquids. Source references are included, together with estimated uncertainties, which are all under 5 percent. Several years later Filippov [154] published another table for 83 substances with some small changes, made in the light of subsequent work, in the tabulated values for the thermal conductivity at 30 C and the temperature coefficients.

Robbins and Kingrea's [447] paper is primarily intended to present the derivation of an empirical formula for the estimation of liquid thermal conductivities. A table is however included of thermal conductivities for about 70 organic liquids, some for a range of temperatures, in which their estimated values are compared with the available experimental values. The average differences were below 4 percent with extreme differences of -108 and $+18.2$ percent.

Tsederberg's [600] Russian book dealing with the thermal conductivities of gases and liquids is available in English translation [601]. In the latter edition, pp. 199–204, two tables are presented. The first contains the thermal conductivity of some 150 organic liquids at 30 C, together with source references. The reliability is thought to be within 5 percent for all the data. The second table lists the temperature coefficients, applicable to temperature ranges within 15 to 90 C for about one third of these liquids.

A subsequent table, pp. 220–221, relates to the thermal conductivities of liquid mixtures (solutions). Comparison is made between experimental values (all Russian, e.g., Filippov and Tsederberg) and calculations made using the additive rule. The original Tsederberg's account should be consulted for details, particularly regarding those solutions for which this rule can and cannot be used. Another table, p. 231, lists the thermal conductivity of aqueous alkali, acid, and salt solutions, at 20 and 30 C, based mainly on the publications of Riedel [441–3] and of Vargaftik and Os'minin [622], while the final table, p. 233, contains values for aqueous solutions of sulphuric and nitric acids for the full range of concentrations and temperatures from 0 to 90 C (93 C for HNO_3 solutions). More recent contributions in this field are those of Chiquillo [80] and Grassmann *et al.* [189]. (See Section B.g for details.)

Vargaftik [621] edited a handbook (in Russian) which contains tabulated values for the thermophysical properties of liquids and gases including thermal conductivity. The many tables of data are presented without discussion or probable reliability estimates but are useful in that the results of Russian measurements are strongly represented.

Jamieson and Tudhope [248] prepared two comprehensive and most useful reports. The first surveys the data available up to 1963 for the thermal conductivity of about 300 liquids at atmospheric pressure. It excludes liquefied gases, liquid metals, and inorganic salts but does include aqueous solutions of electrolytes and nonelectrolytes, mainly from the aforementioned investigations of Riedel and of Vargaftik and Os'minin. Assessments of the accuracy of the tabulated values are also given, together with short descriptions of the methods used in 62 instances. The second reviews data for some 300 liquid mixtures, and includes work published through 1968.

Missenard's [365] book on heat conduction gives extensive data. Pages 116–177 are devoted to pure liquids and pp. 403–444 to liquid mixtures and to salt solutions. Empirical equations are derived (see also reference 364) and tables are presented which enable most existing data to be compared with calculated values. The thermal conductivities of about 300 liquids are presented. A later paper [366] contains additional data and about a 7 percent reduction in the numerical coefficient of the empirical equation is to be noted (p. 421) from 90 to 84, while, according to a private communication of November 1966, a further reduction from 84 to 82 has since been thought desirable. (This equation is equation (12) of Section 3B.)

B. Special Groups of Fluids

In this subsection it has been thought desirable to list recent papers containing thermal conductivity values for particular classes of liquids for which information may be sought. The classes included are:

   a. Organic liquids
   b. Refrigerants
   c. Oils

d. Biological fluids
   e. Fruit juices and sugar solutions
   f. Aqueous solutions of organic compounds: mixtures of organic liquids
   g. Aqueous solutions of inorganic salts
   h. Metal–ammonia solutions
   i.* Some polyphenyls and mixtures thereof
   j.* Molten salts

No attempt has been made to list all available papers, for which the reader should consult the *TPRC Retrieval Guide* [598], but for each group a recent paper has been included and this will often contain references to earlier publications.

a. Organic Liquids

Filippov [150] has measured the thermal conductivities of 50 organic liquids, many over the range 15 to 90 C, and tabulates values for their thermal conductivities at 30 C and for their temperature coefficients. Values for the thermal conductivities of about 150 and 80 organic liquids are given in two later publications [151, 154].

Dick and McCready [120] used a horizontal-plate apparatus in which the liquid thickness could be varied for thermal conductivity determinations at 20 and 60 C on 19 organic compounds. These included isomeric ethers and some esters of varying structure, but of comparable molecular weights. The main purpose of this work was to study the dependence of thermal conductivity on the structure of high-molecular-weight compounds. The general conclusions reached were that the thermal conductivities increase with increasing chain length and that the effect of side chains is to reduce the thermal conductivity.

Sakiadis and Coates [457, 458] made measurements of the thermal conductivities of 88 organic liquids in another type of plate apparatus, in which sample thickness could also be varied. These measurements were mainly for the range 30 to 75 C and values of thermal conductivity and temperature coefficient are tabulated. Details are also given of two methods for predicting thermal conductivities (see Section 3) in which the data for different homologous series are correlated as a function of the reduced temperature and of the number of effective carbon atoms. The original paper should be consulted for details of their correlation method, according to a modification of the theory of corresponding states, and of the value of this last-mentioned parameter for various homologous series of organic liquids. Several examples give good agreement between predicted and experimental values.

Gudzinowicz, Campbell, and Adams [198] have presented the results of thermal conductivity measurements made using a hot-wire cell on some 60 hydrocarbon fuels at temperatures of 63, 104, and 158 C. Using the same apparatus, measurements are also reported for glycerol, *o*-xylene, toluene, Aroclor 1248, *n*-decane, *n*-nonane, chlorobenzene, and nitrobenzene. Comparisons with existing values obtained by plate, cylinder, and sphere methods give maximum differences of $+3$ and -17.6 percent.

Tufeu, LeNeindre, Bury, and Johannin [604] used a concentric-cylinder method to determine the thermal conductivities of seventeen organic liquids over various temperature ranges between 2.7 and 93.2 C. These included the alcohols $C_mH_{(2m+2)}O$ with the number of carbon atoms, m, increasing from 1 to 10, but omitting $C_9H_{20}O$. For this group of liquids the authors plotted the thermal conductivity at 25 C as a function of m and, as m increased, obtained a smooth rapidly falling, slowly rising curve having a minimum at about $m = 4$ or 5.

Missenard contributed an interesting foreword to this paper. He plots in a similar manner thermal conductivity data at 0 C against m for four groups of organic liquids, (i) the acids, $C_mH_{2m}O_2$, (ii) the alcohols, $C_mH_{2m+2}O$, (iii) the saturated hydrocarbons, C_mH_{2m+2}, and (iv) the iodide derivatives $C_mH_{2m+1}I$. The points for each group conform to a different smooth curve, but there is a strong indication that as m increases to high values the four curves should converge to a common limiting value of from 0.00155 to 0.00160 W cm^{-1} C^{-1}. This is at 0 C; presumably other temperatures will yield different values. Such a convergence to a common value will be shown later to be in general accord with an empirical treatment that had been proposed by Missenard [see Equation (12) and the accompanying discussion in Section 3B]. This information could afford a valuable means for estimating the thermal conductivity of the less-studied higher members of various groups of organic liquids, and, in order to check the validity of Missenard's conclusion, thermal conductivity determinations are urgently required on group members for which m has values of the order of 20 or 30.

---

*These classes of substances belong to Volume 2, which should also be consulted, but are included here as the methods employed are common and will make this volume of the maximum assistance to those interested in liquid thermal conductivities.

b. Refrigerants

Tauscher [189, 579] has been responsible for one of the most comprehensive investigations to be reported for the thermal conductivities of liquid refrigerants. This is a field in which considerable differences have occurred in many of the reported values, and doubt as to the true values still exists. Tauscher's values are obtained by an unsteady-state hot-wire method for 14 of the fluoro-chloro-derivatives of methane and ethane, R10, R11, R12, R13, R14, R20, R21, R22, R23, R112, R113, R114, R115, and R116. They cover various sections of the temperature range -125 to $105\,C$ and graphical comparison is made with the existing data, for which references are included.

c. Oils

Powell and Challoner [413] have been responsible for one of the more recent papers dealing with the thermal conductivities of oils. This paper only contains measurements for five transformer oils but includes references to other work. Indeed one of the main reasons for this investigation stemmed from some measurements by Allen [6], which appeared anomalous in yielding a strong positive temperature coefficient for the thermal conductivity of an oil. These measurements are treated in more detail in Section 2Cd. Whereas Smith [550] in 1936 considered the thermal conductivity of oils to lie within 6 to 13 percent of a constant value of 0.00137 W cm^{-1} C^{-1}, Powell and Challoner find the thermal conductivities of the transformer oils which they studied to vary much less with temperature and to be lower by at least the upper limit of this assumed variance. They suggest that for the range 20 to 60 C a value of 0.00118 ± 0.00003 W cm^{-1} C^{-1} seems to be more appropriate for modern oils.

A paper by Rastorguev [425], that was in course of publication at about the same time and so was not referenced by Powell and Challoner, indicates values at 40 C of from 0.00129 to 0.00142 W cm^{-1} C^{-1} for five oils with low solid-hydrocarbon contents. Those of concentrate and raffinate oils with 10 to 14 percent and 15 to 20 percent paraffins respectively were 0.00149 and 0.00157 W cm^{-1} C^{-1} at 40 C. All seven oils measured by Rastorguev had negative temperature coefficients over the range 0 to 120 C. These coefficients agreed closely and were shown to be proportional to the volumetric expansion coefficients.

d. Biological Fluids

Spells [555] has determined the thermal conductivities, mainly near body temperature, of several biological fluids including blood, blood plasma, blood corpuscles, milks, cream, egg white and yolk, and cod liver oil. A fair correlation with water content is claimed. (See also Riedel [440].)

e. Fruit Juices and Sugar Solutions

Riedel [440] has made determinations of the thermal conductivities of apple, pear, and grape juices and of saccharin and glucose solutions for the range 20 to 80 C. This paper also contains values for several types of milk over the range 1.5 to 80 C.

f. Aqueous Solutions of Organic Compounds: Mixtures of Organic Liquids

Rastorguev and Ganiev [426] used a concentric-cylinder apparatus for thermal conductivity determinations on ethylene glycol, diethylene glycol, and dimethylformamide over the range 20 to 80 C and on a wide range of aqueous solutions of these liquids and of glycerol, formamide, and pyridine at 40 C. On the basis of these results and of the most reliable values of earlier workers, they derived for the dependence of k, the thermal conductivity (W m^{-1} C^{-1}) of the aqueous solutions, both on concentration and temperature, the equation

$$k = k_1 x_1 + k_2 x_2 + 1.4 x_1(x_1 - 1)(\Delta k - 0.2) - 1.4 x_1(x_1 - 1)(t - 20)10^{-3} \quad (1)$$

where x_1 and x_2 are respectively the molecular fractions of the organic liquid and of water in the solution, $\Delta k = k_2 - k_1$ is the difference in thermal conductivity between water and the organic liquid, and t is the temperature (C) to which the values of k_1 and k_2 apply. The tabular results also include those of Riedel [445] for eight solutions, and it is stated that the mean and maximum differences between these calculated and his measured values were respectively 1.5 and less than 6.5 percent. Rodriguez [449], however, found that a logarithmic equation due to Jordan and Coates [255]

$$\ln k = x_1 \ln k_1 + x_2 \ln k_2 + x_1 x_2 \ln[e^{(k_2 - k_1)} - 0.5(k_2 + k_1)] \quad (2)$$

when used with units of Btu ft^{-1} h^{-1} F^{-1} gave values which were all within ± 3 percent for 21 different organic–organic and aqueous–organic binary systems, including the values of Riedel [445]. Subsequent work by Shroff [543] also supported equation (2). Shroff, who did not appear to be aware of the work by Rastorguev and Ganiev [426, 428], found equation (2) to yield an average deviation of ±2.2 percent for his own measurements on ethanol–water, n-propanol–toluene, methanol–toluene, methanol–benzene, and n-butanol–benzene as well as for the data of Riedel [445], Rodriguez [449], McLaughlin [351], and Frontas'ev [159]. Yet another semiempirical equation has been proposed by Venart [628]

$$k = [k_1 + (k_2 - k_1)x_2](1 - x_1 x_2 \Delta e/\bar{k}T) \quad (3)$$

which opens up the possibility of intercomparing data for the diffusion, viscosity, and thermal conductivity of liquid mixtures, since the energy interchange term $\Delta e/\bar{k}$, with $\bar{k}$ the Boltzmann constant and Δe a constant indicating the mean energy change of a molecule of one kind on exchanging all its neighbors for molecules of another kind, occurs in expressions for all three properties and is derived in each instance from experimental values for a particular mole fraction and temperature.

g. Aqueous Solutions of Inorganic Salts

Chiquillo [80, 189] used a transient hot-wire method to obtain thermal conductivity values at temperatures between 20 and 40 C for aqueous solutions of 22 salts, namely: the chlorides of lithium, sodium, potassium, rubidium, cesium, magnesium, calcium, strontium, and zinc, the bromides of lithium, sodium, potassium, rubidium, and cesium, the iodides of lithium, sodium, and potassium, potassium fluoride, zinc and copper sulfates, lead nitrate, and potassium carbonate. An appendix details earlier determinations of the thermal conductivity of aqueous solutions.

h. Metal–Ammonia Solutions

Varlashkin and Thompson [626] have reported measurements that yield the thermal conductivity of solutions of the alkali metals lithium, sodium, and potassium in liquid ammonia. For ammonia solutions approaching saturation, near 15 mole percent for sodium or potassium and 20 mole percent for lithium, thermal conductivity values of the high order of from 0.03 to 0.06 W cm^{-1} C^{-1} are obtained. These values are some 6 to 12 times the thermal conductivity of liquid ammonia at about −50 C [625] and approaching 5 to 10 times that of water at 27 C. The thermal conductivity for a particular concentration is fairly independent of the alkali metal used and the high values appear to be due to the presence of free electrons in the metal–ammonia solutions. The measurements extend down to −100 C, and indicate that systems of alkali metals in liquid ammonia could be of value as efficient low-temperature heat transfer media.

i. Some Polyphenyls and Mixtures Thereof

Reiter [430] has used a steady-state hot-wire method to determine the thermal conductivity of the following liquids, smoothed values being given in W m^{-1} C^{-1} over the indicated temperature ranges:

Diphenyl	100 to 250 C	0.133 to 0.110
o-Terphenyl	100 to 350 C	0.127 to 0.100
m-Terphenyl	100 to 350 C	0.134 to 0.117
p-Terphenyl	250 to 350 C	0.126 to 0.111
OM$_2$	100 to 350 C	0.131 to 0.112
Dowtherm A	50 to 250 C	0.134 to 0.106

OM$_2$ is a mixture of 25.5 percent o-terphenyl, 72.2 percent m-terphenyl, 2.1 percent p-terphenyl, and less than 1 percent diphenyl and higher boiling residues; Dowtherm A is mainly 26.5 percent diphenyl with 73.3 percent diphenyloxide. Graphs are included showing comparisons with existing data.

j. Molten Salts

Turnbull [605] has used a transient hot-wire method to determine the thermal conductivities of the following molten salts: potassium and silver nitrates, zinc chloride, sodium, potassium, and ammonium hydrogen sulfates, and potassium thiocyanate. Near the melting point Turnbull finds the ratio of the thermal conductivity of the liquid to that of the solid, k_L/k_S, to be 0.86 ± 0.13, a finding which, if generally true, would be of value in predicting the thermal conductivities of other molten salts from information given in Volume 1 of this series. Two of the molten salts studied by Turnbull, zinc chloride and ammonium hydrogen sulfate, have negative temperature coefficients whereas those of the others are positive just above the melting point, but in no case is the temperature dependence very appreciable.

White and Davis [660] have since suggested that the electrical conductivity of molten salts is too high for use of the hot-wire method, and they used a concentric-cylinder apparatus. Determinations made on the five molten alkali nitrates, LiNO$_3$, NaNO$_3$, KNO$_3$, RbNO$_3$, and CsNO$_3$, all gave positive temperature coefficients, and for NaNO$_3$ and KNO$_3$ the thermal conductivity values of Turnbull and of White and Davis differed considerably. Since Turnbull [605] had certainly considered the possibility mentioned and appeared to be satisfied that under his experimental conditions no trouble should result and since White and Davis do not appear to have checked their apparatus for any material of comparable and well-established thermal conductivity, it seems that further information is required before these differences can be completely resolved. Bloom, Doroszkowski, and Tricklebank [33] had used basically the same method as White and Davis and for KNO_3 in the range 340 to 400 C their values exceed those of White and Davis by about 16 percent and those of Turnbull by only 11 percent. As was stated by McLaughlin [351], when making an advance report of data for $NaNO_3$ from 313 to 465 C, molten salts belong to another class of fluids for which further thermal conductivity determinations are still needed.

C. Influence of Environmental and Structural Factors

a. Conduction

A primary requirement of the design of an apparatus for the measurement of thermal conductivity is that the whole of the supplied energy should be used to establish the observed temperature distribution within the test fluid. As all materials conduct heat, this requirement is never completely realized and corrections have to be made for any undesired heat conduction, to or from the section of the fluid on which observations are being made. Even when the guarding is perfected to avoid extraneous heat transfers, losses can occur within the cell itself. In the main types of apparatus, the parallel plate, concentric sphere, and cylinder, correction has to be made for heat flow that occurs along paths parallel to the required heat flow across a liquid film such as through spacers, supports, and boundary walls, and also for any disturbance in the required, normal, or radial flows, induced by these other conducting paths. Similarly in methods of the hot-wire type it is necessary to allow both for any loss of heat along the wire toward the electrodes as well as for any axial heat flow component within the liquid that might arise because of this end cooling.

b. Effect of Convection

In order to determine thermal conductivity it is necessary for a temperature difference to be established in the test sample. Now it is well known that differences in temperature cause density changes, and, when the test sample is a fluid, these density changes can lead to the establishment of convection currents which influence the heat transfer. In designing and operating thermal conductivity apparatus the necessity of avoiding convection has become well appreciated and, among others, has been dealt with by McLaughlin [351]. The onset of convection is determined by critical values of the Rayleigh number, which is the product of the Grashof and Prandtl numbers. It has been shown that in general this product R should be less than 1000, where $R = (\eta C_p/k)(g\alpha d^2 \Delta T S^3/\eta^2)$, η being the dynamic viscosity, C_p the specific heat at constant pressure, g the gravitational force per unit mass of fluid, α the volumetric expansion coefficient, d the density, ΔT the temperature difference, and S a characteristic length such as the liquid film thickness. While this critical value of R applies to the test section proper of the apparatus, care in the design of any connecting volumes is also of importance since fluid motion set up in these regions may cause fluid movement in the main test section even when R is less than 1000.

When fluid thermal conductivity determinations have been attempted by the guarded hot-plate method close to the critical point of the fluid, special precautions have been necessary to ensure that the marked changes observed are not due to convection. The work of Michels and Sengers [356] is of particular interest in this connection.

With variable-state methods of the hot-wire type the restriction on sample thickness is no longer necessary. For a fixed energy input, the initial rate of heating of the wire remains constant until convection is set up in the fluid. McLaughlin [351] is able to show that the observed elapsed time before the onset of convection agrees with a calculated time that is obtained when a value of 1000 for the Rayleigh criterion is assumed. In this way an independent check for the validity of this criterion is obtained.

c. Effect of Radiation

Poltz and collaborators have been responsible for a series of four papers [158, 405–7] which shows that, unless allowance is made for the heat transmitted by radiation across the test-fluid layer, measurements of thermal conductivity made on liquid layers can be too large by a few percent even at normal temperature, whilst the temperature coefficient is likely to be still more in error.

It is of interest to note that two relevant contributions to this subject had been made in 1954. One was an earlier theoretical treatment by Filippov [149], the other was a comment made by Bonilla when discussing the work of Dick and McCready

[120] which now assumes greater significance. These workers had obtained thermal conductivity increases of from 6 to 8 percent with sample thickness increase from 0.6 to 2.2 mm for several organic liquids when tested at 60 C. The relevant part of Bonilla's contribution states "It is the writer's theory that extrapolation to $\Delta S = 0$ (i.e., to zero sample thickness) has two effects: (a) conduction in the fluid sample increases so that any wall-to-wall radiation becomes negligible, and (b) radiation within the fluid also becomes negligible. Thus the k extrapolated to $\Delta S = 0$ is the true value of k by molecular conduction alone." However, in these particular experiments, Dick and McCready had an alternative explanation that was peculiar to their apparatus.

While the papers of Poltz et al. have been appearing, Leidenfrost [294] has also devoted attention to the errors likely from this cause. Most workers have considered the liquids to be perfectly transparent and have regarded radiation effects to be negligibly small when making thermal conductivity measurements on liquids, at least near room temperature and so long as the adjacent metal surfaces are of low emissivity. Challoner and Powell [73], however, did make determinations on light and heavy water and on carbon tetrachloride using liquid thicknesses of 2 and 3 mm, which revealed no significant differences up to temperatures of about 80 C for the two forms of water and of 56 C for carbon tetrachloride. Also, their measurements on transformer oil over the range 28 to 60 C for 2 and 3 mm thicknesses, showed no differences [413].

The measurements of Fritz and Poltz [158] confirm that for water and also for methanol, liquids which strongly absorb thermal radiation, a thermal conductivity that is independent of sample thickness is obtained for small liquid thicknesses but that an increase, which is attributed to convection, occurs when the thickness exceeds 1 or 2 mm. The later measurements of Poltz and Jugel [407] for liquids that absorb thermal radiation only weakly show a definite dependence on thickness for sample thicknesses of from 0.46 to 1.93 mm, which are thought to be too small for convection to be present. These last include benzene, carbon tetrachloride, toluene, *m*-xylene, nitrobenzene, and liquid paraffin. The anticipated differences between the effective measured thermal conductivities for near zero and large liquid thicknesses, yet for conditions that are considered to be well within the Rayleigh criterion for no convection, are of the order of 5 percent, and,

for carbon tetrachloride Poltz and Jugel show that about four-fifths of this difference occurs as the layer is increased to 2 mm thickness. Thus, according to Poltz and Jugel, although the measurements of Challoner and Powell [73] for 2 and 3 mm thicknesses would be expected to agree to within the limits of accuracy claimed for their method, the measured values at both thicknesses could have included a radiation component of about 4 percent.

The measurements of Poltz and Jugel [407] on toluene are of particular interest. A few years ago Ziebland [675] suggested that toluene might serve as a thermal conductivity standard. The equation proposed was

$$k = (3.36 - 0.0067t)10^{-4} \quad (4)$$

k being in cal cm^{-1} s^{-1} C^{-1} and t in C. In proposing this equation it was pointed out that all determinations from five independent investigations agreed with it to within 3 percent and 85 percent of them to within 1.5 percent. This equation yields at 25 C a thermal conductivity of 3.1925×10^{-4} cal cm^{-1} s^{-1} C^{-1}, or 0.001336 W cm^{-1} C^{-1}, which is 3.5 percent higher than the value of 0.00129 W cm^{-1} C^{-1} given by Poltz and Jugel for the true thermal conductivity corresponding to a zero film thickness, but less than 1 percent greater than their value for a 1 mm thickness. At least three further papers containing data for the thermal conductivity of toluene have appeared since Ziebland proposed equation (4). Of these, the values at 25 and 60 C of Horrocks and McLaughlin [240] and of Venart [627] for the range 0 to 80 C agree with Ziebland's equation to within 1 percent, but those of Tufeu, LeNeindre and Johannin [603, 604] are greater by approaching 5 percent. Those last mentioned values would appear to be too high.

Poltz and Jugel [407] conclude that many of the differences between recent measured values can be attributed to the inclusion of varying radiative components, and their paper includes a useful supporting table for the six liquids which they had studied. Since Tufeu et al. [603, 604] used a concentric cylinder apparatus with a liquid thickness of the order of only 0.5 mm some other reason would presumably need to be found to explain the high values which they reported. Rather surprisingly, in the table of Poltz and Jugel the values attributed to Tufeu et al. and quoted for benzene, toluene, xylene, and carbon tetrachloride have all been reduced by nearly 5 percent. Hence these values no longer appear unduly high, but no reason for the change is given, moreover no change in value appears to have been made by the authors in their papers which preceded [603] and followed [604] that of Poltz and Jugel [407].

It has been considered desirable to include this rather detailed account of recent work on toluene to emphasize something of the uncertainties that are still to be found in the subject of liquid thermal conductivities and which in this case relate to a liquid which had been recommended as a thermal conductivity standard in 1961. These and other uncertainties should certainly receive further consideration.

The subject of the effect of radiation that has been highlighted by the work of Poltz and his collaborators still requires further confirmation. It also demands consideration in connection with other thermal conductivity methods, particularly hot-wire methods in which the observed wire temperature is presumably dependent on the degree of radiation absorption in the surrounding fluid. This additional uncertainty has developed since most of the data sheets that follow have been prepared. Hence no allowance has so far been made for any radiation component. This means that the recommended values for some of the liquids may ultimately prove to be high by a few percent.

A contribution by Ewing, Spann, and Miller [143] also makes definite reference to complicating effects due to radiation, although this work was at much higher temperatures and for materials that belong to Volume 3 of this series. These workers made measurements on molten boric oxide and on molten "Flinak," a mixture, expressed in mole percent, of 11.5 NaF, 42 KF, and 46.5 LiF. With boric oxide, changes were observed in the heat transfer, and consequently in the apparent thermal conductivity, due to changes in the emissivities of the adjacent hot and cold surfaces; with Flinak, observed apparent thermal conductivity changes were attributed to an increased impurity concentration which occurred on heating and which caused increased reradiation to occur within the liquid.

The few examples quoted above indicate some complicating factors which can arise when thermal conductivity determinations are made on liquids. Similar factors can also occur in practical installations and demand consideration when heat transfer estimations are being made.

d. *Possible Effect of an Applied Electrostatic Potential*

When Allen [6] used the variable-state hot-wire method to determine the thermal conductivity of transformer oil over the range 20 to 80 C he obtained a large positive temperature coefficient of the order of 60×10^{-4} C^{-1}, corresponding to an increase in thermal conductivity of nearly 40 percent over the range studied. As most oils had been found by the earlier, more conventional methods to have small negative temperature coefficients, Allen, without checking his method with any other liquid, proceeded to suggest that these earlier measurements might have been subject to error. Evidence was produced, but for rather different conditions, showing that an electrostatic potential could develop across the plates of a steady-state thermal conductivity apparatus when tests are made on a dielectric fluid, and that this could inhibit the heat transfer to a greater extent at high than at low temperatures. Under such conditions the observed temperature coefficient could be too low and the sign could even be reversed. Allen concluded that errors from this cause could have frequently occurred and have probably masked the true variations of thermal conductivity with temperature. He recommends the provision of an electrical connection between the hot and cold plates.

In view of the magnitude of the difference indicated by Allen's data and of his suggestion, which, if found to be true, could have affected the results of most of the usual thermal conductivity methods, Powell and Challoner [413] made a series of careful checks on the guarded hot-plate apparatus in use at the National Physical Laboratory, Teddington. These tests were made with both medicinal paraffin and transformer oil samples of 3 mm thickness and covered mean temperatures of from 24 to 60 C. When operated under normal conditions the voltage gradient between the plates was found to vary from 2 V cm^{-1} at 24 C to 23 V cm^{-1} at 60 C. The plates were then electrically connected and further measurements made. Subsequently, tests at a mean temperature of 41.7 C in the case of medicinal paraffin and 60.7 C in the case of the transformer oil were made with an applied voltage of ± 90 V cm^{-1}. In no instance were the changes of potential across the plates accompanied by any significant change in the observed thermal conductivity.

As a final check the observations were repeated on a sample of the transformer oil that Allen had used. For two different sample thicknesses of 2 and 3 mm and throughout the range 28 to 58.8 C a thermal conductivity of 0.001200 W cm^{-1} C^{-1} was obtained. Hence these tests afforded no evidence for any radiation effect of the type discussed in the previous section. Electrically connecting the hot and cold plates only gave an increase in thermal conductivity of 3 parts in 1200, which was well within the normal experimental variation. The straight line drawn through Allen's results gives this value at about 41 C but is some 7.3 percent lower at 28 C and 9 percent higher at 59 C.

The final conclusion reached was that Allen's measurements would appear to have been subject to some as yet unexplained error and that thermal conductivity determinations made on liquids by means of conventional methods are not likely to be subject to any serious error due to the small electrostatic potentials that can occur during normal operation.

It is clear, however, from the observations of Schmidt and Leidenfrost [529, 530] that an enhanced heat transfer, which is greater at high temperatures, does occur with polar liquids when the applied voltage is sufficiently great, say 0.5 kV cm^{-1} and above. For transformer oil at 21.3 C an applied potential of 10 kV cm^{-1} led to about a 10 percent increase in the apparent thermal conductivity and at 50 C the increase was about doubled. These increases were regarded as due to the setting up of convection.

Rastorguev and Ganiev [427] have more recently studied the influence of an applied electric field on the thermal conductivity of a liquid as measured in a concentric cylinder apparatus. They also observe an increase to occur when the field strength is sufficiently great, but this is attributed to ionic flow that arises from ionization of dipolar molecules. Furthermore, the values obtained are stated to be in striking disagreement with those of Schmidt and Leidenfrost. This appears, therefore, to be another subject requiring further investigation. It is clear that estimates of the heat transfer through certain fluids in the presence of an electric field may need to make allowances for an augmented heat flow and that precise values will need to be determined for specific operational conditions.

e. The Soret Effect

The Soret effect is a measure of the change in concentration of a solution which can occur in the presence of a temperature gradient. Thus, in the case of solutions, it would seem that density changes can arise from two independent causes, namely: the normal thermal expansion and the concentration changes of the Soret effect. Hence this additional factor could presumably lead to complications, not only by modifying the composition of the sample but also due to the fact that the Rayleigh criterion may no longer be adequate to ensure freedom from convection owing to the additional energy transport provided by the migrating molecules.

So far as is known, no definite instance has been reported in which difficulties have arisen due to the Soret effect when thermal conductivity determinations on solutions have been made, but this does not necessarily mean that troubles from this cause can never arise. Indeed, Powell and Tye [418] do refer to two earlier sets of measurements on calcium chloride brines for which unacceptable results had been obtained. The reasons remain uncertain, but may be connected with concentration changes and associated density changes. The Soret effect was also mentioned by Tyrrell [606] as a possible contributing factor that might help to explain the rather different values obtained for liquid mixtures by Frontas'ev [159]. Clearly this is another possible source of error for which more conclusive experimental information is required.

f. Possible Effect of Molecular Orientation

Sutherland, Davis, and Seyer [575] made measurements of the thermal conductivity of *n*-octadecane in an unguarded plate apparatus constructed of copper. By varying the liquid thickness from 6.9 to 0.1 mm they obtained thermal conductivity values for *n*-octadecane that decreased from about 0.00152 to 0.00019 $\text{W cm}^{-1} \text{ C}^{-1}$. The authors did not seem to be at all perturbed by the fact that a thermal conductivity of 0.00019 $\text{W cm}^{-1} \text{ C}^{-1}$ is lower than that of any liquid yet measured by a factor of about 3. Yet, from these results they were led to make two very surprising statements: that "The most reliable method for determining the heat conductivity of a liquid is the thick film method" and that "Under quiescent conditions the orientating forces on the copper surface can extend several millimeters deep into liquid octadecane, and this depth appears to be dependent on the size and shape of the molecule."

The first of these statements is certainly contrary to general belief, and the second casts severe doubt both on the usual thin-film method and on the results of earlier thermal conductivity measurements for long-chain hydrocarbons. Fortunately these observations by Sutherland, Davis, and Seyer have not been supported by subsequent measurements on *n*-octadecane, made first by Powell and Challoner [415] and later by Ziebland and Patient [679]. It now seems clear that should there be any orientation of the kind suggested, its influence on thermal conductivity determinations by the usual methods involving thin liquid films must be small and within the other uncertainties normally associated with these methods.

An instance where application of a magnetic field led to orientation of molecules of *p*-azoxyanisole producing a 25 percent increase in thermal conductivity has been reported by Bereskin and Stewart [29]. When long-chain fluids are flowing through a narrow space, molecular orientation seems likely to occur and such special conditions may introduce some degree of anisotropy into the thermal conduction component required for heat transfer estimations.

3. THEORY AND EMPIRICAL EQUATIONS

In this section an account will be given of the various theories and empirical equations that have been proposed for the conduction of heat in liquids. Owing to the wide range of values obtained experimentally for the thermal conductivity of most liquids, satisfactory comparison is only really possible once values that are known to be reliable have been obtained. Thus some proposed theories will need to be reexamined, whilst empirical equations having a parameter adjusted to yield good agreement with sets of data now known to involve experimental errors, should be revised before comparison with subsequent more accurate data is attempted. The whole subject appears to be in a somewhat unsatisfactory condition and will be due for a renewed appraisal at some future time when thermal conductivity values for a wider range of liquids are known with greater certainty. In view of the uncertainties that have been mentioned such reappraisal will not be attempted at the present time: also whenever reference is made to a comparison between theory and experiment, these uncertainties should be borne in mind.

A. Theories for Electrically Nonconducting Liquids

A fully satisfactory theory for the thermal conductivity of liquids is not available as yet; nor does any one of the many empirical relationships that have been proposed appear to be entirely satisfactory.

The theoretical position can well be appreciated by referring to the previously mentioned book by Tyrrell [606], another by Bondi [37], or a survey paper by McLaughlin [351]. This paper presents brief accounts of the various statistical theories which have been proposed. Enskog [135, 136] extended the dilute gas theory for hard-sphere molecules to liquids by allowing for a change in density. Later approaches through use of this model include Longuet-Higgins and Pople [312] and alternative treatments by Horrocks and McLaughlin [238] and Rice, Kirkwood, Ross, and Zwanzig [435]. Other theories based on a model in which molecules interact with a square-well potential have been proposed by Longuet-Higgins and Valleau [314] and by Davis, Rice, and Sengers [110]. Then follow the more complex treatment of Zwanzig, Kirkwood, Oppenheim, and Alder [681] using nonequilibrium statistical mechanics, an approach that is hampered by lack of knowledge of a molecular friction constant term, the rather approximate but simpler theory of Rice and Kirkwood [434], and others by Collins and Raffel [90], Rice and Allnatt [432, 433], and Helfand [218].

Where comparisons with experimental values have been possible, the agreement is seldom close and the temperature coefficients often disagree. Rather better agreement, together with the correct temperature dependence, appears to result from the treatment by Horrocks and McLaughlin [238], who assumed the liquid to have a quasicrystalline face-centered-cubic lattice-type structure through which the excess energy due to the temperature gradient is transferred with a frequency determined by the molecular mass and the intermolecular forces.

An interesting point emerges from the fact that by differentiating their thermal conductivity expression with respect to temperature at constant pressure, Horrocks and McLaughlin [241] were able to show that the temperature coefficient of thermal conductivity is negative and is a linear function of the coefficient of thermal expansion, as Rastorguev [425] had found (see Section 2Bc). As the simplest liquids usually expand more than complex ones this finding agrees with the smaller negative temperature coefficients usually found for liquids of higher complexity. Furthermore, the line that is obtained when available experimental data are plotted does not pass through the origin but intersects the expansion coefficient axis at a positive value. This means that a liquid having an expansion coefficient below this value should have a thermal conductivity that increases with increase in temperature. Bridgman [44] had noted that the sign of the temperature dependence of most common liquids changed at pressures above about 3000 atm. This change now seems to arise from the considerable reduction of the expansion coefficient that occurs at high pressure, and is consistent with the theory of Horrocks and McLaughlin.

In a very similar manner, but by differentiating the equation of Horrocks and McLaughlin [238] with respect to pressure at constant temperature, Kamal and McLaughlin [257] showed the dependence of thermal conductivity on pressure to be a linear function of the isothermal compressibility. In this instance the line passed through the origin indicating that the thermal conductivity would always be expected to increase with increase in pressure.

Whereas Horrocks and McLaughlin [238] have obtained these results by treating a liquid structure as somewhat like that of a solid, another approach by Kanitkar and Thodos [259] assumes that the modes of energy transfer of the gaseous state still hold for the liquid. Kanitkar and Thodos obtained a simple correlating expression for the thermal conductivities of several monatomic liquids. In these liquids only translational atomic motions can occur, so this same relationship is assumed to hold for the translational contribution to the thermal conductivity of polyatomic liquids and any difference between the measured and the calculated thermal conductivity is then attributed to the net contribution of other forms of motion. Separate correlations are then made for this additional contribution and the total thermal conductivity is obtained as the sum of the two contributions derived in this manner. Kanitkar and Thodos have applied this form of treatment to the simpler liquids of low molecular weight, but the results seem sufficiently encouraging to suggest that further analysis along similar lines could be rewarding. Bondi [37] considers longer molecules and polymer melts.

Concerning the thermal conductivities of binary mixtures of liquids, Thorne [75] has generalized the Enskog theory, Longuet-Higgins, Pople, and Valleau [315] have extended the earlier treatment [312] to apply to a binary mixture of hard-sphere molecules of the same molecular diameter, and Bearman and Kirkwood [27] have extended the theory of Zwanzig et al. [681], their equations being further reduced by Horrocks and McLaughlin [239], whilst Rice and Allnatt [431] have extended that of Rice and Kirkwood [434]. Bondi [37] also has a useful section on liquid mixtures.

B. Empirical Equations for Electrically Nonconducting Liquids

An approach to the problem of predicting the thermal conductivity of liquids that appeared more rewarding than the purely theoretical approach was the use of semiempirical relations, in which what seemed to be reasonable qualitative theoretical equations were formulated and the constants adjusted to fit the available experimental data.

A convenient starting point for the consideration of these contributions to the subject is with the account of Reid and Sherwood [429a] published in 1958, and then to mention their revised account of 1966 [429b], some of the subsequent proposals and a few others taken from other sources.

Reid and Sherwood [429a] tabulated* experimental values for thirty liquids, some over a limited range of temperature, and compared them with values derived from six of the suggested equations. Maintaining the form and the units, these equations were:

Due to Weber [645]

$$k = 359 \times 10^{-3} C_p d(d/M)^{1/3} \quad (5)$$

where M is the molecular weight, d the density, and C_p the specific heat at constant pressure.

Due to Smith [550]

$$k = 430 \times 10^{-3} C_p d(d/M)^{1/3} \quad (6)$$

Due to Palmer [397]

$$k = 94,700 \times 10^{-6} C_p d(d/M)^{1/3} (\Delta S_v)^{-1} \quad (7)$$

with ΔS_v the entropy of vaporization at the normal boiling point, cal (g mole)$^{-1}$ C^{-1}.

Due to Smith [550]

$$10^6 k = 11 + 6450(C_p - 0.45)^3 + 1250(d/M)^{1/3} + 100(\eta/d)^{1/9} \quad (8)$$

with η the viscosity.

Due to Sakiadis and Coates [457, 458] (corresponding-state method)

$$k_r = 2.26\theta_1 - 1.26\theta_2 - 2.10(\theta_1 - \theta_2)T_r \quad (9)$$

k_r being the thermal conductivity at the reduced temperature $T_r = T/T_c$, the ratio of the observed temperature (K) to the critical temperature (K) and θ_1 and θ_2 are constants, with values depending on

---

*This table is in the 1st (1958) edition of their book which is included here since it is referred to by Missenard whose theory will be considered later. The 2nd (1966) edition [429b] has a more extensive table where comparison is given with two subsequent empirical equations.

the structure of the liquid and the homologous series to which it belongs. Also,

$$k = C_p d U L \qquad (10)$$

with U the velocity of sound in the liquid and L a molecular dimension.

The works of Sakiadis and Coates [457, 458] and of Reid and Sherwood [429] should be consulted for details regarding the appropriate values of the quantities θ_1, θ_2, U, and L for the various types of organic liquids. Alternative means are given for deriving L based on liquid densities at the freezing and critical temperatures or at the normal boiling and critical temperatures. This explains the entries of Table 1 under equation 10. The derivation of U also involves the use of a parameter, the values of which depend on the basic molecular structure [429, 455].

This last equation somewhat resembles that of Bridgman [43] but whereas Bridgman assumed

$$k = 2RU\delta^{-2} \qquad (11)$$

with R the gas constant and with δ the mean distance of separation between centers of molecules and equated to $(M/d)^{1/3}$, Sakiadis and Coates, as mentioned above, again derive several different values for the L of equation (10).

The results of the comparison made by Reid and Sherwood [429a] are indicated in Table 1.

This analysis would appear to show none of these equations offer a very satisfactory means for calculating the thermal conductivity of a liquid. Because of the quite large uncertainties which could have been associated with some of the chosen experimental values, the results of this analysis are also indicated by the italicized entries of Table 1 when restricted to the twelve liquids of Reid and Sherwood's table for which recommended values are given in the present work. For all but one of these liquids the present recommended values differ from those used by Reid and Sherwood by less than ±4 percent, for ammonia the present recommended value at −20 C is greater by 16 percent, and at +30 C it is lower by 5 percent. From Table 1, the percentage differences between calculated and experimental values, even for this selected group of liquids, still show equations (5) to (10) to be far from satisfactory.

Missenard [365] in 1965 proposed the equation

$$k_0 = Y(T_b d_0)^{1/2} C_{p0} n^{-1/4} 10^{-6} \qquad (12)$$

k_0 is the thermal conductivity in cal cm^{-1} s^{-1} C^{-1} at 0 C and means for estimating appropriate values of C_{p0} the corresponding specific heat at constant pressure are given by Missenard. T_b is the boiling point of the liquid, d_0 its 0 C density, and n the total number of atoms in each molecule. The constant Y was first allocated a value of 90 but later [366] this was reduced to 84.

In order to allow for the variation of thermal conductivity k with temperature Missenard considers that the equation

$$k_t = k_0 \left[1 + \frac{t}{100} \left(\frac{T_b^{1/2}}{23.5} - 1 \right) \right] \qquad (13)$$

holds for most liquids, other than water and glycerine, and for temperatures from −50 to +50 C.

Table 1. Summary of Comparison between Calculated and Experimental Thermal Conductivities, according to Reid and Sherwood [429a] (Italic entries relate to the 12 liquids which feature in the present work, the others to all 32 liquids included by Reid and Sherwood [429a]).

Calculated by	Number of values	Number with difference*			Extreme difference %		Mean difference, %		
		minus	plus	zero	minus	plus	minus	plus	total
Eq. (5)	79, 35	53, 25	26, 10	0, 0	−49, −33	36, 36	14, 15	11, 10	−6, −8
Eq. (6)	79, 35	18, 11	60, 24	1, 0	−39, −20	63, 63	13, 9	20, 19	13, 10
Eq. (7)	79, 35	14, 10	60, 24	1, 0	−43, −24	70, 38	15, 17	23, 17	16, 7
Eq. (8)	79, 35	4, 0	74, 34	1, 1	−53, 0	130, 130	18, 0	20, 27	18, 27
Eq. (9)	20, 9	12, 7	7, 2	1, 0	−20, −20	34, 3	11, 13	13, 2	−2, −10
Eq. (10)†	54, 18	25, 7	27, 10	2, 1	−61, −24	58, 31	16, 8	18, 9	2, 2
Eq. (10)‡	54, 18	15, 3	38, 15	1, 0	−61, −15	72, 42	14, 9	21, 13	11, 9

*The difference is expressed as 100 (calculated − experimental)/experimental, [429a] but it should be noted that later in the present work such departures and departure plots are given as 100 (experimental − tabulated)/tabulated values.
†Based on d at freezing point.
‡Based on d at boiling point.

Furthermore, equation (14) is recommended for the thermal conductivity $(k_p)_t$ at pressure P kg cm^{-2} and temperature tC for most liquids other than water;

$$(k_p)_t = (k_0)\left[1 + \frac{1}{(T_b d_0)^{1/2}}\left(\frac{P}{144 - 0.3t}\right)^{2/3}\right] \quad (14)$$

Missenard considered equations (12, 13, and 14) to give values that are correct to 10 to 15 percent for most liquids, and, in general, to give results that are superior to those of equations (5) to (10). The large number of liquids to which Missenard [365] applied equation (12) included 25 of those which Reid and Sherwood had studied when making their comparison. For these liquids Missenard claims that equation (12) leads to a mean difference of -5 percent for the 7 liquids for which it yields low values, a mean of $+5$ percent for the 17 liquids for which it yields high values and gives exact agreement for the remaining liquid.

It should be added, that as noted in Section 2A, Missenard [366] reduced the numerical coefficient of equation (12) from 90 to 84 and subsequently to 82 which will have the effect of reducing the above derived values by nearly 9 percent, and will change the foregoing percentages from -5, 5, and 0 to about -14, -4, and -9. This change appears to give consistently low values but the agreement with the experimental values for the majority of these liquids is still fair and superior to that of the equations considered by Reid and Sherwood.

Mention should also be made of the derivation given by Missenard [604] of a common limiting value for certain organic liquids containing very large numbers of carbon atoms (see Section 2A). As n of equation (12) tends toward large values, the ratio $T_b^{1/2}/n^{1/4}$ tends toward about 9 or 9.05, whilst ρ_0 and C_{p0} are assumed to tend toward 0.865 and 0.535 respectively, the values corresponding to the radical CH$_2$. Hence, assuming the value of the numerical coefficient to still be 84 when n is large, the equation indicates a limiting thermal conductivity of 0.000375 to 0.000380 cal cm^{-1} s^{-1} C^{-1}, or 0.00155 to 0.00160 W cm^{-1} C^{-1}.

The various equations that have been proposed for the evaluation of liquid thermal conductivities are also treated in the book by the Russian worker Tsederberg [600] of which there is an English translation by Cess [71, 601]. For many of the additional empirical equations that are discussed in Tsederberg's book considerable departures from the experimental values are again evident.

The equation due to Borovik [38]

$$k = A(\bar{C}_v + 9R/4)U/r^2 \quad (15)$$

with $\bar{C}_v$ the molar heat capacity at constant volume, r the molecular radius, and A a constant, yielded departures of from -21 to $+23$ percent.

The equation due to Kardos [267]

$$k = dC_p UL/2r \quad (16)$$

is similar to that later proposed by Sakiadis and Coates as equation (10) but with L now defined as the average distance between centers of molecules. When L was taken as 0.95×10^{-8} for all liquids, equation (16) gave an average difference from the experimental value of 41 percent, but with L calculated for each liquid, differences of from 5.9 to 17.9 percent were obtained.

Osida [395] and Rao [424] regarded the liquid state as approximating more closely that of the solid and assumed that the energy exchange occurs when two molecules approach each other while vibrating about mean positions with a frequency ν given by the Lindemann equation

$$\nu = fT_m^{1/2}M^{-1/2}V_m^{-1/3} \quad (17)$$

where f is a function of the density of each liquid rather than a numerical constant, T_m is the melting point (K), and V_m the corresponding molecular volume. Both Osida and Rao start by deriving the same expression for the thermal conductivity in terms of the vibrational frequency and the molecular spacing and apparently use the same relationship for ν. They both derive final expressions for the thermal conductivity of the form:

$$k = FT_m^{1/2}M^{-1/2}V_m^{-2/3} \quad (18)$$

but the numerical constant quoted in the two cases has quite different values. Furthermore, from their tabulated data it is seen that the four liquids common to the two treatments have calculated values yielding ratios which are far from constant, indicating the presence of some variable that is not readily apparent. These four fluids are acetic acid, aniline, benzene, and chloroform for which Osida finds percentage differences, (experimental $-$ calculated)/calculated, of 4.4, 65.5, 10.0, and 31.8 whilst Rao gives 3.7, -0.6, -15.7, and -13.3. The inconsistency of these two sets of data which have resulted from such similar treatments is indeed puzzling.

An equation due to Predvoditelev [419]

$$k = AC_p M^{-1/3} d^{4/3} \quad (19)$$

where the constant A is a function of temperature but is approximated as equal to 3.58×10^{-3} yielded values departing by from -18 to $+18.5$ percent at 30 C.

Vargaftik [618, 620] introduced into equation (19) an additional coefficient β, which was 1 for nonassociated fluids, and which took account of the degree of association of the fluid and gave the relation

$$k = \frac{A}{\beta} C_p M^{-1/3} d^{4/3} \qquad (20)$$

The association factor β is given by $D/21$, D being Trouton's constant. He showed that when $A = 4.28 \times 10^{-3}$ the departures for seven of his measured liquids at 30 C did not exceed 3.8 percent. Furthermore, from his experimental data for the range 30 to 140 C for six liquids and 30 to 80 C for another five, Vargaftik found the product AC_p to be independent of temperature.

Hence he proposed that

$$k = \frac{B}{\beta} d^{4/3} \qquad (21)$$

where $B = A_0 C_{p0} M^{-1/3} = $ constant. He recommended that the value for $A_0 C_{p0}$ be taken at a temperature of $0.5 T_c$ and that A have the aforementioned value of 4.28×10^{-3}.

It is interesting to note that Vargaftik's equation (20) only differs from that derived last century by Weber [645] by the inclusion of the coefficient β.

Tsederberg's book also treats the thermal conductivity of liquid solutions and of aqueous electrolytic solutions.

A useful summary of both the theoretical and empirical position regarding the thermal conductivity of liquids was published in 1958 by Scheffy [526]. Scheffy's measurements had been made to fairly high temperatures by means of a concentric cylinder method and he draws attention to a possible limitation of those equations which contain C_p since, at the critical point, C_p tends to infinity whereas the thermal conductivity is often regarded as decreasing toward the gas thermal conductivity. Another difficulty of any of the usual theories is emphasized by Riedel's [439] measurements on two pairs of hydrocarbon isomers for which the branched compounds have thermal conductivity values which are about 20 percent below those of their straight-chain isomers. Further supporting measurements were made later by Dick and McCready [120].

Scheffy's compromise was to obtain a very approximate equation that would fit the linear portion of the curve and would be easy to use. By accepting the parameters used by Osida and Rao he assumed that

$$k_m = f T_m^a M^b V_m^c \qquad (22)$$

He then extrapolated the collected data of Sakiadis and Coates [454, 456] to obtain k_m for 24 liquids, and, by using the least squares method to evaluate the constants a, b, and c, obtained

$$k_m = 4790 \times 10^{-6} T_m^{-0.245} M^{-0.319} V_m^{0.048} \qquad (23)$$

or

$$k_m = 4790 \times 10^{-6} T_m^{-0.245} M^{-0.271} d_m^{-0.048} \qquad (24)$$

Since $d_m^{-0.048}$ approximates to unity, another least-squares fit was made in terms of T_m and M only, and then by using the data of Sakiadis and Coates for the linear range of temperature, Scheffy proposed the equation

$$k = 4.66 \times 10^{-3} M^{-0.300} T_m^{-0.216}$$
$$\times [1 - 0.00126(T - T_m)] \qquad (25)$$

This is the simpler of the two methods recommended by Reid and Sherwood [429b] for the approximate calculation of liquid thermal conductivities, who compare and tabulate the results of these calculations with 150 experimental values for 70 liquids and obtain a mean difference of 13 percent with extreme differences of $+111$ and -58 percent.

In the paper by Robbins and Kingrea [447] another variant of the Weber–Vargaftik equation is proposed, namely,

$$k = G C_p d^{4/3} T^{-\phi} \qquad (26)$$

where G is $(86.0 - 4.83H)(0.55T_c)^{\phi}(1000\Delta S^* M^{1/3})^{-1}$. Equation (26) was considered useful for the determination of the thermal conductivity of organic liquids, but appears somewhat complicated since the expression for the constant G involves several quantities including what was termed a "structural hindrance factor" H for which values ranging from -1 to 6 were allocated for different molecular groupings. ϕ was given values of 1 or zero according to whether the liquid had a density below or above that of water at 20 C.

Calculated and experimental thermal conductivity values are tabulated by Robbins and Kingrea in Btu h^{-1} ft^{-1} F^{-1} for 70 organic liquids. The overall average deviation is ± 3.8 percent with extremes of -10.8 and $+18.2$ percent. This same equation, as converted to give k in cal cm^{-1} s^{-1} C^{-1} by Reid and Sherwood [429b] is

$$k = \{[(88.0 - 4.94H)10^{-3}] \times [\Delta S^*]^{-1}\}(0.55 T_r^{-1})^\phi \bar{C}_p \bar{d}^{4/3} \quad (27)$$

where ΔS^* is the modified Everett entropy of vaporization $[\Delta S^* = \Delta H_{vb} T_b^{-1} + R \ln(273/T_b)]$, with ΔH_{vb} the molar heat of vaporization at the normal boiling point T_b, T_r is the reduced temperature T/T_c, and $\bar{C}_p$ and $\bar{d}$ are the molar heat capacity and density of the liquid. For the liquids that had been listed by Robbins and Kingrea, experimental and calculated comparisons are tabulated and the equation is recommended by Reid and Sherwood [429b] as giving more accurate values than that of Scheffy for the range of reduced temperatures between 0.4 and 0.8.

Yet another correlation for the thermal conductivities of organic liquids is one based on dimensional analysis due to Pachiyappan, Ibrahim, and Kuloor [396]. They examined earlier proposals and combined the parameters that most frequently appear to influence the thermal conductivity to obtain a dimensionless equation of the form

$$M = f(k, V_m, C_p, \Delta H_v, \sigma) \quad (28)$$

In equation (28) the first three terms in the parenthesis stand for previously defined quantities, ΔH_v is the latent heat of vaporization, and σ the Lennard-Jones potential constant. From these considerations they derived the equation

$$M = f\left[\frac{kV_m}{C_p \Delta H_v^{1/2} \sigma}\right] \quad (29)$$

From the experimental data for 51 organic liquids Pachaiyappan et al. obtained

$$k = \frac{5.6 \times 10^{-3}(M)^{1.26} C_p \Delta H_v^{1/2} \sigma}{V_m} \quad (30)$$

with k expressed in cal cm^{-1} s^{-1} C^{-1}. This equation gave an average departure of 11 percent, with a standard deviation of 17 percent.

From a further consideration of an equation of the Bridgman form, namely, $k = dC_p UL$, Viswanath [630] has obtained the equation

$$k = \frac{3.6 \times 10^{-4} \Delta H_b}{C V_m^{2/3} M^{1/2} T^{1/2}} \frac{(1 - T_r)}{(1 - T_{rb})} 0.38 \quad (31)$$

where ΔH_b is the latent heat of vaporization at the boiling point and T_{rb} is the reduced boiling point T_b/T_c. For the sixteen organic substances considered by Reid and Sherwood [429a] (fifty data points), Viswanath finds this equation to give an average deviation of 9.5 percent. Furthermore, on the basis of data for lead, mercury, tin, sodium, and bismuth, he considers that it should hold to ±20 percent for liquid metals

The foregoing account has presented the data seeker with a considerable choice of possible equations, for which an independent assessment, based on well-substantiated experimental data, is still required. In due course this may become possible, as more recommended values become available, not only for the thermal conductivities of liquids, but for the other associated parameters of specific heat, density, etc. It is hoped that some such assessment can be attempted in a later edition of this work. Until this further work can be undertaken, estimates made according to Missenard by equations (12) and (13) and to Vargaftik by equation (20) seem to be relatively simple and to hold sufficient promise to suggest that they be used for predicting the thermal conductivity of a liquid for which no reliable values are available. Equation (25) of Scheffy [526] might also be used.

Before concluding this section, some further thought will be given to the variation of thermal conductivity of a liquid with temperature and pressure and to how the complete curve can best be fitted to observations made over a restricted temperature range, or to an estimated value at one temperature.

As indicated in Section 3A, Horrocks and McLaughlin [241], by regarding a liquid as a quasi-lattice structure, find that

$$\frac{1}{k}\left(\frac{dk}{dT}\right)_p = -\alpha\left[\frac{1}{3} - \gamma\right] \quad (32)$$

where α is the coefficient of expansion and γ is the Grüneisen constant. This equation indicates that there should be a close connection between the expansion coefficient and the temperature coefficient of heat conduction. This result is shown to have experimental support at temperatures that are well below the critical temperature, say up to $0.6T_c$. As T_c is approached, the thermal conductivity no longer obeys a linear relation but decreases at an increasing rate. At T_c the rapidly decreasing thermal conductivity of the liquid phase should presumably merge smoothly with the increasing thermal conductivity of the vapor phase.

Riedel [444], on the basis of the principle of corresponding states, concluded that k/k_c, the ratio of the thermal conductivities at temperatures T and T_c, should be a universal function of T/T_c. Since k_c is seldom known, Riedel plotted k as a function of T/T_c and was satisfied to find that many sets of data, up to $T/T_c = 0.9$, fitted curves of the same family, at least for the unassociated liquids. Data that did not conform to this general pattern were regarded as suspect and in need of confirmation. This would appear to be a sound procedure that can be adopted, either for testing the probable reliability of a particular set of values at $T_r > 0.6$ or for extrapolation from observed values at lower temperatures to this high-temperature range.

The equation that Riedel found offered the best fit to the results that had been first examined in this way was

$$k = k_c[1 + 6.7(1 - T_r)^{2/3}] \quad (33)$$

Riedel stated that further work is needed to determine whether this equation holds right up to the critical temperature. Up to $0.9T_c$ this equation should hold for most liquids, but not for water, glycerine, propylene, glycol, and the alcohols.

In a later publication Riedel [446] carried his analysis a stage further by indicating a method for deriving the thermal conductivity of a liquid at a temperature for which T_r is 0.6.

Although the thermal conductivities of liquids are relatively insensitive to changes in pressure it is desirable to be able to predict the change in thermal conductivity with pressure. Lenoir [299], from an examination of the limited available experimental data, has prepared a set of curves based on the relationship

$$(k_2/k_1)_T = (\omega_2/\omega_1)^p = \epsilon_2/\epsilon_1 \quad (34)$$

where

$$p = -2.94T_r + 3.77 \quad (35)$$

and k_1 and k_2 are the thermal conductivities at two pressures but at the same temperature and ω_1 and ω_2 the corresponding expansion factors as given by Watson [644]. Values of the conductivity factor ϵ are plotted for reduced temperatures of from 0.4 to 1.0 and reduced pressures of from 0 to 12. Calculated values agree to within extremes of 2.3 and -5.3 percent with experimental determinations by three workers on 12 liquids.

In the vicinity of the critical point an anomalous increase in thermal conductivity has been found to occur. This seems first to have been reported by Kardos [268] for carbon dioxide; he directed attention to the close parallel between the temperature variation of thermal conductivity and of the specific heat at constant pressure. For a time this observed thermal conductivity increase was regarded as a consequence of convection that might arise from the rapid change of density with pressure near the critical point, but the careful investigation of Sengers [535] and of Michels and Sengers [356], under conditions for which any such density variations would be minimized, still yielded a sharp maximum for carbon dioxide. Needham and Ziebland [389] have since studied ammonia, and a maximum is again found to occur in the vicinity of the critical point, but, whereas for carbon dioxide Guildner [200] and Sengers [535] had found the location of the maximum to coincide with the specific heat at constant volume, C_v, for ammonia Needham and Ziebland found it to coincide with the specific heat at constant pressure, C_p. On the other hand, Gerts and Filippov [180] when using quite a different method, but one which Filippov claims [154] to be particularly useful for determinations in the critical region, found no anomaly to occur in the thermal conductivity of several liquid mixtures. This form of the hot-wire method does not appear to have been applied by Filippov to the pure liquids, for which the anomaly has been found when using other methods, yet such measurements would seem to be most desirable in providing further essential information for this region.

C. Electrically Conducting Liquids: Molten Metals

Although experimental data for molten metals have been presented in Volume 1 along with the values for the solid phase, it seems appropriate to add a few paragraphs concerning the methods that can be used to estimate the thermal conductivity of metals and alloys when in the liquid state.

Two correlating equations have been suggested. The first was by Ewing, Walker, Grand, and Miller [145] who proposed as a general correlating equation

$$k = 2.61 \times 10^{-8}\sigma T - 2 \times 10^{-17}(\sigma T)^2(C_p d)^{-1} + 97 C_p d^2 (MT)^{-1} \quad (36)$$

the thermal conductivity k being expressed in W cm^{-1} C^{-1} units and σ representing the corresponding electrical conductivity.

This equation is of interest since for electrically nonconducting liquids the terms involving σ vanish and only the last term remains. A small adjustment was later suggested by Ewing, Spann, and Miller [144] giving the relation

$$k = 97 C_p d^2 / \beta M T \qquad (37)$$

where β is an association factor that can vary from unity for liquids that are nonassociated, or only slightly associated, to about 4 for water at 0 C. This adjustment is identical to Vargaftik's as given in equations (20) and (21).

The other equation suggested by Powell [412] is restricted to molten metals and alloys. This has the simpler form

$$k = 2.32 \times 10^{-8} \sigma T + 0.012 \qquad (38)$$

and is thought to yield values for most metals that can be relied on to within about 6 percent. Further measurements on copper are considered necessary, however, since all experimentally determined values [147, 317, 348] appear to be from 13 to 37 percent below the values indicated by equation (38). Indeed, low values for the Lorenz function of molten metals are not confined to copper, and Filippov [154] considers an increasing deviation from the Wiedemann–Franz law in the form of the decrease of Lorenz numbers at high temperature to be a characteristic feature of heat transfer in molten metals. This conclusion seems to be supported by his own measurements [153] on lead to 1000 C and on tin to about 1300 C. The need for the removal of any doubt as to the behavior of the Lorenz function is emphasized by the recent predictions made by Grosse [194–196].

Grosse has proposed means for estimating the thermal conductivity of molten metals from the melting to critical points (T_m to T_c). He has pointed out that mercury is the only metal for which σ has been measured to T_c. From these measurements made by Birch [31] it was deduced by Grosse that

$$a = (\sigma_R + b)(T_R + b) \qquad (39)$$

where a and b are the constants of a hyperbola, $\sigma_R = \sigma_T / \sigma_m$, and $T_R = (T - T_m)/(T_c - T_m)$, with σ_m the electrical conductivity at the melting point and σ_T that at any temperature T.

Grosse makes the suggestion that although the electrical conductivity of other metals is not yet known for the full temperature range the available data can be fitted by an equation of this form and then by means of these derived values for σ and the Wiedemann–Franz–Lorenz relationship

$$k = 2.443 \times 10^{-8} \sigma T \qquad (40)$$

it becomes possible to predict the thermal conductivity of a molten metal over the full range of the liquid phase. He has applied this treatment to the following metals: copper, silver, gold, iron, lithium, aluminum, potassium, cesium, sodium, rubidium, and tungsten. The critical temperature of tungsten is thought to be 23,000 K, so this last estimate involves a tremendous degree of approximation, and it is clear that much further work is required, even for lithium, for which T_c is much less, to determine if equation (39) really holds (Powell and Ho [416]). More specifically there is still an urgent need to investigate further the true behavior of the Lorenz function for molten metals.

4. EXPERIMENTAL METHODS

The methods used for the determination of the thermal conductivity of liquids are often of the same basic type as those used for determinations on gases. References are given in this section to some of the more recent papers that should be consulted for details of the main methods that have been used for thermal conductivity measurements on liquids. A complete listing has not been attempted but the referenced papers will often be found to contain references to earlier papers. No less than nine methods in use for liquids at Moscow University have been described by Filippov [154]. Tyrrell's book [606] contains a useful section on the thermal conductivity of liquids with accounts of the main methods of measurement.

A. Electrically Nonconducting Liquids

a. Horizontal Plane Layer

In the absolute versions, this method includes apparatus of the guarded hot-plate type as used by Challoner and Powell [73, 414] and by Fritz and Poltz [158]. The series of papers by Poltz and his collaborators [158, 405–407] form a most important contribution, as detailed in Section 2Cc, but which should again be mentioned as these results could apply to many of the methods now under consideration.

The influence of radiation heat transfer and energy absorption in the thermal conductivity test fluid, factors that have been considered negligible by most workers, are shown [407] to make both the observed thermal conductivity and its temperature variation a function of sample thickness. Metal surfaces in contact with the test fluid are invariably polished and of low emissivity. This reduces direct radiation transfer, and with liquids such as methanol and water that strongly absorb infrared radiation, any absorption effect would only extend to very short distances and no dependence on thickness would be apparent for normal film thicknesses of 1 to 2 mm. On the other hand a weakly absorbing liquid such as toluene is shown to yield effective conductivity values at 25 C that increase by some 4 to 5 percent as the thickness of the liquid layer is increased from 0.5 to 3 mm. At 80 C the influence of the radiation component would be expected to be about twice as great. Poltz and Jugel [407] showed that the temperature coefficient can be more seriously affected. They estimate that at 25 C the temperature coefficient would decrease from -2.9×10^{-6} to -2.1×10^{-6} W cm^{-1} C^{-2} as the liquid thickness is increased from about zero to infinity (assuming no convection) whereas for liquid paraffin the corresponding change will be from -0.74×10^{-6} to $+0.12 \times 10^{-6}$ W cm^{-1} C^{-2}. They consider that some of the scatter in experimental values can be attributed to failure to allow for the contribution which can arise from thermal radiation in liquids for which the infrared absorption is small. The uncertainty about the effect of thermal radiation on the thermal conductivity of toluene urgently requires further investigation, since toluene had been recommended as a possible liquid for use as a thermal conductivity standard [675], either for testing new equipment or when comparative methods are used.

In the guarded-plate method, as well as in the methods described in Sections 4Ab and 4Ac which follow, errors can arise in the determination of the temperature difference across the test layer of fluid. This difference is seldom measured directly but is estimated from observations made within the good-conducting boundary media.

To minimize convection, the fluid thickness should be small, as indicated in Section 2Cb. Hence uncertainties in the observed thermal conductivity can easily result from the determination of this small spacing and from lack of parallelism between the plates, or of eccentricity, in the methods of Sections 4Ab and 4Ac.

Sengers [535] used a guarded parallel horizontal plate apparatus for his very careful investigation of the thermal conductivity of carbon dioxide in the region of the critical point. This apparatus was chosen as being least likely to be troubled by convection. Precautions were taken to ensure a truly horizontal fluid layer and measurements were conducted with a plate spacing of only 0.04 cm and minimum temperature difference of 0 006 C. However, even with these precautions the thermal conductivity still exhibited a sharp maximum at the critical density.

Methods involving comparative versions of the horizontal layer method have been described by Filippov [148, 154] and by Sakiadis and Coates [458].

b. Concentric Cylinder

Forms of the concentric-cylinder method for the absolute determination of liquid thermal conductivities have often been described. Descriptions can be found in papers by Ziebland and Burton [678] and by Leidenfrost [294].

The apparatus described by Leidenfrost is very ambitious. It has been designed to operate at temperatures of from -180 to 500 C, from vacuum conditions to pressures of 500 atm, and not only to achieve the maximum possible accuracy, about 0.1 percent is suggested for thermal conductivity, but also to serve for concurrent determinations for liquids, gases, and vapors of many other properties including dielectric constant, electrical conductivity, thermal expansion, compressibility, vapor pressure, and the specific heats at constant pressure and constant volume. This preliminary account of a continuing investigation serves to detail the precautions and the corrections that will be necessary before the anticipated high order of accuracy can be obtained. Much more modest apparatus can certainly yield results of sufficient accuracy for most practical purposes, but it is often from careful investigations such as proposed by Leidenfrost that fundamental limitations emerge and real advances are made. The work of Poltz et al. [158, 405–407] and the clearer insight furnished into the radiation errors that have so long been ignored is an example. Many liquid thermal conductivity measurements made previously and claiming 1 to 2 percent accuracy have yielded results differing by much more, and, if modern developments yield an explanation and enable the attainment of reproducible values consistently, this will certainly be a big advance. Although Leidenfrost has chosen the concentric cylinder method, similar careful attention to detail could doubtless also lead to corresponding improvements in the other standard methods.

Comparative radial heat flow methods for liquid thermal conductivity determinations have been described by Scheffy and Johnson [527] and by Filippov [152, 154].

c. Concentric Sphere

With the concentric-cylinder apparatus, increase

in accuracy has been obtained by fitting hemispherical ends to the cylinders. This suggests that a concentric spherical apparatus would be ideal and simpler. It would, apart from the care required to ensure that the supports to the inner sphere and the presence of a filling tube, heater leads, and thermocouple wire do not introduce any indeterminate thermally conducting paths in parallel with the test fluid. Riedel [444] made absolute determinations on a few liquids from −80 to 80 C in three apparatus of the plate, cylinder, and spherical form and obtained values in mutual agreement. Other forms of the spherical apparatus have been used since including those of Schrock and Starkman [532] and of Vanderkooi, Hildenbrand, and Stull [614].

d. Hot-Wire Methods

(i) *Steady-state.* Many of the Russian determinations have been made by means of the hot-wire steady-state method, such as the low-temperature and high-pressure measurements on liquid oxygen by Tsederberg and Timrot [602]. Other papers by Cecil and Munch [70] and by Gudzinowicz, Campbell, and Adams [198] describe measurements by this method at temperatures above normal.

The influence of radiation on determinations made by hot-wire methods appears to be a subject that still requires investigation.

A modification of the hot-wire technique was claimed [154] to be particularly useful for determinations close to the critical point. This consisted of two identical measuring tubes connected as two arms of a Wheatstone bridge. The bridge is balanced when both tubes are filled with the same liquid and are at the same temperature and can be used to measure small differences in thermal conductivity with very small temperature differences established across the liquid layer. In this way measurements of thermal conductivity have been obtained in the critical region of binary liquid mixtures [180]. No anomaly such as the sharp increase reported for pure liquids [200, 389, 535] was found in the thermal conductivity at the critical point for the system nitrobenzene–hexane, nor for the other system tested.

(ii) *Variable-state.* The paper by Horrocks and McLaughlin [240] is a useful example of the use of the transient hot-wire method for determinations to moderately high temperatures.

The various hot-wire methods for thermal conductivity determinations on gases will be detailed later since these involve several modifications and since end losses and temperature discontinuities at the wire surface require more careful consideration.

e. Thermal-Comparator Method

A simple and relatively rapid comparative method has been described by Powell and Tye [418]. This uses the thermal comparator methods developed by Powell at the National Physical Laboratory, Teddington, and used in the first instance for determination on a large variety of solids [411]. To extend these methods to include liquids (and gases) it is only necessary to contain the test fluid in a shallow dish that is covered with a thin tautly stretched sheet of teflon or similar material. With the liquid (or gas) in contact with the underside of this membrane, the warmed metal probe of the thermal comparator is brought into contact with its upper surface. By ensuring that the initial temperature differential and the load and geometry of contact are kept constant, the temperature of the probe at the point of contact and the rate of cooling of the probe are functions of the thermal conductivity of the liquid (or gas) in contact with the underside of the membrane. Arrangements are simply made to measure the temperature of the probe near to this point of contact, or to measure the rate of cooling, for a few liquids of known thermal conductivity. These measurements provide a calibration curve from which values for unknown test liquids can be interpolated so long as consistent conditions of test are maintained. With tests on liquids, the membrane serves two purposes: it prevents evaporation and the accompanying surface cooling, and it ensures that the probe and surface make a reproducible contact. It is of course essential for the membrane to be chemically inert with regard to the test liquid. The thermal comparator method has since been applied by Shroff [543] for determinations of the thermal conductivities at 25 C of several binary liquid systems.

B. Methods Applicable to Molten Metals

a. Longitudinal Heat Flow (*Comparative*)

The comparative longitudinal heat flow method, similar to that often employed for good conducting solids, has been used among others by Powell and Tye [417] and by Cooke [94], with the molten metal held within a thin-walled cavity formed within a metal bar of known thermal conductivity.

b. Direct Electrical Heating Methods

Direct electrical heating methods that rather resemble the necked-down-sample methods used for solids, first by Holm and Störmer [235, 236] and later by Flynn and O'Hagan [156], were extended to temperatures of the order of 2000 K in the course of determination on molten platinum by Hopkins [237]. This method requires the passage of an electric current through a constricted area of a rod-shaped sample, and yields the ratio of the two conductivities k/σ. Hence an independent measurement of the electrical conductivity has to be made before k can be evaluated. As used by Hopkins, the constriction was sufficiently short and thin to be retained in position by surface tension when in a molten state. Furthermore, the small surface area rendered any radiation loss of negligible importance.

Mallon and Cutler [322, 323] used a rather similar method for measurements on semiconducting molten thallium–tellurium solutions and on other electrically conducting liquids including mercury.

In conclusion, it is strongly recommended that, irrespective of the method it is decided to use for thermal conductivity determinations on a particular liquid or group of liquids, any new apparatus should first be tested out with a liquid for which values of the thermal conductivity are well established and are known to within the required accuracy. Appropriate values will be found in the data section of this volume.

Thermal Conductivity of Gases and Gas Mixtures

1. INTRODUCTION

The knowledge of thermal conductivity is essential for a variety of design calculations involving heat transfer. The transport of thermal energy is not simple when polyatomic molecules are involved, for both translational and internal (rotational and vibrational) degrees of freedom participate, and this has aroused not only deep interest among scientists but has also presented to them rather complicated and challenging problems. The last decade has seen considerable development in such directions and today our understanding of this transport property is reasonably sound and remarkably improved. The basic task still remains of providing the most reliable values for engineers and scientists. This task is not straightforward and judicious doses of theory and experiment will have to be mixed to generate this information. There will still be regions and gaps of temperature and pressure left where thermal conductivity estimation will involve new tools of semi-theoretical intuition and indirect guidance from the available experimental data.

In Section 2 of this text is presented a brief account of the different experimental techniques developed and used with varying success to provide thermal conductivity values in different environments. This is followed in Section 3 by a mention of the rigorous nonequilibrium kinetic theory for gases and mixtures of gases of increasing molecular complexity. As the calculation of thermal conductivity from theoretical expressions is tedious and requires a lot of initial information as input, efforts have been made to develop reliable but simple semi-theoretical and empirical procedures which are enumerated in Section 4 and further elaborated to some extent in Section 5. As the basic idea of this volume is to present synthesized knowledge of thermal conductivity of gases and gaseous mixtures, we list major earlier efforts motivated in this direction in Section 6 which have dealt with (i) compilation and recommendation of experimental data on gases and mixtures of gases, (ii) assessment of the various experimental methods of measurement, and (iii) evaluation of the procedures of calculation. The values tabulated in the *Numerical Data* section of this book ignore recommendations given in any of these works and are a fresh evaluation.

2. EXPERIMENTAL METHODS

During the last century, a number of methods have been evolved for the measurement of thermal conductivity of gases under widely different conditions of temperature and pressure. The methods consequently differ and a particular method may be preferable over the others for a measurement covering a given temperature and pressure range. The appropriateness of the method is further governed by such considerations as the time and expense one wants to entail in his apparatus and the accuracy he wishes to associate with his measurement. The various methods fall in one of the two categories, namely: steady-state and unsteady-state methods. The different methods of the first category have one common characteristic in that the test gas is subjected to a temperature profile which is time invariant and hence the general name of steady-state methods. In the unsteady-state methods, on the other hand, the gas experiences a continuously changing temperature field and the thermal conductivity is obtained from measurements of temperature as a function of time. At present the steady-state methods are well developed, but these, in general, are difficult, if not impossible, for measurements above about 1300 K where unsteady-state methods hold better promise. The latter have yet to see a lot of development. A brief description of the various methods, their major developments, and the extent of their applicability and appropriateness for thermal conductivity measurements is given below.

A. The Cooling Method

This method, of historical importance, was the first to receive major attention during the period 1860–1900. The temperature of a warm body (usually a thermometer), immersed in the gas maintained at a constant desired temperature, is determined as a function of time. The thermometer loses heat to the gas by conduction, convection, and radiation as well as through the stem of the thermometer. The rate of fall of the excess temperature of the thermometer also depends upon the geometry of the apparatus. The determination of all these heat losses is not straightforward and hence the experiment is conducted with at least two other gases of known thermal conductivities. This makes the method both relative and approximate in nature. Newton [581], Dalton, Leslie [400], Dulong and Petit [128], Magnus [321], Narr [387], Stefan [570], and Kundt and Warburg [288], etc., are usually credited for the use of this method in the early days. Winkelman [662–4], Graetz [188], Muller [380], Compan [92], and Eckerlein [129] further developed this method. Wassiljewa [643] employed this technique to determine the thermal conductivity of the hydrogen–oxygen system, using the available thermal conductivity data on hydrogen, oxygen, and carbon dioxide to calibrate and check her apparatus. This paper also describes the theory of the method and different corrections. Curie and Lapape [100, 101], employed this method to determine the thermal conductivity of helium relative to air. The known conductivity data of the latter was used to standardize the apparatus. Davidson and Music [109] used this principle as late as 1953 to determine the thermal conductivities of gases and gaseous mixtures. This unsteady-state method has been very little used in this century because of its relative nature, limited accuracy, and difficulty in its extension to either high temperatures or pressures.

B. The Hot-Wire Method

The introduction of this method appears to be in the pioneer work of Andrews [11], who measured the heat transfer through a gas from an electrically heated wire and determined relative thermal conductivity values of gases. In this method an electrically conducting material (usually in wire form) is located axially within a glass or metal cylinder, the space between the wire and the cylinder being filled with the gas for which the thermal conductivity is required. The whole assembly is immersed in a cryostat or furnace maintained at a constant temperature. The wire is heated electrically and acts as a heater as well as a thermometer. The heat input is determined by measurement of the current flowing through the wire and the potential drop across the wire (or a portion of it, see below). The simplest equation which relates the heat input to the wire temperature T_1, surrounding temperature T_2, corresponding distances from the wire axis r_1 and r_2, and the thermal conductivity of the gas k, is the Fourier equation

$$Q = \frac{4\pi k l (T_1 - T_2)}{\ln(r_2/r_1)} \quad (1)$$

for an amount of heat conducted from the wire of length $2l$ in unit time. Various complications arise which necessitate more complex experimental configuration and analysis of the measured values. The various variants of this method used since the work of Andrews [11] differ primarily in the procedure adopted to eliminate, reduce, or correct for that portion of the thermal energy which escapes by conduction along the ends of the wire to the remainder of the system. This heat loss creates a nonisothermal temperature profile along the wire with the result that the temperature of only the central portion of the wire is uniform and decreases toward the ends. The thermal flux through the gas is consequently radial only for a limited central portion of the wire and for which equation (1) applies. It is thus necessary to properly account for end conduction if an accurate determination of thermal conductivity is desired.

Schliermacher [528], in an effort to determine the absolute values of thermal conductivity, avoided end losses by adding two fine potential leads and thus confining the measurement to a small central section of the wire where presumably the temperature is constant along the entire length. This type of thermal conductivity cell is usually referred to as the "potential lead type." Eucken [223], following Goldschmidt [185], corrected for the end losses by taking differential measurements on two identical conductivity hot-wire cells except for their lengths. This variant is designated as the "compensating tube method" and the measurements here refer to the central portion of the longer cell whose length is equal to the difference of the lengths of two individual cells. Both these variants are further called by the common name of the thin hot-wire method as contrasted with the thick wire variety of the hot-wire method first employed by Kannuluik [260] for the determination of the thermal conductivity of wires and by Kannuluik and Martin [265] for powders. The theory and the procedure for measuring conductivity of gases were developed by Kannuluik and Martin [266]. In this innovation of the hot-wire method, the end losses are calculated and the latter is facilitated by maintaining the ends of the wire exactly at the same temperature as the wall of the conductivity cell. The energy supplied to the wire is dissipated by conduction through the gas, along the wire, and by radiation. Under these conditions and with the assumption of the radial heat flow, the differential equation for heat flow is

$$\pi r_1^2 k_1 \frac{d^2\theta}{dz^2} - 2\pi r_1 h\theta + I^2 \rho_0 (1 + \alpha\theta) = 0 \qquad (2)$$

where

$$h = \frac{k_2}{r_1 \ln(r_2/r_1)} \qquad (3)$$

Here k_1 and k_2 are respectively the wire and gas conductivities, r_1 and r_2 the radii of the wire and tube (inner) respectively, ρ_0 the resistance per unit length of the wire at the bath temperature, α the temperature coefficient of resistance of the wire at the bath temperature, θ the excess of the average wire temperature (determined from the average electrical resistance of the wire) over the bath, and I the electrical current. The effect of thermal radiation in the above equation and hence on gas conductivity is accounted for as a correction (described later). Kannuluik and Martin [265, 266] have also given the exact solution of the above heat balance equation by considering the fact that the flow through the gas is not radial and the heat flow lines are slightly curved, particularly at the ends of the tube. This is called the nonradial flow correction. Sherratt and Griffiths [541] attempted to correct for the end conduction by introducing auxiliary heaters to raise the ends of the wire to the same temperature as the central portion.

Some workers [563, 608] have preferred to make the hot-wire cell as one arm of a Wheatstone bridge. The second arm is an adjustable compensating resistor having the same resistance as the cell wire at the temperature desired. The two other arms are fixed resistors and these may be equal. Two variants of this arrangement have been used: (a) constant resistance and (b) constant current. In the former, the bridge balance after the introduction of the gas is restored by passing a suitable current, while in the latter the bridge current is not altered but instead the changed resistance of the cell wire is balanced by adjusting the compensating resistance.

A two-wire-type cell can also be used to obtain relative values of thermal conductivity of gases [62]. One of the wires is heated by a known current and the other acts as a resistance thermometer and gives the temperature rise at a point between the heating wire and the wall. By adopting this arrangement the temperature jump effect (described later) is cleverly avoided but the end effects present complication. The method is not very widely used [190] and a final assessment will have to await additional careful work.

The thin hot-wire method, though used earlier by some workers [533, 568, 636], was later developed and thoroughly discussed by Weber [646-8]. He particularly suggested the use of thin vertical conductivity cells to keep the effect of convection at a minimum. He also employed two cells (compensating tube method) and established through his measurements on air that the conductivity obtained with the two techniques are in good agreement. This showed that these two variants of the hot-wire method were reliable and would lead to conductivity values in good agreement with each other. Schneider [531], then Taylor and Johnston [582], followed by Raw and his collaborators,* developed this type of cell.

The work of Eucken [140-1] on the compensating tube method was followed by Weber [649], Gruss and Schmick [197], and Kornfeld and Hilferding [283]. Gregory and Archer [191-2] undertook a very thorough investigation of this method and measured the thermal conductivity of a number of systems. Their work was followed by Mann and Dickens [324], Dickens [121], Milverton [361], Lambert et al. [290], and Vines [629].

The thick hot-wire method was used extensively by Kannuluik and Martin [265-6], Kannuluik and Law [264], Kannuluik and Donald [263], Kannuluik and Carman [261-2], Srivastava and Saxena [565], Gambhir, Gandhi, and Saxena [167], and Srivastava and Das Gupta [561]. Based on the designs of these conductivity cells, a large number of data have been

―――――――――――
*Pereira, A. N. G. and Raw, C. J. G., "Heat Conductivities of Polyatomic Gases and Their Binary Mixtures," *Phys. Fluids* 6 (8), 1091-6, 1963; Choy, P. and Raw, C. J. G., "Thermal Conductivities of Some Polyatomic Gases at Moderately High Temperature," *J. Chem. Phys.* 45 (5), 1413-7, 1966; Gutweiler, J. and Raw, C. J. G., "Transport Properties of Polar Gas Mixtures. II. Heat Conductivities of Ammonia–Methylamine Mixtures," *ibid.* 48 (6), 2413-5, 1968.

produced on mixtures of gases and these are referred to in the data section of this volume. The hot wire used in all the cells, other than those of Kannuluik and collaborators, was relatively thinner so that end conduction was minimized and the temperature jump effect could not be detected experimentally. The fact that the end conduction is calculable with accuracy in this variant and the jump effect is negligible led quite early [289] to the recognition of its general superiority over other methods.

Hot-wire cells have been used in some other forms, too, for the measurements of either relative or absolute values of thermal conductivity. A single thin hot-wire cell with no auxiliary mountings has been used by Von Ubisch [610], Zaitseva [673], Schafer and Reiter [525], and Timrot and Vargaftik [590, 623–4], among others. In such an arrangement the end conduction is small as compared to the energy radially conducted and radiated so that an approximate calculation is sufficient for an overall accurate determination of thermal conductivity. Stops [573] used the thin hot-wire cell to obtain directly the apparent conductivities which, by calculation, lead to real conductivity values.

The principle of operation of the hot-wire cell is used in the fabrication of katharometers [111] and the latter have been used for the measurement of thermal conductivity by Ibbs and Hirst [244], Thornton [584], Neal, Greenway, and Coutts [388], Minter [362], etc. Hansen, Frost, and Murphy [217] on the other hand used a conventional chromatograph thermal conductivity detector to determine the thermal conductivity values. None of these methods is attractive for the absolute and precise determination of thermal conductivity though they are convenient and simple techniques to obtain relative and approximate values. This limitation is inherent in the instrument by the nature of its design, mode of operation, and the need of calibration [484, 563, and 565].

a. Effect of Convection

Whatever form and arrangement be used for thermal conductivity hot-wire cells, a portion of the energy is lost by other means than radial and end conduction. These are convection and radiation and have been discussed in connection with the thermal conductivity of liquids in Section 2C and in subsections b and c, respectively. Convection is avoided by choosing a narrow-bore tube for cell fabrication and keeping the annulus size at a minimum. If the temperature difference between the hot wire and the cold wall and also the density of the gas are small the Rayleigh number is much below the critical value of 1000 and convected energy is quite small [601]. All these requirements are easily met in this form of conductivity cell and indeed experiments have indicated almost insignificant contribution of convection on conductivity [203]. In the critical region or at reasonably high pressures, extra care and individual assessment is necessary.

b. Effect of Radiation

Most gases are transparent to thermal radiation and very little has been done to determine the effect of energy absorption in the gas on thermal conductivity [294]. In a gas which is transparent to radiation the calculation of radiation energy is simple though it may be approximate. In order to keep the radiation at a minimum the surfaces must be highly polished and properly treated so that their radiation characteristics are time invariant. An estimate of the radiated energy is then possible by performing the experiment in vacuum. Thus, the measurements in vacuum give h_r, i.e., the radiation loss per unit area of the wire for unit difference in the temperature of the wire and surroundings. This is subtracted from the h determined in the presence of the gas to get h_k which gives the conductivity value directly in conjunction with equation (3). This procedure of calculation requires the knowledge of the wire conductivity k_1 at the temperature of measurement. If the emissivity of the hot wire, e, is known, h_r can be computed assuming Stefan's law and is given by

$$h_r = 4e\sigma T^3 \qquad (4)$$

where σ is Stefan–Boltzmann constant and T the absolute temperature of the hot wire.

c. Effect of Temperature Jump

The temperature of a wall bounding a gas which is heated unequally may not be uniform and there may exist an apparent temperature discontinuity at the solid–gas interface. The effect is represented in terms of the accommodation coefficient a which is defined as

$$a \simeq (T_i - T_r)/(T_i - T_w) \qquad (5)$$

where T_i and T_r are the temperatures of the incident and reflected gas streams from a wall at the temperature T_w.

The theory of the temperature jump effect is well described in the books by Kennard [271] and

Present [420]. This treatment shows that thermal conductivity is dependent on pressure in the presence of a jump effect. The value of thermal conductivity k_p at a pressure P is related to its actual value k corresponding to infinite pressure by the relation

$$k \simeq k_p(1 + \delta) \qquad (6)$$

where for $r_2 > r_1$ we have

$$\delta = \frac{2-a}{a} \frac{(2\pi MRT)^{1/2}}{r_1 \ln(r_2/r_1)} \frac{k}{(\gamma + 1)C_v P} \qquad (7)$$

Here M is the molecular weight of the gas, R the gas constant per mole, T the absolute temperature of the gas at the wire, γ the ratio of two specific heats of the gas, and C_v the specific heat of the gas at constant volume. This expression for δ is approximate as its derivation is based on many assumptions. From equations (6) and (7) we get for the pressure dependence of k in the presence of the jump effect

$$\frac{1}{k_p} = \frac{1}{k} + \frac{A}{P} \qquad (8)$$

where A is a constant independent of pressure.

It is clear from the above relations that the jump effect is greater for thin wires, and it increases with decreasing gas pressure. Thus, the thin hot-wire cell has more jump correction than the thick hot-wire cell. The dependence of k_p on P, equation (8), is most often used to determine the correct value of k. Experiments are performed at several pressures and k_p determined as a function of P. A plot of $(1/k_p)$ versus $(1/P)$ on extrapolation gives $(1/k)$ at $(1/P) = 0$.

In this treatment we have assumed that the jump effect is negligible at the cold wall. If this is not valid we still get the same qualitative results except the constant A is now given by a more complicated expression.

The jump effect is minimized by performing experiments at sufficiently high pressures, while convection is minimized by selecting a low gas operating pressure. To minimize the two corrections at the same time, the requirements are thus conflicting and the best choice is determined for an apparatus by actual measurements [203].

d. Effect of Finite Thickness of the Cell Tube

In writing the energy balance equation we have assumed that the inside and outside temperatures of the walls of the conductivity cell tube are at the same temperature while actually there must be a small temperature drop θ_w. This is called "wall effect" and the correction is easily calculated by considering the energy transfer as a conduction through composite cylinders of gas and wall. The final result is that h of equation (3) is now given by

$$h = \frac{k_2}{r_1 \ln(r_2/r_1)} \left[1 - \frac{k_2}{k_w} \frac{\ln(r_3/r_2)}{\ln(r_2/r_1)} \right] \qquad (9)$$

Here r_3 is the outer radius of the cell tube and k_w is the conductivity of its wall material. If k_w is large the wall correction is small and consequently a metal with high conductivity value is preferred over glass as the material for cell construction.

e. Effect of Nonaxiality of the Wire

If the wire is not exactly axial in the tube but instead is displaced by an amount δ from the axis, then the isotherms will be a set of nonconcentric circles whose centers will gradually shift from the center of the wire to the center of the tube. The correction for this nonaxiality of the wire has been given in different form by Kannuluik and Martin [266] and by Vargaftik [616]. The Vargaftik equation is

$$\frac{2\pi k l \Delta T\, dt}{dQ} = \ln[\{\sqrt{(r_2 + r_1)^2 - e^2} + \sqrt{(r_2 - r_1)^2 - e^2}\}/\{\sqrt{(r_2 + r_1)^2 - e^2} - \sqrt{(r_2 - r_1)^2 - e^2}\}] \qquad (10)$$

where e is eccentricity. This equation is also cited in [600-1]. It may be presumed, also from Venart's* discussion, that the Vargaftik form is correct.

In the thin-wire potential lead type of conductivity cell an additional correction arises because of the heat leak from the cell wire to the potential leads. This latter can be calculated only approximately [582].

There is another uncertainty which is bound to appear in any thermal conductivity measurement that lies in the assignment of absolute temperature to which it refers. This is due to the fact that the property thermal conductivity is measured under a temperature gradient. By taking measurements for various decreasing values of the temperature difference between the wire and wall and by extrapolating to zero temperature difference, one can simplify the problem for then the property refers to the wall temperature. This approach will also correct for any unavoidable convection present in the cell. In

---

*Venart, J. E. S., "The Thermal Conductivity of Fluids. A Survey," Univ. Glasgow Mech. Eng. Dept., Rept. TR4, 1961.

actual practice these advantages are offset by the decreasing accuracy of the measurement as the temperature difference between the wire and wall is reduced. For gas mixtures this approach is still more desirable since the various components will separate under a temperature gradient because of thermal diffusion. This correction is likely to be quite small whenever the temperature difference is small and is invariably neglected. Hence, conductivity measurements will be most accurate if the smallest possible value for the temperature difference is chosen consistent with accurate potentiometric measurements.

As an overall assessment of the hot-wire method and its various variants, this method may be regarded as being capable of high precision in the hands of an experienced operator sufficiently patient to disentangle and eliminate the various corrections. Probably, the thick-wire variant of the hot-wire method is most preferable and convenient for investigation of the temperature dependence of thermal conductivity at such pressure where convection is absent. The other methods will probably be preferable for work at increasing pressures (described later).

C. The Parallel-Plate Method

The parallel-plate type of conductivity cell was first used by Christiansen [82] in 1881. This work, though forty-one years later than that of Andrews [11], is still a couple of years ahead of the famous experiments of Schliermacher [528]. The parallel-plate method in its simplest form employs a thin horizontal layer of gas enclosed between two perfectly plane surfaces maintained at two different temperatures. To avoid convection, the upper plate is kept at a higher temperature and guard rings are used to ensure the unidirectional flow of heat in the vertically downward direction. The heat lost by radiation is again determined experimentally as in the case of the hot-wire method and then the thermal conductivity is computed according to the simple relation given by Fourier, i.e.,

$$Q = kA\frac{(T_1 - T_2)}{d} \qquad (11)$$

where A and d are respectively the cross-sectional area and thickness of the gas layer through which the thermal energy Q is being conducted in unit time, T_1 and T_2 are the temperatures of the hot and cold plates respectively, and these enclose the test gas.

Almost the next thirty years record no progress on the use of this method. Todd [591] in 1909 made a very detailed investigation. He modified the method to eliminate the radiation loss by using two different thicknesses of the gas. Hercus and Laby [219–20] and Hercus and Sutherland [221] introduced many other improvements regarding the heat losses from the edges and top surface of the upper plate, refinements in the apparatus design, such as the flatness of the two plates, their exact horizontal mounting, proper thermocouple installations, correct estimation of the real temperature of the gas, and effective area of the upper plate. The latter was estimated on the basis of the analogy of this problem with electrostatics as discussed by Maxwell [347].

The work of Ubbink and deHaas [607] needs special mention. In contrast with other previous workers, they measured thermal resistances of the various thicknesses of the gas layers and estimated the thermal conductivity of the gas through a graphical process which involved the knowledge of the temperature of the plates, guard ring, and wall. One of the disadvantages which is obvious in this method lies in the requirement of varying the distance between the plates.

Michels and Botzen [354–5] and Michels, Botzen, Friedman, and Sengers [357] designed and fabricated a parallel-plate apparatus which could be operated up to 2500 atmosphere pressure and at temperatures between 0 and 75 C. Nuttall and Ginnings [393] on the other hand made measurements up to 500 C but in the pressure range 1–100 atmospheres.

In 1962, Michels and collaborators [356, 358, 359] had nearly perfected the design and operation of a parallel-plate apparatus by a very careful and systematic experimentation with theoretical bias. They thus succeeded in determining thermal conductivity as a function of density of the gas including the critical region. It will be enlightening to include a brief reference to some of the special features of their apparatus [358–9]. The relevant surfaces of the two plates made of electrolytic copper were machined flat within a micron so that gas layers as thin as 0.4 mm could be used. These were highly polished to decrease radiation and coated with a 0.1-micron-thick layer of silicon oxide to avoid any change in emissivity with time. The temperature difference between the two plates was varied between 0.006 to 0.4 C. The temperature of the guard relative to the upper plate was monitored by a thermocouple within a sensitivity of 2×10^{-4} C. They could thus keep the convective energy loss very low so that thermal conductivity was determined within an estimated accuracy of one percent.

In conclusion, this simple-looking apparatus is relatively difficult to set up and is laborious in the measurement of the heat input etc., compared to the hot-wire apparatus. However, for measurements as a function of the density of the gas the parallel-plate apparatus seems to be the most appropriate one because this geometry is found to be the most successful in avoiding losses due to convection [356].

D. The Concentric-Cylinder Method

In 1872, Stefan [569, see also 570] employed the concentric cylinder type of thermal conductivity cell to measure the thermal conductivity of gases. However, hardly any work seems to have been done since then until 1950 when Keyes at M.I.T. revived it. Ideally here, two concentric cylinders are mounted in a perfectly vertical position with a very small annular gap which, during the course of experiment, is charged with the gas. If the thickness of the gap is small compared to the length of the cylinders, the thermal flow will be fairly radial and Fourier's equation (1) will govern the conduction of heat. This cell thus looks like a logical extension of the hot-wire cell though it has a lot in common with the parallel-plate apparatus, particularly as far as experimentation is concerned.

Keyes and Sandell's [276] conductivity cell consisted of two concentric cylinders of silver with an annulus of about 0.6 mm. The inner cylinder or emitter was fitted with an axial heater wire. The bottom end of the cell was closed with a silver piece adjustable in its distance from the cylinders, and this distance was kept the same as the gap between the cylinders. A small correction due to the additional flow of heat from the corners of the disc of the emitter was applied. At the top end there was a heat guard to prevent the flow of heat from or to the emitter. The heat losses by radiation from the emitter surface, conducted away by the centering supports, etc., were evaluated by taking measurements on a highly evacuated cell. The measurements for steam extended up to a maximum temperature of about 350 C at a maximum pressure of about 150 atmospheres.

This cell design was further developed at Berkeley by Bromley and coworkers [77, 78, 451, 452] so that measurements could be extended up to 800 C at ordinary pressures.

Lenoir and Comings [300] designed a conductivity cell of four horizontal concentric steel tubes with appropriate annular gaps between them and measured the thermal conductivity of a number of gases between 1 and about 200 atmospheres. The cell needs to be calibrated with gases of known thermal conductivity. Modifications of this conductivity cell and new designs have been developed by Comings and his collaborators [91, 182, 284, 297] and thermal conductivity measurements reported up to a maximum pressure of 3000 atmospheres at 75 C for a number of pure gases and binary mixtures.

Glassman and Bonilla [183], in an effort to keep thermal radiation small and prevent its rapid increase in magnitude with increasing temperature, used a nonradiant and nonabsorbent cylinder of transparent fused silica. Among others who have used the concentric-cylinder method in thermal conductivity measurements are Ziebland and Burton [676-7], Johannin and Vodar [251-3], Waelbroeck et al. [421, 637], and Misic and Thodos [367].

The merits of this type of conductivity cell are better described in relation to the other methods described above. Thus, this cell is likely to be more free from jump effect and convection in reference to the hot-wire cell. Consequently, for operation with increasing pressure, concentric cylinder would be preferable over hot wire, though again it is likely that these features are still better achieved in the parallel-plate apparatus. However, measurements at ordinary pressures but with increasing temperatures will be satisfactorily performed with a thick hot-wire cell. The latter has also the advantage of simplicity of installation and operation against both of the other types of conductivity cells. At sufficiently high pressures, whether convection is least in the coaxial cylinder type of conductivity cell or parallel-plate type of cell still needs to be resolved by careful experimentation. Of course, this distinction is meaningful only for either very precise work or for measurements near the critical point. This form of experimentation is certainly quite difficult but nevertheless very necessary and useful.

E. The Concentric-Sphere Method

This method was used by Nusselt [392] some sixty years back for asbestos, powdered cork, charcoal, etc., and since then has been used for liquids as pointed out before in Section 4Ac. This configuration was also employed for gases by Sage and coworkers [436-7] in recent years. These workers also used such a cell for liquids. This geometry is theoretically preferable as it avoids the inevitable thermal leakage around the peripheries of the plane surfaces of the parallel-plate apparatus or at the ends of the cylindrical section of the concentric cylinder cell. In order to bring out the relevant features of this method we describe briefly the work of Richter and Sage [436–7].

In the first apparatus [436–7] the innermost sphere was surrounded by two carefully machined spherical shells. The former was fitted with an electric heater and can thus be raised to any desired temperature. The test gas was introduced in the annulus between the two spherical shells, and the latter were properly spaced by six pins. The surfaces were found to be spherical within a maximum deviation of 0.125 mm and an average deviation of 0.005 mm. The effective path length between the two spherical surfaces was 0.5 mm at 23 C. The shells were pressure compensated so that the spacing between them could be determined as a function of pressure and temperature. Thermocouples were installed to determine the temperatures of the relevant surfaces at necessary places. The energy added to the innermost sphere was determined by conventional calorimetric technique within an uncertainty of 0.05 percent. The heat loss occurred through centering pins and the stem containing the leads. The pins were constructed of low-thermal-conductivity steel and at the point of contact were less than 0.25 mm in diameter. The maximum energy loss through the pins and stem was 0.6 and 0.01 percent of the energy supplied to the heater. If one neglects the heat losses through the stem, pins, radiation, and convection, the simple Fourier's equation for such a geometry is

$$Q = \frac{4\pi r_i r_o k (T_i - T_o)}{(r_o - r_i)} \quad (12)$$

where r_o, T_o and r_i, T_i are respectively the radius and temperature of the outer and inner spherical surfaces enclosing the gas.

Convection was found to be present in the apparatus even when the Rayleigh number was less than 1000. The measurements were consequently made at several different temperature gradients and the apparent thermal conductivity was extrapolated to zero temperature gradient to obtain a value free from convection effect. The thermal radiation was determined by taking measurements on the cell evacuated to a pressure less than 1 micron. In a later design, the two spherical shells were eliminated from the high-pressure region and replaced by a new inner sphere. The two spherical surfaces were gold plated to reduce energy transport by radiation.

To give an idea of the accuracy of this type of arrangement, Richter and Sage [436–7] found the combined magnitude of corrections, conduction through pins and radiation, to be 3 and 15 percent at about 4 and 200 C, respectively. The conductivity values were estimated by those authors to be accurate within a probable error of less than 4 percent. It thus seems that the theoretical idea of achieving the best radial flow condition in this type of configuration is offset to a large extent by the failure of achieving it in practice.

F. The Concentric Sphero-Cylinder Method

Schmidt and Leidenfrost [529] realized that the disadvantages of the concentric cylinder and concentric sphere types of conductivity cells can be minimized to a large extent by adopting a spherocylinder geometry, i.e., a vertical cylinder with hemispherical ends. Two such concentric surfaces with suitable size annulus for the introduction of test gas constitute the conductivity cell. Cylindrical heater elements are placed inside the inner surface, and the outer surface acts as the cold surface and heat sink. The details of the apparatus, method of measurement, and calculation of various corrections are described by Leidenfrost [294–6]. Limited measurements [229–30] seem to suggest that this configuration may be among the most precise methods for the absolute measurement of thermal conductivity as a function of temperature and pressure.

G. The Line-Source Flow Method

This technique has been developed only since about 1958 for measurements of thermal conductivity of gases in the moderate temperature range 300–1200 K by Westenberg and his collaborators [640, 657, 658], and was further extended to 1350 K by Krauss and Ferron [285]. Unlike the above-mentioned methods it is not a static method (though it is a steady-state method). A line source of heat maintained at a constant temperature is stretched across the jet exit of the test gas under laminar flow. The former is obtained by passing a fixed direct current in a fine wire and the latter by letting the gas flow through a furnace and a series of precision screens. The gas velocity u is determined with adequate precision by a low-speed anemometer [639].

The gas stream gets heated by interaction with the line source and the temperature at any point downstream rises above its free stream value. By measuring the transverse or axial or both temperature decay profiles downstream the thermal conductivity of the gas can be calculated. However, measurements of the half width at half maximum, r_h, of the thermal wake at a distance z downstream of the line source is considered most advantageous [656]. The thermal conductivity is then calculated from the equation,

$$k = \frac{uC_p\rho(r_h - z)}{\ln(4z/r_h)} \quad (13)$$

Here C_p is the specific heat at constant pressure and ρ the gas density. Thus, absolute temperature differences and heat input rates are not required, and this constitutes the major advantage of this technique as compared to others discussed before. The accuracy of this absolute method is estimated at 2 to 3 percent [656].

H. The Hot-Wire Thermal Diffusion Column Method

Blais and Mann [32] pioneered the use of a hot-wire thermal diffusion column for the measurement of thermal conductivity of light gases in the temperature range 1200–2000 K. The apparatus, though apparently similar to a hot-wire cell, differs both in basic design and operation. Here the temperature difference between the hot wire and the cold wall is of the order of thousands of degrees and the annular gap is wide enough to promote convection. The design has been in lengths varying from a few centimeters to a few meters so that the suggested name [656] of "the long hot-wire method" appears inappropriate.

As discussed by Blais and Mann [32] the heat flow equation, though quite complicated in principle, gets simplified if the axial temperature gradient is much smaller than the radial temperature gradient so that the heat losses by convection and along the axis are small and may be even negligible as compared to the radial heat loss by conduction. It has been shown [32, 459, 462, 588] that the thermal conductivity can then be calculated at the temperature of the hot wire from the knowledge of the electrical power required to heat the wire to different temperatures in the presence of the gas and in vacuum. The difference of the two electrical powers at a fixed temperature of the wire, W_c, is a very good estimate of the energy conducted through the gas. The thermal conductivity is calculated from the expression

$$k = \frac{\ln(r_2/r_1)}{2\pi}\left(\frac{dW_c}{d\theta}\right)_{r=r} \quad (14)$$

Here W_c refers to the unit length of the wire and θ is its excess temperature over any arbitrary chosen reference datum.

The principal corrections which need to be considered are (i) nonuniform distribution of temperature along the length of the wire due to end conduction, (ii) convection, (iii) temperature jump, (iv) radiation, and (v) wall effects. In references [32, 459, 462, 524, 588, 589] detailed descriptions are available for the computation of these corrections and their consequences on the determination of thermal conductivity. Saxena and Saxena [520-2] have reported conductivity data on helium, neon, argon, and nitrogen in the temperature range 350–1500 K with an estimated accuracy of ±2 percent. This technique has also been extended to mixtures with success [201].

In brief, this technique can be used with surprising success from almost room temperature to about 3000 K for the determination of thermal conductivity with varying accuracy depending upon temperature, but errors usually can be kept below possibly 5 percent even at the highest temperature with careful design and experimentation.

I. The Shock-Tube Method

Smiley [546] introduced the use of shock waves [39] for the measurement of thermal conductivity of gases. Since then many attempts have been made [63–5, 88–9, 450, 545] and the shock tube has emerged as a potential unsteady-state method for the determination of thermal conductivity up to as high as 8580 K in argon [291-2]. Matula [346] has determined the thermal conductivity of Ar–He mixtures in the temperature range 650–5000 K by this method.

The principle of measurement is simple and involves the measurement of the heat transfer rates from shock-heated gases to the end wall of a shock tube. The temperature is monitored with time by means of a thin film resistance thermometer gauge located at the end wall of the shock tube. The test gas can be heated up to very high temperatures (1500 K is readily obtained) and in a very small time interval (10^{-9} sec). Convection and radiation effects can be neglected to a first approximation. At high temperatures, the dissociation of the gas can complicate the interpretation [215]. The rise of the end wall temperature depends upon many factors including the thermal conductivity of the gas. Relatively simple model calculations for the gas boundary layer near the end wall allow the determination of the constants used in the representation of the temperature dependence of thermal conductivity. The existence of a large temperature gradient necessitates the knowledge of nonisothermal models and this is one of the greatest disadvantages of this technique.

Both the unique adaptability of the shock tube method for high-temperature measurement of thermal conductivity and the fact that practically no other alternative method is possible has led to the development of this method considerably in recent years from the viewpoints of technique and the interpretation of experimental information [5, 12, 30, 42, 64, 65, 69, 127, 146, 157, 316, 360, 401, 408-10, 423, 544, 583].

J. The Arc Method

An electric arc [554] burning in the test gas was used by Burhorn [59] to determine the thermal conductivity of nitrogen up to 13,000 K. For such a free-burning arc the net energy loss by radiation is negligible in comparison to that by conduction. The mathematical analysis is simple under such an assumption because the electrical energy supplied to a volume element is then completely withdrawn by conduction. The highest temperature yet measured in thermal conductivity research has been by this method. This has been the incentive for the development of this method in the early 1960's [15, 134, 186, 282, 398, 422, 661, 674]. Wienecke [661] measured the total thermal conductivity for the plasma of a high-current carbon arc in the temperature range of 4000 to 10,000 K. This was achieved by observing the cooling trend of a currentless plasma. The evaluation of thermal conductivity by such a study of the heat conduction in a plasma has several coupled theoretical and inherent experimental complications, and efforts are being made to overcome some of these limitations so that reliable data may be generated [40, 41, 287, 331, 403, 652, 668].

K. The Ultrasonic Method

An ultrasonic pulse method has been used by Carnevale et al. [63-69] to determine thermal conductivity at high temperatures. The measurement of sound velocity gives the temperature of the test gas while the absorption is related to the transport properties. They [66-8] have thus determined the thermal conductivity of helium up to 1300 K and argon up to 8000 K. For heating the gases up to such high temperatures several high-temperature sources, muffle-tube ovens, ac (rf-induction-type) plasma generators, dc (arc-type) plasma jets, and shock tubes, were used. The sound absorption was measured in argon and nitrogen up to as high a temperature as 17,000 K. These authors [65-8] also describe an optical technique (Schlieren experiment) to determine thermal conductivity of gases in the temperature range 200 to 11,000 K and for pressures up to 300 atmosphere. Here one measures density and density gradients in the end wall boundary layer of the shock tube and this information coupled with the energy equation for the latter leads to the determination of thermal conductivity.

These techniques need sophisticated instrumentation and have several unresolved complications. However, in view of their great promise the group at Parametrics has been making persistent efforts during the last several years and indeed significant success has been achieved.

L. The Prandtl Number Method

Eckert and Irvine [130-3] suggested a method which directly measures the Prandtl number Pr, and if η and C_p are known for the test gas, k can be determined from the relation

$$\text{Pr} = (C_p \eta / k) \qquad (15)$$

Pr can be measured with fair precision and this method therefore gives an accuracy for k which is limited by those of Pr, η, and C_p. The uncertainty of Pr depends upon the gas and detailed error analysis is given for different systems [130-3, 245, 391]. In general, Pr for a pure gas can be measured with a systematic error of 1 percent and a random error of 0.5 percent. The method is also used for gas mixtures and typical estimates for maximum and random errors are 1.5 and 0.5 percent respectively. The method is capable of extension to high temperatures although present measurements extend only up to 450 K [130-3, 245, 391].

The principle of the method is relatively simple. The test gas is expanded through a convergent nozzle with a subsonic velocity. Along the nozzle axis and parallel to the flow direction is suspended a butt-welded thermocouple with one junction located upstream of the nozzle where the velocity is low and the other junction at the nozzle exit plane where the velocity is maximum. Under the proper conditions, these junctions record the total temperature of the gas stream and the recovery of adiabatic wall temperature respectively. By proper selection of geometry and flow conditions, the boundary layer over the downstream junction approaches a steady two-dimensional laminar-incompressible boundary layer over a flat plate. For such a case the recovery factor bears a simple relation to the Prandtl number [130–3, 245, 391]. The assignment of temperature to the measured Pr between the static and recovery temperatures poses little problem because of the usually feeble temperature dependence of Pr over the range of parameters hitherto studied.

M. The Dilatometric Method

Timrot and Totskii [587] developed a dilatometric method for determining the thermal conductivity of corrosive gases. In principle, this method is similar to that of the concentric cylinder method and necessary changes emerge from the requirement that the test gas can be in contact with only such materials which are resistant to its action.

Timrot and Totskii [587] use two concentric cylinders welded together at the upper end and attached through a thin-walled bellows at the lower end. The test gas is introduced in the annulus and the whole assembly immersed in a furnace maintained at the desired temperature of measurement. A long heater, whose temperature and heat loss per unit length is constant along its length, is inserted inside the inner cylinder and can be adjusted at any desired position. By taking measurements for appropriate and different lengths of insertion of the heater the end effects are accounted. The temperature difference of the two cylindrical surfaces is determined from the relative thermal expansions of the two surfaces. The thermal conductivity is then easily calculated from equation (1).

N. The Thermal-Comparator Method

Powell and Tye [418] proposed a thermal comparator method for the rapid comparative determination of thermal conductivity of gases which is already described in Section 4Ae. It is surprising that this method, so far as is known, has never been used for gases since the initial suggestion [418].

O. The Frequency-Response Technique

Peterson and Bonilla [402] developed an unsteady-state method for the determination of thermal conductivity of gases in which alternating current rather than the direct current is passed in the thin hot-wire cell. The principle of the method is simple. If the fine hot wire, with a large temperature coefficient of resistance, of the conductivity cell is heated with an alternating current, the magnitude of the third harmonic of the fundamental heating current is related to the thermal conductivity of the gas. The basis of the frequency-response analysis technique lies in the measurement of this third harmonic. For this reason, the method is complex mathematically and requires sophisticated electrical equipment. The requirement for the hot wire to be very fine makes it difficult to get absolute values, and Peterson and Bonilla [402] determined conductivity values relative to that at 0 C. However, this geometry helps in keeping the radiation correction small though the temperature jump correction becomes significant, 2 to 3 percent. There are many advantages of this method over the steady-state hot-wire cell method. Besides the radiation correction for the unsteady-state method being relatively small, the conductivity cell need not be held at a constant temperature, no differential temperature measurements are necessary, and cell design can be much simpler. Another requirement of the method lies in the adjustment of the pressure so that capacity-density ratio remains constant as the temperature of measurement is changed. The error in the conductivity measurements was estimated at 2 and 3 percent for He and H_2, respectively. Lee and Bonilla [293] reported measurements for alkali metal vapors and argon. For the latter the highest temperature reached was 1085 C and the error was estimated as 1.7 percent.

Another variant of a somewhat similar setup is described by Bomelburg [34–6]. The method is based on a theoretical analysis which predicts that the phase difference between fluctuating temperature and current in an ac-heated hot wire depends upon the thermal conductivity of the surrounding gas and the mean absolute temperature of the wire. In the actual experiment Bomelburg [34–6] used a thin hot-wire conductivity cell and a modulated rf signal with a lower-frequency current. The ratio of length to diameter has to be large enough so that the end losses are negligible and further the wire diameter should be sufficiently small such that the temperature is uniform over the entire wire cross section. The method was devised with the prospect of its easy extension to high temperatures. Actual experimentation revealed that the method provides conveniently only the relative values of thermal conductivity. Further, the experimental conductivity values for air above 300 C became increasingly smaller than the correct values. The discrepancy was conjectured [34–6] to have resulted due to the decrease in the density of the gas with increasing temperature. In conclusion, the method still remains to be established as a probable variant for obtaining relative values of thermal conductivity.

P. The Transient Hot-Wire Method

Briggs, Goldstein, and Ibele [45-6] have used a transient method to determine thermal conductivity of gases at 39 and 95 C. The cell is again basically a vertical hot-wire type, but the fact that the geometrical constants need not be known very precisely is an obvious advantage. Though the measurements are not quite simple, all that is needed for the determination of thermal conductivity is the knowledge of the change in resistance of the wire with time and rate of heat generation within the wire. The conductivity value thus obtained is required to be corrected for temperature jump if the pressure is not sufficiently large. Free convection is readily avoided by proper operation and design, and the radiation correction is small. Briggs et al. [45-6] have given a detailed error analysis of this method and found their measurements to have a fixed error of less than or equal to 1.0 percent. Random error and precision is given as ± 0.3 percent. However, more experimentation is needed to establish this method and its particular promise for high temperature and pressure work.

In 1968, Burge and Robinson [57] used a line source transient heat transfer technique to determine thermal conductivities of gases and gas mixtures at 297 K. The corresponding steady-state technique was developed by Westenberg and deHaas [640, 657-8] as described before. In principle, this is similar to the Briggs et al. [45-6] work except a different procedure is adopted to compute thermal conductivity of the test gas. A line source, immersed in the test gas, is supplied with a constant power for an appropriate period of time and its temperature is measured as a function of time. This temperature at a given time is compared with the theoretical estimates for various assumed values of thermal conductivity. A value which leads to a reasonable agreement between theory and experiment is taken as the correct conductivity value. The method has many advantages and is especially suited for mixtures because the measurement time is so short that hardly any appreciable thermal diffusion separation occurs. The precision of this work is always better than one percent, though the prospect of its high-temperature operation still remains unresolved.

Q. The Unsteady-State Method of Lindsay and Bromley

Lindsay and Bromley [308] described and demonstrated an unsteady-state method for the measurement of thermal diffusivity of gases and gas mixtures. The test sample is enclosed in a stainless steel tube and is heated instantaneously by passing a charge of electricity. The heat is conducted and radiated inside the gas. The change in the pressure, which is proportional to the average gas temperature, is followed with time. The temperature at any point inside the cylinder depends only on the radius and time and is independent of height and azimuthal angle. Convection is assumed to be absent (though it can be approximately corrected for) and then thermal diffusivity and hence conductivity is determined from the knowledge of the average temperature of the gas as a function of time. The measured conductivity values are reasonable in accuracy. The method on the whole tends to be complicated and is not well suited for high-temperature work.

R. The Flow Methods

Flow methods have been developed for a twin-cell katharometer, forming part of a Wheatstone bridge, which monitors the concentration profile of a gas mixture as a function of time. It is assumed that the katharometer records are dependent on conductivity and that the recorder trace is a linear representation of the thermal conductivity difference. In this way a plot of the thermal conductivity of the mixture ranging from one pure gas to the other gas is obtained as a function of time. The concentration profile as a function of time is calculated from theory. Correlation of these two leads to the continuation of conductivity as a function of composition. Evans and Kenney [142] in this way determined the thermal conductivity of the argon-helium system. Waldmann [638] suggested an unsteady-state method and derived the theoretical relations characterizing the process. The method still remains to be given an experimental trial.

3. THEORETICAL METHODS

A. Introduction

Good reviews of the older simple kinetic mean-free-path theory methods to describe the phenomenon of thermal conduction in gases can be found in the texts of Guggenheim [199], Jeans [249, 250], Kauzman [269], Kennard [271], Loeb [311], Present [420], Saha and Srivastava [453], etc. The more rigorous and mathematically complicated theories as developed by Enskog and Chapman are adequately described in the books of Chapman and Cowling [75], and Hirschfelder, Curtiss, and Bird [228]. The final results of Chapman and Enskog formulation represent thermal conductivity as an infinite series. This is not an unmanageable result because the series is convergent and its first few terms are always sufficient. The only drawback of this theory is its mathematical complexity and any effort [113–6] to find a simpler derivation has met with little success. The mean-free-path theory leads to very poor quantitative results and thus, though attractive from the viewpoint of simplicity and physical fact, has received only limited attention. Monchick [369, 371] has successfully attempted to provide an interconnection between the two theories and their equivalence. Many other notable developments have extended the basic framework of this theory of Chapman and Enskog to take into account such complications which result when this theory is either extended to low and high temperatures and pressures or to more complicated polyatomic and polar molecules. Here, we will very briefly refer to those final results which may be useful for the calculation of thermal conductivity of pure gases and multicomponent mixtures. It will also be pointed out how results for monatomic gases get modified when polyatomic and chemically reacting systems are considered. Several review articles [301–5, 656] include references to many of these works, and some of the pertinent ones will be specifically referred in the text that follows below.

B. The Mean-Free-Path Theories

By considering energy transfer in a homogeneous gas which is dilute [so that only binary collisions between the gas molecules occur and the mean-free-path expression $n_x = n_0 \exp - (x/\lambda)$ gives the variation in concentration n with x, λ being the mean free path], has a gas density that is not too low (so that the gas-to-wall collisions are negligible compared to gas-to-gas collisions), subject to only a small temperature gradient [so that $T_{x+dx} = T_x + (\partial T/\partial x)\Delta x$ accurately describes the temperature variation over Δx], monatomic (so that the specific heat is constant and the perfect gas equation of state applies), classical (so that no quantum effects occur), un-ionized, undissociated, and chemically inert, kinetic theory predicts that

$$k = \tfrac{1}{3}\rho\bar{v}\lambda C_v = \eta C_v \qquad (16)$$

$\bar{v}$ being the mean speed of the molecules. A more rigorous calculation gives

$$k = \frac{25\pi}{64}\rho\bar{v}\lambda C_v = \frac{5}{2}\eta C_v \qquad (17)$$

or more precisely

$$k = (1 + \alpha)\frac{25\pi}{64}\rho\bar{v}\lambda C_v = \frac{5}{2}(1 + \delta)\eta C_v \qquad (18)$$

where α and δ are very small numbers and are exactly equal to zero for a Maxwellian gas.

For a polyatomic gas Eucken [140–1] derived

$$k = f_E \eta C_v \qquad (19)$$

where f_E is called the Eucken factor and is

$$f_E = \tfrac{1}{4}(9\gamma - 5) = 1 + 9R/4C_v \qquad (20)$$

For a binary mixture Wassiljewa [643] derived from a simple mixing rule criterion that

$$k_{\text{mix}} = \frac{k_1}{1 + \Phi_{12}(x_2/x_1)} + \frac{k_2}{1 + \Phi_{21}(x_1/x_2)} \qquad (21)$$

where

$$\Phi_{12} = \frac{1}{\sqrt{2}}\left(\frac{\sigma_1 + \sigma_2}{2\sigma_1}\right)^2 \left(\frac{M_1 + M_2}{M_2}\right)^{1/2} \qquad (22a)$$

and

$$\Phi_{21} = \frac{1}{\sqrt{2}}\left(\frac{\sigma_1 + \sigma_2}{2\sigma_2}\right)^2 \left(\frac{M_1 + M_2}{M_1}\right)^{1/2} \qquad (22b)$$

Here x_1, σ_1 and x_2, σ_2 are the mole fractions and diameters of the components 1 and 2, respectively, in the mixture.

All these relations, when tested against experimental data on even simple systems, did not lead to very satisfactory results. The major deficiencies of these relations lie in neglecting the intermolecular forces between the molecules. In the rigorous statistical treatment of Chapman and Enskog this limitation is overcome and the detailed nature of molecular collision and its dynamics is systematically considered in deriving the expression for k.

C. The Rigorous (Chapman–Enskog) Theories

a. Pure Monatomic Gases

The rigorous statistical theory gives, for a monatomic gas under the same assumptions as listed above, the result [228]

$$[k]_1 = \frac{a(T/M)^{1/2}}{\sigma^2 \Omega^{(2,2)*}(T^*)} = \frac{15}{4}\frac{R}{M}[\eta]_1 \qquad (23)$$

Here the subscript 1 signifies that the formula is represented to the first approximation, T^* is the reduced temperature $T^* = kT/\epsilon$, ϵ and σ are the reduced potential parameters of the intermolecular potential function, in the units of degree K and $A(10^{-8}$ cm), respectively. The quantity a is a numerical factor and if k be represented in mW cm^{-1} K^{-1} its value is 0.8328, $\Omega^{(2,2)*}(T^*)$ is the reduced collision integral and is a function of T^* and thus implicitly of T also.

The higher approximations to k are represented in terms of $[k]_1$ and the nth approximation is

$$[k]_n = [k]_1 f_k^{(n)} = \frac{15}{4} \frac{[\eta]_1 R}{M} \frac{f_k^{(n)}}{f_\eta^{(n)}} \qquad (24)$$

Both $f_k^{(n)}$ and $f_\eta^{(n)}$ are slowly varying functions of T^* and are not much different from unity and are also very feebly dependent on the nature of the intermolecular potential [228], for moderate temperature ranges.

The calculation of k thus requires the knowledge of potential and the collision integrals. For many semitheoretically assumed potential functions the necessary collision integrals have been evaluated and potential parameters determined, so that the calculation of k is a straightforward task. The spherically symmetric potentials for which $\Omega^{(2,2)*}$ and $f_k^{(n)}$ have been determined are: Lennard-Jones (12-6) [228, 483]; Sutherland $(\infty, 6)$ [228]; square-well [228]; simple inverse power [277]; exponential repulsive [368]; exponential attractive [50, 383]; Morse [470, 547]; Lennard-Jones (9-6) and (28-7) [548]; repulsive and attractive screened Coulomb [333]; Lennard-Jones (12-6-5) [549]; modified Buckingham (exp-six) [325]; etc. Collision integrals are also computed for light gases under conditions where quantum effects may be important [122, 246, 372, 384]. For polar gases, Stockmayer type (12-6-3) potential is used to compute the collision integrals [247, 370]. Important additional complications [334] arise as the temperature is sufficiently raised and the gas gets dissociated and ionized. Special techniques are developed to compute the thermal conductivity of such individual systems and will be described in discussing the mixtures below. However, equation (24) still gives the frozen part k^0 of the total thermal conductivity k.

The potential parameters of the intermolecular potentials need to be determined and a number of methods have been tried with varying success. Good reviews are available [228, 328] and we omit any detailed discussion of this topic. Experimental data on molecular beam scattering have proved very valuable in calculating conductivity values at high temperatures [7-10, 258]. Potential parameters are also well estimated on the basis of semitheoretical relations in terms of boiling and critical point constants [228, 460]. A large number of papers written on the determination of the potential parameters from experimental data are referred in a series of review articles by Liley [301-5].

It may be pointed out that the prediction of the simple kinetic theory result that k should be proportional to $\sqrt{T}$ is no longer confirmed because both $\Omega^{(2,2)*}$ and f_k vary with temperature of the gas. Both theories, however, predict that the conductivity of the gas is independent of pressure as long as the assumptions listed above remain valid. This result is indeed confirmed repeatedly up to a high degree of experimental accuracy. If the gas is either in a rarefied state (collisions of the gas molecules with the wall surface are important) or in a dense state (so that many body collisions take place) this formulation breaks down and indeed conductivity will be dependent then on the density (pressure) of the gas.

b. Pure Polyatomic and Polar Gases

A formal semiclassical kinetic theory of polyatomic gases which takes into account the contribution of inelastic collisions is developed by Wang Chang and Uhlenbeck [641-2]. Taxman [580] has developed a classical formulation as an extension to the Chapman-Enskog kinetic theory. However, the theoretical expressions have not been reduced by these workers [580, 641-2] to a form where numerical computation of thermal conductivity will be a straightforward procedure.

In recent years, some efforts have been made to simplify this rigorous theory and derive such expressions which may be handled successfully to generate numerical estimates of conductivity for real gas systems. Here we simply reproduce the final results of these complicated and involved calculations. Hirschfelder [226], after making such assumptions as: the various energy states are in local thermodynamical equilibrium implying thereby that all reactions including rotational–translational and vibrational–translational transfers are fast; for molecules which are not electronically excited, the diffusion coefficients are identical for different molecular quantum states even though they are in different states of excitation, etc., derived a modified Eucken-type correction factor, viz.,

$$f_H = \frac{\rho D}{\eta} + \frac{3R}{2C_v}\left(\frac{5}{2} - \frac{\rho D}{\eta}\right) \qquad (25)$$

If $\rho D/\eta$ is put equal to unity $f_H = f_E$. By giving a theoretically sound value to the factor $(\rho D/\eta)$,

Hirschfelder [226] transformed the above relation to a more practical form which is readily adaptable to numerical computation, viz.,

$$k = k^0[0.115 + 0.354(C_p/R)]$$
$$= k^0[0.115 + 0.354\gamma/(\gamma - 1)] \quad (26)$$

Here k^0 is the "frozen" thermal conductivity of the gas and would be obtained if all internal degrees of freedom were frozen. Consequently, k^0 is not a directly measurable quantity and is obtained from theoretical expressions for k corresponding to the monatomic gases, viz., equations (23) and (24).

Eucken in deriving f_E assumed that there is no interaction between translational and internal degrees of freedom so that the transport of translational energy is unaffected by the presence of internal energy. Hirschfelder's derivation implies a rapid adjustment of the internal degrees of freedom with translational motion. The actual evidence supports the view that the internal energy transport is slow and is characteristic of the molecule [222]. This indeed is confirmed by experimental data on thermal conductivity, which lead to f values lying between f_E and f_H [24-5, 556, 566].

Saxena, Saksena, and Gambhir [517] corrected for the relaxing nature of internal and translational energy exchange [96, 222], on a simple physical model and using the base of Hirschfelder's theory. The final result of their formulation, which is further discussed by Cowling [98], is as follows:

$$f_S = f_H - \frac{3R}{2C_v}\left(\frac{5}{2} - \frac{\rho D}{\eta}\right)\left(1 - \frac{3R}{2C_v}\right)$$
$$\times \left\{1 - \exp\left(-\frac{2C_v}{3RZ}\right)\right\} \quad (27)$$

Here Z represents the number of collisions required for translational–internal energy equilibration. Equation (27) gets simplified if Z is large so that

$$f_S \simeq f_H - \frac{1}{Z}\left(\frac{5}{2} - \frac{\rho D}{\eta}\right)\frac{C_{v\,\text{int}}}{C_v} \quad (28)$$

If τ_c be used to represent the average time between two successive collisions and τ for the relaxation time characterizing the translational–internal energy exchange, we have

$$\tau = Z\tau_c \quad (29)$$

Equation (28) can therefore be written in the following alternative form:

$$f_S = f_H - \frac{\pi}{4}\left(\frac{5}{2} - \frac{\rho D}{\eta}\right)\frac{\eta}{p\tau}\frac{C_{v\,\text{int}}}{C_v} \quad (30)$$

$C_{v\,\text{int}}$ represents the contribution of the internal degrees to C_v.

Mason and Monchick [327] simplified the formal theory [641-2] and, after neglecting those terms which are small and through ingenious approximations, derived the modified Eucken correction as:

$$f_M = f_H - \frac{1}{2}\left(\frac{5}{2} - \frac{\rho D}{\eta}\right)^2 \frac{\eta}{p\tau}\frac{C_{v\,\text{int}}}{C_v} \quad (31)$$

Thus for pure polyatomic gases the thermal conductivity in all formulations is given by equation (19) and f_E is successively changed to f_H, f_M, and f_S. For η, experimental or theoretical values may be used. Mason and Monchick [327] also extended their formulation to include polar gases by considering resonant exchange of rotational energy. This hypothesis has been put to an experimental check by Baker and Brokaw [22-3] and Baker [21]. Systematic experimentation will be of considerable help in developing an adequate theory for such systems.

A number of other formal formulations have appeared for the thermal conductivity of polyatomic gases and some of these are solved with rigor for special molecular models. We quote only some of these: Curtiss and associates [102, 104, 105, 309, 376]; Dahler, Sandler, and Sather [108, 471, 473, 474]; Mueller and Curtiss [378-9]; Dahler [107]; O'Toole and Dahler [596]; Dahler and coworkers [93, 352, 472]; McCourt and Snider [349]; and Monchick, Yun, and Mason [374].

c. Multicomponent Systems of Monatomic Gases

Though the theoretical expression for the binary mixtures of rare gases has been known for some time [75], the correct expression for the multicomponent mixtures was derived somewhat later [377]. For a mixture of n components the final result is,

$$[k_{\text{mix}}]_1 = 4 \begin{vmatrix} L_{11} & \cdots & L_{1n} & x_1 \\ \vdots & & \vdots & \vdots \\ L_{n1} & \cdots & L_{nn} & x_n \\ x_1 & \cdots & x_n & 0 \end{vmatrix} \div \begin{vmatrix} L_{11} & \cdots & L_{1n} \\ \vdots & & \vdots \\ L_{n1} & \cdots & L_{nn} \end{vmatrix} \quad (32)$$

where

$$L_{ii} = -\frac{4x_i^2}{[k_i]_1} - \frac{16T}{25p} \sum_{\substack{j=1 \\ j \neq i}}^{n} \frac{x_i x_j M_i M_j [(15/2)M_i^2 + (25/4)M_j^2 - 3M_j^2 B_{ij}^* + 4M_i M_j A_{ij}^*]}{(M_i + M_j)^2 [D_{ij}]_1} \quad (33)$$

$$L_{ij(i \neq j)} = \frac{16T}{25p} \frac{x_i x_j M_i M_j (55/4 - 3B_{ij}^* - 4A_{ij}^*)}{(M_i + M_j)^2 [D_{ij}]_1} \quad (34)$$

Here A_{ij}^* and B_{ij}^* are functions of collision integrals and are available in tabulated form [228] and elsewhere for various intermolecular potentials as a function of the reduced temperature. $[D_{ij}]_1$, the first approximation to the diffusion coefficient, is given by [75]

$$[D_{ij}]_1 = \frac{3RT}{16NP} \left[\frac{2RT(M_i + M_j)}{M_i M_j} \right]^{1/2} \frac{1}{\sigma_{ij}^2 \Omega_{(1,1)^*}^{ij}} \quad (35)$$

It has been shown by Mason and Saxena [330] that the accuracy of the formula of equation (32) is improved if the first approximation diffusion coefficients are replaced by their accurate values.

Dahler and his collaborators [475, 476, 597] have developed the theory of molecular friction in dilute gases which reveals that the energy exchange between light and heavy molecules becomes more difficult as the difference in the masses of the two colliding molecules increases. This results in thermal relaxation of translational degrees of freedom. Saksena, Saxena, and Mathur [469] applied this idea to the thermal conduction in mixtures of monatomic gases and derived the modified expression for thermal conductivity. We do not reproduce their final result here which gives an appreciable correction for k_{mix} as obtained from equation (32) only when M_1 and M_2 are sufficiently different. Thus, for He-Xe system the correction varies between 1.0 to 1.4 percent depending upon the temperature and composition of the mixture. This correction must therefore be taken into account for proper theoretical interpretation of accurate experimental conductivity data as also for a precise estimation.

d. *Multicomponent Systems of Polyatomic (and Polar) Gases*

Hirschfelder [227] derived a theoretical expression for a multicomponent mixture of nonpolar polyatomic gases by making assumptions similar to those associated with his formulation for pure nonpolar polyatomic gases [226]. His final result is [227],

$$k_{mix} = k_{mix}^0 + \sum_{i=1}^{n} (k_i - k_i^0) \left[1 + \sum_{\substack{j=1 \\ j \neq i}}^{n} \frac{D_{ii} x_j}{D_{ij} x_i} \right]^{-1} \quad (36)$$

k_{mix}^0 is the conductivity of the mixture with frozen internal degrees of freedom and is given by equation (32). The rest of the terms are as defined before.

Saxena, Saksena, Gambhir, and Gandhi [518] improved the theory of Hirschfelder [227] by considering relaxation and energy exchange between translational and rotational degrees of freedom. This phenomenological approach is parallel to their work on pure gases [517] and assumes a simple physical model for the delayed energy exchange between translational and internal modes of polyatomic molecules. The final expression is [518]

$$k_{mix} = k_{mix}(H) - k_{mix}^0 \sum_{i=1}^{n} \frac{[C_i']_{mix}}{[C_{vi}]_{mix}}$$

$$\times \left\{ 1 - \exp\left(-\frac{[C_{vi}]_{mix}}{[Z_i]_{mix} C_{v\,tr}}\right) \right\}$$

$$+ C_{v\,tr} \sum_{i=1}^{n} n[D_i]_{mix} \frac{[C_i]_{mix}}{[C_{vi}]_{mix}}$$

$$\times \left\{ 1 - \exp\left(-\frac{[C_{vi}]_{mix}}{[Z_i]_{mix} C_{v\,tr}}\right) \right\} \quad (37)$$

where

$$[C_{vi}]_{mix} = C_{v\,tr} + [C_i']_{mix}, \quad (38)$$

$$[C_i']_{mix} = x_i C_i', \quad (39)$$

$$[D_i]_{mix} = \frac{D_{ii}}{x_i + \sum_{\substack{j=1 \\ j \neq i}}^{n} x_j (D_{ii}/D_{ij})} \quad (40)$$

Here $k_{mix}(H)$ is the value of k_{mix} as obtained from Hirschfelder's theory [227], equation (36), C_i is shorthand for C_{int}, n is the number of moles of the gas mixture per cc, C_i' is obtained as the difference of the experimental and the translational specific heat, and $[Z_i]_{mix}$ is the number of necessary collisions in the mixture for the ith component to reach equilibrium with the translational energy of the hypothetical pure gas. This is computed from the following expression given by Saxena et al. [518],

$$\frac{1}{[Z_i]_{mix}} = \left(\frac{x_i}{Z_i} + \sum_{\substack{j=1 \\ j \neq i}}^{n} \frac{x_j}{Z_{ij}} \frac{\tau_{ci}}{\tau_{cij}}\right) \frac{[\tau_{ci}]_{mix}}{\tau_{ci}} \quad (41)$$

In this equation $[\tau_{c\,i}]_{\text{mix}}$ is the mean time between successive collisions of the ith component in the mixture, while its value in the pure gas i is $\tau_{c\,i}$, and $\tau_{c\,ij}$ is the collision time of a molecule of ith component in otherwise pure j. Similarly Z_i collisions are required for energy balance between the internal and external degrees of the pure ith gas, Z_{ij} is the number of necessary collisions when a molecule of species i is dispersed in pure j. The various τ are defined as follows:

$$\tau_{c\,i} = \frac{\pi}{4}\frac{n_i}{P}$$

$$\tau_{c\,ij} = \left[\frac{P}{\pi}\left(\frac{M_i+M_j}{2}\right)^{1/2}\left\{\left(\frac{1}{\eta_i}\right)^{1/2}\left(\frac{1}{M_j}\right)^{1/4}\right.\right.$$
$$\left.\left.+\left(\frac{1}{\eta_j}\right)^{1/2}\left(\frac{1}{M_i}\right)^{1/4}\right\}^2\right]^{-1} \quad (42)$$

and

$$[\tau_{c\,i}]_{\text{mix}} = \left[\frac{4P}{\pi}\left\{\frac{x_i}{\eta_i} + \frac{1}{4}\sum_{\substack{j=1\\j\neq i}}^{n} x_j\left(\frac{M_i+M_j}{2}\right)^{1/2}\right.\right.$$
$$\left.\left.\left[\left(\frac{1}{\eta_i}\right)^{1/2}\left(\frac{1}{M_j}\right)^{1/4} + \left(\frac{1}{\eta_j}\right)^{1/2}\left(\frac{1}{M_i}\right)^{1/4}\right]^2\right\}\right]^{-1} \quad (43)$$

Z_{ij} has been determined experimentally for a few systems but otherwise these must be computed [394, 399, 468]. Expressions are available for homonuclear diatomic gases [394, 399] and their binary mixtures [468].

Monchick, Yun, and Mason [375] developed a formal kinetic theory for mixtures of polyatomic gas. Monchick, Pereira, and Mason [373] simplified the formal theory [375] and after making several assumptions derived an expression for the thermal conductivity of polyatomic and polar gas mixtures. Their final result is similar to equation (37) except the correction term giving the departure from Hirschfelder's expression [227] is different and still more complicated. Numerical calculations have confirmed that the correction term is usually small and very often the uncertainty in the experimental data is of the same order. Another discouraging feature of the correction term is that a number of molecular constants are needed and very seldom all of these are available with enough certainty and accuracy. These facts suggest the use of simple equation (36), due to Hirschfelder [227], for all practical needs of moderate accuracy. This indeed is confirmed by a number of investigators who have shown good agreement between the measured and calculated values of thermal conductivity on Hirschfelder's theory. Mathur and Saxena [386] have confirmed this conclusion by adopting a more formal approach. The use of more sophisticated theories [207, 373, 518] may become essential with the availability of accurate data and such constants as occur in the theoretical expressions.

e. Multicomponent Systems of Dissociating Gases

At high temperatures most of the gases of practical interest undergo dissociation and hence a number of chemical reactions are possible. The thermal conductivity of a chemically reacting gas mixture is considerably greater than a nonreacting mixture. The reason for this is that in reacting gases heat is transported not only by molecular collisions as in nonreacting gases but also as chemical enthalpy of molecules which diffuse because of concentration gradients. The latter arise due to the change in composition with temperature. For example, consider a gas which absorbs heat during dissociation. As its temperature is raised, heat is transported by the dissociated fragments which diffuse down the temperature gradient, because in the cooler region there is a lower concentration of dissociated molecules. In the low-temperature region the fragments recombine and release the heat absorbed previously.

Assuming that the reaction rates are fast so that local chemical equilibrium exists throughout the gas mixture, Hirschfelder [225] and Butler and Brokaw [61] have presented a theory for the calculation of thermal conductivity of reacting mixtures. The latter formulation [61], further simplified by Brokaw [48, 53, 54], gives a general expression for the increase in thermal conductivity due to chemical reactions, k_r, and is valid for mixtures involving any number of reactants, inert diluents, and chemical equilibria. The final expression for a system of ν independent chemical reactions involving μ chemical species (both reactants and diluents) is,

$$k_r = -\frac{1}{RT^2}\begin{vmatrix} A_{11} & \cdots & A_{1\nu} & \Delta H_1 \\ \vdots & & \vdots & \vdots \\ A_{\nu 1} & \cdots & A_{\nu\nu} & \Delta H_\nu \\ \Delta H_1 & \cdots & \Delta H_\nu & 0 \end{vmatrix}$$
$$\div \begin{vmatrix} A_{11} & \cdots & A_{1\nu} \\ \vdots & & \vdots \\ A_{\nu 1} & \cdots & A_{\nu\nu} \end{vmatrix} \quad (44)$$

where

$$A_{ij} = A_{ji} = \sum_{k=1}^{\mu-1} \sum_{l=k+1}^{\mu} (RT/D_{kl}P)x_k x_l$$
$$\times [(n_{ik}/x_k) - (n_{il}/x_l)]$$
$$\times [(n_{jk}/x_k) - (n_{jl}/x_l)] \quad i,j = 1, 2, \ldots, \nu \quad (45)$$

$$\Delta H_i = \sum_{k=1}^{\mu} n_{ik} H_k \quad i = 1, 2, \ldots, j, \ldots, \nu \quad (46)$$

Here D_{kl} is the binary diffusion coefficient between components k and l, R the gas constant in pressure volume units, T the absolute temperature, P the pressure, x_k the mole fraction of component k, ΔH the heat of reaction, H_k the enthalpies of the species referred to a common base, and n_{ik} is the stoichiometric coefficient for species k in reaction i. Any gas inert to reaction i or j must be included, but it has a zero stoichiometric coefficient for the reaction in which it does not participate.

For single dissociation reactions of the type $A \rightleftarrows nB$, the increase in thermal conductivity due to chemical reactions has a very simple form as obtained from equation (44), viz.,

$$k_r = \frac{D_{AB}P}{RT} \frac{\Delta H^2}{RT^2} \frac{x_A x_B}{(nx_A + x_B)^2} \quad (47)$$

Usually the chemical reactions are not rapid enough so that the chemical composition will come into equilibrium with the local temperature. This invalidates the applicability of the above expressions to an actual system which may not accord to the requirement of a local chemical equilibrium. If the reaction rates are very slow, the various chemical species diffuse evenly throughout the cell and the chemical composition is sensibly the same everywhere. In between these two extremes there is a range of moderately fast reactions for which the composition as a function of position in the conductivity cell depends sensitively on the reaction rate and other physical parameters. Secrest and Hirschfelder [534] have developed the theory for moderately fast and slow reactions and for an arrangement in which hot and cold plates are parallel, since extended to a cell having cylindrical geometry [567].

Brokaw [49] has derived a general expression for the apparent thermal conductivity of reacting mixtures in which a single reaction proceeds at a finite rate. Because it is essential to assume the constancy of the chemical reaction rate throughout the conductivity cell the theory is valid only for small temperature difference between the hot and cold walls. To consider large temperature differences numerical techniques [534] will have to be used. However, for these intermediate reaction rates the effective conductivity depends on the geometry, scale of the system, and catalytic activity of the surfaces, in contrast to systems where reaction rates are either very high or very low. We do not reproduce the actual expressions here which may be found in references [49, 534, 567]. Only a limited amount of experimental work has been done on reacting systems [26, 51, 55, 72, 557–560, 562] and these do confirm the basic framework of theory mentioned above to be adequate.

The theoretical calculation of the thermal conductivity of dissociating gases is possible in principle from the classical Chapman–Enskog kinetic theory expressions, reproduced above, provided the collision integrals are calculated giving proper weight and averaging over all the possible potential energy curves [334]. The calculation actually is quite involved because the possible number of interactions is quite large even for simple molecules and in many cases these are known only with poor accuracy. The thermal conductivity calculations are available for such relatively simple dissociating gases as hydrogen [615], nitrogen [672], oxygen [672], air [17, 671], etc.

f. Multicomponent Systems of Partially or Fully Ionized Gases

As the temperature of the gas is increased, it will be either partially or fully ionized. The thermal conductivity of such a multicomponent system (plasma) is difficult to measure accurately and so a theoretical calculation with enough sophistication is essential. The increase in thermal conductivity due to ionization, k_{ion}, is difficult to calculate because of many new complications which need not be considered for nonreacting gases. The three interactions involving ions and electrons (ion–ion, electron–electron, and ion–electron), are extremely strong and long range because these follow a Coulomb potential, and hence difficulties arise in evaluating the cross sections. The presence of electrons and their small mass necessitates more elaborate consideration of the Chapman–Enskog formulation of the transport theory.

Most of the methods used for computing the thermal conductivity of partially or fully ionized gases have been reviewed by Ahtye [1–4]. He [1–2] has critically examined the shortcomings and emphasized the importance of higher approximations in Chapman–Enskog theory while dealing with ionized gases. Ahtye [3] has also derived an expression for the total thermal conductivity of partially or fully ionized gases. A more formal and detailed formulation with numerical results of practical value and comparison with the experimental data was also developed by Devoto [117-9]. Attempts to compute thermal conductivity as a function of temperature and for various pressures and their comparison with experimental data have received reasonable success [13, 48, 53, 54, 286, 320, 669].

For partially ionized gas, the contribution to thermal conductivity, k_{ion}, is easily obtained to a fair degree of approximation by a parallel to the theory of Butler and Brokaw [48, 53, 54, 61] as shown by Meador and Staton.* The equation (47) now becomes,

$$k_{ion} = \frac{D_{ia} P}{RT} \frac{\Delta H_I^2}{RT^2} \frac{x_a x_i}{(x_a + x_i)} \quad (48)$$

Here D_{ia} is the diffusion coefficient for ion-atom, H_I the heat of ionization, and x_a and x_i the mole fractions of atoms and ions in the mixture, respectively. The other expressions for k_{ion} are too complicated and will not be reproduced here but will be found in the references of this section.

g. Conclusions

It thus follows that the calculation of thermal conductivity is not simple and becomes increasingly difficult if the gas is composed of molecules which are nonspherical, polyatomic, and polar. Further, complications creep in as the temperature of the gas is increased or decreased and other effects such as dissociation, ionization, or quantum effects set in. The total thermal conductivity k being the sum total of such contributions as arising from molecular collisions (k^0), internal degrees of freedom (k_{int}), dissociation (k_r), ionization (k_{ion}), etc., i.e.,

$$k = k^0 + k_{int} + k_r + k_{ion} \quad (49)$$

The expressions available for calculation of k^0, k_{int}, k_r, and k_{ion} are discussed above in sections (a and c), (b and d), (e), and (f), respectively. Additional difficulties arise if the pressure is too low so that the molecular collisions are scarce or if the pressure is too large so that collisions of order higher than binary are significant. A brief discussion of this density (or pressure) effect is given in the next section.

h. Effect of Density (Pressure)

If the pressure of the test gas in the conductivity

*Meador, W. E. and Staton, L. D., *Phys. Fluids* **8** (9), 1694-1703, 1965.

cell is continuously reduced, the mean free path of the gas molecules will increase because of the reduction in density and will become comparable to the dimensions of the cell. The mechanism of energy transfer now in this rarefied gas is very different because the gas molecules collide more with the walls of the cell than among themselves. Under these conditions the gas molecules leave the interface after surrendering to the walls, on the average, only a fraction of their incident relative momentum, i.e., only an incomplete energy exchange takes place. This is expressed as a coefficient of accommodation, briefly referred before while dealing with the hot-wire method of measuring thermal conductivity and jump effect, and is discussed in almost all texts on kinetic theory of gases [see, e.g., 271, 420]. More recent developments in this area may be found in and through the articles of Wachman and his associates [632-5]. We defer any detailed discussion of this point here with the comment that our knowledge and understanding in this rarefied gas region is far from being adequate.

On the other hand, if the gas pressure is continuously increased the mean free path decreases and both the molecular dynamics and mode of energy transfer becomes much more complicated. It is necessary to consider collisional transfer of energy as well as collisions which involve three and more molecules. Good accounts of work done dealing with the thermal conductivity of such dense gases are available [75, 228] and a brief review of more recent developments is given by Curtiss [103]. Following the pioneer work of Enskog [135] many studies have been made though reliable expressions are still not available and sometimes even a crude estimate of thermal conductivity for large values of gas densities is not possible. We will briefly refer to some of these developments.

Assuming that the spatial pair distribution function depends only on the temperature and density and not on the rate of strain, and the velocity distribution function of a single particle is locally Maxwellian, Longuet-Higgins and Pople [312] developed a theory for a hard-sphere fluid at high densities. A simple theory of hard-sphere fluids is given by Longuet-Higgins and Valleau [313] and for perfectly rough elastic spheres by Valleau [612]. The theory for molecules interacting according to square-well potential is developed by Longuet-Higgins and Valleau [314] and extended to mixtures by Valleau [613] and McLaughlin and Davis [353]. Efforts have also been made to improve the theory due to Enskog [135, 136] so as to apply to polyatomic fluids with internal degrees of freedom [110, 228, 350, 536]. Sather and Dahler [477] considered polyatomic fluids in which molecules interact with impulsive forces. The calculations were based on a rough spherical model with rigid convex molecular "cores" surrounded by potential staircases. Stogryn and Hirschfelder [572] attempted to calculate the initial pressure dependence of thermal conductivity, i.e., the coefficient a in the virial expansion in density (or pressure) of k,

$$k = k_0 + ad + bd^2 + cd^3 + \ldots \quad (50)$$

Here k_0 is the hypothetical limiting value of k as the density d approaches zero. They [572] neglected clusters larger than dimers and the contribution of collisional transfer was obtained by a semiempirical modification of Enskog's theory [135, 136]. They arrived at an important conclusion that at low temperatures molecular association is more important than collisional transfer, while at high temperatures the latter predominates. Whalley [659] has also discussed the thermal conductivity of associating gases. Kim, Flynn, and Ross [280] determined, like Stogryn and Hirschfelder [572], the coefficient a of equation (50). They [280] neglected the contribution of collisional transfer in monomer–monomer collisions but used a broader definition for dimer by including bound, metastable, and orbiting pairs.

The development of kinetic theory of dense gases has received a very systematic and detailed attention in the hands of Curtiss and his associates. Snider and Curtiss [551] developed a theory for moderately dense pure gases made up of spherical molecules and included the effect of collisional transfer but neglected the one arising from multiple collisions. In the limit, when the molecules are rigid spheres, results are obtained [310, 552] which are identical to the one derived by Enskog [135, 136]. Snider and McCourt [553] simplified the expressions [551] and approximately evaluated them for the inverse power potential. Curtiss, McElroy and Hoffman [106] performed the numerical evaluation for molecules obeying Lennard-Jones (12-6) potential. In an effort to derive an expression for a dense gas, Hoffman and Curtiss [232-4] considered the contribution of three-body collisions and derived a Boltzmann equation which is a soft potential generalization of the Enskog dense-gas equation for rigid spheres [232], and obtained the expression for thermal conductivity [233] which, when calculated explicitly for a rigid-sphere gas, is identical with the classic expression of Enskog. These authors [234] have also computed numerically the thermal conductivity second virial a for the case of a Lennard-Jones (12-6) gas but neglecting the effect of bound states.

Another approach used in the general formulation of the transport coefficients for moderately dense gases consists in expressing them in terms of time correlation functions [208, 680]. The expression for thermal conductivity is derived to lowest and first order in density by Ernst, Dorfman, and Cohen [139] and to general order in density by Ernst [137]. Ernst [138] has also specialized his expression [137] for the case of rigid-sphere gas and found it to agree with that of Enskog [135, 136].

The traditional distribution function method is also used by Choh [81] and Garcia-Colin, Green, and Chaos [179] to obtain the theoretical expression for the first density correction. These results are in general agreement with those obtained from the time correlation function, Ernst [137]. These two approaches are compared with each other by Garcia-Colin and Flores [178] and Chaos and Garcia-Colin [74].

Sengers [538-9], following the general distribution function approach, presented an analysis of the triple-collision integrals which occur in the expression for the first density correction of a gas consisting of rigid spheres. He found results which differ from those of Enskog, and he attributed this to the neglect of certain type of recollisions and cyclic collisions in the treatment of Enskog. Dorfman and Cohen [125-6], Weinstock [653], and Goldman and Frieman [184] found that the coefficient b of d^2 in equation (50) is divergent. This causes terms containing logarithm of the density to occur in the virial expansion for the transport coefficient [124-6, 184, 270]. Equation (50) then modifies to

$$k = k_0 + ad + bd^2 + cd^3$$
$$+ \ldots + \phi d^2 \ln d + \ldots \quad (51)$$

Sengers [537-40] has explicitly evaluated the logarithmic term in the special case of a two-dimensional gas of rigid disks. Haines, Dorfman, and Ernst [209] have also considered a similar case and by adopting the alternative approach of time correlation function method derived identical results.

In considering equations (50) and (51) from the viewpoint of a correlator of experimental data, an additional difficulty arises as to the choice to be made for k_0. Many studies have used k_1, the value of thermal conductivity at atmospheric pressure, on the assumption that $k_1 - k_0$ is negligibly small. The latter implies that the pressure coefficient of conductivity dk/dP is negligible below a few atmospheres pressure for most "normal" fluids. However, small experimental uncertainty in thermal conductivity for pressures below a few atmospheres could result in a finite dk/dP resulting in an "apparent" zero pressure thermal conductivity k_0' differing from the value k_1. In addition, a realistic analysis should result in the thermal conductivity being zero at zero pressure. In the tables which follow, no account has been paid to possible pressure effects except for fluids which do not fall in the "normal" category, due to association etc. For general estimation needs, the assumption of invariance in conductivity below a few atmospheres should prove satisfactory. However, for evaluation of conductivities to better than 0.5 percent, pressure effects are probably significant. Completely neglected in the above is the possibility of convection producing a small effect at these low pressures, which may lead to spurious interpretation of experimental values.

4. ESTIMATION METHODS

A. Introduction

There are no reliable estimation procedures available in general for thermal conductivity of either pure gases or mixtures of gases. For monatomic and simple gases several equations derived with guidance from theory have been used with varying success to correlate the existing experimental data. These in turn have also been used as a basis for extrapolation, though the accuracy and reliability of the numbers so generated remain highly questionable. In the absence of data from direct experiments, estimated values of even moderate accuracy receive valuable use in the hands of engineers. It would therefore be highly desirable if reliable and simple procedures could be developed for the estimation of thermal conductivity of pure gases and multicomponent mixtures. In this section, we will outline such procedures and also quote its statistical accuracy on the basis of its ability to reproduce a limited body of experimental data. This may provide some idea of the relative footing of each method and the selection of a particular method to match a particular need when a certain amount of initial information is available.

B. Pure Gases

Many estimation procedures for the thermal conductivity of monatomic (and even other simple molecule) gases have been developed in parallel with the methods used for the viscosity of gases. All such methods are based on the expressions of thermal conductivity derived from theory and the latter are mentioned in the previous section, *Theoretical Methods*. The various forms used so far may be expressed as:

$$k = aT^s \tag{52}$$

$$k = \frac{aT^n}{1 + (S/T)} \tag{53}$$

$$k = \frac{a(1 + bT)T^{1/2}}{1 + (S/T)} \tag{54}$$

$$k = \frac{aT^{1/2}}{1 + \{(S/T)10^{(c/T)}\}} \tag{55}$$

$$k = k_1 + aT + bT^2 + \ldots \tag{56}$$

where the a, b, c, n, s, and S are arbitrary constants and are determined in each case by forcing a particular equation to best fit the experimental thermal conductivity data as a function of temperature. The more flexible the relation is, that is, the more the number of adjustable constants it has, the better it can represent the conductivity data and over a wide temperature range. These equations, in general, are derived on the basis of assuming a simple potential field between two neighboring molecules. Thus equations (52) and (53) are obtained if the intermolecular potential is approximated by inverse power and by placing $n = \frac{1}{2}$ (the Sutherland model) respectively. Equation (55) has been most widely used by Keyes [272, 275]. However, it has been found [306] that while equation (55) may give a fair representation over a few hundred degrees interval of temperature, extrapolated values may be quite in error.

More ambitious efforts [272, 275, 279] attempt to curve fit the collision integral $\Omega^{(2,2)*}(T)^*$ for the Lennard-Jones (12-6) potential by the equations which involve T^* and hence T as the only variable. Thus, if the potential parameters σ and ϵ/k are known, conductivity may be calculated as a function of T, but here again any extrapolation will be uncertain depending upon how realistic is the choice of the intermolecular potential form and how accurately the parameters of the potential are determined. Detailed accounts of the agreement between theory and experiment have been given by many workers [see, e.g., 4, 18, 79, 122, 160-2, 210-2, 228, 246, 256, 372, 384, 464, 466, 467, 480, 494, 495, 504-7, 509, 515]. Some of these have determined the parameters for different potentials from different properties and these thus may be used for prediction purposes. The degree of reliance to be associated with the numbers so generated will have to be judiciously evaluated for the individual cases and it may not be even possible in every case to pinpoint such an uncertainty [213, 281]. The only incentive then behind such an approach is that, in many cases, unless large extrapolation is involved, estimates are possible to meet engineering requirements. All the above-mentioned discussion pertains to dilute gases, i.e., gases at pressures around one atmosphere.

Estimation of thermal conductivity for dense gases is difficult because even the kinetic theory for such systems is not well developed and checked against experimental data. Keyes [272, 273, 275] has discussed a few empirical relations including the formula derived by Enskog [135, 136]. In view of the present inadequacy of theory concerning the pressure effect, attempts have been made to represent thermal conductivity by a generalized chart. The most extensive efforts on this approach have been made by Thodos and his associates [571]. Charts give the reduced thermal conductivity, $k_r = k/k_c$, as a function of reduced temperature, $T_r = T/T_c$, and pressure, $P_r = P/P_c$. This approach suffers from some disadvantages. Firstly, often the values of the critical parameters k_c, P_c, and T_c are unknown and must first be estimated. Especially concerning k_c the experimental uncertainties make the subsequent analysis of doubtful value. Secondly, substances of similar structure should be considered together. Different reduced charts have been found necessary for monatomic, diatomic, etc., fluids. Presumably, the only satisfactory representation by this technique will require the addition of further reducing parameters, such as the critical compressibility coefficient Z_c [318], the acentric factor ω [404], etc.

C. Multicomponent Gas Mixtures

A number of methods have been developed to predict the thermal conductivity of binary and multicomponent mixtures of gases at ordinary pressure, i.e., around one atmosphere. Different procedures require varying amounts of input information but all need the thermal conductivity of the pure gases involved in the mixture to be known at the temperature of interest. Some methods also require knowledge of the thermal conductivity of the binary mixture either at one composition or at two compositions. In the case of multicomponent systems such values must be known for all of the binary combinations involved. As for pure gases, the estimation of the thermal conductivity of mixtures becomes more difficult as the molecular structure of the molecules involved becomes more complicated and consequently the uncertainty in the result increases. In general, the prediction of thermal conductivity for monatomic and simple polyatomic gas mixtures is possible with reasonable accuracy.

The rigorous kinetic theory expressions for thermal conductivity have been mentioned before and for their use the one basic and common information needed is the knowledge of intermolecular potential and evaluated necessary collision integrals for the chosen potential. As the knowledge of the correct potential form and its parameters is not always uniquely possible for the chosen system and, further, in many cases the theory applies only to a limited extent, the theoretical computation leads to only a rough estimate for the thermal conductivity. Information regarding the appropriateness of theory to describe actual systems and the different types of intermolecular models and their parameters for different systems is scattered in a large number of sources [see, e.g., 20, 228, 256, 340-3, 385-6, 467, 478, 479, 486, 488, 498, 505, 506, 508, 510, 523, 654-5]. We briefly describe the various procedures used for estimation purposes and will refer where calculations have been made and with what success. These estimates almost always have considered data at face value and no thorough critical evaluations are available in general for the data on mixtures. Later we will refer to the available sources and the data section of this book does this partially for a number of binary systems. It is expected that at a later date the completed critically evaluated data will be available from TPRC when it will be possible to make a more thorough assessment of the methods described below.

a. Methods Based on the Linear and Reciprocal Mixing Rules

One of the simplest and most straightforward procedures for estimating the thermal conductivity of mixtures is the linear molar mixing rule. The thermal conductivity of a mixture of n components will then be [271]

$$k_{\text{mix}} = \sum_{i=1}^{n} x_i k_i \qquad (57)$$

Here k_i and x_i are the coefficient of thermal conductivity and mole fraction of the component i in the mixture. Invariably, conductivity values obtained from equation (57) are greater than the experimental values [47, 58, 171]. On the other hand, the computed values from the reciprocal mixing rule, i.e.,

$$k_{\text{mix}}^{-1} = \sum_{i=1}^{n} (x_i/k_i) \qquad (58)$$

are always smaller than the experimental values [171]. Burgoyne and Weinberg [58], consequently suggested that a better procedure of estimation would be to combine equations (57) and (58) so that

$$k_{\text{mix}} = 0.5 \left[\sum_{i=1}^{n} x_i k_i + \left\{ \sum_{i=1}^{n} (x_i/k_i) \right\}^{-1} \right] \qquad (59)$$

Brokaw [47] suggested a more general combination of equations (57) and (58), viz.,

$$k_{\text{mix}} = a \sum_{i=1}^{n} x_i k_i + (1-a) \left\{ \sum_{i=1}^{n} (x_i/k_i) \right\}^{-1} \qquad (60)$$

Brokaw [47] determined a of equation (60) from experimental data for 11 gas systems comprising 69 mixtures and found it to vary somewhat systematically with the proportion of one of the components in the mixture. He [47] tabulated these values of a as a function of composition of the lighter component. Extension of equation (60) to multicomponent mixtures is difficult in practice because of the lack of knowledge of the nature of variation of a with composition and so we prefer an approximate value for a of 0.5 in equation (59).

Gandhi and Saxena [171] tested equations (59) and (60) for rare gas mixtures. They found that equation (59) could predict the experimental values for all ten binary systems of five stable rare gases within an average absolute deviation of 7.1 percent at 29 C and 6.2 percent at 520 C. Equation (59) was also tested against the experimental data for the ternary system He–Kr–Xe system [171]. The agreement was relatively poor, average absolute deviation being about 16 percent [171]. Somewhat good success for this method was reported by Mathur, Tondon, and Saxena who [344] found that for 44 binary, 16 ternary, and six quaternary rare gas mixtures, equation (59) could reproduce experimental results within an average absolute deviation of 3.3, 2.6, and 7.3 percent respectively. This method has been tested for binary, ternary, and quaternary mixtures of rare gases and diatomic gases by Mathur, Tondon, and Saxena [345], and Saxena and Gupta [503], for polyatomic gases by Gupta, Mathur, and Saxena [206], and for Ar–He system from 1000 to 5000 K by Saxena [489]. In brief, the simplicity of the method is attractive for determining thermal conductivity of multicomponent systems where one can tolerate moderate accuracy.

b. Methods Based on Empirical Functions for Binary Systems

A quadratic relation was thought to be a more practical choice for representing the data of binary mixtures [271], viz.,

$$k_{\text{mix}} = k_1 x_1^2 + k_2 x_2^2 + k_{12} x_1 x_2 \qquad (61)$$

Here the constant k_{12} has to be determined empirically. Gandhi and Saxena [171] checked the adequacy of this method for results on rare gas mixtures and determined k_{12} by using one of the experimental values close to the middle composition. The average absolute deviations for ten systems were found to be 7.8 percent at 29 C and 6.3 percent at 520 C. The maximum deviations were relatively large, 24.2 percent (29 C) and 18.8 percent (520 C). Mathur, Tondon, and Saxena [345] and Gupta, Mathur, and Saxena [206] examined this method in conjunction with data on polyatomic gases. The method, in general, is not attractive because of determination of the unknown constant k_{12} requires knowledge of the thermal conductivity for one mixture composition, and also large uncertainties are associated with the calculated values [171, 206, 345].

Enskog [135] suggested the following equation for binary systems which indeed can be generalized for multicomponent mixtures [273]:

$$k_{\text{mix}} = \cfrac{\cfrac{k_1}{1 + g_{12}(x_2/x_1)} + \cfrac{k_2}{1 + g_{21}(x_1/x_2)} + \cfrac{2l(k_1 k_2)^{1/2}}{[1 + g_{12}(x_2/x_1)][1 + g_{21}(x_1/x_2)]}}{1 - \cfrac{l^2}{[1 + g_{12}(x_2/x_1)][1 + g_{21}(x_1/x_2)]}} \qquad (62)$$

Here

$$\frac{1}{g_{12}} = \left(\frac{\sigma_1}{\sigma_{12}}\right)^2 \left(\frac{M_1 + M_2}{2M_2}\right)^{1/2} \frac{8(M_1 + M_2)^2}{30M_1^2 + 16M_1 M_2 + 13M_2^2}$$

$$\frac{1}{g_{21}} = \left(\frac{\sigma_2}{\sigma_{12}}\right)^2 \left(\frac{M_1 + M_2}{2M_1}\right)^{1/2}$$

$$\times \frac{8(M_1 + M_2)^2}{30M_2^2 + 16M_1M_2 + 13M_1^2}$$

$$l = 27ab(g_{12}g_{21})$$

$$a = M_1(30M_2^2 + 16M_1M_2 + 13M_1^2)^{-1/2}$$

$$b = M_2(30M_1^2 + 16M_1M_2 + 13M_2^2)^{-1/2}$$

σ_1 and σ_2 are the molecular diameters and σ_{12} may be assumed as the arithmetical mean of σ_1 and σ_2. This equation has not been checked widely. Keyes [273] examined and found tolerable reproduction of the experimental values. However, the method is quite tedious, molecular diameters are needed, and even then the accuracy of the final results is somewhat less than those by other methods to be described later.

Davidson and Music [109] correlated the thermal conductivity data at 0 C for He–CO_2, He–Ne, Ne–CO_2, and N_2–CO_2 systems by the following relation

$$k_{mix} = e^{(a+bx)} \quad (63)$$

within average absolute deviations of 2, 1.2, 0.7, and 1.4 percent respectively. x stands for the mole fraction of the lighter component in the mixture. Because no explicit relations are given for the calculation of a and b of equation (63), its use for estimation of k_{mix} is more limited.

Hirschfelder, Curtiss, and Bird [228], from analysis of the rigorous theoretical expression for thermal conductivity, proposed the following relation for mixtures of heavy isotopes:

$$(k_{mix})^{-1/2} = x_1(k_1)^{-1/2} + x_2(k_2)^{-1/2} \quad (64)$$

This relation is of some practical importance but it has never been tested against measurements.

Minter and Schuldiner [362, 363] used the following equation to correlate data on binary systems:

$$k_{mix} = k_1(1 + a \ln x_1)x_1 + k_2(1 + b \ln x_2)x_2 \quad (64a)$$

Here a and b are empirical constants to be determined from experimental data.

c. Methods Due to Lindsay and Bromley

Lindsay and Bromley [307], paralleling the treatment of Buddenberg and Wilke [56] for viscosity, suggested:

$$k_{mix} = \frac{k_1}{1 + \dfrac{1.114\alpha_1}{D_{12}}\dfrac{x_2}{x_1}} + \frac{k_2}{1 + \dfrac{1.114\alpha_2}{D_{21}}\dfrac{x_1}{x_2}} \quad (65)$$

Here α is the thermal diffusivity and the numerical constant of 1.114 was obtained by considering 49 experimental data points. They [307] found that the observed values were reproduced within an average deviation of 3.5 percent and a maximum deviation of 11.7 percent. The difficulty involved in using equation (65) lies in the requirement of α and D values, which are seldom available.

Lindsay and Bromley [307] also modified the expression given by Sutherland [576] and finally suggested that for a multicomponent mixture,

$$k_{mix} = \sum_{i=1}^{n} \frac{k_i}{1 + \sum_{\substack{j=1 \\ j \neq i}}^{n} \phi_{ij} \dfrac{x_j}{x_i}} \quad (66)$$

where

$$\phi_{ij} = \frac{1}{4}\left[1 + \left\{\frac{\eta_1}{\eta_2}\left(\frac{M_2}{M_1}\right)^{3/4}\frac{(1 + S_i/T)}{(1 + S_j/T)}\right\}^{1/2}\right]^2$$

$$\times \frac{(1 + S_{ij}/T)}{(1 + S_i/T)} \quad (67)$$

The Sutherland constants S_i were obtained by them either from viscosity data or from the empirical relation in terms of the boiling point at one atmosphere pressure, T_B, i.e.,

$$S = 1.5T_B \quad (68)$$

For nonpolar gases

$$S_{ij} = (S_1 S_2)^{1/2} \quad (69)$$

and when one of the constituents is strongly polar

$$S_{ij} = 0.733(S_1 S_2)^{1/2} \quad (70)$$

Lindsay and Bromley [307] could reproduce the experimental k_{mix} values of 85 different compositions of 16 gas pairs within 1.9 percent. This equation has also been tested by a number of other workers. Srivastava and Saxena [565], Saxena [487], and Saxena and Tondon [513] have examined it for rare gas mixtures. They [513] found that the average deviation ranges between 0.6 percent for Kr–Xe system to 8.1 for Ne–Xe system. Still poorer agreement was reported by Gandhi and Saxena [171]. It has been also examined for mixtures involving polyatomic [206, 503] and polar [594] gases with moderate success. A detailed comparison of the available experimental data on thermal conductivity of multicomponent mixtures with the computed values according to this procedure has been reported by Cheung, Bromley, and Wilke [76]. A modification to this formula has been suggested by Srivastava and Saxena [565]. Equation (66) is then written for a binary system as

$$k_{mix} = \frac{k_1}{1 + \phi_{12}(x_2/x_1)} + \frac{k_2}{1 + \phi_{21}(x_1/x_2)} + \frac{c(k_1 k_2)}{\{1 + \phi_{12}(x_2/x_1)\}\{1 + \phi_{21}(x_1/x_2)\}} \quad (71)$$

The constant C is to be determined empirically from one known value of k_{mix}. The reproduction thus improves but at the expense of more initial information. The basic handicap of equations (67) and (71) consists of the large amount of initial data needed for the computation and even then only fair uncertainty, about 5 percent, is assured in k_{mix} values.

d. Method Due to Saxena and Tondon

On the basis of a treatment parallel to that of Strunk, Custead, and Stevenson [574] for viscosity, Saxena and Tondon [513] suggested that thermal conductivity for mixtures of rare gases could be calculated from the following relation:

$$k_{mix} = \frac{A(T/M_{mix})^{1/2}}{\sigma_{mix}^2 \Omega^{(2,2)*}(kT/\epsilon_{mix})} \quad (72)$$

Here

$$\sigma_{mix} = \sum_{i=1}^{n} x_i \sigma_i$$

$$M_{mix} = \sum_{i=1}^{n} x_i M_i$$

and

$$\frac{\epsilon_{mix}}{k} = \frac{\sum_{i=1}^{n} x_i (\epsilon_i/k) \sigma_i^3}{\sigma_{mix}^3}$$

A was found to vary from system to system as determined from k_{mix} data. The procedure thus is somewhat discouraging and the average absolute deviation between theory and experiment varied between 1.1 percent for Kr–Xe to 11.4 percent for He–Xe [513]. This compares unfavorably with the procedure of Lindsay and Bromley [307].

e. Methods Due to Hirschfelder, Curtiss, and Bird

Hirschfelder, Curtiss, and Bird [228] suggested some empirical procedures for computing thermal conductivity on nonreacting polyatomic gas mixtures,

$$k_{mix} = k_{mix}^0 [E_1 x_1 + E_2 x_2] \quad (73)$$

where

$$E_1 = \frac{k_1(\text{exptl})}{k_1^0} = \frac{k_1(\text{exptl})}{\eta_1} \frac{4M}{15R}$$

Equation (73) may be generalized for a multicomponent mixture [206] so that

$$k_{mix} = k_{mix}^0 \sum_{i=1}^{n} x_i E_i \quad (74)$$

These authors [228] have also suggested an alternative procedure in which an adjusted value of the collision diameter σ_i' is computed for each of the pure gases by forcing the theory of monatomic gases and experiment to agree and using the value of ϵ/k as determined from viscosity. The thermal conductivity of the mixture is then obtained from the theoretical expression for monatomic gases using experimental k values for pure components and adjusted σ_i'. For unlike interactions geometric mean rule may be used for ϵ/k and arithmetic mean rule for σ_i'.

If the viscosity of the corresponding mixture is known, these authors [228] suggested the following empirical relation for computing k_{mix}:

$$\frac{1}{k_{mix}} = \frac{1}{\eta_{mix}(\text{exptl})} \left(\frac{x_1^2}{\alpha_1} + \frac{2x_1 x_2}{(\alpha_1 \alpha_2)^{1/2}} + \frac{x_2^2}{\alpha_2} \right) \quad (75)$$

where

$$\alpha_i = \frac{k_i(\text{exptl})}{\eta_i(\text{exptl})}$$

Actual calculations made for a number of binary systems of nonpolar polyatomic gases by these authors [228] using equations (73) and (75) revealed preference for the former. Gupta, Mathur, and Saxena [206] reexamined the method based on relation (73) and its alternative described above and found that though in many cases the k_{mix} values are well reproduced yet for some systems the disagreement between theory and experiment could be well above 10 percent. The methods thus on the whole tend to be less attractive for practical use.

f. Method Due to Ulybin, Bugrov, and Il'in

Ulybin, Bugrov, and Il'in [611] suggested an empirical procedure for computing thermal conductivity of multicomponent mixtures of chemically nonreacting gases. They [611] related the thermal conductivity of a mixture at a temperature t, $k_{\text{mix}}(t)$, to its thermal conductivity at some known temperature t_0, such that

$$k_{\text{mix}}(t) = k_{\text{mix}}(t_0) \sum_{i=1}^{n} \frac{x_i k_i(t)}{k_i(t_0)} \qquad (76)$$

Here $k_i(t)$ and $k_i(t_0)$ are the thermal conductivity values of the ith component at temperatures t and t_0 respectively. These workers [611] assessed the suitability of equation (76) by performing calculations for binary and ternary mixtures of rare gases, nitrogen–carbon dioxide, and oxygen–steam systems. Except for the latter case, good reproduction is found of the conductivity values. The method seems promising and calls for more checks of the procedure. Its big drawback lies in the requirement of the mixture conductivity value at some temperature as well as at the temperature of interest. However, in most cases it may be possible to fulfill these requirements.

g. Methods Due to Cheung, Bromley, and Wilke

Cheung [76] and Cheung, Bromley, and Wilke [77, 78] from semitheoretical arguments derived

$$k_{\text{mix}} = \sum_{i=1}^{n} \frac{k_i}{1 + \sum_{\substack{j=1 \\ i \neq j}}^{n} \frac{D_{ii}}{D_{ij}} \frac{x_j}{x_i}} \qquad (77)$$

The factor D_{ii}/D_{ij} is seldom known from experiments and with enough accuracy, and these workers [78] suggested [336, 461] empirical methods to be used for its evaluation. However, their calculations [78] revealed that equation (77) overestimates the conductivity values, and consequently suggested the following modified form:

$$k_{\text{mix}} = \sum_{i=1}^{n} \frac{k_{ci}}{1 + \sum_{\substack{j=1 \\ i \neq j}}^{n} \frac{D_{ii}}{D_{ij}}\left(\frac{M_{ij}}{M_i}\right)^{1/8}\left(\frac{x_j}{x_i}\right)}$$

$$+ \sum_{i=1}^{n} \frac{k_{di}}{1 + \sum_{\substack{j=1 \\ i \neq j}}^{n} \frac{D_{ii}}{D_{ij}} \frac{x_j}{x_i}} \qquad (78)$$

where

$$k_{ci} = \frac{2.5 C_{vt} + 1.0 C_r}{2.5 C_{vt} + 1.0 C_r + 1.32 (C_v + C_{ir} + \ldots)} k_i$$

$$k_{di} = k_i - k_{ci}$$

$(C_v + C_{ir})$ may be obtained by subtracting $(C_{vt} + C_r)$ from the total heat capacity of component i at constant volume, $C_{vt} = (3/2)R$ per mole, and $C_r = 0$, R, and $(3/2)R$ for monatomic, linear, and nonlinear molecules respectively.

Detailed comparison of the calculated values from equation (78) and experimental data have been made by these authors [78]. In brief, they found for 177 mixtures from thirty different nonpolar binary systems in the temperature range 0–$774\ C$ an average deviation of 2.1 percent and a maximum deviation of 10.5 percent. Thus, though reasonably accurate, the method is quite involved and needs a large amount of initial information.

h. Method Due to Mason and Saxena

Mason and Saxena [329] derived a simple formula for the thermal conductivity of mixtures of monatomic gases by starting from the rigorous kinetic theory result and making some well defined approximations. Their [329] final result is similar to that of Sutherland [576], equation (66), except ϕ_{ij} are now given by the following relation:

$$\phi_{ij} = \frac{1.065}{2\sqrt{2}}\left(1 + \frac{M_i}{M_j}\right)^{-1/2}$$

$$\times \left[1 + \left(\frac{k_i}{k_j}\right)^{1/2}\left(\frac{M_i}{M_j}\right)^{1/4}\right]^2 \qquad (79)$$

For mixtures involving polyatomic gases, again k_{mix} is given by equation (66) and now ϕ_{ij} is defined such that

$$\phi_{ij} = \frac{1.065}{2\sqrt{2}}\left(1 + \frac{M_i}{M_j}\right)^{-1/2}$$

$$\times \left[1 + \left(\frac{k_i^0}{k_j^0}\right)^{1/2}\left(\frac{M_i}{M_j}\right)^{1/4}\right]^2 \qquad (80)$$

They [329] also suggest that the conductivity ratio

$$\frac{k_i^0}{k_j^0} = \frac{\eta_i M_j}{\eta_j M_i} \qquad (81)$$

may be obtained from equation (81) if viscosity data are available or from the relation

$$k_i^0 = k_i/E \qquad (82)$$

where

$$E = 0.115 + 0.354(C_p/R)$$
$$= 0.115 + 0.354\gamma/(\gamma - 1)$$

Tondon and Saxena [594] found that the thermal conductivity of mixtures of nonpolar and polar gases is well correlated again by an equation of the Sutherland [576] type except

$$\phi_{ij} = \frac{0.85}{2\sqrt{2}}\left(1 + \frac{M_i}{M_j}\right)^{-1/2} \times \left[1 + \left(\frac{k_i^0}{k_j^0}\right)^{1/2}\left(\frac{M_i}{M_j}\right)^{1/4}\right]^2 \quad (83)$$

This method has been tested by a number of workers for mixtures of monatomic gases [64, 169, 174, 326, 329, 344, 489, 500, 513], polyatomic gases [165, 173, 202, 326, 329, 338, 339, 345, 503], and polar gases [594]. In general, the method is found to be quite accurate and in many cases as good as the rigorous theory results. What makes it more attractive is its simplicity of calculation and the need of very little initial information. It thus meets the need of general engineering requirements. It may, however, be emphasized that the simple Sutherland form [576] is not rigorously valid for mixtures of polyatomic gases [465].

i. Method Due to Srivastava and Saxena

Srivastava and Saxena [565] suggested the use of equation (66) with ϕ_{ij} determined from the measured values of k_{mix}. Thus if the thermal conductivity of a binary system is known at two compositions referring to the same temperature, ϕ_{12} and ϕ_{21} may be determined empirically if k values for the two pure components are also known. This method has since been tested for a large number of systems of monatomic, polyatomic, and polar gases with success [164, 169, 174, 202, 338, 339, 344, 482, 487, 500, 513, 564, 565, 594]. It has also been demonstrated that while computing k_{mix} at a temperature, ϕ_{ij} determined at some lower temperature may be used with comparable success [164, 169, 174, 202, 338, 344, 513, 594]. The major drawback of the method consists in the requirement of two k_{mix} values in addition to k_1 and k_2.

j. Method Due to Mason and von Ubisch

Mason and von Ubisch [332] developed a method based on equation (66) which avoids the use of absolute expressions for ϕ_{ij} as in the methods in Sections 4Cc, 4Cg, and 4Ch above, and also their empirical determination from k_{mix} values as in Section 4Ci. They preferred to use a relation for the ratio of ϕ_{ij} to ϕ_{ji} as given by the theoretical analysis of Mason and Saxena [329], viz.,

$$\frac{\phi_{ij}}{\phi_{ji}} = \frac{k_i}{k_j} \quad (84)$$

Thus, if the pure conductivity values are available as also one mixture value, equations (66) and (84) will enable the determination of ϕ_{ij} and hence of k_{mix} at any desired composition.

Mason and von Ubisch [332] examined this by performing calculations on binary and ternary mixtures of rare gases with satisfactory results. The success of this method is further demonstrated by the detailed numerical experimentation of Saxena and Gandhi [500], Gambhir and Saxena [164], and Gandhi and Saxena [174]. It is also shown that ϕ_{ij} determined at a lower temperature can be used for prediction of k_{mix} at high temperatures with reasonable success [164, 169, 174].

This method has been extended to mixtures of polyatomic gases by Mathur and Saxena [338]. They showed that relation (84) modifies to

$$\frac{\phi_{ij}}{\phi_{ji}} = \frac{k_i^0}{k_j^0} = \frac{M_j}{M_i} \cdot \frac{\eta_i}{\eta_j} \quad (85)$$

Relations of equations (66) and (85) have been used with success for mixtures of polyatomic gases by Mathur and Saxena [338–9]; Gupta and Saxena [202]; Mathur, Tondon, and Saxena [345]; and Saxena and Gupta [503]. Validity of applying the low temperature determined ϕ_{ij} at higher temperatures has also been demonstrated [202, 338, 345, 503]. The same procedure has been examined for mixtures of polar and nonpolar gases by Tondon and Saxena [594] who found that for 85 mixtures of 12 different binary systems k_{mix} values were reproduced within an average absolute deviation of 1.9 percent. The corresponding deviations for the procedures of equations (66) and (67), (83), and of Srivastava and Saxena, method j, are 2.2, 2.2, and 2.7 percent respectively. The method in general is quite successful and the requirement of one mixture thermal conductivity value is easily met in most of the cases when this value does not necessarily have to refer to the exact temperature of interest.

k. Method Due to Wright and Gray

The success of Sutherland's form in effectively correlating the data on thermal conductivity of multicomponent mixtures of gases has inspired a number of theoretical studies. Brokaw [52] analyzed the theoretical expression of the thermal conductivity of monatomic gases and derived expressions for ϕ_{ij} to different orders of approximations. Burnett [60] presented a critical evaluation and described conditions under which a Sutherland type of equation can approximate to the result derived from rigorous theory. Cowling [97] has given a physical picture which can form the theoretical basis for the derivation of a Sutherland-type expression for thermal conductivity. This model also [99] provides the physical significance to the coefficients ϕ_{ij} in equation (66).

Wright and Gray [667] presented a theoretical analysis of the Chapman–Enskog kinetic theory thermal conductivity expression and suggested the following relation for the Sutherland coefficients:

$$\frac{\phi_{ij}}{\phi_{ji}} = \frac{\eta_i}{\eta_j}\left(\frac{M_j}{M_i}\right)^\nu = \frac{k_i^0}{k_j^0}\left(\frac{M_i}{M_j}\right)^{1-\nu} \quad (86)$$

Saxena and Gambhir [499] analyzed the ϕ_{ij} determined by Saxena and Narayanan [511] and suggested $\nu = 0.85$. This choice of ν and hence the relation in the following form:

$$\frac{\phi_{ij}}{\phi_{ji}} = \frac{k_i^0}{k_j^0}\left(\frac{M_i}{M_j}\right)^{0.15} \quad (87)$$

has been extensively used by Saxena and Gambhir [497] for correlating the experimental data on binary and ternary mixtures of rare gases. Equations (66) and (87) are found to be completely satisfactory in reproducing k_{mix} data within the limits of experimental uncertainties and also low temperature ϕ_{ij} can be used with confidence to predict thermal conductivities at high temperatures [497]. The procedure has indeed been further checked successfully for multicomponent mixtures of rare gases [164, 169, 174].

l. Method Due to Saxena and Gambhir

Saxena and Gambhir [496] have shown that it is possible to use equation (66) for approximately calculating k_{mix} even if ϕ_{ij} are determined by any of the various procedures outlined above and using viscosity data only. This makes it possible to adopt many of the above procedures for evaluating ϕ_{ij} and use viscosity data instead of thermal conductivity, the latter being scarcely available. The promise of this procedure is demonstrated even for multicomponent mixtures [168].

m. Method Due to Gambhir and Saxena

Gambhir and Saxena [163] from the analysis of the theoretical expression for the thermal conductivity of rare gases showed that under well-defined approximations the ratio of the ϕ_{ij} occurring in the Sutherland equation (66) is given by

$$\frac{\phi_{ij}}{\phi_{ji}} = \frac{k_i}{k_j}\frac{59M^2 + 88M + 150}{150M^2 + 88M + 59} \quad (88)$$

where $M = (M_j/M_i)$. Extensive calculations of the thermal conductivity of binary and ternary mixtures of rare gases have been made on the basis of equations (66) and (88) and are found to be in good agreement with the directly measured values [463, 501]. Here also it has been shown, as in many other methods given above, that the low temperature ϕ_{ij} can be used with confidence for predicting k_{mix} values at high temperatures [164, 169, 174, 344, 489, 513].

n. Method Based on Kinetic Theory and Other Data

The Chapman–Enskog kinetic theory itself can be used to calculate thermal conductivity if the various unknown factors are replaced by known and experimental quantities. Such relations are very useful to dispense with the requirement about the detailed nature of intermolecular potentials except what is inherent in theory, viz., that the forces are central. Saxena and Agrawal [493] computed k_{mix} from viscosity data, and Gandhi and Saxena [173] from diffusion data for monatomic gases and their mixtures. Good agreement is found between these indirectly generated and directly observed values. Mathur and Saxena [337] developed such relations for mixtures of polyatomic gases and demonstrated their success against available experimental data. In conclusion, this approach is very useful in generating conductivity data if the reliable base of theory and other needed experimental data are available.

The estimation procedures for thermal conductivity of mixtures of gases at higher pressure (i.e., greater than about one atmosphere) are nonexistent. Any attempt in this direction will await the development of theory for dense gas mixtures and the availability of experimental data. It is hoped that much-needed results on both of these areas will be forthcoming in the near future.

D. Summarizing Remarks

The above-mentioned methods for computing thermal conductivity of mixtures have been checked by many workers and such references have been quoted above while describing a particular method. Invariably, some discussion is also available in these references regarding the comparative success of some of the other methods. One common shortcoming of all these approaches lies in considering the experimental thermal conductivity data at their face value. It is necessary that the data must be critically examined and most reliable values generated before any precise assessment about the relative success of the methods be attempted. However, before such an effort is forthcoming, the papers cited above will be valuable guides to select a particular method for generating conductivity values. Some of the general conclusions may be stated as follows:

a. The method combining the linear and reciprocal mixing rule is attractive in view of its simplicity and moderate reliability. It is therefore recommended for approximate engineering calculations.

b. The method suggested by Mason and Saxena [329] also requires almost the same initial information as in "a" above. It involves some additional computational effort but then it is likely to yield better accuracy. It is appropriate for multicomponent mixtures of monatomic, polyatomic, and polar gases.

c. A better estimation of thermal conductivity is possible if binary mixture thermal conductivity values are available. A more abundant choice of the methods is then probable and efforts made so far cannot discriminate in many cases amongst the various possibilities. If k_{mix} value is available at one composition, methods (j), (k), and (m) are recommended. Detailed calculations have shown the applicability of method (i) for all sorts of mixtures involving nonpolar and polar gases. If the molecular weights involved of the various components are sufficiently different, method (m) should be preferred. The detailed calculations made here on these methods for a large number of binary systems indicate that one method cannot be preferred over the other two. If the criterion of best reproduction of the experimental data is used, sometimes the method (j) is best while for some systems method (k) and for the remaining method (l) is the best. This point will be further discussed and investigated in the next section.

d. The empirical method of Srivastava and Saxena [565] requires that k_{mix} values be known at two compositions. The method of Ulybin, Bugrov, and Il'in [611] makes use of the conductivity value of the same mixture at a lower temperature. Methods (d) and (e) involve the knowledge of intermolecular potentials for all the components present in the mixture. Method (c) needs the Sutherland constant and viscosity data for all the constituents of the mixture. Method (g) assumes the availability of self- and mutual-diffusion coefficients. The amount of information needed for these methods places them below the semitheoretical methods (j), (k),

and (m), found as good as or even better than the above methods in predicting the thermal conductivity of mixtures.

e. Another feature which makes the semitheoretical methods attractive is their easy extension to multicomponent and high-temperature calculations. This involves the assumption of the Sutherland coefficients ϕ_{ij} to be independent of temperature and composition. This point will be further discussed and investigated in the next section.

5. SUTHERLAND COEFFICIENTS

In this section discussion is confined to the determination of Sutherland coefficients ϕ_{ij} of equation (66) by three semitheoretical methods [(j), (k), and (l)] and considering variation of the ϕ_{ij} with temperature and composition for actual systems. For convenience we specify these three semitheoretical methods as follows:

First method: Equations (66) and (84)
Second method: Equations (66) and (88)
Third method: Equations (66) and (87)

A further convenient modification is introduced here while using equations (84) and (87) for mixtures of polyatomic gases in that only k values were used and not k^0. Similarly equation (88) was also used for polyatomic gases. This notation of first, second, and third method will now be used.

In the data section of this book thermal conductivity data on a number of binary systems are reported. We have computed ϕ_{ij} according to the procedure outlined above for the binary systems at the temperature of measurement and for the exact composition of the mixture. These three sets of ϕ_{ij} are reported in Table 1. It will be noted that ϕ_{ij} for the three methods are considerably different though for each method their variation with composition may be regarded as trivial. Next for each system and at each temperature and for all three methods the best set of ϕ_{ij} was chosen. This was done by successively choosing the ϕ_{ij} at different compositions and computing k_{mix} at the remaining compositions and picking that set which best reproduced the experimental data. The criterion used for assessing the best reproduction was to consider the three types of deviations between the computed and experimental conductivity values, viz., average absolute deviation, root-mean-square deviation, and maximum deviation. These ϕ_{ij} are reproduced in Table 2 for each system and for different temperatures. The column 3 of this table

TABLE 1. COMPOSITION AND TEMPERATURE DEPENDENCE OF Φ_{ij} ON DIFFERENT SCHEMES OF COMPUTATION

Gas Pair [Reference]	Temp. (K)	Mole Fraction of Heavier Component	First Method Φ_{12}	Φ_{21}	Second Method Φ_{12}	Φ_{21}	Third Method Φ_{12}	Φ_{21}
Ar-He [610, 332]	302.2	0.1060	0.3099	2.621	0.1497	2.911	0.4215	2.524
		0.2760	0.3271	2.766	0.1544	2.003	0.4454	2.667
		0.5410	0.3189	2.696	0.1483	2.883	0.4342	2.600
		0.7100	0.3359	2.841	0.1562	3.037	0.4564	2.733
Ar-He [610, 332]	793.2	0.1060	0.3167	2.551	0.1538	2.848	0.4299	2.452
		0.2760	0.3184	2.564	0.1508	2.792	0.4328	2.468
		0.5410	0.3081	2.481	0.1431	2.650	0.4198	2.394
		0.7100	0.3286	2.646	0.1526	2.825	0.4467	2.547
Ar-He [586]	291.2	0.0610	0.2029	1.736	0.1045	2.056	0.2690	1.630
		0.2080	0.2731	2.337	0.1297	2.552	0.3713	2.250
		0.2990	0.2721	2.329	0.1275	2.509	0.3709	2.248
		0.4380	0.2921	2.500	0.1357	2.669	0.3987	2.416
		0.5200	0.2989	2.558	0.1384	2.724	0.4080	2.472
		0.5740	0.3130	2.678	0.1451	2.855	0.4266	2.585
		0.6450	0.3140	2.687	0.1453	2.858	0.4279	2.593
		0.7200	0.3075	2.631	0.1417	2.788	0.4195	2.542
		0.7820	0.3023	2.587	0.1389	2.733	0.4128	2.502
		0.8440	0.3040	2.602	0.1396	2.747	0.4152	2.516
		0.9140	0.2797	2.393	0.1273	2.504	0.3836	2.325
Ar-He [486]	311.2	0.1412	0.3353	2.873	0.1601	3.155	0.4573	2.775
		0.2302	0.3283	2.813	0.1555	3.063	0.4474	2.714
		0.4164	0.3319	2.844	0.1554	3.062	0.4517	2.741
		0.6084	0.3408	2.920	0.1590	3.133	0.4630	2.809
		0.8398	0.3199	2.741	0.1476	2.909	0.4356	2.643
Ar-He [164]	308.2	0.2280	0.3514	2.880	0.1670	3.148	0.4782	2.776
		0.4160	0.3359	2.753	0.1578	2.973	0.4566	2.650
		0.7480	0.3402	2.788	0.1583	2.984	0.4617	2.680
Ar-He [164]	323.2	0.2280	0.3461	2.814	0.1647	3.079	0.4709	2.712
		0.4160	0.3342	2.718	0.1570	2.935	0.4543	2.616
		0.7480	0.3376	2.745	0.1570	2.936	0.4583	2.639
Ar-He [164]	343.2	0.2280	0.3466	2.792	0.1650	3.057	0.4714	2.689
		0.4160	0.3359	2.706	0.1579	2.924	0.4565	2.604
		0.7480	0.3349	2.697	0.1557	2.883	0.4548	2.594
Ar-He [164]	363.2	0.2280	0.3493	2.782	0.1665	3.049	0.4749	2.678
		0.4160	0.3439	2.739	0.1620	2.966	0.4670	2.634
		0.7480	0.3324	2.647	0.1544	2.828	0.4516	2.547
Ar-He [636]	273.2	0.0539	0.3509	3.051	0.1685	3.370	0.4803	2.958
		0.1532	0.3485	3.031	0.1658	3.314	0.4760	2.931
		0.5463	0.3244	2.820	0.1509	3.017	0.4417	2.720
		0.7296	0.3230	2.809	0.1495	2.989	0.4397	2.708
Ar-He [77]	373.2	0.2200	0.3524	2.922	0.1675	3.193	0.4799	2.818
		0.4750	0.3263	2.706	0.1525	2.908	0.4440	2.607
		0.7240	0.3424	2.839	0.1595	3.042	0.4646	2.728
Ar-He [77]	589.2	0.2260	0.3793	2.907	0.1814	3.197	0.5153	2.797
		0.4270	0.3622	2.776	0.1714	3.021	0.4908	2.664
		0.6940	0.3916	3.001	0.1856	3.271	0.5273	2.862
Ar-He [308]	295	0.2020	0.3123	2.859	0.1474	3.104	0.4263	2.764
		0.4120	0.3669	3.360	0.1724	3.631	0.4991	3.236
		0.6030	0.3061	2.803	0.1413	2.975	0.4180	2.710
		0.7930	0.3134	2.870	0.1444	3.040	0.4274	2.772
Ar-He [57]	297	0.2500	0.3375	2.853	0.1598	3.105	0.4596	2.751
		0.5000	0.3197	2.702	0.1489	2.894	0.4353	2.606
		0.7500	0.4236	3.580	0.2027	3.939	0.5676	3.397

TABLE 1. COMPOSITION AND TEMPERATURE DEPENDENCE OF Φ_{ij} ON DIFFERENT SCHEMES OF COMPUTATION (continued)

Gas Pair [Reference]	Temp. (K)	Mole Fraction of Heavier Component	First Method Φ_{12}	Φ_{21}	Second Method Φ_{12}	Φ_{21}	Third Method Φ_{12}	Φ_{21}
Ar-Ne [565]	311.2	0.1183	0.6120	1.653	0.4618	1.844	0.6598	1.609
		0.1370	0.6173	1.667	0.4653	1.858	0.6657	1.623
		0.3124	0.6488	1.752	0.4877	1.947	0.6995	1.706
		0.3472	0.6169	1.666	0.4638	1.852	0.6648	1.621
		0.4215	0.6223	1.681	0.4676	1.867	0.6705	1.635
		0.6688	0.6159	1.664	0.4617	1.843	0.6634	1.617
		0.8286	0.6329	1.709	0.4756	1.899	0.6809	1.660
		0.8381	0.6099	1.647	0.4561	1.821	0.6570	1.602
		0.8660	0.6042	1.632	0.4512	1.801	0.6510	1.587
Ar-Ne [610, 332]	302.2	0.2370	0.6078	1.732	0.4555	1.919	0.6560	1.688
		0.4230	0.5850	1.667	0.4374	1.843	0.6311	1.624
		0.6420	0.6007	1.712	0.4486	1.890	0.6477	1.666
		0.8420	0.5743	1.637	0.4259	1.794	0.6199	1.595
Ar-Ne [610, 332]	793.2	0.2370	0.6384	1.648	0.4820	1.840	0.6877	1.603
		0.4230	0.6207	1.603	0.4679	1.786	0.6683	1.558
		0.6420	0.6390	1.650	0.4815	1.838	0.6875	1.602
		0.8420	0.6996	1.807	0.5334	2.036	0.7501	1.748
Ar-Ne [586]	291.2	0.1570	0.5853	1.632	0.4412	1.819	0.6310	1.588
		0.2210	0.5831	1.626	0.4388	1.809	0.6286	1.582
		0.3280	0.5776	1.611	0.4334	1.786	0.6227	1.567
		0.4360	0.5699	1.589	0.4263	1.757	0.6146	1.547
		0.5410	0.5766	1.608	0.4304	1.774	0.6220	1.566
		0.6380	0.5691	1.587	0.4235	1.745	0.6141	1.546
		0.7260	0.5582	1.556	0.4138	1.705	0.6027	1.517
		0.8030	0.5702	1.590	0.4230	1.743	0.6155	1.549
		0.9000	0.5523	1.540	0.4073	1.679	0.5969	1.503
Ar-Ne [344]	313.2	0.2000	0.6171	1.715	0.4634	1.903	0.6658	1.670
		0.4000	0.5861	1.628	0.4392	1.804	0.6320	1.585
		0.6000	0.5718	1.589	0.4261	1.750	0.6169	1.547
		0.8000	0.5621	1.562	0.4163	1.710	0.6070	1.522
Ar-Ne [344]	338.2	0.2000	0.6241	1.665	0.4704	1.855	0.6728	1.620
		0.4000	0.6230	1.662	0.4686	1.848	0.6711	1.616
		0.6000	0.6209	1.656	0.4662	1.838	0.6686	1.610
		0.8000	0.6240	1.664	0.4682	1.846	0.6716	1.617
Ar-Ne [344]	363.2	0.2000	0.6204	1.643	0.4681	1.833	0.6686	1.599
		0.4000	0.6192	1.640	0.4660	1.824	0.6670	1.595
		0.6000	0.6104	1.617	0.4579	1.793	0.6574	1.572
		0.8000	0.6048	1.602	0.4522	1.770	0.6516	1.558
Ar-Ne [563]	273.2	0.2406	0.6179	1.566	0.4681	1.753	0.6649	1.521
		0.5740	0.6850	1.736	0.5191	1.944	0.7361	1.684
		0.7900	0.7720	1.956	0.5960	2.232	0.8253	1.888
Ar-Kr [565]	311.2	0.0866	0.9249	1.731	0.6859	1.952	1.003	1.681
		0.2338	0.7957	1.489	0.6033	1.717	0.8572	1.436
		0.3795	0.7811	1.462	0.5941	1.691	0.8401	1.407
		0.4840	0.7595	1.422	0.5780	1.645	0.8164	1.367
		0.6683	0.7971	1.492	0.6099	1.736	0.8551	1.432
		0.8115	0.8015	1.500	0.6155	1.752	0.8586	1.438
Ar-Kr [610, 332]	302.2	0.2980	0.7208	1.348	0.5495	1.563	0.7749	1.297
		0.5360	0.7418	1.388	0.5639	1.604	0.7974	1.335
		0.7640	0.7111	1.330	0.5369	1.527	0.7650	1.281
Ar-Kr [610, 332]	793.2	0.2980	0.7743	1.325	0.5947	1.548	0.8309	1.273
		0.5360	0.8059	1.379	0.6193	1.612	0.8639	1.323
		0.7640	0.8015	1.372	0.6169	1.606	0.8583	1.315
Ar-Kr [164]	308.2	0.2560	0.7361	1.408	0.5594	1.627	0.7923	1.356
		0.5580	0.7622	1.458	0.5794	1.685	0.8192	1.402
		0.7420	0.7795	1.491	0.5951	1.731	0.8364	1.432

TABLE 1. COMPOSITION AND TEMPERATURE DEPENDENCE OF Φ_{ij} ON DIFFERENT SCHEMES OF COMPUTATION (continued)

Gas Pair [Reference]	Temp. (K)	Mole Fraction of Heavier Composition	First Method		Second Method		Third Method	
			Φ_{12}	Φ_{21}	Φ_{12}	Φ_{21}	Φ_{12}	Φ_{21}
Ar-Kr [164]	323.2	0.2560	0.7460	1.397	0.5680	1.617	0.8025	1.345
		0.5580	0.7858	1.471	0.5994	1.707	0.8440	1.414
		0.7420	0.7776	1.456	0.5939	1.691	0.8342	1.398
Ar-Kr [164]	343.2	0.2560	0.7551	1.410	0.5747	1.632	0.8124	1.357
		0.5580	0.7922	1.479	0.6047	1.717	0.8507	1.421
		0.7420	0.7636	1.426	0.5820	1.653	0.8197	1.370
Ar-Kr [164]	363.2	0.2560	0.7626	1.436	0.5794	1.660	0.8208	1.383
		0.5580	0.7796	1.468	1.5942	1.702	0.8375	1.412
		0.7420	0.7376	1.389	0.5595	1.603	0.7926	1.336
Ar-Kr [585]	291.2	0.1090	0.7344	1.389	0.5599	1.610	0.7904	1.337
		0.2280	0.7931	1.500	0.6006	1.727	0.8547	1.446
		0.3330	0.7772	1.470	0.5902	1.697	0.8364	1.415
		0.4430	0.7825	1.480	0.5952	1.712	0.8414	1.424
		0.5460	0.8146	1.540	0.6221	1.789	0.8748	1.480
		0.6730	0.8239	1.558	0.6324	1.819	0.8833	1.495
		0.7770	0.8299	1.569	0.6399	1.840	0.8884	1.503
		0.8650	0.8212	1.553	0.6339	1.823	0.8785	1.487
Ar-Xe [610, 332]	302.2	0.2710	0.5879	1.788	0.3737	2.078	0.6729	1.712
		0.5040	0.6128	1.864	0.3887	2.161	0.6998	1.780
		0.7500	0.6406	1.948	0.4093	2.276	0.7283	1.853
Ar-Xe [610, 332]	793.2	0.2710	0.6299	1.724	0.4052	2.027	0.7187	1.645
		0.5040	0.6329	1.732	0.4059	2.031	0.7205	1.649
		0.7500	0.6607	1.808	0.4262	2.133	0.7492	1.715
Ar-Xe [584]	291.2	0.1090	0.6163	1.942	0.3892	2.243	0.7090	1.869
		0.2130	0.6262	1.973	0.3953	2.278	0.7193	1.896
		0.3000	0.6216	1.959	0.3929	2.264	0.7127	1.879
		0.4050	0.6047	1.906	0.3822	2.202	0.6921	1.825
		0.4980	0.6044	1.905	0.3819	2.201	0.6910	1.822
		0.5980	0.6066	1.912	0.3834	2.209	0.6927	1.826
		0.7010	0.6121	1.929	0.3875	2.233	0.6980	1.840
		0.7920	0.6180	1.948	0.3920	2.259	0.7037	1.855
		0.9050	0.6232	1.964	0.3965	2.285	0.7084	1.868
Ar-Xe [486]	311.2	0.1757	0.5895	1.913	0.3721	2.208	0.6774	1.838
		0.3231	0.5787	1.878	0.3647	2.164	0.6636	1.801
		0.5023	0.5927	1.903	0.3731	2.214	0.6783	1.841
		0.6727	0.6005	1.948	0.3785	2.246	0.6857	1.861
		0.7517	0.6101	1.979	0.3857	2.288	0.6956	1.888
		0.8339	0.5793	1.880	0.3628	2.152	0.6618	1.796
Ar-Xe [344]	311.2	0.2000	0.4723	1.410	0.3067	1.675	0.5370	1.342
		0.4000	0.4704	1.405	0.2965	1.619	0.5381	1.344
		0.6000	0.4817	1.438	0.2978	1.626	0.5530	1.382
		0.8000	0.4996	1.492	0.3060	1.671	0.5744	1.435
Ar-Xe [344]	366.2	0.2000	0.5307	1.518	0.3441	1.799	0.6042	1.445
		0.4000	0.5311	1.519	0.3376	1.765	0.6063	1.450
		0.6000	0.5376	1.537	0.3371	1.763	0.6149	1.471
		0.8000	0.5591	1.599	0.3492	1.826	0.6392	1.529
Ar-Xe [593]	313.2	0.2410	0.5595	1.748	0.3555	2.031	0.6406	1.674
		0.7580	0.5352	1.672	0.3315	1.894	0.6135	1.603
Ar-Xe [593]	338.2	0.2410	0.5296	1.650	0.3375	1.922	0.6057	1.578
		0.7580	0.4884	1.521	0.2985	1.700	0.5620	1.464
Ar-Xe [593]	366.2	0.2410	0.5547	1.652	0.3550	1.934	0.6337	1.579
		0.7580	0.5010	1.492	0.3078	1.676	0.5756	1.434

TABLE 1. COMPOSITION AND TEMPERATURE DEPENDENCE OF Φ_{ij} ON DIFFERENT SCHEMES OF COMPUTATION (continued)

Gas Pair [Reference]	Temp. (K)	Mole Fraction of Heavier Component	First Method Φ_{12}	Φ_{21}	Second Method Φ_{12}	Φ_{21}	Third Method Φ_{12}	Φ_{21}
He-Ne [586]	291.2	0.1580	0.4887	1.500	0.2912	1.842	0.5832	1.404
		0.2500	0.4904	1.505	0.2873	1.817	0.5871	1.413
		0.3930	0.5308	1.629	0.3053	1.932	0.6375	1.535
		0.5650	0.5184	1.591	0.2934	1.856	0.6239	1.502
		0.6550	0.5224	1.603	0.2940	1.860	0.6290	1.514
		0.7830	0.5297	1.626	0.2966	1.877	0.6378	1.536
		0.8940	0.5361	1.645	0.2993	1.894	0.6452	1.554
He-Ne [610, 332]	302.2	0.1190	0.5735	1.701	0.3388	2.072	0.6883	1.602
		0.1300	0.6192	1.837	0.3616	2.212	0.7464	1.737
		0.3820	0.6174	1.832	0.3588	2.195	0.7403	1.723
		0.7550	0.6299	1.869	0.3662	2.240	0.7502	1.746
He-Ne [610, 332]	793.2	0.1190	0.5005	1.561	0.2985	1.919	0.5982	1.464
		0.1300	0.4784	1.492	0.2866	1.842	0.5704	1.396
		0.3820	0.5687	1.774	0.3276	2.106	0.6834	1.672
		0.7550	0.5515	1.720	0.3114	2.002	0.6626	1.621
He-Ne [109]	273.2	0.2500	0.6113	1.876	0.3544	2.242	0.7360	1.772
		0.5100	0.5881	1.805	0.3384	2.142	0.7052	1.698
		0.7500	0.6017	1.847	0.3460	2.189	0.7190	1.731
He-Ne [174]	303.2	0.2566	0.5808	1.757	0.3382	2.109	0.6978	1.656
		0.4560	0.5976	1.808	0.3454	2.154	0.7165	1.701
		0.7552	0.6329	1.915	0.3680	2.295	0.7537	1.789
He-Ne [174]	323.2	0.2566	0.5781	1.739	0.3369	2.090	0.6942	1.638
		0.4560	0.6068	1.825	0.3513	2.178	0.7272	1.716
		0.7552	0.6229	1.874	0.3611	2.239	0.7425	1.752
He-Ne [174]	343.2	0.2566	0.5736	1.739	0.3340	2.088	0.6890	1.639
		0.4560	0.6099	1.849	0.3530	2.206	0.7311	1.739
		0.7552	0.6014	1.823	0.3459	2.162	0.7185	1.709
He-Ne [174]	363.2	0.2566	0.5727	1.731	0.3337	2.079	0.6877	1.631
		0.4560	0.6184	1.869	0.3584	2.233	0.7411	1.757
		0.7552	0.5940	1.796	0.3408	2.124	0.7102	1.684
He-Ne [57]	297.0	0.2500	0.6955	2.180	0.3996	2.582	0.8410	2.068
		0.5000	0.7967	2.497	0.4723	3.051	0.9495	2.334
		0.7500	0.8878	2.782	0.5602	3.619	1.034	2.542
He-Kr [585]	291.2	0.0690	0.1620	2.621	0.07256	2.852	0.2461	2.524
		0.1510	0.1706	2.760	0.07415	2.914	0.2613	2.680
		0.2720	0.1806	2.923	0.07727	3.037	0.2780	2.851
		0.3530	0.1821	2.947	0.07751	3.046	0.2808	2.879
		0.4390	0.1930	3.124	0.08199	3.223	0.2977	3.052
		0.6000	0.1916	3.101	0.08102	3.184	0.2959	3.034
		0.6980	0.1902	3.078	0.08026	3.155	0.2940	3.015
		0.7970	0.1916	3.101	0.08077	3.174	0.2962	3.037
		0.8910	0.2022	3.272	0.08534	3.354	0.3121	3.200
He-Kr [610, 332]	302.2	0.1200	0.2118	3.350	0.09253	3.555	0.3249	3.257
		0.2500	0.2003	3.168	0.08608	3.307	0.3078	3.085
		0.4230	0.2020	3.195	0.08600	3.304	0.3110	3.117
		0.5100	0.2058	3.256	0.08746	3.360	0.3169	3.177
		0.5780	0.2009	3.178	0.08515	3.272	0.3097	3.104
		0.7600	0.2067	3.270	0.08747	3.361	0.3186	3.194
He-Kr [610, 332]	793.2	0.1200	0.2071	2.854	0.09156	3.065	0.3157	2.757
		0.2500	0.2294	3.161	0.09945	3.329	0.3511	3.066
		0.4230	0.2232	3.077	0.09572	3.204	0.3423	2.990
		0.5100	0.2142	2.952	0.09138	3.059	0.3290	2.874
		0.5780	0.2093	2.884	0.08900	2.979	0.3218	2.811
		0.7600	0.2146	2.958	0.09106	3.048	0.3302	2.884
He-Kr [164]	308.2	0.0790	0.2024	3.174	0.08951	3.408	0.3098	3.077
		0.2470	0.1983	3.109	0.08529	3.248	0.3047	3.027
		0.5410	0.1979	3.102	0.08392	3.195	0.3051	3.031
		0.8980	0.1915	3.002	0.08063	3.070	0.2962	2.942

TABLE 1. COMPOSITION AND TEMPERATURE DEPENDENCE OF Φ_{ij} ON DIFFERENT SCHEMES OF COMPUTATION (continued)

Gas Pair [Reference]	Temp. (K)	Mole Fraction of Heavier Component	First Method Φ_{12}	Φ_{21}	Second Method Φ_{12}	Φ_{21}	Third Method Φ_{12}	Φ_{21}
He-Kr [164]	323.2	0.0790	0.2113	3.217	0.09349	3.457	0.3233	3.119
		0.2470	0.2045	3.114	0.08811	3.258	0.3140	3.029
		0.5410	0.2043	3.111	0.08678	3.209	0.3146	3.036
		0.8980	0.1980	3.015	0.08349	3.088	0.3058	2.951
He-Kr [164]	343.2	0.0790	0.2111	3.174	0.09352	3.416	0.3228	3.076
		0.2470	0.2094	3.149	0.09030	3.298	0.3213	3.062
		0.5410	0.2047	3.078	0.08696	3.176	0.3152	3.004
		0.8980	0.1937	2.913	0.08159	2.980	0.2994	2.853
He-Kr [164]	363.2	0.0790	0.2095	3.143	0.09290	3.384	0.3203	3.044
		0.2470	0.2070	3.104	0.08924	3.251	0.3176	3.019
		0.5410	0.2048	3.072	0.08703	3.171	0.3154	2.998
		0.8980	0.1879	2.819	0.07907	2.880	0.2908	2.764
He-Xe [486]	311.2	0.1139	0.1375	3.821	0.05754	3.952	0.2275	3.746
		0.2603	0.1380	3.836	0.05704	3.918	0.2291	3.773
		0.3460	0.1423	3.955	0.05866	4.029	0.2364	3.893
		0.4963	0.1392	3.870	0.05721	3.929	0.2316	3.815
		0.6333	0.1384	3.847	0.05677	3.899	0.2304	3.794
		0.8991	0.1278	3.553	0.05226	3.589	0.2134	3.514
He-Xe [610, 332]	302.2	0.2130	0.1508	3.879	0.06268	3.983	0.2498	3.806
		0.2830	0.1548	3.980	0.06411	4.074	0.2565	3.908
		0.5820	0.1464	3.765	0.06019	3.825	0.2434	3.708
		0.7980	0.1546	3.976	0.06353	4.037	0.2568	3.913
He-Xe [610, 332]	793.2	0.2130	0.1613	3.555	0.06740	3.670	0.2664	3.477
		0.2830	0.1639	3.612	0.06820	3.713	0.2709	3.537
		0.5820	0.1686	3.716	0.06966	3.792	0.2792	3.644
		0.7980	0.1763	3.885	0.07277	3.962	0.2916	3.806
He-Xe [584]	291.3	0.0630	0.1831	4.939	0.07705	5.134	0.3038	4.854
		0.1390	0.1573	4.241	0.06568	4.376	0.2605	4.161
		0.2010	0.1579	4.258	0.06562	4.372	0.2616	4.180
		0.3040	0.1406	3.792	0.05806	3.869	0.2335	3.730
		0.4010	0.1416	3.818	0.05831	3.885	0.2353	3.759
		0.4940	0.1334	3.598	0.05479	3.651	0.2221	3.548
		0.5940	0.1360	3.668	0.05580	3.718	0.2265	3.618
		0.6870	0.1359	3.664	0.05570	3.711	0.2263	3.616
		0.7920	0.1418	3.825	0.05815	3.875	0.2361	3.772
		0.8980	0.1426	3.847	0.05846	3.895	0.2375	3.795
He-Xe [174]	303.2	0.1203	0.1387	3.615	0.05817	3.745	0.2292	3.539
		0.3011	0.1303	3.394	0.05375	3.461	0.2164	3.340
		0.4810	0.1341	3.494	0.05510	3.547	0.2231	3.444
		0.7837	0.1261	3.286	0.05159	3.321	0.2105	3.249
He-Xe [174]	323.2	0.1203	0.1396	3.537	0.05860	3.669	0.2305	3.460
		0.3011	0.1322	3.350	0.05459	3.418	0.2195	3.295
		0.4810	0.1377	3.489	0.05662	3.545	0.2290	3.438
		0.7837	0.1267	3.210	0.05182	3.245	0.2114	3.173
He-Xe [174]	343.2	0.1203	0.1416	3.516	0.05949	3.651	0.2337	3.438
		0.3011	0.1344	3.339	0.05554	3.408	0.2231	3.282
		0.4810	0.1399	3.474	0.05754	3.531	0.2325	3.421
		0.7837	0.1285	3.193	0.05261	3.228	0.2145	3.156
He-Xe [174]	363.2	0.1203	0.1418	3.470	0.05961	3.605	0.2339	3.391
		0.3011	0.1352	3.309	0.05588	3.379	0.2243	3.252
		0.4810	0.1408	3.446	0.05792	3.503	0.2339	3.392
		0.7837	0.1295	3.169	0.05300	3.205	0.2160	3.132
Kr-Xe [610, 332]	302.2	0.2150	0.8183	1.328	0.6927	1.468	0.8537	1.295
		0.4900	0.7603	1.234	0.6440	1.365	0.7927	1.202
		0.7240	0.8532	1.384	0.7269	1.540	0.8882	1.347
		0.8420	0.8947	1.452	0.7675	1.626	0.9298	1.410
		0.8900	0.9564	1.552	0.8278	1.754	0.9918	1.504

TABLE 1. COMPOSITION AND TEMPERATURE DEPENDENCE OF Φ_{ij} ON DIFFERENT SCHEMES OF COMPUTATION (continued)

Gas Pair [Reference]	Temp. (K)	Mole Fraction of Heavier Component	First Method Φ_{12}	First Method Φ_{21}	Second Method Φ_{21}	Second Method Φ_{21}	Third Method Φ_{12}	Third Method Φ_{21}
Kr-Xe [610, 332]	793.2	0.2150	0.8080	1.292	0.6854	1.431	0.8425	1.259
		0.4900	0.8068	1.290	0.6847	1.430	0.8409	1.257
		0.7240	0.9151	1.463	0.7841	1.637	0.9514	1.422
		0.8420	0.9619	1.538	0.8316	1.736	0.9979	1.492
		0.8900	0.9652	1.543	0.8364	1.746	1.001	1.496
Kr-Xe [344]	313.2	0.2000	0.7830	1.307	0.6628	1.445	0.8168	1.274
		0.4000	0.7756	1.294	0.6564	1.431	0.8090	1.262
		0.6000	0.7714	1.287	0.6523	1.422	0.8046	1.255
		0.8000	0.7748	1.293	0.6548	1.427	0.8081	1.261
Kr-Xe [344]	338.2	0.2000	0.8003	1.315	0.6775	1.454	0.8349	1.283
		0.4000	0.7914	1.301	0.6704	1.439	0.8253	1.268
		0.6000	0.7819	1.285	0.6619	1.421	0.8153	1.253
		0.8000	0.7910	1.300	0.6699	1.438	0.8245	1.267
Kr-Xe [344]	363.2	0.2000	0.7740	1.245	0.6578	1.382	0.8066	1.213
		0.4000	0.7774	1.251	0.6595	1.386	0.8104	1.219
		0.6000	0.7770	1.250	0.6582	1.383	0.8100	1.219
		0.8000	0.7814	1.257	0.6613	1.390	0.8146	1.226
Kr-Xe [584]	291.2	0.1150	0.7810	1.302	0.6615	1.440	0.8148	1.270
		0.2010	0.7917	1.320	0.6699	1.458	0.8261	1.287
		0.2960	0.7945	1.324	0.6722	1.463	0.8289	1.292
		0.3970	0.7886	1.314	0.6674	1.453	0.8226	1.282
		0.4910	0.8036	1.339	0.6803	1.481	0.8380	1.306
		0.5950	0.7981	1.330	0.6759	1.471	0.8322	1.297
		0.6930	0.8039	1.340	0.6813	1.483	0.8380	1.306
		0.7860	0.8112	1.352	0.6883	1.498	0.8452	1.317
		0.8960	0.8123	1.354	0.6899	1.502	0.8461	1.318
Ne-Kr [565]	311.2	0.0712	0.4662	2.357	0.2675	2.656	0.5591	2.283
		0.2076	0.4450	2.250	0.2541	2.523	0.5322	2.173
		0.3092	0.4588	2.319	0.2606	2.588	0.5484	2.239
		0.4277	0.4408	2.229	0.2489	2.472	0.5267	2.151
		0.5070	0.4515	2.283	0.2548	2.530	0.5391	2.201
		0.6707	0.4580	2.315	0.2579	2.561	0.5461	2.230
		0.8556	0.4140	2.093	0.2291	2.275	0.4956	2.024
Ne-Kr [585]	291.2	0.0650	0.4371	2.305	0.2512	2.602	0.5239	2.231
		0.1110	0.4185	2.207	0.2405	2.491	0.5007	2.132
		0.2290	0.4340	2.288	0.2465	2.553	0.5195	2.213
		0.3390	0.4376	2.307	0.2472	2.560	0.5236	2.230
		0.4380	0.4207	2.218	0.2361	2.446	0.5034	2.144
		0.5330	0.4317	2.276	0.2420	2.506	0.5163	2.199
		0.6470	0.4364	2.301	0.2442	2.529	0.5216	2.221
		0.7970	0.4684	2.470	0.2640	2.734	0.5576	2.375
		0.8890	0.5210	2.747	0.2992	3.099	0.6159	2.623
Ne-Kr [610, 332]	302.2	0.3080	0.4470	2.383	0.2527	2.646	0.5351	2.305
		0.4600	0.4367	2.328	0.2454	2.570	0.5223	2.250
		0.7500	0.4379	2.335	0.2445	2.561	0.5231	2.253
Ne-Kr [610, 332]	793.2	0.3080	0.4805	2.124	0.2766	2.402	0.5720	2.042
		0.4600	0.4745	2.097	0.2713	2.355	0.5644	2.015
		0.7500	0.4812	2.126	0.2734	2.374	0.5715	2.040
Ne-Kr [344]	313.2	0.2000	0.4340	2.158	0.2486	2.428	0.5184	2.082
		0.4000	0.4504	2.239	0.2552	2.492	0.5379	2.160
		0.6000	0.4516	2.245	0.2544	2.484	0.5387	2.163
		0.8000	0.4549	2.262	0.2555	2.495	0.5422	2.177
Ne-Kr [344]	338.2	0.2000	0.4714	2.323	0.2695	2.609	0.5638	2.244
		0.4000	0.4596	2.265	0.2609	2.525	0.5486	2.184
		0.6000	0.4597	2.265	0.2595	2.512	0.5480	2.181
		0.8000	0.4653	2.293	0.2623	2.539	0.5539	2.205

TABLE 1. COMPOSITION AND TEMPERATURE DEPENDENCE OF Φ_{ij} ON DIFFERENT SCHEMES OF COMPUTATION (continued)

Gas Pair [Reference]	Temp. (K)	Mole Fraction of Heavier Component	First Method		Second Method		Third Method	
			Φ_{12}	Φ_{21}	Φ_{12}	Φ_{21}	Φ_{12}	Φ_{21}
Ne-Kr [344]	363.2	0.2000	0.4749	2.264	0.2725	2.552	0.5674	2.185
		0.4000	0.4611	2.198	0.2623	2.457	0.5499	2.118
		0.6000	0.4592	2.189	0.2596	2.431	0.5472	2.107
		0.8000	0.4623	2.204	0.2604	2.439	0.5503	2.119
Ne-Xe [584]	291.2	0.1030	0.3026	2.659	0.1534	2.923	0.3888	2.580
		0.1990	0.3085	2.711	0.1542	2.938	0.3969	2.634
		0.2850	0.3162	2.779	0.1569	0.990	0.4069	2.700
		0.3930	0.3195	2.808	0.1577	3.004	0.4112	2.729
		0.5040	0.3241	2.848	0.1594	3.037	0.4169	2.766
		0.5940	0.3333	2.929	0.1639	3.123	0.4283	2.842
		0.6730	0.3368	2.960	0.1656	3.154	0.4326	2.871
		0.7940	0.3435	3.019	0.1689	3.218	0.4406	2.924
		0.9030	0.3530	3.102	0.1739	3.314	0.4520	2.999
Ne-Xe [610, 332]	302.2	0.3300	0.3337	2.887	0.1656	3.106	0.4292	2.803
		0.4300	0.3364	2.910	0.1664	3.120	0.4324	2.825
		0.7040	0.3576	2.527	0.1774	3.155	0.4387	2.866
Ne-Xe [610, 332]	793.2	0.3300	0.3603	2.546	0.1815	2.781	0.4608	2.459
		0.4300	0.3568	2.521	0.1787	2.737	0.4564	2.435
		0.7040	0.3576	2.527	0.1774	2.718	0.4574	2.440
Ne-Xe [174]	303.2	0.1537	0.3039	2.618	0.1531	2.859	0.3905	2.540
		0.5586	0.2938	2.530	0.1434	2.677	0.3788	2.464
		0.7715	0.3004	2.588	0.1459	2.725	0.3875	2.520
Ne-Xe [174]	323.2	0.1537	0.3082	2.596	0.1556	2.841	0.3957	2.518
		0.5586	0.3049	2.569	0.4193	2.726	0.3927	2.498
		0.7715	0.3026	2.550	0.1471	2.687	0.3902	2.482
Ne-Xe [174]	343.2	0.1537	0.3084	2.527	0.1561	2.773	0.3955	2.447
		0.5586	0.3160	2.589	0.1552	2.757	0.4064	2.514
		0.7715	0.3021	2.475	0.1469	2.609	0.3894	2.409
Ne-Xe [174]	363.2	0.1537	0.3108	2.517	0.1575	2.765	0.3986	2.437
		0.5586	0.3231	2.617	0.1590	2.792	0.4153	2.539
		0.7715	0.3003	2.432	0.1460	2.563	0.3872	2.368
Ar-CO_2 [77, 521]	593.2	0.5065	0.1081	0.8176	1.054	0.8457	1.088	0.8108
Ar-D_2 [165]	308.2	0.1010	0.3700	2.737	0.1799	3.057	0.5024	2.635
		0.2420	0.3215	2.378	0.1541	2.619	0.4350	2.281
		0.4370	0.3367	2.491	0.1589	2.701	0.4562	2.392
		0.7960	0.2965	2.194	0.1363	2.316	0.4046	2.122
Ar-D_2 [165]	323.2	0.1010	0.3728	2.736	0.1813	3.058	0.5061	2.634
		0.2420	0.3220	2.364	0.1545	2.605	0.4357	2.267
		0.4370	0.3257	2.391	0.1534	2.588	0.4415	2.298
		0.7960	0.2937	2.156	0.1349	2.275	0.4009	2.086
Ar-D_2 [165]	343.2	0.1010	0.3663	2.667	0.1786	2.988	0.4967	2.564
		0.2420	0.3248	2.364	0.1559	2.608	0.4393	2.267
		0.4370	0.3192	2.324	0.1502	2.513	0.4327	2.233
		0.7960	0.2905	2.115	0.1333	2.229	0.3967	2.048
Ar-D_2 [165]	363.2	0.1010	0.3425	2.468	0.1683	2.786	0.4630	2.365
		0.2420	0.3263	2.351	0.1568	2.596	0.4411	2.253
		0.4370	0.3196	2.303	0.1505	2.492	0.4332	2.213
		0.7960	0.2875	2.072	0.1318	2.182	0.3928	2.007

TABLE 1. COMPOSITION AND TEMPERATURE DEPENDENCE OF Φ_{ij} ON DIFFERENT SCHEMES OF COMPUTATION (continued)

Gas Pair [Reference]	Temp. (K)	Mole Fraction of Heavier Component	First Method		Second Method		Third Method	
			Φ_{12}	Φ_{21}	Φ_{12}	Φ_{21}	Φ_{12}	Φ_{21}
Ar-C_2H_5OH	369	0.3295	1.188	1.012	1.139	1.059	1.200	1.001
		0.5123	1.160	0.9887	1.115	1.036	1.172	0.9773
		0.6834	1.204	1.026	1.159	1.077	1.216	1.014
Ar-H_2 [566]	311.2	0.0770	0.3026	3.078	0.1374	3.385	0.4561	2.964
		0.2760	0.2432	2.474	0.1073	2.644	0.3659	2.378
		0.4940	0.2795	2.844	0.1222	3.010	0.4207	2.734
		0.6710	0.2168	2.206	0.09272	2.284	0.3299	2.144
Ar-H_2 [503, 201]	313.2	0.2020	0.2345	2.307	0.1050	2.502	0.3513	2.208
		0.3380	0.2263	2.227	0.09863	2.350	0.3416	2.147
		0.6140	0.2302	2.265	0.09904	2.360	0.3491	2.194
Ar-H_2 [503, 201]	338.2	0.2020	0.2300	2.312	0.1028	2.502	0.3447	2.214
		0.3880	0.2183	2.194	0.09489	2.310	0.3298	2.118
		0.6140	0.2263	2.275	0.09722	2.367	0.3435	2.206
Ar-H_2 [503, 201]	366.2	0.2020	0.2380	2.312	0.1067	2.511	0.3564	2.212
		0.3880	0.2184	2.122	0.09511	2.238	0.3298	2.047
		0.6140	0.2248	2.184	0.09659	2.273	0.3411	2.117
Ar-H_2 [244]	273.2	0.1980	0.2403	2.489	0.1072	2.689	0.3608	2.388
		0.4000	0.2258	2.339	0.09807	2.461	0.3413	2.259
		0.6000	0.2184	2.262	0.09364	2.350	0.3319	2.197
		0.8200	0.2181	2.259	0.09288	2.330	0.3324	2.200
		0.9100	0.2196	2.275	0.09341	2.344	0.3350	2.217
Ar-H_2 [308]	296	0.2090	0.2333	2.589	0.1033	2.776	0.3515	2.492
		0.3960	0.2008	2.229	0.08666	2.329	0.3048	2.161
		0.6180	0.2275	2.525	0.09761	2.624	0.3457	2.451
		0.8030	0.2064	2.291	0.08767	2.356	0.3154	2.236
Ar-CH_4 [77,78]	811	0.5233	0.4442	1.250	0.3008	1.392	0.4933	1.211
Ar-N_2 [565]	311.2	0.1610	0.9324	1.299	0.8191	1.413	0.9633	1.273
		0.2380	0.9096	1.267	0.8011	1.382	0.9391	1.241
		0.5400	0.8611	1.200	0.7606	1.312	0.8881	1.173
		0.6590	0.8428	1.174	0.7439	1.283	0.8693	1.148
		0.8480	0.8554	1.192	0.7556	1.303	0.8822	1.165
Ar-N_2 [77]	593	0.5034	0.8319	1.216	0.7328	1.326	0.8586	1.190
Ar-N_2 [503, 201]	313.2	0.1880	0.7777	1.127	0.6883	1.234	0.8018	1.101
		0.4920	0.8042	1.165	0.7085	1.270	0.8299	1.140
		0.7670	0.7731	1.120	0.6778	1.215	0.7987	1.097
Ar-N_2 [503, 201]	338.2	0.1880	0.8115	1.219	0.7145	1.329	0.8378	1.194
		0.4920	0.7913	1.189	0.6958	1.294	0.8171	1.164
		0.7670	0.7382	1.109	0.6448	1.199	0.7632	1.087
Ar-N_2 [503, 201]	366.2	0.1880	0.7983	1.186	0.7041	1.295	0.8238	1.160
		0.4920	0.8382	1.245	0.7379	1.357	0.8653	1.219
		0.7670	0.7854	1.167	0.6889	1.267	0.8112	1.143
Ar-N_2 [648]	273.2	0.1296	0.8084	1.189	0.7130	1.298	0.8343	1.163
		0.3892	0.8157	1.200	0.7186	1.308	0.8420	1.174
		0.6413	0.8185	1.204	0.7204	1.311	0.8449	1.178
		0.7962	0.8196	1.205	0.7210	1.312	0.8461	1.180
Ar-O_2 [566]	311.2	0.1780	0.9671	1.436	0.8871	1.508	0.9880	1.419
		0.3500	0.9297	1.381	0.8564	1.456	0.9488	1.363
		0.4640	0.9056	1.345	0.8357	1.421	0.9238	1.327
		0.8220	0.8679	1.289	0.8026	1.364	0.8848	1.271
		0.8900	0.8610	1.279	0.7961	1.353	0.8778	1.261

TABLE 1. COMPOSITION AND TEMPERATURE DEPENDENCE OF Φ_{ij} ON DIFFERENT SCHEMES OF COMPUTATION (continued)

Gas Pair [Reference]	Temp. (K)	Mole Fraction of Heavier Component	First Method		Second Method		Third Method	
			Φ_{12}	Φ_{21}	Φ_{12}	Φ_{21}	Φ_{12}	Φ_{21}
Ar-O_2 [503, 201]	313.2	0.2490	0.8212	1.247	0.7569	1.316	0.8379	1.231
		0.4840	0.8222	1.249	0.7579	1.318	0.8388	1.232
		0.7530	0.7824	1.188	0.7201	1.252	0.7985	1.173
Ar-O_2 [503, 201]	338.2	0.2490	0.7410	1.117	0.6851	1.182	0.7555	1.102
		0.4840	0.7719	1.164	0.7115	1.228	0.7875	1.149
		0.7530	0.7438	1.121	0.6833	1.179	0.7594	1.108
Ar-O_2 [503, 201]	366.2	0.2490	0.7552	1.126	0.6982	1.191	0.7700	1.110
		0.4840	0.7944	1.184	0.7326	1.250	0.8104	1.168
		0.7530	0.7989	1.191	0.7360	1.255	0.8151	1.175
Ar-C_3H_8 [521,519]	591.2	0.5288	1.100	0.5366	1.079	0.5593	1.105	0.5312
Ar-C_3H_8 [521,519]	811.2	0.5282	0.9818	0.3512	0.9684	0.3680	0.9851	0.3472
He-Air [95]	328.3	0.0295	0.4649	2.706	0.2378	3.053	0.6049	2.617
		0.0970	0.4469	2.601	0.2289	2.940	0.5790	2.505
		0.1814	0.4421	2.573	0.2256	2.897	0.5712	2.471
		0.3345	0.4335	2.523	0.2197	2.821	0.5583	2.415
		0.7147	0.4636	2.698	0.2351	3.019	0.5919	2.560
He-n-C_4H_{10} [95]	328.4	0.0903	0.5532	4.782	0.2495	5.125	0.8079	4.674
		0.1941	0.5021	4.340	0.2304	4.733	0.7245	4.192
		0.3560	0.5438	4.700	0.2545	5.230	0.7730	4.472
		0.7220	0.3751	3.242	0.1711	3.516	0.5318	3.077
He-CO_2 [109]	273.2	0.2600	0.3113	3.049	0.1442	3.281	0.4314	2.950
		0.4800	0.3131	3.067	0.1437	3.269	0.4333	2.962
		0.7500	0.3107	3.044	0.1417	3.223	0.4296	2.937
He-CO_2 [109, 78]	590	0.3900	0.5592	3.401	0.2742	3.874	0.7568	3.213
He-C_3H_6 [95]	328.4	0.1145	0.4397	3.908	0.2058	4.230	0.6095	3.806
		0.1963	0.4285	3.808	0.2014	4.139	0.5907	3.689
		0.3890	0.4229	3.759	0.1998	4.106	0.5776	3.608
		0.7161	0.4616	4.103	0.2224	4.571	0.6199	3.872
		0.8072	0.4301	3.823	0.2053	4.219	0.5787	3.614
		0.8957	0.4449	3.954	0.2140	4.397	0.5955	3.719
He-D_2 [173]	303.2	0.2965	1.012	1.126	1.010	1.128	1.013	1.125
		0.4777	0.9937	1.105	0.9913	1.107	0.9942	1.105
		0.7205	1.135	1.262	1.132	1.265	1.136	1.262
He-D_2 [173]	323.2	0.2965	0.9951	1.102	0.9927	1.105	0.9957	1.102
		0.4777	0.9878	1.094	0.9855	1.097	0.9884	1.094
		0.7205	1.078	1.194	1.076	1.197	1.079	1.194
He-D_2 [173]	343.2	0.2965	0.9908	1.096	0.9884	1.098	0.9914	1.096
		0.4777	0.9908	1.096	0.9885	1.098	0.9914	1.096
		0.7205	1.036	1.146	1.034	1.149	1.037	1.146
He-D_2 [173]	363.2	0.2965	0.9855	1.089	0.9832	1.092	0.9861	1.089
		0.4777	0.9878	1.092	0.9855	1.094	0.9884	1.091
		0.7205	1.052	1.163	1.050	1.165	1.053	1.162
He-C_2H_6 [95]	328.4	0.0957	0.5224	3.405	0.2587	3.741	0.6876	3.312
		0.1964	0.5237	3.414	0.2614	3.779	0.6849	3.299
		0.3465	0.4978	3.245	0.2504	3.621	0.6456	3.110
		0.7184	0.4995	3.256	0.2545	3.681	0.6385	3.076

TABLE 1. COMPOSITION AND TEMPERATURE DEPENDENCE OF Φ_{ij} ON DIFFERENT SCHEMES OF COMPUTATION (continued)

Gas Pair [Reference]	Temp. (K)	Mole Fraction of Heavier Component	First Method Φ_{12}	Φ_{21}	Second Method Φ_{12}	Φ_{21}	Third Method Φ_{12}	Φ_{21}
He-CH$_2$CH$_2$ [95]	328.4	0.0942	0.4599	3.136	0.2310	3.456	0.5983	3.046
		0.1857	0.4678	3.190	0.2352	3.520	0.6063	3.087
		0.3487	0.4676	3.188	0.2360	3.532	0.6020	3.065
		0.7106	0.4604	3.139	0.2334	3.493	0.5867	2.987
He-H$_2$ [25]	303.2	0.1464	0.8866	1.055	0.7201	1.268	0.9366	1.005
		0.2891	0.9481	1.128	0.7649	1.347	1.003	1.077
		0.4562	0.9625	1.145	0.7778	1.370	1.018	1.093
		0.5328	0.9661	1.149	0.7817	1.376	1.021	1.096
		0.7136	0.9776	1.163	0.7945	1.399	1.032	1.108
		0.8713	0.9748	1.159	0.7946	1.399	1.028	1.104
He-H$_2$ [25]	318.2	0.1469	0.8755	1.050	0.7112	1.263	0.9247	1.001
		0.2786	0.9232	1.107	0.7453	1.323	0.9768	1.057
		0.4136	0.9646	1.157	0.7783	1.382	1.021	1.105
		0.5462	0.9726	1.167	0.7868	1.397	1.028	1.113
		0.7321	0.9658	1.158	0.7838	1.392	1.020	1.104
		0.8675	0.9732	1.167	0.7928	1.408	1.027	1.111
He-H$_2$ [381]	90.2	0.1140	1.453	1.251	1.162	1.481	1.543	1.199
		0.1930	1.475	1.270	1.187	1.513	1.563	1.214
		0.3890	1.234	1.063	1.024	1.305	1.297	1.008
		0.4920	1.176	1.012	0.9806	1.250	1.234	0.9582
		0.6270	1.152	0.9922	0.9649	1.230	1.208	0.9382
		0.8100	1.139	0.9809	0.9579	1.221	1.193	0.9267
		0.8620	1.165	1.003	0.9842	1.255	1.219	0.9466
		0.9090	1.154	0.9938	0.9752	1.243	1.207	0.9378
		0.9520	1.100	0.9475	0.9227	1.176	1.153	0.8958
He-H$_2$ [381]	258.3	0.1530	0.9894	1.134	0.7976	1.353	1.048	1.083
		0.3030	1.009	1.156	0.8150	1.382	1.068	1.104
		0.6060	1.018	1.167	0.8298	1.408	1.075	1.111
		0.7010	1.036	1.187	0.8480	1.438	1.092	1.129
		0.8070	1.124	1.288	0.9342	1.585	1.180	1.220
		0.8550	1.371	1.570	1.178	1.999	1.428	1.476
		0.9100	1.267	1.452	1.084	1.838	1.322	1.367
		0.9480	1.176	1.347	0.9956	1.689	1.229	1.271
He-H$_2$ [381]	273.3	0.1530	1.009	1.171	0.8101	1.392	1.070	1.120
		0.3030	1.018	1.181	0.8203	1.409	1.078	1.128
		0.6060	1.024	1.189	0.8340	1.433	1.081	1.132
		0.7010	1.033	1.199	0.8451	1.452	1.090	1.141
		0.8070	1.104	1.282	0.9154	1.573	1.161	1.215
		0.8550	1.306	1.516	1.115	1.915	1.364	1.428
		0.9100	1.250	1.450	1.066	1.831	1.304	1.366
		0.9480	1.173	1.361	0.9931	1.706	1.227	1.284
He-H$_2$ [381]	293.3	0.1530	0.9722	1.120	0.7847	1.339	1.029	1.070
		0.3030	0.9960	1.148	0.8044	1.372	1.054	1.096
		0.6060	1.010	1.164	0.8225	1.403	1.067	1.109
		0.7010	1.016	1.170	0.8296	1.415	1.071	1.114
		0.8070	1.128	1.300	0.9381	1.600	1.185	1.232
		0.8550	1.359	1.567	1.167	1.991	1.417	1.473
		0.9100	1.217	1.402	1.033	1.763	1.272	1.322
		0.9480	1.121	1.291	0.9404	1.604	1.174	1.221
He-H$_2$ [381]	393.3	0.1300	0.9946	1.162	0.7985	1.381	1.054	1.112
		0.3960	1.002	1.171	0.8096	1.401	1.060	1.117
		0.8040	1.114	1.302	0.9244	1.599	1.171	1.234
		0.8750	1.328	1.551	1.139	1.970	1.384	1.459
		0.9500	1.187	1.388	1.008	1.743	1.241	1.308
He-H$_2$ [381]	473.3	0.1300	0.9672	1.125	0.7798	1.343	1.024	1.075
		0.3960	1.008	1.172	0.8144	1.403	1.066	1.119
		0.8040	1.059	1.232	0.8727	1.503	1.115	1.170
		0.8750	1.127	1.311	0.9414	1.621	1.182	1.241
		0.9500	1.102	1.282	0.9214	1.587	1.155	1.212

TABLE 1. COMPOSITION AND TEMPERATURE DEPENDENCE OF Φ_{ij} ON DIFFERENT SCHEMES OF COMPUTATION (continued)

Gas Pair [Reference]	Temp. (K)	Mole Fraction of Heavier Component	First Method		Second Method		Third Method	
			Φ_{12}	Φ_{21}	Φ_{12}	Φ_{21}	Φ_{12}	Φ_{21}
He-H$_2$ [95]	303.2	0.8014	1.019	1.222	0.8343	1.481	1.074	1.162
		0.8276	1.017	1.220	0.8337	1.480	1.072	1.160
		0.8779	1.017	1.219	0.8346	1.482	1.071	1.159
		0.9131	1.030	1.236	0.4890	1.507	1.084	1.173
		0.9615	0.9780	1.173	0.7992	1.419	1.031	1.116
He-H$_2$ [95]	328.2	0.0702	0.9650	1.159	0.7737	1.375	1.023	1.109
		0.2856	0.9842	1.182	0.7911	1.406	1.043	1.130
		0.4961	0.9927	1.192	0.8024	1.426	1.050	1.137
		0.7150	1.011	1.214	0.8243	1.465	1.067	1.156
		0.7999	1.023	1.228	0.8376	1.489	1.078	1.168
		0.8907	1.029	1.235	0.8466	1.505	1.083	1.173
		0.9375	1.036	1.244	0.8559	1.521	1.090	1.181
He-H$_2$ [95]	378.3	0.2867	0.9700	1.172	0.7798	1.395	1.028	1.120
		0.5092	0.9828	1.188	0.7942	1.421	1.040	1.133
		0.7192	1.003	1.211	0.8164	1.460	1.058	1.154
		0.8393	1.023	1.236	0.8387	1.500	1.077	1.175
		0.8988	1.038	1.254	0.8557	1.531	1.092	1.191
		0.9515	1.017	1.229	0.8372	1.498	1.070	1.167
He-H$_2$ [95]	353.4	0.2837	0.9812	1.187	0.7880	1.412	1.040	1.135
		0.4956	0.9967	1.206	0.8052	1.443	1.054	1.151
		0.7160	1.012	1.225	0.8251	1.478	1.068	1.167
		0.8630	1.020	1.234	0.8371	1.500	1.074	1.173
		0.9077	1.020	1.235	0.8391	1.503	1.074	1.173
		0.9509	1.031	1.247	0.8508	1.524	1.084	1.184
He-H$_2$ [95]	398.2	0.7116	0.9912	1.201	0.8059	1.445	1.047	1.144
		0.8184	1.024	1.240	0.8392	1.505	1.079	1.179
		0.8568	1.023	1.239	0.8395	1.505	1.077	1.177
		0.8920	1.031	1.249	0.8491	1.523	1.086	1.186
		0.9215	1.020	1.235	0.8389	1.504	1.073	1.173
		0.9603	1.021	1.237	0.8417	1.509	1.074	1.174
He-CH$_4$ [78]	588.2	0.2540	0.7312	2.111	0.4431	2.489	0.8582	2.012
		0.4500	0.8029	2.318	0.4970	2.791	0.9338	2.189
		0.7010	0.7823	2.259	0.4963	2.788	0.8995	2.109
He-CH$_4$ [77]	589.2	0.2540	0.6861	1.880	0.4201	2.239	0.8021	1.785
		0.4500	0.7832	2.146	0.7864	2.592	0.9097	2.024
		0.7010	0.7762	2.127	0.4930	2.628	0.8922	1.985
He-N$_2$ [109]	273.2	0.2400	0.4929	2.837	0.2521	3.183	0.6338	2.725
		0.4500	0.4574	2.633	0.2332	2.945	0.5847	2.514
		0.7400	0.4853	2.793	0.2492	3.148	0.6151	2.644
He-N$_2$ [24]	303.2	0.1136	0.4498	2.673	0.2304	3.005	0.5806	2.578
		0.2568	0.4512	0.682	0.2300	3.000	0.5800	2.575
		0.3959	0.4424	2.692	0.2247	2.931	0.5669	2.517
		0.5319	0.4709	2.799	0.2402	3.133	0.6008	2.667
		0.7107	0.4561	2.711	0.2317	3.022	0.5807	2.578
		0.8472	0.4346	2.583	0.2189	2.854	0.5540	2.459
He-N$_2$ [24]	318.2	0.1349	0.4517	2.664	0.2313	2.994	0.5826	2.566
		0.2638	0.4525	2.669	0.2309	2.988	0.5815	2.561
		0.3759	0.4525	2.669	0.2304	2.981	0.5798	2.554
		0.5019	0.4644	2.739	0.2366	3.063	0.5931	2.612
		0.7038	0.4838	2.854	0.2481	3.211	0.6140	2.705
		0.8438	0.4514	2.662	0.2288	2.962	0.5740	2.528
He-N$_2$ [77]	377.2	0.1630	0.4453	2.524	0.2291	2.850	0.5725	2.424
		0.5910	0.4632	2.626	0.2363	2.940	0.5901	2.498
		0.7810	0.4589	2.601	0.2335	2.905	0.5832	2.469

TABLE 1. COMPOSITION AND TEMPERATURE DEPENDENCE OF Φ_{ij} ON DIFFERENT SCHEMES OF COMPUTATION (continued)

Gas Pair [Reference]	Temp. (K)	Mole Fraction of Heavier Component	First Method Φ_{12}	Φ_{21}	Second Method Φ_{12}	Φ_{21}	Third Method Φ_{12}	Φ_{21}
He-N$_2$ [77]	589.2	0.2610	0.4955	2.589	0.2560	2.935	0.6343	2.476
		0.3630	0.4928	2.575	0.2544	2.917	0.6288	2.454
		0.6950	0.5318	2.779	0.2780	3.188	0.6702	2.616
He-O$_2$ [556]	303.2	0.1238	0.4447	2.511	0.2258	2.853	0.5822	2.406
		0.3159	0.4752	2.683	0.2393	3.024	0.6193	2.560
		0.4319	0.4673	2.638	0.2349	2.968	0.6071	2.509
		0.6619	0.4588	2.590	0.2296	2.902	0.5936	2.454
		0.7941	0.4451	2.513	0.2214	2.798	0.5760	2.381
		0.8616	0.5270	2.976	0.2711	3.426	0.6722	2.778
He-O$_2$ [556]	318.2	0.1636	0.4415	2.488	0.2238	2.821	0.5773	2.381
		0.2879	0.4962	2.796	0.2502	3.155	0.6474	2.670
		0.4537	0.4575	2.578	0.2295	2.894	0.5943	2.451
		0.6016	0.4680	2.637	0.2350	2.963	0.6056	2.498
		0.7416	0.4852	2.733	0.2450	3.089	0.6247	2.577
		0.8539	0.4700	2.648	0.2360	2.976	0.6053	2.497
He-C$_3$H$_8$ [95]	328.4	0.1093	0.5294	4.083	0.2475	4.434	0.7387	3.975
		0.2283	0.5237	4.039	0.2482	4.446	0.7233	3.892
		0.3618	0.5179	3.995	0.2483	4.449	0.7081	3.811
		0.7165	0.5016	3.869	0.2448	4.385	0.6725	3.619
He-CH$_2$CHCH$_3$ [95]	328.4	0.0881	0.4939	3.935	0.2316	4.266	0.6854	3.837
		0.1904	0.4785	3.812	0.2265	4.173	0.6585	3.687
		0.3243	0.4761	3.793	0.2273	4.186	0.6497	3.637
		0.7131	0.4773	3.803	0.2317	4.267	0.6389	3.576
Kr-D$_2$ [165]	308.2	0.0840	0.2156	3.051	0.09593	3.296	0.3287	2.950
		0.2220	0.1970	2.788	0.08539	2.933	0.3014	2.705
		0.4460	0.1898	2.685	0.08078	2.775	0.2920	2.620
		0.8220	0.1696	2.400	0.07121	2.446	0.2629	2.360
Kr-D$_2$ [165]	323.2	0.0840	0.2152	2.958	0.09604	3.205	0.3275	2.856
		0.2220	0.1988	2.732	0.08629	2.880	0.3038	2.649
		0.4460	0.1883	2.588	0.08017	2.675	0.2896	2.525
		0.8220	0.1743	2.396	0.07325	2.445	0.2699	2.354
Kr-D$_2$ [165]	343.2	0.0840	0.2153	2.927	0.09623	3.176	0.3276	2.825
		0.2220	0.1957	2.661	0.08500	2.805	0.2990	2.578
		0.4460	0.1847	2.511	0.07860	2.594	0.2841	2.450
		0.8220	0.1737	2.362	0.07300	2.409	0.2690	2.320
Kr-D$_2$ [165]	363.2	0.0840	0.2142	2.907	0.09576	3.155	0.3257	2.804
		0.2220	0.1960	2.660	0.08513	2.805	0.2994	2.577
		0.4460	0.1809	2.455	0.07693	2.535	0.2784	2.397
		0.8220	0.1690	2.294	0.07095	2.338	0.2619	2.255
Kr-H$_2$ [593]	313.2	0.2530	0.1507	2.791	0.06264	2.884	0.2566	2.717
		0.4690	0.1323	2.450	0.05419	2.495	0.2271	2.405
		0.6530	0.1294	2.396	0.05274	2.428	0.2229	2.360
Kr-H$_2$ [593]	338.2	0.2530	0.1461	2.627	0.06076	2.716	0.2487	2.556
		0.4690	0.1398	2.514	0.05736	2.564	0.2397	2.465
		0.6530	0.1373	2.469	0.05607	2.506	0.2362	2.429
Kr-H$_2$ [593]	366.2	0.2530	0.1411	2.515	0.05867	2.599	0.2402	2.447
		0.4690	0.1318	2.349	0.05400	2.392	0.2262	2.305
		0.6530	0.1341	2.389	0.05472	2.424	0.2308	2.351

TABLE 1. COMPOSITION AND TEMPERATURE DEPENDENCE OF Φ_{ij} ON DIFFERENT SCHEMES OF COMPUTATION (continued)

Gas Pair [Reference]	Temp. (K)	Mole Fraction of Heavier Component	First Method		Second Method		Third Method	
			Φ_{12}	Φ_{21}	Φ_{12}	Φ_{21}	Φ_{12}	Φ_{21}
$Kr-H_2$ [25]	303.2	0.1363	0.1348	2.624	0.05689	2.753	0.2282	2.539
		0.2584	0.1369	2.664	0.05670	2.743	0.2336	2.600
		0.4462	0.1378	2.681	0 05649	2.743	0.2364	2.631
		0.5139	0.1400	2.724	0.05730	2.773	0.2404	2.675
		0.7326	0.1424	2.772	0.05811	2.812	0.2451	2.727
		0.8862	0.3350	6.521	0.1438	6.956	0.5492	6.112
$Kr-H_2$ [25]	318.2	0.1542	0.1200	2.291	0.05051	2.396	0.2031	2.216
		0.2431	0.1362	2.599	0.05652	2.682	0.2322	2.533
		0.3864	0.1790	3.416	0.07408	3.514	0.3053	3.331
		0.5639	0.1333	2.544	0.05446	2.584	0.2293	2.502
		0.7241	0.1298	2.477	0.05286	2.508	0.2238	2.442
		0.8562	0.1331	2.540	0.05416	2.569	0.2296	2.505
$Kr-N_2$ [24]	303.2	0.1371	0.6331	1.684	0.4199	1.975	0.7148	1.613
		0.2756	0.6369	1.694	0.4211	1.981	0.7184	1.621
		0.3728	0.6192	1.647	0.4087	1.923	0.6979	1.575
		0.5364	0.6440	1.713	0.4249	1.999	0.7248	1.636
		0.7139	0.6632	1.764	0.4392	2.066	0.7446	1.680
		0.8084	0.6558	1.744	0.4337	2.040	0.7360	1.661
		0.8914	0.7778	2.069	0.5358	2.520	0.8634	1.948
$Kr-N_2$ [24]	318.2	0.1545	0.5753	1.569	0.3837	1.851	0.6481	1.500
		0.1872	0.5627	1.535	0.3751	1.809	0.6335	1.466
		0.3641	0.5990	1.634	0.3943	1.902	0.6756	1.563
		0.6089	0.4867	1.327	0.3110	1.500	0.5517	1.277
		0.6451	0.6122	1.670	0.4005	1.932	0.6899	1.596
		0.8882	0.6991	1.907	0.4690	2.262	0.7812	1.808
$Kr-O_2$ [556]	303.2	0.1021	0.4926	1.414	0.3459	1.667	0.5446	1.353
		0.2631	0.5927	1.701	0.4044	1.948	0.6599	1.639
		0.3631	0.5860	1.682	0.3987	1.921	0.6523	1.620
		0.4978	0.5595	1.605	0.3781	1.822	0.6230	1.547
		0.6215	0.6265	1.798	0.4265	2.055	0.6960	1.729
		0.7410	0.6074	1.743	0.4116	1.983	0.6749	1.676
$Kr-O_2$ [556]	318.2	0.1545	0.5667	1.630	0.3894	1.881	0.6303	1.569
		0.4751	0.7882	2.267	0.5434	2.625	0.8744	2.177
		0.6059	0.6274	1.805	0.4271	2.063	0.6970	1.735
		0.7384	0.5699	1.639	0.3827	1.849	0.6347	1.580
		0.8914	0.5511	1.585	0.3667	1.771	0.6149	1.531
$Ne-CO_2$ [109]	273.2	0.3100	0.5928	1.892	0.4248	2.100	0.6483	1.841
		0.4000	0.4204	1.342	0.2990	1.478	0.4600	1.306
		0.5300	0.5553	1.772	0.3963	1.959	0.6069	1.723
		0.7400	0.6023	1.922	0.4325	2.138	0.6567	1.865
$Ne-D_2$ [173]	303.2	0.0949	0.6591	1.793	0.3889	2.177	0.7924	1.693
		0.2558	0.6019	1.637	0.3565	1.996	0.7186	1.535
		0.4547	0.5960	1.621	0.3489	1.953	0.7111	1.519
		0.6534	0.6086	1.655	0.3541	1.983	0.7251	1.549
$Ne-D_2$ [173]	323.2	0.0949	0.6727	1.827	0.3957	2.211	0.8097	1.727
		0.2558	0.6040	1.640	0.3578	2.000	0.7211	1.538
		0.4547	0.6040	1.640	0.3540	1.978	0.7205	1.537
		0.6534	0.6103	1.657	0.3553	1.986	0.7271	1.551
$Ne-D_2$ [173]	343.2	0.0949	0.6898	1.890	0.4033	2.274	0.8321	1.791
		0.2558	0.6028	1.652	0.3566	2.011	0.7200	1.550
		0.4547	0.6075	1.665	0.3558	2.007	0.7248	1.560
		0.6534	0.6036	1.654	0.3506	1.977	0.7196	1.549

TABLE 1. COMPOSITION AND TEMPERATURE DEPENDENCE OF Φ_{ij} ON DIFFERENT SCHEMES OF COMPUTATION (continued)

Gas Pair [Reference]	Temp. (K)	Mole Fraction of Heavier Component	First Method Φ_{12}	Φ_{21}	Second Method Φ_{12}	Φ_{21}	Third Method Φ_{12}	Φ_{21}
Ne-D$_2$ [173]	363.2	0.0949	0.7016	1.919	0.4092	2.303	0.8471	1.820
		0.2558	0.6047	1.654	0.3578	2.014	0.7224	1.552
		0.4547	0.6167	1.687	0.3617	2.036	0.7356	1.580
		0.6534	0.6108	1.671	0.3555	2.001	0.7278	1.563
Ne-H$_2$ [25]	303.2	0.1482	0.4525	1.676	0.2417	2.060	0.5906	1.549
		0.2893	0.4375	1.621	0.2271	1.935	0.5730	1.502
		0.4893	0.4809	1.782	0.2442	2.081	0.6319	1.657
		0.6131	0.4910	1.819	0.2477	2.111	0.6449	1.691
		0.7283	0.5009	1.856	0.2519	2.146	0.6569	1.723
		0.8561	0.6156	2.281	0.3246	2.766	0.7907	2.073
Ne-H$_2$ [25]	318.2	0.1462	0.6008	2.245	0.3103	2.667	0.7985	2.112
		0.2983	0.4650	1.738	0.2407	2.069	0.6103	1.614
		0.4654	0.4691	1.753	0.2380	2.045	0.6169	1.632
		0.6041	0.4786	1.788	0.2405	2.067	0.6295	1.665
		0.7486	0.4863	1.817	0.2428	2.087	0.6392	1.691
		0.8671	0.4919	1.838	0.2448	2.104	0.6460	1.709
Ne-H$_2$ [593]	313.2	0.1500	0.4697	1.725	0.2502	2.113	0.6141	1.596
		0.4050	0.4505	1.654	0.2299	1.942	0.5917	1.538
		0.6630	0.4266	1.567	0.2100	1.774	0.5647	1.468
Ne-H$_2$ [593]	338.2	0.1500	0.4820	1.749	0.2566	2.142	0.6305	1.619
		0.4050	0.4472	1.623	0.2285	1.907	0.5871	1.508
		0.6630	0.4446	1.613	0.2204	1.839	0.5871	1.508
Ne-H$_2$ [593]	366.2	0.1500	0.4736	1.714	0.2528	2.105	0.6186	1.585
		0.4050	0.4460	1.614	0.2279	1.897	0.5854	1.500
		0.6630	0.4342	1.572	0.2144	1.785	0.5740	1.471
Ne-H$_2$ [593]	368.2	0.2720	0.4963	1.838	0.2580	2.197	0.6517	1.708
		0.2930	0.5256	1.946	0.2725	2.321	0.6907	1.810
Ne-H$_2$ [593]	408.2	0.2720	0.4214	1.586	0.2190	1.897	0.5517	1.470
		0.2930	0.5213	1.962	0.2697	2.335	0.6858	1.827
Ne-H$_2$ [593]	448.2	0.2720	0.4044	1.494	0.2110	1.794	0.5282	1.381
		0.2930	0.4277	1.580	0.2220	0.887	0.5600	1.464
Ne-N$_2$ [24]	303.2	0.0974	0.8173	1.557	0.7133	1.656	0.8458	1.534
		0.2246	0.7928	1.510	0.6946	1.612	0.8195	1.486
		0.3046	0.6621	1.261	0.5828	1.353	0.6834	1.239
		0.5504	0.7631	1.454	0.6710	1.557	0.7879	1.429
		0.6723	0.7798	1.485	0.6868	1.594	0.8047	1.459
		0.8714	0.7780	1.482	0.6864	1.593	0.8025	1.455
Ne-N$_2$ [24]	318.2	0.1000	0.8635	1.620	0.7520	1.720	0.8941	1.597
		0.2388	0.8133	1.526	0.7128	1.630	0.8407	1.502
		0.3496	0.7704	1.446	0.6769	1.548	0.7957	1.422
		0.5063	0.7862	1.475	0.6918	1.582	0.8116	1.450
		0.6962	0.7733	1.451	0.6812	1.558	0.7979	1.426
		0.8520	0.7228	1.356	0.6342	1.450	0.7464	1.334
Ne-N$_2$ [593]	313.2	0.2030	0.7462	1.380	0.6564	1.479	0.7706	1.356
		0.5110	0.7743	1.431	0.6815	1.535	0.7992	1.407
		0.8050	0.7332	1.355	0.6442	1.451	0.7569	1.332
Ne-N$_2$ [593]	338.2	0.2030	0.7900	1.455	0.6935	1.556	0.8163	1.431
		0.5110	0.8034	1.479	0.7077	1.588	0.8291	1.453
		0.8050	0.7863	1.448	0.6941	1.557	0.8109	1.422
Ne-N$_2$ [593]	366.2	0.2030	0.8071	1.453	0.7089	1.555	0.8338	1.429
		0.5110	0.7973	1.436	0.7028	1.542	0.8227	1.410
		0.8050	0.8055	1.450	0.7124	1.563	0.8303	1.423

TABLE 1. COMPOSITION AND TEMPERATURE DEPENDENCE OF Φ_{ij} ON DIFFERENT SCHEMES OF COMPUTATION (continued)

Gas Pair [Reference]	Temp. (K)	Mole Fraction of Heavier Component	First Method Φ_{12}	Φ_{21}	Second Method Φ_{12}	Φ_{21}	Third Method Φ_{12}	Φ_{21}
Ne-N_2 [593]	368.2	0.2560	1.023	1.914	0.8909	2.031	1.059	1.886
		0.7350	0.9053	1.694	0.8049	1.835	0.9321	1.660
Ne-N_2 [593]	408.2	0.2560	0.8665	1.570	0.7597	1.677	0.8956	1.544
		0.7350	0.8635	1.564	0.7661	1.691	0.8895	1.534
Ne-N_2 [593]	448.2	0.2560	0.8639	1.530	0.7584	1.637	0.8926	1.505
		0.7350	0.9016	1.597	0.8022	1.731	0.9281	1.565
Ne-O_2 [556]	303.2	0.2251	0.7368	1.328	0.6189	1.466	0.7702	1.295
		0.4236	0.7576	1.365	0.6354	1.505	0.7921	1.332
		0.5904	0.7548	1.360	0.6329	1.499	0.7890	1.327
		0.7634	0.7369	1.328	0.6165	1.461	0.7705	1.296
		0.8602	0.7060	1.272	0.5876	1.392	0.7389	1.243
Ne-O_2 [556]	318.2	0.1597	0.7903	1.424	0.6611	1.566	0.8273	1.391
		0.2580	0.7987	1.439	0.6685	1.583	0.8357	1.405
		0.3485	0.8320	1.499	0.6967	1.650	0.8703	1.463
		0.5156	0.8236	1.484	1.6922	1.640	0.8606	1.447
		0.7007	0.7942	1.431	0.6682	1.583	0.8294	1.395
		0.8848	0.6474	1.166	0.5340	1.265	0.6791	1.142
Ne-O_2 [593]	313.2	0.3400	0.7335	1.294	0.6165	1.430	0.7664	1.262
		0.4960	0.7479	1.320	0.6277	1.456	0.7817	1.287
		0.7390	0.7837	1.383	1.6592	1.529	0.8183	1.348
Ne-O_2 [593]	338.2	0.3400	0.7637	1.404	0.6395	1.546	0.7988	1.371
		0.4960	0.7380	1.357	0.6179	1.493	0.7717	1.324
		0.7390	0.7358	1.353	0.6153	1.487	0.7695	1.320
Ne-O_2 [593]	366.2	0.3400	0.7776	1.396	0.6520	1.539	0.8131	1.363
		0.4960	0.7395	1.328	0.6200	1.463	0.7731	1.295
		0.7390	0.7609	1.366	0.6383	1.507	0.7951	1.332
Ne-O_2 [593]	368.2	0.2290	0.9799	1.711	0.8138	1.869	1.028	1.675
		0.4920	0.9657	1.687	0.8152	1.872	1.008	1.643
		0.7440	0.8068	1.409	0.6805	1.562	0.8420	1.372
Ne-O_2 [593]	408.2	0.2290	0.8364	1.462	0.7000	1.608	0.8753	1.428
		0.4920	0.8248	1.441	0.6940	1.594	0.8616	1.405
		0.7440	0.7651	1.337	0.6426	1.476	0.7993	1.304
Ne-O_2 [593]	448.2	0.2290	0.8531	1.449	0.7149	1.596	0.8925	1.415
		0.4920	0.7379	1.253	0.6205	1.385	0.7708	1.222
		0.7440	0.8712	1.480	0.7400	1.652	0.9077	1.439
Xe-D_2 [345, 593]	311.2	0.2000	0.1199	2.585	0.05003	2.665	0.1976	2.526
		0.4000	0.1195	2.577	0.04919	2.620	0.1984	2.538
		0.6000	0.1168	2.518	0.04782	2.547	0.1946	2.488
		0.8000	0.1195	2.578	0.04887	2.603	0.1994	2.551
Xe-D_2 [345, 593]	366.2	0.2000	0.1290	2.573	0.05404	2.662	0.2122	2.509
		0.4000	0.1192	2.377	0.04910	2.419	0.1977	2.338
		0.6000	0.1222	2.437	0.05010	2.468	0.2034	2.405
		0.8000	0.1309	2.610	0.05360	2.640	0.2179	2.577
Xe-D_2 [593]	313.2	0.2550	0.1232	2.803	0.05106	2.871	0.2038	2.750
		0.4960	0.1165	2.652	0.04780	2.687	0.1940	2.618
		0.7590	0.1117	2.542	0.04562	2.564	0.1865	2.517
Xe-D_2 [593]	338.2	0.2550	0.1147	2.639	0.04749	2.699	0.1899	2.590
		0.4960	0.1146	2.636	0.04698	2.670	0.1908	2.603
		0.7590	0.1086	2.499	0.04434	2.520	0.1815	2.476
Xe-D_2 [593]	366.2	0.2550	0.1246	2.637	0.05176	2.706	0.2059	2.583
		0.4960	0.1185	2.508	0.04865	2.543	0.1971	2.474
		0.7590	0.1201	2.541	0.04912	2.568	0.2003	2.514

TABLE 1. COMPOSITION AND TEMPERATURE DEPENDENCE OF Φ_{ij} ON DIFFERENT SCHEMES OF COMPUTATION (continued)

Gas Pair [Reference]	Temp. (K)	Mole Fraction of Heavier Component	First Method Φ_{12}	Φ_{21}	Second Method Φ_{12}	Φ_{21}	Third Method Φ_{12}	Φ_{21}
Xe-H$_2$ [25]	303.2	0.1431	0.07875	2.695	0.03210	2.754	0.1445	2.643
		0.2568	0.09105	3.116	0.03680	3.158	0.1680	3.073
		0.4379	0.08454	2.894	0.03399	2.916	0.1567	2.866
		0.5462	0.08083	2.766	0.03244	2.783	0.1500	2.745
		0.7431	0.07749	2.652	0.03105	2.664	0.1441	2.636
		0.8624	0.08976	3.072	0.03600	3.088	0.1667	3.050
Xe-H$_2$ [25]	318.2	0.1286	0.07225	2.447	0.02954	2.507	0.1323	2.394
		0.2261	0.08163	2.765	0.03302	2.803	0.1505	2.725
		0.4039	0.07901	2.676	0.03176	2.696	0.1464	2.640
		0.5762	0.07337	2.485	0.02942	2.497	0.1364	2.468
		0.7231	0.07811	2.645	0.03131	2.658	0.1452	2.629
		0.8754	0.09093	3.079	0.03647	3.096	0.1689	3.057
Xe-H$_2$ [593]	313.2	0.1600	0.09610	2.952	0.03924	3.021	0.1761	2.891
		0.4340	0.08833	2.713	0.03555	2.737	0.1635	2.684
		0.6080	0.09302	2.857	0.03738	2.877	0.1724	2.830
Xe-H$_2$ [593]	338.2	0.1600	0.08952	2.801	0.03652	2.864	0.1641	2.745
		0.4340	0.08556	2.677	0.03442	2.699	0.1584	2.649
		0.6080	0.08316	2.602	0.03338	2.617	0.1544	2.581
Xe-H$_2$ [593]	366.2	0.1600	0.09082	2.619	0.03714	2.685	0.1662	2.561
		0.4340	0.08902	2.567	0.03584	2.591	0.1647	2.539
		0.6080	0.09325	2.689	0.03748	2.709	0.1728	2.663
Xe-N$_2$ [24]	303.2	0.1313	0.4211	2.052	0.2381	2.351	0.5097	1.970
		0.2683	0.4366	2.127	0.2429	2.399	0.5292	2.045
		0.4274	0.4471	2.178	0.2466	2.435	0.5417	2.093
		0.5889	0.4100	1.998	0.2228	2.200	0.4979	1.924
		0.7265	0.4300	2.095	0.2336	2.306	0.5214	2.015
		0.8821	0.4240	2.066	0.2288	2.260	0.5147	1.989
Xe-N$_2$ [24]	318.2	0.1424	0.4298	2.060	0.2429	2.359	0.5202	1.977
		0.2752	0.4163	1.995	0.2319	2.252	0.5040	1.916
		0.4475	0.4161	1.994	0.2285	2.219	0.5045	1.918
		0.5863	0.4037	1.935	0.2192	2.129	0.4904	1.864
		0.7021	0.4087	1.959	0.2209	2.146	0.4966	1.888
		0.8345	0.3832	1.836	0.2043	1.984	0.4673	1.776
Xe-O$_2$ [556]	303.2	0.1217	0.4343	2.162	0.2512	2.449	0.5181	2.087
		0.2638	0.4126	2.054	0.2359	2.300	0.4921	1.982
		0.4241	0.3984	1.983	0.2246	2.189	0.4757	1.916
		0.5879	0.3750	1.867	0.2082	2.029	0.4492	1.809
		0.7265	0.3976	1.979	0.2205	2.149	0.4758	1.917
		0.8713	0.4695	2.337	0.2655	2.588	0.5576	2.246
Xe-O$_2$ [556]	318.2	0.1281	0.4041	2.001	0.2353	2.280	0.4809	1.926
		0.2564	0.4082	2.021	0.2337	2.266	0.4866	1.949
		0.4138	0.4411	2.184	0.2503	2.426	0.5260	2.107
		0.5772	0.3696	1.830	0.2051	1.988	0.4428	1.774
		0.7238	0.3849	1.905	0.2127	2.062	0.4611	1.847
		0.8543	0.4022	1.991	0.2224	2.156	0.4814	1.928
CHCH-Air [197]	293.2	0.1410	0.8845	1.017	0.8552	1.049	0.8919	1.009
		0.3200	0.8973	1.032	0.8669	1.064	0.9049	1.024
		0.5360	0.9018	1.037	0.8708	1.069	0.9096	1.029
		0.6300	0.9132	1.050	0.8817	1.082	0.9210	1.042
		0.9000	0.9276	1.066	0.8957	1.099	0.9356	1.059
CHCH-Air [197]	338.2	0.2110	0.9036	0.9636	0.8748	0.995	0.9108	0.9559
		0.4640	0.9240	0.9854	0.8935	1.017	0.9317	0.9778
		0.6460	0.9206	0.9818	0.8897	1.013	0.9283	0.9743
		0.8210	0.9359	0.9981	0.9043	1.029	0.9439	0.9906

TABLE 1. COMPOSITION AND TEMPERATURE DEPENDENCE OF Φ_{ij} ON DIFFERENT SCHEMES OF COMPUTATION (continued)

Gas Pair [Reference]	Temp. (K)	Mole Fraction of Heavier Component	First Method Φ_{12}	Φ_{21}	Second Method Φ_{12}	Φ_{21}	Third Method Φ_{12}	Φ_{21}
Air-CO [197]	291.2	0.1080	0.9652	1.015	0.9551	1.025	0.9676	1.013
		0.3210	0.9670	1.017	0.9569	1.027	0.9695	1.015
		0.5620	0.9676	1.018	0.9575	1.028	0.9701	1.015
		0.9780	0.9673	1.017	0.9571	1.027	0.9698	1.015
Air-CH_4 [197]	295.2	0.1200	0.9085	1.087	0.7546	1.273	0.9533	1.044
		0.3000	0.9056	1.084	0.7516	1.268	0.9502	1.041
		0.6100	0.9079	1.086	0.7523	1.269	0.9529	1.044
		0.9240	0.9058	1.084	0.7491	1.264	0.9508	1.041
CO-CH_2CH_2 [77, 78]	591.2	0.4992	0.7287	1.155	0.6173	1.278	0.7676	1.137
CO_2-C_3H_8 [77]	369	0.3646	1.101	0.9032	1.101	0.9038	1.101	0.9030
		0.5510	1.128	0.9250	1.127	0.9256	1.128	0.9248
		0.7088	1.127	0.9244	1.127	0.9250	1.127	0.9243
D_2-N_2 [201, 205]	313.2	0.2220	0.4336	2.184	0.2261	2.496	0.5530	2.083
		0.6010	0.3951	1.990	0.1990	2.197	0.5050	1.902
D_2-N_2 [201, 205]	338.2	0.2220	0.4328	2.134	0.2263	2.446	0.5513	2.033
		0.6010	0.4145	2.044	0.2101	2.271	0.5286	1.949
D_2-N_2 [201, 205]	366.2	0.2220	0.4310	2.058	0.2263	2.369	0.5480	1.957
		0.6010	0.4119	1.967	0.2089	2.187	0.5251	1.875
D_2-N_2 [593]	368.2	0.3320	0.4037	2.037	0.2080	2.300	0.5147	1.941
		0.4960	0.4017	2.027	0.2040	2.256	0.5128	1.934
D_2-N_2 [593]	408.2	0.3320	0.4016	1.960	0.2076	2.221	0.5113	1.866
		0.4960	0.4112	2.007	0.2097	2.243	0.5241	1.912
D_2-N_2 [593]	448.2	0.3320	0.4164	2.008	0.2158	2.280	0.5299	1.910
		0.4960	0.4025	1.941	0.2051	2.167	0.5132	1.850
C_2H_5OH-$C_2H_4O_2$ [77, 78]	374	0.2752	0.8735	1.246	0.7940	1.330	0.8944	1.226
		0.5159	0.8643	1.233	0.7866	1.318	0.8847	1.213
		0.6941	0.8351	1.191	0.7595	1.272	0.8548	1.172
C_2H_5OH-C_3H_8 [77]	368	0.5017	1.064	0.9849	1.051	0.9987	1.068	0.9815
		0.6851	1.126	1.042	1.112	1.057	1.129	1.038
C_2H_5OH-C_3H_8 [77]	591	0.4966	1.062	0.9397	1.049	0.9531	1.081	0.9214
H_2-CO_2 [308]	296	0.2160	0.2543	2.716	0.1123	2.920	0.3879	2.609
		0.4150	0.2503	2.673	0.1085	2.820	0.3831	2.576
		0.5950	0.2813	3.004	0.1219	3.168	0.4291	2.886
		0.8110	0.3238	3.459	0.1416	3.680	0.4893	3.290
H_2-CO_2 [382]	273.2	0.0570	0.2448	3.006	0.1105	3.303	0.3745	2.896
		0.1654	0.2360	2.898	0.1040	3.107	0.3615	2.795
		0.3932	0.2293	2.816	0.09871	2.950	0.3527	2.727
		0.6302	0.2231	2.740	0.09503	2.840	0.3442	2.662
		0.8299	0.2287	2.808	0.09711	2.902	0.3529	2.729
		0.9060	0.2325	2.855	0.09872	2.950	0.3586	2.773
		0.9247	0.2328	2.859	0.09882	2.953	0.3590	2.776

TABLE 1. COMPOSITION AND TEMPERATURE DEPENDENCE OF Φ_{ij} ON DIFFERENT SCHEMES OF COMPUTATION (continued)

Gas Pair [Reference]	Temp. (K)	Mole Fraction of Heavier Component	First Method Φ_{12}	Φ_{21}	Second Method Φ_{12}	Φ_{21}	Third Method Φ_{12}	Φ_{21}
H_2-CO_2 [382]	273.2	0.0500	0.2665	2.990	0.1211	3.308	0.4071	2.877
		0.0990	0.2652	2.976	0.1190	3.249	0.4052	2.863
		0.2500	0.2554	2.866	0.1120	3.058	0.3905	2.760
		0.5000	0.2499	2.804	0.1077	2.941	0.3831	2.707
		0.6450	0.2414	2.709	0.1033	2.822	0.3710	2.621
		0.7500	0.2448	2.747	0.1046	2.857	0.3762	2.659
		0.8580	0.2508	2.815	0.1071	2.926	0.3852	2.722
		0.9000	0.2411	2.705	0.1026	2.803	0.3710	2.622
		0.9510	0.2163	2.428	0.09142	2.497	0.3348	2.366
H_2-CO_2 [283]	298.0	0.0362	0.2529	2.711	0.1172	3.057	0.3837	2.590
		0.0941	0.2621	2.810	0.1185	3.091	0.3992	2.695
		0.5040	0.2471	2.649	0.1065	2.779	0.3787	2.556
		0.8070	0.2401	2.574	0.1024	2.671	0.3693	2.493
		0.9530	0.4523	4.848	0.2086	5.440	0.6588	4.447
H_2-CO [244]	273.2	0.2060	0.3234	2.465	0.1511	2.730	0.4582	2.354
		0.3660	0.3000	2.287	0.1373	2.479	0.4260	2.188
		0.4340	0.3143	2.396	0.1433	2.589	0.4463	2.292
		0.7280	0.3139	2.393	0.1411	2.549	0.4466	2.294
		0.8370	0.3175	2.420	0.1424	2.573	0.4517	2.320
H_2-D_2 [14]	273.2	0.1980	0.9389	1.275	0.7422	1.497	0.9995	1.223
		0.3450	0.9175	1.245	0.7290	1.470	0.9750	1.193
		0.5040	0.8976	1.218	0.7151	1.442	0.9528	1.166
		0.6050	0.8934	1.213	0.7124	1.437	0.9480	1.160
		0.8130	0.8851	1.202	0.7063	1.425	0.9387	1.148
H_2-D_2 [593]	313.2	0.2530	0.9637	1.300	0.7621	1.528	1.026	1.247
		0.4970	0.9125	1.231	0.7275	1.459	0.9684	1.178
		0.7620	1.029	1.388	0.8366	1.677	1.086	1.321
H_2-D_2 [593]	338.2	0.2530	0.9828	1.336	0.7755	1.567	1.047	1.283
		0.4970	0.9499	1.292	0.7577	1.531	1.008	1.236
		0.7620	1.015	1.380	0.8236	1.664	1.072	1.314
H_2-D_2 [593]	366.2	0.2530	0.8759	1.194	0.6967	1.411	0.9306	1.143
		0.4970	0.9208	1.255	0.7337	1.486	0.9775	1.201
		0.7620	0.9560	1.303	0.7699	1.559	1.012	1.243
H_2-D_2 [593]	368.2	0.2430	0.8962	1.231	0.7109	1.450	0.9530	1.179
		0.4880	0.9462	1.299	0.7537	1.538	1.005	1.243
		0.7620	0.9667	1.327	0.7794	1.590	1.023	1.266
		0.9360	0.8144	1.118	0.6402	1.306	0.8665	1.072
H_2-D_2 [593]	408.2	0.2430	0.9793	1.368	0.7700	1.598	1.044	1.315
		0.4880	0.8970	1.253	0.7127	1.479	0.9529	1.200
		0.7620	0.9993	1.396	0.8087	1.678	1.057	1.330
		0.9360	0.8643	1.207	0.6869	1.426	0.9173	1.155
H_2-D_2 [593]	448.2	0.2430	0.9167	1.244	0.7270	1.466	0.9749	1.193
		0.4880	0.9059	1.230	0.7217	1.455	0.9617	1.176
		0.7620	0.9879	1.341	0.7991	1.612	1.045	1.278
		0.9360	0.9040	1.227	0.7249	1.462	0.9574	1.171
H_2-C_2H_4 [283]	298.2	0.1351	0.3114	2.582	0.1460	2.869	0.4426	2.472
		0.3890	0.2973	2.465	0.1350	2.653	0.4235	2.366
		0.4863	0.3038	2.519	0.1373	2.697	0.4330	2.419
		0.6860	0.2974	2.466	0.1330	2.614	0.4247	2.372
		0.8302	0.2732	2.265	0.1208	2.373	0.3922	2.191
H_2-N_2 [201, 295]	313.2	0.1470	0.3230	2.193	0.1550	2.494	0.4539	2.077
		0.3380	0.3203	2.175	0.1486	2.392	0.4523	2.070
		0.5920	0.3041	2.065	0.1375	2.213	0.4320	1.977

TABLE 1. COMPOSITION AND TEMPERATURE DEPENDENCE OF Φ_{ij} ON DIFFERENT SCHEMES OF COMPUTATION (continued)

Gas Pair [Reference]	Temp. (K)	Mole Fraction of Heavier Component	First Method Φ_{12}	Φ_{21}	Second Method Φ_{12}	Φ_{21}	Third Method Φ_{12}	Φ_{21}
H_2-N_2 [201, 205]	338.2	0.1470	0.3365	2.251	0.1615	2.559	0.4732	2.133
		0.3380	0.3361	2.249	0.1565	2.480	0.4742	2.138
		0.5920	0.3161	2.115	0.1435	2.275	0.4482	2.020
H_2-N_2 [201, 205]	366.2	0.1470	0.3329	2.176	0.1604	2.485	0.4671	2.058
		0.3380	0.3295	2.155	0.1536	2.379	0.4645	2.046
		0.5920	0.3163	2.068	0.1437	2.227	0.4483	1.975
H_2-N_2 [593]	368.2	0.2600	0.3246	2.249	0.1520	2.495	0.4581	2.138
		0.5130	0.3093	2.143	0.1407	2.310	0.4388	2.048
		0.8800	0.2742	1.899	0.1212	1.990	0.3934	1.836
H_2-N_2 [593]	408.2	0.2600	0.3407	2.323	0.1599	2.583	0.4806	2.208
		0.5130	0.3067	2.091	0.1396	2.254	0.4351	1.999
		0.8800	0.3139	2.140	0.1406	2.271	0.4468	2.053
H_2-N_2 [593]	448.2	0.2600	0.3346	2.190	0.1577	2.444	0.4710	2.077
		0.5130	0.3245	2.123	0.1486	2.303	0.4589	2.023
		0.8800	0.3181	2.082	0.1427	2.213	0.4523	1.994
H_2-N_2 [214]	273.2	0.1970	0.3953	2.904	0.1851	3.222	0.5612	2.778
		0.2050	0.3979	2.923	0.1862	3.241	0.5648	2.796
		0.3480	0.3813	2.801	0.1772	3.083	0.5392	2.669
		0.6100	0.3424	2.515	0.1560	2.716	0.4847	2.399
		0.8410	0.3413	2.507	0.1544	2.687	0.4831	2.391
H_2-N_2 [190]	298.5	0.1000	0.3371	2.289	0.1633	2.627	0.4741	2.170
		0.1990	0.3332	2.262	0.1579	2.541	0.4693	2.147
		0.2985	0.3312	2.249	0.1546	2.487	0.4672	2.138
		0.3350	0.3067	2.082	0.1421	2.287	0.4332	1.982
		0.4970	0.3097	2.103	0.1412	2.272	0.4389	2.008
		0.5940	0.3261	2.214	0.1484	2.387	0.4619	2.114
		0.6900	0.3145	2.136	0.1419	2.283	0.4467	2.044
		0.7700	0.3548	2.409	0.1617	2.602	0.5006	2.291
		0.8900	0.3200	2.173	0.1436	2.310	0.4548	2.081
H_2-N_2 [190]	348.0	0.1415	0.3509	2.490	0.1667	2.803	0.4962	2.372
		0.3104	0.3296	2.338	0.1530	2.573	0.4659	2.228
		0.5040	0.3150	2.235	0.1434	2.412	0.4467	2.136
		0.7110	0.3253	2.308	0.1470	2.472	0.4616	2.207
		0.8530	0.3313	2.351	0.1494	2.511	0.4699	2.247
H_2-N_2 [190]	372.3	0.0820	0.3164	2.125	0.1556	2.475	0.4426	2.003
		0.1875	0.3156	2.120	0.1504	2.393	0.4435	2.007
		0.3520	0.3053	2.050	0.1413	2.248	0.4312	1.952
		0.6045	0.3271	2.197	0.1488	2.369	0.4632	2.097
		0.8140	0.3466	2.328	0.1573	2.504	0.4898	2.217
H_2-N_2 [190]	422.5	0.0820	0.3170	2.103	0.1562	2.455	0.4430	1.980
		0.1415	0.3461	2.296	0.1662	2.612	0.4869	2.177
		0.3105	0.3256	2.160	0.1520	2.389	0.4589	2.052
		0.5040	0.3273	2.171	0.1499	2.357	0.4628	2.069
		0.6045	0.3306	2.193	0.1506	2.368	0.4678	2.091
		0.7110	0.3273	2.171	0.1482	2.330	0.4638	2.073
		0.8140	0.3283	2.178	0.1481	2.328	0.4655	2.081
		0.8530	0.2961	1.964	0.1320	2.075	0.4228	1.890
H_2-N_2 [308]	298.0	0.2140	0.1974	1.384	0.09347	1.553	0.2758	1.303
		0.4100	0.3219	2.258	0.1480	2.459	0.4557	2.153
		0.6800	0.3659	2.566	0.1677	2.787	0.5156	2.437
		0.8240	0.6374	4.470	0.3270	5.433	0.8508	4.021
H_2-N_2O [244]	273.2	0.1880	0.2414	2.567	0.1072	2.773	0.3678	2.462
		0.4010	0.2577	2.739	0.1120	2.897	0.3940	2.638
		0.6140	0.2621	2.786	0.1130	2.923	0.4010	2.685
		0.7910	0.2585	2.749	0.1108	2.868	0.3962	2.652
		0.9250	0.2768	2.942	0.1190	3.080	0.4225	2.828

TABLE 1. COMPOSITION AND TEMPERATURE DEPENDENCE OF Φ_{ij} ON DIFFERENT SCHEMES OF COMPUTATION (continued)

Gas Pair [Reference]	Temp. (K)	Mole Fraction of Heavier Component	First Method Φ_{12}	First Method Φ_{21}	Second Method Φ_{12}	Second Method Φ_{21}	Third Method Φ_{12}	Third Method Φ_{21}
H_2-O_2 [201, 205]	313.2	0.2090	0.3355	2.173	0.1584	2.453	0.4796	2.052
		0.4910	0.3241	2.099	0.1474	2.284	0.4661	1.994
		0.7960	0.3351	2.170	0.1504	2.329	0.4827	2.065
H_2-O_2 [201, 205]	338.2	0.2090	0.3171	2.114	0.1493	2.380	0.4535	1.997
		0.4910	0.3102	2.068	0.1404	2.239	0.4471	1.969
		0.7960	0.3022	2.015	0.1341	2.138	0.4383	1.930
H_2-O_2 [201, 205]	366.2	0.2090	0.3205	2.089	0.1513	2.358	0.4578	1.971
		0.4910	0.3104	2.023	0.1407	2.193	0.4471	1.925
		0.7960	0.3001	1.956	0.1331	2.074	0.4353	1.874
H_2-O_2 [643]	295.2	0.0526	0.3791	2.536	0.1822	2.913	0.5460	2.412
		0.2500	0.3226	2.158	0.1506	2.409	0.4622	2.042
		0.5000	0.3131	2.094	0.1417	2.267	0.4513	1.994
		0.7500	0.3297	2.205	0.1479	2.365	0.4755	2.100
		0.9664	0.5658	3.784	0.2813	4.499	0.7699	3.401
CH_4-C_2H_4 [77]	590	0.5106	0.9559	1.271	0.7946	1.465	1.002	1.225
CH_4-C_3H_8 [77]	368	0.3208	0.8264	1.348	0.5915	1.651	0.9090	1.274
		0.5145	0.9059	1.477	0.6531	1.824	0.9938	1.393
		0.6870	0.9198	1.500	0.6714	1.875	1.005	1.408
N_2-CO_2 [109]	273.2	0.2500	0.7387	1.257	0.6253	1.392	0.7706	1.226
		0.5200	0.7089	1.207	0.5971	1.329	0.7401	1.177
		0.6600	0.8284	1.410	0.7021	1.563	0.8636	1.374
N_2-CO_2 [274]	323.2	0.3350	0.8360	1.279	0.7104	1.421	0.8712	1.246
		0.4712	0.8276	1.266	0.7037	1.408	0.8622	1.233
		0.6594	0.8183	1.252	0.6956	1.392	0.8523	1.219
N_2-CO_2 [274]	423.2	0.3350	0.8486	1.125	0.7290	1.264	0.8819	1.092
		0.4712	0.8444	1.119	0.7244	1.256	0.8777	1.087
		0.6594	0.8363	1.108	0.7158	1.241	0.8697	1.077
N_2-CO_2 [274]	523.2	0.3350	0.9113	1.072	0.7887	1.213	0.9452	1.039
		0.4712	0.9017	1.060	0.7799	1.199	0.9353	1.028
		0.6594	0.9048	1.064	0.7816	1.202	0.9388	1.032
N_2-CO_2 [274]	623.2	0.3350	0.9137	0.9689	0.7976	1.106	0.9455	0.9370
		0.4712	0.9095	0.9645	0.7922	1.099	0.9417	0.9332
		0.6594	0.9230	0.9788	0.8018	1.112	0.9563	0.9477
N_2-CO_2 [77]	642.2	0.2500	0.9512	1.020	0.8287	1.162	0.9849	0.9868
		0.4700	0.9446	1.013	0.8222	1.153	0.9783	0.9802
		0.5000	1.025	1.099	0.8921	1.251	1.061	1.063
N_2-CO_2 [77]	645.2	0.1700	0.9137	0.9806	0.7988	1.121	0.9452	0.9479
		0.3300	0.9241	0.9918	0.8056	1.131	0.9566	0.9594
		0.5000	0.9400	1.009	0.8179	1.148	0.9736	0.9764
		0.6700	0.9578	1.028	0.8331	1.169	0.9921	0.9950
N_2-CO_2 [77]	648.2	0.1700	0.8793	0.9328	0.7722	1.071	0.9085	0.9006
		0.3300	0.9055	0.9606	0.7909	1.097	0.9369	0.9288
		0.5000	0.9289	0.9854	0.8088	1.122	0.9619	0.9535
		0.6700	0.9481	1.006	0.8246	1.144	0.9820	0.9735
N_2-CO_2 [77]	745.2	0.4400	0.9425	0.9557	0.8239	1.093	0.9750	0.9239
N_2-CO_2 [77]	842.2	0.3300	0.9110	0.8776	0.8023	1.011	0.9405	0.8467
		0.5000	0.9437	0.9090	0.8270	1.042	0.9755	0.8781
		0.6700	0.9644	0.9290	0.8431	1.062	0.9976	0.8980
N_2-CO_2 [77]	846.2	0.3300	0.9141	0.8825	0.8047	1.016	0.9438	0.8514
		0.5000	0.9399	0.9074	0.8236	1.040	0.9717	0.8766
		0.6700	0.9654	0.9320	0.8440	1.066	0.9986	0.9009

TABLE 1. COMPOSITION AND TEMPERATURE DEPENDENCE OF Φ_{ij} ON DIFFERENT SCHEMES OF COMPUTATION (continued)

Gas Pair [Reference]	Temp. (K)	Mole Fraction of Heavier Component	First Method Φ_{12}	Φ_{21}	Second Method Φ_{12}	Φ_{21}	Third Method Φ_{12}	Φ_{21}
N_2-CO_2 [77]	950.2	0.2500	0.9060	0.8574	0.8014	0.9919	0.9343	0.8262
		0.5000	0.9357	0.8855	0.8210	1.016	0.9670	0.8551
		0.7500	0.9533	0.9021	0.8323	1.030	0.9864	0.8723
N_2-CO_2 [77]	961.2	0.5000	0.9472	0.8896	0.8316	1.021	0.9788	0.8590
N_2-CO_2 [77]	1047.2	0.5000	0.9636	0.8951	0.8466	1.029	0.9955	0.8641
N_2-C_2H_4 [77]	591	0.4980	0.7807	1.120	0.7803	1.120	0.7808	1.119
		0.7558	0.7627	1.094	0.7622	1.094	0.7628	1.094
N_2-O_2 [201, 205]	313.2	0.2490	1.016	0.9687	0.9764	1.010	1.026	0.9589
		0.5290	1.053	1.004	1.012	1.047	1.063	0.9939
		0.7620	1.085	1.035	1.044	1.080	1.095	1.024
N_2-O_2 [201, 205]	338.2	0.2490	0.9948	0.9914	0.9554	1.033	1.005	0.9815
		0.5290	0.9985	0.9950	0.9588	1.036	1.009	0.9851
		0.7620	0.9623	0.9590	0.9232	0.9979	0.9721	0.9496
N_2-O_2 [201, 205]	366.2	0.2490	0.9952	0.9920	0.9557	1.033	1.005	0.9821
		0.5290	0.9986	0.9954	0.9589	1.037	1.009	0.9855
		0.7620	1.026	1.023	0.9860	1.066	1.036	1.013
N_2-O_2 [593]	368.2	0.2270	0.9979	0.9315	0.9603	0.9722	1.007	0.9217
		0.5140	0.9440	0.8812	0.9076	0.9189	0.9532	0.8721
		0.7820	1.009	0.9420	0.9698	0.9819	1.019	0.9324
N_2-O_2 [593]	408.2	0.2270	0.9808	0.9462	0.9431	0.9870	0.9902	0.9365
		0.5140	0.9239	0.8914	0.8877	0.9290	0.9331	0.8824
		0.7820	0.9227	0.8902	0.8846	0.9257	0.9323	0.8817
N_2-O_2 [593]	448.2	0.2270	1.077	1.033	1.034	1.075	1.088	1.023
		0.5140	0.9979	0.9570	0.9589	0.9975	1.008	0.9473
		0.7820	0.9715	0.9316	0.9325	0.9700	0.9813	0.9224
N_2-O_2 [78, 514]	592.2	0.6098	1.054	0.9702	0.7680	1.374	1.164	0.8706
N_2-C_3H_8 [77, 78]	591.2	0.4747	1.055	0.7690	0.9428	0.8998	1.085	0.7389
N_2-C_3H_8 [514, 519]	811.2	0.5239	1.052	0.5496	0.9611	0.6576	1.075	0.5250
O_2-CO_2 [78]	370	0.2240	0.8592	1.246	0.9648	1.155	0.8350	1.270
		0.4644	0.8669	1.257	0.9714	1.162	0.8427	1.282
		0.6847	0.8728	1.266	0.9758	1.168	0.8489	1.292
		0.7301	0.8732	1.267	0.9758	1.168	0.8494	1.292
C_3H_8-$C_2H_4O_2$ [77, 78]	369.2	0.5300	1.182	0.7668	0.9566	1.063	1.309	0.6534
NH_3-Air [197]	293.2	0.1950	0.9533	0.8732	0.8276	1.036	0.9880	0.8357
		0.3920	0.9006	0.8250	0.7793	0.9752	0.9342	0.7902
		0.6340	0.8785	0.8047	0.7510	0.9398	0.9141	0.7732
		0.7540	0.8541	0.7823	0.7235	0.9054	0.8907	0.7534
NH_3-Air [197]	353.2	0.2580	0.8678	0.9113	0.7448	1.069	0.9022	0.8748
		0.4240	0.8492	0.8917	0.7254	1.041	0.8839	0.8570
		0.5900	0.8310	0.8725	0.7041	1.010	0.8666	0.8403
		0.7840	0.8093	0.8498	0.6772	0.9715	0.8466	0.8209
NH_3-CO [197]	295.2	0.2100	0.8969	0.8672	0.7839	1.017	0.9278	0.8326
		0.3800	0.9023	0.8724	0.7838	1.017	0.9349	0.8390
		0.6620	0.8859	0.8566	0.7613	0.9875	0.9205	0.8260
		0.7800	0.8860	0.8567	0.7576	0.9827	0.9217	0.8271

TABLE 1. COMPOSITION AND TEMPERATURE DEPENDENCE OF Φ_{ij} ON DIFFERENT SCHEMES OF COMPUTATION (continued)

Gas Pair [Reference]	Temp. (K)	Mole Fraction of Heavier Component	First Method Φ_{12}	Φ_{21}	Second Method Φ_{12}	Φ_{21}	Third Method Φ_{12}	Φ_{21}
NH_3-C_2H_4 [283]	298.2	0.2268	0.8299	0.9911	0.7120	1.142	0.8628	0.9560
		0.4121	0.8330	0.9948	0.7106	1.139	0.8673	0.9610
		0.7360	0.8241	0.9841	0.6958	1.116	0.8600	0.9529
NH_3-H_2 [308]	299	0.2060	0.1973	1.357	0.09825	1.523	0.2582	1.289
		0.4220	0.2353	1.618	0.1114	1.727	0.3135	1.565
		0.6400	0.2369	1.629	0.1100	1.705	0.3180	1.588
		0.7980	0.2366	1.627	0.1090	1.689	0.3186	1.591
NH_3-H_2 [190]	298.5	0.0900	0.2678	1.937	0.1374	2.240	0.3503	1.840
		0.1450	0.2662	1.926	0.1335	2.178	0.3498	1.837
		0.2080	0.2877	2.081	0.1414	2.307	0.3800	1.996
		0.3240	0.2693	1.948	0.1295	2.113	0.3573	1.876
		0.4160	0.2706	1.958	0.1287	2.099	0.3602	1.892
		0.5230	0.2640	1.909	0.1242	2.025	0.3525	1.851
		0.5990	0.2489	1.800	0.1161	1.893	0.3335	1.751
		0.6660	0.2504	1.811	0.1163	1.897	0.3359	1.764
		0.7690	0.2750	1.989	0.1280	2.088	0.3682	1.934
		0.8870	0.2483	1.796	0.1143	1.864	0.3343	1.756
NH_3-H_2 [190]	348.0	0.1750	0.2751	1.823	0.1381	2.065	0.3606	1.736
		0.3875	0.2640	1.750	0.1265	1.891	0.3502	1.686
		0.4880	0.2437	1.616	0.1149	1.717	0.3251	1.565
		0.7510	0.2448	1.623	0.1133	1.693	0.3290	1.583
		0.8315	0.2630	1.744	0.1219	1.821	0.3530	1.699
NH_3-H_2 [190]	372.3	0.0945	0.2970	1.942	0.1532	2.258	0.3878	1.841
		0.1650	0.2690	1.759	0.1360	2.005	0.3518	1.670
		0.4230	0.2645	1.729	0.1263	1.862	0.3512	1.668
		0.5980	0.2877	1.882	0.1359	2.004	0.3833	1.820
		0.8200	0.2959	1.935	0.1385	2.042	0.3951	1.876
NH_3-H_2 [190]	422.5	0.0780	0.3257	1.976	0.1696	2.321	0.4247	1.871
		0.2960	0.2977	1.807	0.1466	2.005	0.3918	1.726
		0.4620	0.2972	1.803	0.1428	1.953	0.3936	1.734
		0.6850	0.3035	1.842	0.1435	1.963	0.4037	1.779
		0.8675	0.2732	1.657	0.1268	1.735	0.3661	1.613
NH_3-N_2 [308]	300	0.1770	0.3079	0.3141	0.2890	0.3955	0.3125	0.2958
		0.3950	0.4257	0.4342	0.3743	0.5122	0.4393	0.4158
		0.5970	0.6923	0.7062	0.5895	0.8067	0.7207	0.6823
		0.7890	0.3408	0.3476	0.2708	0.3706	0.3607	0.3415
NH_3-N_2 [190]	298.5	0.0975	0.8397	0.7883	0.7447	0.9379	0.8652	0.7537
		0.1830	0.8997	0.8446	0.7896	0.9944	0.9297	0.8100
		0.2450	0.9169	0.8607	0.8018	1.010	0.9484	0.8263
		0.3730	0.8841	0.8300	0.7708	0.9708	0.9152	0.7973
		0.5040	0.9181	0.8618	0.7959	1.002	0.9518	0.8292
		0.5810	0.8908	0.8362	0.7696	0.9692	0.9243	0.8052
		0.6565	0.8905	0.8359	0.7668	0.9657	0.9247	0.8056
		0.7260	0.8689	0.8156	0.7446	0.9378	0.9034	0.7870
		0.9080	0.8534	0.8011	0.7227	0.9101	0.8898	0.7752
NH_3-N_2 [190]	348.0	0.1500	0.7461	0.7552	0.6612	0.8978	0.7689	0.7223
		0.3410	0.7532	0.7624	0.6570	0.8922	0.7794	0.7322
		0.5000	0.7492	0.7583	0.6452	0.8761	0.7778	0.7306
		0.5125	0.7619	0.7712	0.6556	0.8902	0.7912	0.7432
		0.6670	0.7458	0.7549	0.6332	0.8598	0.7770	0.7299
		0.8555	0.6441	0.6519	0.5302	0.7199	0.6761	0.6351
NH_3-N_2 [190]	372.3	0.0975	0.9021	0.9264	0.7850	1.082	0.9343	0.8905
		0.3480	0.7697	0.7905	0.6695	0.9225	0.7971	0.7598
		0.6565	0.8450	0.8678	0.7221	0.9950	0.8791	0.8379
		0.8240	0.9485	0.9742	0.8139	1.121	0.9860	0.9398

TABLE 1. COMPOSITION AND TEMPERATURE DEPENDENCE OF Φ_{ij} ON DIFFERENT SCHEMES OF COMPUTATION (continued)

Gas Pair [Reference]	Temp. (K)	Mole Fraction of Heavier Component	First Method		Second Method		Third Method	
			Φ_{12}	Φ_{21}	Φ_{12}	Φ_{21}	Φ_{12}	Φ_{21}
NH_3-N_2 [190]	422.5	0.1120	0.6764	0.7395	0.6014	0.8822	0.6964	0.7066
		0.7090	0.8804	0.9625	0.7505	1.101	0.9165	0.9300
		0.8910	0.7369	0.8057	0.6122	0.8980	0.7719	0.7832
Steam-Air [197]	353.2	0.4810	0.8853	0.6756	0.7857	0.7944	0.9121	0.6482
		0.5560	0.8727	0.6659	0.7704	0.7789	0.9003	0.6398
		0.6940	0.8595	0.6559	0.7503	0.7585	0.8893	0.6320
		0.8030	0.8331	0.6357	0.7185	0.7264	0.8645	0.6143

TABLE 2. RECOMMENDED SETS OF Φ_{ij} FOR THE THERMAL CONDUCTIVITY DATA

Gas Pair	Temp. (K)	Mole Fraction of Heavier Component	First Method Φ_{12}	Φ_{21}	Second Method Φ_{12}	Φ_{21}	Third Method Φ_{12}	Φ_{21}
Ar-He								
	302.2	0.2760	0.3271	2.766	0.1544	3.003	0.4454	2.667
	793.2	0.2760	0.3184	2.564	0.1508	2.792	0.4328	2.468
	291.2	0.4380	0.2921	2.500	0.1357	2.669	0.3987	2.416
	311.2	0.4164	0.3319	2.844	0.1554	3.062	0.4517	2.741
	308.2	0.7480	0.3402	2.788	0.1583	2.984	0.4617	2.680
	323.2	0.7480	0.3376	2.745	0.1570	2.936	0.4583	2.639
	343.2	0.4160	0.3359	2.706	0.1579	2.924	0.4565	2.604
	363.2	0.4160	0.3439	2.739	0.1620	2.966	0.4670	2.634
	273.2	0.5463	0.3244	2.820	0.1509	3.017	0.4417	2.720
	373.2	0.7240	0.3424	2.839	0.1595	3.042	0.4646	2.728
	589.2	0.2260	0.3793	2.907	0.1814	3.197	0.5153	2.797
	295	0.7930	0.3134	2.870	0.1444	3.040	0.4274	2.772
Ar-Ne								
	311.2	0.4215	0.6223	1.681	0.4676	1.867	0.6705	1.635
	302.2	0.6420	0.6007	1.712	0.4486	1.890	0.6477	1.666
	793.2	0.6420	0.6390	1.650	0.4815	1.838	0.6875	1.602
	291.2	0.4360	0.5699	1.589	0.4263	1.757	0.6146	1.547
	313.2	0.4000	0.5861	1.628	0.4392	1.804	0.6320	1.585
	338.2	0.4000	0.6230	1.662	0.4686	1.848	0.6711	1.616
	363.2	0.6000	0.6104	1.617	0.4579	1.793	0.6574	1.572
	273.2	0.5740	0.6850	1.736	0.5191	1.944	0.7361	1.684
Ar-Kr								
	311.2	0.2338	0.7957	1.489	0.6033	1.717	0.8572	1.436
	302.2	0.2980	0.7208	1.348	0.5495	1.563	0.7748	1.297
	793.2	0.7640	0.8015	1.372	0.6169	1.606	0.8583	1.315
	291.2	0.2280	0.7931	1.500	0.6006	1.727	0.8547	1.446
	308.2	0.5580	0.7622	1.458	0.5794	1.685	0.8192	1.402
	323.2	0.7420	0.7776	1.456	0.5939	1.691	0.8342	1.398
	343.2	0.7420	0.7636	1.426	0.5820	1.653	0.8197	1.370
	363.2	0.2560	0.7626	1.436	0.5794	1.660	0.8208	1.383
Ar-Xe								
	302.2	0.5040	0.6128	1.864	0.3887	2.161	0.6998	1.780
	793.2	0.5040	0.6329	1.732	0.4059	2.031	0.7205	1.649
	291.2	0.7010	0.6121	1.929	0.3875	2.233	0.6980	1.840
	311.2	0.5023	0.5927	1.923	0.3731	2.214	0.6783	1.841
	311.2	0.6000	0.4817	1.438	0.2978	1.626	0.5530	1.382
	366.2	0.6000	0.5376	1.537	0.3371	1.763	0.6149	1.471
	313.2	0.2410	0.5595	1.748	0.3555	2.031	0.6406	1.674
	338.2	0.7580	0.4884	1.521			0.5620	1.464
	338.2	0.2410			0.3375	1.922		
	366.2	0.7580	0.5010	1.492			0.5756	1.434
	366.2	0.2410			0.3550	1.934		
He-Ne								
	291.2	0.5650	0.5184	1.591	0.2934	1.856	0.6239	1.502
	302.2	0.3820	0.6174	1.832	0.3588	2.195	0.7403	1.723
	793.2	0.7550	0.5515	1.720	0.3114	2.002	0.6626	1.621
	273.2	0.7500	0.6017	1.847	0.3460	2.189	0.7190	1.731
	303.2	0.4560	0.5976	1.808	0.3454	2.154	0.7165	1.701
	323.2	0.4560	0.6068	1.825	0.3513	2.178	0.7272	1.716
	343.2	0.7552	0.6014	1.823	0.3459	2.162	0.7185	1.709
	363.2	0.7552	0.5940	1.796	0.3408	2.124	0.7102	1.684
He-Kr								
	291.2	0.6980	0.1902	3.078	0.08026	3.155	0.2940	3.015
	302.2	0.4230	0.2020	3.195	0.08600	3.304	0.3110	3.117
	793.2	0.7600	0.2146	2.958	0.09106	3.048	0.3302	2.884
	308.2	0.5410	0.1979	3.102	0.08392	3.195	0.3051	3.031
	323.2	0.5410	0.2043	3.111	0.08678	3.209	0.3146	3.036
	343.2	0.5410	0.2047	3.078	0.08696	3.176	0.3152	3.004
	363.2	0.5410	0.2048	3.072	0.08703	3.171	0.3154	2.998

TABLE 2. RECOMMENDED SETS OF Φ_{ij} FOR THE THERMAL CONDUCTIVITY DATA (continued)

Gas Pair	Temp. (K)	Mole Fraction of Heavier Component	First Method Φ_{12}	Φ_{21}	Second Method Φ_{12}	Φ_{21}	Third Method Φ_{12}	Φ_{21}
He-Xe								
	311.2	0.6333	0.1384	3.847	0.05677	3.899	0.2304	3.794
	302.2	0.2130	0.1508	3.879	0.06268	3.983	0.2498	3.806
	793.2	0.5820	0.1686	3.716	0.06966	3.792	0.2792	3.644
	291.3	0.4010	0.1416	3.818	0.05831	3.885	0.2353	3.759
	303.2	0.3011	0.1303	3.394	0.05375	3.461	0.2164	3.340
	323.2	0.3011	0.1322	3.350	0.05459	3.418	0.2195	3.295
	343.2	0.3011	0.1344	3.339	0.05554	3.408	0.2231	3.282
	363.2	0.3011	0.1352	3.309	0.05588	3.379	0.2243	3.252
Kr-Xe								
	302.2	0.2150	0.8183	1.328	0.6927	1.468	0.8537	1.295
	793.2	0.7240	0.9151	1.463	0.7841	1.637	0.9514	1.422
	313.2	0.4000	0.7756	1.294	0.6564	1.431	0.8090	1.262
	338.2	0.8000	0.7910	1.300	0.6699	1.438	0.8245	1.267
	363.2	0.4000	0.7774	1.251	0.6595	1.386	0.8104	1.219
	291.2	0.5950	0.7981	1.330	0.6759	1.471	0.8322	1.297
Ne-Kr								
	311.2	0.5070	0.4515	2.283	0.2548	2.530	0.5391	2.201
	291.2	0.3390	0.4376	2.307	0.2472	2.560	0.5236	2.230
	302.2	0.7500	0.4379	2.335	0.2445	2.561	0.5231	2.253
	793.2	0.7500	0.4812	2.127	0.2734	2.374	0.5715	2.040
	313.2	0.4000	0.4504	2.239	0.2552	2.492	0.5379	2.160
	338.2	0.8000	0.4653	2.293	0.2623	2.539	0.5539	2.205
	363.2	0.8000	0.4623	2.204	0.2604	2.439	0.5503	2.119
Ne-Xe								
	291.2	0.5040	0.3241	2.848	0.1594	3.037	0.4169	2.766
	302.2	0.4300	0.3364	2.910	0.1664	3.120	0.4324	2.825
	793.2	0.7040	0.3576	2.527	0.1774	2.718	0.4574	2.440
	303.2	0.7715	0.3004	2.588	0.1459	2.725	0.3875	2.520
	323.2	0.5586	0.3049	2.569	0.1493	2.726	0.3927	2.498
	343.2	0.1537	0.3084	2.527	0.1561	2.773	0.3955	2.447
	363.2	0.1537	0.3108	2.517	0.1575	2.765	0.3986	2.437
Ar-D_2								
	308.2	0.2420	0.3215	2.378	0.1541	2.619	0.4350	2.281
	323.2	0.2420	0.3220	2.364	0.1545	2.605	0.4357	2.267
	343.2	0.4370	0.3192	2.324	0.1502	2.513	0.4327	2.233
	363.2	0.4370	0.3196	2.303	0.1505	2.492	0.4332	2.213
Ar-C_2H_5OH								
	369	0.3295	1.188	1.012	1.139	1.059	1.200	1.001
Ar-H_2								
	311.2	0.2760	0.2432	2.474	0.1073	2.644	0.3659	2.378
	313.2	0.6140	0.2302	2.265	0.09904	2.360	0.3491	2.194
	338.2	0.6140	0.2263	2.275	0.09722	2.367	0.3435	2.206
	366.2	0.6140	0.2248	2.184	0.09659	2.273	0.3411	2.117
	273.2	0.4000	0.2258	2.339	0.09807	2.461	0.3413	2.259
	296	0.8030	0.2064	2.291	0.08767	2.357	0.3154	2.236
Ar-N_2								
	311.2	0.5400	0.8611	1.200	0.7606	1.312	0.8810	1.173
	313.2	0.1880	0.7777	1.127	0.6883	1.234	0.8018	1.101
	338.2	0.4920	0.7913	1.189	0.6958	1.294	0.8171	1.164
	366.2	0.1880	0.7983	1.186	0.7041	1.295	0.8238	1.160
	273.2	0.3892	0.8157	1.200	0.7186	1.308	0.8420	1.174
Ar-O_2								
	311.2	0.4640	0.9056	1.345	0.8357	1.421	0.9238	1.327
	313.2	0.2490	0.8212	1.247	0.7569	1.316	0.8379	1.231
	338.2	0.7530	0.7438	1.122	0.6833	1.179	0.7594	1.108
	366.2	0.4840	0.7944	1.184	0.7326	1.250	0.8104	1.168

TABLE 2. RECOMMENDED SETS OF Φ_{ij} FOR THE THERMAL CONDUCTIVITY DATA (continued)

Gas Pair	Temp. (K)	Mole Fraction of Heavier Component	First Method Φ_{12}	Φ_{21}	Second Method Φ_{12}	Φ_{21}	Third Method Φ_{12}	Φ_{21}
He-Air	328.3	0.1814	0.4211	2.573	0.2256	2.897	0.5712	2.471
He-n-C_4H_{10}	328.4	0.1941	0.5021	4.340	0.2304	4.733	0.7245	4.192
He-CO_2	273.2	0.2600	0.3113	3.049	0.1442	3.281	0.4314	2.950
He-C_3H_6	328.4	0.1145	0.4397	3.908	0.2058	4.230		
	328.4	0.8957					0.5955	3.719
He-D_2	303.2	0.2965	1.012	1.126	1.010	1.128	1.013	1.125
	323.2	0.2965	0.9951	1.102	0.9927	1.105	0.9957	1.102
	343.2	0.4777	0.9908	1.096	0.9885	1.099	0.9914	1.096
	363.2	0.4777	0.9878	1.092	0.9855	1.094	0.9884	1.091
He-C_2H_6	328.4	0.7184	0.4995	3.256	0.2545	3.681	0.6385	3.076
He-CH_2CH_2	328.4	0.3487	0.4676	3.188	0.2360	3.532	0.6020	3.065
He-H_2	303.2	0.4562	0.9625	1.145	0.7778	1.370	1.018	1.093
	318.2	0.4136	0.9646	1.157	0.7783	1.382	1.021	1.105
	90.2	0.3890	1.235	1.063	1.024	1.305	1.297	1.008
	258.3	0.7010	1.036	1.187	0.8480	1.438	1.092	1.129
	273.3	0.7010	1.033	1.199	0.8451	1.452	1.090	1.141
	293.3	0.7010	1.016	1.170	0.8296	1.415	1.071	1.114
	393.2	0.8040	1.114	1.302	0.9244	1.599	1.296	1.108
	473.3	0.3960	1.008	1.172	0.8144	1.403	1.066	1.119
	303.3	0.8014	1.019	1.222	0.8343	1.481	1.074	1.162
	328.2	0.4961	0.9927	1.192	0.8024	1.426	1.050	1.137
	378.3	0.5092	0.9828	1.188	0.7942	1.421	1.040	1.133
	353.4	0.4956	0.9967	1.206	0.8052	1.443	1.054	1.151
	398.2	0.9215	1.020	1.235	0.8389	1.504	1.073	1.173
He-CH_4	589.2	0.7010	0.7762	2.127			0.8922	1.985
	589.2	0.4500			0.4864	2.592		
He-N_2	273.2	0.7400	0.4853	2.793	0.2492	3.148	0.6151	2.644
	303.2	0.1136	0.4498	2.673	0.2304	3.005	0.5806	2.578
	318.2	0.5019	0.4644	2.739	0.2366	3.063	0.5931	2.612
	377.2	0.7810	0.4589	2.601	0.2335	2.905	0.5832	2.469
	589.2	0.2610	0.4955	2.589	0.2560	2.935	0.6343	2.476
He-O_2	303.2	0.4319	0.4673	2.638	0.2349	2.968	0.6071	2.509
	318.2	0.8539	0.4700	2.648	0.2360	2.976	0.6053	2.497
He-C_3H_8	328.4	0.3618	0.5179	3.995	0.2484	4.449	0.7081	3.811
He-CH_2CHCH_3	328.4	0.3243	0.4761	3.793	0.2272	4.186	0.6497	3.637
Kr-D_2	308.2	0.4460	0.1898	2.685	0.08078	2.775	0.2920	2.620
	323.2	0.4460	0.1883	2.588	0.08017	2.675	0.2896	2.525
	343.2	0.4460	0.1847	2.511	0.07860	2.594	0.2841	2.450
	363.2	0.4460	0.1809	2.455	0.07693	2.535	0.2784	2.397

TABLE 2. RECOMMENDED SETS OF Φ_{ij} FOR THE THERMAL CONDUCTIVITY DATA (continued)

Gas Pair	Temp. (K)	Mole Fraction of Heavier Component	First Method Φ_{12}	First Method Φ_{21}	Second Method Φ_{12}	Second Method Φ_{21}	Third Method Φ_{12}	Third Method Φ_{21}
Kr-H$_2$								
	313.2	0.4690	0.1323	2.450	0.05419	2.495	0.2271	2.405
	338.2	0.4690	0.1398	2.514	0.05736	2.564	0.2397	2.465
	366.2	0.6530	0.1341	2.389	0.05472	2.424	0.2308	2.351
	303.2	0.7326	0.1424	2.772	0.05811	2.812	0.2451	2.727
	318.2	0.2431	0.1362	2.599	0.05652	2.682	0.2322	2.533
Kr-N$_2$								
	303.2	0.5364	0.6440	1.713	0.4249	1.999	0.7248	1.636
	318.2	0.1872	0.5627	1.535	0.3751	1.809	0.6335	1.466
Kr-O$_2$								
	303.2	0.3631	0.5860	1.682	0.3987	1.921	0.6523	1.620
	318.2	0.6059	0.6274	1.805	0.4271	2.063	0.6970	1.735
Ne-CO$_2$								
	273.2	0.5300	0.5553	1.772	0.3963	1.959	0.6069	1.723
Ne-D$_2$								
	303.2	0.2558	0.6019	1.637	0.3565	1.996	0.7186	1.535
	323.2	0.6534	0.6103	1.657	0.3553	1.986	0.7271	1.551
	343.2	0.4547	0.6075	1.665	0.3558	2.007	0.7248	1.560
	363.2	0.4547	0.6167	1.687	0.3617	2.036	0.7356	1.580
Ne-H$_2$								
	303.2	0.7283	0.5009	1.856	0.2519	2.146	0.6569	1.723
	318.2	0.6041	0.4786	1.788	0.2405	2.067	0.7420	1.590
	313.2	0.7486	0.4863	1.817	0.2428	2.087	0.6392	1.691
	338.2	0.4050	0.4472	1.623	0.2285	1.907	0.5871	1.508
	366.2	0.4050	0.4460	1.614	0.2279	1.897	0.5854	1.500
	368.2	0.2930	0.5256	1.946	0.2725	2.321	0.6907	1.810
	408.2	0.2930	0.5213	1.962	0.2697	2.335	0.6858	1.827
	448.2	0.2930	0.4277	1.580	0.2220	1.887	0.5600	1.464
Ne-N$_2$								
	303.2	0.5504	0.7631	1.454	0.6710	1.557	0.7879	1.429
	318.2	0.5063	0.7862	1.475	0.6918	1.582	0.8116	1.450
	313.2	0.5110	0.7743	1.431	0.6815	1.535	0.7992	1.407
	338.2	0.5110	0.8034	1.479	0.7077	1.588	0.8291	1.453
	366.2	0.5110	0.7973	1.436	0.7028	1.542	0.8227	1.410
	368.2	0.2560	1.023	1.914	0.8909	2.031	1.059	1.886
	408.2	0.2560	0.8665	1.570	0.7597	1.677	0.8956	1.544
	448.2	0.7350	0.9016	1.597	0.8022	1.731	0.9281	1.565
Ne-O$_2$								
	303.2	0.5904	0.7548	1.360	0.6329	1.499	0.7890	1.327
	318.2	0.2580	0.7987	1.439	0.6685	1.583	0.8357	1.405
	313.2	0.4960	0.7479	1.320	0.6277	1.456	0.7817	1.287
	338.2	0.4960	0.7380	1.357	0.6179	1.493	0.7717	1.324
	366.2	0.7390	0.7609	1.366	0.6383	1.507	0.7951	1.332
	368.2	0.4920	0.9657	1.687	0.8152	1.872	1.008	1.643
	408.2	0.4920	0.8248	1.441	0.6940	1.594	0.8616	1.405
	448.2	0.2290	0.8531	1.449	0.7149	1.596	0.8925	1.415
Xe-D$_2$								
	311.2	0.4000	0.1195	2.577	0.04919	2.620	0.1984	2.538
	366.2	0.6000	0.1222	2.437	0.05010	2.468	0.2034	2.405
	313.2	0.4960	0.1165	2.652	0.04780	2.687	0.1940	2.618
	338.2	0.4960	0.1146	2.636	0.04698	2.670	0.1908	2.603
	366.2	0.7590	0.1201	2.541	0.04912	2.568	0.2003	2.514
Xe-H$_2$								
	303.2	0.4379	0.08454	2.894	0.03399	2.916	0.1567	2.866
	318.2	0.4039	0.07901	2.676	0.03176	2.696	0.1464	2.650
	313.2	0.6080	0.09302	2.857	0.03738	2.877	0.1724	2.830
	338.2	0.4340	0.08556	2.677	0.03442	2.699	0.1584	2.649
	366.2	0.1600	0.09082	2.619	0.03714	2.685	0.1662	2.561

TABLE 2. RECOMMENDED SETS OF Φ_{ij} FOR THE THERMAL CONDUCTIVITY DATA (continued)

Gas Pair	Temp. (K)	Molecular Fraction of Heavier Component	First Method Φ_{12}	Φ_{21}	Second Method Φ_{12}	Φ_{21}	Third Method Φ_{12}	Φ_{21}
Xe-N_2								
	303.2	0.7265	0.4300	2.095	0.2336	2.306	0.5214	2.015
	318.2	0.7021	0.4087	1.959	0.2209	2.146	0.4966	1.888
Xe-O_2								
	303.2	0.4241	0.3984	1.983	0.2246	2.189	0.4757	1.916
	318.2	0.8543	0.4022	1.991	0.2224	2.156	0.4814	1.928
CHCH-Air								
	293.2	0.5360	0.9018	1.037	0.8708	1.069	0.9096	1.029
	338.2	0.6460	0.9206	0.9818	0.8897	1.013	0.9283	0.9743
Air-CO								
	291.2	0.3210	0.9670	1.017	0.9569	1.027	0.9695	1.015
Air-CH_4								
	295.2	0.6100	0.9079	1.086	0.7523	1.269	0.9529	1.044
CO_2-C_3H_8								
	369	0.7088	1.127	0.9244	1.127	0.9250	1.127	0.9243
D_2-N_2								
	313.2	0.6010	0.3951	1.990	0.1990	2.197	0.5050	1.902
	338.2	0.6010	0.4145	2.044	0.2101	2.271	0.5286	1.949
	366.2	0.6010	0.4119	1.967	0.2089	2.187	0.5251	1.875
	368.2	0.4960	0.4017	2.027	0.2040	2.256	0.5128	1.934
	408.2	0.4960	0.4112	2.007	0.2097	2.243	0.5241	1.912
	448.2	0.4960	0.4025	1.941	0.2051	2.167	0.5132	1.850
C_2H_5OH-$C_2H_4O_2$								
	374	0.5159	0.8643	1.233	0.7866	1.318	0.8847	1.213
C_2H_5OH-C_3H_8								
	368	0.5017	1.064	0.9849	1.051	0.9987	1.068	0.9815
H_2-CO_2								
	296	0.5950	0.2813	3.004	0.1219	3.168	0.4291	2.886
	273.2	0.8299	0.2287	2.808	0.09711	2.902	0.3529	2.729
	273.2	0.5000	0.2499	2.804	0.1077	2.941	0.3831	2.707
	298.0	0.0941	0.2621	2.810	0.1185	3.091	0.3992	2.695
H_2-CO								
	273.2	0.4340	0.3143	2.396	0.1433	2.589	0.4463	2.292
H_2-D_2								
	273.2	0.5040	0.8976	1.218	0.7151	1.442	0.9528	1.166
	313.2	0.2530	0.9637	1.300	0.7621	1.528	1.026	1.247
	338.2	0.2530	0.9828	1.336	0.7755	1.567	1.047	1.283
	366.2	0.4970	0.9208	1.255	0.7337	1.486	0.9775	1.201
	368.2	0.4880	0.9462	1.299	0.7537	1.538	1.005	1.243
	408.2	0.2430	0.9793	1.368	0.7700	1.598	1.044	1.315
	448.2	0.2430	0.9167	1.244	0.7270	1.466	0.9750	1.193
H_2-C_2H_4								
	298.2	0.3890	0.2973	2.465	0.1350	2.653	0.4235	2.366
H_2-N_2								
	313.2	0.5920	0.3041	2.065	0.1375	2.213	0.4320	1.977
	338.2	0.5920	0.3161	2.115	0.1435	2.275	0.4482	2.020
	366.2	0.3380	0.3295	2.155	0.1536	2.379	0.4645	2.046
	368.2	0.5130	0.3093	2.143	0.1407	2.310	0.4388	2.048
	408.2	0.8800	0.3139	2.140	0.1406	2.271	0.4468	2.053
	448.2	0.5130	0.3245	2.123	0.1486	2.303	0.4589	2.023
	273.2	0.3480	0.3813	2.801	0.1772	3.083	0.5392	2.669
	298.5	0.5940	0.3261	2.214	0.1484	2.387	0.4619	2.114
	348.0	0.7110	0.3253	2.308	0.1470	2.472	0.4616	2.207
	372.3	0.6045	0.3271	2.197	0.1488	2.369	0.4632	2.097
	422.5	0.3105	0.3256	2.160	0.1520	2.389	0.4589	2.052
	298.0	0.6800	0.3659	2.566	0.1677	2.787	0.5156	2.437

TABLE 2. RECOMMENDED SETS OF Φ_{ij} FOR THE THERMAL CONDUCTIVITY DATA (continued)

Gas Pair	Temp. (K)	Mole Fraction of Heavier Component	First Method Φ_{12}	Φ_{21}	Second Method Φ_{12}	Φ_{21}	Third Method Φ_{12}	Φ_{21}
H_2-N_2O	273.2	0.4010	0.2577	2.739	0.1120	2.897	0.3940	2.638
H_2-O_2	313.2	0.4910	0.3241	2.099	0.1474	2.284	0.4661	1.994
	338.2	0.4910	0.3102	2.068	0.1404	2.239	0.4471	1.969
	366.2	0.4910	0.3104	2.023	0.1407	2.193	0.4471	1.925
	295.2	0.2500	0.3226	2.158	0.1506	2.409	0.4622	2.042
CH_4-C_3H_8	368	0.5145	0.9059	1.477	0.6531	1.824	0.9938	1.393
N_2-CO_2	273.2	0.2500	0.7387	1.257	0.6253	1.392	0.7706	1.226
	323.2	0.4712	0.8276	1.266	0.7037	1.408	0.8622	1.233
	423.2	0.4712	0.8444	1.119	0.7244	1.256	0.8777	1.087
	523.2	0.6594	0.9048	1.064	0.7816	1.202	0.9388	1.032
	623.2	0.3350	0.9137	0.9689	0.7976	1.106	0.9455	0.9370
	642.2	0.2500	0.9512	1.020	0.8287	1.162	0.9849	0.9868
	645.2	0.5000	0.9400	1.009	0.8179	1.148	0.9736	0.9764
	648.2	0.5000	0.9289	0.9854	0.8088	1.122	0.9619	0.9535
	842.2	0.5000	0.9437	0.9090	0.8270	1.042	0.9755	0.8781
	846.2	0.5000	0.9399	0.9074	0.8236	1.040	0.9717	0.8766
	961.2	0.5000	0.9472	0.8896	0.8316	1.021	0.9788	0.8590
N_2-C_2H_4	591	0.4980	0.7807	1.120	0.7803	1.120	0.7808	1.119
N_2-O_2	313.2	0.5290	1.053	1.004	1.012	1.047	1.063	0.9939
	338.2	0.2490	0.9948	0.9914	0.9554	1.033	1.005	0.9815
	366.2	0.5290	0.9986	0.9954	0.9589	1.037	1.009	0.9855
	368.2	0.2270	0.9979	0.9315	0.9603	0.9722	1.007	0.9217
	408.2	0.5140	0.9239	0.8914	0.8877	0.9290	0.9331	0.8824
	448.2	0.5140	0.9979	0.9570	0.9589	0.9975	1.008	0.9473
O_2-CO_2	370	0.4644	0.8669	1.257	0.9714	1.162	0.8427	1.282
NH_3-Air	293.2	0.3920	0.9006	0.8250	0.7793	0.9752	0.9342	0.7902
	353.2	0.4240	0.8492	0.8917	0.7254	1.041	0.8839	0.8570
NH_3-CO	295.2	0.2100	0.8969	0.8672	0.7839	1.017	0.9278	0.8326
NH_3-C_2H_4	298.2	0.2268	0.8299	0.9911	0.7120	1.142	0.8628	0.9560
NH_3-H_2	299	0.4220	0.2353	1.618			0.3135	1.565
	299	0.7980			0.1090	1.689		
	298.5	0.5230	0.2640	1.909	0.1242	2.025	0.3525	1.851
	348.0	0.8315	0.2630	1.744	0.1219	1.821		
	348.0	0.7510					0.3290	1.583
	372.3	0.5980	0.2877	1.882	0.1359	2.004	0.3833	1.820
	422.5	0.4620	0.2972	1.803	0.1428	1.953	0.3936	1.734
NH_3-N_2	300	0.3950	0.4257	0.4342	0.3743	0.5122	0.4393	0.4158
	298.5	0.5810	0.8908	0.8362	0.7696	0.9692	0.9243	0.8052
	348.0	0.6670	0.7458	0.7549	0.6332	0.8598	0.7770	0.7299
	372.3	0.6565	0.8450	0.8678	0.7221	0.9950	0.8791	0.8379
	422.5	0.7090	0.8804	0.9625	0.7505	1.101	0.9165	0.9300
Steam-Air	353.2	0.5560	0.8727	0.6659	0.7704	0.7789	0.9003	0.6398

gives the composition of the mixture and the corresponding ϕ_{ij} were indeed the best set to reproduce the data of all the ones listed in Table 1. Here, again it is interesting to note that for each method ϕ_{ij} does not vary much with temperature, though here again for different methods ϕ_{ij} are sufficiently different. These ϕ_{ij} may be used for calculation of thermal conductivity at high temperatures and for multicomponent systems.

It may be interesting to point out that Gray and Wright [190] and Wright [666] have described methods of finding out ϕ_{ij} from the experimental data. Huck and Thornton [242-3] on the other hand determined the whole range of coupled ϕ_{ij} values which will reproduce the conductivity data within a prescribed uncertainty. They [242-3] then attempt to develop, on empirical grounds, such relations for the product and ratio of ϕ_{ij} which may lead directly to an acceptable set. This as explained elsewhere [491] has an inherent limitation of employing an empirical scheme to determine that set which is theoretically acceptable. We have therefore preferred a theoretical approach, though a weak one unfortunately, and used the experimental data as guidance to pinpoint the best set.

6. SOURCES OF FURTHER INFORMATION

In this section we cite other available publications which deal, though less extensively than attempted here, with the collection and evaluation of thermal conductivity data, techniques of measurement, calculation, or estimation of thermal conductivity of gases and gaseous mixtures. Sources which refer extensively to thermal conductivity literature are also mentioned.

Lenoir [298] compiled recommended values for 29 inorganic and 50 organic gases at atmospheric pressure as well as tabulating Wassiljewa coefficients for binary gas mixtures. While the sources used for experimental values of the inorganic gases are listed, it is particularly unfortunate that no index for the organics is given. For some ten years after publication, many of the values cited in this publication, especially for the organic gases, were unique.

Hilsenrath et al. [223-4] compiled recommended values for air, argon, carbon dioxide, carbon monoxide, helium, hydrogen, nitrogen, nitric oxide, oxygen, and steam at atmospheric pressure. This was one of the earliest critical compilations over an extended temperature range and also about the pioneering publication of the departure plot concept used in the present work.

Johnson, V. J. (ed.), et al. [254] compiled the properties of materials at low temperatures. Included were values for the thermal conductivity in solid, liquid, and gaseous phases. Substances considered were air, argon, carbon monoxide, fluorine, helium, hydrogen, methane, neon, nitrogen, and oxygen. Not all the phases were considered for all these substances. The information is presented mainly as graphical or tabular citations of the original works cited and, for the most part, little critical analysis was made of the varying data.

Vargaftik, N. B. [617, 619, 621], published a collection of thermophysical properties of substances in the liquid and gaseous state. The values included are reproductions from original data sources. The work is valuable in that it probably gives the most comprehensive coverage presently available for many properties of many substances, also for making data available from many unobtainable Russian works. Its principal drawbacks are variable coverage of works foreign to the USSR and the absence of explanation of the various table sources. In several instances, the tabular material is unique. An up-to-date English language publication would undoubtedly receive great attention.

Tsederberg [71, 601] has written a monograph on the thermal conductivity of real gases (and liquids) and gas mixtures. This work is of Russian origin [600] and therefore cites much material which is not readily available. Its theoretical coverage for dilute gases is rather surprising. Almost no work since 1954 is cited. Experimental details appear much more adequate, as do techniques and results for high pressure. For such information the book is recommended.

Galloway and Sage [160-1] extensively analyzed the normal paraffin hydrocarbon series from methane to *n*-decane and have tested several intermolecular potentials. This publication appeared after the main work in this volume was completed.

Hanley et al. [79, 210-2] have analyzed data for some selected pure gases using intermolecular potentials. A preliminary comparison with the values appearing here yields reasonable agreement.

Saxena and coworkers [166, 170, 175, 516] have analyzed the available experimental data of rare gases and their binary mixtures, nine polyatomic gases, and thirty-two binary systems and recommended, critically evaluated, and graphically smoothed values at round temperatures and compositions. This group has also described the various methods very often employed for thermal conductivity measurement [490, 502] and estimation [171, 338, 463, 500-1, 513].

Svehla [577] has reported the estimated values of thermal conductivity of about 200 gases in the ground state at a pressure of one atmosphere and at suitable spaced temperatures in the range 100 to 5000 K.

Todheide, Hensel, and Franck [592] have reported thermal conductivity of pure gases at round temperatures and for some at round pressures and for binary systems at five compositions (mole fraction 0.1, 0.25, 0.50, 0.75, and 0.90). The survey includes data published till 1963.

Cheung [76, see also 77, 78] has collected data for 226 binary mixtures, *Westenberg* [656] has briefly described the major methods of measurement and calculation of thermal conductivity, and *Curtiss* [103] has written a very useful review of the theoretical work done to understand the complicated process of heat conduction in dilute and dense gases.

Reid and Sherwood [429] and *Missenard* [365] in their books refer to various thermal conductivity data and estimation methods.

Other sources to references may be cited. In addition to the well-known abstracting journals, the *Retrieval Guide* [598] provides a convenient index by substance for gaseous thermal conductivity. A more specialized survey for cryogenic fluids has appeared [83, 84] which presents graphical illustrations of available ranges of pressure and temperature for which experimental or correlated data are available. Some 16 pure fluids were considered as well as some binary mixtures. Earlier collections of references include [181, 301-6].

References to Text

1. Ahtye, W. F., "A Critical Evaluation of Methods for Calculating Transport Coefficients of a Partially Ionized Gas," *Proc. Heat Transfer Fluid Mech. Inst.*, 211–25, 1964.
2. Ahtye, W. F., "A Critical Evaluation of Methods for Calculating Transport Coefficients of Partially and Fully Ionized Gases," NASA TN D-2611, 110 pp., 1965.
3. Ahtye, W. F., "Total Thermal Conductivity of Partially and Fully Ionized Gases," *Phys. Fluids* **8**(10), 1918–9, 1965; **9**(1), 224, 1966.
4. Ahtye, W. F., "Calculation of the Total Thermal Conductivity of Ionized Gases," *Thermal Conductivity—Proceedings of the 7th Conference*, NBS Special Publication 302, 551–60, 1968; also NASA-TM-X-60857, 15 pp., 1967.
5. Ahtye, W. F. and Peng, T. C., "Approximation for the Thermodynamic and Transport Properties of High Temperature Nitrogen with Shock Tube Applications," NASA TN D-1303, 108 pp., 1962.
6. Allen, P. H. G., "Fluid Thermal Conductivity by a Transient Method," *Thermodynamic and Transport Properties of Gases, Liquids and Solids*, ASME, Symposium on Thermal Properties, 350–357, 1959.
7. Amdur, I., "High Temperature Transport Properties of Gases; Limitations of Current Calculating Methods in the Light of Recent Experimental Data," *Am. Inst. Chem. Eng.* **8**(4), 521–6, 1962.
8. Amdur, I., "Intermolecular Potentials from Scattering Experiments: Results, Applications, and Limitations," *Progress in International Research on Thermodynamic and Transport Properties*, ASME, 369–77, 1962.
9. Amdur, I. and Mason, E. A., "Properties of Gases at Very High Temperatures," *Phys. Fluids* **1**(5), 370–83, 1958.
10. Amdur, I. and Ross, J., "On the Calculation of Properties of Gases at Elevated Temperatures," *Combust. Flame* **2**, 412–20, 1958.
11. Andrews, T., *Proc. Roy. Irish Acad.* **1**, 465, 1840.
12. Anisimov, S. I., Kuznetsov, N. M., and Nogotov, E. F., "The Structure of Shock Waves in Air with Kinetics of Chemical Reactions and Excitation of Molecular Vibrations in the Nitrogen Taken into Account," *High Temperature* **2**(3), 304–10, 1964.
13. Anon., Avco Corporation Report, "Evaluation of High Temperature Gas Transport Properties," Final Report AVSSD-0414-67-RR, Space Systems Division, Aerophysics Laboratory, Wilmington, Massachusetts, 87 pp., 1966–67.
14. Archer, C. T., "Thermal Conduction in Hydrogen–Deuterium Mixtures," *Proc. Roy. Soc.* **A165**, 474–85, 1938.
15. Asinovskii, E. I. and Kirillin, A. V., "Experimental Determination of the Thermal Conductivities of Argon Plasma," *High Temperature* **3**(5), 633, 1965.
16. Asinovskii, E. I. and Kirillin, A. V., "Experimental Determination of the Thermal Conductivities of Argon Plasmas," *Intern. Chem. Eng.* **7**(2), 281–8, 1967.
17. Bade, W. L., Mason, E. A., and Yun, K. S., "Transport Properties of Dissociated Air," *Am. Rocket Soc. J.* **31**, 1151–3, 1961. Many earlier references are cited.
18. Bahethi, O. P. and Saxena, S. C., "Morse Potential Parameters for Hydrogen," *Indian J. Pure Appl. Phys.* **2**(8), 267–9, 1964.
19. Bahethi, O. P. and Saxena, S. C., "Intermolecular Potentials for Krypton," *Indian J. Pure Appl. Phys.* **3**(1), 12–15, 1965.
20. Bahethi, O. P., Gambhir, R. S., and Saxena, S. C., "Properties of Gases and Gaseous Mixtures with a Morse Potential," *Zeit. Naturforsch.* **19a**, 1478–85, 1964.
21. Baker, C. E., "Thermal Conductivities of Gaseous HF, DF, and the Equimolar HF–DF Mixture," *J. Chem. Phys.* **46**(7), 2846–8, 1967.
22. Baker, C. E. and Brokaw, R. S., "Thermal Conductivities of Gaseous H_2O, D_2O and the Equimolar H_2O–D_2O Mixtures," *J. Chem. Phys.* **40**(6), 1523–8, 1964.
23. Baker, C. E. and Brokaw, R. S., "Thermal Conductivities of Ordinary and Isotopically Substituted Polar Gases and Their Equimolar Mixtures," *J. Chem. Phys.* **43**(10), 3519–28, 1965.
24. Barua, A. K., "Thermal Conductivity and Eucken Type Correction for Binary Mixtures of N_2 with Some Rare Gases," *Physica* **25**, 1275–86, 1959.
25. Barua, A. K., "Thermal Conductivity and Eucken Type Factor for the Binary Mixtures H_2–He, H_2–Ne, H_2–Kr and H_2–Xe," *Indian J. Phys.* **34**(4), 169–83, 1960.
26. Barua, A. K. and Chakraborti, P. K., "Measurement of the Thermal Conductivity of Singly and Doubly Dissociating Nitrogen Peroxide," *J. Chem. Phys.* **36**(11), 2817–20, 1962.
27. Bearman, R. J. and Kirkwood, J. G., "Statistical Mechanics of Transport Processes. XI. Equations of Transport in Multicomponent Systems," *J. Chem. Phys.* **28**(1), 136–45, 1958.
28. Bennett, L. A. and Vines, R. G., "Thermal Conductivity of Organic Vapor Mixtures," *J. Chem. Phys.* **23**, 1587–91, 1955.
29. Bereskin, S. and Stewart, G. W., "Interesting Case where the Heat Flow Decreases the Heat Conductivity of Fluid," *Proc. Iowa Acad. Sci.* **48**, 305, 1941. [Abstract only.]
30. Biberman, L. M. and Takubov, I. T., "The State of a Gas behind a Strong Shock-Wave Front," *High Temperature* **3**, 309–320, 1965.

31. Birch, F., "The Electrical Resistance and the Critical Point of Mercury," *Phys. Rev.* **41**, 641–8, 1932.
32. Blais, N. B. and Mann, J. B., "Thermal Conductivity of Helium and Hydrogen at High Temperatures," *J. Chem. Phys.* **32**, 1459–65, 1960.
33. Bloom, H., Doroszkowski, A., and Tricklebank, S. B., "Molten Salt Mixtures: IX. The Thermal Conductivities of Molten Nitrate Systems," *Australian J. Chem.* **18**(8), 1171–76, 1965.
34. Bomelburg, H. J., "A Direct Method to Measure Thermal Diffusivity of Gases," BRL Rept. 1058, 28 pp., 1958. [AD 209 438]
35. Bomelburg, H. J., "The Heat Loss from Very Thin Heated Wires in Rarefied Gases," *Phys. Fluids* **2**, 717–8, 1959.
36. Bomelburg, H. J., "A Method for Determining the Heat Conductivity Coefficient of a Gas with a Single AC Heated Hot Wire," Ballistic Research Laboratories (Aberdeen Proving Ground, Maryland) Rept. No. 1107, May 1960.
37. Bondi, A., "Physical Properties of Molecular Crystals, Liquids and Glasses," Chapter 10, *Thermal Conductivity of Non-Associated Liquids*, 298–313, John Wiley and Sons, Inc., New York, 1968.
38. Borovik, E., "Formula for the Heat Conductivity of Liquids" (in Russian), *J. Exptl. Theoret. Phys. USSR* **18**, 48–51, 1948.
39. Bradley, J. N., *Shock Waves in Chemistry and Physics*, John Wiley and Sons, Inc., New York, 1962.
40. Braun, E., "Transport Coefficients in a Plasma. I. Pure Gas," *Phys. Fluids* **10**(4), 731–6, 1967.
41. Braun, E., "Transport Coefficients in a Plasma. II. Binary Mixtures," *Phys. Fluids* **10**(4), 737–40, 1967.
42. Breed, B. R., "Impossibility of Three Confluent Shocks in Two-Dimensional Irrotational Flow," *Phys. Fluids* **10**(1), 21–3, 1967.
43. Bridgman, P. W., "The Thermal Conductivity of Liquids under Pressure," *Proc. Am. Acad. Arts Sci.* **59**, 141–69, 1923.
44. Bridgman, P. W., *The Physics of High Pressure*, Bell and Sons, London, p. 315, 1952.
45. Briggs, D. G., "The Measurement of the Thermal Conductivity of Gases by a Transient Method," Ph.D. thesis, University of Minnesota, 137 pp., UM 66–8862, 1965.
46. Briggs, D. G., Goldstein, R. J., and Ibele, W. E., "Precision Measurement of the Thermal Conductivity of Gases in a Transient Hot-Wire Cell," *Proc. 4th Symposium Thermophysical Properties*, ASME, N.Y., 452 61, 1968.
47. Brokaw, R. S., "Estimating Thermal Conductivities for Nonpolar Gas Mixtures Simple Empirical Method," *Ind. Eng. Chem.* **47**, 2398–2400, 1955.
48. Brokaw, R. S., "Thermal Conductivity of Gas Mixtures in Chemical Equilibrium. II," *J. Chem. Phys.* **32**(4), 1005–6, 1960.
49. Brokaw, R. S., "Thermal Conductivity and Chemical Kinetics," *J. Chem. Phys.* **35**(5), 1569–80, 1961.
50. Brokaw, R. S., "Estimated Collision Integrals for the Exponential Attractive Potential," *Phys. Fluids* **4**(8), 944–6, 1961.
51. Brokaw, R. S., "Energy Transport in High Temperature and Reacting Gases," *Planetary Space Sci.* **3**, 238–52, 1961.
52. Brokaw, R. S., "Approximate Formulas for the Viscosity and Thermal Conductivity of Gas Mixtures," Pt. I, *J. Chem. Phys.* **29**(2), 391–7, 1958; ibid., Pt. II **42**(4): 1140–6, 1965.
53. Brokaw, R. S., "Transport Properties of High Temperature Gases," NASA TM X-52315, 15 pp. and 12 figures, 1967.
54. Brokaw, R. S., *Symposium on High Temperature Technology*, Stanford Research Institute, California, Sept. 17–20, 1967.
55. Brokaw, R. S. and Svehla, R. A., "Viscosity and Thermal Conductivity of the $N_2O_4 \rightleftharpoons 2NO_2$ System," *J. Chem. Phys.* **44**, 4643–5, 1966.
56. Buddenberg, J. W. and Wilke, C. R., "Calculation of Gas-Mixture Viscosities," *Ind. Eng. Chem.* **41**, 1345–7, 1949.
57. Burge, H. L. and Robinson, L. B., "Thermal Conductivities of He, Ne, Ar and of Their Mixtures," *J. Appl. Phys.* **39**(1), 51–4, 1968.
58. Burgoyne, J. H. and Weinberg, F., *Fourth Symposium on Combustion*, Williams and Wilkins Co., Baltimore, 294–302, 1953.
59. Burhorn, F., "Calculation and Measurement of the Thermal Conductivity of Nitrogen up to 13 000 K," *Z. Phys.* **155**, 42–58, 1959.
60. Burnett, D., "Viscosity and Thermal Conductivity of Gas Mixtures. Accuracy of Some Empirical Formulas," *J. Chem. Phys.* **42**(7), 2533–40, 1965.
61. Butler, J. N. and Brokaw, R. S., "Thermal Conductivity of Gas Mixtures in Chemical Equilibrium," *J. Chem. Phys.* **26**(6), 1636–43, 1957.
62. Callear, A. B. and Robb, J. C., "An Experimental Method of Measuring the Thermal Conductivity of Gases," *Trans. Faraday Soc.* **51**, 630–8, 1955.
63. Carey, C. A., Carnevale, E. H., and Marshall, T., "Heat Transfer and Ultrasonic Methods for the Determination of High Temperature Gas Transport Properties," *Proc. 6th Conf. Thermal Conductivity*, Dayton, Ohio, 177–234, AFML, 1966.
64. Carey, C. A., Carnevale, E. H., and Marshall, T., "Experimental Determination of the Transport Properties of Gases," Part II*, Parametrics, Technical Rept. AFML-TR-65-141, August 1964, 96 pp. [AD 804 604]
65. Carnevale, E. H., Carey, C. A., and Larson, G. S., "Experimental Determination of the Transport Properties of Gases," AFML-TR-65-141, 1–65, 1965. [AD 474 535]
66. Carnevale, E. H., Carey, C., Marshall, T., and Uva, S., "Experimental Determination of Gas Properties at High Temperatures and/or Pressures," AEDC-TR-68-105, June 1968, 121 pp. [AD 670 192]
67. Carnevale, E. H., Larson, G., Lynnworth, L. C., Carey, C., Panaro, M., and Marshall, T., "Experimental Determination of Transport Properties of High Temperature Gases," NASA CR-789, 113 pp., June 1967.
68. Carnevale, E. H., Lynnworth, L. C., and Larson, G. S., "Ultrasonic Determination of Transport Properties of Monatomic Gases at High Temperatures," *J. Chem. Phys.* **46**(8), 3040–7, 1967.
69. Carnevale, E. H., Wolnik, S., Larson, G., Carey, C., and Waves, G. W., "Simultaneous Ultrasonic and Line

*Heat Transfer and Ultrasonic Measurements, September 1966.

Reversal Temperature Determination in a Shock Tube," *Phys. Fluids* **10**(7), 1459–67, 1967.

70. Cecil, O. B. and Munch, R. H., "Thermal Conductivity of Some Organic Liquids," *Ind. Eng. Chem.* **48**(3), 437–40, 1956.

71. Cess, R. D. (ed.), *Thermal Conductivity of Gases and Liquids*, English translation, the MIT Press, Massachusetts Institute of Technology, Cambridge, Mass., 246 pp., 1965. (See reference 600.)

72. Chakraborti, P. K., "Thermal Conductivity of Dissociating Phosphorous Pentachloride," *J. Chem. Phys.* **36**(3), 575–7, 1963.

73. Challoner, A. R. and Powell, R. W., "Thermal Conductivities of Liquids: New Determinations for Seven Liquids and Appraisal of Existing Values," *Proc. Roy. Soc.* (*London*) **A238**, 90–106, 1956.

74. Chaos, F. and Garcia-Colin, L. S., "Density Expansions of the Transport Coefficients for a Moderately Dense Gas," *Phys. Fluids* **9**, 382–9, 1966.

75. Chapman, S. and Cowling, T. G., *The Mathematical Theory of Non-Uniform Gases*, Cambridge University Press, London, 431 pp., 1953.

76. Cheung, H., "Thermal Conductivity and Viscosity of Gas Mixtures," UCRL-8230, 146 pp., 1958.

77. Cheung, H., Bromley, L. A., and Wilke, C. R., "Thermal Conductivity and Viscosity of Gas Mixtures," UCRL-8230 Rev., 64 pp., 1959.

78. Cheung, H., Bromley, L. A., and Wilke, C. R., "Thermal Conductivity of Gas Mixtures," *Am. Inst. Chem. Eng. J.* **8**(2), 221–8, 1962.

79. Childs, G. E. and Hanley, H. J. M., "The Viscosity and Thermal Conductivity Coefficients of Dilute Nitrogen and Oxygen," N.B.S. TN 350, 1–27, 1966; NASA-CR-81126, 33 pp., 1966.

80. Chiquillo, A., "Measurement of the Relative Thermal Conductivities of Aqueous Salt Solutions by an Unsteady State Hot-Wire Method" (in German), Technical High School, Zurich, Dissertation, No. 3955, 1967.

81. Choh, S. T., "The Kinetic Theory Phenomena in Dense Gases," Ph.D. Thesis, University of Michigan, 1958.

82. Christiansen, C., "Einige Versuche über die Warmeleitung," *Ann. Physik.* **14**: 23–33, 1881.

83. Clark, R. G., Hyman, F. L., and Wilson, G. M., "Literature Survey on Refrigerants Essential to Cryocooler Technology," AFML-TR-66-136, 1–486, 1966. [AD488257L]

84. Clark, R. G., Kuebler, G. P., Weimer, R. F., and Bailey, B., "Research on Materials Essential to Cryocooler Technology," AFML-TR-67-229, 1–226, 1967.

85. Clingman, W. H., Brokaw, R. S., and Pease, R. N., "Burning Velocities of Methane with Nitrogen, Oxygen, Argon–Oxygen and Helium–Oxygen Mixtures," *4th Symp.* (*Int.*) *Combustion*, Williams and Wilkins Co., Baltimore, Maryland, 310–3, 1953.

86. Cohen, E. G. D., "On the Statistical Mechanics of Moderately Dense Gases Not In Equilibrium," *Lect. Theo. Phys.* **8A**, 145–86, University Colorado Press, Boulder, Colorado, 1966.

87. Cohen, E. G. D., "Kinetic Theory of Dense Gases," *Lect. Theo. Phys.* **9C**, 279–333, Gordon and Breach, 1967. See also references cited by Cohen and others in the 1967 volume.

88. Collins, D. J. and Menard, W. A., "Measurement of the Thermal Conductivity of Noble Gases in the Temperature Range 1500 to 5000 deg Kelvin," *J. Heat Transfer, Trans. ASME, Series C* **88**, 52–6, 1966.

89. Collins, D. J., Greif, R., and Bryson, A. E., Jr., "Measurements of the Thermal Conductivity of Helium in the Temperature Range 1600–6700 K," *Intern. J. Heat Mass Transfer* **8**, 1209–16, 1965.

90. Collins, F. C. and Raffel, H., "Statistical Mechanical Theory of Transport Processes in Liquids," *J. Chem. Phys.* **29**(4), 699–710, 1958.

91. Comings, E. W., Lee, W., and Kramer, F. R., "A Cylindrical Thermal Conductivity Cell for Gases at Pressures to 3000 Atmospheres," *Proc. Joint Conf. on Thermodynamics and Transport Properties of Fluids, London, July 1957*, 188–92, (Publ.) Inst. Mech. Engrs., London, 1958.

92. Compan, M., "Radiation Laws at Low Temperatures," *Compt. Rend.* **133**, 813–5, 1901.

93. Condiff, D. W., Lu, W. K., and Dahler, J. S., "Transport Properties of Polyatomic Fluids; A Dilute Gas of Perfectly Rough Spheres," *J. Chem. Phys.* **42**(10), 3445–75, 1965.

94. Cooke, J. W., "Experimental Determination of the Thermal Conductivity of Molten Lithium from 320 to 830 C," *J. Chem. Phys.* **40**(7), 1902–9, 1964.

95. Cotton, J. E., "Thermal Conductivity of Binary Mixtures of Gases," Ph.D. Thesis, Chemistry Dept., Univ. of Oregon, Eugene, Oregon, 88 pp., 1962.

96. Cottrell, T. L. and McCoubrey, J. C., *Molecular Energy Transfer in Gases*, Butterworths Scientific Publications Ltd., London, 1962.

97. Cowling, T. G., "Appendix: The Theoretical Basis for Wassiljewa's Equation," *Proc. Roy. Soc.* (*London*) **A263**, 186–8, 1961.

98. Cowling, T. G., "Heat Conductivity of Polyatomic Gases," *Brit. J. Appl. Phys.* **15**, 959–62, 1964.

99. Cowling, T. G., Gray, P., and Wright, P. G., "The Physical Significance of Formulae for the Thermal Conductivity and Viscosity of Gaseous Mixtures," *Proc. Roy. Soc.* **A276**, 69–82, 1963.

100. Curie, M. and Lepape, A., "Thermal Conductivity of the Rare Gases," *Compt. Rend.* **193**, 842–3, 1931.

101. Curie, M. and Lepape, A., "Thermal Conductivity of Rare Gases," *J. Phys. Radium* **2**, 392–7, 1931.

102. Curtiss, C. F., "Kinetic Theory of Nonspherical Molecules," *J. Chem. Phys.* **24**(2), 225–41, 1956.

103. Curtiss, C. F., "Transport Phenomena in Gases," *Ann. Rev. Phys. Chem.* **18**, 125–34, 1967.

104. Curtiss, C. F. and Dahler, J. S., "Kinetic Theory of Nonspherical Molecules V," *J. Chem. Phys.* **38**(10), 2352–62, 1963.

105. Curtiss, C. F. and Muckenfuss, C., "Kinetic Theory of Nonspherical Molecules II," *J. Chem. Phys.* **26**(6), 1619–36, 1957.

106. Curtiss, C. F., McElroy, M. B., and Hoffman, D. K., "The Transport Properties of a Moderately Dense Lennard-Jones Gas," *Intern. J. Eng. Sci.* **3**, 269–83, 1965.

107. Dahler, J. S., "Transport Phenomena in a Fluid Composed of Diatomic Molecules," *J. Chem. Phys.* **30**(6), 1447–75, 1959.

108. Dahler, J. S. and Sather, N. F., "Kinetic Theory of

Loaded Spheres, I," *J. Chem. Phys.* **38**(10), 2362–82, 1963.

109. Davidson, J. M. and Music, J. F., "Experimental Thermal Conductivities of Gases and Gaseous Mixtures at Zero Centigrade," Rept. HW 29021, 7–30, 1953. [AD18 359]
110. Davis, H. T., Rice, S. A., and Singers, J. V., "On the Kinetic Theory of Dense Fluids. IX. The Fluids of Rigid Spheres with a Square-Well Attraction," *J. Chem. Phys.* **35**, 2210–33, 1961.
111. Daynes, H. A., *Gas Analysis by Measurement of Thermal Conductivity*, Cambridge University Press, London, 1933.
112. Delaplace, R., "Thermal Conductivities of Saturated Gaseous Hydrocarbons at Low Pressures," *Compt. Rend.* **203**, 1505–7, 1936.
113. Desloge, E. A., "Collision Term in the Boltzmann Transport Equation," *Am. J. Phys.* **28**(1), 1–11, 1960.
114. Desloge, E. A., "Coefficients of Diffusion, Viscosity, and Thermal Conductivity of a Gas," *Am. J. Phys.* **30**(12), 911–20, 1962.
115. Desloge, E. A., "Transport Properties of a Simple Gas," *Am. J. Phys.* **32**(10), 733–42, 1964.
116. Desloge, E. A., "Transport Properties of a Gas Mixture," *Am. J. Phys.* **32**(10), 742–8, 1964.
117. Devoto, R. S., "Transport Coefficients of Partially Ionized Monatomic Gases," *Phys. Fluids* **9**(6), 1230–40, 1966.
118. Devoto, R. S., "Transport Coefficients of Partially Ionized Argon," *Phys. Fluids* **10**(2), 354–64, 1967.
119. Devoto, R. S., "Tables of the Compositions and Transport Coefficients of Partially Ionized Argon," A Dept. of Aeron. Astron. Rept., Stanford University, undated.
120. Dick, M. F. and McCready, D. W., "The Thermal Conductivities of Some Organic Liquids," *Trans. ASME* **76**, 831–9, 1954.
121. Dickins, B. G., "The Effect of Accommodation on Heat Conduction through Gases," *Proc. Roy. Soc.* **A143**, 517–40, 1934.
122. Diller, D. E. and Mason, E. A., "Low-Temperature Transport Properties of Gaseous H_2, D_2 and HD," *J. Chem. Phys.* **44**, 2604–9, 1966. [AD634 900]
123. Dimiduk, P., "A Bibliography of References for the Thermophysical Properties of Helium-4, Hydrogen, Deuterium, Hydrogen Deuteride, Neon, Argon, Nitrogen, Oxygen, Carbon Dioxide, Methane, Ethane, Krypton and Refrigerants 13, 14 and 23," NBS Rept. 8808, 1965, AFML-TR-65-338, 1965. [AD467 519 and AD480 173]
124. Dorfman, J. R., "Transport Coefficients for Dense Gases," *Dynamics of Fluids and Plasmas*, 199–212, Academic Press, New York, 1966.
125. Dorfman, J. R. and Cohen, E. G. D., "On the Density Expansion of the Pair Distribution Functions for a Dense Gas Not in Equilibrium," *Phys. Letters* **16**, 124–5, 1965.
126. Dorfman, J. R. and Cohen, E. G. D., "Difficulties in the Kinetic Theory of Dense Gases," *Phys. Fluids* **8**(2), 282–97, 1967.
127. Drudzhaliev, E. A., "On the Theory of Shock Waves in the Dynamics of a Real Gas," *Intern. J. Heat Mass Transfer* **6**, 935–40, 1963.
128. Dulong, P. L. and Petit, A. T., "On the measurement of Temperature and the Law of the Transfer of Heat," *Ann. Chem. Phys.* **7**, 113–54, 225–64, 337–67, 1817.
129. Eckerlein, P. A., "Ueber die Wärmeleitungsfähigkeit der Gase und ihre Abhängigkeit von der Temperatur (bei tiefen Temperaturen)," *Ann. Physik* **3**, 120–54, 1900.
130. Eckert, E. R. G. and Irvine, T. F., Jr., "Ein neues Verfahren zum Messen der Prandtlzahl und der Wärmeleitzahl von Gasen," *Forsch. Gebiete Ingenieurw.* **23**, 91–94, 1957.
131. Eckert, E. R. G. and Irvine, T. F., Jr., "A New Method to Measure Prandtl Number and Thermal Conductivity of Fluids," *J. Appl. Mechs.* **24**, 25–8, 1957.
132. Eckert, E. R. G., Ibele, W. E., and Irvine, T. F., Jr., "Thermal Conductivity of Helium–Air Mixtures," *Thermodynamic and Transport Properties of Gases, Liquids and Solids*, Symp., Lafayette, Indiana, 295–300, 1959.
133. Eckert, E. R. G., Ibele, W. E., and Irvine, T. F., Jr., "Prandtl Number, Thermal Conductivity and Viscosity of Air–Helium Mixtures," NASA TN D533, 1–39, 1960. [AD241 819]
134. Emmons, H. W., "Arc Measurement of High Temperature Gas Transport Properties," *Phys. Fluids* **10**, 1125–36, 1967.
135. Enskog, D., "Kinetic Theory of Processes in Moderately Low Pressure Gases," Inaugural Dissertation, Uppsala, Sweden, 1917. As quoted in ref. 273.
136. Enskog, D., "Kinetic Theory of Heat Conductivity, Viscosity and Diffusion in Certain Dense Gases and Liquids," (In German) *Kungl. Svenska Velenskapsakademiens Handlingar* **63**(4), 5–44, 1922.
137. Ernst, M. H., "Formal Theory of Transport Coefficients to General Order in the Density," *Physica* **32**, 209–43, 1966.
138. Ernst, M. H., "Hard Sphere Transport Coefficients from Time Correlation Functions," *Physica* **32**, 273–88, 1966.
139. Ernst, M. H., Dorfman, J. R., and Cohen, E. G. D., "Transport Coefficients in Dense Gases. I. The Dilute and Moderately Dense Gas," *Physica* **31**, 493–521, 1965.
140. Eucken, A., "Ueber die Temperaturabhängigkeit der Wärmeleitfähigkeit einiger Gase," *Physik. Z.* **12**, 1101–7, 1911.
141. Eucken, A., "On the Thermal Conductivity, Specific Heat and Viscosity of Gas," *Physik Z.* **14**, 324–32, 1913.
142. Evans, E. V. and Kenney, C. N., "A Flow Method for Determining the Thermal Conductivity of Gas Mixtures," *Nature* **203**, 184–5, 1964.
143. Ewing, C. T., Spann, J. R., and Miller, R. R., "Radiant Transfer of Heat in Molten Inorganic Compounds at High Temperatures," *J. Chem. Eng. Data* **7**(2), 246–50, 1962.
144. Ewing, C. T., Spann, J. R., and Miller, R. R., "Semitheoretical Relations for Heat Transfer in Dielectric Ceramics and Inorganic Liquids," *Proceedings of the 2nd Conference on Thermal Conductivity*, National Research Council, Ottawa, 202–14, October 1962.
145. Ewing, C. T., Walker, B. E., Grand, J. A., and Miller, R. R., "Thermal Conductivity of Metals," *Chem. Eng. Progress Symposium Series* **53**(20), 19–24, 1957.
146. Fay, J. A. and Kemp, N. H., "Theory of Heat Transfer to a Shock-Tube End Wall from an Ionized Monatomic Gas," *J. Fluid Mech.* **21**(4), 659–72, 1965.

147. Fieldhouse, I. B., Hedge, J. C., Lang, J. I., and Waterman, T. E., "Measurements of Thermal Properties," WADC Tech. Rept. 55-495, Part II, 1-18, 1956. [AD 110 510]
148. Filippov, L. P., "Measurement of the Thermal Conductivity of Liquids and Gases. Development of a Comparative Method Using a Flat Layer," (In Russian) *Vest. Moskov Univ. Ser. Fiz. Mat. i Est. Nauk* **8**(9), 109–14, 1953. For English Translation, SLA 61-10145 = TPRC Transl. 2, 1960.
149. Filippov, L. P., "The Influence of Radiation and Absorption by a Medium on the Process of Heat Transmission," (In Russian) *Vest. Mosk. Univ. Ser. Fiz. Mat. i Est. Nauk*. No. 2, 51–6, 1954.
150. Filippov, L. P., "Thermal Conductivity of Fifty Organic Liquids," (In Russian), *Vest. Mosk. Univ. Ser. Fiz. Mat. i Est. Nauk* **9**(12), 45–8, 1954.
151. Filippov, L. P., "Thermal Conductivity of Organic Liquids," (In Russian) *Vest. Mosk. Univ. Ser. Fiz. Astron.* **15**(3), 61–8, 1960.
152. Filippov, L. P., "Relative Methods of Measuring the Thermal Properties of Fluids," *Intern. Chem. Eng.* **2**(2), 182–4, 1962.
153. Filippov, L. P., "High-Temperature Investigations of Solid and Liquid Metal Properties," (In Russian) *Vest. Mosk. Gos. Univ. Ser. Fiz. Astron.*, No. 4, 90–4, 1964.
154. Filippov, L. P. "Liquid Thermal Conductivity Research at Moscow University," *Intern. J. Heat Mass Transfer* **11**, 331–45, 1968.
155. Fischer, J., "The Measurement of Thermal Conductivity and the Temperature Dependence from the Compensation Process with Schleiermacher's Measuring Tube Method and with the Plate Method," *Ann. Phys.* **34**, 669–88, 1939.
156. Flynn, D. R. and O'Hagan, M. E., "Measurements of the Thermal Conductivity and Electrical Resistivity of Platinum from 100 to 900 C," *J. Res. Natl. Bur. Stand., C. Engineering and Instrumentation* **71C**(4), 255–84, 1967.
157. Freidman, H. S. and Fay, J. A., "Heat Transfer from Argon and Xenon to the End Wall of a Shock Tube," *Phys. Fluids* **8**(11), 1968–75, 1965.
158. Fritz, W. and Poltz, H., "Absolute Determination of the Thermal Conductivity of Liquids. I. Critical Investigation of a New Parallel Plate Apparatus," *Intern. J. Heat Mass Transfer* **5**(2), 307–16, 1962.
159. Frontas'ev, V. P., "Heat Conductivity as a Method for Physical-Chemical Analysis of Binary Liquid Systems," (In Russian) *Zh. Fiz. Khim.* **20**(1), 91–104, 1946.
160. Galloway, T. R. and Sage, B. H., "Prediction of the Transport Properties of Paraffin Hydrocarbons," *Chem. Eng. Sci.* **22**, 979–95, 1967.
161. Galloway, T. R. and Sage, B. H., "Transport Properties of the Normal Paraffins at Attenuation," *J. Chem. Eng. Data* **12**, 59–65, 1967; Errata **13**(4), 598, 1968.
162. Gambhir, R. S. and Saxena, S. C., "Zero Pressure Joule-Thomson Coefficient for a Few Non-Polar Gases on the Morse Potential," *Indian J. Phys.* **37**(10), 540–2, 1963.
163. Gambhir, R. S. and Saxena, S. C., "Translational Thermal Conductivity and Viscosity of Multicomponent Gas Mixtures," *Trans. Faraday Soc.* **60**, 38–44, 1964.
164. Gambhir, R. S. and Saxena, S. C., "Thermal Conductivity of Binary and Ternary Mixtures of Krypton, Argon and Helium," *Mol. Phys.* **11**(3), 233–41, 1966.
165. Gambhir, R. S. and Saxena, S. C., "Thermal Conductivity of the Gas Mixtures: Ar–D_2, Kr–D_2, and Ar–Kr–D_2," *Physica* **32**, 2037–43, 1966.
166. Gambhir, R. S. and Saxena, S. C., "Thermal Conductivity of Common Non-Polar Polyatomic Gases," *Suppl. Defence Sci. J. (India)* **17**(2A), 35–46, 1967.
167. Gambhir, R. S., Gandhi, J. M., and Saxena, S. C., "Thermal Conductivity of Rare Gases, 'Deuterium and Air'," *Indian J. Pure Appl. Phys.* **5**, 457–63, 1967.
168. Gandhi, J. M. and Saxena, S. C., "An Approximate Method for the Simultaneous Prediction of Thermal Conductivity and Viscosity of Gas Mixtures," *Indian J. Pure Appl. Phys.* **2**(3), 83–5, 1964.
169. Gandhi, J. M. and Saxena, S. C., "Calculation of Translational Thermal Conductivity of Gas Mixtures at High Temperatures," *Indian J. Pure Appl. Phys.* **3**(8), 312–3, 1965.
170. Gandhi, J. M. and Saxena, S. C., "Thermal Conductivity of Monatomic Gases and Binary Gas Mixtures," *U. Rajasthan Studies* **1**, 7–24, 1965.
171. Gandhi, J. M. and Saxena, S. C., "Thermal Conductivity of Multicomponent Mixtures of Inert Gases: Part III—Some Simpler Methods of Computation," *Indian J. Pure Appl. Phys.* **4**(12), 461–6, 1966.
172. Gandhi, J. M. and Saxena, S. C., "Correlation between Thermal Conductivity and Diffusion of Gases and Gas Mixtures," *Proc. Phys. Soc. (London)* **87**, 273–9, 1966.
173. Gandhi, J. M. and Saxena, S. C., "Thermal Conductivities of the Gas Mixtures D_2–He, D_2–Ne and D_2–He–Ne," *Brit. J. Appl. Phys.* **18**, 807–12, 1967.
174. Gandhi, J. M. and Saxena, S. C., "Thermal Conductivity of Binary and Ternary Mixtures of Helium, Neon and Xenon," *Mol. Phys.* **12**(1), 57–68, 1967.
175. Gandhi, J. M. and Saxena, S. C., "Correlated Thermal Conductivity Data of Rare Gases and Their Binary Mixtures at Ordinary Pressures," *J. Chem. Eng. Data* **13**(3), 357–61, 1968.
176. Garcia-Colin, L. S. and Flores, A., "The Generalization of Choh–Uhlenbeck's Method in the Kinetic Theory of Dense Gases," *J. Math. Phys.* **7**, 254–9, 1966.
177. Garcia-Colin, L. S. and Flores, A., "On the Transport Coefficients of Moderately Dense Gas," *Physica* **32**, 289–303, 1966.
178. Garcia-Colin, L. S. and Flores, A., "Note on the Transport Coefficient of a Moderately Dense Gas," *Physica* **32**, 444–9, 1966.
179. Garcia-Colin, L. S., Green, M. S., and Chaos, F., "The Chapman–Enskog Solution of the Generalized Boltzmann Equation," *Physica* **32**, 450–78, 1966.
180. Gerts, I. G. and Filippov, L. P., "Critical Point Thermal Conductivity of Binary Systems," (In Russian), *Zh. Fiz. Khim.* **30** 2424–7, 1956.
181. Gillum, T. L., "Cryogenics and Low Temperature Research. An ASTIA Report Bibliography," 1–65, 1962, [AD 271 000]; Thompson, E. E., "Cryogenics and Low Temperature Research," 1–311, 1963. [AD 419 460]
182. Gilmore, T. F. and Comings, E. W., "Thermal Conductivity of Binary Mixtures of Carbon Dioxide, Nitrogen, and Ethane at High Pressures: Comparison

with Correlation and Theory," *Am. Inst. Chem. Eng. J.* **12**, 1172–8, 1966.
183. Glassman, I. and Bonilla, C. F., "Thermal Conductivity and Prandtl Number of Air at High Temperatures," *Chem. Eng. Progr. Sym. Ser.* **49**, 153–62, 1953.
184. Goldman, R. and Frieman, E. A., "Higher Order Considerations in the BBGKY Hierarchy for a Boltzmann Gas," *Bull. Am. Phys. Soc.* **10**, 531, 1965.
185. Goldschmidt, R., "On the Thermal Conductivity of Liquids," *Physik. Z.* **12**, 417–24, 1911.
186. Golubev, U. S., Kasabov, G. A., and Konakh, V. F., "Study of a Steady-State Argon–Cesium Plasma with Non-Equilibrium Conductivity," *High Temperature* **2**(4), 445–59, 1964.
187. Goodman, F. O. and Wachman, H. Y., "Formula for Thermal Accommodation Coefficients," *J. Chem. Phys.* **46**(6), 2376–86, 1967.
188. Graetz, L., "Ueber die Wärmeleitungsfähigkeit von Gasen und ihre Abhängigkeit von der Temperatur," *Wied. Ann.* **14**, 232–60, 1881.
189. Grassmann, P., Tauscher, W., and Chiquillo, A., "Measurement of the Thermal Conductivity of Refrigerants and Salt Solutions," in *Proceedings of the Fourth Symposium on Thermophysical Properties*, The American Society of Mechanical Eng., New York, 282–5, 1968.
190. Gray, P. and Wright, P. G., "The Thermal Conductivity of Mixtures of Nitrogen, Ammonia and Hydrogen," *Proc. Roy. Soc. (London)* **A263**, 161–88, 1961.
191. Gregory, H. S. and Archer, C. T., "The Variation of the Thermal Conductivity of Gases with Pressure," *Phil. Mag.* **1**, 593–606, 1926.
192. Gregory, H. S. and Archer, C. T., "Experimental Determination of the Thermal Conductivities of Gases," *Proc. Roy. Soc. (London)* **A110**, 91–122, 1926.
193. Gribkova, S. I., "Thermal Conductivity of the Vapors of Some Ethers," *J. Exp. Theoret. Phys. USSR* **11**, 364–71, 1941.
194. Grosse, A. V., "Thermal Conductivity of Liquid Metals over Their Entire Liquid Range, i.e., from Melting Point to Critical Point, in Relation to Their Electrical Conductivities and the Fallacy of Dividing Metals into 'Normal' and 'Abnormal' Thermally Conducting Ones," The Res. Inst. of Temple University, Philadelphia, Pa., USAEC TID-21737, 1–71, Report dated Dec. 7, 1964.
195. Grosse, A. V., "An Empirical Relationship between the Electrical Conductivity of Mercury and Temperature, over its Entire Liquid Range, also Its Thermal Conductivity and the Latter's Regular Behaviour," *J. Inorg. Nucl. Chem.* **28**, 803–11, 1966.
196. Grosse, A. V., "Electrical and Thermal Conductivities of Metals over Their Entire Liquid Range, i.e., from Melting Point to Critical Point, and the Containment of Metallic Substances up to 5000 K for Substantial Periods of Time," *Rev. Intern. Hautes Temper. Refract.* **3**(2), 115–46, 1966.
197. Gruss, H. and Schmick, H., "Ueber die Wärmeleitfähigkeit von Gasgemisch," *Wiss. Veröffentl. Siemens-Konzern* **7**, 202–24, 1928.
198. Gudzinowicz, B. J., Campbell, R. H., and Adams, J. S., Jr., "Thermal Conductivity Measurements of Complex Saturated Hydrocarbons," *J. Chem. Eng. Data* **9**(1), 79–82, 1964.
199. Guggenheim, E. A., "The International Encyclopedia of Physical Chemistry and Chemical Physics," Topic 6. "The Kinetic Theory of Gases," Volume 1, *Elements of the Kinetic Theory of Gases*, 92 pp., Pergamon Press, Oxford, 1960.
200. Guildner, L. A., "Thermal Conductivity of Gases. II. Thermal Conductivity of Carbon Dioxide Near the Critical Point," *J. Res. Natl. Bur. Std.* **66A**, 341–8, 1962.
201. Gupta, G. P. "Studies on Thermal Conductivity and Other Properties of Gases," Ph.D. Thesis, University of Rajasthan, Jaipur, India, Chapter 8, 243–57, 1968. [Available in microfiche form from TPRC, Purdue University, Lafayette, Indiana.]
202. Gupta, G. P. and Saxena, S. C., "Calculation of Thermal Conductivity of Polyatomic Gas Mixtures at High Temperatures," *Defence Sci. J.* **16**(3), 165–76, 1966.
203. Gupta, G. P. and Saxena, S. C., "Pressure Dependence of Thermal Conductivity in a Hot Wire Type of Cell," *Can. J. Phys.* **45**, 1418–20, 1967.
204. Gupta, G. P. and Saxena, S. C. "Prediction of Thermal Conductivity of Pure Gases and Mixtures," *Defence Sci. J.* **17**(2A), 21–34, 1967.
205. Gupta, G. P. and Saxena, S. C., "Thermal Conductivity of Multicomponent Mixtures of Monatomic and Diatomic Gases," to be published. Also contained in 201.
206. Gupta, G. P., Mathur, S., and Saxena, S. C., "Certain Methods for Calculating Thermal Conductivity of Multicomponent Mixtures Involving Polyatomic Gases," *Defence Sci. J.* **18**(4), 195–204, 1968.
207. Gutweiler, J. and Raw, C. J. G., "Transport Properties of Polar Gas Mixtures. II. Heat Conductivities of Ammonia–Methylamine Mixtures," *J. Chem. Phys.* **48**, 2413–5, 1968.
208. Haines, L. K., "On the Theory of Transport Coefficients for Moderately Dense Gases," Ph.D. Thesis, Technical Note BN-460, AROD-5960-1, 226 pp., June 1966, Univ. of Maryland, College Park, Maryland. [AD 637 798]
209. Haines, L. K., Dorfman, J. R., and Ernst, M. H., "Divergent Transport Coefficients and the Binary Collision Expansion," *Phys. Rev.* **144**, 207–15, 1966.
210. Hanley, H. J. M., "Comparison of the Lennard-Jones, exp-6, and Kihara Potential Functions from Viscosity Data of Dilute Argon," *J. Chem. Phys.* **44**(11), 4219–22, 1966.
211. Hanley, H. J. M., "The Viscosity and Thermal Conductivity Coefficients of Dilute Argon between 100 and 2000 K," NBS TN 333, 1–23, 1966; also NASA-CR-76397.
212. Hanley, H. J. M. and Childs, G. E., "The Viscosity and Thermal Conductivity Coefficients of Dilute Neon, Krypton and Xenon," NBS TN 352, 1–24, 1967; also NASA-CR-83899 and NASA-CR-87476.
213. Hanley, H. J. M. and Klein, M., "On the Selection of the Intermolecular Potential Function. Application of Statistical Mechanical Theory to Experiment," NBS Tech. Note 360, 82 pp., 1967; also NASA-CR-93126.
214. Hanley, H. J. M., McCarty, R. D., and Sengers, J. V., "Density Dependence of Experimental Transport Coefficients of Gases," *J. Chem. Phys.* **50**, 857–70, 1969.
215. Hansen, C. F., Early, R. A., Alzofon, F. E., and Witteborn, F. C., "Theoretical and Experimental Investigation

of Heat Conduction in Air, Including Effects of Oxygen Dissociation," NASA TR-R-27, 1959.
216. Hansen, C. F., "Interpretation of Linear Approximations for the Viscosity of Gas Mixtures," *Phys. Fluids* **4**(7), 926–7, 1961.
217. Hansen, R. S., Frost, R. R., and Murphy, J. A., "The Thermal Conductivity of Hydrogen–Helium Mixtures," *J. Phys. Chem.* **68**, 2028–9, 1964.
218. Helfand, E., "Transport Coefficients from Dissipation in a Canonical Ensemble," *Phys. Rev.* **119**, 1–9, 1960.
219. Hercus, E. O. and Laby, T. H., "Thermal Conductivity of Gases," *Phil. Mag.* **3**, 1061–4, 1927.
220. Hercus, E. O. and Laby, T. H., "The Thermal Conductivity of Air," *Proc. Roy. Soc. (London)* **A95**, 190–210, 1919.
221. Hercus, E. O. and Sutherland, D. M., "The Thermal Conductivity of Air by a Parallel Plate Method," *Proc. Roy. Soc. (London)* **A145**, 599–611, 1934.
222. Herzfeld, K. F. and Litovitz, T. A., *Absorption and Dispersion of Ultrasonic Waves*, Academic Press, Inc., New York, 1959.
223. Hilsenrath, J. and Touloukian, Y. S., "The Viscosity, Thermal Conductivity, and Prandtl Number for Air, O_2, N_2, NO, H_2, CO, CO_2, H_2O, He and A," *Trans. ASME* **76**, 967–85, 1954.
224. Hilsenrath, J., Beckett, C. W., Benedict, W. S., Fano, L., Hoge, H. J., Masi, J. F., Nuttall, R. L., Touloukian, Y. S., and Woolley, H. W., *Tables of Thermal Properties of Gases*, NBS Circular 564, 478 pp., 1955; *Tables of Thermodynamic and Transport Properties of Air, Argon, Carbon Dioxide, Carbon Monoxide, Hydrogen, Nitrogen, Oxygen, and Steam*, Pergamon Press, 478 pp., 1960.
225. Hirschfelder, J. O., "Heat Transfer in Chemically Reacting Mixtures. I," *J. Chem. Phys.* **26**(2), 274–81, 1957.
226. Hirschfelder, J. O., "Heat Conductivity in Polyatomic or Electronically Excited Gases. III," *J. Chem. Phys.* **26**(2), 282–5, 1957.
227. Hirschfelder, J. O., "Heat Conductivity in Polyatomic, Electronically Excited, or Chemically Reacting Mixtures. III," *Sixth International Combustion Symposium*, Reinhold Publishing Corporation, New York, 351–66, 1957.
228. Hirschfelder, J. O., Curtiss, C. F., and Bird, R. B., *Molecular Theory of Gases and Liquids*, Wiley, N. Y., 1219 pp., 1954; reprinted with notes added, 1249 pp., 1964.
229. Ho, C. Y., "Precise Determination of the Thermal Conductivity of Helium Gas at High Pressures and Moderate Temperatures," Ph.D. Thesis, Purdue University, Lafayette, Indiana, 1964.
230. Ho, C. Y. and Leidenfrost, W., "Precise Determination of the Thermal Conductivity of Helium Gas at High Pressures and Moderate Temperatures," *Proc. Sixth Conf. on Thermal Conductivity*, 105–58, 1966; also in *Progress in Heat and Mass Transfer*, Monograph Series of the International Journal of Heat and Mass Transfer, Vol. 1, Chap. 2, Pergamon Press, Oxford, 55–98, 1969.
231. Hoffman, D. K., "The Density Expansions of the Transport Coefficients," Univ. Wisconsin Theoretical Chemistry Institute Report WIS-TCI-48, 1–131, 1964, NASA-CR-56919.
232. Hoffman, D. K. and Curtiss, C. F., "Kinetic Theory of Dense Gases. III. The Generalized Enskog Equation," *Phys. Fluids* **7**, 1887–97, 1964.
233. Hoffman, D. K. and Curtiss, C. F., "Kinetic Theory of Dense Gases. IV. Transport Virial Coefficients," *Phys. Fluids* **8**, 667–82, 1965.
234. Hoffman, D. K. and Curtiss, C. F., "Kinetic Theory of Dense Gases. V. Evaluation of the Second Transport Virial Coefficient," *Phys. Fluids* **8**, 890–95, 1965.
235. Holm, R., "Thermal Conductivity Measurements at High Temperature by a Variant of the Kohlrausch Method," (In German) *Z. Tech. Phys.* **10**, 621–3, 1929.
236. Holm, R. and Störmer, R., "Measurements of the Thermal Conductivity of a Platinum Probe in the Temperature Range 19–1020 C," (In German) *Wiss. Veröffentl. Siemens-Konzern* **9**(2), 312–22, 1930.
237. Hopkins, M. R., "The Thermal and Electrical Conductivities of Metals at High Temperatures," (In English) *Zeit. Phys.* **147**(2), 148–60, 1957.
238. Horrocks, J. K. and McLaughlin, E., "Thermal Conductivity of Simple Molecules in the Condensed State," *Trans. Faraday Soc.* **56**(2), 206–12, 1960.
239. Horrocks, J. K. and McLaughlin, E., "Transport Properties of Binary Regular Solutions," *Trans. Faraday Soc.* **58**, 1357–62, 1962.
240. Horrocks, J. K. and McLaughlin, E., "Nonsteady-State Measurements of the Thermal Conductivities of Liquid Polyphenyls," *Proc. Roy. Soc. (London)* **A273**, 259–74, 1963.
241. Horrocks, J. K. and McLaughlin, E., "Temperature Dependence of the Thermal Conductivity of Liquids," *Trans. Faraday Soc.* **59**, 1709–16, 1963.
242. Huck, R. J. and Thornton, E., "Sutherland–Wassiljewa Coefficients for the Viscosities of Binary Rare Gas Mixtures," *Proc. Phys. Soc.* **92**, 244–52, 1967.
243. Huck, R. J. and Thornton, E., "Sutherland–Wassiljewa Coefficients for the Thermal Conductivities of Binary Rare Gas Mixtures," *Proc. Fourth Sym. Thermophysical Properties*, ASME, N. Y., 366–71, 1968.
244. Ibbs, T. L. and Hirst, A. A., "The Thermal Conductivity of Gas Mixtures," *Proc. Roy. Soc. (London)* **A123**, 134–42, 1929.
245. Ibele, W. E. and Briggs, D. G., "Prandtl Number Measurements and Transport Property Calculations for N_2–CO_2 Mixtures," *Proc. 4th Symposium Thermophysical Properties*, ASME, N. Y., 392–7, 1968.
246. Iman-Rahajoe, S., Curtiss, C. F., and Bernstein, R. B., "Numerical Evaluation of Quantum Effects on Transport Cross Sections," *J. Chem. Phys.* **42**, 530–6, 1965.
247. Itean, E. C., Glueck, A. R., and Svehla, R. A., "Collision Integrals for a Modified Stockmayer Potential," NASA Tech. Note D-481, 29 pp., Jan. 1961. [AD 248 768]
248. Jamieson, D. T. and Tudhope, J. S., (a) "The Thermal Conductivity of Liquids: A Survey to 1963," NEL Report No. 137, 67 pp., 1964; (b) "The Thermal Conductivity of Binary Liquid Mixtures: A Survey to 1968," NEL Rept. No. 435, 65 pp., 1969, National Engineering Lab., East Kilbride, Glasgow, Scotland.
249. Jeans, J. H., *Dynamical Theory of Gases*, Cambridge Univ. Press, 444 pp., 1925; Dover Publ. reprint, 444 pp., 1954.

250. Jeans, J. H., *An Introduction to the Kinetic Theory of Gases*, Cambridge Univ. Press, 311 pp., 1946.
251. Johannin, P., "Thermal Conductivity of Nitrogen between 75 and 700 Degrees and About 1000 to 1600 Atmospheres," *Compt. Rend.* **244**, 2700–3, 1957.
252. Johannin, P. and Vodar, B., "Thermal Conductivity of Nitrogen at High Temperatures and Pressures," *Ind. Eng. Chem.* **49**, 2040–1, 1957.
253. Johannin, P., Wilson, M., Jr., and Vodar, B., "Heat Conductivity of Compressed Helium at Elevated Temperatures," *Second Symp. on Thermophysical Properties*, ASME, Academic Press, 418–33, 1962.
254. Johnson, V. J. (ed), "A Compendium of the Properties of Materials at Low Temperatures (Phase I), Part I. Properties of Fluids," WADD-TR-60-56 (Part I), 1–489, 1960. [AD 249 644]
255. Jordan, H. B., "Prediction of the Thermal Conductivity of Miscible Binary Liquid Mixtures from the Pure Component Values," Master's Thesis, Louisiana State University, 1961.
256. Joshi, K. M. and Saxena, S. C., "Viscosity of Polar Gases," *Physica* **27**, 329–36, 1961; *ibid.*, **27**, 1101, 1961.
257. Kamal, I. and McLaughlin, E., "Pressure and Volume Dependence of the Thermal Conductivity of Liquids," *Trans. Faraday Soc.* **60**(497), 809–16, 1964.
258. Kamnev, A. B. and Leonas, V. B., "Kinetic Coefficients for Inert Gases at High Temperatures," *Teplofiz. Vysokikh Temperatur* **4**(2), 288–9, 1966. Transl. in *High Temperature* **4**(2), 283–4, 1966.
259. Kanitkar, D. and Thodos, G., "Thermal Conductivities of Normal Liquids," *Proceedings of the Fourth Symposium on Thermophysical Properties*, ASME, New York, 286–91, 1968.
260. Kannuluik, W. G., "On the Thermal Conductivity of Some Metal Wires," *Proc. Roy. Soc. (London)* **A131**, 320–35, 1931.
261. Kannuluik, W. G. and Carman, E. H., "The Temperature Dependence of the Thermal Conductivity of Air," *Australian J. Sci. Res.* **A4**, 305–14, 1951.
262. Kannuluik, W. G. and Carman, E. H., "The Thermal Conductivity of Rare Gases" *Proc. Phys. Soc. (London)* **B65**, 701–9, 1952.
263. Kannuluik, W. G. and Donald, H. B., "The Pressure Dependence of the Thermal Conductivity of Polyatomic Gases at 0 C," *Australian J. Sci. Res.* **A3**, 417–27, 1950.
264. Kannuluik, W. G. and Law, P. G., "The Thermal Conductivity of Carbon Dioxide Between −78.5 C and 100 C," *Proc. Roy. Soc. Victoria* **58**, 142–56, 1947.
265. Kannuluik, W. G. and Martin, L. H., "Conduction of Heat in Powders," *Proc. Roy. Soc.* **A141**, 144–58, 1933.
266. Kannuluik, W. G. and Martin, L. H., "The Thermal Conductivity of Some Gases at 0 Degrees," *Proc. Roy. Soc.* **A144**, 496–513, 1934.
267. Kardos, A., "Theory of Thermal Conductivity of Liquids," (In German) *Forsch. Gebiete Ingenieurw.* **5B**, 14–24, 1934.
268. Kardos, A., "The Thermal Conductivity of Various Liquids," (In German) *Z. Ges. Kälte-Ind.* **41**, 1–6 and 29–35, 1934.
269. Kauzman, W., *Kinetic Theory of Gases*, Benjamin, N. Y., 248 pp., 1966.
270. Kawasaki, K. and Oppenheim, I., "Logarithmic Term in the Density Expansion of Transport Coefficients," *Phys. Rev.* **139A**, 1763–8, 1965.
271. Kennard, E. H., *Kinetic Theory of Gases, with an Introduction to Statistical Mechanics*, McGraw-Hill, N. Y., 483 pp., 1938.
272. Keyes, F. G., "A Summary of Viscosity and Heat-Conduction Data for He, Ar, H_2, O_2, N_2, CO, CO_2, H_2O and Air," *Trans. ASME* **73**, 589–96, 1951.
273. Keyes, F. G., "Measurement of the Heat Conductivity of Nitrogen–Carbon Dioxide Mixtures," *Trans. ASME* **73**, 597–603, 1951.
274. Keyes, F. G., "Additional Measurements of Heat Conductivity of Nitrogen, Carbon Dioxide and Mixtures," *Trans. ASME* **74**, 1303–6, 1952.
275. Keyes, F. G., "Thermal Conductivity of Gases," *Trans. ASME* **76**, 809–16, 1954.
276. Keyes, F. G. and Sandell, D. J., Jr., "New Measurements of the Heat Conductivity of Steam and Nitrogen," *Trans. ASME* **72**, 767–78, 1950.
277. Kihara, T., Taylor, M. H., and Hirschfelder, J. O., "Transport Properties for Gases Assuming Inverse Power Intermolecular Potentials," *Phys. Fluids* **3**(5), 715–20, 1960.
278. Kim, S. K. and Ross, J., "Viscosity of Moderately Dense Gases," *J. Chem. Phys.* **42**(1), 263–71, 1965; also BRN-13-P, PB 166 949, AD 448 076, 29 pp., 1964.
279. Kim, S. K. and Ross, J., "On the Determination of Potential Parameters from Transport Coefficients," *J. Chem. Phys.* **46**(2), 818, 1967.
280. Kim, S. K., Flynn, G. P., and Ross, J., "Thermal Conductivity of Moderately Dense Gases," *J. Chem. Phys* **43**(11), 4166–9, 1965.
281. Klein, M., "Determination of Intermolecular Potential Functions from Macroscopic Measurements," *J. Res. Natl. Bur. Std.* **70A**(3), 259–69, 1966.
282. Knopp, C. F. and Cambel, A. B., "Experimental Determination of the Thermal Conductivity of Atmospheric Argon Plasma," *Phys. Fluids* **9**(5), 989–6, 1966.
283. Kornfeld, G. and Hilferding, K., "Energy Exchange in Gas Mixtures," *Bedenstein-Festband* (Supplement to *Z. Physik. Chem. Leipzig*), 792–800, 1931.
284. Kramer, F. R. and Comings, E. W., "Thermal Conductivity of Butane at High Pressure: Correlation With Other Gases," *J. Chem. Eng. Data* **5**, 462–7, 1960.
285. Krauss, B. J. and Ferron, J. R., "Thermal Conductivity of Carbon Dioxide at One Atmosphere," Proj. Squid TR DEL-16-P, 1966.
286. Krinberg, I. A., "Calculation of the Thermal Conductivity of Some Gases at Temperatures of 1000–20,000 K at Atmospheric Pressure," *Teplofiz. Vysokikh Temperatur* **3**(4), 654–7, 1965; *High Temperature* **3**(4), 606–9, 1965.
287. Krinberg, I. A., "Effects of Ionization Reactions on the Thermal Conductivity of a Plasma," *High Temperature* **3**(6), 782–7, 1965.
288. Kundt, A. and Warburg, E., "On the Viscosity and Heat Conduction in Rarified Gases," *Ann. Physik* **156**, 177–211, 1875.
289. Laby, T. H. "Measurement of the Thermal Conductivity of Gases," *Nature* **137**, 741, 1936.
290. Lambert, J. D., Staines, E. N., and Woods, S. D.,

"Thermal Conductivities of Organic Vapors," *Proc. Roy. Soc. (London)* **A200**, 262–71, 1950.

291. Lauver, M. R., "Evaluation of Shock Tube Heat Transfer Experiments to Measure Thermal Conductivity of Argon from 700 to 8600 K," NASA TN D-2117, 18 pp., 1964.
292. Lauver, M. R., "Shock-Tube Thermal Conductivity," *Phys. Fluids* **7**(4), 611–2, 1964.
293. Lee, C. S. and Bonilla, C. F., "Thermal Conductivity of the Alkali Metals Vapors and Argon," *Proc. Seventh Conference on Thermal Conductivity*, NBS SP 302, 561–78, 1968.
294. Leidenfrost, W., "An Attempt to Measure the Thermal Conductivity of Liquids, Gases and Vapors with a High Degree of Accuracy over Wide Ranges of Temperature (-180 to 500 C) and Pressure (Vacuum to 500 Atm.)," *Intern. J. Heat Mass Transfer* **7**, 477–78, 1964.
295. Leidenfrost, W., "A Critical Analysis of the Experimental Determination of the Thermal Conductivity of Steam," Proc. 7th Intl. Conf. Props. Steam, Tokyo, 1968.
296. Leidenfrost, W., "Theory and Design Considerations in Developing a Multipurpose Instrument for the Determination of Twelve Properties," *Intern. J. Heat Mass Transfer*, in press.
297. Leng, D. E. and Comings, E. W., "Thermal Conductivity of Propane," *Ind. Eng. Chem.* **49**, 2042–5, 1957.
298. Lenoir, J. M., "Thermal Conductivity of Gases at Atmospheric Pressure," *Univ. Arkansas, Eng. Exp. Sta. Bull.* **18**, 1–48, 1953.
299. Lenoir, J. M., "Effect of Pressure on Thermal Conductivity of Liquids," *Petroleum Refiner* **36**(8), 162–4, 1957.
300. Lenoir, J. M. and Comings, E. W., "Thermal Conductivity of Gases, Measurement at High Pressure," *Chem. Eng. Progr.* **47**, 223–31, 1951.
301. Liley, P. E., "Survey of Recent Work on the Viscosity, Thermal Conductivity and Diffusion of Gases and Gas Mixtures," *Thermodynamic and Transport Properties of Gases, Liquids and Solids*, Symposium ASME, N. Y., 40–69, 1959.
302. Liley, P. E., "Review of Work on the Transport Properties of Gases and Gas Mixtures," Purdue University, TPRC Report 10, 1–57, 1959.
303. Liley, P. E., "Review of Work on the Transport Properties of Gases and Gas Mixtures, Supplement 1," Purdue University, TPRC Report 12, 1–14, 1961.
304. Liley, P. E., "Survey of Recent Work on the Viscosity, Thermal Conductivity and Diffusion of Gases and Liquefied Gases Below 500 K," Purdue University, TPRC Report 13, 1–33, 1961.
305. Liley, P. E., "Survey of Recent Work on the Viscosity, Thermal Conductivity and Diffusion of Gases and Liquefied Gases below 500 K," *Progress in International Research on Thermodynamic and Transport Properties*, Academic Press, N. Y., 313–30, 1962.
306. Liley, P. E., "Some Remarks on the Thermal Conductivity of Gases and Liquids," *Proc. 6th Conf. Thermal Conductivity*, Dayton, Ohio, 3–14, 1966.
307. Lindsay, A. L. and Bromley, L. A., "Thermal Conductivity of Gas Mixtures," *Ind. Eng. Chem.* **42**, 1508–11, 1950.
308. Lindsay, A. L. and Bromley, L. A., "Use of the Unsteady State Method for the Determination of the Thermal Conductivity of Gases," AEC R. and D. Rept. UCRL 1128, 1–49, 1951.
309. Livingston, P. M. and Curtiss, C. F., "Kinetic Theory of Nonspherical Molecules, IV. Angular Momentum Transport Coefficient," *J. Chem. Phys.* **31**(6), 1643–5, 1959.
310. Livingston, P. M. and Curtiss, C. F., "Kinetic Theory of Moderately Dense Rigid Sphere Gases," *Phys. Fluids* **4**(7), 816–33, 1961.
311. Loeb, L. B., *The Kinetic Theory of Gases*, McGraw-Hill, N. Y., 2nd Edition, 687 pp., 1934.
312. Longuet-Higgins, H. C. and Pople, J. A., "Transport Properties of a Dense Fluid of Hard Spheres," *J. Chem. Phys.* **25**, 884–9, 1956.
313. Longuet-Higgins, H. C. and Valleau, J. P., "Transport of Energy and Momentum in a Dense Fluid of Hard Spheres," *Disc. Faraday Soc.* No. 22, 47–53, 1956.
314. Longuet-Higgins, H. C. and Valleau, J. P., "Transport Coefficients of Dense Fluids of Molecules Interacting According to a Square-Well Potential," *Mol. Physics* **1**, 284–94, 1958.
315. Longuet-Higgins, H. C., Pople, J. A., and Valleau, J. P., *International Symposium on Transport Processes in Statistical Mechanics, Brussels, 1956*, I. Prigogine (ed.), Interscience Publishers, Inc., New York, p. 73, 1958.
316. Lubin, M. J. and Resler, E. L., "Precurser Studies in an Electromagnetically Driven Shock Tube," *Phys. Fluids* **10**(1), 1–8, 1967.
317. Lucks, C. F. and Deem, H. W., "Thermal Properties of Thirteen Metals," Am. Soc. for Testing Materials, Special Tech. Publ. No. 227, 1–29, 1958.
318. Lydersen, A. L., Greenkorn, R. A., and Hougen, O. A., "Generalized Thermodynamic Properties of Pure Fluids," University of Wisconsin, Eng. Exp. Sta. Rept. 4, 1–99, 1955.
319. Maecker, H., "Measurement and Evaluation of Arc Characteristics (Ar, N_2)," *Z. Phys.* **158**, 392–404, 1960.
320. Maecker, H., Chapters 18 (Theory of Thermal Plasma and Application to Observed Phenomena), 19 (Arc Measurements and Results), and 20 (Different Types of Arcs), *Introduction to Discharge and Plasma Physics*, 245–89, 1963. Published as a proceeding of the Summer School held at the University of New England, Jan. 25–Feb. 3, 1963; Department of University of Extension, The University of New England, Armidale, N.S.W., Australia.
321. Magnus, G., "On the Conduction of Heat by Gases," *Phil. Mag.* **20**, 510–2, 1860.
322. Mallon, C. E. and Cutler, M., "The Thermal Conductivity of Liquid Semi-conducting Thallium–Tellurium Solutions," *Phil. Mag.* **11**(112), 667–72, 1965.
323. Mallon, C. E. and Cutler, M., "Thermal Conductivity of Electrically Conducting Liquids," *Rev. Sci. Instr.* **36**(7), 1036–40, 1965.
324. Mann, W. B. and Dickins, B. G., "Thermal Conductivities of the Saturated Hydrocarbons in the Gaseous State," *Proc. Roy. Soc. (London)* **A134**, 77–96, 1931.
325. Mason, E. A., "Transport Properties of Gases Obeying a Modified Buckingham (exp-six) Potential," *J. Chem. Phys.* **22**(2), 169–86, 1954.
326. Mason, E. A., "Thermal Conductivity of Multicomponent Gas Mixtures," *J. Chem. Phys.* **28**(5), 1000–1, 1958.

327. Mason, E. A. and Monchick, L., "Heat Conductivity of Polyatomic and Polar Gases," *J. Chem. Phys.* **36**, 1622–39, 1962.
328. Mason, E. A. and Monchick, L., "Methods for the Determination of Intermolecular Forces," *Advan. Chem. Phys.* **12**, 329–87, 1967.
329. Mason, E. A. and Saxena, S. C., "Approximate Formula for the Thermal Conductivity of Gas Mixtures," *Phys. Fluids* **1**(5), 361–9, 1958.
330. Mason, E. A. and Saxena, S. C., "Thermal Conductivity of Multicomponent Gas Mixtures. II," *J. Chem. Phys.* **31**(2), 511–4, 1959.
331. Mason, E. A. and Sherman, M. P., "Effect of Resonant Charge Exchange on Heat Conduction in Plasmas," *Phys. Fluids* **9**(10), 1989–91, 1966.
332. Mason, E. A. and von Ubisch, H., "Thermal Conductivities of Rare Gas Mixtures," *Phys. Fluids* **3**(3), 355–61, 1960.
333. Mason, E. A., Munn, R. J., and Smith, F. J., "Transport Coefficients of Ionized Gases," *Phys. Fluids* **10**(8), 1827–32, 1967.
334. Mason, E. A., Vanderslice, J. T., and Yos, J. M., "Transport Properties of High Temperature Multicomponent Gas Mixtures," *Phys. Fluids* **2**(6), 688–94, 1959.
335. Mathur, S., "A Few Properties of Gases and Solids," Ph.D. Thesis, Physics Dept., Rajathan, Univ., India, 441 pp., 1966.
336. Mathur, B. P. and Saxena, S. C., "A New Method for the Calculation of Diffusion Coefficients of Multicomponent Gas Mixtures," *Indian J. Pure Appl. Phys.* **4**(7), 266–8, 1966. (Many earlier works are referenced.)
337. Mathur, S. and Saxena, S. C., "Relations Between Thermal Conductivity and Diffusion Coefficients of Pure and Mixed Polyatomic Gases," *Proc. Phys. Soc. (London)* **89**, 753–64, 1966.
338. Mathur, S. and Saxena, S. C., "Methods of Calculating Thermal Conductivity of Binary Mixtures Involving Polyatomic Gases," *Appl. Sci. Res.* **17**, 155–68, 1967.
339. Mathur, S. and Saxena, S. C., "Calculation of Thermal Conductivity of Ternary Mixtures of Polyatomic Gases," *Indian J. Pure Appl. Phys.* **5**(4), 114–6, 1967.
340. Mathur, B. P. and Saxena, S. C., "Composition Dependence of the Thermal Diffusion Factor in Binary Gas Mixtures," *Z. Naturforsch.* **22a**, 164–9, 1967.
341. Mathur, B. P. and Saxena, S. C., "Note on the Composition Dependence of the Thermal Diffusion Factor of Ar–He System," *Z. Naturforsch.* **22a**, 840, 1967.
342. Mathur, B. P. and Saxena, S. C., "Measurement of the Concentration Diffusion Coefficient for He–Ar and Ne–Kr by a Two Bulb Method," *Appl. Sci. Res.* **18**, 325–35, 1968.
343. Mathur, B. P., Joshi, R. K., and Saxena, S. C., "Thermal Diffusion Factors from the Measurements on a Trennschaukel: Ar–He and Kr–Ne," *J. Chem. Phys.* **46**(12), 4601–3, 1967.
344. Mathur, S., Tondon, P. K., and Saxena, S. C., "Thermal Conductivity of Binary, Ternary, and Quaternary Mixtures of Rare Gases," *Mol. Phys.* **12**(6), 569–79, 1967.
345. Mathur, S., Tondon, P. K., and Saxena, S. C., "Thermal Conductivity of the Gas Mixtures: D_2–Xe, D_2–Ne–Kr, D_2–Ne–Ar and D_2–Ar–Kr–Xe," *J. Phys. Soc. Japan* **25**(2), 530–5, 1968.
346. Matula, R. A., "High Temperature Thermal Conductivity of Rare Gases and Gas Mixtures," *J. Heat Transfer, Trans. ASME, Series C* **90**(3), 319–27, 1968.
347. Maxwell, J. C., *Electricity and Magnetism*, Clarendon Press, Oxford, 1881, Second Edition, Vol. 1, 280.
348. McClelland, J. D., Rasor, N. S., Dahleen, R. C., and Zehms, E. H., "Thermal Conductivity of Liquid Copper," WADC Tech. Rept. 56-400, Part II, 1–6, 1957. [AD 118 243]
349. McCourt, F. R. and Snider, R. F., "Thermal Conductivity of a Gas with Rotational States," *J. Chem. Phys.* **41**(10), 3185–94, 1964; "Transport Properties with Rotational States," *J. Chem. Phys.* **43**(7), 2276–83, 1965.
350. McCoy, B. J., Sandler, S. I., and Dahler, J. S., "Transport Properties of Polyatomic Fluids. IV. The Kinetic Theory of a Dense Gas of Perfectly Rough Spheres," *J. Chem. Phys.* **45**(10), 3485–512, 1966.
351. McLaughlin, E., "The Thermal Conductivities of Liquids and Dense Gases," *Chem. Rev.* **64**(4), 390–428, 1964.
352. McLaughlin, I. L. and Dahler, J. S., "Transport Properties of Polyatomic Fluids, III. The Transport-Relaxation Equations for a Dilute Gas of Rough Spheres," *J. Chem. Phys.* **44**(12), 4453–9, 1966.
353. McLaughlin, I. L. and Davis, H. T., "Kinetic Theory of Dense Fluid Mixtures. I. Square-Well Model," *J. Chem. Phys.* **45**(6), 2020–31, 1966.
354. Michels, A. and Botzen, A., "A Method for the Determination of the Thermal Conductivity of Gases at High Pressures," *Physica* **18**, 605–12, 1952.
355. Michels, A. and Botzen, A., "The Thermal Conductivity of Nitrogen at Pressures up to 2500 Atmospheres," *Physica* **19**, 585–98, 1953.
356. Michels, A. and Sengers, J. V., "The Thermal Conductivity of Carbon Dioxide in the Critical Region. 3. Verification of the Absence of Convection," *Physica* **28**, 1238–64, 1962.
357. Michels, A., Botzen, A., Friedman, A. S., and Sengers, J. V., "The Thermal Conductivity of Argon between 0 and 75 C at Pressures up to 2500 Atmospheres," *Physica* **22**, 121–8, 1956.
358. Michels, A., Sengers, J. V., and Van Der Gulik, P. S., "The Thermal Conductivity of Carbon Dioxide in the Critical Region, I. The Thermal Conductivity Apparatus," *Physica* **28**, 1201–15, 1962.
359. Michels, A., Sengers, J. V., and Van Der Gulik, P. S., "The Thermal Conductivity of Carbon Dioxide in the Critical Region, II. Measurements and Conclusion," *Physica* **28**, 1216–37, 1962.
360. Miller, B., "Experimental Study of Normal Ionizing Shock Waves," *Phys. Fluids* **10**(1), 9–16, 1967.
361. Milverton, S. W., "Experimental Determination of the Thermal Conductivity of Air Between 0 and 100 Degrees," *Phil. Mag.* **17**, 397–422, 1934.
362. Minter, C. C., "Thermal Conductivity of Binary Mixtures of Gases. I. Hydrogen–Helium Mixtures," *J. Phys. Chem.* **72**(6), 1924–6, 1968.
363. Minter, C. C. and Schuldiner, S., "Thermal Conductivity of Equilibrated Mixtures of H_2, D_2, and HD," *J. Chem. Eng. Data* **4**, 223–6, 1959.

364. Missenard, F. A., "New Simple Expressions, Theoretical and Experimental for the Thermal Conductivity of Pure Liquids at Different Temperatures and Pressures," (In French), *Revue Générale Thermique* 3(35), 1403–29, 1964.
365. Missenard, F. A., *Thermal Conductivity of Solids, Liquids, Gases and Their Mixtures*, (In French), 554 pp., Editions Eyrolles, Paris, 1965.
366. Missenard, F. A., "Precise Values of the Thermal Conductivity of Liquids at Different Temperatures," (In French), *Revue Générale Thermique* 4(40), 409–28, 1965.
367. Misic, D. and Thodos, G., "Thermal Conductivity Measurements for Nitrogen in Dense Gaseous State," *Am. Inst. Chem. Eng. J.* **11**, 650–6, 1965.
368. Monchick, L., "Collision Integrals for the Exponential Repulsive Potential," *Phys. Fluids* **2**(6), 695–700, 1959.
369. Monchick, L., "Equivalence of the Chapman–Enskog and the Mean-Free-Path Theory of Gases," *Phys. Fluids* **5**, 1393–8, 1962.
370. Monchick, L. and Mason, E. A., "Transport Properties of Polar Gases," *J. Chem. Phys.* **35**(5), 1676–97, 1961.
371. Monchick, L. and Mason, E. A., "Free-Flight Theory of Gas Mixtures," *Phys. Fluids* **10**(7), 1377–90, 1967.
372. Monchick, L., Mason, E. A., Munn, R. J., and Smith, F. J., "Transport Properties of Gaseous He³ and He⁴," *Phys. Rev.* **139**, A1076–82, 1965.
373. Monchick, L., Pereira, A. N. G., and Mason, E. A., "Heat Conductivity of Polyatomic and Polar Gases and Gas Mixtures," *J. Chem. Phys.* **42**, 3241–56, 1965.
374. Monchick, L., Yun, K. S., and Mason, E. A., "Relaxation Effects in the Transport Properties of a Gas of Rough Spheres," *J. Chem. Phys.* **38**(6), 1282–7, 1963.
375. Monchick, L., Yun, K. S., and Mason, E. A., "Formal Kinetic Theory of Transport Phenomena in Polyatomic Gas Mixtures," *J. Chem. Phys.* **39**(3), 654–69, 1963.
376. Muckenfuss, C. and Curtiss, C. F., "Kinetic Theory of Nonspherical Molecules. III," *J. Chem. Phys.* **29**(6), 1257–72, 1958.
377. Muckenfuss, C. and Curtiss, C. F., "Thermal Conductivity of Multicomponent Gas Mixtures," *J. Chem. Phys.* **29**(6), 1273–7, 1958.
378. Mueller, J. J. and Curtiss, C. F., "Quantum-Mechanical Kinetic Theory of Loaded Spheres," *J. Chem. Phys.* **46**(1), 283–302, 1967.
379. Mueller, J. J. and Curtiss, C. F., "Quantum-Mechanical Kinetic Theory of Loaded Spheres. II. The Classical Limit," *J. Chem. Phys.* **46**(4), 1252–64, 1967.
380. Muller, E., "Experimentelle Untersuchungen über die absolute Wärmeleitungsconstante der Luft," *Ann. Physik.* **60**, 82–118, 1897.
381. Mukhopadhyay, P. and Barua, A. K., "Thermal Conductivity of Hydrogen–Helium Gas Mixtures," *Brit. J. Appl. Phys.* **18**, 635–40, 1967.
382. Mukhopadhyay, P., das Gupta, A., and Barua, A. K., "Thermal Conductivity of Hydrogen–Nitrogen and Hydrogen–Carbon Dioxide Mixtures," *Brit. J. Appl. Phys.* **18**, 1301–6, 1967, see also **18**, 1307, 1967.
383. Munn, R. J., Mason, E. A., and Smith, F. J., "Collision Integrals for the Exponential Attractive Potential," *Phys. Fluids* **8**(6), 1103–5, 1965.
384. Munn, R. J., Smith, F. J., and Mason, E. A., "Transport Collision Integrals for Quantum Gases Obeying a 12-6 Potential," *J. Chem. Phys.* **42**, 537–9, 1965.
385. Nain, V. P. S. and Saxena, S. C., "On the Appropriateness of Dymond, Rigby and Smith Intermolecular Potential," *Chem. Phys. Letters* **1**, 46–7, 1967.
386. Nain, V. P. S. and Saxena, S. C., "Second Virial Coefficient of Non-Polar Gases and Gas Mixtures and Buckingham–Carra–Konowalow Potential," *Indian J. Phys.* **41**, 199–208, 1967.
387. Narr, F., *Pogg. Ann.* **142**, 123–33, 1871.
388. Neal, W. E. J., Greenway, J. E., and Coutts, P. W., "Thermal Conduction in Hydrogen–Helium Gas Mixtures," *Proc. Phys. Soc. (London)* **87**, 577–9, 1966.
389. Needham, D. P. and Ziebland, H., "The Thermal Conductivity of Liquid and Gaseous Ammonia and Its Anomalous Behaviour in the Vicinity of the Critical Point," Explosives Research and Development Establishment, Waltham Abbey, England, ERDE-14/R/63, 1–41, 1964. [AD 445 673]
390. Nelson, R. D., Lide, D. R., Jr., and Maryott, A. A., "Selected Values of Electric Dipole Moments for Molecules in the Gas Phase," NSRDS-NBS 10, 49 pp., Sept. 1967.
391. Novotny, J. L. and Irvine, T. F., Jr., "Thermal Conductivity and Prandtl Number of Carbon Dioxide and Carbon Dioxide–Air Mixtures at One Atmosphere," *J. Heat Transfer* **83C**, 125–32, 1961.
392. Nusselt, W., "Thermal Conductivity of Thermal Insulators," *Z. Ver. Dt. Ing.* **52**, 901–12, 1909.
393. Nuttall, R. L. and Ginnings, D. C., "Thermal Conductivity of Nitrogen from 50 to 500 C and 1 to 100 Atmospheres," *J. Res. Natl. Bur. Stds.* **58**, 271–8, 1957.
394. Nyeland, C., "Rotational Relaxation of Homonuclear Diatomic Molecules," *J. Chem. Phys.* **46**(1), 63–7, 1967.
395. Osida, I., "The Thermal Conductivity of Liquids," *Proc. Phys.-Math. Soc. Japan* **21**, 353–6, 1939.
396. Pachaiyappan, V., Ibrahim, S. H., and Kuloor, N. R., "Thermal Conductivities of Organic Liquids—A New Correlation," *J. Chem. Eng. Data* **11**(1), 73–6, 1966.
397. Palmer, G., "Thermal Conductivity of Liquids," *Ind. Eng. Chem.* **40**, 89–92, 1948.
398. Panevin, I. G. and Kulik, P. P., "Experimental Determination of the Coefficient of Thermal Conductivity of High-Temperature Gas," *Teplofiz. Vysokikh Temperatur* **1**(3), 394–8, 1963; *High Temperature* **1**(3), 353–7, 1963.
399. Parker, J. G., "Rotational and Vibrational Relaxation in Diatomic Gases," *Phys. Fluids* **2**(4), 449–62, 1959.
400. Partington, J. R., *An Advanced Treatise on Physical Chemistry. I. Fundamental Principles. The Properties of Gases*, Longmans, Green and Co., 888 pp., 1949.
401. Peng, T. C. and Ahtye, W. F., "Experimental and Theoretical Study of Heat Conduction for Air up to 500 K," NASA TN-D 687, 1961.
402. Peterson, J. R. and Bonilla, C. F., "The Development of a Frequency Response Analysis Technique for Thermal Conductivity Measurement and Its Application to Gases at High Temperatures," *Advan. Thermophys. Properties at Extreme Temperatures and Pressures*, ASME, N.Y., 264–76, 1965.
403. Pfender, E., Eckert, E. R. G., and Raithby, G. D., "Energy Transfer Studies in a Wall-Stabilized, Cascaded

Arc," *Proc. Seventh International Conf. on Phenomena in Ionized Gases*, **1**, 691–6, 1966.
404. Pitzer, K. S., "The Volumetric and Thermodynamic Properties of Fluids. II. Compressibility Factor, Vapor Pressure and Entropy of Vaporization," *J. Am. Chem. Soc.* **77**, 3433–40, 1955.
405. Poltz, H., "The Thermal Conductivity of Liquids. II. The Radiation Component of the Effective Thermal Conductivity," *Intern. J. Heat Mass Transfer* **8**(4), 515–27, 1965.
406. Poltz, H., "The Thermal Conductivity of Liquids. III. Dependence of the Thermal Conductivity of Organic Liquids on the Film Thickness," *Intern. J. Heat Mass Transfer* **8**(4), 609–20, 1965.
407. Poltz, H. and Jugel, R., "Thermal Conductivity of Liquids. IV. Temperature Dependence of Thermal Conductivity," *Intern. J. Heat Mass Transfer* **10**(8), 1075–88, 1967.
408. Polyakov, Y. A., "Pulse Probe for Measuring Heat Transfer in Ionized Gas Flow," *High Temperature* **3**(5), 695–8, 1965.
409. Polyakov, Y. A., "Investigation of Heat Transfer When a Shock Wave Is Reflected," *High Temperature* **3**(6), 818–26, 1965.
410. Polyakov, Y. A., Naboko, I. M., and Makanov, Y. V., "Measurement of the Working Period of a Shock Tube with a Thermal Probe," *High Temperature* **3**, 411–5, 1965.
411. Powell, R. W., "Experiments Using a Simple Thermal Comparator for Measurements of Thermal Conductivity, Surface Roughness and Thickness of Foils or of Surface Deposits," *J. Sci. Instrum.* **34**, 485–92, 1957.
412. Powell, R. W., "Correlation of Metallic Thermal and Electrical Conductivities for Both Solid and Liquid Phases," *Intern. J. Heat Mass Transfer* **8**, 1033–45, 1965.
413. Powell, R. W. and Challoner, A. R., "Thermal Conductivity Measurements on Oils," *J. Inst. Petroleum* **46**, 267–71, 1960.
414. Powell, R. W. and Challoner, A. R., "New Measurements of the Thermal Conductivities of Several Liquid Refrigerants of the Fluorochloro-Derivative Type," *Proc. 10th International Congress of Refrigeration* **1**, 382–7, 1960.
415. Powell, R. W. and Challoner, A. R., "Measurement of Thermal Conductivity of *n*-Octadecane; Does Molecular Orientation Affect the Data?" *Ind. Eng. Chem.* **53**, 581–2, 1961.
416. Powell, R. W. and Ho, C. Y., "The State of Knowledge Regarding the Thermal Conductivity of the Metallic Elements," *Thermal Conductivity—Proc. 7th Conf.* (Eds. D. R. Flynn and B. A. Peavy, Jr.), NBS Spec. Publ. 302, 1–31, Sept. 1968, Washington, D.C.
417. Powell, R. W. and Tye, R. P., "The Thermal and Electrical Conductivity of Liquid Mercury," *International Developments in Heat Transfer*, ASME, 856–62, 1961.
418. Powell, R. W. and Tye, R. P., "Thermal-Comparator Methods for the Rapid Measurements of Thermal Conductivities, with Particular Reference to Fluids," *Proc. Intern. Meas. Conf.*, Acta IMEKO, Stockholm, 397–413, 1964.
419. Predvoditelev, A. S., "Some Invariant Quantities in the Theories of Heat Conductance and of Viscosity of Liquids," (In Russian), *J. Phys. Chem. USSR*, **22**, 339–48, 1948.
420. Present, R. D., *Kinetic Theory of Gases*, McGraw-Hill, N. Y., 1958.
421. Prigogine, I. and Waelbroeck, F. G., "Heat Conductivity and Chemical Reactions in Gases, *Proc. Joint Conf. Thermodynamics and Transport Properties of Fluids, London, 1957*, Inst. Mech. Engrs. (London), 128–32, 1958.
422. Pustogarov, A. V., "Calculation of the Parameters of an Arc in Argon," *High Temperature* **3**, 22–5, 1965.
423. Radin, S. H. and Mintzer, D., "Orthogonal Polynomial Solution of the Boltzmann Equation for a Strong Shock Wave," *Phys. Fluids* **9**(9), 1621–33, 1966.
424. Rao, M. R., "The Thermal Conductivity of Liquids," *Indian J. Phys.* **16**, 161–7, 1942.
425. Rastorguev, Yu. L., "Experimental Study of the Thermal Conductivity of Oils and Other Petroleum Products." (English translation of *Izv. Vysshikh Uchebn. Zavedenii, Neft i Gaz.* **2**(8), 51–3, 1959; Associated Technical Services Inc., East Orange, N.J., ATS-14P63R, 4 pp., 1962.)
426. Rastorguev, Yu. L. and Ganiev, Yu. A., "Thermal Conductivity of Aqueous Solutions of Organic Liquids," *Zh. Fiz. Khim.* **40**(7), 1608–12, 1966; *Russian J. Phys. Chem.* **40**(7), 869–71, 1966.
427. Rastorguev, Yu. L. and Ganiev, Yu. A., "Thermal Conductivity of Liquids in a Constant Electric Field," *Elektronaya Obrabotka Mater. Akad. Nauk. Mold. SSR*, No. 1, 64–72, 1967.
428. Rastorguev, Yu. L. and Ganiev, Yu. A., "Thermal Conductivity of Non-Electrolyte Solutions," *Zh. Fiz. Khim.* **41**, 1352–6, 1967; *Russian J. Phys. Chem.* **41**, 717–20, 1967.
429. Reid, R. C. and Sherwood, R. K., *The Properties of Gases and Liquids, Their Estimation and Correlation*, McGraw-Hill Book Co., Inc., New York, Toronto and London, (a) 1st edition, 386 pp., 1958; (b) 2nd edition, 646 pp., 1966.
430. Reiter, F. W., "Thermal Conductivity of Some Polyphenyls and Mixtures Thereof," (In German), U.S. Atomic Energy Commission, Brussels, Belgium, EUR-582D, N64-19006, 1–32, 1964. (Obtainable from OTS, and The Belgium American Bank and Trust Co., New York.)
431. Rice, S. A. and Allnatt, A. R., "Approximate Theory of Transport in Simple Dense Fluid Mixtures," *J. Chem. Phys.* **34**(2), 409–20, 1961.
432. Rice, S. A. and Allnatt, A. R., "On the Kinetic Theory of Dense Fluids. VI. Single Distribution Function for Rigid Spheres with an Attractive Potential," *J. Chem. Phys.* **34**, 2144–55, 1961.
433. Rice, S. A. and Allnatt, A. R., "On the Kinetic Theory of Dense Fluids. VII. The Doublet Distribution Function for Rigid Spheres with an Attractive Potential," *J. Chem. Phys.* **34**, 2156–65, 1961.
434. Rice, S. A. and Kirkwood, J. G., "An Approximate Theory of Transport in Dense Media," *J. Chem. Phys.* **31**, 901–8, 1959.
435. Rice, S. A., Kirkwood, J. G., Ross, J., and Zwanzig, R. W., "Statistical Mechanical Theory of Transport Properties, XII. Dense Rigid Sphere Fluids," *J. Chem. Phys.* **31**, 575–83, 1959.

436. Richter, G. N. and Sage, B. H., "Thermal Conductivity of Fluids, Nitrogen Dioxide in the Liquid Phase," *Ind. Eng. Chem. J. Chem. Eng. Data, Ser. 2*, No. 1, 61–6, 1957.
437. Richter, G. N. and Sage, B. H., "Thermal Conductivity of Fluids. Nitric Oxide," *J. Chem. Eng. Data* **4**, 36–40, 1959.
438. Richter, G. N. and Sage, B. H., "Thermal Conductivity of Fluids. Nitrous Oxide," *J. Chem. Eng. Data* **8**, 221–5, 1963.
439. Riedel, L., "Thermal Conductivity of Liquids," (In German) *Mitt. Kältetech. Inst. Karlsruhe* (2), 1–48, 1948; SLA TT-64-14402, 83 pp.
440. Riedel, L., "Thermal Conductivity Measurements on Sugar Solutions, Fruit Juices and Milk," (In German) *Chem. Ing. Tech.* **21**(9), 340–1, 1949.
441. Riedel, L., "Thermal Conductivity Measurements on Sodium and Potassium Hydroxide Solutions at Different Concentrations and Temperatures," (In German) *Chem. Ing. Tech.* **22**(3), 54–6, 1950; AEC-TR-2503, 12 pp., 1962.
442. Riedel, L., "Thermal Conductivity Determinations on Important Refrigerating Salt Solutions," (In German) *Kältetechnik* **2**(4), 99–101, 1950.
443. Riedel, L., "The Thermal Conductivities of Aqueous Solutions of Strong Electrolytes," (In German) *Chem. Ing. Tech.* **23**(3), 59–64, 1951; AEC-TR-2501, 23 pp., 1962.
444. Riedel, L., "New Thermal Conductivity Measurements on Organic Liquids," (In German), *Chem. Ing. Tech.* **23**(13), 321–4, 1951; AEC-TR-1822, 18 pp., 1962.
445. Riedel, L., "Thermal Conductivity Determinations on Mixtures of Different Organic Compounds with Water," (In German) *Chem. Ing. Tech.* **23**, 465–9, 1951; DDC RSIC-40, 15 pp., 1963. [AD 418 164]
446. Riedel, L., "Compressibility, Surface Tension and Thermal Conductivity in the Liquid State, Investigations Relating to an Extension of the Theorem of Corresponding States, Part IV," (In German) *Chem. Ing. Tech.* **27**, 209–13, 1955; SLA TT-64-16119, 23 pp., 1963, and TT-65-11850, 1963.
447. Robbins, L. A. and Kingrea, C. L., "Estimation of Thermal Conductivities of Organic Liquids over Useful Temperature Ranges," *Petroleum Refiner* **41**(5), 133–41, 1962; *Proc. Am. Petrol. Inst. Sect. III* **42**, 52–61, 1962.
448. Robin, J., Dewasnes, P., and Mabboux, C., "Dynamic Determination of the Thermal Conductivity of Carbon Dioxide at Different Temperatures," *Compt. Rend.* **250**, 3003–5, 1960.
449. Rodriguez, H. V., "Molecular Field Relationships to Liquid Viscosity, Compressibility, and Prediction of Thermal Conductivity of Binary Liquid Mixtures," University Microfilms Publ., 62-3663, 1–121, 1962.
450. Rolinski, E. J. and Zakanycz, S., "Thermal Conductivity of Gases at High Temperatures Using Shocktube Techniques," *Proc. 6th Conf. Thermal Conductivity*, Dayton, Ohio, AFML, 159–76, 1966.
451. Rothman, A. J., "Thermal Conductivity of Gases at High Temperatures," Ph.D. Thesis, USAEC UCRL-2339, 115 pp., 1954.
452. Rothman, A. J. and Bromley, L. A., "High Temperature Thermal Conductivity of Gases, Measurements on Nitrogen, Carbon Dioxide, Argon, and Nitrogen Carbon Dioxide Mixtures at Temperatures up to 775 C," *Ind. Eng. Chem.* **47**, 899–906, 1955.
453. Saha, M. N. and Srivastava, B. N., *A Treatise on Heat*, Indian Press Ltd., Allahabad, 1958.
454. Sakiadis, B. C. and Coates, J., "A Literature Survey of the Thermal Conductivity of Liquids," *Louisiana State University, Engineering Expt. Sta. Bull.*, No. 34, 1–70, 1952.
455. Sakiadis, B. C. and Coates, J., "Studies of Thermal Conductivity of Liquids, Part IV. A Literature Survey of Ultrasonic Velocities in Liquids and Solutions," *Louisiana State University, Eng. Expt. Sta. Bull.*, No. 46, 1–61, 1954.
456. Sakiadis, B. C. and Coates, J., "Studies of Thermal Conductivity of Liquids: Supplement to Part 1, A Literature Survey of the Thermal Conductivities of Liquids," *Louisiana State University, Eng. Expt. Sta. Bull.* No. 48, 1–24, 1954.
457. Sakiadis, B. C. and Coates, J., "Studies of Thermal Conductivity of Liquids, Parts I and II," *Am. Inst. Chem. Eng. J.* **1**(3), 275–88, 1955.
458. Sakiadis, B. C. and Coates, J., "Studies of Thermal Conductivity of Liquids, Part III," *Am. Inst. Chem. Eng. J.* **3**(1), 121–6, 1957.
459. Saksena, M. P. and Saxena, S. C., "Measurement of Thermal Conductivity of Gases Using Thermal Diffusion Columns: Air and Helium," *Defence Sci. J. (India)* **16**, 235–46, 1966.
460. Saksena, M. P. and Saxena, S. C., "On Possible Correlation between Potential Parameters and Critical or Boiling Point Constants," *Indian J. Pure Appl. Phys.* **4**(2), 86–7, 1966.
461. Saksena, M. P. and Saxena, S. C., "Calculation of Diffusion Coefficients of Binary Gas Mixtures," *Indian J. Pure Appl. Phys.* **4**(3), 109–16, 1966.
462. Saksena, M. P. and Saxena, S. C., "Measurement of Thermal Conductivity of Gases Using Thermal Diffusion Columns," *Phys. Fluids* **9**, 1595–9, 1966.
463. Saksena, M. P. and Saxena, S. C., "Thermal Conductivity of Mixtures of Monatomic Gases," *Proc. Natl. Inst. Sci. (India)* **31A**, 26–32, 1966.
464. Saksena, M. P. and Saxena, S. C., "Equilibrium Properties of Gases and Gaseous Mixtures," *Proc. Natl. Inst. Sci. (India)* **32A**, 177–95, 1966.
465. Saksena, M. P. and Saxena, S. C., "Thermal Conductivity of Polyatomic Gas Mixtures and Wassiljewa Form," *Appl. Sci. Res.* **17**, 326–30, 1967.
466. Saksena, M. P. and Saxena, S. C., "Certain Equilibrium Properties of Gases and Gas Mixtures on Steeper Lennard-Jones and Stockmayer Type Potentials," *Def. Sci. J.* **17**(2), 79–94, 1967.
467. Saksena, M. P., Nain, V. P. S., and Saxena, S. C., "Second Virial and Zero-Pressure Joule–Thomson Coefficients of Polar and Nonpolar Gases and Gas Mixtures," *Indian J. Phys.* **41**, 123–33, 1967.
468. Saksena, M. P., Saxena, S. C., and Mathur, S. N., "Rotational Relaxation in Binary Gas Mixtures," *Z. Phys. Chem. (Leipzig)* **232**, 253–8, 1966.
469. Saksena, M. P., Saxena, S. C., and Mathur, S. N., "Thermal Conductivity of Mixtures of Monatomic Gases and Translational Relaxation," *Trans. Faraday Soc.* **63**(3), 591–5, 1967.

470. Samoilov, E. V. and Tsitelauri, N. N., "Collision Integrals for the Morse Potential," *Teplofiz. Vysokikh Temperatur* **2**(4), 565–72, 1964; *High Temperature* **2**(4), 509–515, 1964.
471. Sandler, S. I. and Dahler, J. S., "Kinetic Theory of Loaded Spheres. II," *J. Chem. Phys.* **43**(5), 1750–9, 1965.
472. Sandler, S. I. and Dahler, J. S., "Transport Properties of Polyatomic Fluids. II. A Dilute Gas of Spherocylinders," *J. Chem. Phys.* **44**(3), 1229–37, 1966.
473. Sandler, S. I. and Dahler, J. S., "Kinetic Theory of Loaded Spheres. III. Transport Coefficients for the Dense Gas," *J. Chem. Phys.* **46**(9), 3520–31, 1967.
474. Sandler, S. I. and Dahler, J. S., "Kinetic Theory of Loaded Spheres. IV. Thermal Diffusion in a Dilute-Gas Mixture of D_2 and HT," *J. Chem. Phys.* **47**(8), 2621–30, 1967.
475. Sather, N. F. and Dahler, J. S., "Molecular Friction in Dilute Gases. II. Thermal Relaxation of Translational and Rotational Degrees of Freedom," *J. Chem. Phys.* **35**(6), 2029–37, 1961.
476. Sather, N. F. and Dahler, J. S., "Molecular Friction in Dilute Gases. III. Rotational Relaxation in Polyatomic Fluids," *J. Chem. Phys.* **37**(9), 1947–51, 1962.
477. Sather, N. F. and Dahler, J. S., "Approximate Theory of Viscosity and Thermal Conductivity in Dense Polyatomic Fluids," *Phys. Fluids* **5**(7), 754–68, 1962.
478. Saxena, S. C., "Thermal Diffusion of Gas Mixtures and Determination of Force Constants," *Indian J. Phys.* **29**(3), 131–40, 1955.
479. Saxena, S. C., "Higher Approximations to Diffusion Coefficients and Determination of Force Constants," *Indian J. Phys.* **29**(10), 453–60, 1955.
480. Saxena, S. C., "Thermal Conductivity and Force Between Like Molecules," *Indian J. Phys.* **29**(12), 587–602, 1955.
481. Saxena, S. C., "Evaluation of Collision Integrals Occurring in Higher Approximations to Diffusion Coefficients," *J. Chem. Phys.* **24**(6), 1209–10, 1956.
482. Saxena, S. C., "Thermal Conductivity of He–Ar–Xe Ternary Mixture," *J. Chem. Phys.* **25**(2), 360–1, 1956.
483. Saxena, S. C., "On the Two Schemes of Approximating the Transport Coefficients (Chapman–Cowling and Kihara), *J. Phys. Soc. (Japan)* **11**(4), 367–9, 1956.
484. Saxena, S. C., "Thermal Conduction and Gas Analysis," *Nature* **178**, 1462, 1956.
485. Saxena, S. C., "Generalized Relations for the Three Elementary Transport Coefficients of Inert Gases," *Physica* **22**, 1242–6, 1956.
486. Saxena, S. C., "Transport Coefficients and Force between Unlike Molecules," *Indian J. Phys.* **31**(3), 146–55, 1957.
487. Saxena, S. C., "Thermal Conductivity of Binary and Ternary Mixtures of Helium, Argon, and Xenon," *Indian J. Phys.* **31**(12), 597–606, 1957.
488. Saxena, S. C., "Determination of Unlike Interactions of Polyatomic Gases from Properties of Mixtures," *Rajasthan University Studies, Physical Sciences Section*, 1–5, 1963.
489. Saxena, S. C., Discussion on the paper "High Temperature Thermal Conductivity of Rare Gases and Gas Mixtures," *J. Heat Transfer, Trans. ASME* **90C**, 324–7, 1968.
490. Saxena, S. C., "Measurement of Thermal Conductivity of Gases," *Proc. 7th Conf. Thermal Conductivity*, NBS, Washington, D.C., 1967. (Publ. 1968.)
491. Saxena, S. C., "Thermal Conductivity of Binary Mixtures of Rare Gases and Wassiljewa Coefficients," *Thermal Conductivity, Proceedings of the 8th Conference*, Plenum Press, N. Y., 265–80, 1969.
492. Saxena, S. C. and Agrawal, J. P., "Thermal Conductivity of Polyatomic Gases and Relaxation Phenomena," *J. Chem. Phys.* **35**(6), 2107–13, 1961.
493. Saxena, S. C. and Agrawal, J. P., "Interrelation of Thermal Conductivity and Viscosity of Binary Gas Mixtures," *Proc. Phys. Soc. (London)* **80**, 313–5, 1962.
494. Saxena, S. C. and Bahethi, O. P., "Transport Properties of Some Simple Nonpolar Gases on the Morse Potential," *Mol. Phys.* **7**(2), 133–9, 1963–4.
495. Saxena, S. C. and Bahethi, O. P., "Intermolecular Potentials for Krypton," *Indian J. Pure Appl. Phys.* **3**(1), 12–15, 1964.
496. Saxena, S. C. and Gambhir, R. S., "The Viscosity and Translational Thermal Conductivity of Gas Mixtures," *Brit. J. Appl. Phys.* **14**, 436–8, 1963.
497. Saxena, S. C. and Gambhir, R. S., "Semi-empirical Formula for the Translational Thermal Conductivity of Gas Mixtures," *Indian J. Pure Appl. Phys.* **1**(9), 318–21, 1963.
498. Saxena, S. C. and Gambhir, R. S., "Second Virial Coefficient of Gases and Gaseous Mixtures on the Morse Potential," *Mol. Phys.* **6**(6), 577–83, 1963.
499. Saxena, S. C. and Gambhir, R. S., "Semi-empirical Formulae for the Viscosity and Translational Thermal Conductivity of Gas Mixtures," *Proc. Phys. Soc.* **81**(4), 788–9, 1963.
500. Saxena, S. C. and Gandhi, J. M., "Thermal Conductivity of Multicomponent Mixtures of Inert Gases," *Rev. Mod. Phys.* **35**(4), 1022–32, 1963.
501. Saxena, S. C. and Gandhi, J. M., "Thermal Conductivity of Multicomponent Mixtures of Inert Gases, II," University of Rajasthan Studies (Physics), 6 pp., 1965.
502. Saxena, S. C. and Gandhi, J. M., "Methods of Measuring Thermal Conductivity of Gases," *J. Sci. Indian Res.* **26**, 458–65, 1967.
503. Saxena, S. C. and Gupta, G. P., "Thermal Conductivity of Binary, Ternary and Quaternary Mixtures of Polyatomic Gases," *Proc. 7th Conf. Thermal Conductivity*, NBS, Washington, D.C., 1967.
504. Saxena, S. C. and Joshi, K. M., "Second Virial and Zero Pressure Joule–Thomson Coefficients of Nonpolar Quasi-Spherical Molecules," *Indian J. Phys.* **36**(8), 422–30, 1962.
505. Saxena, S. C. and Joshi, K. M., "Second Virial Coefficient of Polar Gases," *Phys. Fluids* **5**(10), 1217–22, 1962.
506. Saxena, S. C. and Joshi, K. M., "Zero Pressure Joule–Thomson Coefficient of Polar Gases," *Indian J. Pure Appl. Phys.* **1**(12), 420–6, 1963.
507. Saxena, S. C. and Joshi, K. M., "Thermal Diffusion Factors for Krypton and Xenon," *Physica* **29**, 257–60, 1963.
508. Saxena, S. C. and Mathur, B. P., "Thermal Diffusion in Binary Gas Mixtures and Intermolecular Forces," *Rev. Mod. Phys.* **37**(2), 316–25, 1965.
509. Saxena, S. C. and Mathur, B. P., "Thermal Diffusion

in Isotopic Gas Mixtures and Intermolecular Forces," *Rev. Mod. Phys.* **38**(2), 380–90, 1966.

510. Saxena, S. C. and Mathur, B. P., "Central Molecular Potentials, Combination Rules and Properties of Gases and Gas Mixtures," *Chem. Phys. Letters* **1**, 224–6, 1967.
511. Saxena, S. C. and Narayanan, T. K. S., "Multicomponent Viscosities of Gaseous Mixtures at High Temperatures," *Ind. Eng. Chem. Fundamentals* **1**(3), 191–5, 1962.
512. Saxena, S. C. and Srivastava, B. N., "Second Approximation to the Thermal Diffusion Factor on the Lennard-Jones 12:6 Model," *J. Chem. Phys.* **23**(9), 1571–4, 1955.
513. Saxena, S. C. and Tondon, P. K., "Thermal Conductivity of Multicomponent Mixtures of Rare Gases," *Proc. Fourth Symp. Thermophysical Properties*, ASME, N. Y., 398–404, 1968.
514. Saxena, S. C., Gupta, G. P., and Saxena, V. K., "Measurement of the Thermal Conductivity of Nitrogen (350 to 1500 K) by the Column Method," *Thermal Conductivity, Proceedings of the 8th Conference*, Plenum Press, N. Y., 125–39, 1969.
515. Saxena, S. C., Kelley, J. G., and Watson, W. W., "Temperature Dependence of the Thermal Diffusion Factor for Helium, Neon and Argon," *Phys. Fluids* **4**(10), 1216–25, 1961.
516. Saxena, S. C., Mathur, S., and Gupta, G. P., "The Thermal Conductivity of Some Binary Gas Mixtures including Non-Polar Polyatomic Gases," *Suppl. Defence Sci. J. (India)* **16**(4), 99–112, 1966.
517. Saxena, S. C., Saksena, M. P., and Gambhir, R. S., "The Thermal Conductivity of Non-polar Polyatomic Gases," *Brit. J. Appl. Physics* **15**, 843–9, 1964.
518. Saxena, S. C., Saksena, M. P., Gambhir, R. S., and Gandhi, J. M., "The Thermal Conductivity of Nonpolar Polyatomic Gas Mixtures," *Physica* **31**, 333–41, 1965.
519. Saxena, V. K., "Studies on the Transport Properties of Gases," Ph.D. Thesis, Phys. Dept., Rajasthan Univ., India, 448 pp., 1967.
520. Saxena, V. K. and Saxena, S. C., "Measurement of the Thermal Conductivity of Helium Using a Hot Wire Type of Thermal Diffusion Column," *Brit. J. Appl. Phys. (J. Phys. D) Ser. 2* **1**, 1341–51, 1968.
521. Saxena, V. K. and Saxena, S. C., "Measurement of the Thermal Conductivity of Argon Using Hot Wire Type Thermal Diffusion Columns," *Chem. Phys. Letters* **2**, 44–6, 1968.
522. Saxena, V. K. and Saxena, S. C., "Measurement of the Thermal Conductivity of Neon Using Hot-Wire Type of Thermal Diffusion Columns," *J. Chem. Phys.* **48**(12), 5662–7, 1968.
523. Saxena, V. K., Nain, V. P. S., and Saxena, S. C., "Thermal-Diffusion Factors from the Measurements on a Trennschaukel: Ne-Ar and Ne-Xe," *J. Chem. Phys.* **48**(8), 3681–5, 1968.
524. Saxena, V. K., Saksena, M. P., and Saxena, S. C., "Measurements of Thermal Conductivity of Gases Using Thermal Diffusion Column: Neon," *Indian J. Phys.* **40**, 597–604, 1966.
525. Schafer, Kl. and Reiter, F. W., "A Measuring Method for the Determination of Thermal Conductivity at 1100 C," *Z. Elektrochem.* **61**, 1230–5, 1957; SLA TT-64-30306, 10 pp., 1961.
526. Scheffy, W. J., "Thermal Conduction in Liquids," Tech. Rept. PR-85-R, Princeton Univ., Princeton, N.J., 1–99, 1958. [AD 204 891]
527. Scheffy, W. J. and Johnson, E. F., "Thermal Conductivities of Liquids at High Temperature," *J. Chem. Eng. Data* **6**(2), 245–9, 1961.
528. Schliermacher, A., "Ueber die Wärmeleitung der Gase," *Wied. Ann.* **34**, 623–46, 1888.
529. Schmidt, E. and Leidenfrost, W., "The Influence of Electric Fields on the Heat Transfer in Liquid Electrical Nonconductors," (In German) *Forschung Gebiete Ingenieurw.* **19**(3), 65–80, 1953.
530. Schmidt, E. and Leidenfrost, W., "Heat Transport in Liquid Non-conductors under the Influence of Electric Fields," (In German) *Chem. Ing. Tech.* **26**, 35–8, 1954.
531. Schneider, E., "The Thermal Conductivity of Air and Hydrogen," *Ann. Physik.* **79**, 177–203, 1926; ibid. **80**, 215–6, 1927.
532. Schrock, V. E. and Starkman, E. S., "Spherical Apparatus for Measuring the Thermal Conductivity of Liquids," *Rev. Sci. Instr.* **29**(7), 625–9, 1958.
533. Schwarze, W., "Bestimmung der Wärmeleitfähigkeit von Argon and Helium nach der Methode von Schliermacher," *Ann. Physik.* **11**, 303–30 and 1144, 1903.
534. Secrest, D. and Hirschfelder, J. O., "Slowly Reacting Gas Mixture in a Heat Conductivity Cell," *Phys. Fluids* **4**(1), 61–73, 1961.
535. Sengers, J. V., *Thermal Conductivity Measurements at Elevated Gas Densities Including the Critical Region*, Van Gorcum and Co., Amsterdam, Netherlands, 1–126, 1962.
536. Sengers, J. V., "Thermal Conductivity and Viscosity of Simple Fluids," *Intern. J. Heat Mass Transfer* **8**, 1103–16, 1965.
537. Sengers, J. V., "Density Expansion of the Viscosity of a Moderately Dense Gas," *Phys. Rev. Letters* **15**(12), 515–7, 1965.
538. Sengers, J. V., "Triple Collision Contribution to the Transport Coefficients of Gases," *Boulder Lectures in Theoretical Physics*, Univ. of Colorado Press, 1966.
539. Sengers, J. V., "Triple Collision Contribution to the Transport Coefficients of a Rigid Sphere Gas," *Phys. Fluids* **9**(7), 1333–47, 1966.
540. Sengers, J. V., "Divergence in the Density Expansion of the Transport Coefficients of Two-Dimensional Gas," *Phys. Fluids* **9**(9), 1685–96, 1966.
541. Sherratt, G. G. and Griffiths, E., "Hot Wire Method for the Thermal Conductivities of Gases," *Phil. Mag.* **27**, 68–75, 1939.
542. Shister, A. R., "On the Conductivity of a Nonisothermal Plasma," *High Temperature* **3**(6), 855–7, 1965.
543. Shroff, G. H., "Measurement and Correlation of Transport Properties in Non-Ideal Binary Liquid Systems," Ph.D. Thesis, Dept. of Chem. Eng., Univ. of New Brunswick, 1–167, 1968.
544. Slepicvka, F., "Attenuation in Shock Tubes," *Phys. Fluids* **9**(9), 1865–6, 1966.
545. Smeets, G., "Determination of the Thermal Conductivity of Hot Gases from the Temperature Boundary Layer in the Shocktube," *Z. Naturforsch.* **20a**, 683–9, 1965.
546. Smiley, E. F., Jr., "The Measurement of the Thermal Conductivity of Gases at High Temperatures with a Shocktube; Experimental Results in Argon at

Temperatures between 1000 K and 3000 K," Ph.D. Thesis, Catholic Univ. America, 1–32, 1957.
547. Smith, F. J. and Munn, R. J., "Automatic Calculation of the Transport Collision Integrals with Tables for the Morse Potential," *J. Chem. Phys.* **41**(11), 3560–8, 1964.
548. Smith, F. J., Mason, E. A., and Munn, R. J., "Transport Collision Integrals for Gases Obeying (9-6) and (28-7) Potentials," *J. Chem. Phys.* **42**(4), 1334–9, 1965.
549. Smith, F. J., Munn, R. J., and Mason, E. A., "Transport Properties of Quadrupolar Gases," *J. Chem. Phys.* **46**(1), 317–21, 1967.
550. Smith, J. F. D., "The Thermal Conductivity of Liquids," Trans. ASME **58**, 719–25, 1936.
551. Snider, R. F. and Curtiss, C. F., "Kinetic Theory of Moderately Dense Gases," *Phys. Fluids* **1**(2), 122–38, 1958.
552. Snider, R. F. and Curtiss, C. F., "Kinetic Theory of Moderately Dense Gases: Rigid Sphere Limit," *Phys. Fluids* **3**(6), 903–4, 1960.
553. Snider, R. F. and McCourt, F. R., "Kinetic Theory of Moderately Dense Gases: Inverse Power Potentials," *Phys. Fluids* **6**(7), 1020–5, 1963.
554. Somerville, J. M., *The Electric Arc*, Wiley, New York, 1959.
555. Spells, K. E., "Thermal Conductivities of Some Biological Fluids," *Phys. Med. Biol.* **5**(2), 139–53, 1960.
556. Srivastava, B. N. and Barua, A. K., "Thermal Conductivity of Binary Mixtures of Diatomic and Monatomic Gases," *J. Chem. Phys.* **32**(2), 427–35, 1960.
557. Srivastava, B. N. and Barua, A. K., "Heat Transfer in Polyatomic Gases and Chemically Reacting Gas Mixtures," *Proc. Natl. Inst. Sci. India* **26A**, 143–56, 1960.
558. Srivastava, B. N. and Barua, A. K., "Thermal Conductivity and Equilibrium Constant of the System $N_2O_4 \rightleftharpoons 2NO_2$," *J. Chem. Phys.* **35**(1), 329–34, 1961.
559. Srivastava, B. N. and Barua, A. K., "Effect of Relaxation of Chemical Energy on the Thermal Conductivity of the System $N_2O_4 \rightleftharpoons 2NO_2$," *J. Chem. Phys.* **35**(2), 649–51, 1961.
560. Srivastava, B. N. and Chakraborti, P. K., "Thermal Conductivity of the Slowly Reacting System $2H \rightleftharpoons H_2 + I_2$," *Indian J. Phys.* **38**(1), 1–6, 1964.
561. Srivastava, B. N. and Das Gupta, A., "Thermal Conductivity of Gaseous Ammonia at Different Temperatures," *Phys. Fluids* **9**, 722–5, 1966.
562. Srivastava, B. N. and Das Gupta, A., "Thermal Conductivity of Ammonia–Diethyl Ether Mixtures at Different Temperatures," *J. Chem. Phys.* **46**, 3592–4, 1967.
563. Srivastava, B. N. and Madan, M. P., "Thermal Conductivity of Argon–Neon Mixtures," *Proc. Natl. Inst. Sci. (India)* **20**, 587–97, 1954.
564. Srivastava, B. N. and Saxena, S. C., "Formulas for Thermal Conductivity of Ternary Gas Mixtures," *J. Chem. Phys.* **27**(2), 583–4, 1957.
565. Srivastava, B. N. and Saxena, S. C., "Thermal Conductivity of Binary and Ternary Rare Gas Mixtures," *Proc. Phys. Soc. (London)* **B70**, 369–78, 1957.
566. Srivastava, B. N. and Srivastava, R. C., "Thermal Conductivity and Eucken Correction for Diatomic Gases and Binary Gas Mixtures," *J. Chem. Phys.* **30**, 1200–5, 1959.
567. Srivastava, B. N., Barua, A. K., and Chakraborti, P. K., "Thermal Conductivity of Slowly Reacting Systems," *Trans. Faraday Soc.* **59**, 2522–7, 1963.
568. Stafford, O. J., *Z. Phys. Chem.* **77**, 66 pp., 1911.
569. Stefan, J., "Investigations on the Thermal Conductivity of Gases," *Sitzber. Akad.-Wiss. Wien. Math. Naturw. Kl. IIA* **65**, 45–69, 1872.
570. Stefan, J., "Untersuchungen über die Wärmeleitung in Gasen. Relative Bestimmung der Wärmeleitungsvermögen verschiedener Gase," *Sitzber. Akad. Wiss. Wien. Math. Naturw. Kl.*, **72**, 69–101, 1875.
571. Stiel, L. I. and Thodos, G., "The Prediction of the Transport Properties of Pure Gaseous and Liquid Substances," *Progr. Int. Res. Thermodynamic and Transport Properties*, Academic Press, New York, 352–65, 1962, and references cited therein.
572. Stogryn, D. E. and Hirschfelder, J. O., "Initial Pressure Dependence of Thermal Conductivity and Viscosity," *J. Chem. Phys.* **31**(6), 1545–54, 1959; ibid., **33**(3), 942–3, 1960.
573. Stops, D. W., "Effect of Temperature on the Thermal Conductivity of Gases," *Nature* **164**, 966–7, 1949.
574. Strunk, M. R., Custead, W. G., and Stevenson, G. L., "The Prediction of the Viscosity of Nonpolar Binary Gaseous Mixtures at Atmospheric Pressure," *Am. Inst. Chem. Eng. J.* **10**, 483–6, 1964.
575. Sutherland, R. D., Davis, R. S., and Seyer, W. F., "Heat Transfer Effects: Molecular Orientation of Octadecane," *Ind. Eng. Chem.* **51**, 585–8, 1959.
576. Sutherland, W., "The Viscosity of Mixed Gases," *Phil. Mag.* **40**, 421–31, 1895.
577. Svehla, R. A., "Estimated Viscosities and Thermal Conductivities of Gases at High Temperatures," NASA TR R-132, 140 pp., 1962. [AD 272 963]
578. Svehla, R. A. and Brokaw, R. S., "Thermodynamic and Transport Properties for the $N_2O_4 \rightleftharpoons 2NO_2 \rightleftharpoons 2NO + O_2$ System," NASA TN D-3327, 57 pp., 1966.
579. Tauscher, W., "Thermal Conductivity of Liquid Refrigerants Measured by an Unsteady State Hot-Wire Method," (In German) *Kältetechnik-Klimatisierung* **19**(9), 288–92, 1967.
580. Taxman, N., "Classical Theory of Transport Phenomena in Dilute Polyatomic Gases," *Phys. Rev.* **110**, 1235–9, 1958.
581. Taylor, H. S. and Glasstone, S., "A Treatise on Physical Chemistry," *States of Matter*, Vol. 2, 3rd Edition, Van Nostrand, p. 126, 1951.
582. Taylor, W. J. and Johnston, H. L., "An Improved Hot-Wire Cell for Accurate Measurements of Thermal Conductivities of Gases over a Wide Temperature Range. Results with Air between 87 and 375 K," *J. Chem. Phys.* **14**, 219–33, 1946.
583. Thompson, W. P. and Emrich, R. J., "Turbulent Spots and Wall Roughness Effects in Shock Tube Boundary Layer Transition," *Phys. Fluids* **10**(1), 17–20, 1967.
584. Thornton, E., "Viscosity and Thermal Conductivity of Binary Gas Mixtures: Xenon–Krypton, Xenon–Argon, Xenon–Neon, Xenon–Helium," *Proc. Phys. Soc. (London)* **76**, 104–12, 1960.
585. Thornton, E., "Viscosity and Thermal Conductivity of Binary Gas Mixtures: Krypton–Argon, Krypton–Neon and Krypton–Helium," *Proc. Phys. Soc. (London)* **77**, 1166–9, 1961.

586. Thornton, E. and Baker, W. A. D., "Viscosity and Thermal Conductivity of Binary Gas Mixtures: Argon-Neon, Argon-Helium and Neon-Helium," *Proc. Phys. Soc. (London)* **80**, 1171-5, 1962.
587. Timrot, D. L. and Totskii, E. E., "Dilatometric Method for the Experimental Determination of the Thermal Conductivity of Corrosive Gases, and Vapors at High Temperatures," *High Temperature* **3**, 685-90, 1965; *Teplofiz. Vysokikh Temperatur* **3**(5), 740-6, 1965.
588. Timrot, D. L. and Umanskii, A. S., "Investigation of the Thermal Conductivity of Helium in the Temperature Range 400-2400 K," *High Temperature* **3**, 345-51, 1965; *Teplofiz. Vysokikh Temperatur* **3**(3), 381-8, 1965.
589. Timrot, D. L. and Umanskii, A. S., "Thermal Conductivity of Hydrogen and Argon," *High Temperature* **4**(2), 285-7, 1966; *Teplofiz. Vysokikh Temperatur* **4**(2), 289-92, 1966.
590. Timrot, D. L. and Vargaftik, N. B., "Heat Conductivity, Viscosity and Thermodynamical Properties of Steam at High Temperatures and Pressures," *Trans. Fourth World Power Conference*, III, 1642-66, 1950.
591. Todd, G. W., "Thermal Conductivity of Air and Other Gases," *Proc. Roy. Soc. (London)* **A83**, 19-39, 1909.
592. Todheide, K., Hensel, F., and Franck, E. U., "Thermal Conductivity of Gases," *Landolt-Bornstein*, Ser. II, 39-71 and 385-92, Springer-Verlag, 1968.
593. Tondon, P. K., "Studies on Non-equilibrium Properties of Gases," Ph.D. Thesis, Physics Dept., Rajasthan Univ., India, 268 pp., 1968.
594. Tondon, P. K. and Saxena, S. C., "Calculation of Thermal Conductivity of Polar-Nonpolar Gas Mixtures," *Appl. Sci. Res.* **19**, 163-70, 1968.
595. Tondon, P. K., Gandhi, J. M., and Saxena, S. C., "Thermal Conductivity of Hydrogen-Helium and Deuterium-Helium Systems," *Proc. Phys. Soc. (London)* **92**, 253-5, 1967.
596. O'Toole, J. T. and Dahler, J. S., "On the Kinetic Theory of a Fluid Composed of Rigid Spheres," *J. Chem. Phys.* **32**(4), 1097-1106, 1960.
597. O'Toole, J. T. and Dahler, J. S., "Molecular Friction in Dilute Gases," *J. Chem. Phys.* **33**(5), 1496-504, 1960.
598. Touloukian, Y. S., Gerritsen, J. K., and Moore, N. Y., *Thermophysical Properties Research Literature Retrieval Guide*, 3 Books, 2759 pp., Plenum Press, New York, 1967.
599. Tree, D. R. and Leidenfrost, W., "Prediction of Minor Heat Losses in a Thermal Conductivity and Other Calorimeter-Type Cells," *Proc. 7th Conf. Thermal Conductivity*, NBS, Washington, D.C., 1967 (published 1968).
600. Tsederberg, N. V., *Teploprovodrost' Gazov i Zhidkostev*, published by Gosenergoizdat, Moscow and Leningrad, 1963.
601. Tsederberg, N. V., *Thermal Conductivity of Gases and Liquids*, M.I.T. Press, Cambridge, Mass., 246 pp., 1965.
602. Tsederberg, N. V. and Timrot, D. L., "Experimental Determination of the Thermal Conductivity of Liquid Oxygen," (In Russian) *Zh. Tekh. Fiz.* **26**(8), 1849-56, 1956; *Sov. Phys. Tech. Phys.* **1**, 1791-7, 1957.
603. Tufeu, R., LeNeindre, B., and Johannin, P., "Thermal Conductivity of Several Liquids," (In French) *C.R. Acad. Sci. Paris, Series B* **262**, 229-31, 1966.
604. Tufeu, R., LeNeindre, B., Bury, P., and Johannin, P., "Measurement of the Thermal Conductivity of Several Liquids," with a foreword by Missenard, F. A., (In French) *Revue Générale Thermique* **7**(76), 365-77, 1968.
605. Turnbull, A. G., "The Thermal Conductivity of Molten Salts: II. Theory and Results for Pure Salts," *Australian J. Appl. Sci.* **12**(3), 324-9, 1961.
606. Tyrrell, H. J. V., "Diffusion and Heat Flow in Liquids," Chapter 11, *Thermal Conductivity of Liquids*, 291-319, Butterworths, London, 1961.
607. Ubbink, J. B. and DeHaas, W. J., "Apparatus to Measure the Specific Thermal Conductivity of Gases at Low Temperatures," *Physica* **10**, 451-64, 1943.
608. Von Ubisch, H., "On the Conduction of Heat in Rarefied Gases and Its Manometric Application. I," *Appl. Sci. Res.* **A2**, 364-402, 1951.
609. Von Ubisch, H., "On the Theory of the Temperature Jump in a Rarefied Gas and Its Experimental Varification," *Arkiv. Fys.* **10**, 157-63, 1956.
610. Von Ubisch, H., "The Thermal Conductivities of Mixtures of Rares Gases at 29 C and 520 C," *Arkiv. Fys.* **16**, 93-100, 1959.
611. Ulybin, S. A., Bugrov, V. P., and Il'in, A. V., "Temperature Dependence of the Thermal Conductivity of Chemically Nonreacting Rarefied Gas Mixtures," *Teplofiz. Vysok. Temp.* **4**(2), 214-7, 1966.
612. Valleau, J. P., "Transport of Energy and Momentum in a Dense Fluid of Rough Spheres," *Mol. Phys.* **1**(1), 63-7, 1958.
613. Valleau, J. P., "Transport in Dense Square-Well Fluid Mixtures," *J. Chem. Phys.* **44**(7), 2626-32, 1966.
614. Vanderkooi, W. N., Hildenbrand, D. L., and Stull, D. R., "Liquid Thermal Conductivities: The Apparatus, Values for Several Glycols and Their Aqueous Solutions, and Five Molecular Weight Hydrocarbons," *J. Chem. Eng. Data* **12**(3), 377-9, 1967.
615. Vanderslice, J. T., Weissman, S., Mason, E. A., and Fallon, R. J., "High Temperature Transport Properties of Dissociating Hydrogen," *Phys. Fluids* **5**(2), 155-64, 1962.
616. Vargaftik, N. B., "Dependence of the Coefficient of Thermal Conductivity of Gases and Vapors on Pressure," *J. Tech. Phys.* **7**, 1199-216, 1937.
617. Vargaftik, N. B., *Teplofizichoskiya Svoystvo Vsehestv*, Moscow and Leningrad, 368 pp., 1956.
618. Vargaftik, N. B., "Thermal Conductivity of Liquids and Condensed Gases," *Proceedings of the Joint Conference on Thermodynamic and Transport Properties of Fluids*, published by Inst. Mech. Engrs., London, 142-9, 1958.
619. Vargaftik, N. B., "Thermophysical Properties of Substances," F-TS-9537/V, 485 pp., 1959.
620. Vargaftik, N. B., "Thermal Conductivities of Liquids," (In Russian), *Izv. Vses. Teplotekhn. Inst.* **18**(8), 6, 1959.
621. Vargaftik, N. B., *Spravochnik po Teplofizicheskim Svoistam Gasov i Zhidkostei*, Gosudarstvennoe Izdatel'stvo Fiziko-Mathematichesko i Literatury, Moscow (M. Fizmatgiz) 708 pp., 1963.
622. Vargaftik, N. B. and Os'minin, Yu. P., "Thermal Conductivity of Aqeuous Solutions of Salts, Acids and Alkalies," (In Russian), *Teploenerg.* **3**(7), 11-16, 1956.
623. Vargaftik, N. B. and Zimina, N. Kh., "Heat Conductivity of Argon at High Temperatures," *Teplofiz. Vysokikh*

Temperatur **2**, 716–24, 1964; *High Temperature* **2**(5), 645–51, 1964.

624. Vargaftik, N. B. and Zimina, N. Kh., "Thermal Conductivity of Nitrogen at High Temperatures," *Teplofiz. Vysokikh Temperatur* **2**, 869–78, 1964; *High Temperature* **2**(6), 782–90, 1964.

625. Varlashkin, P. G. and Thompson, J. C., "Thermal Conductivity of Liquid Ammonia," *J. Chem. Eng. Data* **8**(4), 526, 1963.

626. Varlashkin, P. G. and Thompson, J. C., "Heat Conduction in Metal–Ammonia Solutions," *J. Chem. Phys.* **38**(8), 1974–7, 1963.

627. Venart, J. E. S., "Liquid Thermal Conductivity Measurements," *J. Chem. Eng. Data* **10**(3), 239–41, 1965.

628. Venart, J. E. S., "The Thermal Conductivity of Binary Organic Liquid Mixtures," *Proceedings of the Fourth Symposium on Thermophysical Properties*, ASME, New York, N.Y., 292–5, 1968.

629. Vines, R. G., "The Thermal Conductivity of Organic Vapours: The Influence of Molecular Interaction," *Australian J. Chem.* **6**, 1–26, 1953.

630. Viswanath, D. S., "On Thermal Conductivity of Liquids," *Am. Inst. Chem. Eng. J.* **13**(5), 850–3, 1967.

631. Vodar, B. and Saurel, J., "The Properties of Compressed Gases," Chapter 3, *High Pressure Physics and Chemistry*, I (R. S. Bradley, Ed.), Academic Press, New York, 51–143, 1963.

632. Wachman, H. Y., "The Thermal Accommodation Coefficient: A Critical Survey," *Am. Rocket Soc. J.* **32**(1), 2–12, 1962.

633. Wachman, H. Y., "Method for Determining Accommodation Coefficients from Data in the Temperature Jump Range Without Applying Temperature-Jump Theory," *J, Chem. Phys.* **42**(5), 1850–1, 1965.

634. Wachman, H. Y., "Thermal Accommodation Coefficients of Helium on Tungsten and Hydrogen on Hydrogen-Covered Tungsten at 325, 403 and 473 K," *J. Chem. Phys.* **45**(5), 1532–8, 1966.

635. Wachman, H. Y., "Accommodation Coefficients of Nitrogen and Helium on Nitrogen-Covered Tungsten Between 325–496 K," *Rarefied Gas Dynamics, 5th Symposium* **1**, 173–86, 1967.

636. Wachsmuth, J., "About Heat Conduction in Mixtures of Argon and Helium," *Z. Physik.* **9**, 235–40, 1908.

637. Waelbroeck, F. G. and Zuckerbrodt, P., "Thermal Conductivity of Gases at Low Pressures. I. Monatomic Gases, Helium and Argon," *J. Chem. Phys.* **28**, 523–4, 1958.

638. Waldmann, L., "Ueber eine nichtstationäre Messmethode für die Wärmeleitkonstante von Gasen," *Z. Naturforsch.* **18a**, 1360–1, 1963.

639. Walker, R. E. and Westenberg, A. A., "Absolute Low Speed Anemometer," *Rev. Sci. Instr.* **27**, 844–8, 1956.

640. Walker, R. E., deHaas, N., and Westenberg, A. A., "New Method of Measuring Gas Thermal Conductivity," *Phys. Fluids* **3**, 482–83, 1960.

641. Wang Chang, C. S. and Uhlenbeck, G. E., "Transport Phenomena in Polynomic Gases," Univ. of Michigan, Eng. Res. Rept. No. CM-681, 1951.

642. Wang Chang, C. S., Uhlenbeck, G. E., and deBoer, J., "The Heat Conductivity and Viscosity of Polyatomic Gases," *Studies in Statistical Mechanics*, *II* (Edited by deBoer J. and Uhlenbeck, G. E.) John Wiley and Sons, Inc., New York, 1964.

643. Wassiljewa, A., "Wärmeleitung in Gasgemischen, I," *Z. Physik* **5**, 737–42, 1904.

644. Watson, K. M., "Thermodynamics of the Liquid State. Generalized Prediction of Properties," *Ind. Eng. Chem.* **35**, 398–406, 1943.

645. Weber, H. F., "VII. Investigations into the Thermal Conductivity of Liquids," (In German), *Ann. Physik* **10**(3), 103–29, 1880.

646. Weber, S., "Experimentelle Untersuchungen über die Wärmeleitfähigkeit der Gase, I," *Ann. Physik* **54**, 325–56, 1917.

647. Weber, S., "Untersuchungen über die Wärmeleitfähigkeit der Gase. II," *Ann. Physik* **54**, 437–62, 1917.

648. Weber, S., "Theoretische und experimentelle Untersuchungen über die Wärmeleitfähigkeit von Gasgemischen," *Ann. Physik* **54**, 481–502, 1917.

649. Weber, S., "Uber die Wärmeleitfähigkeit der Gase," *Ann. Physik.* **82**, 479–503, 1927.

650. Wei, C. C. and Davis, H. T., "Kinetic Theory of Dense Fluid Mixtures. II. Solution to the Singlet Distribution Functions for the Rice-Allnatt Model," *J. Chem. Phys.* **45**(7), 2533–44, 1966.

651. Wei, C. C. and Davis, H. T., "Kinetic Theory of Dense Fluid Mixtures. III. The Doublet Distribution Functions of the Rice-Allnatt Model," **46**(9), 3456–67, 1967. [AD 659 832]

652. Weibel, E. S., "Anomalous Skin Effect in a Plasma," *Phys. Fluids* **10**(4), 741–8, 1967.

653. Weinstock, J., "Nonanalyticity of Transport Coefficients and the Complete Density Expansion of Momentum Correlation Functions," *Phys. Rev.* **140A**, 460–5, 1965.

654. Weissman, S., Saxena, S. C., and Mason, E. A., "Intermolecular Forces from Diffusion and Thermal Diffusion Measurements," *Phys. Fluids* **3**(4), 510–18, 1960.

655. Weissman, S., Saxena, S. C., and Mason, E. A., "Diffusion and Thermal Diffusion in Ne–CO_2," *Phys. Fluids* **4**(5), 643–8, 1961.

656. Westenberg, A. A., "A Critical Survey of the Major Methods for Measuring and Calculating Dilute Gas Transport Properties," *Advances in Heat Transfer, Vol. 3* 253–302, Academic Press, N. Y., 1966.

657. Westenberg, A. A. and deHaas, N., "Gas Thermal Conductivity Studies at High Temperature. Line-Source Technique and Results in N_2, CO_2 and N_2–CO_2 Mixtures," *Phys. Fluids* **5**, 266–73, 1962.

658. Westenberg, A. A. and deHaas, N., "High Temperature Gas Thermal Diffusivity with the Line Source Technique," *Progr. Int. Res. Thermodynamic and Transport Properties*, Academic Press, N. Y., 412–7, 1962.

659. Whalley, E., "The Thermal Conductivity of Associating Gases," *Disc. Faraday Soc.*, No. 22, 54–63, 1956.

660. White, L. R. and Davis, H. T., "Thermal Conductivity of Molten Alkali Nitrates," *J. Chem. Phys.* **47**(12), 5433–9, 1967.

661. Wienecke, R., "Experimental and Theoretical Determination of the Thermal Conductivity of the Plasma in a High-Current Carbon Arc," *Z. Physik.* **146**, 39–58, 1956; SCL-T-48, 22 pp., 1960; TPRC Translation No. 9.

662. Winkelmann, A., "Ueber die Wärmeleitung der Gase," *Pogg. Ann.* **156**, 497–531, 1875.

663. Winkelmann, A., "Ueber die Wärmeleitung der Gase," *Wied. Ann.* **44**, 429–56, 1891; **46**, 323, 1892.
664. Winkelmann, A., "Ueber den absoluten Werth der Wärmeleitung der Luft," *Wied. Ann.* **48**, 180–7, 1892.
665. Wood, H. T., "The Quantum Corrections to the Transport Collision Integrals," Univ. Wisconsin Theoretical Chemistry Institute Report WIS-TCI-111, 1–180, 1965.
666. Wright, P. G., "A Method of Obtaining Sutherland–Wassiljewa Coefficients," *Proc. Leeds Philosophical and Literary Society, Scientific Section, IX* Part VIII, 215–21, 1964.
667. Wright, P. G. and Gray, P., "Collisonal Interference Between Unlike Molecules Transporting Momentum or Energy in Gases," *Trans. Faraday Soc.* **58**, 1–16, 1962.
668. Wutzke, S. A., Pfendev, E., and Eckert, E. R. G., "Study of Electric-Arc Behavior with Superimposed Flow," *AIAAJ* **5**(4), 707–14, 1967.
669. Yos, J. M., "Transport Properties of Nitrogen, Hydrogen, Oxygen and Air to 30,000 K," Tech. Memorandum RAD-TM-63-7, 65 pp., 22 March 1963 reprinted April 1966, and a subsequent tabulation, 50 pp., Nov. 1967.
670. Yos, J. M., "Approximate Equations for the Viscosity and Translational Thermal Conductivity of Gas Mixtures," AVCO Missile, Space, and Electronics Group Rept., Wilmington, Massachusetts, 56 pp., April 1967.
671. Yun, K. S. and Mason, E. A., "Collision Integrals for the Transport Properties of Dissociating Air at High Temperature," *Phys. Fluids* **5**(4), 380–6, 1962.
672. Yun, K. S., Weissman, S., and Mason, E. A., "High Temperature Transport Properties of Dissociating Nitrogen and Dissociating Oxygen," *Phys. Fluids* **5**(6), 672–8, 1962.
673. Zaitseva, L. S., "An Experimental Investigation of the Heat Conductivity of Monatomic Gases over Wide Temperature Intervals," *Soviet Phys.—Tech. Phys.* **4**, 444–50, 1959; *Zh. Tekhn. Fiz.* **29**(4), 497–505, 1959.
674. Zarkova, L. P. and Stefanov, B. I., "Experimental Determination of Coefficient of Thermal Conductivity of a Cesium Plasma," *Mezhdunarodnyii Simpozium po Svoistvam i Primeneniyu Nizkotemperaturnoi Plazmy, pri XX. Mezhdunarodnom Kongresse po Teoreticheskoi i Prikladnoi Khimii*, Moscow, 1–16, 1965. [AD 653 555, translation]
675. Ziebland, H., "The Thermal Conductivity of Toluene. New Determinations and an Appraisal of Recent Experimental Work," *Int. J. Heat Mass Transfer* **2**, 273–9, 1961.
676. Ziebland, H. and Burton, J. T. A., "The Thermal Conductivity of Liquid and Gaseous Oxygen," *Brit. J. Appl. Phys.* **6**, 416–20, 1955.
677. Ziebland, H. and Burton, J. T. A., "The Thermal Conductivity of Nitrogen and Argon in the Liquid and Gaseous States," *Brit. J. Appl. Phys.* **9**, 52–9, 1958.
678. Ziebland, H. and Burton, J. T. A., "The Thermal Conductivity of Heavy Water Between 75 and 260 C at Pressures up to 300 Atm.," *Int. J. Heat Mass Transfer*, **1**(2), 242–54, 1960.
679. Ziebland, H. and Patient, J. E., "The Thermal Conductivity of n-Octadecane; New Measurements and a Critical Appraisal of the Paper by Sutherland, Davis and Seyer," *J. Chem. Eng. Data* **7**(1), 530–1, 1961.
680. Zwanzig, R., "Time Correlation Functions and Transport Coefficients in Statistical Mechanics," *Ann. Rev. Phys. Chem.* **16**, 67–102, 1965.
681. Zwanzig, R. W., Kirkwood, J. G., Oppenheim, I., and Alder, B. J., "Statistical Mechanical Theory of Transport Processes. VII. The Coefficient of Thermal Conductivity of Monatomic Liquids," *J. Chem. Phys.* **22**, 783–90, 1954.

Numerical Data

Data Presentation and Related General Information

This volume contains values for the thermal conductivity of nonmetallic elements and compounds which exist in the liquid or gaseous state at normal temperature (25 C or 77 F) and pressure (1 atm or 14.696 lb in.$^{-2}$). The material for a given system is arranged in the sequence solid, saturated liquid, saturated vapor or gaseous physical state. Within each state the sequence consists of text, table, and departure plot(s) (if any) for pure substances and air (here considered as a simple substance and not a mixture).

The available experimental thermal conductivity data on 82 binary, 23 ternary, and seven quaternary systems are tabulated in this volume along with some other details. Graphical representation of the data on binary systems is included together with tabulations of the graphically smoothed values. All these systems are grouped together in different categories depending upon the complexity of the molecule, i.e., whether it is polar or nonpolar and, further, whether monatomic or polyatomic.

1. SCOPE OF COVERAGE

The substances contained in this volume were selected based on consideration of scientific and technological interest and needs. The pure substances were originally selected to match a parallel program for specific heat and viscosity corresponding tables for which will appear in Volumes 6 and 11 respectively.

A good estimate of the available knowledge on the thermal conductivity of mixtures of gases can be found from the retrieval guide to the research literature published by the Thermophysical Properties Research Center [719]. A search revealed that experimental data exist on about two hundred binary and higher order multicomponent systems in the gaseous phase. In this volume, we consider only the mixtures of those substances which exist in the gaseous state at normal temperature and pressure. Further, this survey considers only a limited number of gas systems, viz., 82 binary, 23 ternary, seven quaternary, and one multicomponent. Thus, a little over half the systems are covered here.

Thermal conductivity is strongly and intricately dependent on the shape and structure of the molecules. Consequently, different varieties and complexities of molecules and their different permutations in the mixtures have been selected. It is hoped that such an investigation of classified categories of gas molecules and their combinations will help in elucidating the various ways in which thermal conductivity of gaseous mixtures can vary with changes in such variables as pressure, temperature, and mixture composition. The groupings of the present work are (a) monatomic, (b) nonpolar polyatomic, (c) combination of monatomic and nonpolar polyatomic, (d) polar–nonpolar, and (e) polar–polar systems. Several representative gas mixtures in each category are considered.

Very few data exist on the pressure dependence of thermal conductivity of gas mixtures, and all the measurements considered here refer to a pressure around or below one atmosphere. Thermal conductivity can usually be regarded as independent of pressure for pressures close to atmospheric and the data interpreted in this work may be regarded as referred to one atmosphere. Further, the data analysis is here restricted to the actual temperature of measurement but extends over the entire possible range of composition for binary systems. To achieve the latter, a graphical approach was used. For systems composed of more than two gases only the actual compositions of the mixtures examined were considered.

The second phase of the work on gaseous mixtures, which is not reported here but is in progress, will involve smoothing these data with reference to temperature and thereby determining the most probable thermal conductivity values as functions of temperature and composition. This approach will indeed apply to only a limited number of binary systems where such data exist over an appreciable temperature range. However, some of the developed semitheoretical techniques may be carefully used as the basis for extrapolation of the temperature range. An assessment of such a procedure may be based on the values generated according to the method outlined above.

2. PRESENTATION OF DATA

For the pure substances (and in this work air has been thus regarded) a brief discussion of the available information in the literature and of the procedure adopted in analyzing the values is given for each substance. Recommended values are then presented in tabular form, accompanied by indications of phase transition temperatures, where these fall within the range of the tabulation. Comments are appended concerning the probable accuracy of the recommended values and regions where further experimental or theoretical study are needed are often described. A departure plot, or plots, showing the concordance between the various experimental and/or theoretical values and the recommended values, is given if sufficient original experimental data are available.

In preparing the departure plots,

Percent departure =

$$\frac{\text{Experimental value} - \text{Tabulated value}}{\text{Tabulated value}} 10^2$$

By the above definition, departures are positive if the experimental data are greater than the tabular values and vice versa. Extrapolation of the values beyond the limits of the table is not recommended. If, however, this must be done, the departure plots should be examined to obtain an indication of the probable trend in the values in regions not yet experimentally studied.

A consistent style is adopted in presenting the thermal conductivity data for (a) binary and (b) multicomponent systems. In the former category, the data are classified in the following subheads:

    A. Monatomic–monatomic systems,
    B. Monatomic–nonpolar polyatomic systems,
    C. Nonpolar polyatomic–nonpolar polyatomic systems,
    D. Polar–nonpolar polyatomic systems, and
    E. Polar–polar systems.

The arrangement of different systems in each group may be explained as follows: In Section 4A the two monatomic components of the mixture are arranged in alphabetical order. In Section 4B the monatomic constituent of the mixture is given first (arranged alphabetically) and then the second constituent again arranged in alphabetical order for a common first component. In Section 4C the two polyatomic constituents are arranged in a simple alphabetical order, as are also all the members of this group. In Section 4D the polar component is given first, and these are arranged in alphabetical order. If more than one polyatomic gas permutes with the same polar gas the former are again arranged alphabetically. In Section 4E both the components are polar and their order is again determined alphabetically. In Section 5A all the gases are monatomic and these are arranged in the increasing order of their molecular weights. In Section 5B the three gases of a particular mixture are arranged in the increasing order of their atomicity (rare gases first, then diatomic, and so on). Gases having the same atomicity are again arranged in the order of increasing molecular weight. For each gas system the original data of different workers are reproduced in a tabular form. The tables cite the temperature of measurement, the composition of mixture, and the thermal conductivity, converted from the original units of the author into mW cm^{-1} K^{-1}, where the necessary conversion factors used are given in Section 6. In these tables thermal conductivity values are listed to four significant figures. This is usually not warranted by the probable uncertainties of the measurement and by the absolute accuracy of the data, except where the thermal conductivity is quite large (greater than 1 mW cm^{-1} K^{-1}) and the precision is better than a percent. To facilitate the numerical handling of the original data and to reduce the rounding errors we have consistently reported the k values in four significant figures. Interpretation of the tabular data will first require reducing them to the correct number of significant figures consistent with the experimental errors. In the remarks column, wherever possible, the purity of the gases, the experimental technique, and the precision and accuracy of the measurement are given.

Raw data of different workers for a gas system

are shown plotted for the actual temperature(s) of measurement as a function of the relative proportion of the heavier component in the mixture. These graphs serve to indicate the relative scatter of the experimental points of a single investigation. In some cases where the data of different workers are available either exactly at one or at very closely spaced temperatures, these could be used to analyze the variation of thermal conductivity with composition for selected identical temperatures and of thermal conductivity with temperature for selected compositions. A study of the regularity of the values thus obtained would afford an estimate of the accuracy of the original data and technique. Some of these features have assumed increasing importance and interest in the background of different experimental and theoretical studies made during the last decade.

The thermal conductivity data for each binary mixture were carefully plotted on suitably sized graphs. The entire composition range, 0.00 to 1.00, when 1 refers to the heavier compound mole fraction, was divided into twenty increments of 0.05. These graphically smoothed values at the temperature of actual measurement were read off of the graphs and are reported in tabular form.

For ternary and quaternary systems this graphical smoothing to generate values at rounded compositions is not as straightforward and in this work the original data are therefore reported as such.

3. SYMBOLS AND ABBREVIATIONS USED IN THE FIGURES AND TABLES

Most abbreviations and symbols used are those generally accepted in engineering and scientific practice. The abbreviation NTP, signifies normal temperature and pressure. The notation cgsu, signifying "centimeter gram second unit," is an abbreviation used to indicate that the thermal conductivity has been expressed in calories per cm per sec per degree rather than the Watt unit.

The notations "n.m.p.," "n.b.p.," and "c.p.," are used to indicate phase transition temperatures (see Section 5 below). Curve numbers on the departure plots are surrounded by circles or squares, the latter being used to indicate a single data point or a cluster of points. Solid lines on the plots are used to connect experimental data points and dotted lines indicate major correlated or calculated values. To emphasize the distinction the word "data" is reserved for experimentally determined quantities; all quantities determined by calculation or estimation are referred to as "values." Numbers in parentheses in the text and signified by "Ref." on the departure plot correspond to the references listed at the end of the chapter. The term "mole fraction" is used to denote the ratio of the number of molecules of one kind present in a given mixture to the total number of molecules. Thus, an argon–helium mixture in which the stated mole fraction of argon is 0.20 implies that in the mixture argon is 20 percent by number and hence that $\frac{1}{5}$ the total volume is argon. The mole fraction of a given component will often vary between the extreme limits 0 and 1 referring to complete absence and presence respectively.

4. CONVENTION FOR BIBLIOGRAPHIC CITATION

The numbers in square brackets refer to the bibliographic citations. These follow the same policies adopted in other TPRC publications (see, e.g., [719]) in that the sequence adopted in each citation is author(s), journal title (abbreviated usually as the *Chemical Abstracts* list), volume number (underlined), page(s), and year; or author(s), book title, publisher, relevant page(s), and year. The sequence of the citations follows the sequence used historically at TPRC in that the first 414 citations are arranged alphabetically and then successive entries have been added as additional material was located and used.

5. NAME, FORMULA, MOLECULAR WEIGHT, TRANSITION TEMPERATURES, AND PHYSICAL CONSTANTS OF ELEMENTS AND COMPOUNDS

The gases of which the binary, ternary, and quaternary mixtures we have considered may be divided in such broad categories as (a) spherically symmetric, (b) linear diatomic, (c) nonpolar polyatomic, (d) polar polyatomic, (e) nonpolar organic, and (f) polar organic. Because of the pronounced dependence of thermal conductivity on the molecular structure of gas, one can look for possible correlations only within the mixtures which are constituted out of gases which fall in identical categories. To facilitate such an approach we list in the table on page 106a some such physical constants of the gases considered by us, and which we think are

NAME, FORMULA, MOLECULAR WEIGHT, TRANSITION TEMPERATURES, AND PHYSICAL CONSTANTS OF ELEMENTS AND COMPOUNDS

Name	Formula	Molecular Weight	Density (25 C), g cm^{-3}	Melting (or Triple) Point, K	Normal Boiling Point, K	Critical Temp., K	C_p (25°C), cal g^{-1}K^{-1}	C_v (25°C), cal g^{-1}K^{-1}	Dipole Moment, Debyes
Acetone	C_3H_6O	58.081	0.933 (ℓ)†	178	329	508	0.528 (ℓ)		2.88
Acetylene	C_2H_2	26.039	1.077 -3**	179	191	309	0.407	0.329	0
Air		28.966	1.184 -3	60	79b, 82d	133	0.240	0.172	
Ammonia	NH_3	17.031	0.601 -3	195	240	405	0.515	0.387	1.47
Argon	Ar	39.948	1.634 -3	84	88	151	0.125	0.075	0
Benzene	C_6H_6	78.117	0.876 (ℓ)	279	353	563	0.415 (ℓ)	--	0
Boron Trifluoride	BF_3	67.807		146	172	261	--	--	0
Bromine	Br_2	159.818		266	332	584	0.113	--	0
i-Butane	i-C_4H_{10}	58.126		114	262	408	0.404		0.132
n-Butane	n-C_4H_{10}	58.126	2.491 -3	137	273	426	0.409	0.358	≤0.05
Carbon Dioxide	CO_2	44.010	1.811 -3	216(5 atm)	195	304	0.203	0.158	0
Carbon Monoxide	CO	28.011	1.145 -3	68	81	134	0.249	0.177	0.112
Carbon Tetrachloride	CCl_4	153.824	1.589 (ℓ)	250	350	556	0.204 (ℓ)	--	0
Chlorine	Cl_2	70.906	2.944 -3	172	239	417	0.114*	0.084	0
Chloroform	$CHCl_3$	119.378	1.469 (ℓ)	210	334	535	0.228 (ℓ)	--	1.01
n-Decane	$C_{10}H_{22}$	142.290	0.728 (ℓ)	243	446	619	0.527 (ℓ)	--	
Deuterium	D_2	4.028	0.165 -3	19(.16 at)	23	38	1.731*	1.241	0
Diethylamine	$C_4H_{11}N$	73.143	0.711 (ℓ)	233	329	496	0.516 (ℓ)	--	1.11
Ethane	C_2H_6	30.070	1.243 -3	95±6	185	305	0.422	0.335	0
Ethyl Alcohol	C_2H_6O	46.070	0.789 (ℓ)	159±3	351	516	0.580 (ℓ)	--	1.69
Ethyl Ether	$C_4H_{10}O$	74.125	0.716 (ℓ)	157(α), 150(β)	308	467	0.559 (ℓ)	--	1.15
Ethylene	C_2H_4	28.055	1.155 -3	104	170	283	0.374	0.297	0
Ethylene Glycol	$C_2H_6O_2$	62.070	1.100 (ℓ)	258	471		0.575 (ℓ)		2.28
Fluorine	F_2	37.997	1.553 -3	54	86	144	0.197*	0.152	0
Freon 11	CCl_3F	137.369	5.840 -3	162	297	471	0.136*	0.125	0.45
Freon 12	CCl_2F_2	120.914	5.045 -3	116	243	385	0.146	0.128	0.51
Freon 13	$CClF_3$	104.460	4.388 -3	91	191	302	0.153*	0.138	0.50
Freon 21	$CHCl_2F$	102.924	4.284 -3	133	282	451	0.141*	0.119	1.29
Freon 22	$CHClF_2$	86.469	3.588 -3	113	233	369	0.151	0.133	1.42
Freon 113	$C_2Cl_3F_3$	187.377	1.564 (ℓ)	238	321	481	0.225 (ℓ)		
Freon 114	$C_2Cl_2F_4$	170.922	7.012 -3	179	276	419	0.170	0.157	0.5
Glycerol	$C_3H_8O_3$	92.096	1.263 (ℓ)	291	563		0.567 (ℓ)		
Helium	He	4.003	0.164 -3	3.5	4	5.4	1.240*	0.748	0
n-Heptane	C_7H_{16}	100.208	0.681 (ℓ)	183	371	540	0.536 (ℓ)	--	
n-Hexane	C_6H_{14}	86.181	0.657 (ℓ)	179	342	508	0.543 (ℓ)	--	
Hydrogen	H_2	2.016	0.082 -3	14	20	33	3.420	2.438	0
Hydrogen Chloride	HCl	36.461	1.502 -3	166	188	325	0.191*	0.140	1.08
Hydrogen Iodide	HI	127.913		223	237	423	0.054*		
Hydrogen Sulfide	H_2S	34.080	1.409 -3	190	214	374	0.240*	0.157	0.97
Iodine	I_2	253.809	4.93 (s)	387	458	785	0.052 (s)	--	0
Krypton	Kr	83.80	3.429 -3	117	120	210	0.059*	0.035	0
Methane	CH_4	16.043	0.657 -3	90	112	190	0.533	0.409	0
Methyl Alcohol	CH_4O	32.043	0.789 (ℓ)	175	338	513	0.602 (ℓ)		1.70
Methyl Chloride	CH_3Cl	50.488		175	249	416	0.193		
Methyl Formate	$C_2H_4O_2$	60.054	0.974 (ℓ)	174	305	487	0.516	--	
Neon	Ne	20.183	0.824 -3	25	27	44	0.246*	0.150	0
Nitric Oxide	NO	30.006	1.228 -3	111	121	180	0.238	0.167	0.153
Nitrogen	N_2	28.018	1.146 -3	63	78	126	0.249	0.178	0
Nitrogen Peroxide	NO_2	46.006	1.44 (ℓ)	263	295	431	0.369 (ℓ)		0.316
Nitrous Oxide	N_2O	44.013		176±7	184	310	0.209*	0.170	0.167
n-Nonane	C_9H_{20}	128.262	0.714 (ℓ)	220	424	594	0.529 (ℓ)	--	
n-Octane	C_8H_{18}	114.234	0.701 (ℓ)	216	399	569	0.530 (ℓ)	--	
Oxygen	O_2	31.999	1.310 -3	55	90	155	0.220	0.157	0
n-Pentane	C_5H_{12}	72.154	0.621 (ℓ)	144	309	470	0.561 (ℓ)		
Cyclopropane	C_3H_6	42.098	0.720 (ℓ)	146	240			--	
Propane	C_3H_8	44.098	1.854 -3	86	231	369	0.400	0.350	0.084
Propylene	C_3H_6	42.082	0.514 (ℓ)	87	226	365	0.370	0.320	
Radon	Rn	226		202	211	377			0
Sulfur Dioxide	SO_2	64.063	2.679 -3	198	263	430	0.149*	0.081	1.63
Toluene	C_7H_8	92.144	1.028 (ℓ)	178	384	594	0.410 (ℓ)	--	0.36
Tritium	T_2			21	26	44			0
Water	H_2O	18.015	0.997 (ℓ)	273	373	647	0.998 (ℓ)	--	1.85
Xenon	Xe	131.30	5.397 -3	161	165	289	0.0378*	0.0227*	0

* For ideal gas state.

** The notation -3 signifies 10^{-3}, so that 1.077 -3 means 1.077 x 10^{-3}, etc.

† (ℓ) and (s) designate liquid and solid state, respectively.

Numerical Data

CONVERSION FACTORS FOR UNITS OF THERMAL CONDUCTIVITY

MULTIPLY by appropriate factor to OBTAIN →	$Btu_{IT} hr^{-1} ft^{-1} F^{-1}$	$Btu_{IT} in. hr^{-1} ft^{-2} F^{-1}$	$Btu_{th} hr^{-1} ft^{-1} F^{-1}$	$Btu_{th} in. hr^{-1} ft^{-2} F^{-1}$	$cal_{IT} sec^{-1} cm^{-1} C^{-1}$	$cal_{th} sec^{-1} cm^{-1} C^{-1}$	$kcal_{th} hr^{-1} m^{-1} C^{-1}$	$J sec^{-1} cm^{-1} K^{-1}$	$W cm^{-1} K^{-1}$	$W m^{-1} K^{-1}$	$mW cm^{-1} K^{-1}$
$Btu_{IT} hr^{-1} ft^{-1} F^{-1}$	1	12	1.00067	12.0080	4.13379×10^{-3}	4.13656×10^{-3}	1.48916	1.73073×10^{-2}	1.73073×10^{-2}	1.73073	17.3073
$Btu_{IT} in. hr^{-1} ft^{-2} F^{-1}$	8.33333×10^{-2}	1	8.33891×10^{-2}	1.00067	3.44482×10^{-4}	3.44713×10^{-4}	0.124097	1.44228×10^{-3}	1.44228×10^{-3}	0.144228	1.44228
$Btu_{th} hr^{-1} ft^{-1} F^{-1}$	0.999331	11.9920	1	12	4.13102×10^{-3}	4.13379×10^{-3}	1.48816	1.72958×10^{-2}	1.72958×10^{-2}	1.72958	17.2958
$Btu_{th} in. hr^{-1} ft^{-2} F^{-1}$	8.32776×10^{-2}	0.999331	8.33333×10^{-2}	1	3.44252×10^{-4}	3.44482×10^{-4}	0.124014	1.44131×10^{-3}	1.44131×10^{-3}	0.144131	1.44131
$cal_{IT} sec^{-1} cm^{-1} C^{-1}$	2.41909×10^{2}	2.90291×10^{3}	2.42071×10^{2}	2.90485×10^{3}	1	1.00067	3.60241×10^{2}	4.1868	4.1868	4.1868×10^{2}	4.1868×10^{3}
$cal_{th} sec^{-1} cm^{-1} C^{-1}$	2.41747×10^{2}	2.90096×10^{3}	2.41909×10^{2}	2.90291×10^{3}	0.999331	1	3.6×10^{2}	4.184	4.184	4.184×10^{2}	4.184×10^{3}
$kcal_{th} hr^{-1} m^{-1} C^{-1}$	0.671520	8.05824	0.671969	8.06363	2.77592×10^{-3}	2.77778×10^{-3}	1	1.16222×10^{-2}	1.16222×10^{-2}	1.16222	11.6222
$J sec^{-1} cm^{-1} K^{-1}$	57.7789	6.93347×10^{2}	57.8176	6.93811×10^{2}	0.238846	0.239006	86.0421	1	1	1×10^{2}	1×10^{3}
$W cm^{-1} K^{-1}$	57.7789	6.93347×10^{2}	57.8176	6.93811×10^{2}	0.238846	0.239006	86.0421	1	1	1×10^{2}	1×10^{3}
$W m^{-1} K^{-1}$	0.577789	6.93347	0.578176	6.93811	2.38846×10^{-3}	2.39006×10^{-3}	0.860421	1×10^{-2}	1×10^{-2}	1	10
$mW cm^{-1} K^{-1}$	5.77789×10^{-2}	0.693347	5.78176×10^{-2}	0.693811	2.38846×10^{-4}	2.39006×10^{-4}	8.60421×10^{-2}	1×10^{-3}	1×10^{-3}	0.1	1

very useful. The electric dipole moment values are quoted from the recent compilation of Nelson, Lide, and Maryott [720].

6. CONVERSION FACTORS

The conversion factors given in the table on page 107a are based upon the following basic definitions:

1 in. = 0.0254 (exactly) m*
1 lb = 0.45359237 kg*
1 cal$_{th}$ = 4.184 (exactly) J*
1 cal$_{IT}$ = 4.1868 (exactly) J*
1 Btu$_{th}$lb^{-1} F^{-1} = 1 cal$_{th}$g^{-1} C^{-1}†
1 Btu$_{IT}$lb^{-1} F^{-1} = 1 cal$_{IT}$g^{-1} C^{-1}†

*National Bureau of Standards, "New Values for the Physical Constants Recommended by NAS-NRC," *Natl. Bur. Std. (U.S.), Tech. News Bull.* **47**(10), 175–7, 1963.
†Mueller, E. F. and Rossini, F. D., "The Calory and the Joule in Thermodynamics and Thermochemistry," *Am. J. Phys.* **12**(1), 1–7, 1944.

TABLE 1 THERMAL CONDUCTIVITY OF ARGON

RECOMMENDED VALUES

[Temperature, T, K; Thermal Conductivity, k, mW cm^{-1}K^{-1}]

SOLID

T	k
8	60
9	46
10	37
12	27
14	22
16	18
18	16
20	13.6
25	9.9
30	7.8
35	6.5
40	5.6
45	5.1
50	4.6
60	3.8
70	3.3
80	3.0

DISCUSSION

SOLID

Available data on the thermal conductivity of solid argon includes the work of Dobbs and Jones (675), White and Woods (674, 675), Berne, Boato et al. (680) while some calculations and correlations have appeared (673, 678). Most of the above results have only been presented in graphical form.

Comparison of the available information reveals reasonable agreement above about 10 K and severe disagreement at lower temperatures. In the region of from 5 to 8 K an order of magnitude difference exists between the (674, 675) and (680) data. Such differences are probably produced by structure variations caused by different impurity content, although further high precision experimentation is needed to confirm this supposition.

The recommended values were obtained from a large scale plot of the available information and were not generated for temperatures below 8 K due to the experimental uncertainty. From 8 to 10 K the uncertainty may be as much as fifty percent while the higher temperature values should be accurate to ten percent. Due to the almost complete absence of tabulated data no departure plot is given.

DISCUSSION

SATURATED LIQUID

Three experimental works are available on the thermal conductivity of liquid argon. Keyes (192) made measurements in a coaxial-cylinder apparatus near saturation conditions at three temperatures from 87 K to 112 K. The extensive measurements of Uhlir (353) were made in a coaxial-cylinder apparatus using the gas thermometer, covering temperatures from 86 to 150 K and pressures up to 96 atm. The uncertainty in the measurements was reported to be from 0.5 to 2.5 percent. Other measurements for the liquid and gaseous phases were carried out in a coaxial-cylinder apparatus with an accuracy of two percent, by Ziebland-Burton (57, 413), over the temperature range from 93 up to 151 K for the liquid phase under various pressures up to 120 atm. From the standpoint of the experimental method and procedure, all the above measurements are considered to be reliable.

In the initial analysis, the values under saturated vapor pressures were obtained from the graphical extrapolation of the data of both Uhlir and Ziebland-Burton. No correction was made to the values of Keyes. The three sets of data for the saturated liquid were given equal weight and were fitted to the quadratic equation represented by

$$10^6 k \text{ (cgsu)} = 516.609 - 2.32178\,T - 0.00255768\,T^2 \quad (T \text{ in K}).$$

In arriving at this formula, the values at the critical point were excluded.

The above equation should be valid in the temperature range from 80 to 140 K. This equation was found to fit the above-enumerated values with a mean deviation of 0.61 percent and a maximum of 1.9 percent. The recommended values below 140 K were generated from the above formula, and should be correct within two percent.

Subsequent to the initial analysis, the data of Ikenberry and Rice (672) from 91 to 150 K were examined. Values for saturated vapor pressure were similarly obtained by graphical extrapolation. No significant discord with the above formulation was noted.

Values above 140 K were obtained from a large-scale graph. The experimental difficulties increase considerably in this temperature region and the recommended values at 145, 150 K and the critical point are probably uncertain by as much as five, ten and twenty-five percent, respectively.

TABLE 1 THERMAL CONDUCTIVITY OF ARGON

[Temperature, T, K; Thermal Conductivity, k, mW cm^{-1}K^{-1}]

RECOMMENDED VALUES

SATURATED LIQUID

T	k
80	(1.316)†
85	1.258
90	1.201‡
95	1.142‡
100	1.082‡
105	1.021‡
110	0.963‡
115	0.903‡
120	0.842‡
125	0.780‡
130	0.718‡
135	0.655‡
140	0.592‡
145	0.518‡
150	0.404‡
151.2*	0.25

†Extrapolated for the supercooled liquid. (n.m.p. = 83 K)
‡Under saturated vapor pressures. (n.b.p. = 88 K)
*Critical point.

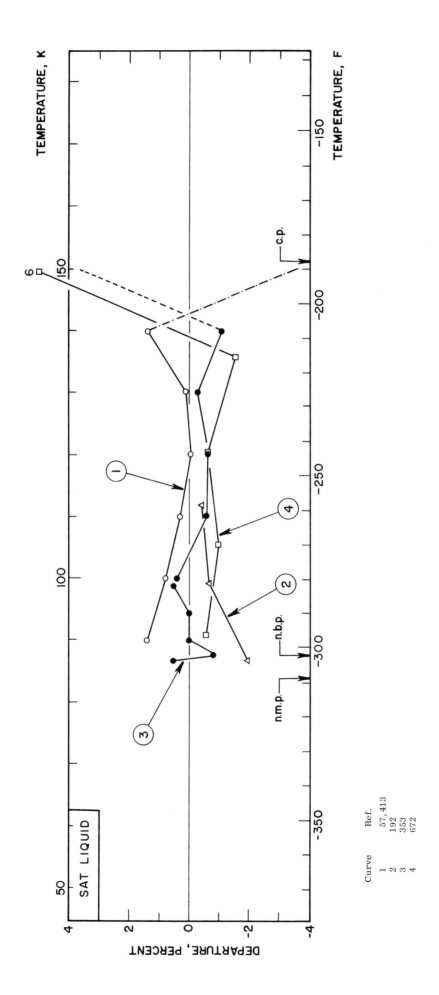

FIGURE 1   DEPARTURE PLOT FOR THERMAL CONDUCTIVITY OF LIQUID ARGON

TABLE 1   THERMAL CONDUCTIVITY OF ARGON

RECOMMENDED VALUES

[Temperature, T, K; Thermal Conductivity, k, mW cm^{-1}K^{-1}]

SATURATED VAPOR

T	k
85	0.055
90	0.059
95	0.064
100	0.068
105	0.072
110	0.077
115	0.082
120	0.088
125	0.095
130	0.103
135	0.109
140	0.120
145	0.140
150	0.19
151.2*	0.25

* Critical temperature

DISCUSSION

SATURATED VAPOR

No experimental data were found for the thermal conductivity of saturated argon vapor. The only information located was estimations of Owens and Thodos (268), Uhlir (353) and Ziebland et al. (413, 542). Below about 140 K the estimates are in fair agreement, the Uhlir values being intermediate. Above 140 K a wide variation in estimates exists.

The values were plotted on a large scale graph in which the Owens and Thodos values were adjusted to agree with the atmospheric pressure value at 88 K. The increase necessary at 88 K was linearly reduced for higher temperatures to zero at the critical temperature. Values obtained in this way were in excellent agreement with the Uhlir values up to 125 K. Above 125 K they were lower than the other estimates.

The recommended values were deduced from the plot of the Owens and Thodos estimates. Based upon the agreement of these with other estimates and upon the uncertainty in the saturated liquid values, they should be accurate to about 2.5 percent below 125 K, fifteen percent at 135 K and twenty-five percent at and above 145 K. Due to the lack of experimental data no departure plot is given. Experimental measurements to confirm these estimates are urgently required.

GAS

DISCUSSION

Experimental measurements have been reported for the thermal conductivity of gaseous argon for temperatures between about 90 and 1173 K and many correlations and calculations have appeared, the more recent extending to temperatures well above 15000 K. At atmosphere pressure the ionization reaches some one, five and ten percent at temperatures of 9400, 10900 and 11750 K and the tabulation of recommended values only extends to 10000 K so that the recommended values may, without serious error, be said to refer to neutral argon.

As shown by the departure plots, most experimental, correlated and calculated values are in reasonable accord and the accuracy of the recommended values, derived by drawing a smooth curve through these sources, can be assessed as about one percent for temperatures between 100 and 500 K, five percent for temperatures below 100 K and between 500 and 1500 K, and ten percent between 1500 and 10000 K.

TABLE 1   THERMAL CONDUCTIVITY OF ARGON

RECOMMENDED VALUES

[Temperature, T, K; Thermal Conductivity, k, mW cm^{-1}K^{-1}]

GAS

T	k	T	k	T	k	T	k
50	(0.0326)*	400	0.2233	750	0.353	1500	0.561
60	(0.0392)*	410	0.2276	760	0.356	1550	0.575
70	(0.0456)*	420	0.2318	770	0.359	1600	0.588
80	(0.0522)*	430	0.2359	780	0.362	1650	0.602
90	0.0587	440	0.2400	790	0.366	1700	0.615
100	0.0652	450	0.2441	800	0.369	1750	0.628
110	0.0716	460	0.2481	810	0.372	1800	0.641
120	0.0779	470	0.2520	820	0.375	1850	0.654
130	0.0839	480	0.2559	830	0.378	1900	0.667
140	0.0898	490	0.2599	840	0.381	1950	0.680
150	0.0957	500	0.2638	850	0.384	2000	0.692
160	0.1016	510	0.268	860	0.387	2100	0.717
170	0.1074	520	0.272	870	0.390	2200	0.741
180	0.1131	530	0.276	880	0.393	2300	0.766
190	0.1188	540	0.280	890	0.396	2400	0.790
200	0.1244	550	0.283	900	0.398	2500	0.815
210	0.1300	560	0.287	910	0.401	2600	0.839
220	0.1355	570	0.290	920	0.404	2700	0.864
230	0.1409	580	0.294	930	0.407	2800	0.888
240	0.1462	590	0.297	940	0.410	2900	0.913
250	0.1515	600	0.301	950	0.413	3000	0.938
260	0.1567	610	0.305	960	0.416	3100	0.962
270	0.1619	620	0.308	970	0.418	3200	0.987
280	0.1671	630	0.311	980	0.421	3300	1.011
290	0.1722	640	0.315	990	0.424	3400	1.036
300	0.1772	650	0.319	1000	0.427	3500	1.060
310	0.1822	660	0.322	1050	0.441	3600	1.084
320	0.1871	670	0.326	1100	0.454	3700	1.109
330	0.1919	680	0.329	1150	0.468	3800	1.133
340	0.1966	690	0.333	1200	0.481	3900	1.158
350	0.2013	700	0.336	1250	0.495	4000	1.182
360	0.2059	710	0.339	1300	0.508	4100	1.207
370	0.2103	720	0.343	1350	0.521	4200	1.231
380	0.2147	730	0.346	1400	0.535	4300	1.256
390	0.2190	740	0.349	1450	0.548	4400	1.281

*Extrapolated for the gas phase ignoring pressure dependence (n.b.p. = 88 K).

TABLE 1   THERMAL CONDUCTIVITY OF ARGON (continued)

RECOMMENDED VALUES

[Temperature, T, K; Thermal Conductivity, k, mW cm^{-1} K^{-1}]

GAS

T	k	T	k
4500	1.305	8000	2.48
4600	1.330	8200	2.63
4700	1.354	8400	2.80
4800	1.379	8600	2.99
4900	1.404	8800	3.18
5000	1.429	9000	3.39
5200	1.478	9200	3.61
5400	1.526	9400	3.83
5600	1.575	9600	4.07
5800	1.624	9800	4.31
6000	1.673	10000	4.56
6200	1.722		
6400	1.771		
6600	1.821		
6800	1.881		
7000	1.950		
7200	2.03		
7400	2.12		
7600	2.23		
7800	2.35		

FIGURE 1   DEPARTURE PLOT FOR THERMAL CONDUCTIVITY OF GASEOUS ARGON

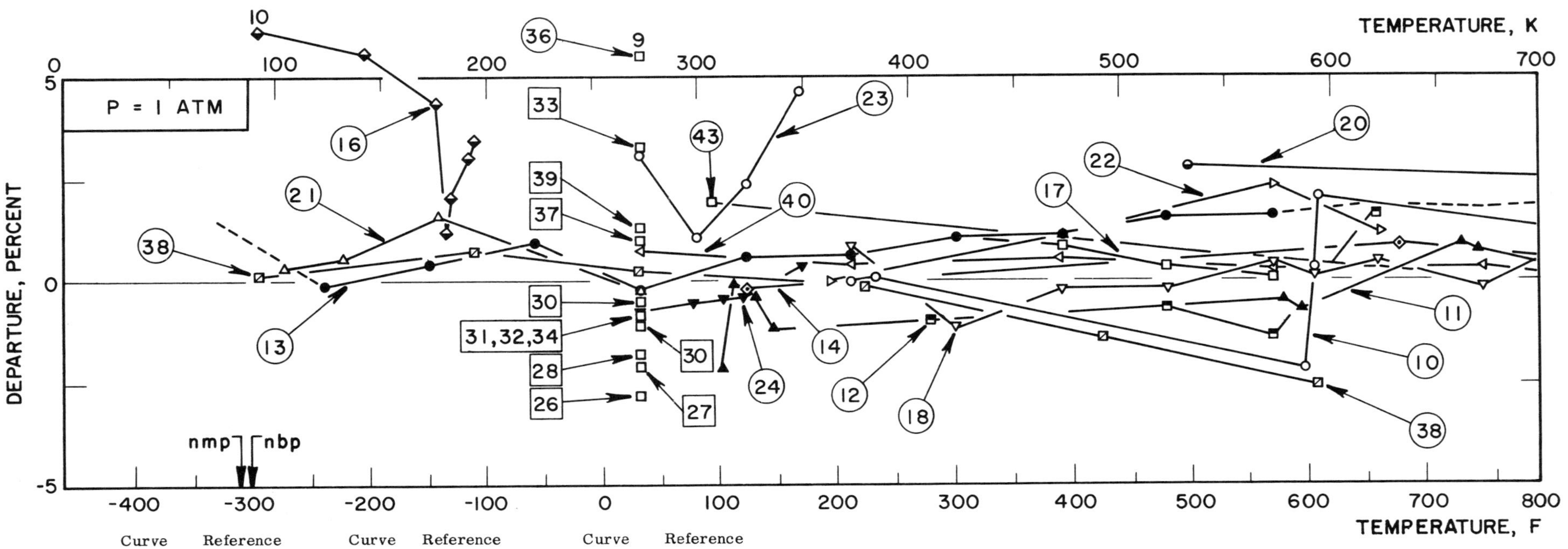

Curve	Reference	Curve	Reference	Curve	Reference
10	64, 65	21	192	32	96
11	617	22	191	33	81, 82
12	618	23	249	34	401
13	187	24	619	36	325
14	280, 281	26	177	37	174
16	412	27	379, 385	38	173
17	112	28	218	39	86
18	305	30	353	40	292, 293
20	369	31	156	43	628

P = 1 ATM

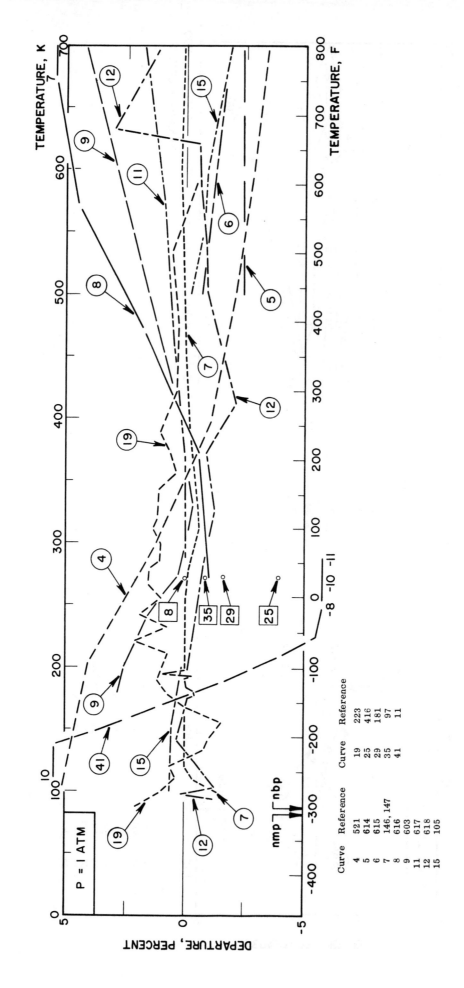

FIGURE 1   DEPARTURE PLOT FOR THERMAL CONDUCTIVITY OF GASEOUS ARGON (continued)

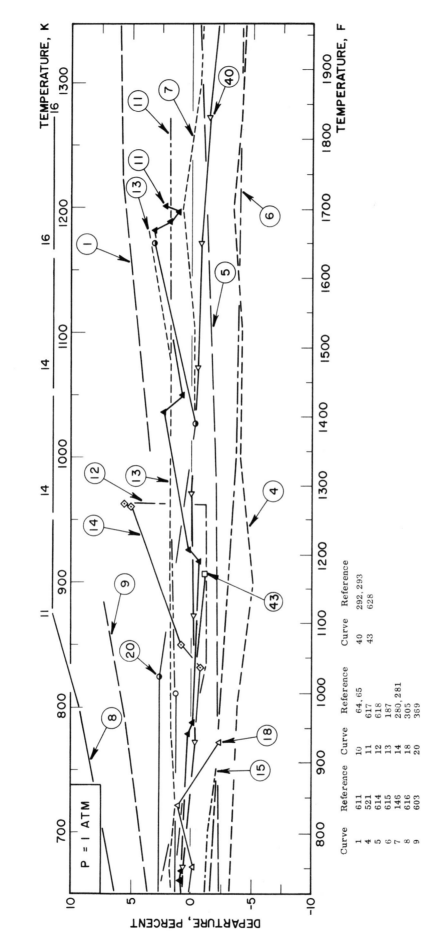

FIGURE 1 DEPARTURE PLOT FOR THERMAL CONDUCTIVITY OF GASEOUS ARGON (continued)

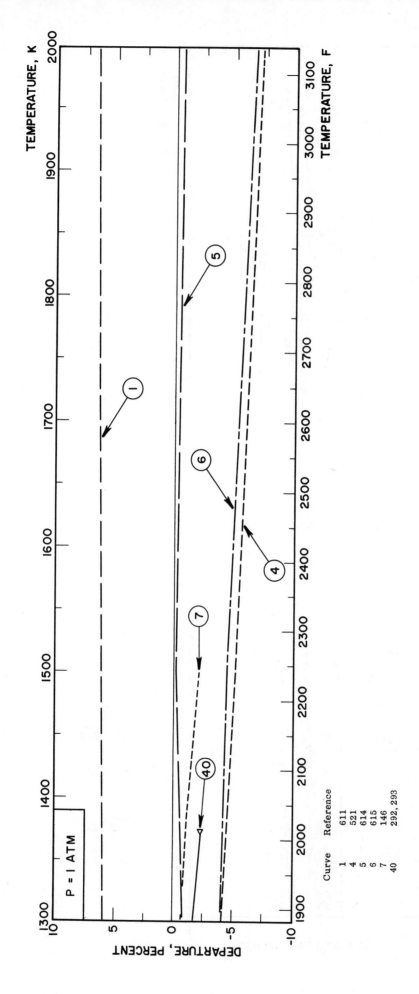

FIGURE 1 DEPARTURE PLOT FOR THERMAL CONDUCTIVITY OF GASEOUS ARGON (continued)

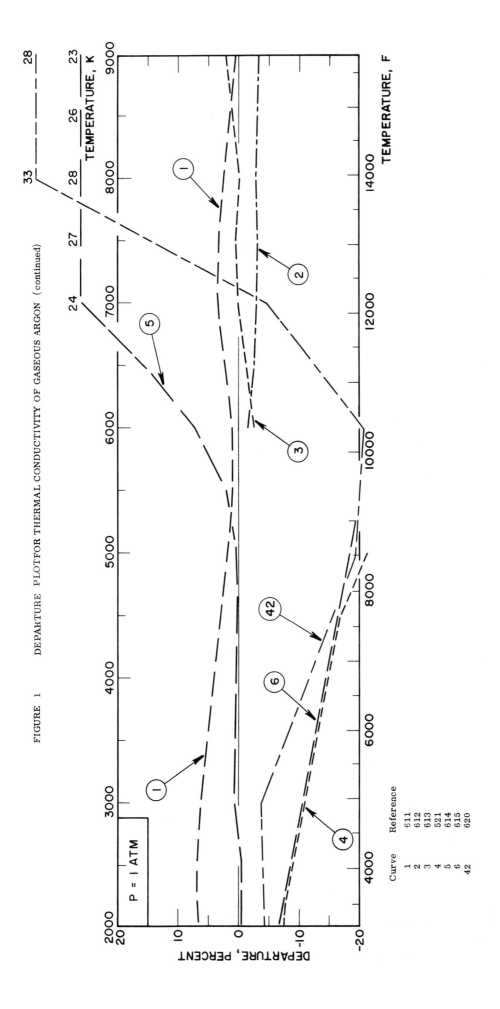

FIGURE 1 DEPARTURE PLOT FOR THERMAL CONDUCTIVITY OF GASEOUS ARGON (continued)

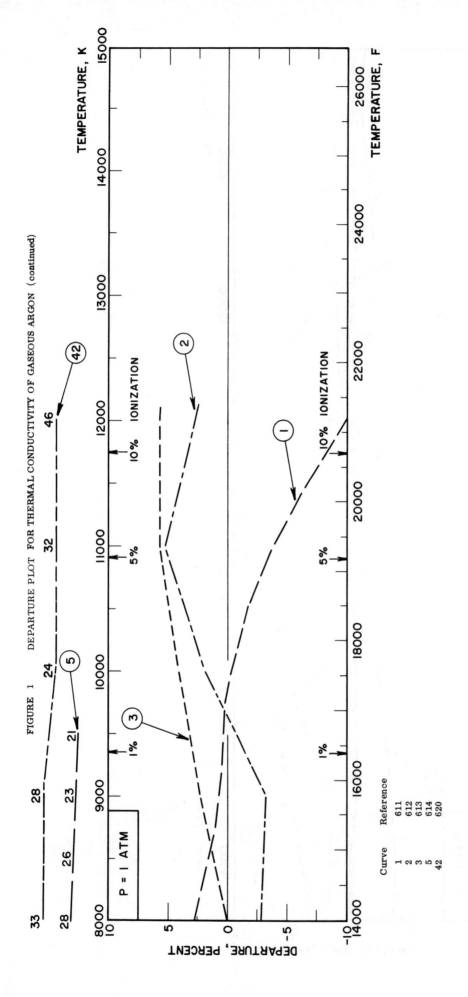

FIGURE 1   DEPARTURE PLOT FOR THERMAL CONDUCTIVITY OF GASEOUS ARGON (continued)

DISCUSSION

SATURATED LIQUID

The only experimental data located for the thermal conductivity of saturated liquid Bromine were some recent experimental measurements from 283 to 323 K (676). In addition a correlation for saturated liquid and vapor diatomic substances (575) was considered.

Intercomparison of the measurements and the correlation showed good agreement below 305 K but progressively increasing disagreement at higher temperatures. The trend of the measured data is to predict a critical temperature far below other estimates. Possibilities to explain the discrepancy are association in the liquid phase, partial vaporization or an experimental effect similar to that observed with Keyes (192) data for liquid krypton. It was decided to base the recommended values on a smooth curve which followed the correlation for temperatures near and at the critical point.

Based upon the disagreement noted above and the errors possible in the critical point thermal conductivity, the recommended values are subject to an uncertainty of some fifteen percent below about 500 K and an unknown amount at higher temperatures.

TABLE 2   THERMAL CONDUCTIVITY OF BROMINE

RECOMMENDED VALUES

[Temperature, T, K; Thermal Conductivity, k, mW cm^{-1}K^{-1}]

SATURATED LIQUID

T	k
260	(1.32)‡
270	1.30
280	1.27
290	1.25
300	1.22
310	1.20
320	1.18
330	1.16
340	1.14
350	1.11
360	1.09
370	1.06
380	1.04
390	1.02
400	0.99
410	0.97
420	0.94
430	0.92
440	0.89
450	0.87
460	0.84
470	0.82
480	0.79
490	0.76
500	0.73
510	0.70
520	0.66
530	0.63
540	0.59
550	0.55
560	0.50
570	0.44
580	0.35
584*	0.28

‡Extrapolated for the supercooled liquid (n.m.p. = 266 K)
*Critical point

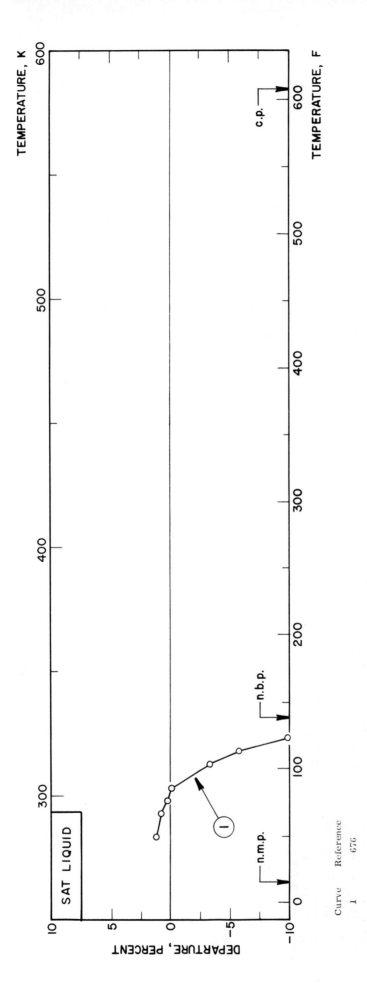

FIGURE 2 DEPARTURE PLOT FOR THERMAL CONDUCTIVITY OF SATURATED LIQUID BROMINE

DISCUSSION

SATURATED VAPOR

No experimental data were located for the thermal conductivity of saturated Bromine vapor. The correlation of Schaefer and Thodos (575) was used to construct a graph of the thermal conductivity as a function of temperature and it was found that this agreed very closely with values for atmospheric pressure gas in the region 310 to 350 K. However, the latter values depend on a single experimental data point of uncertain accuracy. The recommended values, derived from the plot of the correlation (575), must therefore be regarded as tentative. Their accuracy could possibly be within about ten percent below 450 K, fifteen percent from 450 to 550 K and unknown at higher temperatures. Due to the complete lack of experimental data for the saturated vapor, no departure plot is given.

TABLE 2   THERMAL CONDUCTIVITY OF BROMINE

RECOMMENDED VALUES

[Temperature, T, K; Thermal Conductivity, k, mW cm^{-1}K^{-1}]

SATURATED VAPOR

T	k
300	0.047
310	0.049
320	0.051
330	0.053
340	0.055
350	0.057
360	0.059
370	0.061
380	0.063
390	0.065
400	0.068
410	0.070
420	0.072
430	0.075
440	0.078
450	0.080
460	0.084
470	0.087
480	0.091
490	0.095
500	0.099
510	0.104
520	0.109
530	0.116
540	0.128
550	0.144
560	0.16
570	0.19
580	0.23
584*	0.28

*Critical point

TABLE 2 THERMAL CONDUCTIVITY OF BROMINE

RECOMMENDED VALUES

[Temperature, T, K; Thermal Conductivity, k, mW cm^{-1} K^{-1}]

GAS

T	k
250*	0.038
260*	0.040
270*	0.042
280*	0.044
290*	0.046
300*	0.048
310*	0.049
320*	0.051
330*	0.053
340	0.055
350	0.057

DISCUSSION

GAS

Only one experimental data value is available for the thermal conductivity of Bromine, being that of Franck (105) for 276 K, 65 mm. Hg. Lenoir (223) has calculated values at 255, 283 and 311 K. Interpolation of these values at 276 K gives a value ten percent lower than that of Franck. No information is given as to the pressure for which the Lenoir values are valid. The tabulated values were generated by parallel displacement of the Lenoir values to coincide with the single experimental data point. Due to the paucity of experimental data no departure plot or recommendation as to accuracy is possible.

*Ignoring pressure dependence. (n. b. p. = 332 K)

DISCUSSION

SATURATED LIQUID

No experimental or other values were found for the thermal conductivity of saturated liquid chlorine apart from a correlation by Schaefer and Thodos (575) for the thermal conductivity of liquid and gaseous states of diatomic substances. Their correlation was based principally on data for nitrogen and is thus subject to several uncertainties - in the original data, in the correlation process and in the validity of the principle of corresponding staes. The recommended values here presented were derived using the above correlation and must thus be regarded as tentative and of uncertain accuracy. Experimental measurements are urgently required to confirm their correctness. In the absence of such measurements no departure plot appears.

TABLE 3 THERMAL CONDUCTIVITY OF CHLORINE

RECOMMENDED VALUES

[Temperature, T, K; Thermal Conductivity, k, mW cm^{-1}K^{-1}]

SATURATED LIQUID

T	k
170	(1.94)‡
180	1.89
190	1.85
200	1.81
210	1.76
220	1.72
230	1.67
240	1.63
250	1.58
260	1.54
270	1.49
280	1.44
290	1.39
300	1.34
310	1.29
320	1.24
330	1.18
340	1.13
350	1.07
360	1.01
370	0.95
380	0.88
390	0.80
400	0.72
410	0.62
417*	0.40

‡Extrapolated for the supercooled liquid (n.m.p. = 172 K)
*Critical point.

DISCUSSION

SATURATED VAPOR

No experimental or other values were found for the thermal conductivity of saturated vapor chlorine apart from a correlation by Schaefer and Thodos (575) for the thermal conductivity of liquid and gaseous states of diatomic substances. Their correlation was based principally on data for nitrogen and is thus subject to several uncertainties - in the orginal data, in the correlation process and in the validity of the principle of corresponding states. The recommended values here presented were derived using the above correlation and must thus be regarded as tentative and of uncertain accuracy. Experimental measurements are urgently required to confirm their correctness. In the absence of such measurements no departure plot appears.

TABLE 3   THERMAL CONDUCTIVITY OF CHLORINE

RECOMMENDED VALUES

[Temperature, T, K; Thermal Conductivity, k, mW cm^{-1} K^{-1}]

SATURATED VAPOR

T	k
200	0.054
210	0.058
220	0.061
230	0.065
240	0.069
250	0.074
260	0.078
270	0.082
280	0.086
290	0.092
300	0.097
310	0.103
320	0.110
330	0.117
340	0.125
350	0.134
360	0.144
370	0.155
380	0.168
390	0.185
400	0.210
410	0.25
417*	0.40

*Critical point

TABLE 3 THERMAL CONDUCTIVITY OF CHLORINE

RECOMMENDED VALUES

[Temperature, T, K; Thermal Conductivity, k, mW cm^{-1} K^{-1}]

GAS

T	k	T	k
200	(0.054)*	500	0.156
210	(0.057)*	510	0.160
220	(0.061)*	520	0.163
230	(0.064)*	530	0.166
240	0.068	540	0.170
250	0.0711	550	0.173
260	0.0747	560	0.176
270	0.0782	570	0.180
280	0.0818	580	0.183
290	0.0854	590	0.186
300	0.0889	600	0.190
310	0.0925	610	0.192
320	0.0960	620	0.195
330	0.0995	630	0.197
340	0.1031	640	0.200
350	0.1066	650	0.202
360	0.1100	660	0.205
370	0.1135	670	0.207
380	0.1169	680	0.210
390	0.1204	690	0.212
400	0.1238	700	0.215
410	0.127		
420	0.131		
430	0.134		
440	0.137		
450	0.141		
460	0.144		
470	0.147		
480	0.150		
490	0.153		

DISCUSSION

GAS

The only data available for this substance are experimental values of Franck (105) from 198 to 676 K obtained for pressures between 50 and 250 mm Hg upon which a tabulation by Lenoir (223) was based. Examination of the Franck data shows that the variation of thermal conductivity with pressure in this temperature interval is irregular, hence in the analysis it was neglected. The Lenoir values show good agreement with the Franck data above the normal boiling point, and were selected as a basis for the most probable values. Below the normal boiling point the single experimental point was used in the construction of the recommended values. The recommended values should be accurate to within five percent.

*Ignoring pressure dependence.
(n. b. p. = 239 K)

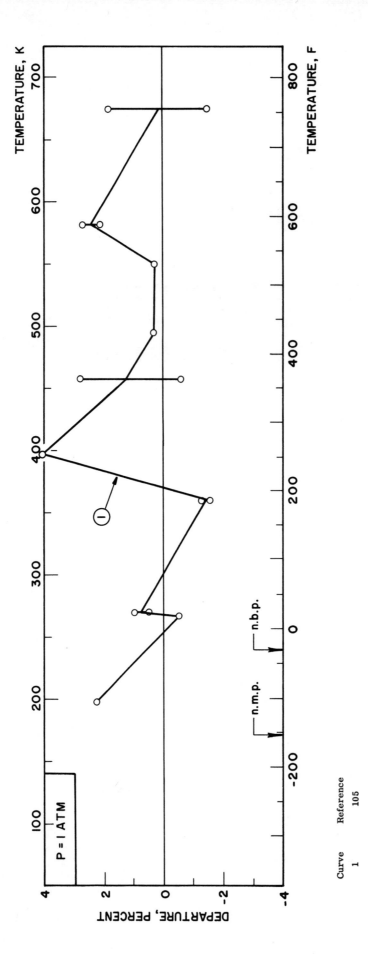

FIGURE 3 DEPARTURE PLOT FOR THERMAL CONDUCTIVITY OF GASEOUS CHLORINE

TABLE 4 THERMAL CONDUCTIVITY OF DEUTERIUM

[Temperature, T, K; Thermal Conductivity, k, mW cm^{-1}K^{-1}]

DISCUSSION

SATURATED LIQUID

Only one set of experimental data were located for the thermal conductivity of liquid deuterium. Powers et al. (533) made measurements in a parallel-plate apparatus for normal and ortho deuterium for temperatures between 20 and 24 K. To within a two percent uncertainty they considered the thermal conductivity to be independent of the ortho-para composition. The values so obtained were fitted to a linear equation in temperature by Friedman and Hilsenrath (524) who then used the equation to generate values from 20 to 30 K.

In this work the correlation of Kerrisk et al. (667) was also compared with the above measurements. As read from the source graph, values some six percent lower than the data (533) were obtained. These were adjusted to coincide with the data values and a curve drawn through the latter to exhibit the same trend as the Kerrisk curve in that the thermal conductivity reaches a maximum just below 28 K and then decreases rapidly to the critical point value of (667). The recommended values were read from this smooth curve.

The accuracy of the recommended values depends on the accuracy of the only set of experimental data (533) and on the (667) estimate for the critical point. It is considered that five percent is a reasonable estimate below 30 K, ten percent from 30 to 35 K and steadily increasing uncertainty to the critical point of about 38.3 K.

RECOMMENDED VALUES

SATURATED LIQUID

T	k
18	1.22†
19	1.24
20	1.26
21	1.28
22	1.30
23	1.33
24	1.34
25	1.36
26	1.37
27	1.38
28	1.38
29	1.38
30	1.37
31	1.35
32	1.32
33	1.29
34	1.25
35	1.21
36	1.15
37	1.07
38	0.91
38.26*	0.83

†Extrapolated for the supercooled liquid (n.m.p. = 19 K)
*Critical temperature

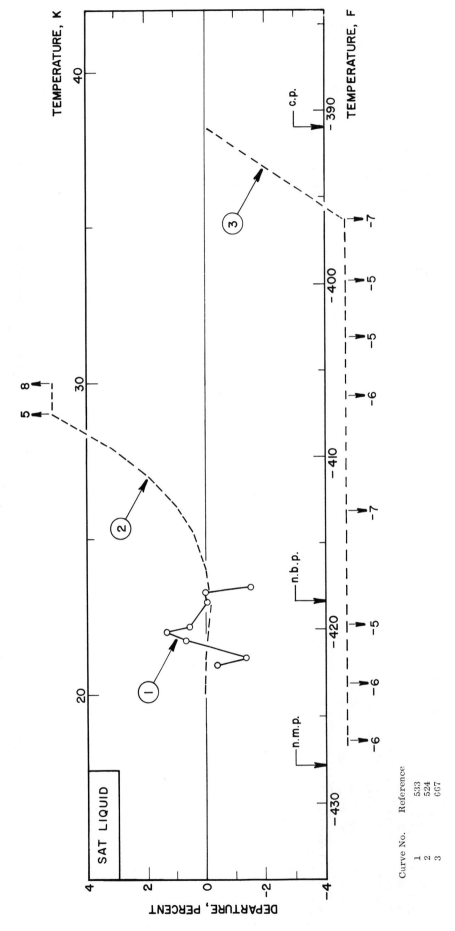

FIGURE 4 DEPARTURE PLOT FOR THERMAL CONDUCTIVITY OF SATURATED LIQUID DEUTERIUM

TABLE 4 THERMAL CONDUCTIVITY OF DEUTERIUM

RECOMMENDED VALUES

[Temperature, T, K; Thermal Conductivity, k, mW cm^{-1}K^{-1}]

DISCUSSION

SATURATED VAPOR

No experimental measurements or estimates were found for the thermal conductivity of saturated deuterium vapor apart from a critical point estimate of Kerrisk et al. (667). An attempt was made in this work to provide a set of estimated values using the above mentioned critical point value, the value obtained from the atmospheric pressure tables for the normal boiling point (23 K) and the fact that the Kerrisk correlation indicated a maximum in the saturated liquid conductivity just below 28 K.

The thermal conductivity-temperature values for the saturated liquid were plotted on a graph together with the atmospheric pressure value at 23 K. Assuming the rectilinear diameter relation to apply for thermal conductivity above about 36 K, values for the saturated liquid were used to obtain values for the saturated vapor. The rectilinear diameter line so obtained clearly did not pass through the point obtained at 23 K so a curve of minimum curvature was drawn which passed through this point and the critical point. Values for a few other temperatures for the saturated vapor were thus obtained and a smooth curve drawn to pass through these and the values at 23 K and the critical point.

The recommended values, obtained in this manner, are of uncertain accuracy. In addition to possible quantum effects and phenomena near the critical point they depend upon the rectilinear diameter and visual smoothing for their accuracy. They must be regarded as extremely tentative and subject to corrections of at least twenty-five percent. Due to the complete lack of experimental data, no departure plot is given.

SATURATED VAPOR

T	k
20	0.084
21	0.096
22	0.109
23	0.123
24	0.138
25	0.155
26	0.174
27	0.193
28	0.215
29	0.238
30	0.26
31	0.29
32	0.32
33	0.36
34	0.40
35	0.45
36	0.51
37	0.58
38	0.70
38.26*	0.83

*Critical Point

DISCUSSION

GAS

Available data for the thermal conductivity of deuterium gas are confined to a set of experimental measurements from 15 to 20 K and from 65 to 89 K (349) and single measurements at the ice point (13, 20, 21, 67, 171, 260, 521). A correlation by Lenoir (223) extends from 16 to 366 K and fits the data very well. After analysis of the data it was decided to base the recommended values on a smooth curve which is almost identical to that through the Lenoir values.

Some evidence exists from the data of (349) that a quantum effect may occur between 50 and 70 K but no confirmation from independent measurements is available. In the preparation of the recommended values the deviation from the smooth curve was regarded as being due to experimental error and therefore neglected. Should new measurements show that the deviation is real then the recommended values in this range will need revision.

Between 20 and 100 K and 260 to 280 K the recommended values are considered accurate to about two percent. Between 100 and 260 K the values may only be accurate to within five percent. Values above 280 K may be in error by as much as ten percent. Further experimental data are to be desired, especially above 30 K.

TABLE 4 THERMAL CONDUCTIVITY OF DEUTERIUM

RECOMMENDED VALUES

[Temperature, T, K; Thermal Conductivity, k, mW cm^{-1}K^{-1}]

GAS

T	k	T	k
20	(0.100)*	200	1.014
25	0.139	210	1.056
30	0.175	220	1.097
35	0.206	230	1.138
40	0.236	240	1.178
45	0.268	250	1.217
50	0.299	260	1.256
60	0.360	270	1.294
70	0.421	280	1.331
80	0.475	290	1.369
90	0.527	300	1.406
100	0.577	310	1.44
110	0.625	320	1.48
120	0.672	330	1.51
130	0.718	340	1.55
140	0.762	350	1.59
150	0.806	360	1.62
160	0.848	370	1.66
170	0.890	380	1.69
180	0.931	390	1.73
190	0.973	400	1.76

*Extrapolated for the gas phase ignoring pressure dependence. (n. b. p. = 23 K)

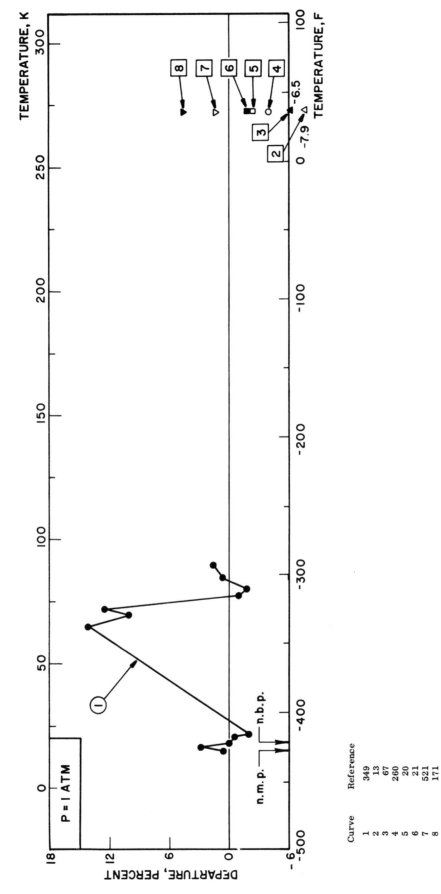

FIGURE 4   DEPARTURE PLOT FOR THERMAL CONDUCTIVITY OF GASEOUS DEUTERIUM

Curve	Reference
1	349
2	13
3	67
4	260
5	20
6	21
7	521
8	171

TABLE 5   THERMAL CONDUCTIVITY OF FLUORINE

RECOMMENDED VALUES
[Temperature, T, K; Thermal Conductivity, k, mW cm^{-1}K^{-1}]

SATURATED LIQUID

T	k
50	(2.1)‡
60	1.95
70	1.80
80	1.65
90	1.50
100	1.35
110	1.21
120	1.06
130	0.90
140	0.66
144*	0.40

‡ Extrapolated for the supercooled liquid (n.m.p. = 54 K)
* Critical temperature

DISCUSSION

SATURATED LIQUID

The values here presented for the thermal conductivity of saturated fluorine liquid are interpolated from a graph drawn in the engineering system (670). No experimental data were located for the thermal conductivity of the saturated liquid in the unclassified literature. In the source cited (670), it was stated that the saturated values were adjusted from those derived from the Stiel and Thodos correlation (671) although no detailed description of the adjustment method was given.

In view of the uncertainty possible in the original correlation (671) plus that introduced in the adjustment procedure (670), no detailed error estimate can be given. An assessment of ten percent below 125 K and unknown above 125 K would appear reasonable. Due to the lack of experimental data no departure plot is given. Experimentation is highly desirable to confirm the estimates here reproduced.

TABLE 5   THERMAL CONDUCTIVITY OF FLUORINE

DISCUSSION

RECOMMENDED VALUES
[Temperature, T, K; Thermal Conductivity, k, mW cm^{-1}K^{-1}]

SATURATED VAPOR

SATURATED VAPOR

T	k
50	0.036
60	0.045
70	0.054
80	0.064
90	0.076
100	0.089
110	0.105
120	0.125
130	0.16
140	0.22
144*	0.40

The values here presented for the thermal conductivity of saturated fluorine vapor are interpolated from a graph drawn in the engineering system (670). No experimental data were located for the thermal conductivity of the saturated vapor in the unclassified literature. In the source cited (670), it was stated that the saturated values were adjusted from those derived from the Stiel and Thodos correlations (671) although no detailed description of the adjustment method was given.

In view of the uncertainty possible in the original correlation (671) plus that introduced in the adjustment procedure (670), no detailed error estimate can be given. An assessment of ten percent below 125 K and unknown above 125 K would appear reasonable. Due to the lack of experimental data no departure plot is given. Experimentation is highly desirable to confirm the estimates here reproduced.

* Critical temperature

TABLE 5 THERMAL CONDUCTIVITY OF FLUORINE

RECOMMENDED VALUES

[Temperature, T, K; Thermal Conductivity, k, mW cm^{-1}K^{-1}]

GAS

T	k	T	k
80	(0.075)*	450	0.413
90	0.082	460	0.421
100	0.090	470	0.430
110	0.100	480	0.438
120	0.109	490	0.446
130	0.118	500	0.455
140	0.128	510	0.463
150	0.137	520	0.471
160	0.146	530	0.479
170	0.156	540	0.486
180	0.165	550	0.493
190	0.174	560	0.500
200	0.184	570	0.507
210	0.193	580	0.514
220	0.202	590	0.520
230	0.212	600	0.527
240	0.221	610	0.534
250	0.231	620	0.541
260	0.241	630	0.547
270	0.251	640	0.552
280	0.260	650	0.557
290	0.269	660	0.563
300	0.279	670	0.568
310	0.288	680	0.573
320	0.298	690	0.579
330	0.307	700	0.583
340	0.316	710	0.588
350	0.326	720	0.592
360	0.335	730	0.596
370	0.344	740	0.599
380	0.354	750	0.603
390	0.363	760	0.607
400	0.371	770	0.610
410	0.378	780	0.613
420	0.388	790	0.616
430	0.397	800	0.618
440	0.405		

* Extrapolated (n.b.p. = 86 K)

DISCUSSION

GAS

The only experimental data reported for the thermal conductivity of fluorine is that due to Franck and co-workers (105, 109) for pressures below atmospheric. These data have been correlated by Franck (105) and Lenoir (223).

Analysis of the experimental data was made difficult by the fact that with one exception, only two pressures were studied by Franck at any one temperature. No information is given be either Franck or Lenoir on the method of obtaining conductivity values for atmospheric pressure. In this analysis a linear variation was assumed. Extrapolation of the data to atmospheric pressure yielded a set of values which appear to be consistently higher by about five percent than either the Lenoir of Franck tabulations. It appears that the Franck smoothed values are not for atmospheric pressure and that Lenoir erroneously assumed this. The recommended values were calculated from the equation

$$10^5 k \text{ (cgsu)} = 0.082526 + 1.980619 \cdot 10^{-2}T + 1.24895 \cdot 10^{-5}T^2 - 1.78771 \cdot 10^{-8}T^3$$
(T in K)

which fitted the extrapolated values to within 2.2 percent.

The accuracy of the recommended data is difficult to estimate accurately due to the scatter in the original experimental values at lower pressures and the possibility of dimerization in the vapor but a value of ten percent should be adequate for all temperatures tabulated.

TABLE 6   THERMAL CONDUCTIVITY OF HELIUM

RECOMMENDED VALUES

[Temperature, T, K; Thermal Conductivity, k, mW cm^{-1}K^{-1}]

SOLID

T	k(He^3)	k(He^4)
0.5	–	415
0.6	250	705
0.7	104	1060
0.8	55	1200
0.9	33	650
1.0	20	245
1.1	14.4	97
1.2	10.8	45
1.3	8.9	27
1.4	7.3	16
1.5	5.7	10.5
1.6	4.6	6.9
1.7	3.8	4.9
1.8	3.0	3.4
1.9	2.5	2.5
2.0	2.1	1.8

DISCUSSION

SOLID

A survey (682) of the thermal conductivity of solid isotopic mixtures of helium considered a solid volume of 20.2 cm^3/mole, at which molar volume pure solid He^4 was considered to exist as a hexagonal close-packed structure and pure solid He^3 as a body-centered cubic structure. For a molar volume of 19.5 cm^3/mole the latter structure is hcp, and the thermal conductivities are in general much larger. In this analysis the experimental data of Bertmann (683), also reproduced by (682), were selected for a volume of 20.2 cm^3/mole, close to the mean of different values investigated by Bertmann. While Bertmann evaluates the accuracy in conductivity values as being from 2-15 percent, which appears to be realistic, some other work on pure He^4 (684) yielded values which usually were comparable with the Bertmann data at the lower limit but which could be greater by a factor of ten at the upper limit. The latter work claims that their technique allows the production of crystals of much higher purity and that the influence of the container has also to be taken into account. Based on a consideration of these effects it would appear that while the above and other measurements are reliable to the individual accuracies assessed the limiting factor has been the characterization of the sample.

The recommended values here presented were based on the Bertmann 20.2 cm^3/mole sets of data (682, 683) and must be regarded as only order of magnitude in absolute accuracy due to sample characterization difficulties. In view of this factor no departure plot is given.

TABLE 6 THERMAL CONDUCTIVITY OF HELIUM

[Temperature, T, K; Thermal Conductivity, k, mW cm^{-1}K^{-1}]

RECOMMENDED VALUES

SATURATED LIQUID

T	k
2.0	0.199
2.5	0.191
3.0	0.203
3.5	0.232
4.0	0.281
4.5	(0.348)‡
5.0	(0.434)‡

DISCUSSION

SATURATED LIQUID

A number of experimental works has been reported on the thermal conductivity of liquid helium from the standpoint of the interest in low temperature physics. As is well known, a thermodynamic transition in the liquid phase of helium takes place at a temperature near 2.17 K, referred to as the "lambda point". At temperatures above the lambda point, the liquid is called helium-I and below this point it is called helium-II. Helium-I is not particularly remarkable, but helium-II has a number of interesting properties especially flow and conduction properties due to the quantum nature of this liquid.

The thermal conductivity of liquid helium-I was first measured by Keesom-Keesom (547), and it was found that the value is of the same order of magnitude as that of gases at ordinary temperatures. Grenier (133, 545, 546) made measurements in a parallel-plate apparatus within the uncertainty of 10 percent, covering the temperature range from 2.2 to 4.2 K, and found that the thermal conductivity of helium-I decreases with decreasing temperature and exhibits a minimum near 2.4 K. He concluded that helium-I behaves more like a gas than a normal liquid. Bowers (48, 49) also measured it in a longitudinal capillary apparatus. Although his measurements were not a precise absolute evaluation of the thermal conductivity, he obtained a linear relation down to the lambda point with considerable scattering. More recently, Fairbank - Lee (544) obtained more accurate values at temperatures from 2.3 to 3.9 K under saturated vapor pressures, using a capillary method. As their results are considered to be the most reliable to date, all their reported points are given equal weight in this analysis and are fitted to a quadratic equation, represented by

$$10^9 k \text{ (cgsu)} = 99.5614 - 43.8934\,T + 8.94877\,T^2 \quad (T \text{ in K}).$$

This equation should be valid at temperatures above 2.2 K. The above equation is found to fit the data of Fairbank - Lee with a mean deviation of 1.7 percent and a maximum of 3.7 percent. The recommended values are generated from this equation, and the values should be substantially correct within two percent.

---

‡ Extrapolated for the liquid under vapor pressures, ignoring pressure dependence. (n.b.p. = 4 K)

TABLE 6 THERMAL CONDUCTIVITY OF HELIUM (continued)

RECOMMENDED VALUES

[Temperature, T, K; Thermal Conductivity, k, mW cm^{-1}K^{-1}]

SATURATED LIQUID

SATURATED LIQUID

DISCUSSION

On the other hand, of all the physical properties of helium-II, the most remarkable is the extraordinarily high transport of heat. Preliminary measurements by Keesom - Keesom (547) at 1.4 and 1.75 K gave values of the thermal conductivity of about 190 cal cm^{-1} sec^{-1}K^{-1}. It may be noted that this value is about 200 times that of copper at ordinary temperatures. In their further measurements (548, 560, 561), it was found that the thermal conductivity as a function of temperature has a very pronounced maximum near 1.92 K. and under some condition, a thermal conductivity as high as 810 cal^{-1} sec^{-1}k^{-1} was observed. Hence, liquid helium-II is by far the best heat conducting substance known.
A number of investigations have been carried out on the super-heat-conduction of liquid helium-II (198, 231, 548-567). The mechanism of heat transport in liquid helium-II is quite different from that in helium-I or other liquids due to its extreme fluidity and the associated transport of energy by virtue of convective currents. Under these circumstances it is not possible to observe a "true" thermal conductivity as a transport property. Therefore, observed thermal conductivity values are found to depend markedly on the conditions of measurement, that is, the heat current density, the temperature gradient, and the dimensions of the test cell used. It is considered to be impossible to treat the heat conduction in helium-II in the same way as in other liquids. Therefore, no correlation is attempted in this analysis.

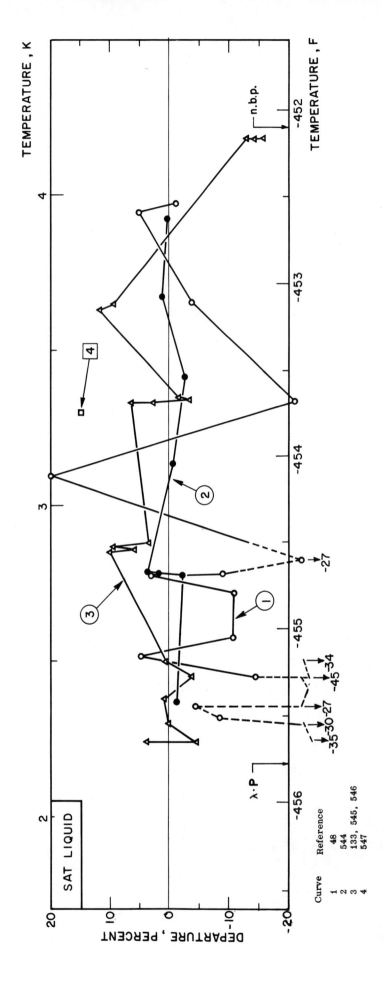

FIGURE 6 DEPARTURE PLOT FOR THERMAL CONDUCTIVITY OF LIQUID HELIUM

TABLE 6 THERMAL CONDUCTIVITY OF HELIUM

RECOMMENDED VALUES

[Temperature, T, K; Thermal Conductivity, k, mW cm^{-1} K^{-1}]

GAS

DISCUSSION

Helium is one of the few gases where quantum effects become significant at low temperatures. In addition to the experimental data of Ubbink and deHaas (251) calculated values have been made for such temperatures and the recommended values for such temperatures were deduced from a curve drawn through both calculated and experimental values. The usual increment of 10 K for the tabulation of recommended values is inadequate for helium at low temperatures and appropriate increments have been chosen in the tabulation.

No experimental data were found below about 2 K and gaps exist between 4 to 14 K, 21 to 73 K and above 2000 K. Some discrepancies exist between the measurements of different workers. The trend of the Johnston and Grilly data (168) and some previous correlations (147, 187, 223, 521, 570, 630, 631) is to produce values lower than the Kannuluik and Carmen (173) data. The values were selected so as to agree with the higher temperature data of the latter workers. This selection agrees with the trend of the considerably higher temperature data of Blais and Mann (569) and of Petersen and Bonilla (628).

Certain conclusions can be reached concerning previous analyses. The calculated values of Amdur (10) and the recommended tables of Chelton and Mann (81) should be disregarded below 100 K. The tables of Lenoir (223) agree to within about one percent between 20 and 450 K while the NBS tables (147) only agree to the same accuracy between about 205 and 415 K.

Many correlations (147, 187, 223, 521, 630, 631) fail above about 600 K. Of the seventeen different values found in the literature for the thermal conductivity at the ice point (273.15 K), thirteen agree to within two percent with the recommended value from this table. Further experimental measurements are desirable for temperatures below 100 K and above 600 K.

Below 100 K the recommended values should be accurate to within five percent, from 100 to 400 K the accuracy should be one percent, from 400 to 700 K five percent, from 700 to 2000 K ten percent and above 2000 K as much as twenty-five percent.

T	k	T	k	T	k
0.08	0.00044	25	0.2962	350	1.649	700	2.78
0.09	0.00053	30	0.3330	360	1.678	710	2.81
0.10	0.00064	35	0.3669	370	1.708	720	2.84
0.15	0.00130	40	0.4000	380	1.737	730	2.87
0.20	0.00231	45	0.4314	390	1.766	740	2.90
0.25	0.0039	50	0.4623	400	1.795	750	2.92
0.30	0.0062	60	0.521	410	1.824	760	2.95
0.35	0.0089	70	0.578	420	1.853	770	2.98
0.40	0.0120	80	0.631	430	1.882	780	3.01
0.45	0.0154	90	0.679	440	1.914	790	3.04
0.5	0.0187	100	0.730	450	1.947	800	3.07
0.6	0.0231	110	0.776	460	1.980	810	3.09
0.7	0.0252	120	0.819	470	2.013	820	3.12
0.8	0.0262	130	0.863	480	2.046	830	3.15
0.9	0.0266	140	0.907	490	2.080	840	3.18
1.0	0.0269	150	0.950	500	2.114	850	3.21
1.25	0.0281	160	0.992	510	2.15	860	3.23
1.5	0.0306	170	1.033	520	2.18	870	3.26
2.0	0.0393	180	1.072	530	2.22	880	3.29
2.5	0.0502	190	1.112	540	2.25	890	3.32
3.0	0.0607	200	1.151	550	2.29	900	3.35
3.5	0.0732	210	1.190	560	2.33	910	3.37
4.0	0.0803	220	1.228	570	2.36	920	3.40
4.5	0.0879	230	1.266	580	2.40	930	3.43
5.0	0.0962	240	1.304	590	2.43	940	3.46
6	0.1113	250	1.338	600	2.47	950	3.49
7	0.1247	260	1.372	610	2.51	960	3.52
8	0.1393	270	1.405	620	2.54	970	3.54
9	0.1523	280	1.437	630	2.58	980	3.57
10	0.1640	290	1.468	640	2.61	990	3.60
12	0.1866	300	1.499	650	2.64	1000	3.63
14	0.2067	310	1.530	660	2.67	1050	3.76
16	0.2435	320	1.560	670	2.69	1100	3.89
18	0.2435	330	1.590	680	2.72	1150	4.03
20	0.2582	340	1.619	690	2.75	1200	4.16

TABLE 6   THERMAL CONDUCTIVITY OF HELIUM (continued)

RECOMMENDED VALUES

[Temperature, T, K; Thermal Conductivity, k, mW cm^{-1} K^{-1}]

GAS

T	k	T	k
1250	4.29	3000	8.51
1300	4.43	3100	8.72
1350	4.55	3200	8.95
1400	4.69	3300	9.16
1450	4.82	3400	9.37
1500	4.94	3500	9.58
1550	5.07	3600	9.79
1600	5.21	3700	10.00
1650	5.33	3800	10.22
1700	5.45	3900	10.43
1750	5.57	4000	10.64
1800	5.70	4100	10.85
1850	5.83	4200	11.06
1900	5.96	4300	11.27
1950	6.08	4400	11.48
2000	6.20	4500	11.69
2100	6.44	4600	11.90
2200	6.69	4700	12.11
2300	6.93	4800	12.31
2400	7.16	4900	12.51
2500	7.39	5000	12.71
2600	7.62		
2700	7.85		
2800	8.07		
2900	8.29		

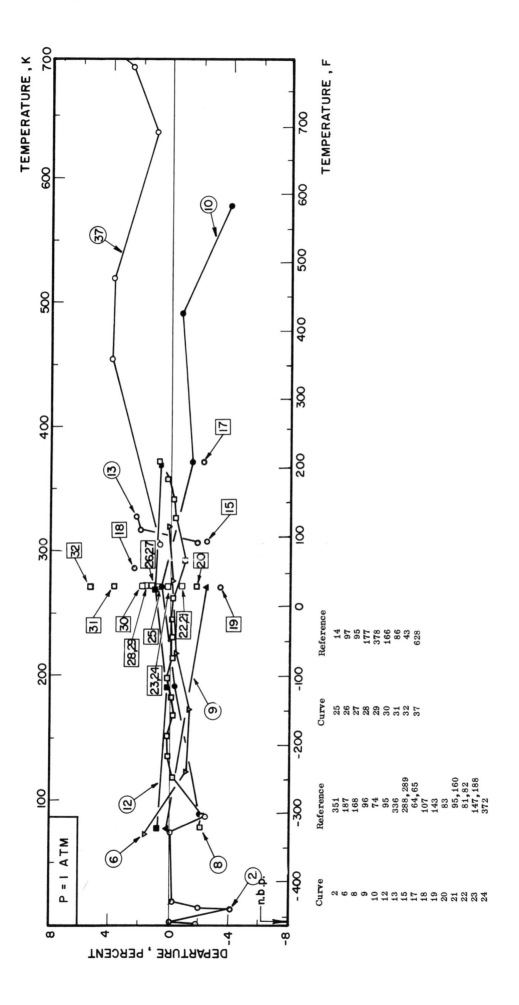

FIGURE 6   DEPARTURE PLOT FOR THERMAL CONDUCTIVITY OF GASEOUS HELIUM

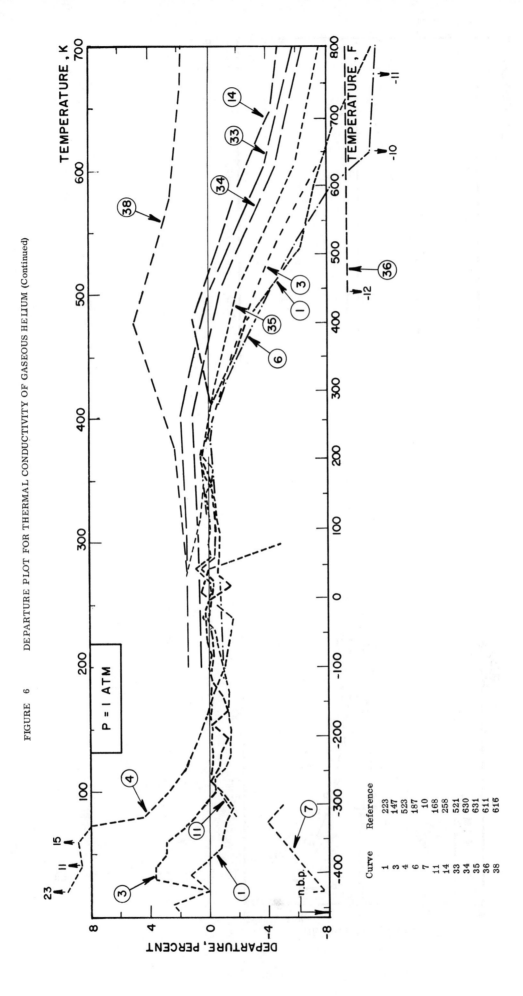

FIGURE 6 DEPARTURE PLOT FOR THERMAL CONDUCTIVITY OF GASEOUS HELIUM (Continued)

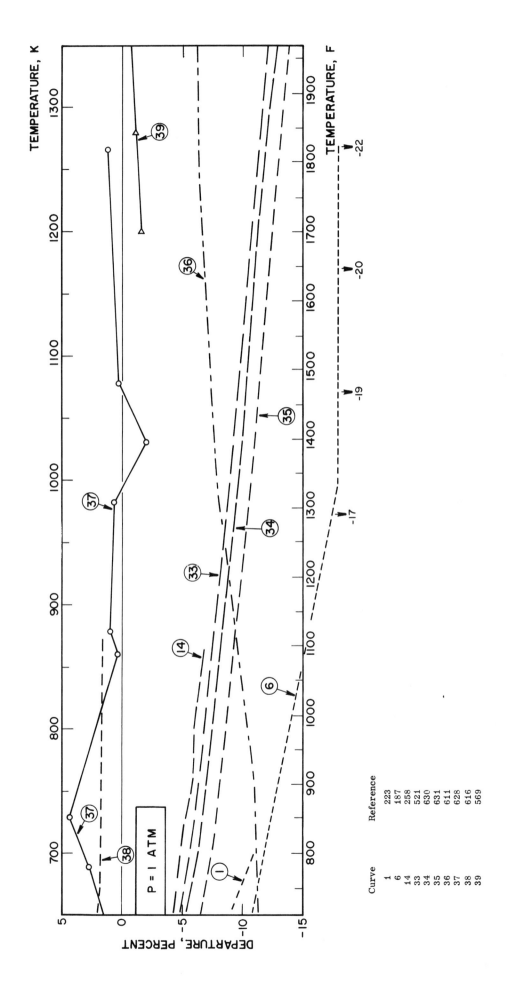

FIGURE 6 DEPARTURE PLOT FOR THERMAL CONDUCTIVITY OF GASEOUS HELIUM (continued)

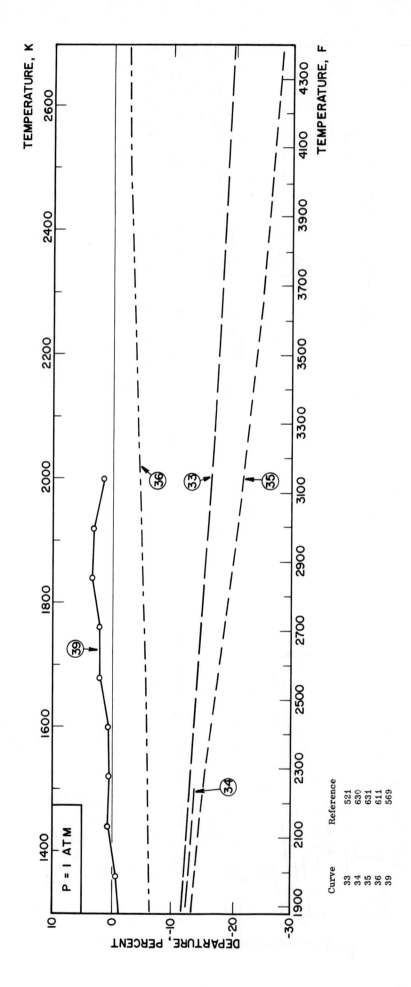

FIGURE 6 DEPARTURE PLOT FOR THERMAL CONDUCTIVITY OF GASEOUS HELIUM (continued)

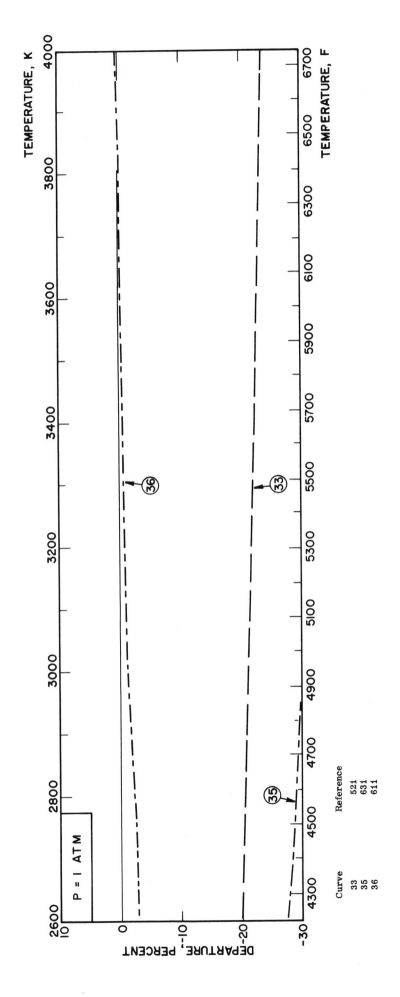

FIGURE 6 DEPARTURE PLOT FOR THERMAL CONDUCTIVITY OF GASEOUS HELIUM (continued)

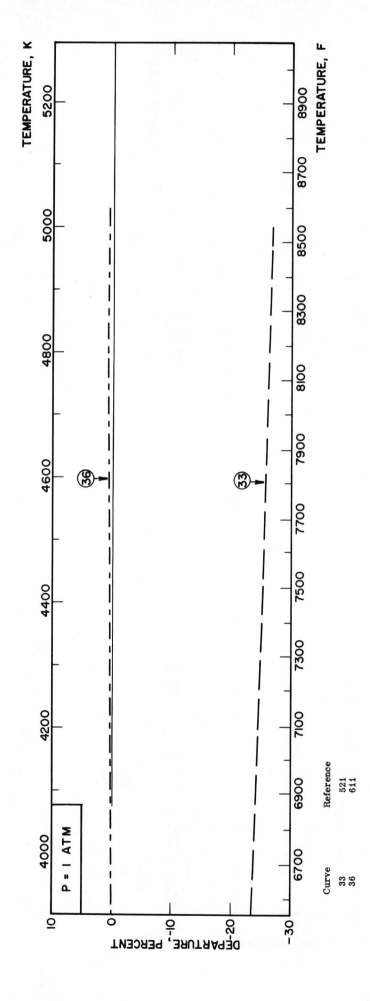

FIGURE 6 DEPARTURE PLOT FOR THERMAL CONDUCTIVITY OF GASEOUS HELIUM (continued)

TABLE 7   THERMAL CONDUCTIVITY OF HYDROGEN

RECOMMENDED VALUES

[Temperature, T, K; Thermal Conductivity, k, mW cm^{-1}K^{-1}]

SOLID

T	k
4	2300
5	550
6	190
7	83
8	43
9	23
10	15.8
11	12.5
12	10.0
13	9.5
14	9.0
15	9.0
16	8.9
17	8.9‡

DISCUSSION

SOLID

Available data for the thermal conductivity of solid hydrogen appears to be confined to two sources (669, 668). The first reports data for 0.5, 1 and 5 percent orthohydrogen in parahydrogen for temperatures below 11 K while the latter contains three experimental data for a one percent orthohydrogen, ninety nine percent parahydrogen mixture for temperatures between 15 and 17 K and pressures from 88 to 201 atmospheres.

The earlier (669) report was not available at the time of this compilation. The information was derived from the published curve (668). The recommended values were obtained by smoothing of values read from a large scale reproduction of the (668) curve. For the lowest temperature datum, 4K, the value cited was obtained by extrapolation of the various orthohydrogen content data to a zero percent orthohydrogen content. Values for temperatures below 4 K appear to be even more sensitive to the orthohydrogen content. While possibly the thermal conductivity of pure parahydrogen reaches a maximum at about 3K, the available information was not sufficiently detailed to enable this to be determined accurately. The tables extend to 17K, a temperature higher than the triple (14 K) or melting (16 K) points. Some uncertainty is introduced into the values due to the pressures of 88 to 201 atmospheres used by (668) but this is considered much less than the uncertainty in the data themselves, which is assessed at eleven percent by (668) and which could be even larger than their estimates.

Based upon the smoothness of the data reported by the two workers, the values from 5 to 10 K should be consistent to about five percent, from 10 to 17 K to about fifteen percent and at 4 K by about twenty percent. The word 'consistent' is here used to emphasise the fact that no overlap between the different measurements exists and that confirmatory measurements are very desirable. As only a graphical reproduction of the (669) values was available at the time of this correlation, no departure plot is given.

‡ n.m.p. = 16 K

TABLE 7 THERMAL CONDUCTIVITY OF HYDROGEN

RECOMMENDED VALUES

[Temperature, T, K; Thermal Conductivity, k, mW cm^{-1}K^{-1}]

SATURATED LIQUID

T	k
14	0.989
15	1.022
16	1.055
17	1.088
18	1.121
19	1.153
20	1.184
21	1.213‡
22	1.238‡
23	1.258‡
24	1.272‡
25	1.269‡
26	1.251‡
27	1.217‡
28	1.168‡
29	1.117‡
30	1.06 ‡
31	1.00 ‡
32	0.91 ‡
33	0.74 ‡
33.18*	0.60

‡Under saturated vapor pressures. (n.b.p. = 20 K)
*Critical point

DISCUSSION

SATURATED LIQUID

Only one set of experimental data is available on the thermal conductivity of liquid hydrogen. The measurement was made in a parallel-plate apparatus for normal and para hydrogen covering the temperature range between 16 and 25 K by Powers et al. (532). The thermal conductivity of liquid hydrogen was found to have a positive temperature coefficient. No significant difference was observed between normal and para hydrogen within the experimental error of 3.5 percent. The authors fitted their data by a linear equation which was used by Friedman and Hilsenrath (524) to tabulate the thermal conductivity between 16 and 30 K. These values have been cited in a number of compendia on cryogenics (61, 167, 523, 529).

Since the above works a number of other studies have been made (664-8). These consist of generalized analyses, based either upon critical properties alone and the principle of corresponding states (664) or upon a more extended analysis which involves the quantum parameter which becomes significant for hydrogen (665-8). However, no new measurements of the thermal conductivity of the saturated or compressed liquid have been located. The entire body of material in the literature would seem to rest upon the experimental measurements of Powers, Mattox, et al. (532).

In several works (665-8), the temperature dependence of the thermal conductivity is shown to change from positive to negative at about 25 K whereas in the Schaefer and Thodos correlation (664) no such effect is exhibited. The latter must therefore be regarded as physically inaccurate below this temperature. The remaining works (665-8) were found to agree in trend, but not in magnitude, as to the variation of thermal conductivity with temperature. The differences between the various estimates increases markedly as the critical point is approached. Somewhat arbitrarily, it was decided to base the recommended values upon the tables of Jones (665) as the latter appear to be the most recent and also the most detailed available.

The recommended values, derived from the tables of Jones (665), should be accurate to about three percent for temperatures below 25 K. Due to the lack of precise knowledge of the critical point parameters, the uncertainty increases rapidly for higher temperatures and can be assessed as about thirty percent at 30 K and even a hundred percent for the critical temperature.

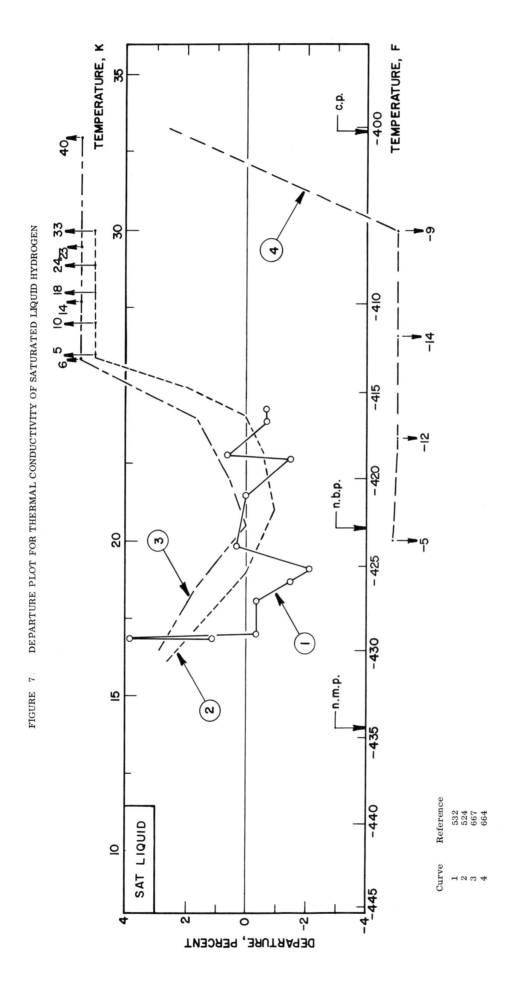

FIGURE 7. DEPARTURE PLOT FOR THERMAL CONDUCTIVITY OF SATURATED LIQUID HYDROGEN

TABLE 7   THERMAL CONDUCTIVITY OF HYDROGEN

RECOMMENDED VALUES

[Temperature, T, K; Thermal Conductivity, k, mW cm^{-1}K^{-1}]

SATURATED VAPOR

T	k
15	0.117
16	0.126
17	0.134
18	0.142
19	0.150
20	0.159
21	0.169
22	0.180
23	0.192
24	0.205
25	0.22
26	0.23
27	0.25
28	0.27
29	0.29
30	0.31
31	0.35
32	0.40
33	0.58
33.2*	0.60

* Critical Temperature

DISCUSSION

SATURATED VAPOR

No experimental data have been located for the thermal conductivity of saturated hydrogen vapor. The correlations of Schaefer and Thodos (664) and of Jones (665) have been examined. The former was found to yield values for the gas at atmospheric pressure which were somewhat lower than those considered most probable. The shape of the envelope enclosing the two-phase mixture of liquid and vapor was found to disagree at the saturated liquid boundary. The correlations of Jones proved much more satisfactory for both cases.

The recommended values were therefore obtained from a large-scale graph of the Jones values. They should be accurate to within three percent below 25 K. As the critical point is approached the uncertainty increases due to the errors possible in the correlated values and due to the difficulty of interpolating these values. Possibly, uncertainties of twenty percent at 30 K and even a hundred percent at the critical point may prove to be reliable error estimates. As noted by Jones, experimentation to confirm the accuracy of these estimates is urgently required. The recommended values must therefore be regarded as tentative pending this work.

GAS

TABLE 7 THERMAL CONDUCTIVITY OF HYDROGEN

RECOMMENDED VALUES

[Temperature, T, K; Thermal Conductivity, k, mW cm^{-1}K^{-1}]

GAS

DISCUSSION

Experimental data for the thermal conductivity of gaseous hydrogen extend from 15 to 2000 K. Calculated values have appeared for the range 10 to 10^{4}K. In this analysis, recommended values were generated to 2000 K only. Values for higher temperatures are not at present thought to be sufficiently accurate to enable a recommendation to be made. Such studies should also consider the effect of dissociation and ionization. At 2000 K about two percent of the hydrogen is dissociated. In view of the effect of dissociation upon the thermal conductivity, these recommended values can adequately be said to refer to diatomic hydrogen.

In preparing the recommended values, little difficulty was found for temperatures from 100 to 250 K. The correlation by the NBS(146,147) was found perfectly adequate. Some other correlations (61, 223, 408) were found to exhibit rather large errors in certain temperature intervals. At temperatures above 250 K, all these correlations yield values which are progressively larger than the recommended values with increasing temperature. In this range the previous correlations by Keyes (187) and Svehla (521) appear most accurate. The recommended values above about 450 K were based on the experimental measurements of Geier and Schafer (587) which extend to 1473 K and which can be seen from the departure plot to be in reasonable accord with theoretical estimates even at 2000 K. The data of Blais and Mann (569) were found to be considerably higher than all other measured or estimated values and were not used in preparing the recommended values. Further experimental measurements are desirable below 100 K and above 500 K.

The accuracy of the recommended values can be estimated as two percent or better below 400 K, within five percent from 400 to 1350 K and fifteen percent from 1350 K to 2000 K.

T	k	T	k	T	k	T	k
		350	2.033	750	3.43	1150	4.78
		360	2.069	760	3.46	1160	4.82
		370	2.106	770	3.50	1170	4.85
		380	2.142	780	3.53	1180	4.88
		390	2.177	790	3.56	1190	4.92
20	0.159	400	2.212	800	3.60	1200	4.95
25	0.193	410	2.248	810	3.63	1210	4.98
30	0.227	420	2.283	820	3.67	1220	5.02
35	0.261	430	2.318	830	3.70	1230	5.05
40	0.294	440	2.354	840	3.74	1240	5.08
45	0.328						
50	0.361	450	2.389	850	3.77	1250	5.12
60	0.426	460	2.424	860	3.80	1260	5.15
70	0.489	470	2.459	870	3.84	1270	5.18
80	0.552	480	2.494	880	3.87	1280	5.21
90	0.614	490	2.529	890	3.91	1290	5.25
100	0.676	500	2.564	900	3.94	1300	5.28
110	0.738	510	2.60	910	3.97	1310	5.31
120	0.801	520	2.64	920	4.01	1320	5.35
130	0.864	530	2.67	930	4.04	1330	5.38
140	0.926	540	2.70	940	4.08	1340	5.41
150	0.986	550	2.74	950	4.11	1350	5.45
160	1.046	560	2.77	960	4.14	1360	5.49
170	1.105	570	2.80	970	4.18	1370	5.52
180	1.164	580	2.84	980	4.21	1380	5.55
190	1.222	590	2.88	990	4.25	1390	5.59
200	1.280	600	2.91	1000	4.28	1400	5.62
210	1.338	610	2.95	1010	4.31	1410	5.65
220	1.395	620	2.98	1020	4.35	1420	5.69
230	1.451	630	3.01	1030	4.38	1430	5.72
240	1.506	640	3.05	1040	4.42	1440	5.76
250	1.560	650	3.08	1050	4.45	1450	5.79
260	1.613	660	3.12	1060	4.48	1460	5.82
270	1.665	670	3.15	1070	4.52	1470	5.86
280	1.717	680	3.19	1080	4.55	1480	5.90
290	1.767	690	3.22	1090	4.59	1490	5.93
300	1.815	700	3.25	1100	4.62	**1500**	5.97
310	1.863	710	3.29	1110	4.65	1510	6.00
320	1.910	720	3.32	1120	4.69	1520	6.04
330	1.954	730	3.36	1130	4.72	1530	6.07
340	1.994	740	3.39	1140	4.75	1540	6.11

TABLE 7 THERMAL CONDUCTIVITY OF HYDROGEN (continued)

RECOMMENDED VALUES
[Temperature, T, K; Thermal Conductivity, k, mW cm^{-1}K^{-1}]

GAS

T	k	T	k
1550	6.14	1800	7.08
1560	6.18	1810	7.13
1570	6.21	1820	7.17
1580	6.25	1830	7.21
1590	6.28	1840	7.25
1600	6.32	1850	7.29
1610	6.36	1860	7.33
1620	6.40	1870	7.37
1630	6.43	1880	7.42
1640	6.47	1890	7.46
1650	6.51	1900	7.50
1660	6.54	1910	7.55
1670	6.58	1920	7.59
1680	6.62	1930	7.64
1690	6.66	1940	7.68
1700	6.69	1950	7.72
1710	6.73	1960	7.77
1720	6.77	1970	7.82
1730	6.81	1980	7.87
1740	6.85	1990	7.92
1750	6.89	2000	7.96
1760	6.93		
1770	6.97		
1780	7.00		
1790	7.04		

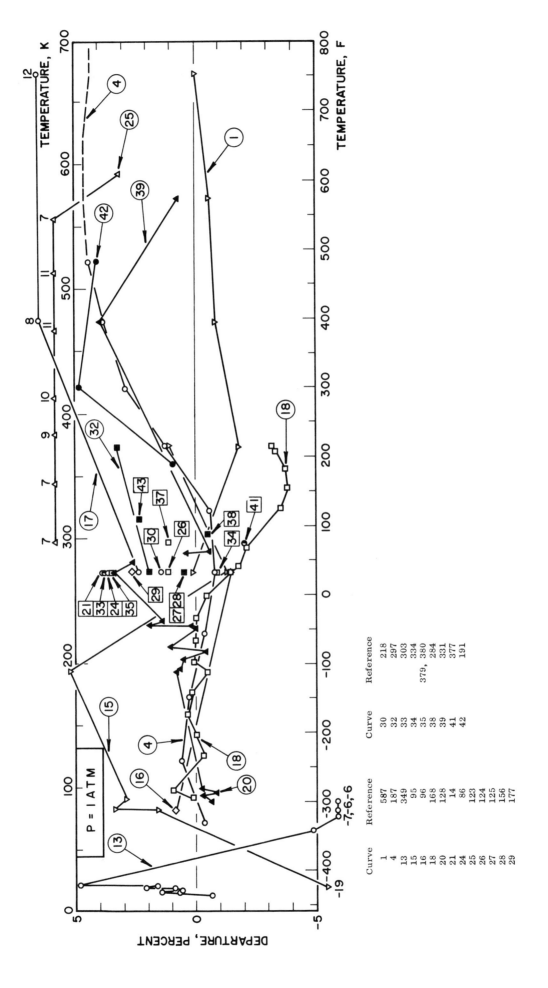

FIGURE 7   DEPARTURE PLOT FOR THERMAL CONDUCTIVITY OF GASEOUS HYDROGEN

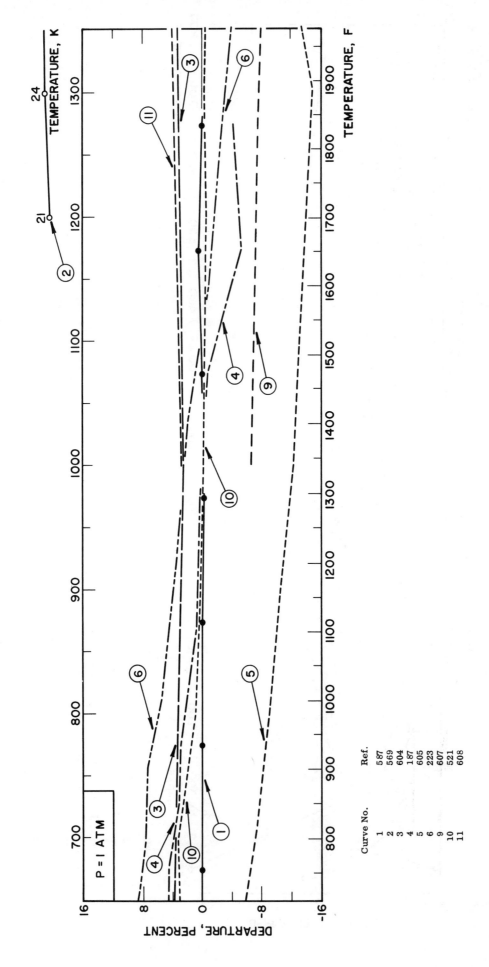

FIGURE 7 DEPARTURE PLOT FOR THE THERMAL CONDUCTIVITY OF GASEOUS HYDROGEN (continued)

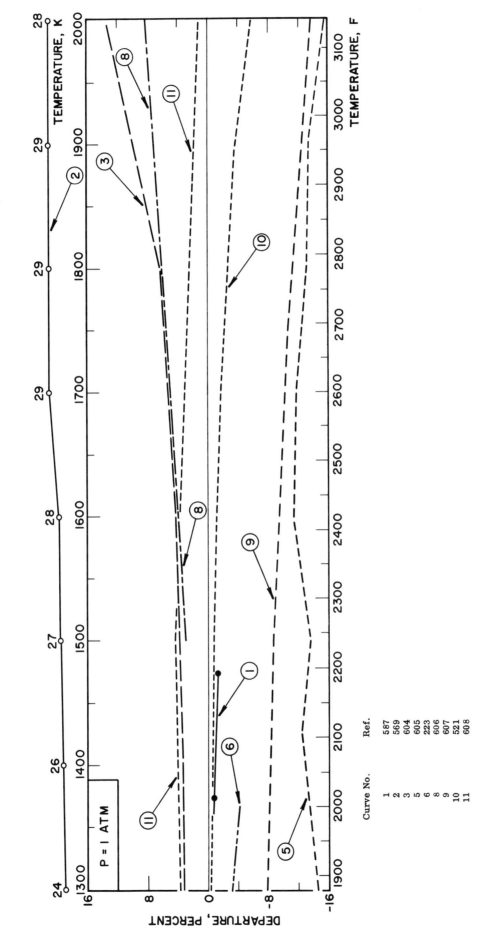

FIGURE 7 DEPARTURE PLOT FOR THE THERMAL CONDUCTIVITY OF GASEOUS HYDROGEN (continued)

TABLE 8 THERMAL CONDUCTIVITY OF KRYPTON

RECOMMENDED VALUES

[Temperature, T, K; Thermal Conductivity, k, mW cm^{-1}K^{-1}]

DISCUSSION

SOLID

The thermal conductivity of solid krypton has been experimentally measured by White and Woods (674) from about 3 to 90 K. Calculations have also been made by Julian (673). The recommended values were obtained from a plot of the theoretical and experimental values cited in these two sources. Where possible, preference was given to the experimental data. Above 90 K, the values were obtained by extrapolation. For temperatures above 8 K the accuracy should be of the order of ten percent. At lower temperatures the accuracy is more difficult to assess and further experimentation is necessary to confirm the correctness of the White and Woods values, both above and below 8 K. As the experimental data were only presented in graphical form no departure plot is given.

SOLID

T	k
1	0.4
1.5	0.8
2	1.3
2.5	2.0
3	2.7
3.5	3.5
4	4.4
4.5	5.4
5	6.5
6	8.9
7	10.7
8	14.4
9	16
10	17
12	16
14	15
16	14
18	13
20	12
25	9.8
30	8.3
35	7.1
40	**6.2**
45	5.6
50	5.1
60	4.3
70	3.8
80	3.4
90	3.1
100	2.8
110	2.6
120	(2.4) ‡

‡ n.m.p. = 117 K

TABLE 8   THERMAL CONDUCTIVITY OF KRYPTON

RECOMMENDED VALUES

[Temperature, T, K; Thermal Conductivity, k, mW cm^{-1}K^{-1}]

SATURATED LIQUID

T	k
115	(0.938)‡
120	0.905
125	0.872
130	0.839
135	0.806
140	0.773
145	0.740
150	0.708
155	0.675
160	0.642
165	0.609
170	0.576
175	0.543
180	0.510
185	0.477
190	0.444
195	0.408
200	0.366
205	0.31
210*	0.21

‡Extrapolated for the supercooled liquid (n.m.p. = 117 K)
*Critical point.

DISCUSSION

SATURATED LIQUID

Data for the thermal conductivity of liquid krypton have been reported by Keyes (192) and by Ikenberry and Rice (672). The former source reports three data points from 123 to 162 K while the latter reports data for temperatures from 126 to 200 K. In both cases, apparently, no values at exactly saturation conditions were obtained and extrapolation to saturation pressure is necessary. In general, this was simple. However, some error could occur with the highest temperature (672) data.

While the accord between the two sets of data is usually good, the highest temperature point of Keyes, for 162 K, is suspect. Keyes has confirmed in private communication that this point is in error. He was unable to supply a revised figure as some of his original notebooks were destroyed by fire. Possibly partial vaporization of the sample could have produced the apparently very low value.

For temperatures below 190 K the majority of the experimental data deviated insignificantly from a straight line relationship and the recommended values were calculated from the equation

k(mw cm^{-1}K^{-1}) = 1.69375 - 6.573·10^{-3} T (T in K)

which fitted the data to about 0.25 percent. However, the accuracy is more probably about two percent. For temperatures above 190 K the critical thermal conductivity estimate of Owens and Thodos (268) was used with a graphical plot to obtain the recommended values. The uncertainty increases rapidly above 190 K and can be assessed at about ten percent to 205 K and possibly twenty percent at the critical point.

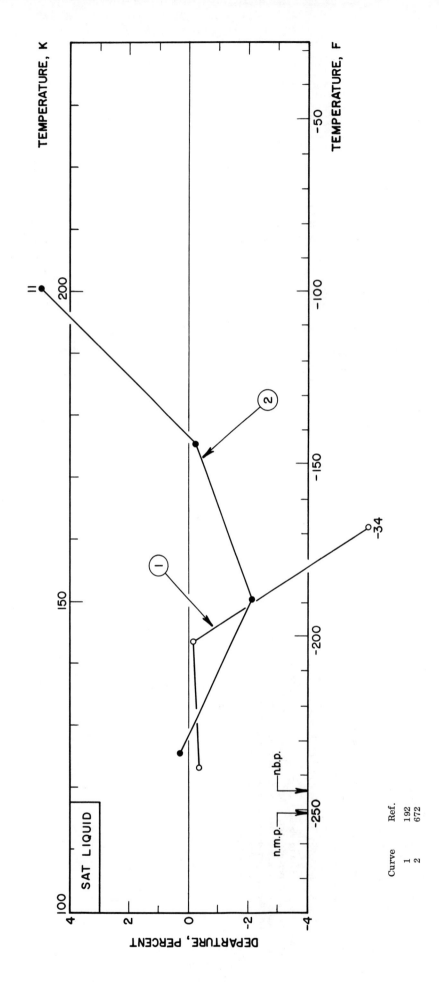

FIG. 8   DEPARTURE PLOT FOR THERMAL CONDUCTIVITY OF LIQUID KRYPTON

DISCUSSION

TABLE 8 THERMAL CONDUCTIVITY OF KRYPTON

RECOMMENDED VALUES

[Temperature, T, K; Thermal Conductivity, k, mW cm^{-1}K^{-1}]

SATURATED VAPOR

No measurements of the thermal conductivity of saturated krypton vapor were located or vapor phase measurements sufficiently close to saturation to enable an estimate of the saturation conditions to be made. The correlation of Owens and Thodos (268) was used and it was found that this gave a value at the normal boiling point some three percent too high. Accordingly, the values so obtained were reduced by percentages which varied linearly with temperature from three percent at the normal boiling point to zero at the critical point. These adjusted values are presented as the recommended values.

In the absence of any experimental values, no departure plot is given. It can be estimated that the probable accuracy of the recommended values is some five percent below 150 K, ten percent at 200 K and possibly twenty percent at the critical point.

SATURATED VAPOR

T	k
120	0.0406
125	0.0429
130	0.0452
135	0.0476
140	0.0501
145	0.0527
150	0.0554
155	0.059
160	0.062
165	0.065
170	0.070
175	0.074
180	0.079
185	0.085
190	0.093
195	0.101
200	0.112
205	0.135
210*	0.21

*Critical point

TABLE 8   THERMAL CONDUCTIVITY OF KRYPTON

RECOMMENDED VALUES

[Temperature, T, K; Thermal Conductivity, k, mW cm^{-1}K^{-1}]

DISCUSSION

GAS

Thirteen experimental measurements were found in the temperature range from 131 to 579 K. Successive fittings of polynomials of orders 2(1)7 showed that little advantage was secured by considering any higher than a second order. Accordingly, the equation

10^5k (cgsu) = 9.06926.10^{-3} + 8.33613.10^{-3}T - 2.8711δ.10^{-6}T^2 (T in K)

was used to compute the recommended values, fitting the original data to ±0.9 percent. Three points, due to Curie and Lepape (81), Waldmann (376), and the Landolt-Bornstein Tables (214), all for 273.2 K all disagree by more than one percent.

Further experimental measurements below 195 K and above 273 K are to be desired in order to verify the single sets of data presently existing for these temperature regions. Pending such studies the recommended values should be accurate to two percent below 195 K, one percent or better between 195 and 273 K and up to five percent at the highest temperature tabulated.

GAS

T	k	T	k
100	(0.034)*	400	0.1207
110	(0.037)*	410	0.1232
120	0.0405	420	0.1257
130	0.0437	430	0.1281
140	0.0469	440	0.1306
150	0.0501	450	0.1330
160	0.0533	460	0.1354
170	0.0562	470	0.1378
180	0.0593	480	0.1401
190	0.0623	490	0.1424
200	0.0653	500	0.1447
210	0.0683	510	0.147
220	0.0713	520	0.149
230	0.0742	530	0.151
240	0.0772	540	0.154
250	0.0801	550	0.156
260	0.0829	560	0.158
270	0.0857	570	0.160
280	0.0886	580	0.162
290	0.0914	590	0.164
300	0.0942	600	0.166
310	0.0970	610	0.168
320	0.0997	620	0.170
330	0.1024	630	0.172
340	0.1051	640	0.174
350	0.1077	650	0.176
360	0.1104	660	0.178
370	0.1130	670	0.180
380	0.1156	680	0.182
390	0.1181	690	0.184
		700	0.186

*Extrapolated for the gas phase ignoring pressure dependence. (m. b. p. = 120 K)

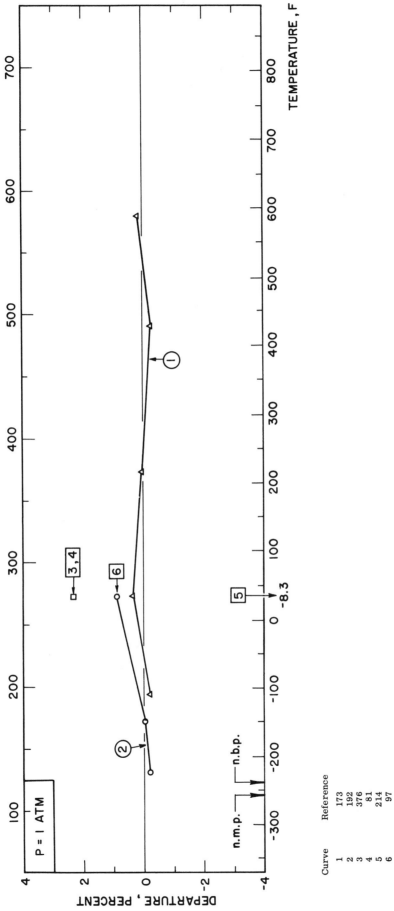

FIGURE 8 DEPARTURE PLOT FOR THERMAL CONDUCTIVITY OF GASEOUS KRYPTON

TABLE 9 THERMAL CONDUCTIVITY OF NEON

DISCUSSION

SOLID

Little information was found to be available on the thermal conductivity of solid neon. A graph of the experimental data was given by White and Woods (674) while calculations were made by Julian (673). The recommended values were derived from a smooth curve drawn through the White and Woods values and, from the scatter in those values, should be accurate to within ten percent for temperatures above 2 K. Further confirmatory measurements are highly desirable to check the accuracy of the White and Woods values. As no tabulation of the original data is given no departure plot appears.

RECOMMENDED VALUES

[Temperature, T, K; Thermal Conductivity, k, mW cm^{-1} K^{-1}]

SOLID

T	k
1.0	7.3
1.5	18.5
2.0	29.5
2.5	39.8
3.0	45.7
3.5	47.1
4.0	44.0
4.5	39.3
5	33.6
6	24.5
7	17.0
8	13.0
9	10.2
10	8.4
12	6.0
14	4.5
16	3.7
18	3.1
20	2.7
22	2.3
24	2.1
26	2.0 ‡

‡ n.m.p. = 25 K

TABLE 9   THERMAL CONDUCTIVITY OF NEON

RECOMMENDED VALUES

[Temperature, T, K; Thermal Conductivity, k, mW cm^{-1}K^{-1}]

DISCUSSION

SATURATED LIQUID

Only one set of experimental data were located for the thermal conductivity of saturated liquid neon, the measurements of Lochtermann (683). These were compared with the correlation of Owens and Thodos (268). Severe disagreement is evident. The correlation predicted values are at least fifteen percent greater than the experimental data. In addition, the trend of the experimental data with temperature would predict a critical point of about 32 K, considerably below the accepted values.

The procedure here adapted was to retain the correlation values for temperatures above 35 K and to fair these into the experimental values below about 27.5 K by drawing a smooth curve. While it is possible that the correlation values above 35 K are in error, no experimental evidence is available. New measurements are urgently required to resolve the anomalous trend in the only set of experimental data presently available and to confirm the correlation for higher temperatures. The uncertainty in the recommended values is probably about twenty percent below 35 K and can be as much as forty percent at the critical point.

SATURATED LIQUID

T	k
25	1.168
26	1.151
27	1.134
28	1.117
29	1.100
30	1.082
31	1.063
32	1.042
33	1.018
34	0.991
35	0.960
36	0.924
37	0.883
38	0.837
39	0.785
40	0.73
41	0.67
42	0.61
43	0.54
44	0.46
44.5*	0.33

*Critical Point

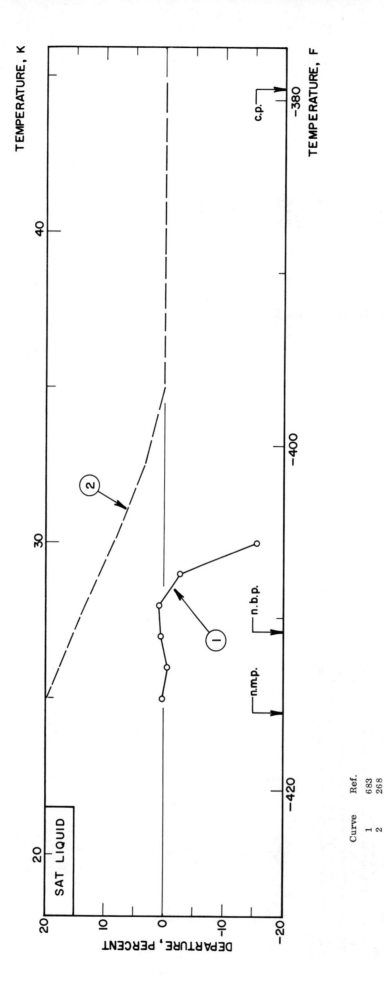

FIGURE 9 DEPARTURE CURVE FOR THERMAL CONDUCTIVITY OF SATURATED LIQUID NEON

TABLE 9   THERMAL CONDUCTIVITY OF NEON

RECOMMENDED VALUES

[Temperature, T, K; Thermal Conductivity, k, mW cm^{-1}K^{-1}]

DISCUSSION

SATURATED VAPOR

No experimental data were located for the thermal conductivity of saturated neon vapor. The correlation of Owens and Thodos (268) was used and it was found that the value at the normal boiling point was eight percent lower than obtained from the atmospheric pressure thermal conductivity correlation. The values were therefore adjusted by amounts varying linearly with temperature, from eight percent at the boiling point to zero at the critical point.

In view of the complete lack of experimental data, no departure plot appears. An estimate of accuracy is difficult as possibly quantum effects could become important at the lower temperatures and the correlation has neglected these. Tabular uncertainties of ten percent below 30 K, twenty percent at 40 K and even forty percent at the critical point are probable. Experimentation to confirm the recommended values and error estimates is urgently required.

SATURATED VAPOR

T	k
27	0.078
28	0.082
29	0.085
30	0.089
31	0.093
32	0.097
33	0.102
34	0.107
35	0.112
36	0.118
37	0.124
38	0.131
39	0.138
40	0.147
41	0.158
42	0.172
43	0.192
44	0.237
44.5*	0.33

*Critical point

TABLE 9 THERMAL CONDUCTIVITY OF NEON

DISCUSSION

GAS

Major determinations of the thermal conductivity of gaseous neon have been reported by Kannuluik and Carman (173) from 90 to 579 K, by Keyes (191) from 91 to 273 K, by Zaitseva (656) from 413 to 800 K, and indirectly by Collins and Menard (657) from about 500 to 5000° K. Other determinations, mainly at single temperatures, have been reported (33, 81, 83, 97, 173, 326, 336, 378-381), while some of these values and other values reproduced without source references are cited in (14, 60, 325, 376). In addition, sets of calculated data from 60 to 200 K (11), 100(100)5000 K (521), 89 to 598 K (223) and to 15000 K (12) have appeared as well as numerous Russian correlations (601-3, 616, 653).

In this analysis the data of Kannuluik and Carman (173) were found to be represented, to within 2.5 percent, by the equation

$$10^5 k(\text{cgsu}) = 0.49159 + 5.47196 \cdot 10^{-2} T - 7.19790 \cdot 10^{-5} T^2 + 5.06172 \cdot 10^{-8} T^3 \quad (T \text{ in K})$$

which was used to generate the recommended values to 500 K. Values for temperatures from 500 to 1000 K were obtained from a smooth curve drawn to pass through the experimental value of Collins and Menard at 1000 K. Values above 1000 K were calculated using an equation given by these authors which closely fitted their experimental data.

Some serious discrepancies occur between the different sets of data. In common with experience for other gases, the Russian high temperature data (656) and correlations (601-3, 616, 653) seem to be significantly higher than other data or values and they were neglected in this analysis. Also, the calculations of Amdur and Mason (12) and of Svehla (521) yield higher values than the data of Collins and Menard. As, usually, the thermal conductivity values of Svehla have been found to be high for other gases, the choice of the Collins and Menard data should be more nearly accurate. The departure plot also shows some values calculated using the Lennard-Jones potential function (with $\epsilon/k = 35.7$ K and $\sigma = 2.789$ Å).

Based upon a consideration of these, and other, factors, the recommended values should be accurate to within two percent for temperatures below 500 K. This uncertainty is considered to increase to about ten percent from about 1000 to 3000 K and can be as much as twenty five percent at 5000 K.

RECOMMENDED VALUES

[Temperature, T, K; Thermal Conductivity, k, mW cm^{-1} K^{-1}]

GAS

T	k	T	k	T	k	T	k
25	0.076*	400	0.590	800	0.914	2000	1.625
30	0.086	410	0.599	810	0.920	2100	1.68
35	0.097	420	0.608	820	0.926	2200	1.73
40	0.107	430	0.616	830	0.932	2300	1.78
45	0.117	440	0.625	840	0.938	2400	1.83
50	0.128	450	0.634	850	0.945	2500	1.88
60	0.148	460	0.643	860	0.951	2600	1.93
70	0.168	470	0.651	870	0.957	2700	1.98
80	0.186	480	0.660	880	0.964	2800	2.02
90	0.204	490	0.669	890	0.971	2900	2.07
100	0.222	500	0.678	900	0.978	3000	2.12
110	0.239	510	0.687	910	0.985	3100	2.16
120	0.256	520	0.696	920	0.992	3200	2.20
130	0.272	530	0.705	930	1.001	3300	2.25
140	0.288	540	0.714	940	1.006	3400	2.29
150	0.303	550	0.723	950	1.012	3500	2.33
160	0.318	560	0.732	960	1.019	3600	2.38
170	0.333	570	0.741	970	1.026	3700	2.42
180	0.347	580	0.750	980	1.033	3800	2.47
190	0.361	590	0.759	990	1.039	3900	2.51
200	0.375	600	0.768	1000	1.046	4000	2.55
210	0.388	610	0.777	1050	1.081	4100	2.59
220	0.401	620	0.786	1100	1.115	4200	2.63
230	0.414	630	0.795	1150	1.148	4300	2.67
240	0.426	640	0.803	1200	1.181	4400	2.71
250	0.438	650	0.811	1250	1.213	4500	2.75
260	0.449	660	0.818	1300	1.244	4600	2.78
270	0.461	670	0.826	1350	1.274	4700	2.82
280	0.472	680	0.832	1400	1.303	4800	2.86
290	0.483	690	0.838	1450	1.332	4900	2.89
300	0.493	700	0.844	1500	1.360	5000	2.93
310	0.504	710	0.851	1550	1.389		
320	0.514	720	0.858	1600	1.418		
330	0.524	730	0.865	1650	1.446		
340	0.534	740	0.872	1700	1.474		
350	0.544	750	0.879	1750	1.500		
360	0.553	760	0.886	1800	1.526		
370	0.563	770	0.893	1850	1.551		
380	0.572	780	0.900	1900	1.576		
390	0.581	790	0.907	1950	1.600		

*n. b. p. = 27 K

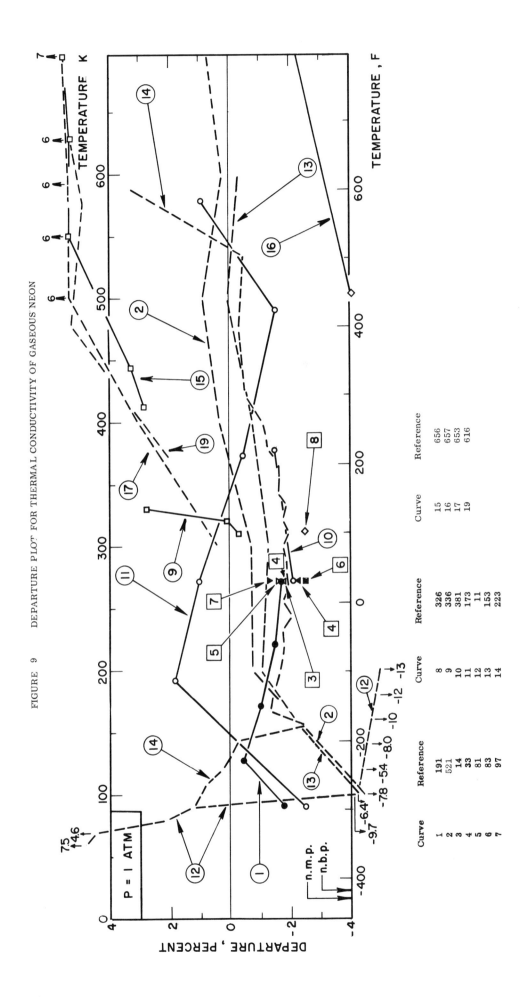

FIGURE 9 DEPARTURE PLOT FOR THERMAL CONDUCTIVITY OF GASEOUS NEON

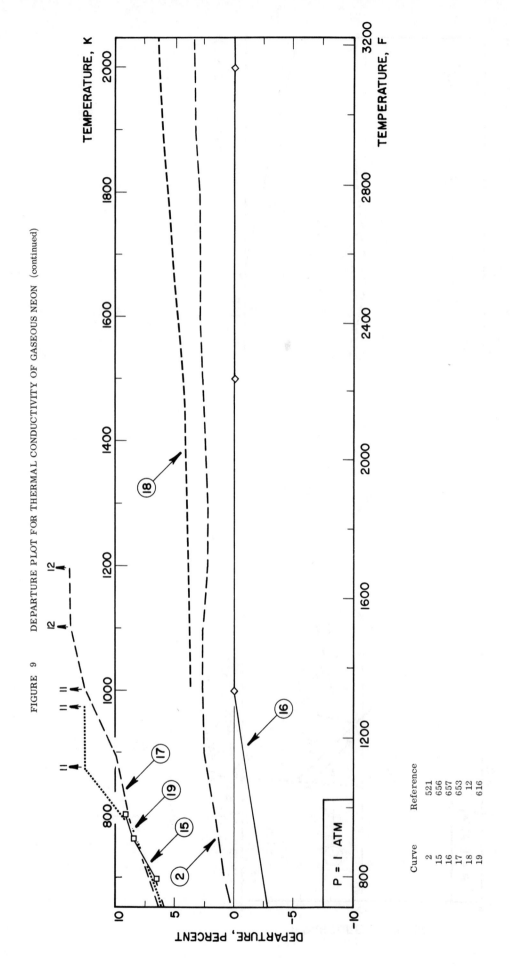

FIGURE 9 DEPARTURE PLOT FOR THERMAL CONDUCTIVITY OF GASEOUS NEON (continued)

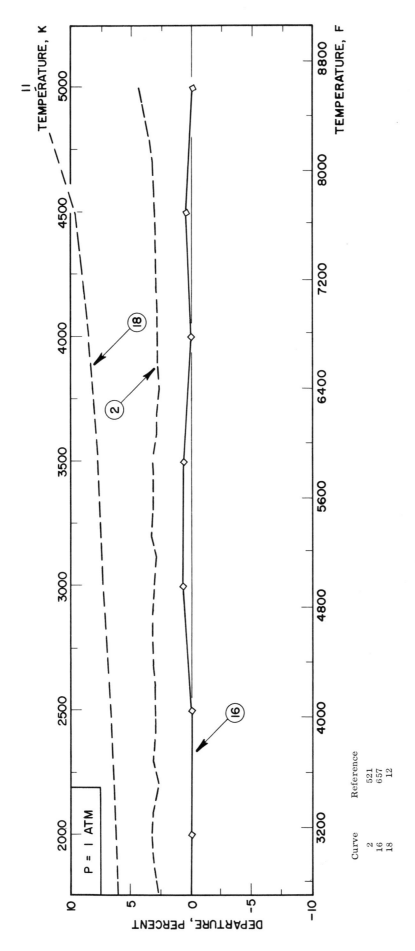

FIGURE 9 DEPARTURE PLOT FOR THERMAL CONDUCTIVITY OF GASEOUS NEON (continued)

Curve	Reference
2	521
16	657
18	12

TABLE 10   THERMAL CONDUCTIVITY OF NITROGEN

RECOMMENDED VALUES

[Temperature, T, K; Thermal Conductivity, k, mW cm^{-1}K^{-1}]

DISCUSSION

SOLID

The only experimental data located for the thermal conductivity of solid nitrogen were values of Roder (677) between 4 and 28 K. Five measurements around 4 K, one at 14 K and fifteen from 20 to 28 K were reported.

To compare these data with theory, some experimental evidence for the temperature at which the thermal conductivity reaches a maximum is required. This could not be obtained from the only source of data (677). Study of the papers of Julian (673) and Keyes (678) reveals no predicted maximum (in disagreement with experiment) while the White and Woods (674) paper only considers the theory at sufficiently high temperatures where the thermal conductivity could be considered to vary inversely with absolute temperature. The experimental evidence was insufficient to determine if the maximum occurred below or above 4 K.

The recommended values were obtained from a double logarithmic plot of thermal conductivity versus temperature and must be regarded as tentative below 12 and above 30 K. Within these temperature limits an uncertainty of ten percent appears probable.

SOLID

T	k
4	56
5	45
6	37
7	30
8	25
9	20
10	17
11	14
12	12
13	10
14	9.2
15	7.6
16	6.5
17	5.6
18	4.9
19	4.5
20	4.0
21	3.8
22	3.6
23	3.5
24	3.3
25	3.2
26	3.1
27	3.0
28	2.9
29	2.8
30	2.7
35	2.3
40	2.0
45	1.8
50	1.6

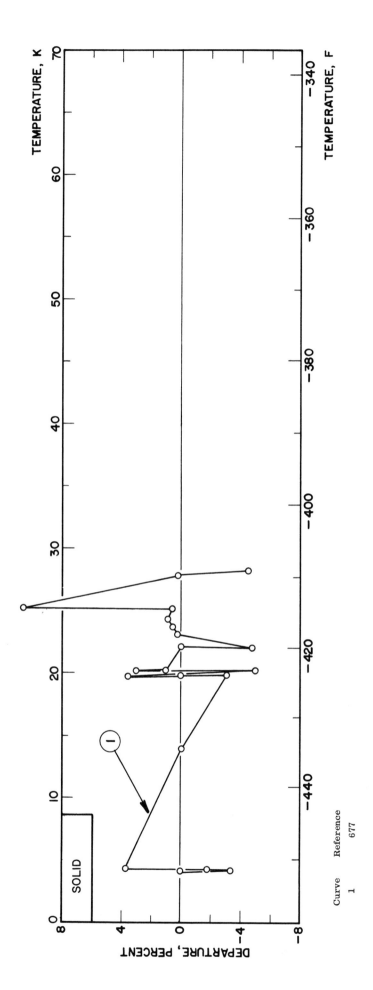

FIGURE 10 DEPARTURE PLOT FOR THERMAL CONDUCTIVITY OF SOLID NITROGEN

TABLE 10 THERMAL CONDUCTIVITY OF NITROGEN

RECOMMENDED VALUES

[Temperature, T, K; Thermal Conductivity, k, mW cm^{-1}K^{-1}]

SATURATED LIQUID

T	k
60	(1.69)†
65	1.60
70	1.51
75	1.413
80	1.322‡
85	1.231‡
90	1.142‡
95	1.053‡
100	0.966‡
105	0.880‡
110	0.795‡
115	0.710‡
120	0.628‡
125	0.520‡
126.25	0.37 *

†Extrapolated for the supercooled liquid (n.m.p. = 63K)
‡Under saturated vapor pressures (n.b.p. = 78K)
*Critical Temperature

DISCUSSION

SATURATED LIQUID

There exist eight available experimental works on the thermal conductivity of liquid nitrogen. Extensive measurements of both Uhlir (353) and Ziebland - Burton (57, 413) were considered reliable from the standpoint of the experimental method and procedure. As they did not give the values for the saturated liquid, graphical extrapolation was used to obtain the values at the saturated vapor pressures. All of the values thus obtained are given equal weight. Another set of recommended values reported by Powers et al. (276, 531) was also partly used in this analysis. On the other hand, two sets of data reported by Borovik (42, 46) deviate considerably, and the values of Hammann (139) and Prosad (535) are too high. Therefore, no weight was given to these sets of data.

The correlation formula was determined from the reliable values described above, excluding those at the critical point considered to be less reliable. The correlation formula was given by

$10^6 k$ (cgsu) = 695.957 - 5.15493 T + 0.00504635 T^2 (T in K)

and should be valid between 60 and 123 K. It was found that this equation fits the above-enumerated values with a mean deviation of 0.8 percent and a maximum of 2.2 percent. The recommended values up to 120 K were calculated from the above equation.

Above 120 K, the recommended values were obtained from a large-scale plot of all the available information. The principal uncertainty introduced is in the thermal conductivity at the critical point. At least a ten percent uncertainty exists in this value. On the departure plot only a part of the results of Hammann (curve 3) is plotted for the sake of clarity. The recommended values are considered accurate to a few percent below 120 K, the uncertainty reaching five percent at 125 K and at least ten percent at the critical temperature.

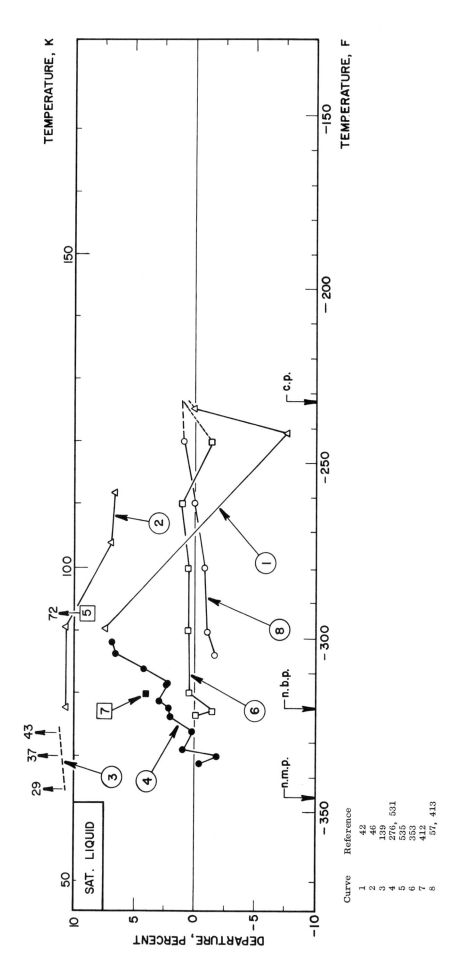

FIGURE 10 DEPARTURE PLOT FOR THERMAL CONDUCTIVITY OF LIQUID NITROGEN

Curve	Reference
1	42
2	46
3	139
4	276, 531
5	535
6	353
7	412
8	57, 413

67

TABLE 10   THERMAL CONDUCTIVITY OF NITROGEN

RECOMMENDED VALUES

[Temperature, T, K; Thermal Conductivity, k, mW cm^{-1}K^{-1}]

SATURATED VAPOR

DISCUSSION

SATURATED VAPOR

No data were located for the thermal conductivity of saturated nitrogen vapor. Various correlations were plotted on a large scale. Between the normal boiling and critical points, that of Schaefer and Thodos (575) appeared lower than those of Petrozzi (670) and Johnson (167). In addition, the Aerojet General values exhibited a somewhat anomalous variation with temperature below 110 K. The values predicted by the two correlations (167) and (575) at the normal boiling were respectively slightly above and below the atmospheric pressure value considered most probable for the gas. The average of these two correlations was selected as the most probable value for temperatures below 125 K. This procedure also resulted in good agreement with a value extrapolated from an isotherm of Keyes (192) at 92 K. The recommended values are thought to have an average error of a few percent for temperatures up to 90 K, the uncertainty then increasing to about ten percent for temperatures from 100 K to the critical point.

SATURATED VAPOR

T	k
60	0.056
65	0.061
70	0.066
75	0.071
80	0.077
85	0.084
90	0.091
95	0.100
100	0.111
105	0.123
110	0.138
115	0.160
120	0.195
125	0.265
126.25	0.37 *

*Critical Temperature

GAS

TABLE 10 THERMAL CONDUCTIVITY OF NITROGEN

RECOMMENDED VALUES

[Temperature, T, K; Thermal Conductivity, k, mW cm^{-1}K^{-1}]

GAS

T	k	T	k	T	k.	T	k
50	(0.0485)*	450	0.3564	850	0.564	2500	1.406
60	(0.0578)*	460	0.3626	860	0.569	2600	1.449
70	(0.0670)*	470	0.3688	870	0.574	2700	1.494
80	0.0762	480	0.3749	880	0.578	2800	1.542
90	0.0852	490	0.3808	890	0.583	2900	1.590
100	0.0941	500	0.3864	900	0.587	3000	1.640
110	0.1030	510	0.392	910	0.592	3100	1.691
120	0.1119	520	0.398	920	0.596	3200	1.743
130	0.1208	530	0.403	930	0.600	3300	1.795
140	0.1296	540	0.408	940	0.605	3400	1.853
150	0.1385	550	0.414	950	0.609	3500	1.915
160	0.1474	560	0.420	960	0.613		
170	0.1562	570	0.425	970	0.618		
180	0.1651	580	0.431	980	0.622		
190	0.1739	590	0.436	990	0.626		
200	0.1826	600	0.441	1000	0.631		
210	0.1908	610	0.446	1050	0.651		
220	0.1989	620	0.452	1100	0.672		
230	0.2067	630	0.457	1150	0.693		
240	0.2145	640	0.462	1200	0.713		
250	0.2222	650	0.467	1250	0.733		
260	0.2298	660	0.472	1300	0.754		
270	0.2374	670	0.478	1350	0.775		
280	0.2449	680	0.483	1400	0.797		
290	0.2524	690	0.488	1450	0.819		
300	0.2598	700	0.493	1500	0.842		
310	0.2671	710	0.498	1550	0.867		
320	0.2741	720	0.503	1600	0.893		
330	0.2808	730	0.508	1650	0.921		
340	0.2874	740	0.513	1700	0.950		
350	0.2939	750	0.517	1750	0.981		
360	0.3002	760	0.522	1800	1.013		
370	0.3065	770	0.526	1850	1.046		
380	0.3127	780	0.531	1900	1.080		
390	0.3189	790	0.536	1950	1.113		
400	0.3252	800	0.541	2000	1.146		
410	0.3314	810	0.546	2100	1.207		
420	0.3376	820	0.551	2200	1.263		
430	0.3438	830	0.555	2300	1.314		
440	0.3501	840	0.559	2400	1.361		

*Extrapolated for the gas phase ignoring pressure dependence. (n.b.p. = 78 K)

DISCUSSION

Many experimental, theoretical and correlated sets of values are available for the thermal conductivity of gaseous nitrogen. In view of this fact, it is surprising that the departure plots show the degree of disagreement between these different values to be larger than would be expected.

As will be observed from the departure plots, the recommended values, obtained by drawing a smooth curve through the experimental data, are somewhat lower than most previous correlations for temperatures between about 250 and 700 K and, for the Keyes (187) and NBS (146) correlations, for higher temperatures. It seems that the more recent measurements justify this change.

While measurements up to about 1200 K appear in reasonable agreement, for higher temperatures the trend of the experimental and theoretical values differs. The recommended values were selected to occur midway between the experimental values at 1200 K and to approach the theoretical estimates at about 2500 K. Theoretical estimates for temperatures above about 3500 K differ according to whether consideration is given to the influence of dissociation on the thermal conductivity. Even supposedly similar calculations differ increasingly at higher temperatures. Due to this reason, the tabulation of recommended values was only undertaken for temperatures to 3500 K, at which temperature the reaction contribution of some two percent is less than the uncertainty in the recommended values. The recommended values can thus be considered as applying to both the equilibrium and the frozen gas.

Further experiments are to be desired for the entire temperature range if accuracy better than two percent is desired. More accurate calculations are also required, possibly for temperatures from 1000 to 4000 K and certainly for higher temperatures. The accuracy of the recommended values can be assessed as two percent for temperatures below about 350 K, five percent for temperatures from 350 to 1200 K and ten percent above 1200 K.

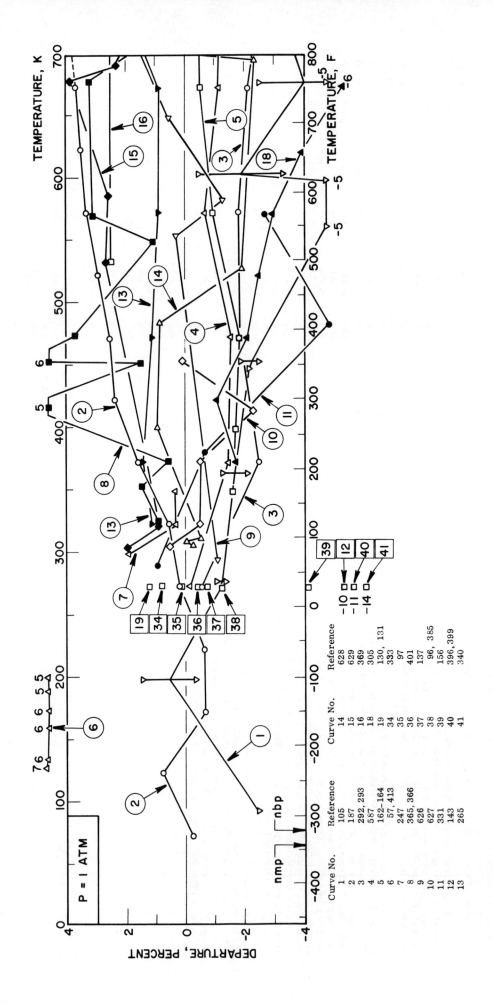

FIGURE 10   DEPARTURE PLOT FOR THERMAL CONDUCTIVITY OF GASEOUS NITROGEN

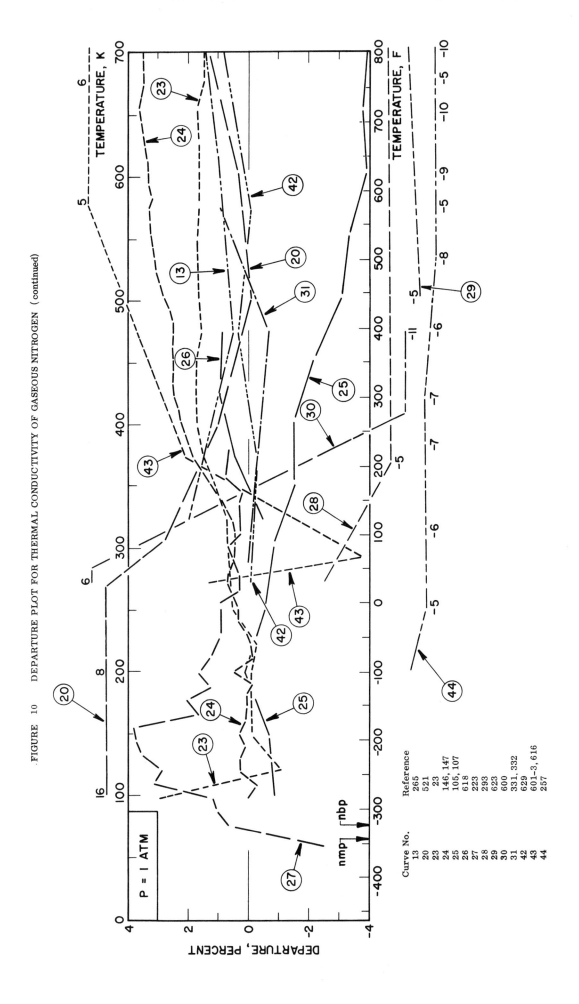

FIGURE 10   DEPARTURE PLOT FOR THERMAL CONDUCTIVITY OF GASEOUS NITROGEN (continued)

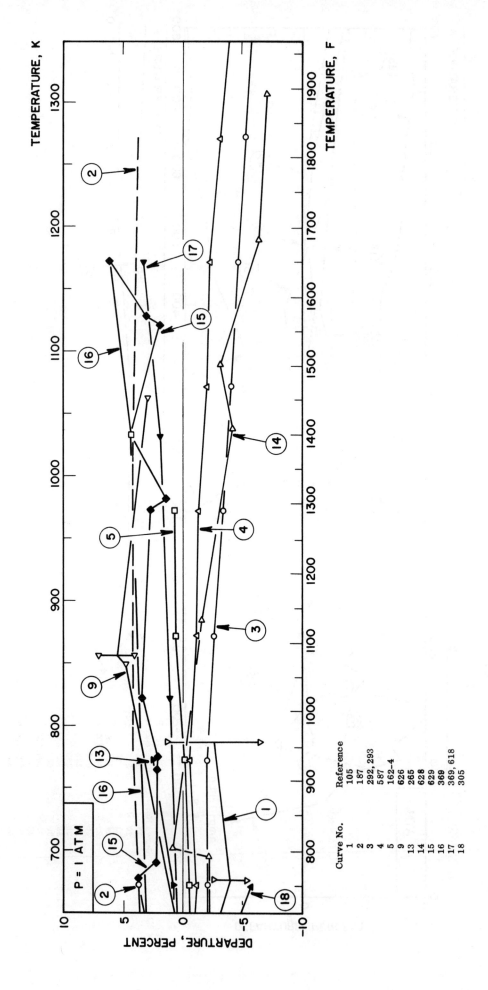

FIGURE 10   DEPARTURE PLOT FOR THERMAL CONDUCTIVITY OF GASEOUS NITROGEN (continued)

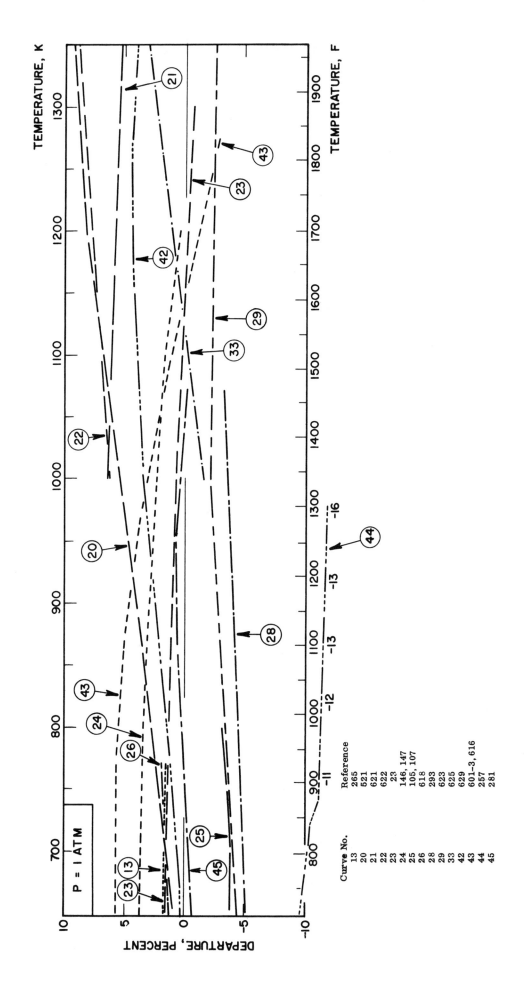

FIGURE 10 DEPARTURE PLOT FOR THERMAL CONDUCTIVITY OF GASEOUS NITROGEN (continued)

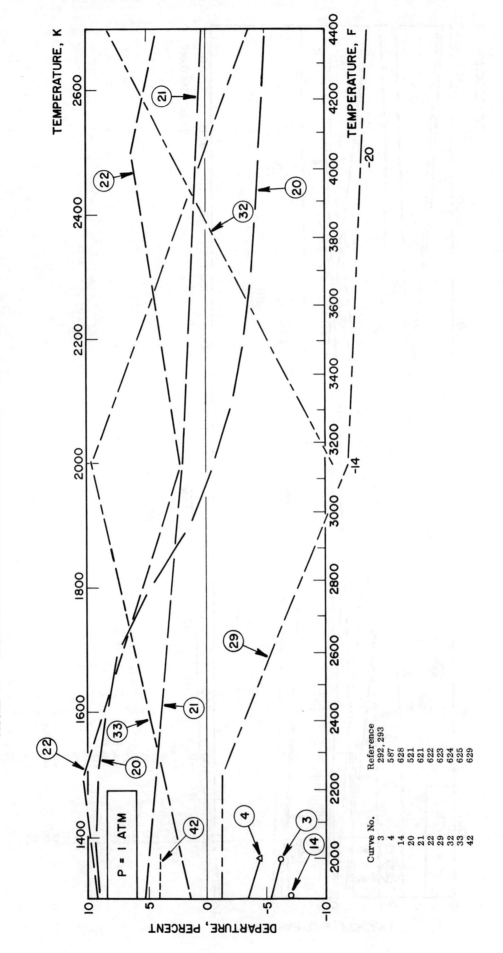

FIGURE 10   DEPARTURE PLOT FOR THERMAL CONDUCTIVITY OF GASEOUS NITROGEN (continued)

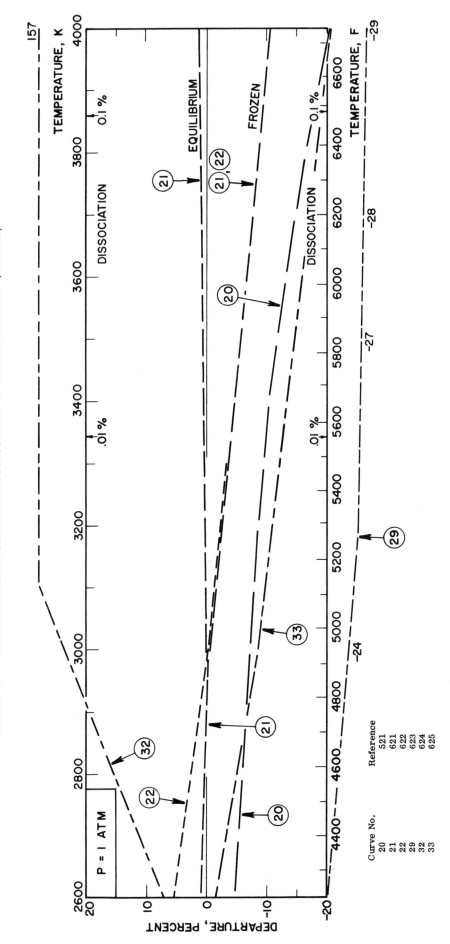

FIGURE 10 DEPARTURE PLOT FOR THERMAL CONDUCTIVITY OF GASEOUS NITROGEN (continued)

TABLE 11 THERMAL CONDUCTIVITY OF OXYGEN

RECOMMENDED VALUES

[Temperature, T, K; Thermal Conductivity, k, mW cm^{-1}K^{-1}]

DISCUSSION

SATURATED LIQUID

Six experimental investigations were located in the literature on the thermal conductivity of liquid oxygen. The extensive measurements of both Ziebland - Burton (56, 412) and Tsederberg - Timrot (356, 539) were considered to be reliable from the standpoint of the experimental method and procedure. Values of the thermal conductivity for the saturated liquid, read from their diagrams, were used and given equal weight in this analysis. The data of Keyes (192), obtained near the saturated vapor pressures, were also used for the estimation of the most probable values. On the other hand, three investigations reported by Hammann (139), Prosad (534) and Waterman (541) give very high values. Therefore, no weight was given to these three sets of data.

The correlation formula obtained for the saturated liquid is given by

$$10^6 k \text{ (cgsu)} = 568.807 - 1.66779\, T - 0.00740052\, T^2. \quad (T \text{ in K})$$

In deriving this formula, values near the critical point were excluded, because the thermal conductivity of the liquid decreases at an extremely rapid rate near the critical point and the experimental accuracy also decreases. Therefore, the equation should be valid in the temperature range from 50 to 140 K and is found to fit the above-enumerated values with a mean deviation of 0.95 percent and a maximum of 2.7 percent. The recommended values up to 140 K were calculated from this equation. Above 140 K they were obtained from a large-scale plot of the available information.

In the departure plot, only a part of the results of Prosad (curve 3) is plotted to aid clarity. The recommended values up to 150 K should be accurate to within about two percent. At the critical point itself an uncertainty of up to fifteen percent is possible.

SATURATED LIQUID

T	k
50	(1.95)†
55	1.90
60	1.85
65	1.80
70	1.74
75	1.682
80	1.623
85	1.563
90	1.501
95	1.437‡
100	1.372‡
105	1.306‡
110	1.237‡
115	1.168‡
120	1.096‡
125	1.023‡
130	0.949‡
135	0.873‡
140	0.796‡
145	0.712‡
150	0.610‡
154.7*	0.41

†Extrapolated for the supercooled liquid. (n.m.p. = 55 K)
‡Under saturated vapor pressures. (n.b.p. = 90 K)
*Critical point.

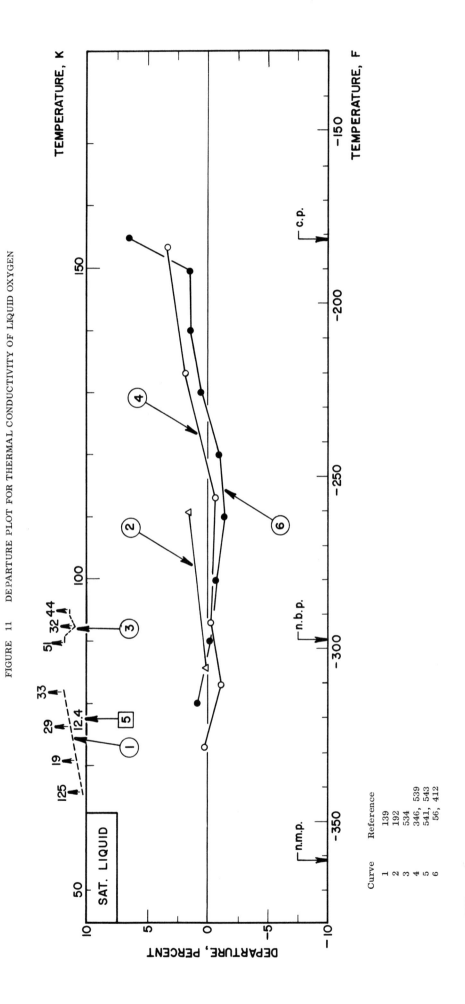

FIGURE 11   DEPARTURE PLOT FOR THERMAL CONDUCTIVITY OF LIQUID OXYGEN

TABLE 11   THERMAL CONDUCTIVITY OF OXYGEN

RECOMMENDED VALUES

[Temperature, T, K; Thermal Conductivity, k, mW cm^{-1}K^{-1}]

SATURATED VAPOR

T	k
90	0.081
95	0.087
100	0.093
105	0.100
110	0.108
115	0.116
120	0.124
125	0.135
130	0.15
135	0.16
140	0.18
145	0.21
150	0.25
154.7*	0.41

* Critical point

DISCUSSION

SATURATED VAPOR

No experimental data were located. Various correlations were plotted on a large scale. Between the normal boiling and critical points, that of Schaefer and Thodos (575) appeared lower than those of Petrozzi (670) and Air Products (679), the latter evidently being based on a monatomic gas correlation (268). The (575) correlation agreed well with the values deduced from the atmospheric pressure correlation at the normal boiling point. Experimental measurements are highly desirable to confirm the various correlations. The recommended values were obtained from the (575) correlation and are thought to have an uncertainty of a few percent below 100 K, the uncertainty then increasing to about ten percent at 125 K and as much as fifteen percent at the critical point.

TABLE 11 THERMAL CONDUCTIVITY OF OXYGEN

DISCUSSION

GAS

Experimental data for the thermal conductivity of gaseous oxygen extend from about 80 K to 1380 K. Above 787 K only the measurements of Geier and Schafer (587) are available.

As shown by the departure plots, most previous experimental and correlated values agree to within one percent between about 80 and 330 K and the recommended values, derived from a smooth curve drawn through the experimental data plotted as a function of temperature, should be accurate to one half percent in this temperature range. In the range 330 to 787 K only two sets of measurements (105, 587) are available for temperatures above 373 K. It was decided to base the recommended values for the higher temperatures upon the Geier and Schafer (587) data as the results of these workers have been found to be accurate for other gases and somewhat superior to those of Franck (105). The departure plot for temperatures from 700 to 1400 K shows that the Geier and Schafer data also fall nicely between previous estimates of the high temperature values.

The accuracy of the recommended values can thus be assessed as within one half percent from 80 to 330 K, two percent from 330 to 600 K, four percent from 600 to 900 K and probably within six percent from 900 to 1500 K. Further experimental measurements are to be desired below the normal boiling point and above 373 K.

RECOMMENDED VALUES

[Temperature, T, K; Thermal Conductivity, k, mW cm^{-1}K^{-1}]

GAS

T	k	T	k	T	k
50	(0.045)*	400	0.3420	750	0.574
60	(0.054)*	410	0.3490	760	0.579
70	(0.063)*	420	0.356	770	0.585
80	(0.072)*	430	0.363	780	0.591
90	0.0813	440	0.370	790	0.597
100	0.0905	450	0.377	800	0.603
110	0.0998	460	0.384	810	0.609
120	0.1092	470	0.391	820	0.615
130	0.1187	480	0.398	830	0.620
140	0.1281	490	0.405	840	0.626
150	0.1376	500	0.412	850	0.632
160	0.1466	510	0.419	860	0.638
170	0.1556	520	0.426	870	0.644
180	0.1646	530	0.433	880	0.650
190	0.1735	540	0.440	890	0.655
200	0.1824	550	0.447	900	0.661
210	0.1911	560	0.453	910	0.667
220	0.1997	570	0.460	920	0.672
230	0.2083	580	0.467	930	0.678
240	0.2168	590	0.474	940	0.684
250	0.2254	600	0.480	950	0.689
260	0.2339	610	0.487	960	0.695
270	0.2424	620	0.493	970	0.701
280	0.2509	630	0.500	980	0.706
290	0.2592	640	0.506	990	0.712
300	0.2674	650	0.513	1000	0.717
310	0.2753	660	0.519	1010	0.723
320	0.2831	670	0.525	1020	0.728
330	0.2907	680	0.532	1030	0.734
340	0.2982	690	0.538	1040	0.739
350	0.3056	700	0.544	1050	0.745
360	0.3130	710	0.550	1060	0.750
370	0.3204	720	0.556	1070	0.755
380	0.3276	730	0.562	1080	0.760
390	0.3348	740	0.568	1090	0.765

*Extrapolated for the gas phase ignoring pressure dependence (n. b. p. = 90 K)

TABLE 11   THERMAL CONDUCTIVITY OF OXYGEN (continued)

RECOMMENDED VALUES

[Temperature, T, K; Thermal Conductivity, k, mW cm^{-1}K^{-1}]

GAS

T	k	T	k	T	k
1100	0.771	1250	0.846	1400	0.921
1110	0.776	1260	0.851	1410	0.926
1120	0.781	1270	0.856	1420	0.931
1130	0.786	1280	0.861	1430	0.936
1140	0.791	1290	0.866	1440	0.941
1150	0.796	1300	0.871	1450	0.946
1160	0.801	1310	0.876	1460	0.951
1170	0.806	1320	0.881	1470	0.956
1180	0.811	1330	0.886	1480	0.960
1190	0.816	1340	0.891	1490	0.965
1200	0.821	1350	0.896	1500	0.970
1210	0.826	1360	0.901		
1220	0.831	1370	0.906		
1230	0.836	1380	0.911		
1240	0.841	1390	0.916		

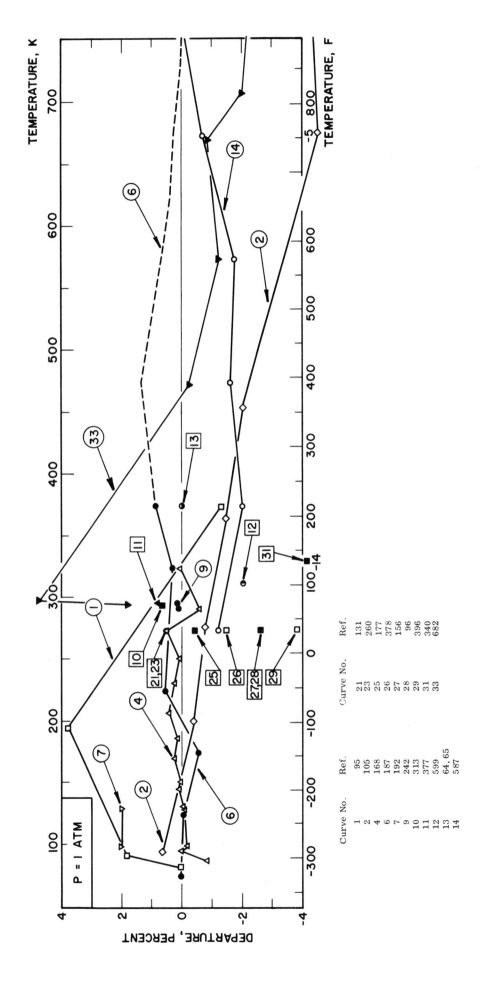

FIGURE 11 DEPARTURE PLOT FOR THERMAL CONDUCTIVITY OF GASEOUS OXYGEN

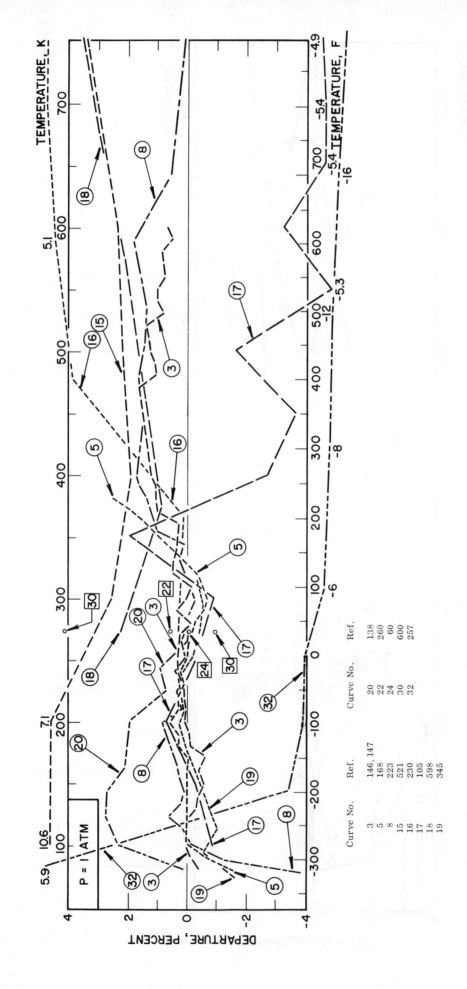

FIGURE 11　DEPARTURE PLOT FOR THERMAL CONDUCTIVITY OF GASEOUS OXYGEN　(continued)

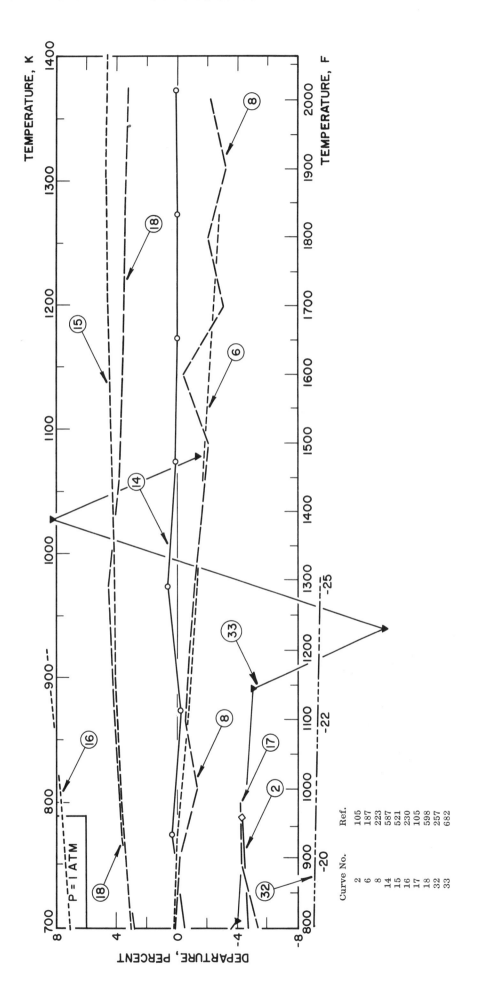

FIGURE 11 DEPARTURE PLOT FOR THERMAL CONDUCTIVITY OF GASEOUS OXYGEN (continued)

TABLE 12   THERMAL CONDUCTIVITY OF RADON

RECOMMENDED VALUES

[Temperature, T, K; Thermal Conductivity, k, mW cm^{-1}K^{-1}]

SATURATED LIQUID

T	k
200	(0.609)‡
210	0.586
220	0.562
230	0.540
240	0.518
250	0.498
260	0.477
270	0.456
280	0.437
290	0.417
300	0.396
310	0.375
320	0.353
330	0.330
340	0.305
350	0.278
360	0.249
370	0.213
377.2*	0.138

‡Extrapolated for the supercooled liquid (n.m.p. = 202 K)
*Critical Temperature

DISCUSSION

SATURATED LIQUID

No experimental data or estimates were found for the thermal conductivity of saturated liquid radon. The recommended values here presented were obtained using the generalized correlation of Owens and Thodos (268) for the thermal conductivity of monatomic liquids and vapors. It was necessary to obtain values of the critical parameters for this substance. While the critical temperature of 377.16 K was located in the literature, no value of the critical thermal conductivity was located. A variety of estimation methods yielded the result (3.30 ± 0.45) 10^{-5} cal/cm sec°K. These critical parameters were then used together with the correlation to obtain the recommended values.

The accuracy of the recommended values is difficult to assess. Based upon the uncertainty in the original correlation, in the estimation of critical data and in the validity of the principle of corresponding states, an uncertainty of about twenty-five percent would seem to be a reasonable estimate. The values here presented must be considered as tentative and in need of experimental verification.

TABLE 12   THERMAL CONDUCTIVITY OF RADON

RECOMMENDED VALUES

[Temperature, T, K; Thermal Conductivity, k, mW cm^{-1}K^{-1}]

DISCUSSION

SATURATED VAPOR

No experimental data or estimates were found for the thermal conductivity of saturated radon vapor. The recommended values here presented were obtained using the generalized correlation of Owens and Thodos (268) for the thermal conductivity of monatomic liquids and vapors. It was necessary to obtain values of the critical parameters for this substance. While the critical temperature of 377.16 K was located in the literature, no value of the critical thermal conductivity was located. A variety of estimation methods yielded the result $(3.30 \pm 0.45) \times 10^{-5}$ cal/cm sec°K. These critical parameters were then used together with the correlation to obtain the recommended values.

The accuracy of the recommended values is difficult to assess. Based upon the uncertainty in the original correlation, in the estimation of critical data and in the validity of the principle of corresponding states, an uncertainty of about twenty-five percent would seem to be a reasonable estimate. The values here presented must be considered as tentative and in need of experimental verification.

SATURATED VAPOR

T	k
200	0.025
210	0.027
220	0.028
230	0.030
240	0.032
250	0.034
260	0.035
270	0.038
280	0.040
290	0.042
300	0.045
310	0.047
320	0.051
330	0.055
340	0.060
350	0.065
360	0.073
370	0.089
377.2*	0.138

*Critical temperature

DISCUSSION

TABLE 12   THERMAL CONDUCTIVITY OF RADON

RECOMMENDED VALUES

[Temperature, T, K; Thermal Conductivity, k, mW cm^{-1}K^{-1}]

GAS

No experimental values were found for the thermal conductivity of radon gas at atmospheric pressure. Values were obtained in two ways: first, by using the generalized correlation of Owens and Thodos (268) with the critical temperature of 377.16 °K and critical thermal conductivity of 3.30 x 10^{-5} cal.cm^{-1}sec^{-1}°K^{-1} and, secondly, by using the Lennard-Jones 6-12 potential function with molecular parameters quoted by Chakraborti (J. Chem. Phys., 44, 3137, 1966). Surprisingly good agreement was obtained in the results of the two methods. At 300 K, the values agreed to 0.5 percent, at 600 K to 0.7 percent and at 1000 K to 2.5 percent. No preference can be assigned for the better method. The generalized correlation approach relies on the principle of corresponding states and the estimated critical thermal conductivity while the Lennard-Jones approach relies upon the molecular parameters estimated from thermal diffusion studies.

The recommended values here presented are the mean of the two sets of values derived above. In view of the complete lack of experimental data they must be regarded as tentative and of uncertain accuracy. Possibly five percent uncertainty below 500 K and ten percent to 1000 K would prove a reasonable error estimate.

GAS

T	k
200	0.0242
210	0.0254
220	0.0266
230	0.0279
240	0.0291
250	0.0303
260	0.0315
270	0.0327
280	0.0339
290	0.0351
300	0.0364
310	0.0376
320	0.0387
330	0.0398
340	0.0410
350	0.0422
360	0.0433
370	0.0445
380	0.0457
390	0.0468
400	0.0480
410	0.0490
420	0.0501
430	0.0512
440	0.0523
450	0.0534
460	0.0544
470	0.0555
480	0.0566
490	0.0576
500	0.0586
550	0.0643
600	0.0690
650	0.0740
700	0.0789
750	0.0832
800	0.0874
850	0.0915
900	0.0964
950	0.0997
1000	0.1042

TABLE 13   THERMAL CONDUCTIVITY OF TRITIUM

DISCUSSION

SATURATED LIQUID

No experimental values of the thermal conductivity of saturated liquid tritium have been located. The values here reproduced were obtained from a large-scale plot of the correlated values of Kerrisk et al. (667) and must thus be regarded as tentative pending experimental verification. Due to the complete absence of experimental data no departure plot or error estimate can be given.

RECOMMENDED VALUES

[Temperature, T, K; Thermal Conductivity, k, mW cm^{-1}K^{-1}]

SATURATED LIQUID

T	k
20	1.22
22	1.28
24	1.32
26	1.36
28	1.37
30	1.34
32	1.30
34	1.25
36	1.18
38	1.10
40	1.00
42	0.89
43.6*	0.68

*Critical point

TABLE 14   THERMAL CONDUCTIVITY OF XENON

RECOMMENDED VALUES

[Temperature, T, K; Thermal Conductivity, k, mW cm^{-1} K^{-1}]

SOLID

T	k
50	14.4
60	12.0
70	10.5
80	9.2
90	8.2
100	7.5
110	6.8
120	6.3
130	5.8
140	5.4
150	5.1
160	4.8
170	(4.5) ‡

DISCUSSION

SOLID

The only information found for the thermal conductivity of solid xenon was a set of calculated values of Julian (673) from 3.8 to 152 K. As noted in that source, and evident from the graphs there presented, severe disagreement exists between theory and experiment at sufficiently low temperatures, where the theory does not predict a maximum in the thermal conductivity to occur. The recommended values, obtained from a plot of the Julian values, have thus been restricted to temperatures of and above 50 K, where this difficulty should not arise. However, even for such temperatures a comparison of theory and experiment for the other substances considered by Julian indicates errors of twenty percent are quite possible in the calculated values for xenon. The values presented here should thus be regarded as tentative and in urgent need of checking by accurate experimental measurements.

‡ n.m.p. = 161 K

TABLE 14   THERMAL CONDUCTIVITY OF XENON

RECOMMENDED VALUES

[Temperature, T, K; Thermal Conductivity, k, mW cm^{-1}K^{-1}]

SATURATED LIQUID

DISCUSSION

In studying the thermal conductivity of saturated liquid xenon, the experimental data of Keyes (192) and of Ikenberry and Rice (672) were compared with the correlation of Owens and Thodos (268). Where necessary, the experimental values of (672) were extrapolated to saturation conditions. The result of the intercomparison was to indicate that the two sets of experimental data agreed to within a few percent. Above 210 K the correlated values appeared too high and the recommended values were derived from a smooth curve which passed through the mean of the two sets of available data at the lowest temperatures, through the Ikenberry and Rice data above 200 K and which passed through a critical point value about five percent lower than the Owens and Thodos value. The maximum difference between the present values and those of the correlation was about nine percent at 260 K.

The recommended values should be accurate to within two percent for the entire tabulated range, except possibly in the immediate vicinity of the critical point.

SATURATED LIQUID

T	k
150	(0.79)†
160	(0.74)†
170	0.70
180	0.66
190	0.62
200	0.58
210	0.54
220	0.50
230	0.46
240	0.42
250	0.38
260	0.34
270	0.31
280	0.27
289*	0.16

† Extrapolated for the supercooled liquid (n. m. p. = 161 K)
* Critical point

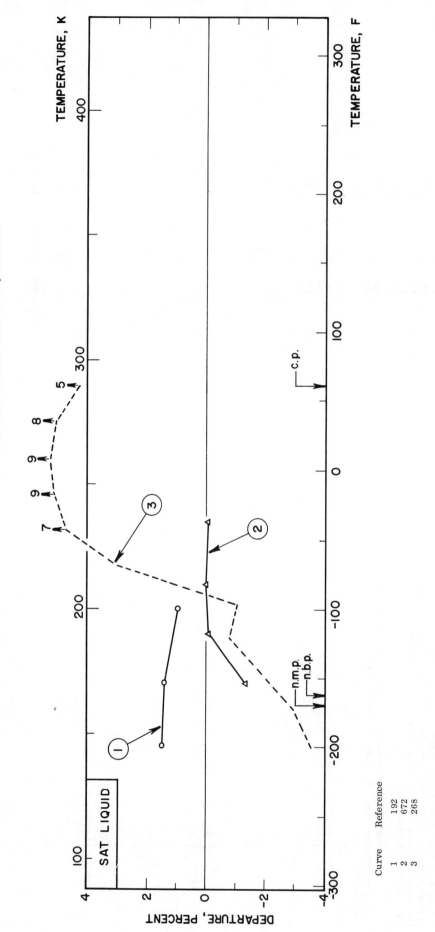

FIGURE 14　DEPARTURE PLOT FOR THERMAL CONDUCTIVITY OF LIQUID XENON

TABLE 14 THERMAL CONDUCTIVITY OF XENON

DISCUSSION

SATURATED VAPOR

No experimental data were located for the thermal conductivity of saturated xenon vapor. The recommended values were derived from a correlation of Owens and Thodos (268) and must be regarded as of uncertain accuracy until experimental measurements are available. Based upon a comparison of such values with atmosphere pressure values at low temperatures a few percent uncertainty below 250 K would appear reasonable, the uncertainty then gradually increasing to the critical temperature, at which a magnitude of about twenty five percent seems a reasonable estimate. Due to the absence of experimental data no departure plot is given.

RECOMMENDED VALUES

[Temperature, T, K; Thermal Conductivity, k, mW cm^{-1}K^{-1}]

SATURATED VAPOR

T	k
150	(0.029)†
160	(0.031)†
170	0.034
180	0.037
190	0.041
200	0.044
210	0.048
220	0.051
230	0.055
240	0.060
250	0.066
260	0.073
270	0.084
280	0.098
289*	0.16

† Extrapolated (n.b.p. = 165 K)
* Critical point

TABLE 14   THERMAL CONDUCTIVITY OF XENON

RECOMMENDED VALUES

[Temperature, T, K; Thermal Conductivity, k, mW cm^{-1} K^{-1}]

DISCUSSION

GAS

The most extensive set of experimental measurements for xenon were reported by Kannuluik and Carmen (173) for 195-579 K. With two exceptions (192, 289), other determinations have been confined to the ice point (14, 81, 214).

All values were given equal weight in the analysis and it was found that the equation

$$10^5 k \text{ (cgsu)} = 5.95280 \cdot 10^{-4} + 4.87582 \cdot 10^{-3} T - 1.32482 \cdot 10^{-6} T^2 \quad (T \text{ in K})$$

fitted the data with maximum deviations of -2.7 and +1.8 percent. The equation was used to generate the recommended values between 150 and 750 K. From 150 to 200 K the uncertainty may be five percent, from 210 to 590 K the values should be accurate to two or three percent while above 590 K the uncertainty may increase to about five percent at the highest temperature tabulated.

GAS

T	k	T	k
150	(0.029)*	500	0.0881
160	(0.031)*	510	0.0896
170	0.033	520	0.0911
180	0.035	530	0.0926
190	0.037	540	0.0940
200	0.0386	550	0.0955
210	0.0404	560	0.0969
220	0.0422	570	0.0983
230	0.0440	580	0.0997
240	0.0458	590	0.1011
250	0.0476	600	0.1025
260	0.0493	610	0.1038
270	0.0511	620	0.1052
280	0.0528	630	0.1065
290	0.0545	640	0.1079
300	0.0562	650	0.1092
310	0.0579	660	0.1105
320	0.0596	670	0.1118
330	0.0613	680	0.1131
340	0.0630	690	0.1144
350	0.0646	700	0.1157
360	0.0662	710	0.1169
370	0.0678	720	0.1183
380	0.0695	730	0.1194
390	0.0712	740	0.1206
400	0.0728	750	0.1218
410	0.0743		
420	0.0759		
430	0.0775		
440	0.0791		
450	0.0806		
460	0.0821		
470	0.0837		
480	0.0852		
490	0.0867		

*Extrapolated for the gas phase ignoring pressure dependence. (n.b.p. = 165 K)

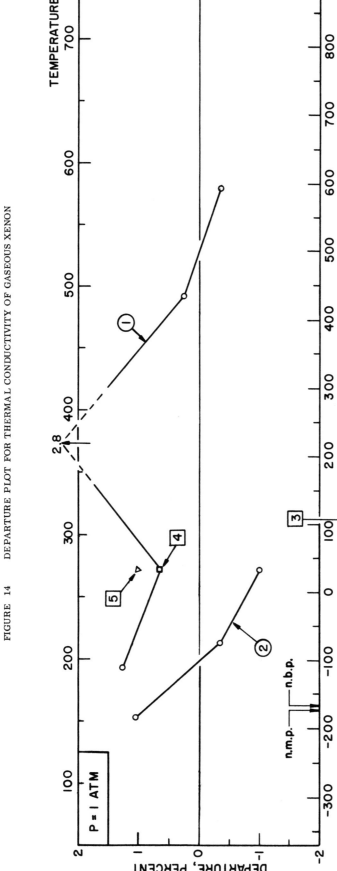

FIGURE 14   DEPARTURE PLOT FOR THERMAL CONDUCTIVITY OF GASEOUS XENON

Curve	Reference
1	173
2	192
3	289
4	81
5	14

TABLE 15 THERMAL CONDUCTIVITY OF AMMONIA

RECOMMENDED VALUES

[Temperature, T, K; Thermal Conductivity, k, mW cm^{-1}K^{-1}]

SATURATED LIQUID

T	k
220	6.60
230	6.37
240	6.14
250	5.92[‡]
260	5.69[‡]
270	5.46[‡]
280	5.24[‡]
290	5.01[‡]
300	4.785[‡]
310	4.559[‡]
320	4.333[‡]
330	4.106[‡]
340	3.880[‡]
350	3.653[‡]
360	3.427[‡]
370	3.200[‡]
380	2.97[‡]
390	2.75[‡]
400	2.52[‡]

[‡] Under saturated vapor pressures. (n.b.p. = 240 K).

SATURATED LIQUID   DISCUSSION

Only two experimental investigations are available in the literature on the thermal conductivity of liquid ammonia. Kardos (462, 526) made measurements in a hot-wire apparatus covering the temperature range from 258 to 303 K. However, because of the high electrical conductivity of liquid ammonia, his results were not of sufficient reliability, and he gave only a mean value over the whole temperature range. On the other hand, Sellschopp (538) used a coaxial-cylinder apparatus and measured the thermal conductivity of saturated liquid ammonia at temperatures from 303 to 373 K. His results were presented by a linear equation. Since these data are considered to be most reliable at present, the correlation equation is based on Sellschopp's measurements and is given by

$$10^6 k \text{ (cgsu)} = 2767.4 - 5.4122 \, T \quad (T \text{ in K}).$$

The recommended values are calculated from the above formula. The values in the temperature range from 300 to 375 K should be substantially correct but outside the range the uncertainty increases.

In the departure plot, the original equation given by Sellschopp is compared with the above formula at four temperatures. Incidentally, Koch (527) calculated the thermal conductivity of saturated liquid ammonia by means of an empirical correlation. However, his values are found to be considerably different from the present correlation, as shown in the departure plot.

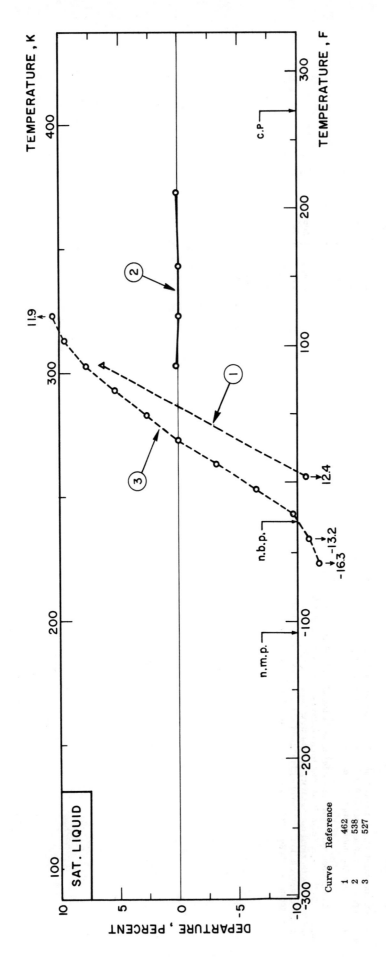

FIGURE 15   DEPARTURE PLOT FOR THERMAL CONDUCTIVITY OF LIQUID AMMONIA

TABLE 15   THERMAL CONDUCTIVITY OF AMMONIA

RECOMMENDED VALUES

[Temperature, T, K; Thermal Conductivity, k, mW cm^{-1}K^{-1}]

GAS

T	k	T	k	T	k
200	0.153*	450	0.433	700	0.811
210	0.162*	460	0.447	710	0.828
220	0.171*	470	0.462	720	0.844
230	0.180*	480	0.476	730	0.861
240	0.188	490	0.491	740	0.877
250	0.197	500	0.506	750	0.894
260	0.206	510	0.520	760	0.910
270	0.215	520	0.535	770	0.927
280	0.225	530	0.550	780	0.944
290	0.235	540	0.565	790	0.962
300	0.246	550	0.580	800	0.977
310	0.256	560	0.595	810	0.996
320	0.267	570	0.610	820	1.013
330	0.279	580	0.625	830	1.029
340	0.290	590	0.640	840	1.046
350	0.302	600	0.656	850	1.063
360	0.314	610	0.671	860	1.080
370	0.327	620	0.686	870	1.096
380	0.339	630	0.702	880	1.113
390	0.352	640	0.717	890	1.130
400	0.364	650	0.733	900	1.146
410	0.377	660	0.749		
420	0.390	670	0.764		
430	0.404	680	0.780		
440	0.418	690	0.795		

*n.b.p. = 240 K.

DISCUSSION

GAS

Several measurements of the thermal conductivity of gaseous ammonia made over moderately large temperature ranges (51, 95, 96, 105, 568, 587-589), for smaller ranges or single temperatures (14, 59, 86, 187, 228) and correlations (105, 187, 223, 521) have been compared with more recent experimental and correlated values (59, 644-647).

The trend with temperature of the Keyes (187) correlation appears to be erroneous above 400 K and was ignored in the preparation of the recommended values. More difficult to explain are the published data of Ziebland et al. (646) which are higher than his preliminary values (589) and which, with the exception of two data points (568, 644) are higher than all other values. The measurements of Baker and Brokaw (644), exhibit a trend at higher temperatures more in agreement with all other work.

The recommended values were therefore chosen to fall near the average of all measurements for temperatures below 400 K and to approach the trend in the Geier and Schafer data (587) for the highest temperatures. The conclusion which can be drawn from the departure plot is that the recommended values should be accurate to about 1.5 percent for temperatures below 400 K and possibly ten percent for the highest temperature tabulated. More precise estimation will require more accurate measurements to be undertaken in order to resolve differences of up to thirteen percent which exist between present data.

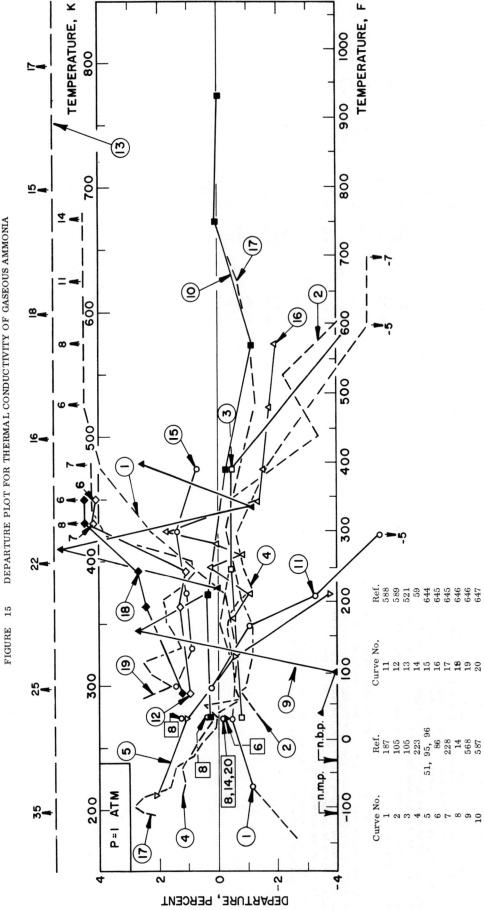

FIGURE 15 DEPARTURE PLOT FOR THERMAL CONDUCTIVITY OF GASEOUS AMMONIA

GAS

TABLE 16   THERMAL CONDUCTIVITY OF BORON TRIFLUORIDE

DISCUSSION

Data for the thermal conductivity of gaseous boron trifluoride have been given for temperatures between 273 and 353 K by McKenzie and Raw (234), and are reported as accurate to within two percent or better. These authors assumed a linear variation with temperature. In the preparation of the recommended values, it was found that a smooth curve gave a better fit to the data and the recommended values were obtained from this curve. The recommended values should be accurate to within two percent.

RECOMMENDED VALUES

[Temperature, T, K; Thermal Conductivity, k, mW cm^{-1}K^{-1}]

GAS

T	k
250	0.157
260	0.164
270	0.171
280	0.178
290	0.184
300	0.190
310	0.197
320	0.203
330	0.208
340	0.214
350	0.220
360	0.225
370	0.230
380	0.236
390	0.241
400	0.246

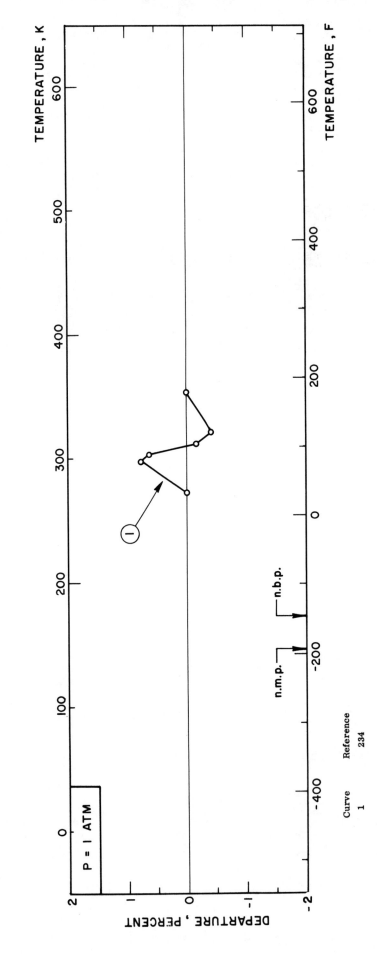

FIGURE 16   DEPARTURE PLOT FOR THERMAL CONDUCTIVITY OF GASEOUS BORON TRIFLUORIDE

TABLE 17   THERMAL CONDUCTIVITY OF HYDROGEN CHLORIDE

RECOMMENDED VALUES

[Temperature, T, K; Thermal Conductivity, k, mW cm^{-1}K^{-1}]

GAS

T	k	T	k
		450	0.218
		460	0.223
		470	0.227
180	(0.080)*	480	0.232
190	0.086	490	0.236
200	0.092	500	0.240
210	0.097	510	0.244
220	0.103	520	0.249
230	0.108	530	0.253
240	0.113	540	0.257
250	0.119	550	0.261
260	0.124	560	0.265
270	0.129	570	0.269
280	0.135	580	0.273
290	0.140	590	0.277
300	0.145	600	0.281
310	0.150	610	0.285
320	0.155	620	0.289
330	0.160	630	0.293
340	0.165	640	0.297
350	0.170	650	0.301
360	0.176	660	0.305
370	0.180	670	0.309
380	0.185	680	0.313
390	0.190	690	0.317
400	0.195	700	0.321
410	0.200		
420	0.205		
430	0.209		
440	0.214		

DISCUSSION

GAS

Only two sets of experimental data are available for the thermal conductivity of gaseous hydrogen chloride, the measurements of Franck (105) from 197 to 577 K at pressures below atmospheric, and of Baker and Brokaw (644) from 300 to 471 K. It was found that, while each of these could be closely approximated by a linear relationship between thermal conductivity and temperature, the difference between the two sets could be as much as two percent. By drawing a smooth curve through the data taken as a whole, agreement to about one percent was obtained. Further experimental measurements will be necessary if it is desired to establish which set of data is the more nearly correct or to reduce the uncertainty below one percent. The Franck data were also used as the basis of correlations by Brokaw (420), Lenoir (223) and Schaefer and Thodos (575) while Andrussow (574) cites a single value.

In the temperature range of 200 to 500 K, the recommended values, read from the smooth curve mentioned above, should be accurate to about two percent. At the highest temperature tabulated the uncertainty may be as much as five percent.

*Extrapolated for the gas phase ignoring pressure dependence.
(n. b. p. = 188 K)

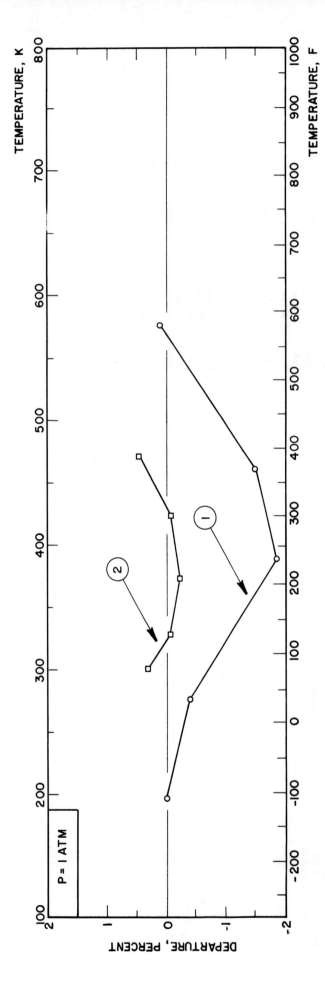

FIGURE 17 DEPARTURE PLOT FOR THERMAL CONDUCTIVITY OF GASEOUS HYDROGEN CHLORIDE

TABLE 18   THERMAL CONDUCTIVITY OF HYDROGEN IODIDE

RECOMMENDED VALUES

[Temperature, T. K; Thermal Conductivity, k, mW cm^{-1}K^{-1}]

GAS

DISCUSSION

GAS

The only thermal conductivity values which have been noted for gaseous hydrogen iodide is the compilation of Svehla (521) for temperatures from 100 to 5000° K. These values were obtained using force constants fitted to the Lennard-Jones 6-12 intermolecular potential function.

The recommended values were obtained by a finite difference smoothing and interpolation of the Svehla tables. In view of the complete lack of experimental data, the tabulation of recommended values has been restricted to a temperature range of from 7 K below the normal boiling point of a temperature of 1000 K. Based upon experience with similar values for other gases in uncertainty of at least ten percent must be assumed in the recommended values. Experimental data are urgently required for this substance.

T	k	T	k	T	k
		500	0.103	800	0.162
		510	0.105	810	0.164
		520	0.107	820	0.166
230	0.047*	530	0.109	830	0.168
240	0.049	540	0.111	840	0.170
250	0.051	550	0.113	850	0.172
260	0.053	560	0.115	860	0.174
270	0.055	570	0.117	870	0.176
280	0.058	580	0.119	880	0.178
290	0.060	590	0.121	890	0.180
300	0.062	600	0.123	900	0.182
310	0.064	610	0.125	910	0.183
320	0.066	620	0.127	920	0.185
330	0.068	630	0.129	930	0.187
340	0.070	640	0.131	940	0.189
350	0.072	650	0.133	950	0.191
360	0.074	660	0.135	960	0.193
370	0.077	670	0.137	970	0.195
380	0.079	680	0.139	980	0.197
390	0.081	690	0.141	990	0.199
400	0.083	700	0.143	1000	0.200
410	0.085	710	0.145		
420	0.087	720	0.147		
430	0.089	730	0.149		
440	0.091	740	0.151		
450	0.093	750	0.153		
460	0.095	760	0.154		
470	0.097	770	0.156		
480	0.099	780	0.158		
490	0.101	790	0.160		

*n.b.p. = 237 K.

TABLE 19 THERMAL CONDUCTIVITY OF HYDROGEN SULFIDE

DISCUSSION

GAS

A set of values for the thermal conductivity of hydrogen sulfide was given by Lenoir (223) for temperatures between 255 and 311 K, based on values given in the International Critical Tables (416) and the Landolt-Börnstein tables (214). It was found that, to within the accuracy of the values, the thermal conductivity could be represented as a linear function of temperature which was used to generate the recommended values between 200 and 400 K. These values should be accurate to within two percent between 250 and 310 K and possibly to within five percent for other temperatures.

RECOMMENDED VALUES

[Temperature, T, K; Thermal Conductivity, k, mW cm^{-1}K^{-1}]

GAS

T	k
200	(0.082)*
210	(0.088)*
220	0.095
230	0.101
240	0.108
250	0.114
260	0.121
270	0.128
280	0.134
290	0.141
300	0.147
310	0.154
320	0.160
330	0.167
340	0.173
350	0.180
360	0.186
370	0.193
380	0.199
390	0.206
400	0.212

*Extrapolated for the gas phase ignoring pressure dependence. (n. b. p. = 214 K)

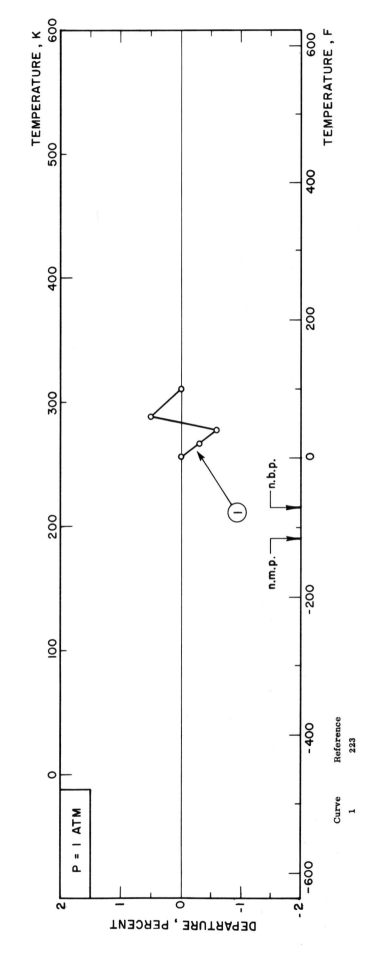

FIGURE 19   DEPARTURE PLOT FOR THERMAL CONDUCTIVITY OF GASEOUS HYDROGEN SULFIDE

TABLE 20   THERMAL CONDUCTIVITY OF NITRIC OXIDE

RECOMMENDED VALUES

[Temperature, T, K; Thermal Conductivity, k, mW cm^{-1}K^{-1}]

DISCUSSION

GAS

The principal experimental measurements of the thermal conductivity of nitric oxide have been made by Choy and Raw (660), Dresvyannikov (661), Gardiner and Schafer (112), Johnston and Grilly (168), Keyes (187) and Richter and Sage (417). In addition, less extensive data have been reported (96, 627). Some of the above data have been requoted (17, 51, 153) and correlated (146, 223), the latter evidently based principally upon the Johnston and Grilly data.

In a preliminary analysis of the data, poor agreement was found between the Richter and Sage data and other work. It must be concluded that some systematic error was equally responsible for the deviation of their data for nitric oxide as was also found with their data for nitrous oxide. Accordingly, these data were not considered further in this analysis. Above 300 K, the trend of the Johnston and Grilly values is to approach the Dresvyannikov data which are considerably higher than other work (112, 627, 660). The high temperature (661) data were likewise ignored in the preparation of the recommended values.

It was therefore decided to base the recommended values for temperatures up to 300 K upon the (168) values. The recommended values above 300 K, read from a large scale plot, were selected to approach the (112, 627, 660) data at about 450 K and are indistinguishable from these to the highest temperature (710 K) reported by these workers. Values above 710 K were obtained by extrapolation. The recommended values should be accurate to about two percent for temperatures below about 500 K, the uncertainty increasing to possibly as much as ten percent for 1000 K. Further experimentation is desirable to confirm the correctness of the choice of values above about 500 K.

GAS

T	k	T	k	T	k
100	(0.090)*	450	0.364	750	0.562
110	(0.099)*	460	0.371	760	0.569
120	(0.1079)*	470	0.377	770	0.575
130	0.1168	480	0.384	780	0.582
140	0.1256	490	0.390	790	0.588
150	0.1345	500	0.396	800	0.595
160	0.1432	510	0.403	810	0.601
170	0.1519	520	0.409	820	0.608
180	0.1606	530	0.415	830	0.614
190	0.1691	540	0.422	840	0.621
200	0.1776	550	0.429	850	0.627
210	0.1860	560	0.435	860	0.634
220	0.1943	570	0.442	870	0.640
230	0.2025	580	0.449	880	0.647
240	0.2107	590	0.456	890	0.653
250	0.2188	600	0.462	900	0.659
260	0.227	610	0.469	910	0.666
270	0.235	620	0.476	920	0.672
280	0.243	630	0.482	930	0.678
290	0.251	640	0.489	940	0.685
300	0.259	650	0.496	950	0.691
310	0.267	660	0.502	960	0.698
320	0.274	670	0.509	970	0.704
330	0.282	680	0.516	980	0.710
340	0.289	690	0.523	990	0.717
350	0.296	700	0.529	1000	0.723
360	0.303	710	0.536		
370	0.310	720	0.542		
380	0.317	730	0.549		
390	0.324	740	0.556		
400	0.331				
410	0.338				
420	0.344				
430	0.351				
440	0.358				

*n. b. p. = 120 K.

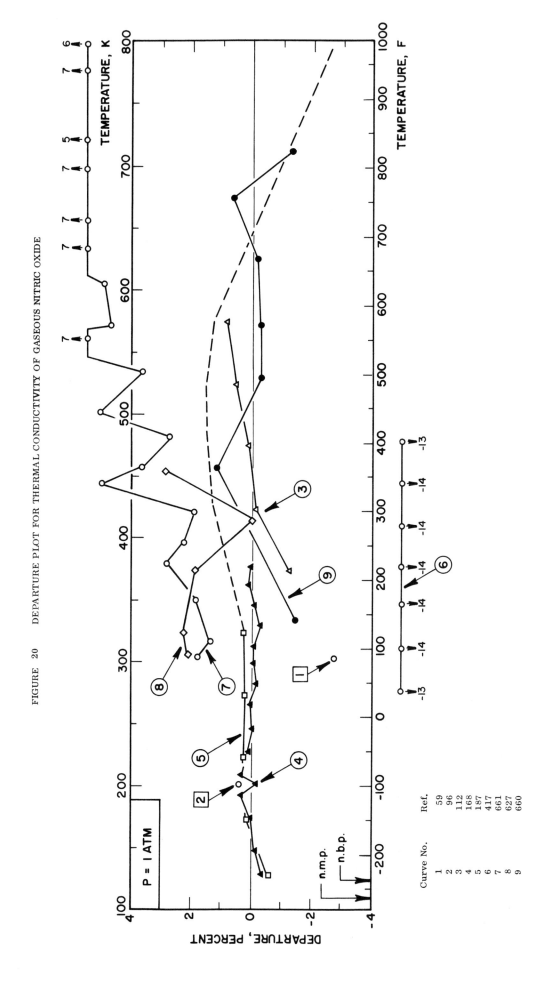

FIGURE 20 DEPARTURE PLOT FOR THERMAL CONDUCTIVITY OF GASEOUS NITRIC OXIDE

TABLE 21   THERMAL CONDUCTIVITY OF NITROGEN PEROXIDE

RECOMMENDED VALUES

[Temperature, T, K; Thermal Conductivity, k, mW cm^{-1}K^{-1}]

SATURATED LIQUID

T	k
260	(1.442) †
270	1.422
280	1.391
290	1.351
300	1.302 ‡
310	1.243 ‡
320	1.174 ‡
330	1.096 ‡
340	1.009 ‡
350	0.911 ‡
360	0.805 ‡
370	0.605 ‡
380	0.581 ‡
390	0.427 ‡
400	0.282 ‡

DISCUSSION

SATURATED LIQUID

There exists only one set of experimental data on the thermal conductivity of liquid nitrogen peroxide. Richter-Sage (536, 537) made measurements in a coaxial-spherical apparatus with an accuracy of two percent, covering various pressures up to 5000 psia at three temperatures from 278 to 344 K. As their results are considered to be reliable from the standpoint of the experimental method and procedure, the values for the saturated liquid at each temperature are used in the present analysis. Equal weight is given to the three data points and the data are fitted to a quadratic equation.

The correlation formula obtained for the saturated liquid is given by

10^{6}k (cgsu) = -329.566 + 5.56691 T - 0.0114368 T^2 (T in K).

This equation should be valid between 260 and 370 K, and is found to fit the above experimental data within 0.07 percent. The recommended values are calculated from the above equation. The tabulated data should be correct within two percent in the temperature range from 280 to 340 K. Outside this range the uncertainty increases.

† Extrapolated for the supercooled liquid.   (n.m.p. = 263 K)
‡ Under saturated vapor pressures.   (n.b.p. = 295 K)

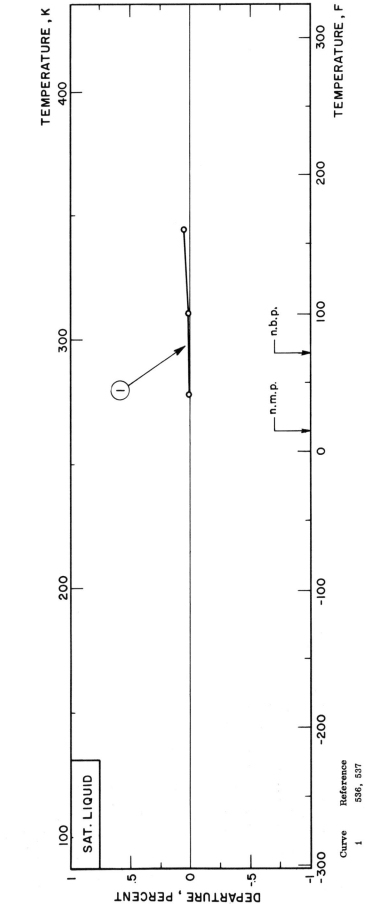

FIGURE 21 DEPARTURE PLOT FOR THERMAL CONDUCTIVITY OF LIQUID NITROGEN PEROXIDE

DISCUSSION

TABLE 21 THERMAL CONDUCTIVITY OF NITROGEN PEROXIDE

RECOMMENDED VALUES

[Temperature, T, K; Thermal Conductivity, k, mW cm^{-1} K^{-1}]

GAS

T	k(a)	k(b)	k(c)
300	0.151	1.180	1.184
310	0.163	1.460	1.456
320	0.176	1.598	1.602
330	0.188	1.611	1.582
340	0.201	1.527	1.540
350	0.213	1.280	1.397
360	0.226	1.042	1.063
370	0.238	0.833	0.843
380	0.247	0.632	0.674
390	0.255	0.502	0.552
400	0.264	0.431	0.502
410	0.272	0.385	0.473
420	0.280	0.356	0.464
430	0.289	0.339	0.485
440	0.297	0.331	0.502
450	0.303	0.327	0.531
460	0.310	0.328	0.586
470	0.318	0.330	0.640
480	0.326	0.335	0.703
490	0.335	0.341	0.778
500	0.343	0.347	0.845
510	0.351	0.354	0.933
520	0.359	0.361	1.021
530	0.366	0.368	1.109
540	0.373	0.375	1.213
550	0.381	0.382	1.297
560	0.389	0.390	1.423
570	0.396	0.397	1.523
580	0.403	0.404	1.636
590	0.410	0.411	1.741
600	0.418	0.418	1.833
610	0.427	0.427	1.916
620	0.435	0.435	2.004
630	0.441	0.441	2.077
640	0.448	0.448	2.100

a. Frozen   b. No NO   c. With NO

GAS

The present tables represent the third consideration of the thermal conductivity of nitrogen peroxide. The original correlation was based upon the work of Brokaw (50) and Coffin and O'Neal (421) for temperatures from 290 to 490 K. The second included the measurements of Barua et al. (580-583) from 305 to 363 and from 373 to 473 K. This analysis also includes the recent detailed discussion by Svehla and Brokaw (654), also briefly reported elsewhere (655).

An analysis of the recent publications (654, 655) reveals that two major uncertainties arise in the interpretation of the experimental measurements; (1), the question as to whether the nitrogen peroxide is to be regarded as being described by the equation

$$N_2O_4 \rightleftharpoons 2NO_2 \text{ or by } N_2O_4 \rightleftharpoons 2NO_2 \rightleftharpoons 2NO + O_2,$$

(2) the question as to what corrections are necessary to the experimental Coffin and O'Neal (421) data. Svehla and Brokaw applied an unspecified correction for error which they considered introduced in the calibration of the apparatus and a second correction, considered negligible above 400 K and always less than five percent for lower temperatures. If the first correction is assumed to be from three to five percent then the total correction can be anywhere from three to ten percent. As Svehla and Brokaw only reproduce their corrected values graphically, two sets of values were considered in the analysis, (a) the original Coffin and O'Neal data (421), (b) the values read from the Svehla and Brokaw graph (654).

Disagreement still occurs between the NASA (50, 421, 654, 655) and Indian (580-583) works due to several reasons. One is the extent to which the dissociation of nitrogen peroxide into nitric oxide and oxygen is considered. While both groups consider that the dissociation is a slow process a more detailed analysis of their works is not possible as the Indian data only extend to a highest pressure of about 0.87 atmospheres and extrapolation to atmospheric pressure is, for some temperatures, questionable. Another disagreement occurs in the values cited for frozen nitrogen peroxide.

In view of the difficulties encountered with the Indian data the recommended values cited here were based upon the most recent NASA publications (654, 655). Three sets of values are given; the frozen conductivity, which assumes no association or dissociation, the equilibrium conductivity, assuming no nitric oxide formation, and the equilibrium conductivity allowing for nitric oxide formation. In application of these recommended values to design problems it appears that the second set should probably be satisfactory to within ten percent for temperatures below about 550 K. For higher temperatures no estimation is presently possible.

TABLE 21 THERMAL CONDUCTIVITY OF NITROGEN PEROXIDE (continued)

RECOMMENDED VALUES

[Temperature, T, K; Thermal Conductivity, k, mW cm^{-1}K^{-1}]

GAS

T	k(a,b)	k(c)	T	k(a,b)	k(c)
650	0.456	2.186	1000	0.695	0.979
660	0.464	2.226	1010	0.701	0.967
670	0.471	2.259	1020	0.706	0.958
680	0.477	2.259	1030	0.711	0.950
690	0.485	2.251	1040	0.716	0.941
700	0.492	2.238	1050	0.722	0.933
710	0.499	2.209	1060	0.728	0.929
720	0.506	2.173	1070	0.734	0.925
730	0.514	2.130	1080	0.741	0.920
740	0.522	2.075	1090	0.747	0.918
750	0.529	2.018	1100	0.753	0.916
760	0.536	1.956	1110	0.758	0.914
770	0.542	1.888	1120	0.763	0.912
780	0.548	1.824	1130	0.768	0.912
790	0.556	1.749	1140	0.773	0.912
800	0.565	1.695	1150	0.779	0.912
810	0.571	1.632	1160	0.786	0.912
820	0.577	1.569	1170	0.792	0.912
830	0.584	1.510	1180	0.798	0.912
840	0.590	1.456	1190	0.803	0.914
850	0.596	1.406	1200	0.808	0.916
860	0.602	1.356	1210	0.814	0.918
870	0.609	1.310	1220	0.820	0.920
880	0.615	1.268	1230	0.826	0.922
890	0.623	1.230	1240	0.831	0.925
900	0.631	1.197	1250	0.836	0.927
910	0.638	1.163	1260	0.841	0.929
920	0.644	1.134	1270	0.847	0.933
930	0.651	1.109	1280	0.853	0.937
940	0.657	1.084	1290	0.859	0.943
950	0.663	1.063	1300	0.865	0.950
960	0.669	1.042			
970	0.676	1.025			
980	0.682	1.008			
990	0.688	0.992			

a. Frozen b. No NO c. With NO

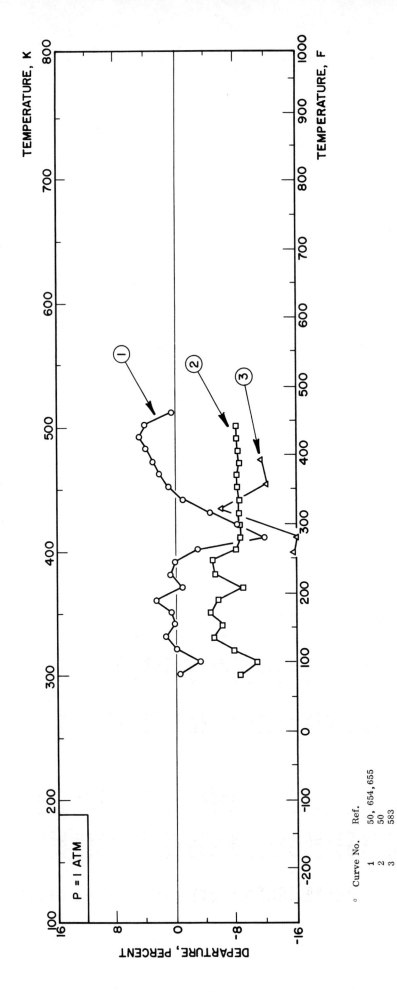

FIGURE 21   DEPARTURE PLOT FOR THERMAL CONDUCTIVITY OF GASEOUS NITROGEN PEROXIDE (EQUILIBRIUM)

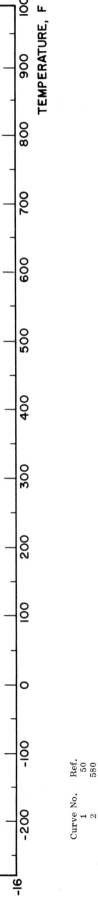

FIGURE 21 DEPARTURE PLOT FOR THERMAL CONDUCTIVITY OF GASEOUS NITROGEN PEROXIDE (FROZEN)

Curve No.	Ref.
1	50
2	580

TABLE 22   THERMAL CONDUCTIVITY OF NITROUS OXIDE

RECOMMENDED VALUES

[Temperature, T, K; Thermal Conductivity, k, mW cm^{-1}K^{-1}]

DISCUSSION

GAS

Measurements of the thermal conductivity of gaseous nitrous oxide have been reported for various temperature ranges by (168, 191, 415, 573, 627, 660) while data for various single temperatures are given in (86, 96, 125, 126, 156, 177, 379, 396, 410). As noted by Richter and Sage in 1958 (415) their values for atmospheric pressure disagree with all other data. Their final publication (573) also shows similar disagreement. No explanation was offered for this discrepancy. The Richter and Sage data were thus disregarded in this analysis. Selections of the available data have also been requoted in (14, 17, 51, 126, 160, 214, 228, 242, 368, 416) and correlations for the ranges 178 to 700 K (223), 123 to 1273 K (187) and 0 to 5000 K (521) have been published.

The data considered most reliable and used as a basis for determining the recommended values were those of Johnston and Grilley (168) for temperatures below about 360 K and Choy and Raw (660) for higher temperatures. The recommended values were read from smooth curves drawn through these data and were checked by differencing. The departure plot shows that, in the range 180 to 360 K, the recommended values should be accurate to one percent or better. Above about 360 K further measurements are desirable to check the accuracy of the (660) data. There would appear to be a possibility that their data are a few percent high at the highest temperatures. Pending such measurements, the recommended values are felt to be accurate to within two percent at 400 K, four percent at 600 K and possibly within ten percent for temperatures from 800 to 1000 K.

GAS

T	k	T	k
180	(0.0843)*	500	0.341
190	0.0909	510	0.349
		520	0.357
200	0.0976	530	0.365
210	0.1045	540	0.373
220	0.1115		
230	0.1187	550	0.380
240	0.1260	560	0.388
		570	0.395
250	0.1335	580	0.403
260	0.1411	590	0.410
270	0.1488		
280	0.1568	600	0.418
290	0.1650	610	0.426
		620	0.433
300	0.1735	630	0.440
310	0.182	640	0.448
320	0.191		
330	0.200	650	0.455
340	0.209	660	0.463
		670	0.470
350	0.218	680	0.477
360	0.227	690	0.485
370	0.236		
380	0.244	700	0.492
390	0.252	710	0.500
		720	0.507
400	0.260	730	0.514
410	0.268	740	0.522
420	0.276		
430	0.285	750	0.529
440	0.293	800	0.566
		850	0.602
450	0.301	900	0.638
460	0.309	950	0.672
470	0.317		
480	0.325	1000	0.705
490	0.333		

*n.b.p. = 183 K. Extrapolated for the gas phase ignoring pressure dependence

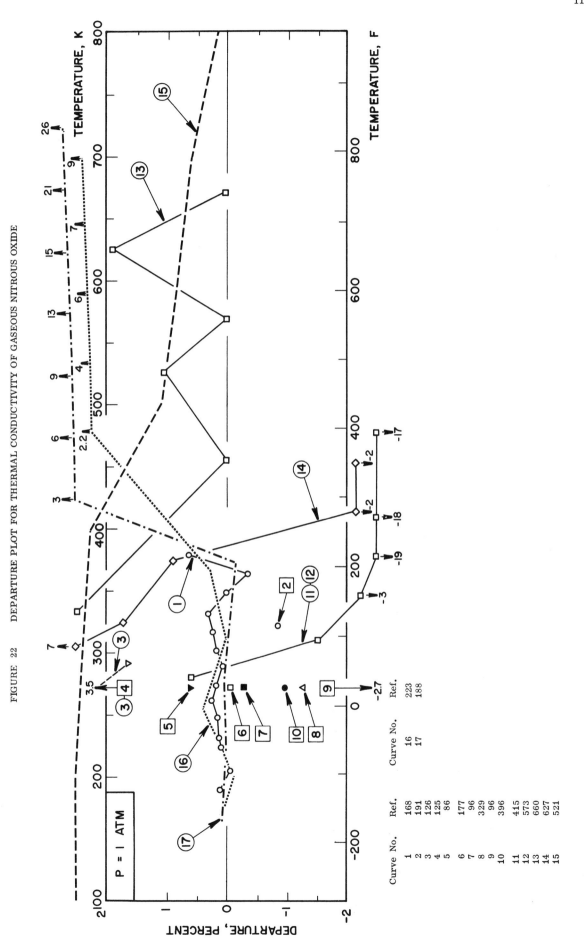

FIGURE 22 DEPARTURE PLOT FOR THERMAL CONDUCTIVITY OF GASEOUS NITROUS OXIDE

TABLE 23   THERMAL CONDUCTIVITY OF SULFUR DIOXIDE

DISCUSSION

SATURATED LIQUID

Only one experimental work is available in the literature for the thermal conductivity of liquid sulfur dioxide, being that of Kardos (462, 526) in the temperature range between 260 and 298 K. The measurements were made in a hot-wire apparatus and the accuracy was reported to be two percent. All of the reported values are given equal weight in this analysis and are fitted to a linear relation, presented by the formula,

$$10^6 k \text{ (cgsu)} = 912.85 - 1.4931\,\overline{T} \quad (\text{T in K}).$$

This equation is found to agree with the experimental data with a mean deviation of 0.74 percent and a maximum of 1.4 percent. The above equation is used to generate the recommended values. The values between 250 and 300 K should be correct within two percent, but outside this temperature range the uncertainty increases.

In the departure plot, a set of the calculated values, which were obtained from an empirical correlation for saturated liquids by Koch (527), is also plotted by a dotted line.

RECOMMENDED VALUES

[Temperature, T, K; Thermal Conductivity, k, mW cm^{-1} K^{-1}]

SATURATED LIQUID

T	k
200	2.57
210	2.51
220	2.45
230	2.38
240	2.32
250	2.258
260	2.195
270	2.133‡
280	2.070‡
290	2.007‡
300	1.945‡
310	1.88 ‡
320	1.82 ‡
330	1.76 ‡
340	1.70 ‡
350	1.63 ‡
360	1.57 ‡
370	1.51 ‡
380	1.45 ‡
390	1.38 ‡
400	1.32 ‡

‡ Under saturated vapor pressures. (n.b.p. = 263 K)

FIGURE 23 DEPARTURE PLOT FOR THERMAL CONDUCTIVITY OF LIQUID SULFUR DIOXIDE

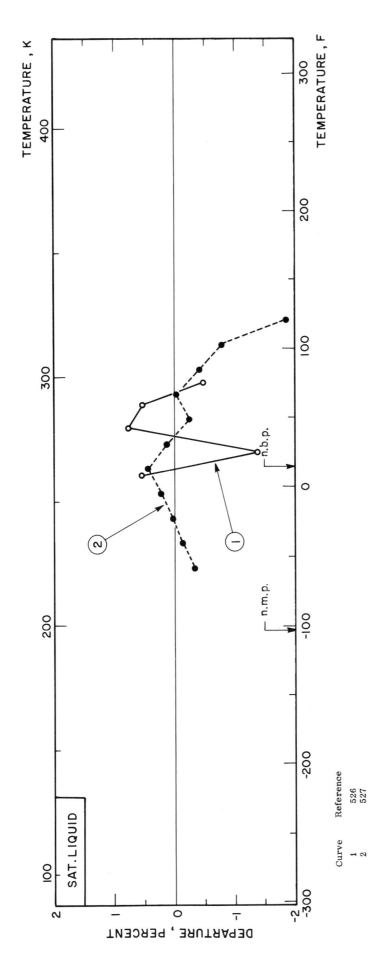

Curve	Reference
1	526
2	527

TABLE 23   THERMAL CONDUCTIVITY OF SULFUR DIOXIDE

RECOMMENDED VALUES

[Temperature, T, K; Thermal Conductivity, k, mW cm^{-1}K^{-1}]

GAS

DISCUSSION

GAS

Data have recently been obtained (579) for temperatures between about 300 and 900 K which considerably extend the range (275 to 315 K) previously available. However, the agreement between the different sets of data is only moderate. The correlated values of Lenoir (223) disagree by a considerable amount at the higher temperatures. The values of thermal conductivity only cited at even hundred degrees Kelvin increments in (579) and used in the previous correlation have been replaced in this analysis by the direct experimental values for various temperatures from 296 to 887 K. The result was to indicate that a fit of the data as a linear function of temperature was inferior to a graphical curve fit. All the recommended values were obtained from a smooth curve.

The recommended values should be accurate to within eight percent between 250 and 360 K and possibly within fifteen percent for the higher temperatures. Further experimental measurements are urgently required to check these values.

T	k	T	k	T	k
250	(0.078)*	500	0.200	750	0.329
260	(0.081)*	510	0.206	760	0.333
270	0.085	520	0.212	770	0.338
280	0.089	530	0.218	780	0.343
290	0.092	540	0.223	790	0.348
300	0.096	550	0.229	800	0.352
310	0.100	560	0.234	810	0.357
320	0.105	570	0.240	820	0.361
330	0.109	580	0.245	830	0.367
340	0.113	590	0.250	840	0.371
350	0.118	600	0.256	850	0.376
360	0.122	610	0.261	860	0.380
370	0.127	620	0.266	870	0.385
380	0.133	630	0.271	880	0.390
390	0.138	640	0.276	890	0.395
400	0.143	650	0.281	900	0.400
410	0.148	660	0.286		
420	0.154	670	0.291		
430	0.159	680	0.295		
440	0.165	690	0.300		
450	0.170	700	0.305		
460	0.176	710	0.310		
470	0.182	720	0.315		
480	0.188	730	0.319		
490	0.194	740	0.324		

*Extrapolated (n. b. p. = 263 K).

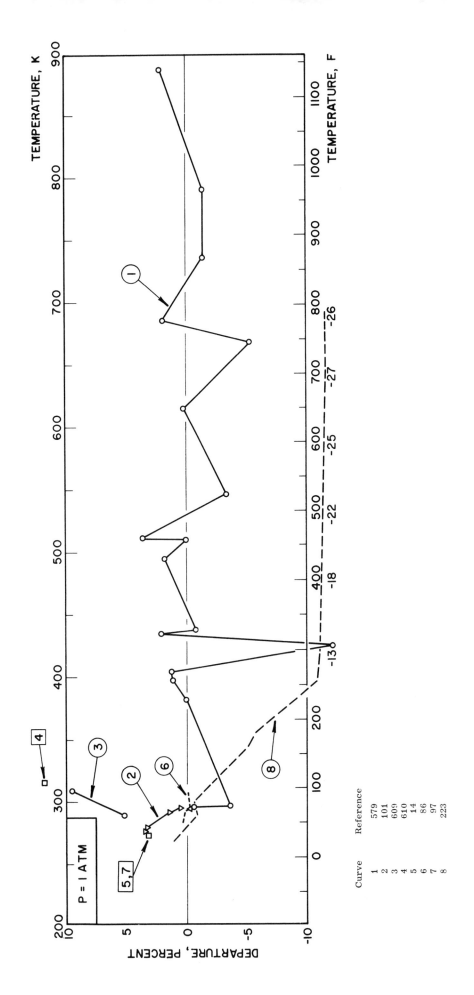

FIGURE 23 DEPARTURE PLOT FOR THERMAL CONDUCTIVITY OF GASEOUS SULFUR DIOXIDE

TABLE 24   THERMAL CONDUCTIVITY OF WATER

RECOMMENDED VALUES

[Temperature, T, K; Thermal Conductivity, k, mW cm^{-1}K^{-1}]

SATURATED LIQUID

T	k	T	k
		400	6.864‡
		410	6.855‡
		420	6.836‡
		430	6.809‡
230	4.899†	440	6.774‡
235	4.980†	450	6.727‡
240	5.062†	460	6.672‡
245	5.143†	470	6.606‡
250	5.225†	480	6.530‡
255	5.306†	490	6.445‡
260	5.388†	500	6.348‡
265	5.470†	510	6.239‡
270	5.551†	520	6.123‡
280	5.818	530	5.995‡
290	5.918	540	5.857‡
300	6.084	550	5.708‡
310	6.233	560	5.55 ‡
320	6.367	570	5.38 ‡
330	6.485	580	5.20 ‡
340	6.587	590	5.01 ‡
350	6.673	600	4.81 ‡
360	6.743		
370	6.797		
380	6.836‡		
390	6.858‡		

cgsu

† Supercooled liquid. (n.m.p. = 273 K)
‡ Under vapor pressure, ignoring pressure dependence. (n.b.p. = 373 K)

DISCUSSION

SATURATED LIQUID

More than sixty experimental works are available on the thermal conductivity of liquid water. With two exceptions (217, 488), experimental results show that the thermal conductivity of water increases with increasing temperature from the normal melting point to the normal boiling point and reaches a maximum near 400 K. Beyond this temperature the thermal conductivity first decreases gradually and at a faster rate near the critical point. The extensive results of Timrot - Vargaftik (339) and Schmidt - Sellschopp (494) have long been considered to be most reliable and were cited in review papers (258, 465, 598, 520) and many handbooks. Subsequently, more careful measurements were reported by Powell-Challoner (434, 480), Riedel (279, 485, 486, 487), Schmidt-Leidenfrost (301, 492, 493), Vargaftik-Oleshchuk (509, 510), Wright (518), and some other investigators. Furthermore, Powell (479) made a study and recommended the most probable values.

This analysis is divided into the following three parts:

(a) The supercooled state below the normal melting point.

There is only one set of data reported by Riedel (485), who extrapolated the values of various salt solutions to zero concentration, and covered temperatures down to 233.16 K. The reported values are exactly linear with temperature. However, his value at the normal melting point is about one percent higher than the most probable value of the normal state at the same temperature. Therefore, Riedel's data were adjusted by a parallel displacement and the final correlation formula is

$$10^6 k \text{ (cgsu)} = 273.778 + 3.90000 \, T \qquad (T \text{ in } K)$$

This equation should be valid in the temperature range from 233.16 to 273.16 K, and should be accurate within one percent. The recommended values from 230 to 270 K were calculated from this equation.

(b) The normal liquid state ——— from the normal melting point to the thermal conductivity maxima.

Seven sets of data (279, 479, 480, 487, 493, 510, 518) were selected as the most reliable and were given equal weight. The correlation formula obtained is

$$10^6 k \text{ (cgsu)} = -1390.53 + 15.1937 \, T - 0.0190398 \, T^2 \qquad (T \text{ in } K).$$

This equation should be valid between 273.16 and 413.16 K. It is found that this equation fits the above-mentioned data with a mean deviation of 0.24 percent and a maximum of 0.82 percent. This equation was used to generate the recommended values from 280 to 410 K.

(c) The higher vapor pressure state ——— from near the thermal conductivity maxima up to the critical point.

The values of Vargaftik-Oleshchuk (509, 510) are considered to be more reliable than the older data of Timrot-Vargaftik (339). Therefore, the weight given in this analysis is two to the former and one to the latter. The correlation formula obtained is

TABLE 24 (continued)

$$10^6 k \text{ (egsu)} = -339.838 + 9.86669 \, T - 0.0123045 \, T^2 \quad (T \text{ in K}).$$

This equation should be valid in the temperature range from 413.16 to 613.16 K, and is found to fit the experimental data of Vargaftik-Oleshchuk with a mean deviation of 0.39 percent and a maximum of 1.4 percent. No further extrapolation is recommended since the deviation becomes extremely large beyond 613.16 K. The tabulated values from 420 to 600 K are calculated from this formula.

For the sake of clarity, the departure curves are presented on three plots. The first plot consists of 16 sets of data up to 380 K. The second plot represents 26 other works in the same temperature range. The third plot depicts three sets of data at vapor pressures higher than one atm.

Seven sets of data (217, 450, 461, 471, 472, 473, 488) which yield departures greater than 10 percent, and older data (429, 430, 453, 454, 455, 456, 511, 512, 513, 514, 516) which were published in the 19th century, are not shown at all.

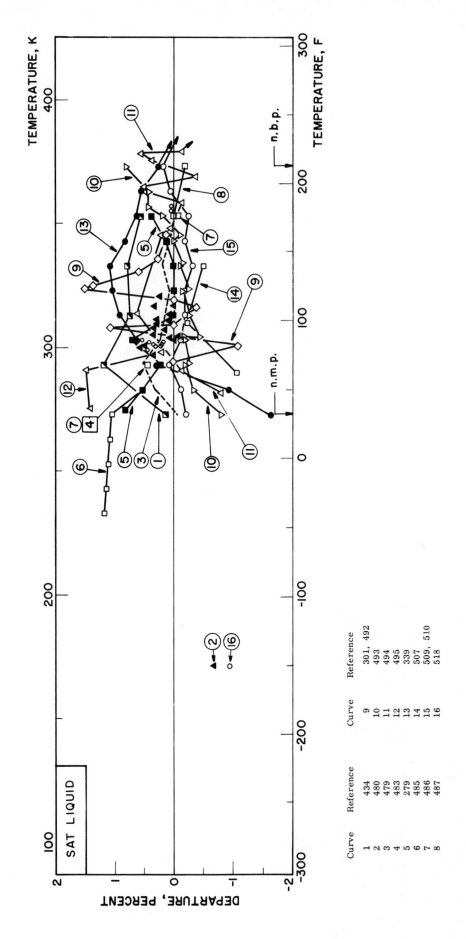

FIGURE 24 DEPARTURE PLOT FOR THERMAL CONDUCTIVITY OF LIQUID WATER (A)

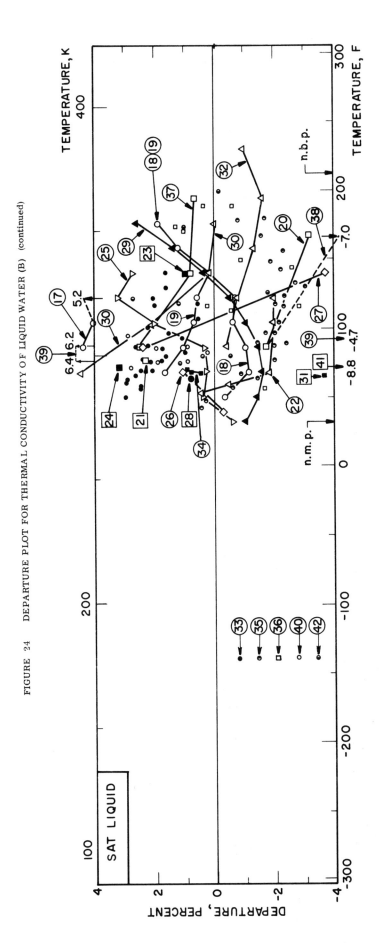

FIGURE 24 DEPARTURE PLOT FOR THERMAL CONDUCTIVITY OF LIQUID WATER (B) (continued)

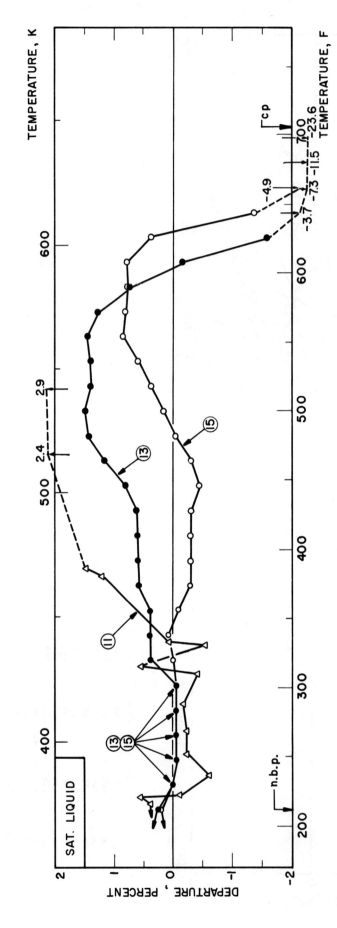

FIGURE 24   DEPARTURE PLOT FOR THERMAL CONDUCTIVITY OF LIQUID WATER (C)   (continued)

Curve	Reference
11	494
13	339
15	509, 510

TABLE 24 THERMAL CONDUCTIVITY OF GASEOUS WATER (STEAM)

RECOMMENDED VALUES

[Temperature, T, K; Thermal Conductivity, k, mW cm^{-1} K^{-1}]

GAS

DISCUSSION

In preparing tables of recommended values for the thermal conductivity of water vapor a more complete collection of tabulations based upon correlating equations, etc. was made than usual. This was done because severe disagreement exists between different sets of data for this substance. Many users of tabulated values have been unaware of the fact that such primary data were subject to large errors. Recent measurements have shown that at least one set of primary data are in considerable error and hence also many tabulations.

Examination of the departure plots reveals that there now exist a large number of experimental data which agree with the recommended values to within some three percent. In severe disagreement are the measurements of Keyes, reported by himself (187) and with Sandell (195). These show a systematic trend with temperature in disagreement with others and also with more recent measurements and should be disregarded. A large number of tabulations have wholly or in part been based upon the Keyes data. Those of Lenoir (223), Keyes (187), Nusselt (263, see also 201), Keenan and Keyes (294), Jakob (593), van Iterson (592) and Grober and Erk (591), are unsatisfactory above about 373 K and only moderately accurate for the few cases (201, 223, 263, 594) where they extend to lower temperatures. In addition to the recommended values, the tabulations of Koch and Fritz (201), the recalculated values of Keyes and Sandell (360, 365, 366) and the Russian data cited by Keyes (596) are reasonably accurate for all temperatures as are those of Keyes and Vines (590) for temperatures above 420 K. Still in severe disagreement are the high temperature data of Geier and Schafer (587) and Vargaftik et. al. (360, 365, 366). For this reason, the tabulation of recommended values has been curtailed to 900 K.

The accuracy of the recommended values can be estimated as being within two percent from 320 to 700 K, and five percent from 250 to 310 K and 710 to 900 K. The uncertainty at the higher temperatures is produced by the problem of estimating the radiation error in the vapor. Due to the high boiling point of water as compared to most fluids, pressure effects are significant to higher temperatures than usual and hence influence the recommended values to about 600 K. More precise recommended values for temperatures below 600 K will require a detailed consideration to be made of the pressure effect and has thus limited the suggested accuracy to two percent rather than a closer tolerance. Further experimentation is to be desired for all temperatures and pressures so that the uncertainties due to pressure, radiation, and accomodation effects can be reduced.

GAS

T	k	T	k
250	(0.140)*	600	0.464
260	(0.148)*	610	0.475
270	(0.156)*	620	0.486
280	0.164	630	0.497
290	0.172	640	0.508
300	0.181	650	0.518
310	0.189	660	0.529
320	0.197	670	0.540
330	0.205	680	0.551
340	0.214	690	0.562
350	0.222	700	0.572
360	0.231	710	0.58
370	0.239	720	0.59
380	0.248	730	0.60
390	0.256	740	0.62
400	0.264	750	0.63
410	0.273	760	0.64
420	0.282	770	0.65
430	0.291	780	0.66
440	0.300	790	0.67
450	0.307	800	0.68
460	0.317	810	0.69
470	0.327	820	0.70
480	0.337	830	0.71
490	0.347	840	0.72
500	0.357	850	0.73
510	0.368	860	0.74
520	0.378	870	0.75
530	0.389	880	0.76
540	0.400	890	0.77
550	0.411	900	0.78
560	0.422		
570	0.432		
580	0.443		
590	0.454		

*Extrapolated

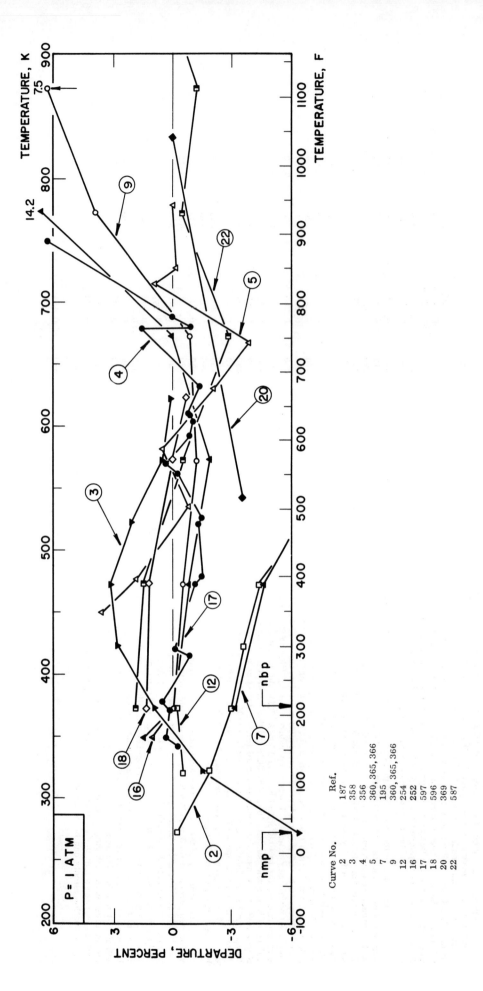

FIGURE 24 DEPARTURE PLOT FOR THERMAL CONDUCTIVITY OF GASEOUS WATER (STEAM)

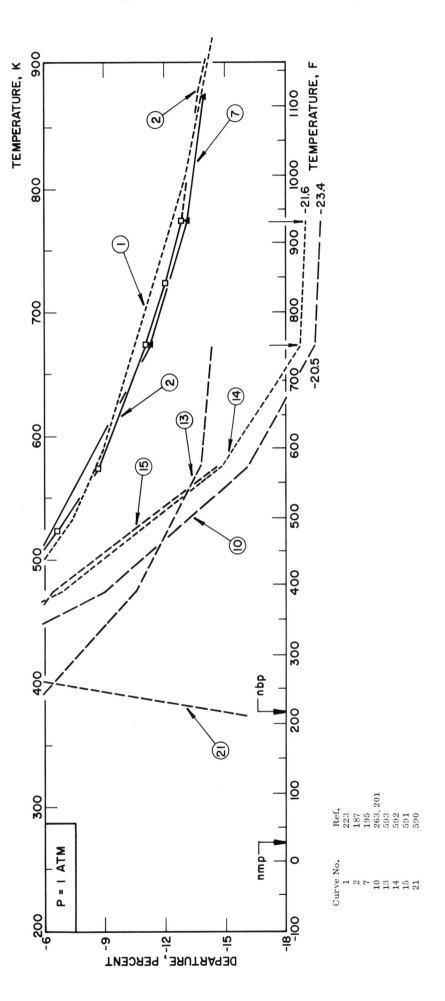

FIGURE 24 DEPARTURE PLOT FOR THERMAL CONDUCTIVITY OF GASEOUS WATER (STEAM) (continued)

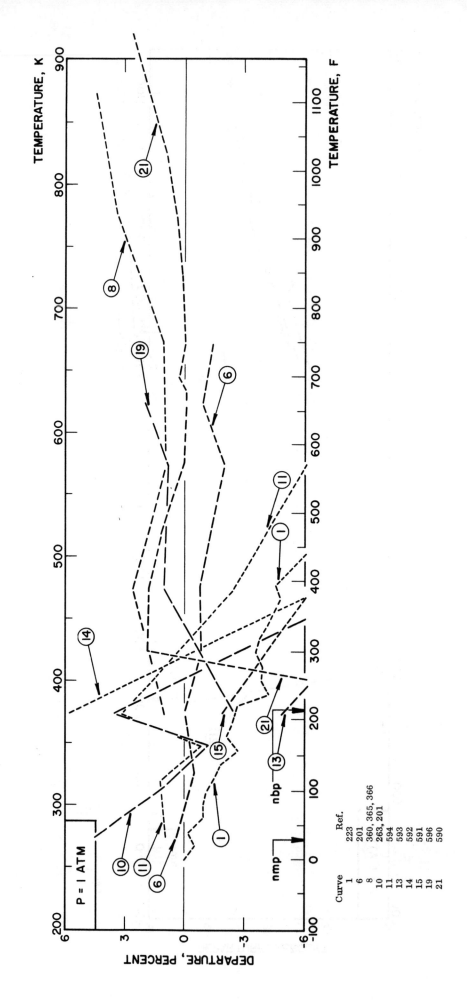

FIGURE 24 DEPARTURE PLOT FOR THERMAL CONDUCTIVITY OF GASEOUS WATER (STEAM) (continued)

TABLE 25   THERMAL CONDUCTIVITY OF ACETONE

RECOMMENDED VALUES

[Temperature, T, K; Thermal Conductivity, k, mW cm^{-1}K^{-1}]

SATURATED LIQUID

T	k
150	(2.15)†
160	(2.11)†
170	(2.08)†
180	2.040
190	2.003
200	1.966
210	1.928
220	1.891
230	1.854
240	1.817
250	1.780
260	1.743
270	1.706
280	1.668
290	1.631
300	1.594
310	1.557
320	1.520
330	1.483‡
340	1.45‡
350	1.41‡
360	1.37‡
370	1.33‡
380	1.30‡
390	1.26‡
400	(1.22)‡

†Extrapolated for the supercooled liquid. (n.m.p. = 178 K)
‡Extrapolated for the liquid under vapor pressures, ignoring pressure dependence. (n.b.p. = 329 K)

DISCUSSION

SATURATED LIQUID

There exist 19 experimental works of the thermal conductivity of liquid acetone. The discrepancy between the reported values of different investigators is rather large. Although the values of Bridgman (431) were thought to be accurate for a long time, recent investigations gave results which were about 10 to 15 percent lower than Bridgman's data. Accordingly, the extensive measurements of Abas-Zade (3), who covered the temperature range from 273 K to the critical point, should be considered to be less reliable, because his values are close to those of Bridgman. In recent careful measurements, the results of Mason (475) and Riedel (483, 484, 279, 486, 487) are considered to be the most reliable from the standpoint of their experimental methods and procedures. Therefore, their values are given heavy weight in this analysis. The values of Filippov (442) and Frontas'ev-Gusakov (447, 448) are also included in the estimation of the most probable formula.

The correlation formula obtained is

$$10^6 k \text{ (cgsu)} = 647.361 - 0.887911\, T \qquad (T \text{ in K}).$$

This equation fits the above-enumerated measurements with a mean deviation of 0.94 percent and a maximum one of 2.4 percent.

The above formula is used to generate the recommended values. The values below the melting point are those for the metastable supercooled liquid, and the data above the boiling point are those for the liquid under vapor pressures, where the pressure dependence on the thermal conductivity is ignored. Therefore, in the range from the melting point to the boiling point, the tabulated values should be substantially correct. Outside this range the uncertainty increases.

Although two measurements (3, 490, 491) were made at higher temperatures and under vapor pressures, their values are less reliable as seen from the departure plot. Therefore, no correlation is attempted in this range. An old value of Guthrie (454, 455) is not plotted.

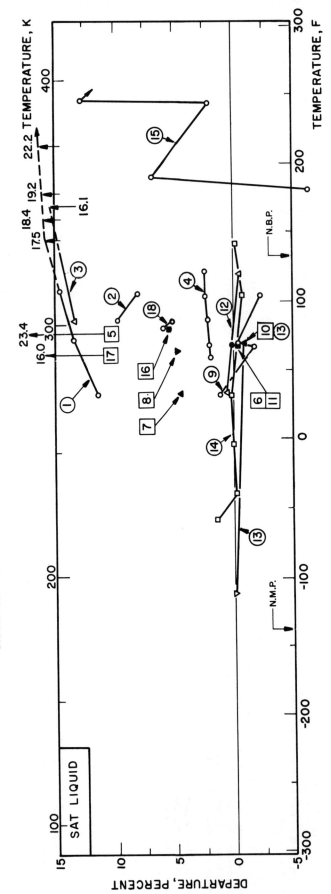

FIGURE 25 DEPARTURE PLOT FOR THERMAL CONDUCTIVITY OF LIQUID ACETONE

Curve	Reference	Curve	Reference
1	3	9	475
2	29	10	483
3	431	11	484
4	442	12	279
5	443	13	486
6	447, 448	14	487
7	450	15	490, 491
8	458	16	497
		17	507
		18	508

TABLE 25 THERMAL CONDUCTIVITY OF ACETONE

RECOMMENDED VALUES

[Temperature, T, K; Thermal Conductivity, k, mW cm^{-1} K^{-1}]

GAS

DISCUSSION

GAS

Most of the experimental data for the thermal conductivity of gaseous acetone are reported from two laboratories (254; 367, 368, 370) while two other values (212) are available. In addition, two correlations have been made (223; 601, 602), the former with no source references and the latter based on (254, 367, 368, 370).

In an analysis of these data, the Vargaftik method of plotting the logarithm of thermal conductivity as a function of the logarithm of absolute temperature was tested and the best curve through the available points was found to differ insignificantly from a straight line. The recommended values were therefore obtained assuming a straight line relationship, the data of (367, 368, 370) being preferred over those of (254). The departure plot shows that all the experimental values except three data points of (254) fall close to such a correlation and that the average deviation is about one percent. The three data points of (254) are some four to six percent lower. More experimental measurements are desirable to confirm the accuracy of the choice made above.

The accuracy of the recommended values can be assessed at within two percent for temperatures below 450 K and possibly as low as eight percent for the highest temperature tabulated.

T	k
250	(0.0803)*
260	(0.0867)*
270	(0.0933)*
280	0.1002
290	0.1073
300	0.1146
310	0.1222
320	0.1300
330	0.1380
340	0.1463
350	0.1548
360	0.1635
370	0.1725
380	0.1817
390	0.1911
400	0.2008
410	0.2106
420	0.2206
430	0.2309
440	0.2412
450	0.252
460	0.263
470	0.275
480	0.286
490	0.298
500	0.310

* Extrapolated (n.b.p. = 329 K)

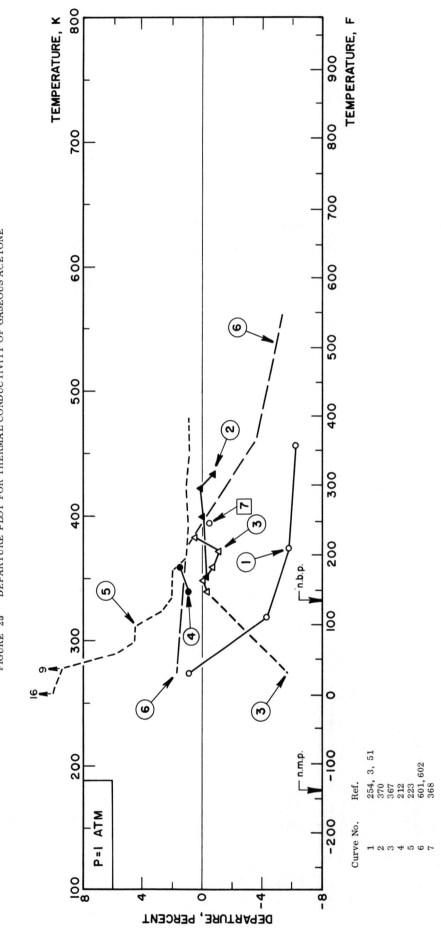

FIGURE 25 DEPARTURE PLOT FOR THERMAL CONDUCTIVITY OF GASEOUS ACETONE

TABLE 26   THERMAL CONDUCTIVITY OF ACETYLENE

DISCUSSION

GAS

The most extensive tabulation of values for this substance is that of Lenoir (223) who quotes nineteen values between 189 and 422 K. These values are cited with no source references. Gardiner and Schafer (112) reported five experimental values between 373 and 573 K. These two sources formed the basis for the correlation of available data.

Polynomials were fitted to these sets of data and it was found that

$10^6 k$ (cgsu) $= -1.19046 + 9.07061 \cdot 10^{-2} T + 2.76330 \cdot 10^{-4} T^2$ (T in K)

$10^5 k$ (cgsu) $= -4.44784 + 3.36669 \cdot 10^{-2} T - 6.32639 \cdot 10^{-6} T^2$ (T in K)

fitted the Lenoir and the Gardiner and Schafer values to within one and 0.2 percent respectively. These equations were accordingly used to generate the recommended values in the ranges 144 to 422 K and 433 to 622 K respectively. Values in the vicinity of 430 K were adjusted to provide a smooth transition between the two equations above.

For temperatures below 350 K the probable error in the recommended values is difficult to estimate due to the paucity of direct measurements but may be estimated as five percent. For temperatures from 350 to 600 K the probable error is one percent and at 650 K may be increased to five percent. New measurements are to be desired in order to verify this tabulation and error estimate.

RECOMMENDED VALUES

[Temperature, T, K; Thermal Conductivity, k, mW cm^{-1} K^{-1}]

GAS

T	k	T	k
190	(0.109)*	450	0.394
		460	0.406
		470	0.418
		480	0.429
		490	0.441
200	0.117	500	0.452
210	0.126	510	0.463
220	0.134	520	0.475
230	0.143	530	0.486
240	0.153	540	0.497
250	0.162	550	0.508
260	0.172	560	0.520
270	0.182	570	0.531
280	0.192	580	0.542
290	0.202	590	0.553
300	0.213	600	0.564
310	0.224	610	0.575
320	0.235	620	0.586
330	0.246	630	0.597
340	0.258	640	0.608
350	0.269	650	0.619
360	0.281		
370	0.294		
380	0.306		
390	0.319		
400	0.332		
410	0.345		
420	0.358		
430	0.371		
440	0.382		

* n. b. p. = 191 K Extrapolated for the gas phase ignoring pressure dependence.

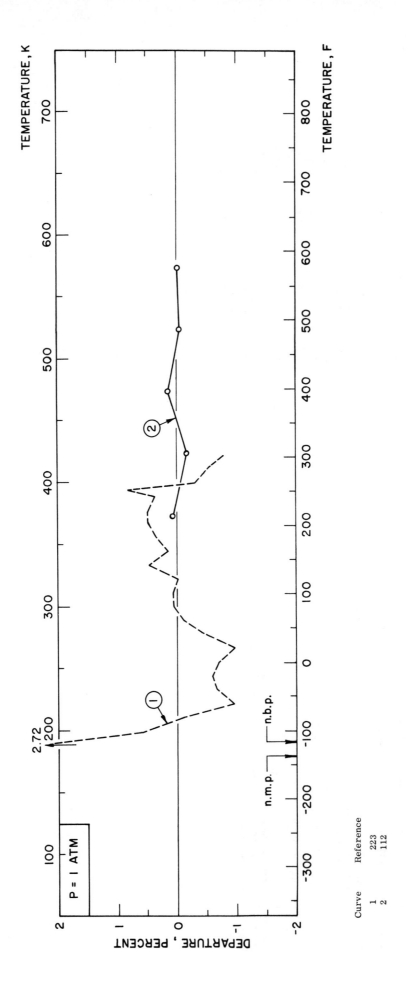

FIGURE 26   DEPARTURE PLOT FOR THERMAL CONDUCTIVITY OF GASEOUS ACETYLENE

TABLE 27   THERMAL CONDUCTIVITY OF BENZENE

DISCUSSION

SATURATED LIQUID

Nineteen experimental works are available on the thermal conductivity of liquid benzene. Excluding two sets of results of Smith (500, 501), who used Bridgman's apparatus (431), a set of data of Goldschmidt (450), and several single point values (443, 497, 507, 512, 513), the discrepancy between the reported values of different investigators is rather small. From the standpoint of the experimental method and procedure, the extensive values of Filippov (442), Riedel (486), Schmidt-Leidenfrost (301, 492), and Vargaftik (508) are considered to be most reliable and are given equal weight in this analysis. Furthermore, the single point values of Frontas'ev-Gusakov (447, 448) and Riedel (483, 484) are also used to estimate the correlation formula.

The correlation equation obtained is

$10^6 k$ (cgsu) $= 525.278 - 0.604093\ T$ (T in K).

This equation is found to fit the above-enumerated values with a mean deviation of 0.55 percent and a maximum of 1.4 percent.

The recommended values are calculated from the above equation. The tabulated data should be substantially correct in the temperature range between 275 and 375 K. The uncertainty increases outside this range.

Abas-Zade (2, 4, 423) made measurements covering the temperature range up to the critical point. Scheffy-Johnson (490, 491) also measured the thermal conductivity of this substance under various vapor pressures up to 491.16 K. Both results are found to agree with the extrapolated values of the above equation up to 420 K within five percent. However, no correlation is attempted at pressures higher than one atm.

Older values of Goldschmidt (450) and Weber (512, 513) are not shown in the departure plot.

RECOMMENDED VALUES

[Temperature, T, K; Thermal Conductivity, k, mW cm^{-1}K^{-1}]

SATURATED LIQUID

T	k
250	(1.566)†
260	(1.541)†
270	(1.515)†
280	1.490
290	1.465
300	1.440
310	1.414
320	1.390
330	1.364
340	1.338
350	1.313
360	(1.288)‡
370	(1.263)‡
380	(1.237)‡
390	(1.212)‡
400	(1.187)‡
410	(1.161)‡
420	(1.136)‡
430	(1.111)‡
440	(1.086)‡
450	(1.060)‡
460	(1.035)‡
470	(1.010)‡

†Extrapolated for the supercooled liquid. (n.m.p. = 279 K)

‡Extrapolated for the liquid under vapor pressures, ignoring pressure dependence. (n.b.p. = 353 K)

FIGURE 27   DEPARTURE PLOT FOR THERMAL CONDUCTIVITY OF LIQUID BENZENE

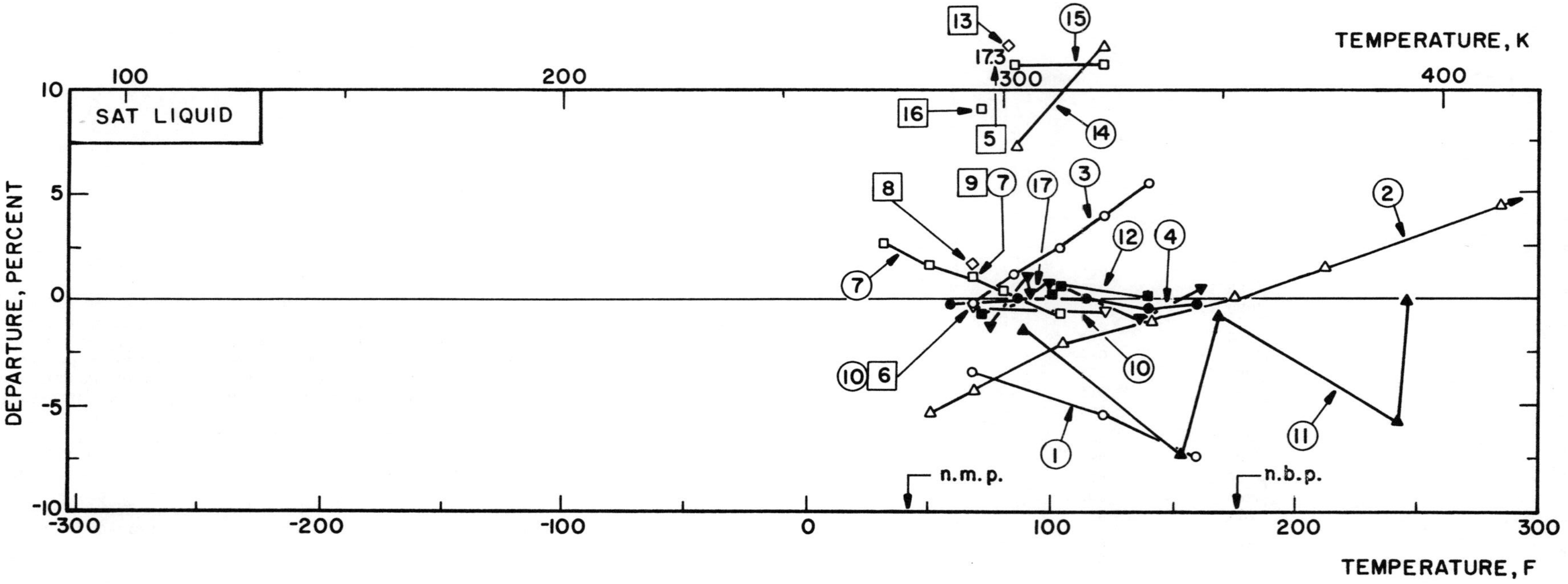

Curve	Reference		Curve	Reference
1	1		10	486
2	2, 4, 423		11	490, 491
3	432		12	301, 492
4	442		13	497
5	443		14	500
6	447, 448		15	501
7	472		16	507
8	483		17	508
9	484			

TABLE 27 THERMAL CONDUCTIVITY OF BENZENE

DISCUSSION

GAS

Extensive measurements on the thermal conductivity of benzene have not yet been made. Vines (367) measured six values in the range 273 to 383 K while Moser (254) obtained four values from 273 to 486 K. Mostly single values were measured by (211, 212, 218, 370) and all these data are also requoted by (51, 211, 223, 368, 416).

All experimental values except that of Vines (367) at 273 K were given equal weight. The formula used to calculate thermal conductivity was

$$10^5 k \text{ (cgsu)} = 3.04565 - 1.94679 \cdot 10^{-2} T + 5.87783 \cdot 10^{-5} T^2 \quad (T \text{ in K}).$$

The departure plot shows that, with one exception, a fit to within five percent is obtained. The value of Vines at 273 K appears unduly low. This conclusion is confirmed by a least mean square fit of the tabulated values (from unspecified sources) of Lenoir (223) which gave a value at 273 K higher than that shown by curves 1, 2, thereby confirming the trend shown on the departure plot.

The recommended values are thought to be accurate to within five percent over the range 280 to 490 K. The quadratic formula was used to extrapolate data to 600 K, the tabulation of (223) being used as a check in this process. Good agreement was found to exist between the recommended values and those tabulated by (223).

RECOMMENDED VALUES

[Temperature, T, K; Thermal Conductivity, k, mW cm^{-1} K^{-1}]

GAS

T	k
250	0.077
260	0.082
270	0.087
280	0.092
290	0.098
300	0.104
310	0.111
320	0.119
330	0.127
340	0.135
350	0.144
360	0.153
370	0.163
380	0.173
390	0.184
400	0.195
410	0.207
420	0.219
430	0.232
440	0.245
450	0.259
460	0.273
470	0.288
480	0.303
490	0.319
500	0.335
510	0.352
520	0.369
530	0.387
540	0.405
550	0.424
560	0.443
570	0.462
580	0.482
590	0.503
600	0.524

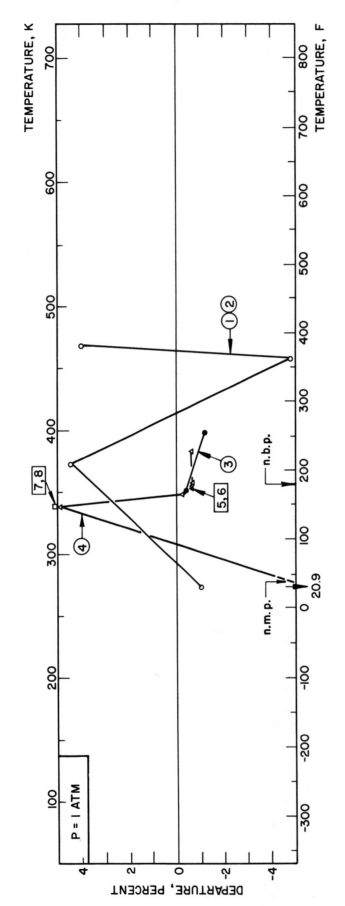

FIGURE 27   DEPARTURE PLOT FOR THERMAL CONDUCTIVITY OF GASEOUS BENZENE

Curve	Reference
1	51
2	254
3	218
4	367
5	368
6	370
7	80
8	211

TABLE 28 THERMAL CONDUCTIVITY OF iso-BUTANE

DISCUSSION

GAS

Few experimental measurements have been reported for the thermal conductivity of iso-butane. Such data as are available - single data values at 273 (237), 339 (211) and 373 K (51) - agree well with a correlation extending from 267 to 505 K (223). The recommended values were read from a smooth curve passing through the correlated and experimental data. Due to the paucity of experimental data an estimate of accuracy of the data is difficult. It is thought that the recommended data are accurate to about five percent below 400 K, the uncertainty possibly reaching 10 percent at the highest temperature.

RECOMMENDED VALUES

[Temperature, T, K; Thermal Conductivity, k, mW cm^{-1}K^{-1}]

GAS

T	k
250	(0.120)*
260	(0.128)*
270	0.137
280	0.145
290	0.154
300	0.163
310	0.173
320	0.184
330	0.194
340	0.205
350	0.216
360	0.227
370	0.238
380	0.250
390	0.261
400	0.272
410	0.283
420	0.294
430	0.305
440	0.316
450	0.328
460	0.339
470	0.351
480	0.362
490	0.374
500	0.385

*Extrapolated for the gas phase ignoring pressure dependence. n. b. p. = 262 K.

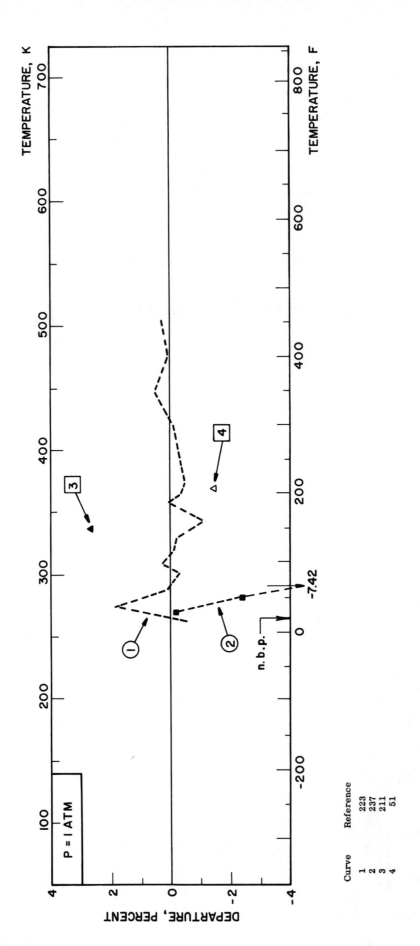

FIGURE 28   DEPARTURE PLOT FOR THERMAL CONDUCTIVITY OF GASEOUS iso-BUTANE

TABLE 29 THERMAL CONDUCTIVITY OF n-BUTANE

RECOMMENDED VALUES

[Temperature, T, K; Thermal Conductivity, k, mW cm^{-1}K^{-1}]

DISCUSSION

SATURATED LIQUID

The only available experimental data reported for the thermal conductivity of liquid n-butane are those of Kramer (467). The measurements were made in a coaxial-cylinder apparatus under various pressures up to 1000 atm along four isotherms from 348 to 437 K, covering both gaseous and liquid phases. Since the author did not give the values of the thermal conductivity for the saturated liquid, in this analysis the values are obtained by the graphical extrapolation of the three isotherms reported for the liquid phase. However, these values obtained would have some uncertainty because each isotherm for the liquid phase has a small hump near the saturation line. These three values of the saturated liquid are given equal weight and the correlation formula obtained is given by

$10^6 k$ (cgsu) $= 471.57 - 0.72393\,T$ (T in K).

This equation fits the three points with a mean deviation of 0.17 percent and a maximum deviation of 0.26 percent. The above correlation formula is used to generate the recommended values. The tabulated values from 350 to 420 K should be reliable within two percent, but the uncertainty below 340 K cannot be ascertained because of the lack of the experimental data.

SATURATED LIQUID

T	k
250	(1.216)
260	(1.186)
270	(1.155)
280	(1.125)
290	(1.095)
300	(1.064) ‡
310	(1.034) ‡
320	(1.004) ‡
330	(0.969) ‡
340	(0.943) ‡
350	0.913 ‡
360	0.883 ‡
370	0.852 ‡
380	0.822 ‡
390	0.792 ‡
400	0.761 ‡
410	0.731 ‡
420	0.701 ‡
430	0.671 ‡
440	0.640 ‡
450	0.610 ‡

( ) Extrapolated.
‡ Under saturated vapor pressures. (n.b.p. = 273 K)

142

FIGURE 29   DEPARTURE PLOT FOR THERMAL CONDUCTIVITY OF LIQUID n-BUTANE

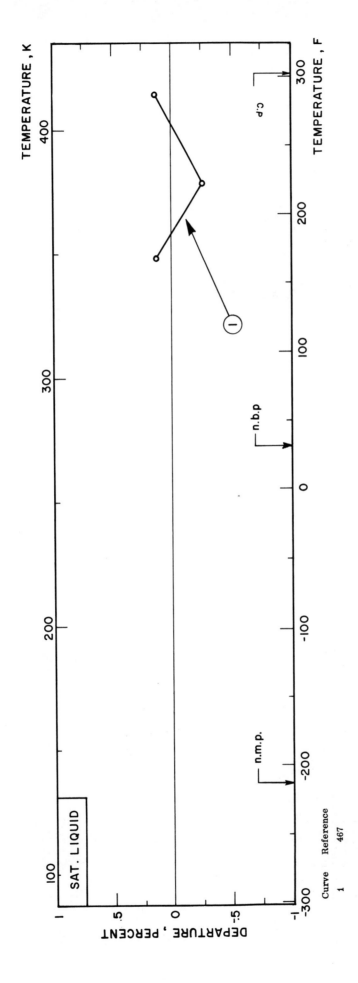

TABLE 29   THERMAL CONDUCTIVITY OF n-BUTANE

RECOMMENDED VALUES

[Temperature, T, K; Thermal Conductivity, k, mW cm^{-1}K^{-1}]

GAS

T	k
250	(0.117)*
260	(0.125)*
270	(0.1329)*
280	0.1416
290	0.1506
300	0.1599
310	0.1694
320	0.1793
330	0.1895
340	0.1997
350	0.2101
360	0.220
370	0.231
380	0.242
390	0.253
400	0.264
410	0.275
420	0.286
430	0.297
440	0.309
450	0.320
460	0.331
470	0.343
480	0.354
490	0.365
500	0.377

DISCUSSION

GAS

The most extensive tabulation of values for n-butane is that given by Lenoir (223), from 266 to 505 K. No source references were given for these values. Actual experimental data have been given by various investigators, mainly in the region 273 - 300 K. In analysis of the data it was found that the ice point value quoted by Mann and Dickens (237) agreed well with the recommended values but that the temperature coefficient quoted by them was considerably low. No attempt was made to evaluate the alignment chart of Brokaw (420) as C_p/R data are required, or the reduced chart of Codegone (70, 73) which involves a knowledge of the thermal conductivity at the critical point.

The recommended values were obtained from a smooth curve drawn through all the available values. In the range 250 to 375 K the recommended values should be accurate to about two percent while the error at the highest temperature tabulated may be five percent.

*Extrapolated for the gas phase ignoring pressure dependence.
(n. b. p. = 273 K)

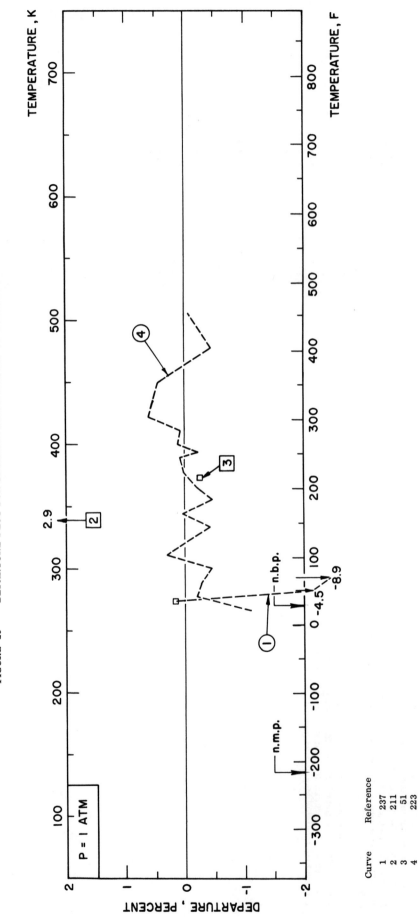

FIGURE 29   DEPARTURE PLOT FOR THERMAL CONDUCTIVITY OF GASEOUS n-BUTANE

TABLE 30   THERMAL CONDUCTIVITY OF CARBON DIOXIDE

RECOMMENDED VALUES

[Temperature, T, K; Thermal Conductivity, k, mW cm^{-1} K^{-1}]

SATURATED LIQUID

T	k
250	1.338 ‡
255	1.278 ‡
260	1.218 ‡
265	1.117 ‡
270	1.099 ‡
275	1.040 ‡
280	0.980 ‡
285	0.920 ‡
290	0.861 ‡
295	0.801 ‡
300	0.741 ‡

DISCUSSION

SATURATED LIQUID

There exist two experimental works on the thermal conductivity of liquid carbon dioxide. Kardos (526) made measurements in a hot-wire apparatus covering the temperature range from 260 to 284 K at a constant pressure of 60 atm. Although the accuracy of his measurements was reported to be within 2.5 percent, his results are greatly distorted by convection, as pointed out by Planck (530), and the values which he obtained are several times greater than the corresponding data of Sellschopp (312). The careful measurements of Sellschopp were carried out in a coaxial cylinder apparatus for both liquid and gaseous phases under various pressures up to 90 atm at temperatures from 283 to 313 K. The estimated accuracy of these measurements was one percent for the liquid phase. However, his values are also partially distorted by convection, as pointed out by Sellschopp himself. Therefore, Borovik (44, 45) attempted to correct Sellschopp's data by means of a theoretical calculation.

Since the results corrected by Borovik are considered to be the most reliable at present, the values reported for the liquid phase are used to obtain those of the saturated liquid by means of a graphical extrapolation used in this analysis. The values thus obtained are given equal weight and are fitted to a linear formula, represented by

$10^6 k$ (cgsu) = 1032.2 - 2.8500 T   (T in K).

As the value at the critical point is excluded in the present estimation, the equation should be valid between 260 and 303 K. It is found that the equation fits the corrected values within 0.15 percent. The recommended values, which are calculated from the above formula, should be correct within two percent.

In the departure plot, a value at the critical point using Borovik's correction is given beyond the limit of validity of the formula. Departures of Kardos' results at 60 atm are also plotted without any adjustment. However, Sellschopp's original data are extrapolated graphically to the saturation line. Incidentally, a set of calculated values, which are obtained from an empirical correlation by Koch (527), is also plotted as a dotted line beyond the lower limit of validity of the above formula.

‡ Under saturated vapor pressures. (n.s.p. = 195 K)

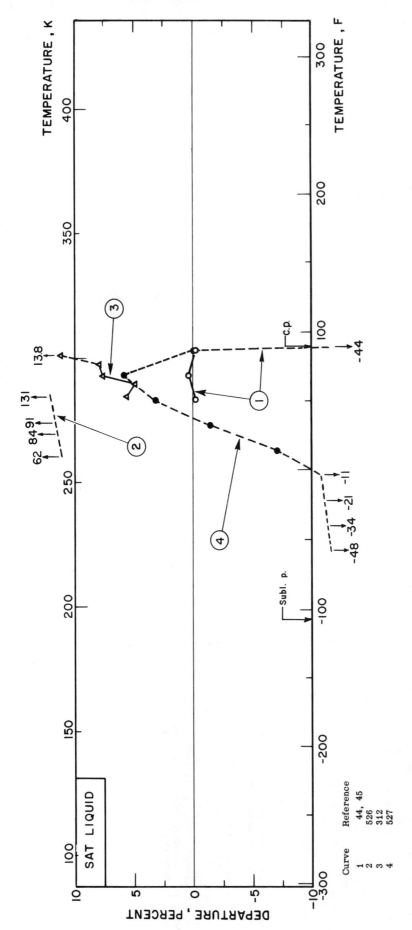

FIGURE 30 DEPARTURE PLOT FOR THERMAL CONDUCTIVITY OF LIQUID CARBON DIOXIDE

TABLE 30   THERMAL CONDUCTIVITY OF CARBON DIOXIDE

RECOMMENDED VALUES

[Temperature, T, K; Thermal Conductivity, k, mW cm^{-1}K^{-1}]

GAS

DISCUSSION

Thirty seven different sources were found of data on the thermal conductivity of gaseous carbon dioxide. The data available were found to be somewhat more accurate than that for many other gases but, surprisingly, not as accurate as that for a few special cases. Only two sets of measurements (192, 331) were found which yielded values differing by more than seven percent from the recommended values for temperatures below 600 K. Estimations and correlations were also examined and proved to be reliable below 400 K, values agreeing with the recommended values to within one percent being the general rule. However, at 600 K these values can differ by as much as eight percent.

The recommended values were obtained from a large scale plot of the available data as a function of temperature, the values so obtained being adjusted by examination of first and second differences of adjacent values. From 273 to 400 K they should be accurate to about two percent, from 400 to 600 K and below 273 K about five percent and above 600 K about ten percent. Further measurements would be desirable below 273 K and above 600 K in order to reduce this uncertainty and to provide data for a critical test of statistical mechanical theories. No attempt was made in this compilation to include values estimated for the plasma temperature region.

GAS

T	k	T	k	T	k
		550	0.363	950	0.651
		560	0.371	960	0.656
		570	0.379	970	0.662
180	(0.0828)*	580	0.387	980	0.668
190	(0.0890)*	590	0.395	990	0.674
200	0.0953	600	0.403	1000	0.680
210	0.1017	610	0.412	1010	0.685
220	0.1083	620	0.421	1020	0.691
230	0.1151	630	0.429	1030	0.696
240	0.1219	640	0.438	1040	0.702
250	0.1289	650	0.446	1050	0.707
260	0.1360	660	0.455	1060	0.713
270	0.1433	670	0.463	1070	0.718
280	0.1508	680	0.472	1080	0.723
290	0.1585	690	0.480	1090	0.728
300	0.1662	700	0.487	1100	0.733
310	0.1740	710	0.495	1110	0.738
320	0.1817	720	0.503	1120	0.743
330	0.1895	730	0.510	1130	0.747
340	0.1973	740	0.518	1140	0.752
350	0.2050	750	0.525	1150	0.757
360	0.2128	760	0.532	1160	0.761
370	0.2206	770	0.540	1170	0.766
380	0.2284	780	0.547	1180	0.771
390	0.2362	790	0.553	1190	0.775
400	0.2441	800	0.560	1200	0.780
410	0.2519	810	0.566	1210	0.785
420	0.2598	820	0.572	1220	0.789
430	0.2677	830	0.578	1230	0.793
440	0.2756	840	0.584	1240	0.798
450	0.2834	850	0.590	1250	0.804
460	0.2912	860	0.597	1300	0.825
470	0.2991	870	0.603	1350	0.846
480	0.3070	880	0.609	1400	0.867
490	0.3149	890	0.615	1450	0.888
500	0.3228	900	0.621	1500	0.909
510	0.331	910	0.627		
520	0.339	920	0.633		
530	0.347	930	0.639		
540	0.355	940	0.645		

*Ignoring pressure dependence.
(n. b. p. = 195 K)

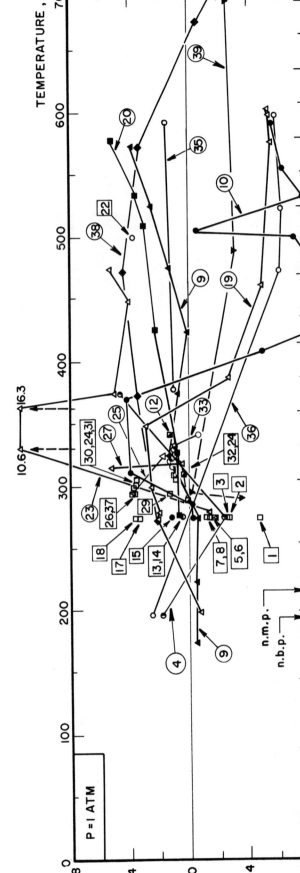

FIGURE 30   DEPARTURE PLOT FOR THERMAL CONDUCTIVITY OF GASEOUS CARBON DIOXIDE

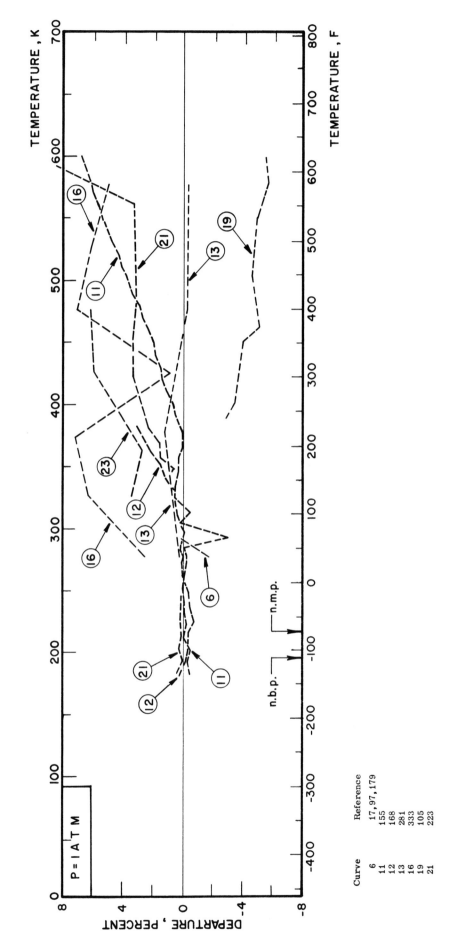

FIGURE 30 DEPARTURE PLOT FOR THERMAL CONDUCTIVITY OF GASEOUS CARBON DIOXIDE (Continued)

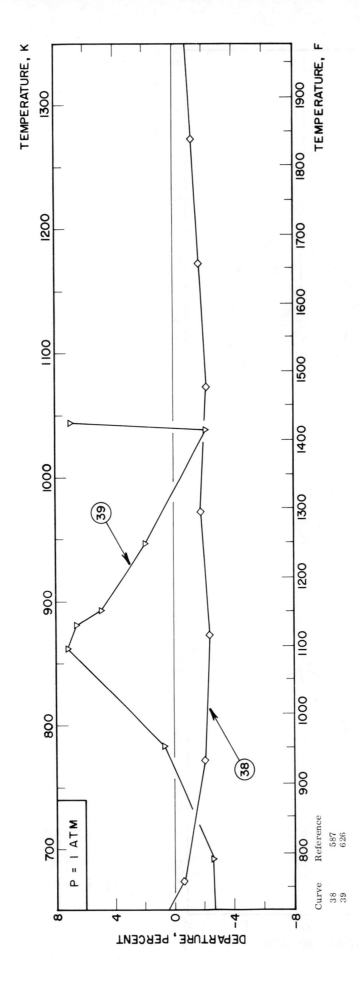

FIGURE 30 DEPARTURE PLOT FOR THERMAL CONDUCTIVITY OF GASEOUS CARBON DIOXIDE (continued)

TABLE 31 THERMAL CONDUCTIVITY OF CARBON MONOXIDE

RECOMMENDED VALUES

[Temperature, T, K; Thermal Conductivity, k, mW cm^{-1}K^{-1}]

DISCUSSION

SATURATED LIQUID

Only one experimental investigation is available on the thermal conductivity of liquid carbon monoxide, being that of Borovik et al. (46) in the temperature range between 78 and 112 K. Using a hot-wire apparatus, they reported four experimental points. The results were cited in a compendium by Johnson (167). In the present analysis, all of the original points are given equal weight and are fitted to a linear equation.

The correlation formula obtained is given by

$10^6 k$ (cgsu) = 682.33 - 4.2638 T (T in K).

This equation fits the experimental results with a mean deviation of 2.7 percent. The maximum deviation is found to be 3.6 percent, as shown in the departure plot. The above formula is used to generate the recommended values. All of the tabulated values are thought to be correct within three percent.

SATURATED LIQUID

T	k
65	1.695 †
70	1.606
75	1.517
80	1.428
85	1.339 ‡
90	1.249 ‡
95	1.160 ‡
100	1.071 ‡
105	0.982 ‡
110	0.893 ‡
115	0.803 ‡
120	0.714 ‡
125	0.625 ‡

† Extrapolated for the supercooled liquid. (n.m.p. = 68 K)
‡ Under saturated vapor pressures. (n.b.p. = 81 K)

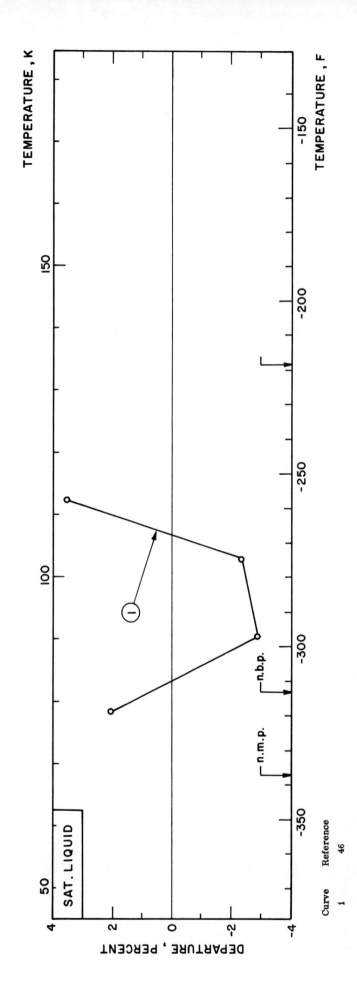

FIGURE 31   DEPARTURE PLOT FOR THERMAL CONDUCTIVITY OF LIQUID CARBON MONOXIDE

TABLE 31 THERMAL CONDUCTIVITY OF CARBON MONOXIDE

RECOMMENDED VALUES

[Temperature, T, K; Thermal Conductivity, k, mW cm^{-1}K^{-1}]

GAS

DISCUSSION

GAS

Measurements of the thermal conductivity of carbon monoxide gas have been reported over appreciable temperature ranges by Eucken (96), Johnston and Grilly (168), Keyes (187), Gardiner and Schafer (112) and Geier and Schafer (587). In addition, data are available from other sources (14, 17, 86, 126, 127, 146, 147, 156, 168, 177, 207, 218, 329, 396) for mainly single temperatures and three correlations (146, 147 and 223, 521) have appeared.

Below 400 K the overall agreement in the data can be assessed as better than one percent and the recommended values should be accurate to one half percent. At higher temperatures the recommended values have been selected to merge with the trend of the Geier and Schafer data which agree remarkably with the Keyes calculated values. The recommended values should be accurate to two percent from 400 to 750 K and five percent for higher temperatures. Experimental measurements to verify the Geier and Schafer high temperature data are desirable.

T	k	T	k	T	k	T	k
80	(0.0693)*	400	0.323	700	0.497	1000	0.644
90	0.0780	410	0.329	710	0.502	1010	0.649
		420	0.336	720	0.508	1020	0.654
100	0.0875	430	0.342	730	0.513	1030	0.659
110	0.0962	440	0.349	740	0.518	1040	0.663
120	0.1049	450	0.355	750	0.523	1050	0.668
130	0.1138	460	0.362	760	0.528	1060	0.673
140	0.1228	470	0.367	770	0.533	1070	0.677
		480	0.374	780	0.538	1080	0.683
		490	0.380	790	0.544	1090	0.687
150	0.1318	500	0.386	800	0.549	1100	0.692
160	0.1406	510	0.392	810	0.553	1110	0.697
170	0.1492	520	0.398	820	0.558	1120	0.701
180	0.1577	530	0.404	830	0.563	1130	0.706
190	0.1661	540	0.410	840	0.568	1140	0.710
200	0.1745	550	0.416	850	0.572	1150	0.715
210	0.1825	560	0.421	860	0.577	1160	0.720
220	0.1904	570	0.427	870	0.582	1170	0.725
230	0.1983	580	0.433	880	0.587	1180	0.729
240	0.2062	590	0.438	890	0.592	1190	0.733
250	0.2141	600	0.444	900	0.596	1200	0.738
260	0.222	610	0.450	910	0.601	1210	0.743
270	0.229	620	0.455	920	0.606	1220	0.747
280	0.237	630	0.460	930	0.611	1230	0.752
290	0.244	640	0.465	940	0.616	1240	0.756
300	0.252	650	0.471	950	0.621	1250	0.761
310	0.259	660	0.476	960	0.625		
320	0.266	670	0.481	970	0.630		
330	0.274	680	0.486	980	0.635		
340	0.281	690	0.491	990	0.639		
350	0.288						
360	0.295						
370	0.302						
380	0.309						
390	0.316						

*Extrapolated for the gas phase ignoring pressure dependence. (m. b. p. = 81 K)

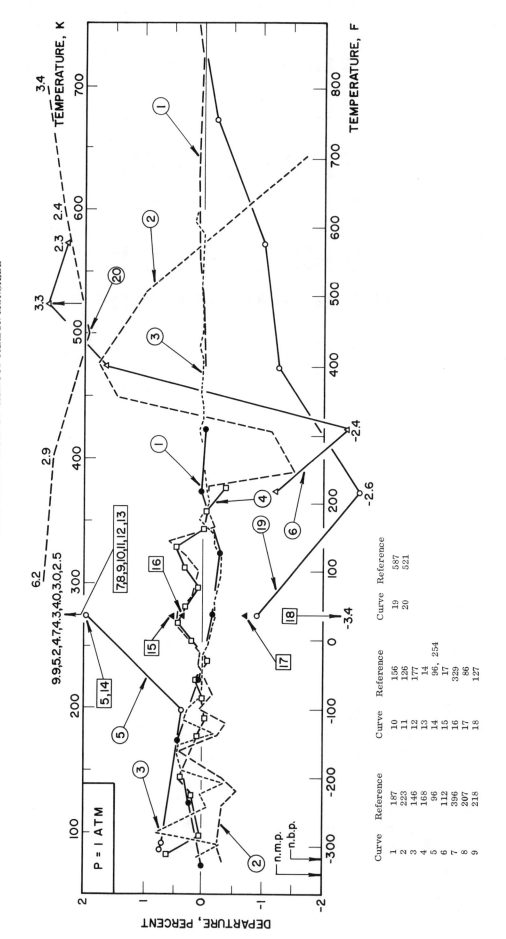

FIGURE 31 THERMAL CONDUCTIVITY OF GASEOUS CARBON MONOXIDE

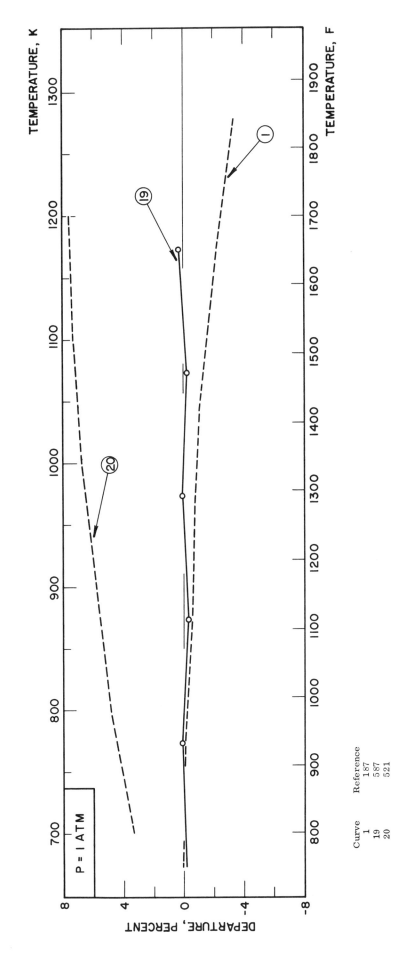

FIGURE 31 DEPARTURE PLOT FOR THERMAL CONDUCTIVITY OF GASEOUS CARBON MONOXIDE (continued)

Curve	Reference
1	187
19	587
20	521

TABLE 32   THERMAL CONDUCTIVITY OF CARBON TETRACHLORIDE

RECOMMENDED VALUES

[Temperature, T, K; Thermal Conductivity, k, mW cm^{-1}K^{-1}]

DISCUSSION

SATURATED LIQUID

There exist 20 experimental measurements in the literature on the thermal conductivity of liquid carbon tetrachloride. The extensive measurements of Challoner-Powell (434), Filippov (442), Mason (475), Riedel (279, 486), Schmidt-Leidenfrost (492) are considered to be reliable from the standpoint of their experimental methods and procedures, and are given heavy weight in this analysis. Furthermore, the single point values of Frontas'ev-Gusakov (447, 448) Riedel (483, 484) and Van der Held-Van Drunen (507) are also reliable and are used for the estimation of the most probable correlation. Although there are several other extensive measurements (427, 29, 437, 517), these are considered to be less reliable, and therefore, they are given no weight in this analysis.

The correlation formula obtained is

10^6k (cgsu) = 384.690 - 0.457184 T   (T in K).

This equation is found to fit the above enumerated measurements with a mean deviation of 1.4 percent and a maximum of 4.9 percent.

The recommended values are calculated from the above equation. The tabulated values are considered to be substantially correct in the range from 250 to 375 K and outside this range the uncertainty would increase.

SATURATED LIQUID

T	k
240	(1.150)†
250	(1.131)†
260	1.112
270	1.093
280	1.074
290	1.055
300	1.036
310	1.017
320	0.997
330	0.978
340	0.959
350	0.940
360	0.921 ‡
370	0.902 ‡
380	0.883 ‡
390	0.864 ‡
400	0.844 ‡
410	0.825 ‡
420	0.806 ‡
430	0.787 ‡
440	0.768 ‡

†Extrapolated for the supercooled liquid. (n.m.p. = 250 K)
‡Extrapolated for the liquid under vapor pressures, ignoring pressure dependence. (n.b.p. = 350 K)

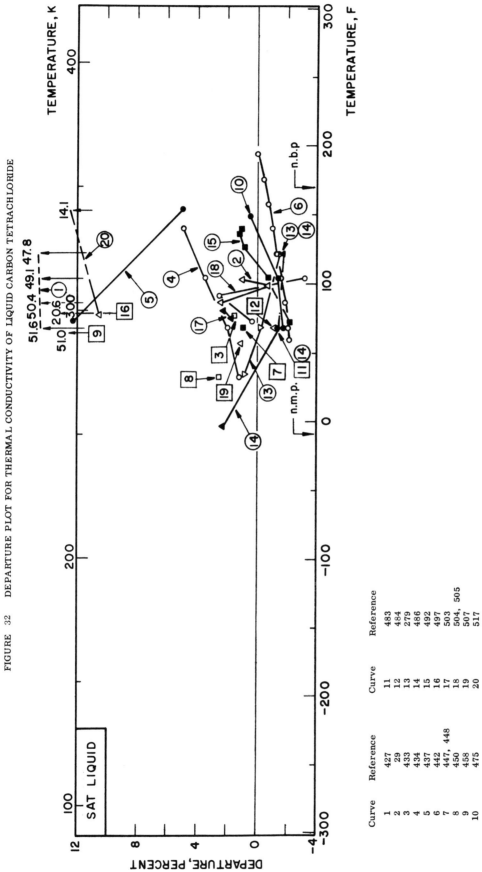

FIGURE 32 DEPARTURE PLOT FOR THERMAL CONDUCTIVITY OF LIQUID CARBON TETRACHLORIDE

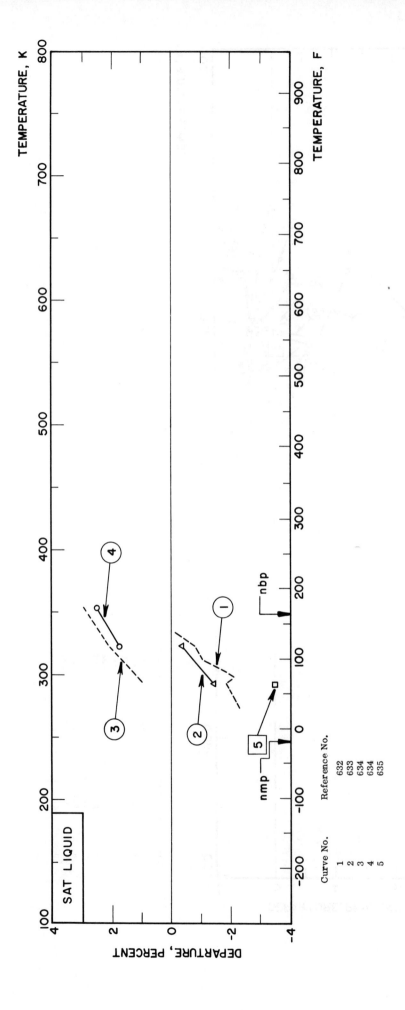

FIGURE 32   DEPARTURE PLOT FOR THERMAL CONDUCTIVITY OF LIQUID CARBON TETRACHLORIDE   (continued)

TABLE 32 THERMAL CONDUCTIVITY OF CARBON TETRACHLORIDE

RECOMMENDED VALUES

[Temperature, T, K; Thermal Conductivity, k, mW cm^{-1}K^{-1}]

DISCUSSION

GAS

The recommended values for the thermal conductivity of gaseous carbon tetrachloride include tabular data by Masia and co-workers (571, 578). Also included are earlier data by Eucken (97) and Moser (254) and two sets of correlated values by Lenoir (223) and Vargaftik (601, 602).

As will be noted from the departure plot, a five percent difference exists between the Masia and Moser values. The Lenoir correlation, given without source references, was evidently based on the Moser data while that of Vargaftik evidently considered only the Eucken and Moser data. The recommended values here presented were based on the Eucken and Masia data for temperatures from 273 to 373 K and the trend for higher temperatures was adjusted to approach the Vargaftik value at about 573 K.

In view of the disagreement between the Masia and Moser data, the tabulation of recommended values only extends from 250 to 500 K, and the values must be regarded as only being accurate to five percent. If further experimentation shows the Masia data to be accurate, this error estimate can be reduced to one or two percent.

GAS

T	k
250	0.0528
260	0.0555
270	0.0583
280	0.0612
290	0.0642
300	0.0673
310	0.0705
320	0.0738
330	0.0770
340	0.0803
350	0.0835
360	0.0866
370	0.0897
380	0.0928
390	0.0959
400	0.0989
410	0.1019
420	0.1049
430	0.1079
440	0.1108
450	0.1136
460	0.1163
470	0.1189
480	0.1214
490	0.1238
500	0.1261

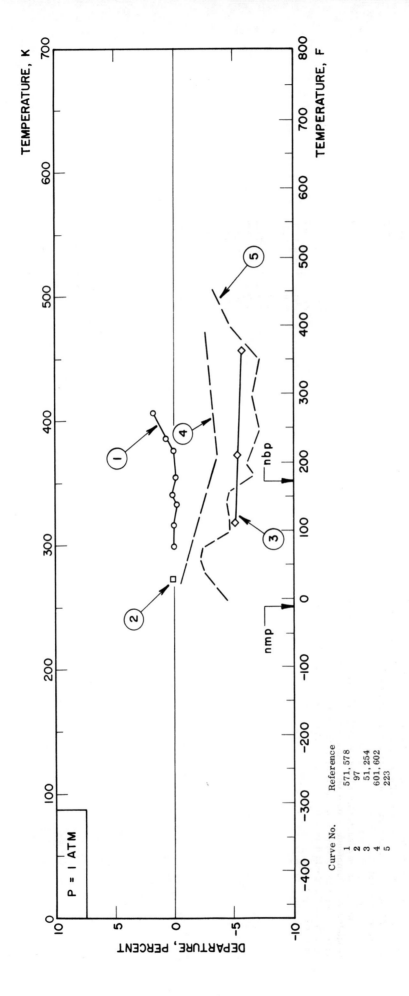

FIGURE 32 DEPARTURE PLOT FOR THERMAL CONDUCTIVITY OF GASEOUS CARBON TETRACHLORIDE

TABLE 33 THERMAL CONDUCTIVITY OF CHLOROFORM

DISCUSSION

SATURATED LIQUID

Fourteen sets of experimental data are available in the literature on the thermal conductivity of liquid chloroform. Considering the experimental methods and procedures, the results of Filippov (442), Mason (475), Riedel (483, 484), and Van der Held-Van Drunen (507) should be reliable, and are given equal weight in this analysis. Although there exist some measurements of Bates-Hazzard-Palmer (427) and Woolf-Sibbitt (517), their values are five to ten percent higher than the above-enumerated measurements, and, accordingly, are given no weight.

The correlation formula obtained is

$10^6 k$ (cgsu) = 423.344 - 0.480210 T   (T in K).

This equation is found to fit the above-mentioned values with a mean deviation of 0.9 percent and a maximum of 1.7 percent. The recommended values are calculated from the above equation. In the table of the recommended values, the data should be substantially correct in the range from 260 to 360 K, and outside this range the uncertainty would increase.

Older data of Guthrie (454, 455) and Weber (512, 513, 514) are not shown in the departure plot.

RECOMMENDED VALUES

[Temperature, T, K; Thermal Conductivity, k, mW cm^{-1}K^{-1}]

SATURATED LIQUID

T	k
200	(1.369)†
210	1.349
220	1.329
230	1.309
240	1.289
250	1.269
260	1.249
270	1.229
280	1.209
290	1.189
300	1.169
310	1.148
320	1.128
330	1.108
340	1.088‡
350	1.068‡
360	1.048‡
370	1.028‡
380	1.008‡
390	0.988‡
400	0.968‡

†Extrapolated for the supercooled liquid. (n.m.p. = 210 K)
‡Extrapolated for the liquid under vapor pressures, ignoring pressure dependence. (n.b.p. = 334 K)

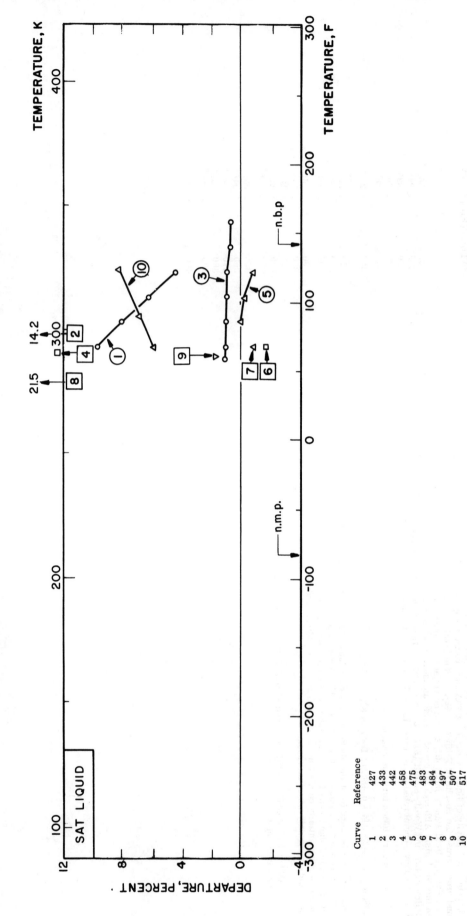

FIGURE 33 DEPARTURE PLOT FOR THERMAL CONDUCTIVITY OF LIQUID CHLOROFORM

TABLE 33 THERMAL CONDUCTIVITY OF CHLOROFORM

DISCUSSION

GAS

Only one set of values for the thermal conductivity of chloroform vapor are available in the literature, values from 0 to 500 F quoted without source reference by Lenoir (223). A request for information concerning the sources of the values was not answered. The values quoted as recommended values are interpolated for 10 K increments between 250 and 550 K from the Lenoir tabulation. No estimate is possible for the accuracy of this tabulation.

RECOMMENDED VALUES

[Temperature, T, K; Thermal Conductivity, k, mW cm^{-1}K^{-1}]

GAS

T	k
250	0.059*
260	0.062*
270	0.065*
280	0.068*
290	0.071*
300	0.075*
310	0.078*
320	0.082*
330	0.085*
340	0.088
350	0.092
360	0.096
370	0.100
380	0.104
390	0.107
400	0.111
410	0.115
420	0.119
430	0.123
440	0.127
450	0.131
460	0.135
470	0.139
480	0.143
490	0.147
500	0.151
510	0.155
520	0.159
530	0.163
540	0.168
550	0.172

*No consideration given to pressure dependence. (n. b. p. = 334 K)

TABLE 34 THERMAL CONDUCTIVITY OF n-DECANE

RECOMMENDED VALUES

[Temperature, T, K; Thermal Conductivity, k, mW cm^{-1}K^{-1}]

SATURATED LIQUID

T	k
230	1.610†
240	1.577†
250	1.545
260	1.512
270	1.479
280	1.447
290	1.414
300	1.381
310	1.349
320	1.316
330	1.283
340	1.250
350	1.218
360	1.185
370	1.152
380	1.120
390	1.087
400	1.054
410	1.022
420	0.989
430	0.956
440	0.923
450	0.891‡

DISCUSSION

SATURATED LIQUID

There are only three available experimental works on the thermal conductivity of liquid n-decane. The discrepancy between the reported values of different investigators is rather large, reaching as much as ten percent even near room temperature. As the results of Smith (501) are considered to be less reliable because he calibrated his apparatus using Bridgman's data (431), no weight is given to his data in this analysis. The data reported by Briggs (432) are close to those of Sakiadis - Coates (489) near room temperature, but the temperature dependence of the former is considerably smaller than that of the latter. This trend can be also seen in the case of some other liquids, and the results of Sakiadis, et al. should be considered to be more reliable from the point of view of the experimental accuracy. Therefore, the weights given in this analysis are two for the data of Sakiadis, et al. and one for Briggs' values.

The correlation formula obtained is

$10^6 k$ (cgsu) = 564.599 - 0.781586 T   (T in K).

This equation is found to fit the considered values with a mean deviation of 3.7 percent and a maximum of 9.0 percent.

The recommended values are calculated from the above equation. The tabulated values in the temperature range between 290 and 365 K should be substantially correct, but the uncertainty increases outside this range.

†Extrapolated for the subcooled liquid. (n. m. p. = 243 K)
‡Extrapolated for the liquid under vapor pressure, ignoring pressure dependence. (n. b. p. = 446 K)

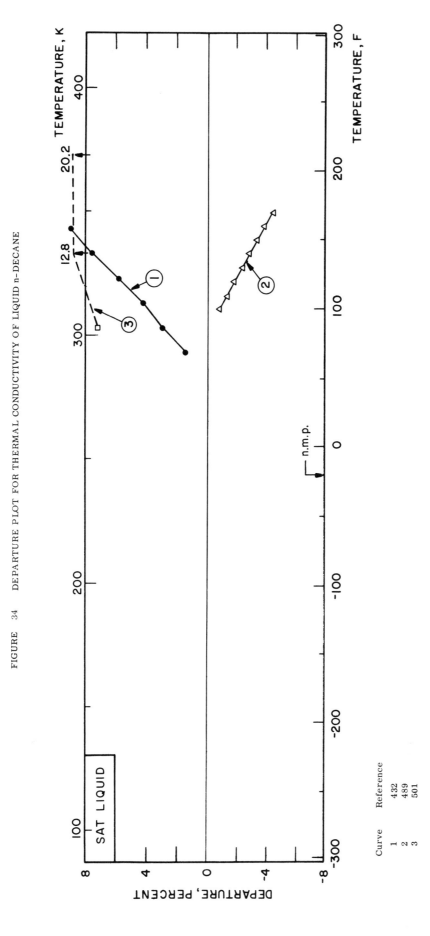

FIGURE 34   DEPARTURE PLOT FOR THERMAL CONDUCTIVITY OF LIQUID n-DECANE

TABLE 34 THERMAL CONDUCTIVITY OF n-DECANE

RECOMMENDED VALUES
[Temperature, T, K; Thermal Conductivity, k, mW cm^{-1}K^{-1}]

DISCUSSION

GAS

No experimental data for n-decane have been found. The values recommended here are based upon the tabulation of Lenoir (223) from 283 to 478K which is given with no source references and were obtained from the equation

$$10^5 k \text{ (cgsu)} = -1.43507 + 9.06309 \cdot 10^{-3}T + 1.78134 \cdot 10^{-5}T^2 \quad (T \text{ in K})$$

which fitted the Lenoir tabulation to within 0.4 percent. No departure plot or recommendation as to accuracy is possible.

GAS

T	k
250	(0.081)*
260	(0.089)*
270	(0.097)*
280	(0.105)*
290	(0.113)*
300	(0.121)*
310	(0.129)*
320	(0.138)*
330	(0.146)*
340	(0.155)*
350	(0.164)*
360	(0.173)*
370	(0.182)*
380	(0.192)*
390	(0.201)*
400	(0.211)*
410	(0.221)*
420	(0.231)*
430	(0.241)*
440	(0.251)*
450	0.262
460	0.272
470	0.283
480	0.294
490	0.305
500	0.316

*Extrapolated for the gas phase ignoring pressure dependence. (n. b. p. = 446K)

TABLE 35 THERMAL CONDUCTIVITY OF ETHANE

RECOMMENDED VALUES

[Temperature, T, K; Thermal Conductivity, k, mW cm^{-1} K^{-1}]

DISCUSSION

GAS

The measurements of Geier and Schafer (587) from 273 to 873 K considerably extend the temperature range for which data have been obtained for the thermal conductivity of gaseous ethane. They have also allowed of a more reliable correlation of the previously available data to be made within the range 273 to 523 K.

The recommended values above 473 K were based upon the Geier and Schafer data. Below 473 K the values were derived from analysis of the different data available. Only a few sets of data (191, 225, 576, 587) are available over any range of temperature, the remaining values being either single temperature data (17, 96, 97, 211, 237, 254, 313, 414) or correlated values (223, 521). The correlation of Lenoir (223) proved to be superior to that of Svehla (521). Further experimental measurements are desirable below 270 K and above about 500 K.

The recommended values should be accurate to within two percent from 250 to 600 K, five percent from 220 to 240 and 610 to 850 K and ten percent from 200 to 210 and 860 to 1000 K. These estimates include consideration of probable errors due to convection, radiation and pressure effects.

GAS

T	k	T	k	T	k
200	0.102	500	0.516	800	1.070
210	0.112	510	0.532	810	1.10
220	0.123	520	0.549	820	1.12
230	0.134	530	0.565	830	1.15
240	0.145	540	0.582	840	1.17
250	0.156	550	0.599	850	1.20
260	0.167	560	0.616	860	1.23
270	0.179	570	0.633	870	1.25
280	0.191	580	0.651	880	1.28
290	0.204	590	0.668	890	1.31
300	0.218	600	0.685	900	1.34
310	0.231	610	0.702	910	1.37
320	0.244	620	0.719	920	1.40
330	0.258	630	0.736	930	1.43
340	0.272	640	0.754	940	1.46
350	0.286	650	0.772	950	1.49
360	0.300	660	0.790	960	1.52
370	0.315	670	0.808	970	1.55
380	0.330	680	0.826	980	1.58
390	0.345	690	0.843	990	1.61
400	0.360	700	0.861	1000	1.64
410	0.375	710	0.880		
420	0.390	720	0.899		
430	0.406	730	0.918		
440	0.422	740	0.937		
450	0.437	750	0.957		
460	0.453	760	0.978		
470	0.469	770	0.999		
480	0.485	780	1.021		
490	0.500	790	1.045		

FIGURE 35   DEPARTURE PLOT FOR THERMAL CONDUCTIVITY OF GASEOUS ETHANE

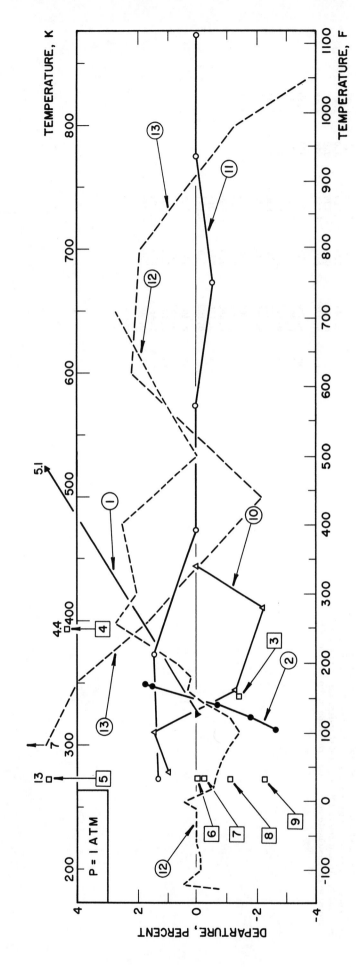

Curve	Reference
1	191
2	225
3	211
4	313
5	414
6	237
7	17, 97
8	254
9	96
10	576
11	587
12	223
13	521

TABLE 36 THERMAL CONDUCTIVITY OF ETHYL ALCOHOL

RECOMMENDED VALUES

[Temperature, T, K; Thermal Conductivity, k, mW cm^{-1}K^{-1}]

DISCUSSION

SATURATED LIQUID

Thirty-one experimental works are available on the thermal conductivity of liquid ethyl alcohol. The discrepancy between the reported values of different investigators is extremely large. The results of several extensive measurements fall into two groups, one group being about ten percent higher than the other. The results of Abas-Zade (3, 5) and Vargaftik (508) who used the hot-wire method, and the values of Bridgman (431), Daniloff (435), Markwood-Benning (238) and Smith (500) who used the coaxial-cylinder method, all fall in the higher set. On the other hand, recent careful measurements of Challoner-Powell (434), Filippov (442), Frontas'ev-Gusakov (447, 448), Mason (475), Riedel (487) and Sakiadis-Coates (489) fall within the lower group. From the point of view of experimental methods and procedures, the latter are more reliable. Therefore, only six sets of data mentioned above are used in this analysis.

The correlation formula obtained is

$10^6 k$ (cgsu) = 609.512 - 0.709240 T (T in K).

This equation is found to fit their values with a mean deviation of 2.4 percent and a maximum one of 8.9 percent.

The recommended values are calculated from the above formula. In the range from 220 to 365 K, the tabulated values should be substantially correct. Outside this range the uncertainty would increase.

The departure plot shows that the values of Riedel are most satisfactory over the whole temperature range. Several older values of Graetz (451, 452, 453), Guthrie (454, 455), Weber (512, 513, 514) and Winkelmann (516)are not shown in this figure. The data of Tsederberg-Timrot (506), who measured 94 percent alcohol solution at temperatures from 193 K to 343 K, are also omitted.

Although several measurements (3, 5, 466, 490, 491) were made at higher temperatures and corresponding vapor pressures, these values are less reliable as seen from the departure plot, and no correlation is attempted in this region.

SATURATED LIQUID

T	k
150	2.105†
160	2.075
170	2.046
180	2.016
190	1.986
200	1.957
210	1.927
220	1.897
230	1.868
240	1.838
250	1.808
260	1.778
270	1.750
280	1.719
290	1.690
300	1.660
310	1.630
320	1.601
330	1.571
340	1.541
350	1.512
360	1.482‡
370	1.452‡
380	1.423‡
390	1.393‡
400	1.363‡
410	1.334‡
420	1.304‡
430	1.274‡
440	1.245‡

†Extrapolated for the supercooled liquid. (n.m.p. = 159±3 K)
‡Extrapolated under vapor pressure. (n.b.p. = 351 K)

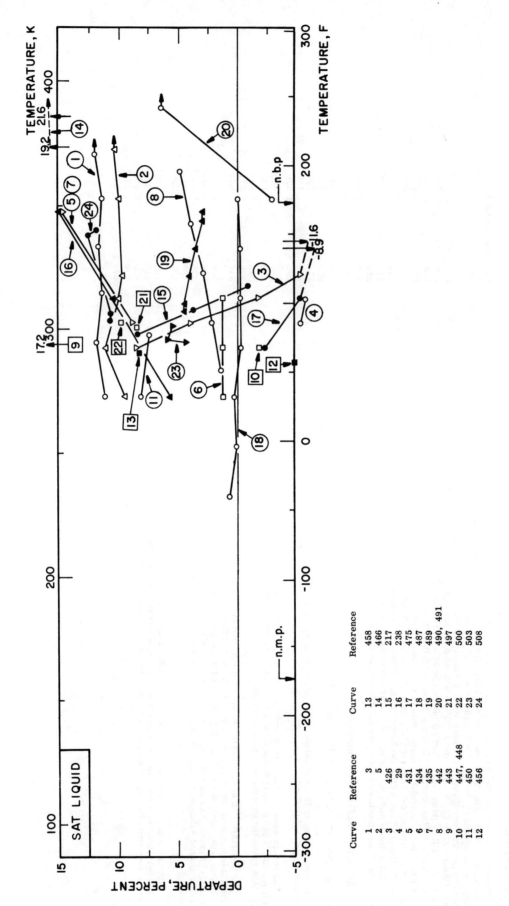

FIGURE 36 DEPARTURE PLOT FOR THERMAL CONDUCTIVITY OF LIQUID ETHYL ALCOHOL

TABLE 36 THERMAL CONDUCTIVITY OF ETHYL ALCOHOL

RECOMMENDED VALUES

[Temperature, T, K; Thermal Conductivity, k, mW cm^{-1}K^{-1}]

DISCUSSION

GAS

The most extensive data available for ethyl alcohol is that of Shushpanov (319) who reported twelve data values in the range 325 to 400K. Single data values are reported by (254) and requoted by (51) while Lenoir (223) lists ten values in the range 20 to 350 F with no source references. Further experimental measurements are to be desired.

The temperature variation of the Shushpanov data could be represented equally well by a quadratic or cubic function; agreement with the experimental values to within 2.6 percent was secured. Accordingly, the equation

$$10^5 k \text{ (cgsu)} = -5.909020 + 3.732944 \cdot 10^{-2} T - 1.975439 \cdot 10^{-5} T^2 \quad (T \text{ in K})$$

was used to compute values at 250(10) 500 K. Examination of these values showed that the Lenoir values in the range 125 to 260 F were consistently low by about five percent.

Extrapolation of the equation below 125 F gave values between 60 and 80 F agreeing with the Lenoir tabulation to within two percent. The deviation of the value of Moser from both tabulations is eight percent. It can therefore be concluded that the recommended values are accurate to about three percent from 320 to 400 K and to within ten percent for all other temperature tabulated.

GAS	
T	k
250	0.092*
260	0.103*
270	0.114*
280	0.125*
290	0.136*
300	0.147*
310	0.158*
320	0.168*
330	0.178*
340	0.188*
350	0.198*
360	0.208
370	0.218
380	0.227
390	0.236
400	0.245
410	0.254
420	0.263
430	0.272
440	0.280
450	0.288
460	0.296
470	0.304
480	0.312
490	0.320
500	0.327

*Extrapolated for the gas phase ignoring pressure dependence (n. b. p. = 351 K)

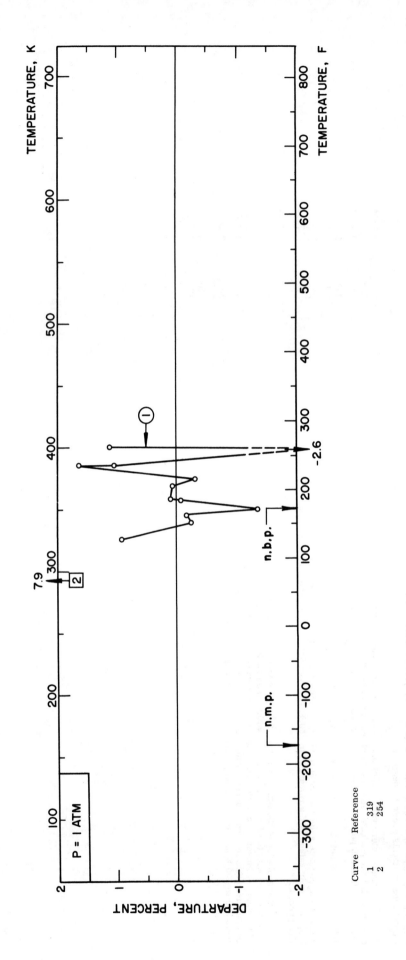

FIGURE 36 DEPARTURE PLOT FOR THERMAL CONDUCTIVITY OF GASEOUS ETHYL ALCOHOL

TABLE 37 THERMAL CONDUCTIVITY OF ETHYLENE

DISCUSSION

SATURATED LIQUID

The only available experimental data reported for the thermal conductivity of liquid ethylene is that due to Borovik-Matveev-Panina (46). They reported six data points over the temperature range from 112.66 K to 273.46 K. In this analysis, all of the points are given equal weight.

The correlation obtained is given by

10^6 (cgsu) = 883.091 - 2.49707 T (T in K).

This equation fits the experimental results with a mean deviation of 3.9 percent. The maximum deviation is found to be +9.7 percent, as shown in the departure plot. This high maximum is inherent to the scattering of the experimental results.

The above correlation formula is used to calculate the recommended values. The tabulated values should be correct within two percent in the temperature range between the melting point and the boiling point. Above the boiling point, the uncertainty would increase, because of the scattering in the original data.

RECOMMENDED VALUES

[Temperature, T, K; Thermal Conductivity, k, mW cm^{-1}K^{-1}]

SATURATED LIQUID

T	k
100	2.65†
110	2.547
120	2.441
130	2.337
140	2.232
150	2.128
160	2.023
170	1.919‡
180	1.81‡
190	1.71‡
200	1.61‡
210	1.50‡
220	1.40‡
230	1.29‡
240	1.19‡
250	1.08‡
260	0.98‡
270	0.87‡
280	0.77‡

†Extrapolated for the supercooled liquid. (n.m.p. = 104 K)
‡Extrapolated under vapor pressures. (n.b.p. = 170 K)

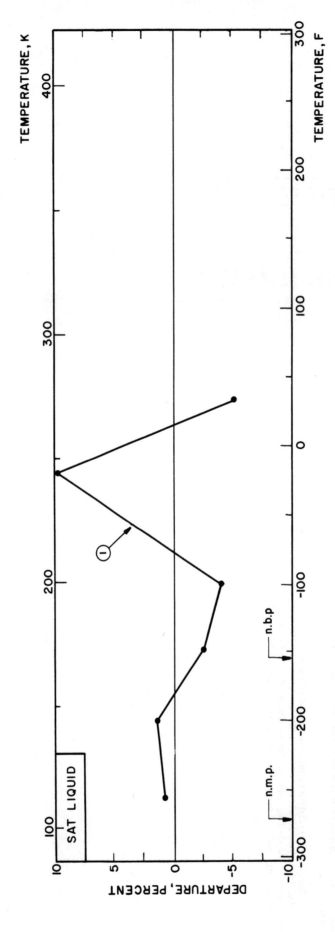

FIGURE 37 DEPARTURE PLOT FOR THERMAL CONDUCTIVITY OF LIQUID ETHYLENE

TABLE 37 THERMAL CONDUCTIVITY OF ETHYLENE

RECOMMENDED VALUES

[Temperature, T, K; Thermal Conductivity, k, mW cm^{-1}K^{-1}]

DISCUSSION

Various experimental data have been reported for ethylene vapor for temperatures between 178 and 426 K (59, 96, 97, 191, 204, 211, 218, 224, 313). In addition, two correlations have appeared (223, 524), the former from 169 to 394 K and the latter at 100 K intervals to 5000 K. Comparison of the correlations with the experimental data indicates the correlation of (223) to be in fair agreement, especially between 270 and 330 K, while the correlation of (524) gives values which are consistently high.

Recommended values were obtained from a smooth curve drawn through the data points and should be accurate to within a few percent between 250 and 400 K. Further experimental measurements are needed so that the variation with temperature can be more accurately determined below 250 and above 400 K.

GAS

T	k
200	0.088
210	0.095
220	0.103
230	0.114
240	0.125
250	0.138
260	0.151
270	0.164
280	0.177
290	0.191
300	0.204
310	0.217
320	0.231
330	0.245
340	0.259
350	0.274
360	0.289
370	0.304
380	0.320
390	0.335
400	0.350
410	0.365
420	0.380
430	0.396
440	0.411
450	0.427

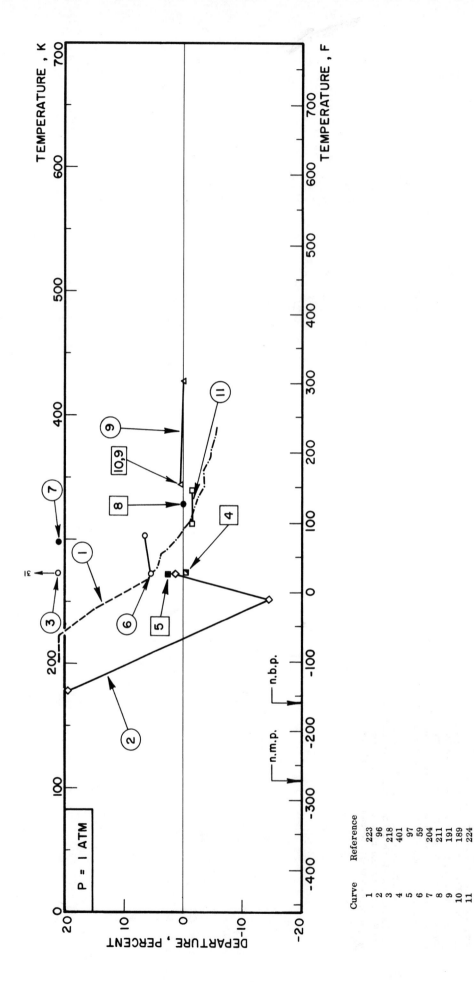

FIGURE 37 DEPARTURE PLOT FOR THERMAL CONDUCTIVITY OF GASEOUS ETHYLENE

TABLE 38 THERMAL CONDUCTIVITY OF ETHYLENE GLYCOL

RECOMMENDED VALUES

[Temperature, T, K; Thermal Conductivity, k, mWcm^{-1}K^{-1}]

SATURATED LIQUID

T	k
250	(2.509)[†]
260	2.522
270	2.536
280	2.549
290	2.563
300	2.576
310	2.590
320	2.603
330	2.616
340	2.630
350	2.643
360	2.657
370	2.670
380	2.684
390	2.697
400	2.710
410	2.724
420	2.737
430	2.751
440	2.764
450	2.78
460	2.79
470	2.80
480	2.82[‡]
490	2.83[‡]
500	2.85[‡]

DISCUSSION

SATURATED LIQUID

Nine experimental works are available in the literature on the thermal conductivity of liquid ethylene glycol. A pronounced anomaly, which produces a sharp depression in the thermal conductivity of about 14 percent between 310 K and 330 K, was found by Kraus (468). However, no confirmation of such an anomaly is found in the recent measurements of Filippov (442) and Riedel (487), whose results are considered to be more reliable from the standpoint of the experimental method and procedure. Therefore, no weight is given to the results of Kraus in this analysis. Furthermore, Bates - Hazzard (428) and Woolf - Sibbitt (517) showed that the thermal conductivity of ethylene glycol decreases with increasing temperature in the same manner as most other normal liquids, but their results are also given no weight, because recent extensive measurements (442, 487, 499) give a positive slope. Concluding that the thermal conductivity of ethylene glycol has a positive slope with temperature and no anomaly at any temperature, equal weight is given to the results of Filippov (442), Riedel (487) and Slawecki - Molstad (499).

The correlation formula obtained is

$10^6 k$ (cgsu) $= 519.442 + 0.320920$ T (T in K).

This equation is found to fit the considered values with a mean deviation of 1.3 percent and a maximum of 2.5 percent, and is used to generate the recommended values. The tabulated data should be substantially correct in the temperature range from 250 to 400 K but outside this range the uncertainty increases.

In the departure plot, only a part of Kraus's data is shown, in order to avoid unnecessary confusion.

[†]Extrapolated for the supercooled liquid. (n.m.p. = 258 K)
[‡]Extrapolated for the liquid under vapor pressures ignoring pressure dependence. (n.b.p. = 471 K)

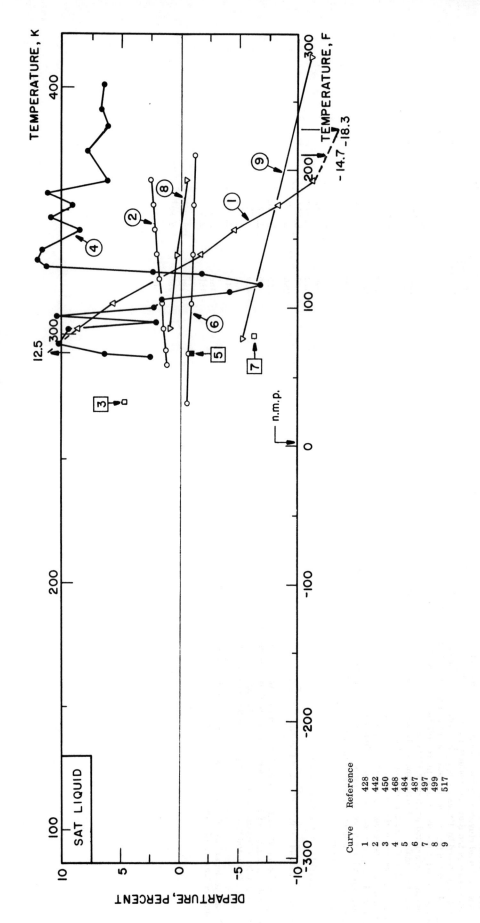

FIGURE 38 DEPARTURE PLOT FOR THERMAL CONDUCTIVITY OF LIQUID ETHYLENE GLYCOL

TABLE 39 THERMAL CONDUCTIVITY OF ETHYL ETHER

RECOMMENDED VALUES

[Temperature, T, K; Thermal Conductivity, k, mW cm^{-1}K^{-1}]

DISCUSSION

SATURATED LIQUID

Twelve sets of experimental data are available in the literature for the thermal conductivity of liquid ethyl ether. Except for some old data (452, 512, 513), the discrepancy between different investigators is rather small. The results of Bridgman (431) are high in the same trend as several other liquids measured by him. In this analysis the values reported by Filippov (442), Mason (475), Riedel (483, 484, 486) and Spencer (504, 505) are given equal weight.

The correlation formula obtained is

$10^6 k$ (cgsu) = 597.554 - 0.964187 T (T in K).

This equation is found to fit the selected data sets with a mean deviation of 2.2 percent and a maximum one of 4.9 percent.

The recommended values are calculated from the above equation. The tabulated data from 190 to 330 K should be substantially correct. Outside this range the uncertainty would increase.

Although an extensive measurement up to the critical point was made by Abas-Zade (3), the correlation is not extended to high saturation vapor pressures. Older data, published in the 19th century, (452, 512, 513, 514) are not shown in the departure plot.

SATURATED LIQUID

T	k
160	1.855
170	1.814
180	1.774
190	1.734
200	1.693
210	1.653
220	1.613
230	1.572
240	1.532
250	1.492
260	1.451
270	1.411
280	1.371
290	1.330
300	1.290
310	1.250‡
320	1.209‡
330	1.169‡
340	1.129‡
350	1.088‡
360	1.048‡
370	1.008‡
380	0.967‡
390	0.927‡
400	0.887‡

‡Extrapolated under vapor pressure. (n.b.p. = 308 K)

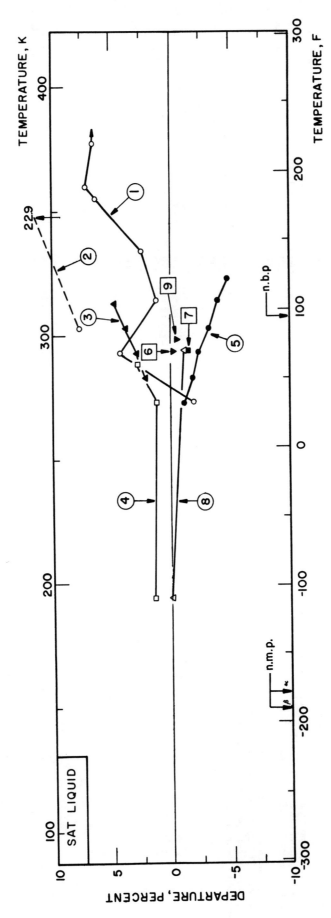

FIGURE 39 DEPARTURE PLOT FOR THERMAL CONDUCTIVITY OF LIQUID ETHYL ETHER

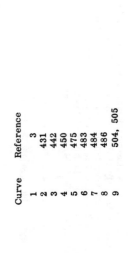

TABLE 39   THERMAL CONDUCTIVITY OF ETHYL ETHER

RECOMMENDED VALUES

[Temperature, T, K; Thermal Conductivity, k, mW cm^{-1}K^{-1}]

GAS

T	k
250	0.111*
260	0.118*
270	0.126*
280	0.134*
290	0.143*
300	0.1514*
310	0.1602
320	0.1692
330	0.1785
340	0.1880
350	0.1977
360	0.2077
370	0.2178
380	0.2282
390	0.2389
400	0.2497
410	0.2608
420	0.2721
430	0.2837
440	0.2954
450	0.3074
460	0.320
470	0.332
480	0.345
490	0.358
500	0.371

*Ignoring pressure dependence. (n. b. p. = 308 K)

DISCUSSION

GAS

Four principal sets of data are available for ethyl ether, being those reported by Gribkova (134) from 325 to 390 K, Moser (254) from 273 to 485 K, Vines (367) from 350 to 383 K and Vines and Bennett (370) from 330 to 422 K. These data have also been requoted in (51). A few other values are also available (85, 212, 272). A tabulation from 0 to 300 F, given with no source references, has also appeared (223).

Analysis of these data showed that those of Moser and of Vines and Bennett were in very good agreement. The Gribkova data were higher than all other measurements while the trend of the Vines data at the higher temperatures was not substantiated by the other measurements. Therefore the equation

$$10^5 k \text{ (cgsu)} = 0.166798 + 4.457152 \cdot 10^{-3} T + 2.720293 \cdot 10^{-5} T^2 \qquad (T \text{ in K})$$

fitting the Moser data to within 0.83 percent was used to generate thermal conductivity values from 250 to 500 K. Comparison with the other data showed that the Gribkova values were, on the average, 6.4 percent high; those of Vines and of Vines and Bennett agreed to within 1.3 percent. Within the range 310 to 420 K the recommended values should be accurate to one percent. For temperatures outside this range the accuracy should be a few percent. The Lenoir data agree with the recommended values to within five percent from 0 to 200 F but at 300 F are 10 percent higher. It appears that Lenoir extrapolated the Vines data to obtain his tabulated values and, as noted above, such an extrapolation is not in agreement with the other data.

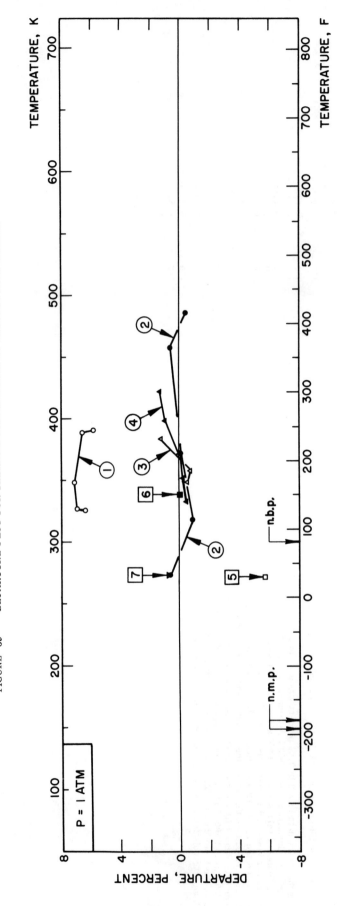

FIGURE 39 DEPARTURE PLOT FOR THERMAL CONDUCTIVITY OF GASEOUS ETHYL ETHER

Curve	Reference
1	134
2	254
3	367
4	370
5	85
6	212
7	272

TABLE 40   THERMAL CONDUCTIVITY OF FREON-11

RECOMMENDED VALUES

[Temperature, T, K; Thermal Conductivity, k, mW cm^{-1}K^{-1}]

SATURATED LIQUID

T	k
150	(1.42)†
160	(1.383)†
170	1.350
180	1.318
190	1.285
200	1.252
210	1.220
220	1.187
230	1.154
240	1.122
250	1.089
260	1.057
270	1.024
280	0.991
290	0.959
300	0.926‡
310	0.894‡
320	0.861‡
330	0.828‡
340	0.796‡
350	0.763‡

DISCUSSION

SATURATED LIQUID

Six experimental works are available in the literature on the thermal conductivity of liquid Freon-11. Excluding the extremely high values of Malhotra (473) and Markwood-Benning (238), the other results are considered to be reliable from the standpoint of their experimental methods and procedures. Therefore, equal weight was given to the data of Cherneyeva (422), Powell-Challoner (481), Riedel (483), and Danilova (436) (except for one value at 243.16 K), for the estimation of the most probable correlation.

The correlation equation obtained is

10^6k (cgsu) = 455.248 - 0.779660 T   (T in K).

This equation is found to fit the above-enumerated measurements with a mean deviation of 2.3 percent and a maximum of 6.5 percent.

The recommended values are calculated from the above formula. In the range from 200 K to 300 K, the tabulated data should be substantially correct. Outside this range the uncertainty would increase, because of the inconclusive nature of the available data.

†Extrapolated for the supercooled liquid.   (n. m. p. = 162 K)
‡Extrapolated under vapor pressures.   (n. b. p. = 297 K)

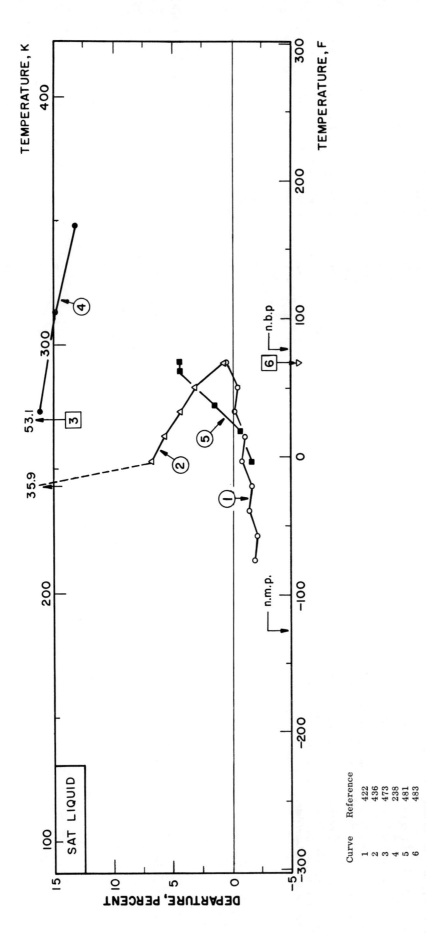

FIGURE 40   DEPARTURE PLOT FOR THERMAL CONDUCTIVITY OF LIQUID FREON-11

TABLE 40   THERMAL CONDUCTIVITY OF FREON-11

RECOMMENDED VALUES

[Temperature, T, K; Thermal Conductivity, k, mW cm^{-1}K^{-1}]

GAS

T	k
250	0.059*
260	0.063*
270	0.067*
280	0.071*
290	0.075*
300	0.079
310	0.083
320	0.087
330	0.091
340	0.095
350	0.100
360	0.104
370	0.108
380	0.112
390	0.116
400	0.120
410	0.124
420	0.128
430	0.132
440	0.136
450	0.140
460	0.144
470	0.148
480	0.152
490	0.156
500	0.160

DISCUSSION

GAS

Eight recent determinations of the thermal conductivity of gaseous Freon-11 have been reported (571) which differ considerably from the Cherneyeva (422) data, the deviation being of the order of fifteen percent. The correlation of Lenoir (223) given without source references, shows a fairly constant departure of about seven percent. It seems probable that Lenoir selected his values about half way between those of Cherneyeva and those of Markwood and Benning (238).

While a smooth curve through the Masia et. al. values gave a slightly better fit than a straight line the difference between the two fittings was smaller than the experimental uncertainty and much smaller than the differences between the different sets of data. The recommended values were obtained from a straight line fit of the Masia data, extrapolated where necessary. The recommended values fit the original data to within 0.40 percent. However, a much greater uncertainty must be ascribed to them in view of the poor agreement between the different data. Further experimental measurements on the substance are urgently required.

*Ignoring pressure dependence.
(n. b. p. = 297 K)

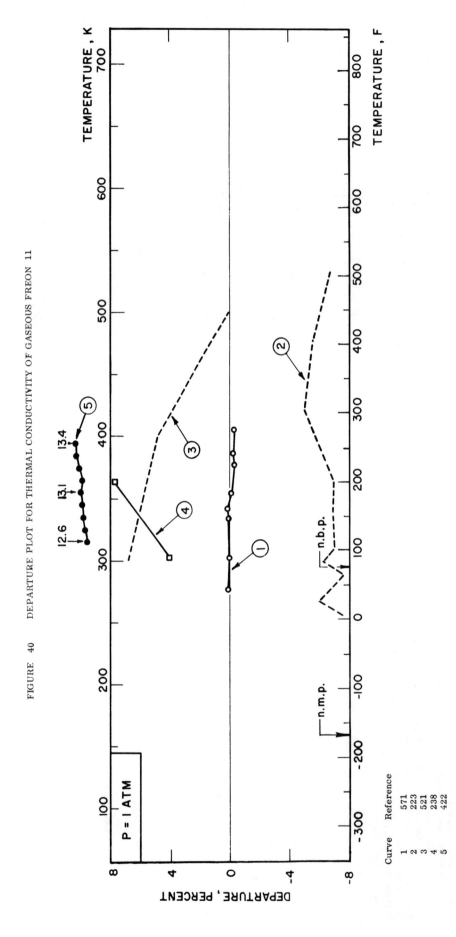

FIGURE 40 DEPARTURE PLOT FOR THERMAL CONDUCTIVITY OF GASEOUS FREON 11

TABLE 41 THERMAL CONDUCTIVITY OF FREON-12

SATURATED LIQUID

DISCUSSION

There exist six available experimental works on the thermal conductivity of liquid Freon-12. The results of Danilova (436) and Markwood-Benning (238) are considerable high, because they used Bridgman's data (431) for the calibration of their coaxial-cylinder apparatus. A set of old measurements of Griffiths et al. (525) is considered to be too low. Although the measurements of Powell-Challoner (481) are seen to be reliable, their results showed considerable scattering. Therefore, these values are excluded in the present analysis. The extensive results of Cherneyeva (422) and the single point value of Riedel (483) are used for the estimation of the most probable values.

The correlation formula obtained is given by

$$10^6 k \text{ (cgsu)} = 493.37 - 1.0945\, T \quad (T \text{ in K}).$$

This equation fits the above enumerated measurements with a mean deviation of 0.21 percent and a maximum of 0.38 percent. The above formula is used to generate the recommended values. The tabulated values should be correct in the temperature range between 230 to 360 K. Outside this range the uncertainty increases. In the departure plot, the recommended values by Powell-Challoner (481) are distinguished from their original experimental data by a dotted line.

RECOMMENDED VALUES

[Temperature, T, K; Thermal Conductivity, k, mW cm^{-1} K^{-1}]

SATURATED LIQUID

T	k
150	1.38
160	1.33
170	1.29
180	1.24
190	1.19
200	1.148
210	1.103
220	1.057
230	1.011
240	0.965
250	0.919‡
260	0.874‡
270	0.830‡
280	0.782‡
290	0.736‡
300	0.690‡
310	0.645‡
320	0.599‡
330	0.553‡
340	0.507‡
350	0.461‡

‡Under saturated vapor pressures. (n.b.p. = 243 K).

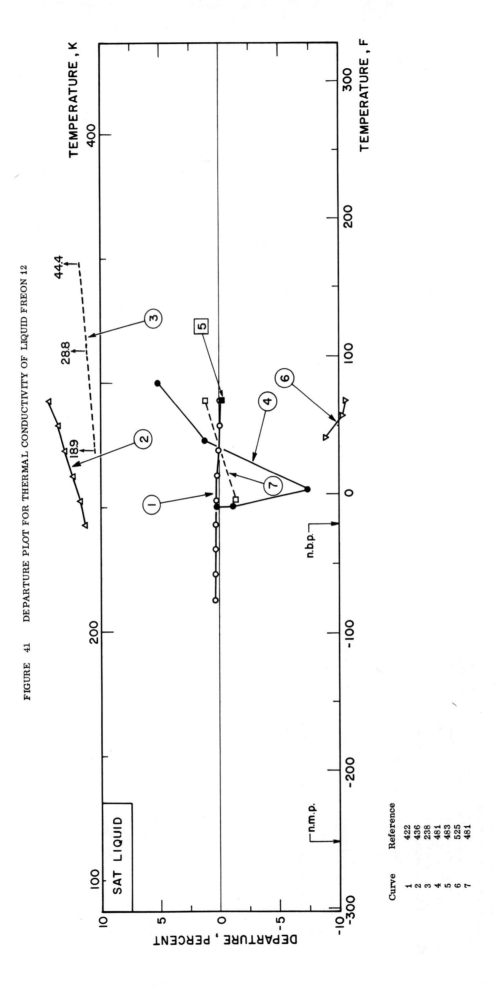

FIGURE 41   DEPARTURE PLOT FOR THERMAL CONDUCTIVITY OF LIQUID FREON 12

TABLE 41   THERMAL CONDUCTIVITY OF FREON 12

RECOMMENDED VALUES

[Temperature, T, K; Thermal Conductivity, k, mW cm^{-1} K^{-1}]

GAS

DISCUSSION

Three sets of experimental measurements have been reported for the thermal conductivity of gaseous Freon 12 which extend over moderate temperature ranges; those of (317, requoted by 51) from 278 to 437 K, those of (191) from 323 to 423 K and those of (570) from 278 to 408 K. A comparison of data in 1956 (275) also included data of (238, 422). Two correlations (233, 264) have appeared, the former covering the range 244 to 477 K and the latter the range 250 to 500 K.

Comparison of the differing data has been made and the recommended values were read from a smooth curve passing near to the data of (191, 317 and 570). The measurements of (570) resolve the difficulty of deciding which data are the more probable and the error estimate can be assessed as about 2.5 percent below and five percent for higher temperatures.

GAS

T	k
250	0.072
260	0.077
270	0.082
280	0.087
290	0.092
300	0.097
310	0.102
320	0.107
330	0.113
340	0.118
350	0.123
360	0.128
370	0.134
380	0.139
390	0.145
400	0.151
410	0.156
420	0.162
430	0.168
440	0.173
450	0.179
460	0.185
470	0.191
480	0.197
490	0.203
500	0.209

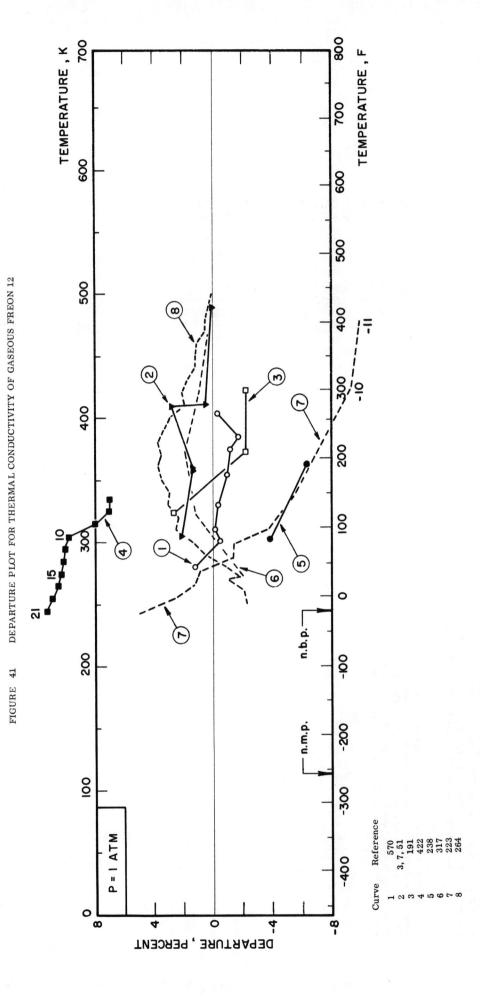

FIGURE 41 DEPARTURE PLOT FOR THERMAL CONDUCTIVITY OF GASEOUS FREON 12

TABLE 42 THERMAL CONDUCTIVITY OF FREON 13

RECOMMENDED VALUES

[Temperature, T, K; Thermal Conductivity, k, mW cm^{-1}K^{-1}]

GAS

T	k
250	0.0911
260	0.0972
270	0.1033
280	0.1094
290	0.1155
300	0.1217
310	0.1279
320	0.1341
330	0.1403
340	0.1466
350	0.1529
360	0.1592
370	0.1655
380	0.1718
390	0.1781
400	0.1844
410	0.191
420	0.197
430	0.204
440	0.210
450	0.216
460	0.223
470	0.229
480	0.236
490	0.242
500	0.249

DISCUSSION

GAS

The only source of experimental data available for the thermal conductivity of gaseous Freon 13 was the thesis of Riendā (570), which reports eight experimental values between 278 and 407 K. The information contained in two other sources was found incomplete. Attempts are being made to determine if the values are appropriate for inclusion. Svehla (521) quotes values at 100 K increments calculated from intermolecular potential values determined from viscosity data. Two values falling within the range of experimental measurements show departures of some 7.8 percent.

The recommended values were read from a smooth curve drawn through the experimental data and agree with these to within one half percent. Further measurements are to be desired to check the accuracy of this single source of data. The accuracy in the recommended values should be within one or two percent from 280 to 400 K and five percent for all other temperatures tabulated.

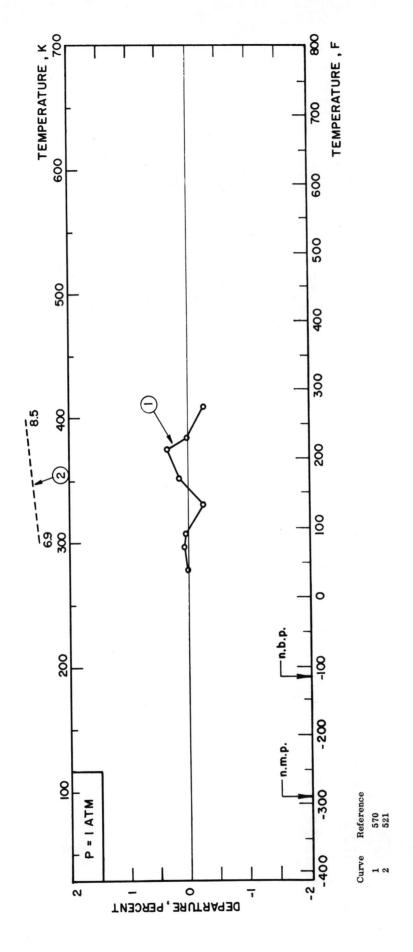

FIGURE 42 DEPARTURE PLOT FOR THERMAL CONDUCTIVITY OF GASEOUS FREON 13

TABLE 43   THERMAL CONDUCTIVITY OF FREON-21

DISCUSSION

SATURATED LIQUID

Only two sets of experimental data are available on the thermal conductivity of liquid Freon-21. These data are in disagreement by as much as 12 to 14 percent. The results of Markwood-Benning (238) are too high, because they calibrated their apparatus using the data of Bridgman (431). Therefore, no weight is given to their work in this analysis. As the recent measurement of Powell-Challoner (481) are considered to be reliable from the standpoint of their experimental methods and procedures, all of the reported points are given equal weight.

The correlation equation obtained is

$$10^6 k \text{ (cgsu)} = 502.497 - 0.815192\, T \quad (T \text{ in K}).$$

This equation is found to agree with the considered values within ± 0.5 percent. On the other hand, the data of Markwood-Benning deviate as much as +11.6 to +13.8 percent, as seen from the departure plot.

The above equation is used to generate the recommended values. Because of the narrow temperature range of experimental data, the reliability of the tabulated values would be unknown at temperatures both below 225 and above 325 K.

RECOMMENDED VALUES

[Temperature, T, K; Thermal Conductivity, k, mW cm^{-1} K^{-1}]

SATURATED LIQUID

T	k
130	(1.659)[†]
140	1.625
150	1.591
160	1.557
170	1.523
180	1.489
190	1.454
200	1.420
210	1.386
220	1.352
230	1.318
240	1.284
250	1.250
260	1.216
270	1.182
280	1.147
290	1.113[‡]
300	1.079[‡]
310	1.045[‡]
320	1.011[‡]
330	0.977[‡]
340	0.943[‡]
350	0.909[‡]
360	0.875[‡]
370	0.840[‡]
380	0.806[‡]
390	0.772[‡]
400	0.738[‡]

[†] Extrapolated for supercooled liquid. (n.m.p. 133 K).

[‡] Extrapolated for the liquid under vapor pressures, ignoring pressure dependence. (n.b.p. 282 K).

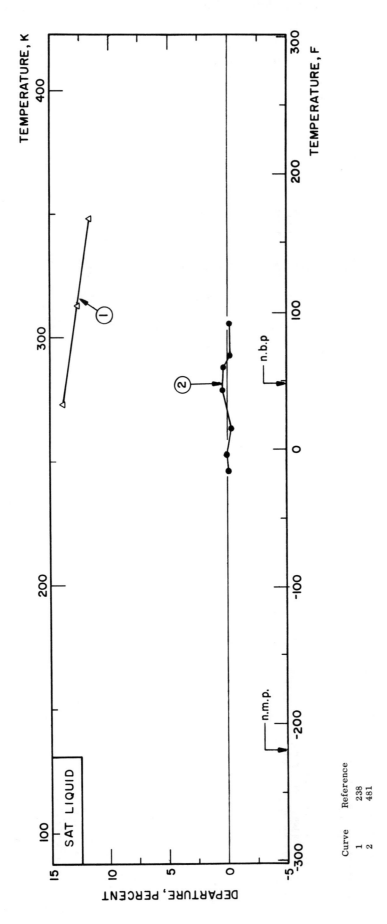

FIGURE 43   DEPARTURE PLOT FOR THERMAL CONDUCTIVITY OF LIQUID FREON-21

TABLE 43   THERMAL CONDUCTIVITY OF FREON-21

RECOMMENDED VALUES

[Temperature, T, K; Thermal Conductivity, k, mW cm^{-1}K^{-1}]

DISCUSSION

GAS

Eight experimental measurements of the thermal conductivity of gaseous Freon-21 for the range 278-408 K have recently been reported (570). These show a rather large disagreement with the two data points of Markwood and Benning (238) which are in substantial agreement with the correlated values of Lenoir (223) given without source references. From experience with the other Freon compounds, it was decided to base the revised recommended values on the recent (570) measurements. The recommended values were obtained from a smooth curve drawn through the experimental data points which agreed with the data to within 0.2 percent. Further experimental measurements are desirable in order that the accuracy of the recent data may be checked. Assuming that these are correct, it is considered that the recommended values are accurate to within two percent between 250 and 330 K and within five percent for the other temperatures tabulated.

GAS

T	k
250	0.066*
260	0.070*
270	0.074*
280	0.078
290	0.082
300	0.086
310	0.090
320	0.095
330	0.100
340	0.105
350	0.109
360	0.115
370	0.120
380	0.125
390	0.131
400	0.138
410	0.144
420	0.151
430	0.157
440	0.164
450	0.172

*Ignoring pressure dependence.
(n. b. p. = 282 K)

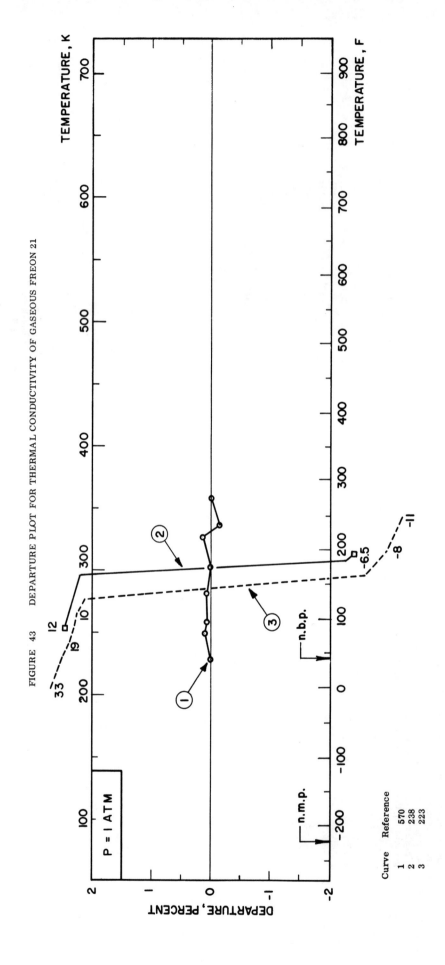

FIGURE 43 DEPARTURE PLOT FOR THERMAL CONDUCTIVITY OF GASEOUS FREON 21

TABLE 44   THERMAL CONDUCTIVITY OF FREON-22

RECOMMENDED VALUES

[Temperature, T, K; Thermal Conductivity, k, mW cm^{-1}K^{-1}]

SATURATED LIQUID

DISCUSSION

Only three sets of experimental works are available on the thermal conductivity of liquid Freon-22. The results of Markwood - Benning (238) are extremely high, because they calibrated their apparatus using Bridgman's data (431). Therefore, they are less reliable and no weight is given to them. The values of Cherneyeva (418), who covered the temperature range from 193 K to 293 K, and the values recommended by Powell - Challoner (481), who measured at temperatures between 253 K and 284 K, both are considered to be reliable, and are given equal weight in this analysis.

The correlation formula obtained is

$10^6 k$ (cgsu) = 521.817 - 1.05375 T   (T in K).

The mean deviation of the considered values from this equation is found to be 2.9 percent with a maximum of 3.7 percent.

This equation is used to generate the recommended values. In this table, the data should be correct within three percent in the temperature range from 190 to 320 K, and outside this range the uncertainty increases.

SATURATED LIQUID

T	k
150	1.52
160	1.48
170	1.43
180	1.39
190	1.35
200	1.301
210	1.257
220	1.213
230	1.170
240	1.125‡
250	1.081‡
260	1.037‡
270	0.993‡
280	0.949‡
290	0.905‡
300	0.861‡
310	0.817‡
320	0.772‡
330	0.728‡
340	0.684‡
350	0.640‡

‡Under saturated vapor pressures.   (n.b.p. = 233 K).

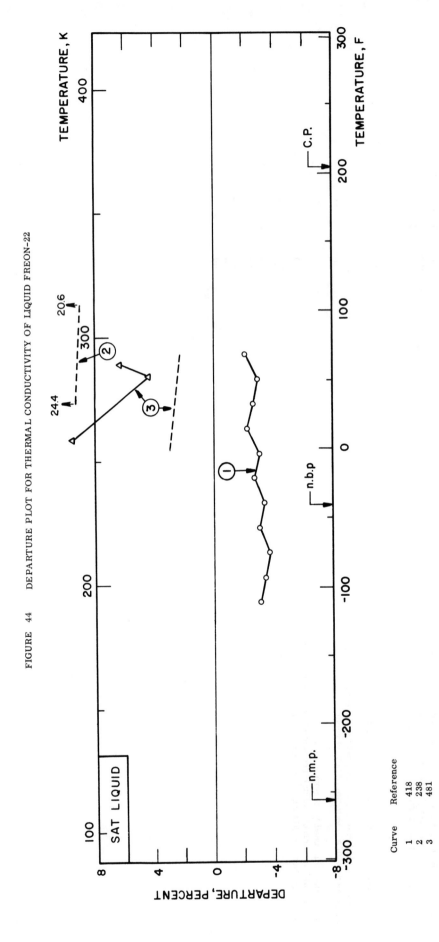

FIGURE 44 DEPARTURE PLOT FOR THERMAL CONDUCTIVITY OF LIQUID FREON-22

Curve	Reference
1	418
2	238
3	481

TABLE 44   THERMAL CONDUCTIVITY OF FREON-22

RECOMMENDED VALUES

[Temperature, T, K; Thermal Conductivity, k, mW cm^{-1}K^{-1}]

DISCUSSION

GAS

Eight data points obtained (570) for the thermal conductivity of gaseous Freon 22 from 278 to 407 K show a larger increase of conductivity with temperature than exhibited by previous experimental values (238, 418, 419) and correlations (223, 653) apparently based on the (238, 418) values respectively. The trend of the most extensive previous data is to approach more closely the recent data at the higher temperatures, however, severe disagreement exists at the lower temperatures. The agreement with the calculated values of Svehla (521) when compared with other gases suggests that the recent (570) data are more probable than the earlier data.

The recommended values were obtained from a smooth curve drawn through the (570) data. New measurements of the conductivity are to be desired, especially below 325 K. Pending such determination, the recommended values should be accurate to about ten percent for the entire range of temperature tabulated.

GAS

T	k
250	0.0801
260	0.0850
270	0.0899
280	0.0951
290	0.1003
300	0.1056
310	0.1112
320	0.1169
330	0.1228
340	0.1289
350	0.1350
360	0.1413
370	0.1477
380	0.1542
390	0.1608
400	0.1678
410	0.174
420	0.181
430	0.188
440	0.195
450	0.203
460	0.210
470	0.218
480	0.225
490	0.233
500	0.241

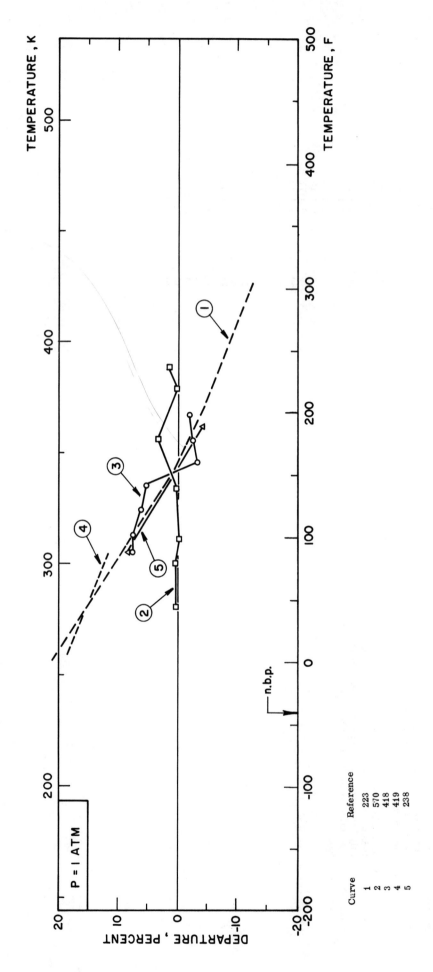

FIGURE 44 DEPARTURE PLOT FOR THERMAL CONDUCTIVITY OF GASEOUS FREON-22

TABLE 45 THERMAL CONDUCTIVITY OF FREON-113

RECOMMENDED VALUES

[Temperature, T, K; Thermal Conductivity, k, mW cm^{-1}K^{-1}]

SATURATED LIQUID

T	k
230	(0.874) †
240	0.854
250	0.833
260	0.812
270	0.791
280	0.771
290	0.750
300	0.729
310	0.708
320	0.688
330	0.667 ‡
340	0.646 ‡
350	0.626 ‡
360	0.605 ‡
370	0.584 ‡
380	0.563 ‡
390	0.543 ‡
400	0.522 ‡

† Extrapolated for the supercooled liquid (n.m.p. = 238 K).
‡ Under saturated vapor pressures. (n.b.p. = 321 K).

DISCUSSION

SATURATED LIQUID

Six experimental investigations are available in the literature on the thermal conductivity of liquid Freon-113. Excluding the extremely high values of Markwood - Benning (238) who used Bridgman's data (431) for the calibration of their apparatus, the other results fall within 10 percent. From the standpoint of the experimental method and procedure, the extensive values of Cherneyeva (63) and Powell - Challoner (481), and the single point value of Riedel (483) are considered to be the most reliable and are used to estimate the most probable correlation in this analysis.

The correlation formula obtained is given by

$10^6 k$ (cgsu) = 322.94 - 0.49552 T (T in K).

This equation is found to fit the experimental values of the above enumerated investigators with a mean deviation of 3.2 percent and a maximum of 6.8 percent. The recommended values are tabulated using the above equation. The values in the temperature range from 240 to 330 K are thought to be correct within the error limits described above, but outside this range the uncertainty increases. In the departure plot, the recommended values by Powell - Challoner (481) are distinguished from their original experimental data by a dotted line.

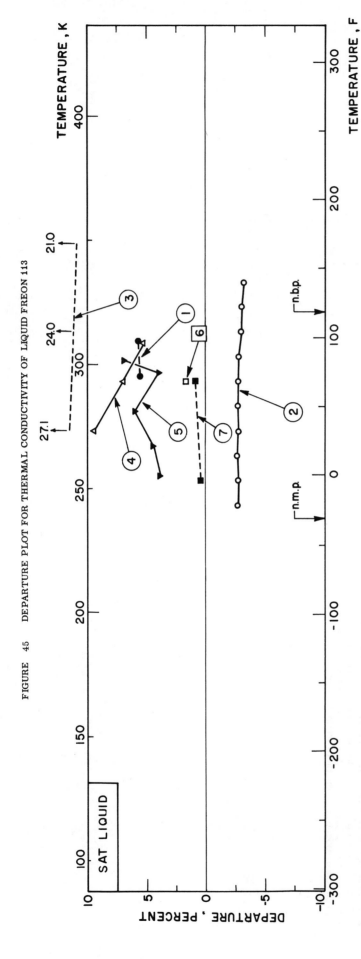

FIGURE 45   DEPARTURE PLOT FOR THERMAL CONDUCTIVITY OF LIQUID FREON 113

Curve	Reference
1	522
2	63
3	238
4	275
5	481
6	483
7	481

TABLE 45   THERMAL CONDUCTIVITY OF FREON 113

RECOMMENDED VALUES

[Temperature, T, K; Thermal Conductivity, k, mW cm^{-1}K^{-1}]

GAS

T	k
250	0.0579*
260	0.0610*
270	0.0644*
280	0.0681*
290	0.0722*
300	0.0766*
310	0.0813*
320	0.0863*
330	0.0915
340	0.0970
350	0.1027
360	0.109
370	0.115
380	0.121
390	0.128
400	0.135

* Ignoring pressure dependence.
(n. b. p. = 321 K)

GAS

DISCUSSION

Twelve experimental values for the thermal conductivity of gaseous Freon 113 are quoted by Cherneyeva (63) for temperatures between 273 and 383 K. Only three other values, for 258, 303 and 348 K are apparently available in the literature (238, 419, 572) and it appears that these are all due to the work of Markwood and Benning (238) although their original publication only cites two of them. In addition, a correlation, given without source references by Lenoir (223), gives values from 266 to 366 K.

Comparison of the available data shows that poor agreement exists between the two sets of experimental data. In view of the assumptions made in analysis of other Cherneyeva data that such data are high, the recommended values were obtained from a graph of the three data values from 258, 303 and 348 K as a function of temperature. It is considered very probable that these values are high for temperatures above 355 and below 266 K and somewhat low for intermediate temperatures. New measurements are urgently required for this substance. The error possible in the recommended values is assessed at ten percent pending such new measurements.

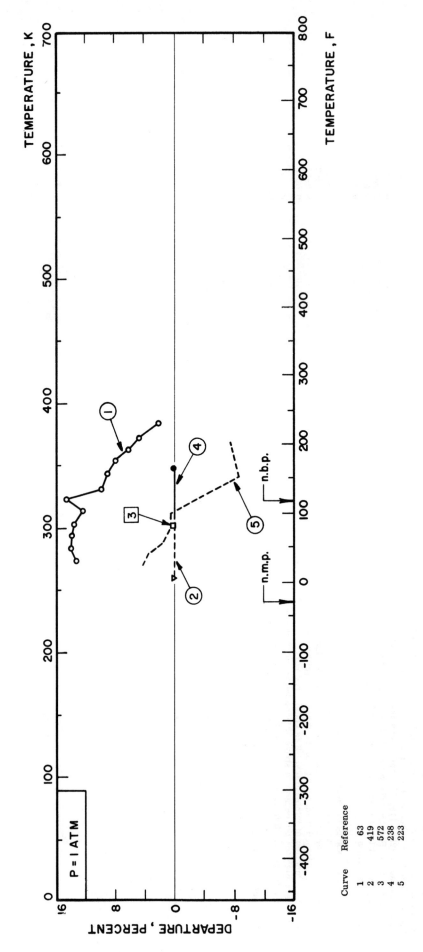

FIGURE 45   DEPARTURE PLOT FOR THERMAL CONDUCTIVITY OF GASEOUS FREON 113

TABLE 46   THERMAL CONDUCTIVITY OF FREON-114

RECOMMENDED VALUES

[Temperature, T, K; Thermal Conductivity, k, mW cm^{-1}K^{-1}]

DISCUSSION

SATURATED LIQUID

Only three sets of experimental data are available on the thermal conductivity of liquid Freon-114. The values of Powell - Challoner (481), covering the temperature range from 250 K to 290 K, are considered to be most reliable. Their measurement agrees closely with the single measurement of Riedel (483) at 293.16 K. Another measurement at three temperatures between 270 K and 350 K was made by Markwood - Benning (238). However, as they calibrated their apparatus using Bridgman's data (431), their results are extremely high. Therefore, no weight is given to their works in this analysis.

All of the reported points of Powell, et al. and Riedel are given equal weight. The correlation formula obtained is

$10^6 k$ cgsu = 335.962 - 0.605000 T   (T in K).

This equation is found to fit the considered values with a mean deviation of 0.42 percent and a maximum deviation of 1.2 percent, and is used to calculate the recommended values.

Because of the narrow temperature range of experimental data, the reliability of the tabulated values would be unknown at temperatures both below 220 and above 310 K.

SATURATED LIQUID

T	k
170	(0.975)†
180	0.950
190	0.925
200	0.899
210	0.874
220	0.849
230	0.823
240	0.798
250	0.773
260	0.748
270	0.722
280	0.697‡
290	0.671‡
300	0.646‡
310	0.621‡
320	(0.596)‡
330	(0.570)‡
340	(0.545)‡
350	(0.520)‡
360	(0.494)‡
370	(0.469)‡
380	(0.444)‡
390	(0.418)‡
400	(0.393)‡

† Extrapolated for the supercooled liquid. (n.m.p. = 179 K).
‡ Under saturated vapor pressures. (n.b.p. = 276 K).

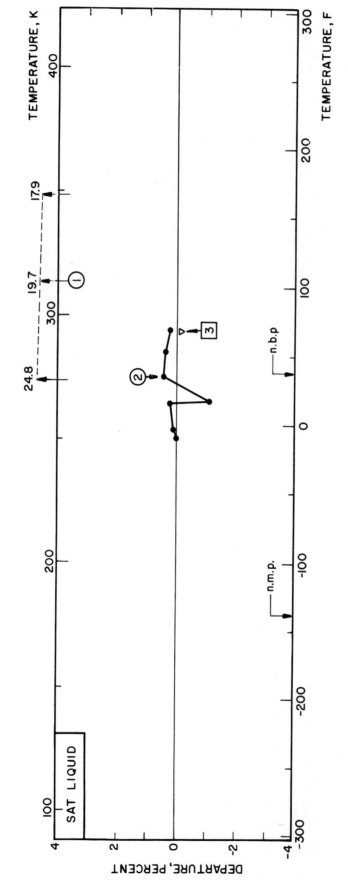

FIGURE 46   DEPARTURE PLOT FOR THERMAL CONDUCTIVITY OF LIQUID FREON-114

Curve	Reference
1	238
2	481
3	483

TABLE 46   THERMAL CONDUCTIVITY OF FREON-114

RECOMMENDED VALUES

[Temperature, T, K; Thermal Conductivity, k, mW cm^{-1}K^{-1}]

DISCUSSION

GAS

Few experimental data are available for this substance. Keyes (191) published three provisional values in 1954 but has never since referred to or amended these data. A tabulation from 266 to 366 K was given by Lenoir (223) without source references. In this work the Lenoir tabulation was analyzed and used as the basis for the tabulation of recommended values. It was found that the equation

$$10^5 k \text{ (cgsu)} = 3.79439 - 1.81602 \cdot 10^{-2}T + 4.76397 \cdot 10^{-5}T^2 \qquad \text{(T in K)}$$

fitted the seven values of Lenoir with a maximum error of 1.5 percent. Accordingly, it was used to tabulate data from 250 to 400 K.

The three data values of Keyes show an average deviation of about 5.5 percent from the Lenoir tabulation. In the absence of any other data no assessment of the most probable set of values can be made. Further experimental data are required for this substance. The recommended values meanwhile are estimated to have a probable error of from about five percent between 260 and 370 K to as much as ten percent at 400 K.

GAS

T	k
250	0.093*
260	0.096*
270	0.099*
280	0.102
290	0.106
300	0.110
310	0.115
320	0.120
330	0.125
340	0.131
350	0.137
360	0.144
370	0.151
380	0.158
390	0.166
400	0.174
410	0.182
420	0.191
430	0.201
440	0.210
450	0.220
460	0.231
470	0.242
480	0.253
490	0.265
500	0.277

*Ignoring pressure dependence. (n. b. p. = 276 K)

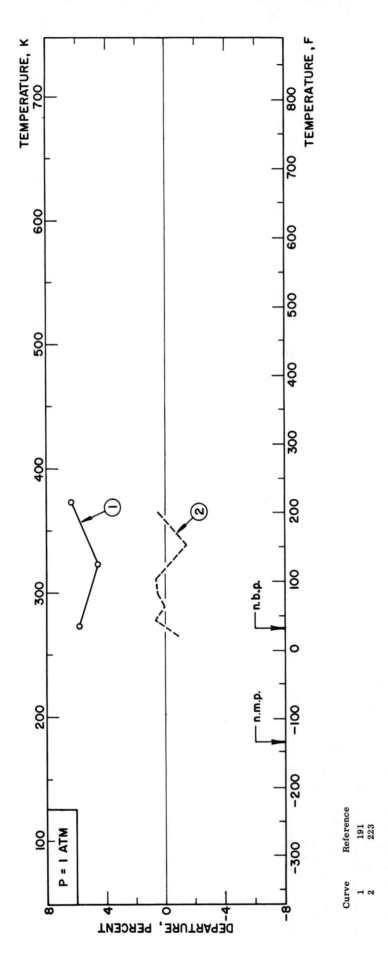

FIGURE 46 DEPARTURE PLOT FOR THERMAL CONDUCTIVITY OF GASEOUS FREON-114

TABLE 47   THERMAL CONDUCTIVITY OF GLYCEROL

RECOMMENDED VALUES

[Temperature, T, K; Thermal Conductivity, k, mW cm^{-1}K^{-1}]

SATURATED LIQUID   DISCUSSION

There exist 24 experimental works on the thermal conductivity of liquid glycerol. The discrepancy between the reported values of reliable investigators is rather small, excluding the results of Lees (217), Scheffy-Johnson (490, 491), who gave a negative slope with temperature, and of several older investigators (452, 454, 455, 512, 513, 516). Although the results of Kurtener-Malyshev (469) deviate positively, the data are considered to be reasonable because they measured 95 percent glycerol - 5 percent of water solution. Bates (425), claims that the thermal conductivity of glycerol is independent of temperature from 283.16 K to 353.16 K. However, several extensive measurements show that the thermal conductivity of this substance increases with increasing temperature, as in the case of water and ethylene glycol. Therefore, the data of Challoner - Powell (434), Filippov (442), Kaye - Higgins (464), McCready (471), Mason (475), Riedel (484, 487) and Vargaftik (508) are considered to be reliable, and are given equal weight in this analysis.

The correlation formula obtained is

10^{6}k (cgsu) = 593,771 + 0.315103 T   (T in K).

This equation is found to fit the above-enumerated results with a mean deviation of 1.7 percent and a maximum of 4.7 percent.

The recommended values are calculated from the above equation. The tabulated values should be substantially correct in the temperature range from 250 to 400 K. Outside this range the uncertainty increases.

An extensive measurement for the thermal conductivity of supercooled liquid glycerol was made by Schulz (496), to 203.16 K. However, no consideration for this range is given in this analysis. Several older values of Graetz (452), Guthrie (454, 455), Weber (512, 513) and Winkelmann (516) are not shown in the departure plot.

SATURATED LIQUID

T	k
250	(2.81)†
260	(2.83)†
270	(2.84)†
280	(2.85)†
290	(2.867)†
300	2.880
310	2.893
320	2.906
330	2.919
340	2.933
350	2.946
360	2.959
370	2.972
380	2.985
390	2.999
400	3.012
410	3.025
420	3.038
430	3.051
440	3.064
450	3.078
460	3.09
470	3.10
480	3.12
490	3.13
500	3.14
510	3.16
520	3.17
530	3.18
540	3.20
550	3.21

†Extrapolated for the supercooled liquid.   (n.m.p. = 291 K)

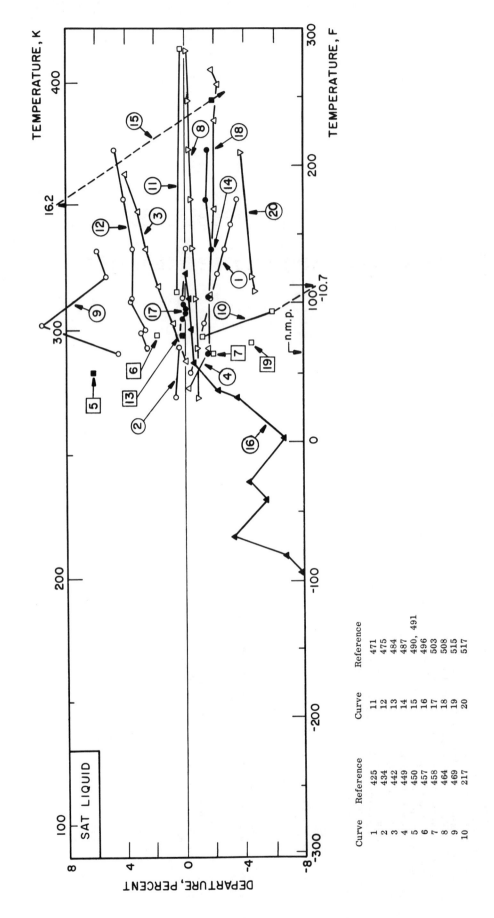

FIGURE 47 DEPARTURE PLOT FOR THERMAL CONDUCTIVITY OF LIQUID GLYCEROL

TABLE 48   THERMAL CONDUCTIVITY OF n-HEPTANE

RECOMMENDED VALUES

[Temperature, T, K; Thermal Conductivity, k, mW cm^{-1}K^{-1}]

SATURATED LIQUID

T	k
250	1.38
260	1.36
270	1.33
280	1.311
290	1.288
300	1.265
310	1.242
320	1.219
330	1.196
340	1.173
350	1.150
360	1.127
370	1.104
380	(1.081) ‡
390	(1.06) ‡
400	(1.04) ‡

DISCUSSION

SATURATED LIQUID

Six experimental works are available in the literature for the thermal conductivity of liquid n-heptane. The extensive measurements of Filippov (442) are considered to be most reliable, and the single point data of both Riedel (484) and Frontas'ev-Gusakov (447, 448) are also thought to be reasonable. Therefore, all of these experimental points are given equal weight in this analysis. On the other hand, although the values reported by Sakiadis-Coates (489) fall in the neighborhood of the above enumerated data, the trend with temperature is seen to be too steep. A set of data by Briggs (432) shows a too flat trend with temperature. Therefore, no weight is given to these two sets of data.

The correlation formula obtained is given by

$$10^6 k \text{ (cgsu)} = 466.94 - 0.54892\, T \quad (T \text{ in K}).$$

This equation is found to fit Filippov's data within 0.23 percent, and deviates -1.3 percent from the value of Frontas'ev-Gusakov and +0.88 percent from Riedel's single point. The above equation is used to generate the recommended values. The data from 290 to 370 K should be substantially correct, and outside this range the uncertainty increases.

The calculated values from a correlation between the thermal conductivity and the molecular weight for the homologous series of the saturated n-hydrocarbons were reported by Smith (501). However, as he used his own experimental values for the correlation, the absolute values reported are found to be very high, as shown in the departure plot.

‡ Extrapolated for the liquid under vapor pressures, ignoring pressure dependence. (n.b.p. = 371 K)

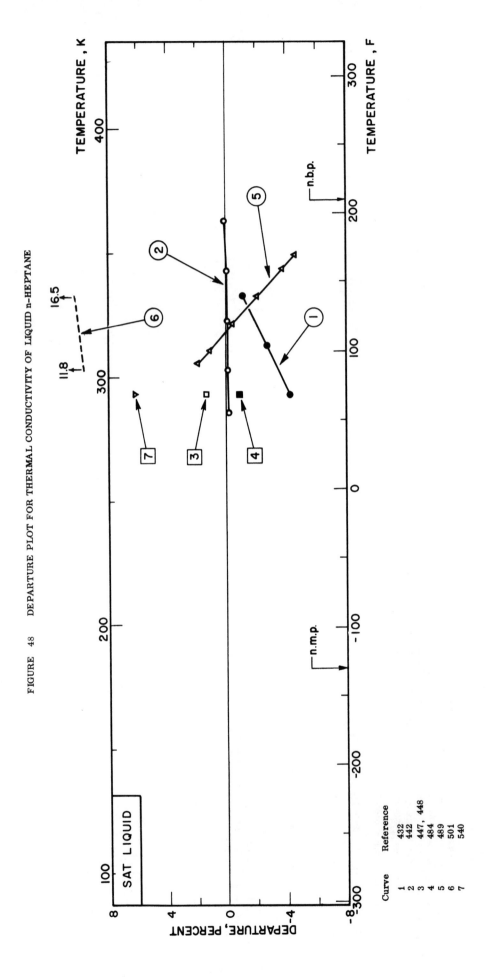

FIGURE 48   DEPARTURE PLOT FOR THERMAL CONDUCTIVITY OF LIQUID n-HEPTANE

DISCUSSION

GAS

In addition to the correlation of Lenoir (223) for the range 266-505 K given without source references and the single experimental data point by Lambert et al. (211) at 339 K, some recent tabulations have been given by Russian workers. Vargaftik gave values at 273 (100) 673 K (602, 603), subsequently (653) repeating these values and citing other values (658) for 373 (20) 633 K. The latter values are substantially similar to a further publication (659).

Graphical plotting of these data revealed close agreement between the various Russian tabulations. Furthermore these tabulations are much closer to the Lambert measurement than the Lenoir tabulation. Accordingly the recommended values were obtained from a smooth curve drawn through the Russian tabulations for temperatures from 373 to 673 K. Below 373 K these tabulations were adjusted so as to pass through the Lambert value. Above 673 K the values were obtained from a logarithm thermal conductivity -- logarithm absolute temperature plot. Some curvature in this plot was evident.

As no experimental data were cited in any of the available Russian publications no definite assessment of accuracy is possible. It would appear probable that the recommended values have an accuracy of within five percent below 500 K, ten percent at 750 K and fifteen percent at 1000 K. Further experimental measurements to confirm the tabulated values are highly desirable. Due to the fact that only one experimental data point is presently available, no departure plot is given.

TABLE 48 THERMAL CONDUCTIVITY OF n-HEPTANE

RECOMMENDED VALUES

[Temperature, T, K; Thermal Conductivity, k, mW cm^{-1}K^{-1}]

GAS

T	k	T	k
250	0.082*	550	0.386
260	0.090*	560	0.398
270	0.097*	570	0.411
280	0.105*	580	0.423
290	0.112*	590	0.435
300	0.120*	600	0.447
310	0.128*	610	0.459
320	0.137*	620	0.472
330	0.146*	630	0.484
340	0.155*	640	0.496
350	0.165*	650	0.508
360	0.174*	660	0.521
370	0.184*	670	0.533
380	0.194	680	0.546
390	0.204	690	0.559
400	0.214	700	0.573
410	0.225	710	0.586
420	0.235	720	0.599
430	0.246	730	0.613
440	0.256	740	0.626
450	0.267	750	0.639
460	0.279	800	0.709
470	0.290	850	0.779
480	0.302	900	0.850
490	0.313	950	0.926
500	0.325	1000	0.970
510	0.337		
520	0.349		
530	0.361		
540	0.373		

*Ignoring pressure dependence.
(n. b. p. = 371 K).

TABLE 49   THERMAL CONDUCTIVITY OF n-HEXANE

RECOMMENDED VALUES

[Temperature, T, K; Thermal Conductivity, k, mW cm^{-1}K^{-1}]

SATURATED LIQUID

DISCUSSION

There exist eight experimental investigations on the thermal conductivity of liquid n-hexane. The extensive measurements of Filippov (442) and the single point values of both Riedel (484) and Frontas' ev - Gusakov (447, 448) are considered to be reliable from the standpoint of the experimental method and procedure. Therefore, equal weight is given to these data in this analysis. The results of Sakiadis - Coates (489) are found to be reasonable in magnitude, however, as their values show a too steep trend with temperature, no weight is given to them.

The correlation formula obtained is given by

$10^6 k$ (cgsu) = 467.25 - 0.57760 T (T in K).

This equation is able to reproduce the values which are used for the estimation of this correlation with a mean deviation of 1.4 percent and a maximum of 3.6 percent. The recommended values tabulated are generated from the above equation. The data should be substantially correct in the temperature range between 290 and 330 K. Outside this range the uncertainty increases.

SATURATED LIQUID

T	k
250	1.35
260	1.33
270	1.30
280	1.28
290	1.25
300	1.230
310	1.206
320	1.182
330	1.157
340	1.133
350	(1.109)‡
360	(1.09)‡
370	(1.06)‡
380	(1.04)‡
390	(1.01)‡
400	(0.99)‡

‡Extrapolated for the liquid under vapor pressures, ignoring pressure dependence. (n.b.p. = 342K)

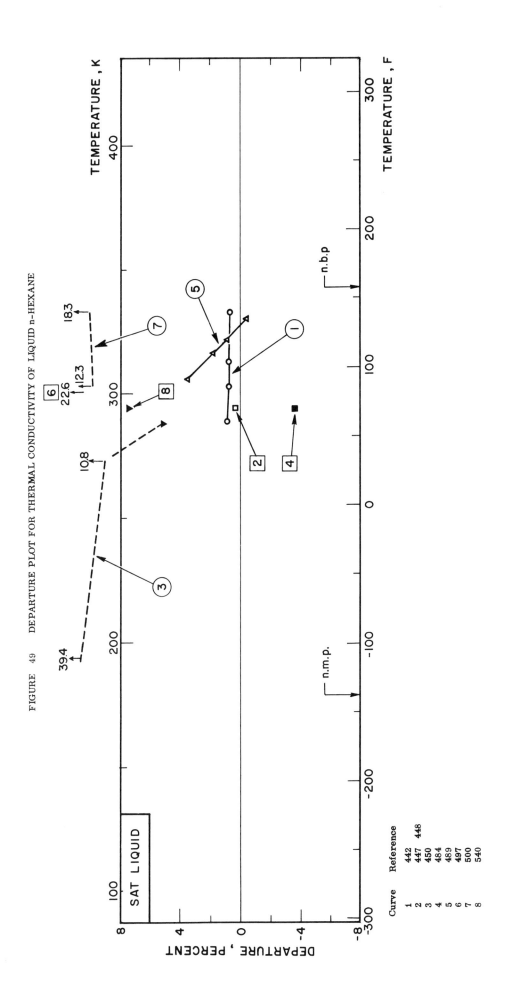

FIGURE 49 DEPARTURE PLOT FOR THERMAL CONDUCTIVITY OF LIQUID n-HEXANE

TABLE 49   THERMAL CONDUCTIVITY OF n-HEXANE

RECOMMENDED VALUES

[Temperature, T, K; Thermal Conductivity, k, mW cm^{-1}K^{-1}]

DISCUSSION

GAS

The available temperature range for which data have been tabulated for the thermal conductivity of gaseous n-hexane has been considerably extended in some Russian publications (601-603) which have recently become available. Little information is contained in these publications as to the source of their values.

The recommended values were obtained from a smooth curve drawn through all the available information. At 273 K an error of fifteen percent may occur. From 300 to 600 K the accuracy should be within a few percent. The higher temperature values are more uncertain and the error may again be as much as fifteen percent for the highest temperature tabulated. Experimental measurements are desirable for all temperatures, particularly those below 300 K or above 600 K.

GAS

T	k	T	k	T	k
250	(0.090)*	550	0.421	800	0.870
260	(0.097)*	560	0.435	810	0.891
270	0.104	570	0.451	820	0.912
280	0.112	580	0.467	830	0.933
290	0.120	590	0.483	840	0.954
300	0.128	600	0.499	850	0.975
310	0.137	610	0.515	860	0.996
320	0.147	620	0.531	870	1.019
330	0.156	630	0.547	880	1.042
340	0.166	640	0.565	890	1.065
350	0.176	650	0.581	900	1.088
360	0.187	660	0.598	910	1.11
370	0.198	670	0.617	920	1.14
380	0.209	680	0.636	930	1.16
390	0.220	690	0.655	940	1.19
400	0.232	700	0.674	950	1.21
410	0.244	710	0.693	960	1.24
420	0.256	720	0.711	970	1.27
430	0.268	730	0.730	980	1.30
440	0.280	740	0.749	990	1.33
450	0.292	750	0.768	1000	1.36
460	0.304	760	0.787		
470	0.316	770	0.807		
480	0.329	780	0.828		
490	0.342	790	0.849		
500	0.355				
510	0.367				
520	0.380				
530	0.393				
540	0.407				

*Extrapolated. (n. b. p. = 342 K)

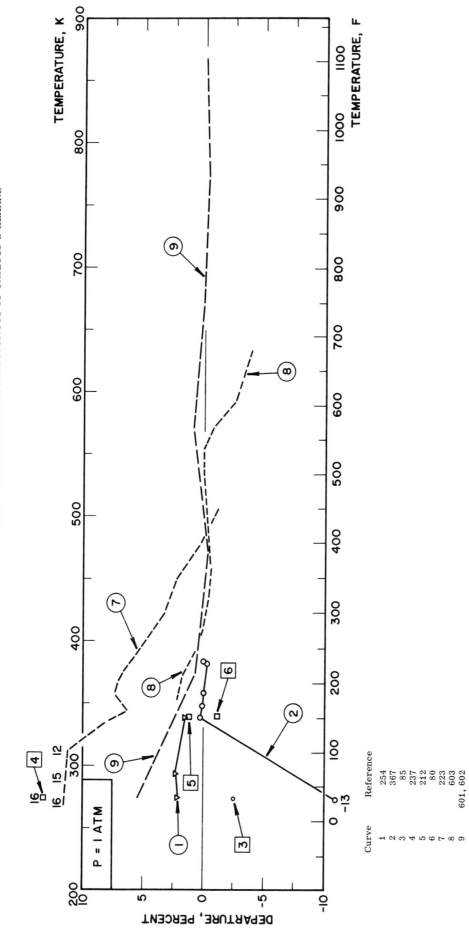

FIGURE 49 DEPARTURE PLOT FOR THERMAL CONDUCTIVITY OF GASEOUS n-HEXANE

TABLE 50   THERMAL CONDUCTIVITY OF METHANE

[Temperature, T. K; Thermal Conductivity, k, mW cm^{-1}K^{-1}]

RECOMMENDED VALUES

SATURATED LIQUID

T	k
90	2.23
100	2.08
110	1.93
120	1.78‡
130	1.63‡
140	1.48‡
150	1.33‡
160	1.18‡
170	1.03‡
180	0.88‡
190	0.73‡

‡ Under saturated vapor pressures. (n.b.p. = 112 K)

DISCUSSION

SATURATED LIQUID

The only available experimental work reported for the thermal conducitivity of liquid methane is that due to Borovik et al. (46). The results are cited in a compendium by Johnson (167). In the measurements, four data points were reported over the temperature range from 103 to 173 K. In this analysis, all of the points are given equal weight, and are fitted to a linear equation.

The correlation formula obtained is given by

$$10^6 k \text{ (cgsu)} = 854.68 - 3.5760\,T \quad (T \text{ in K}).$$

This equation is found to fit the experimental results with a mean deviation of 4.0 percent. The maximum deviation is -7.4 percent, as shown in the departure plot. The high maximum is inherent to the scattering of the original results. The above formula is used to calculate the recommended values. The tabulated values should be correct between 100 and 180 K within the error described above, and outside this temperature range the uncertainty increases.

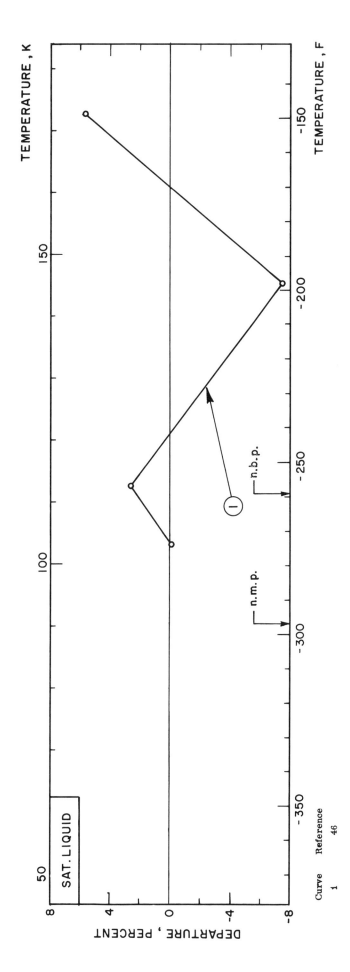

FIGURE 50 DEPARTURE PLOT FOR THERMAL CONDUCTIVITY OF LIQUID METHANE

TABLE 50   THERMAL CONDUCTIVITY OF METHANE

RECOMMENDED VALUES

[Temperature, T, K; Thermal Conductivity, k, mW cm^{-1}K^{-1}]

DISCUSSION

About twenty measurements of the thermal conductivity of gaseous methane at atmospheric pressure have been reported, of which eleven (96, 168, 187, 305, 331, 587, 603, 645, 649, 650, 651) extend over appreciable temperature ranges. A graphical plotting of all the data revealed reasonable agreement below about 300 K and relatively poor agreement above 400 K. After a careful analysis of the differing data it was decided to base the higher temperature values upon the measurements of Geier and Schafer (587) which, as can be ascertained from the departure plots, fall almost exactly midway between the extremes of other measurements for temperatures above about 500 K. Even the Geier and Schafer data appear somewhat uncertain in trend for temperatures above 800 K and this has limited the extent of the extrapolation of the values to higher temperatures.

The recommended values were obtained from a smooth curve drawn through all the data for temperatures below 400 K and through the Geier and Schafer data for higher temperatures. The recommended values are considered accurate to one percent for temperatures below 300 K, two percent for temperatures from 300 to 450 K and six percent for all other temperatures tabulated.

GAS

T	k	T	k	T	k
100	0.106*	450	0.578	750	1.137
110	0.117*	460	0.596	760	1.16
120	0.128	470	0.615	770	1.18
130	0.139	480	0.634	780	1.20
140	0.150	490	0.652	790	1.22
150	0.162	500	0.671	800	1.24
160	0.173	510	0.690	810	1.26
170	0.184	520	0.710	820	1.28
180	0.195	530	0.729	830	1.30
190	0.207	540	0.749	840	1.33
200	0.218	550	0.767	850	1.35
210	0.230	560	0.786	860	1.37
220	0.242	570	0.804	870	1.39
230	0.254	580	0.823	880	1.41
240	0.266	590	0.840	890	1.44
250	0.277	600	0.858	900	1.46
260	0.289	610	0.877	910	1.48
270	0.301	620	0.894	920	1.50
280	0.314	630	0.912	930	1.53
290	0.329	640	0.930	940	1.55
300	0.343	650	0.948	950	1.57
310	0.356	660	0.967	960	1.60
320	0.371	670	0.985	970	1.62
330	0.384	680	1.004	980	1.64
340	0.399	690	1.022	990	1.67
350	0.412	700	1.041	1000	1.69
360	0.426	710	1.060		
370	0.440	720	1.083		
380	0.455	730	1.098		
390	0.469	740	1.126		
400	0.484				
410	0.503				
420	0.522				
430	0.540				
440	0.560				

*n. b. p. = 112 K

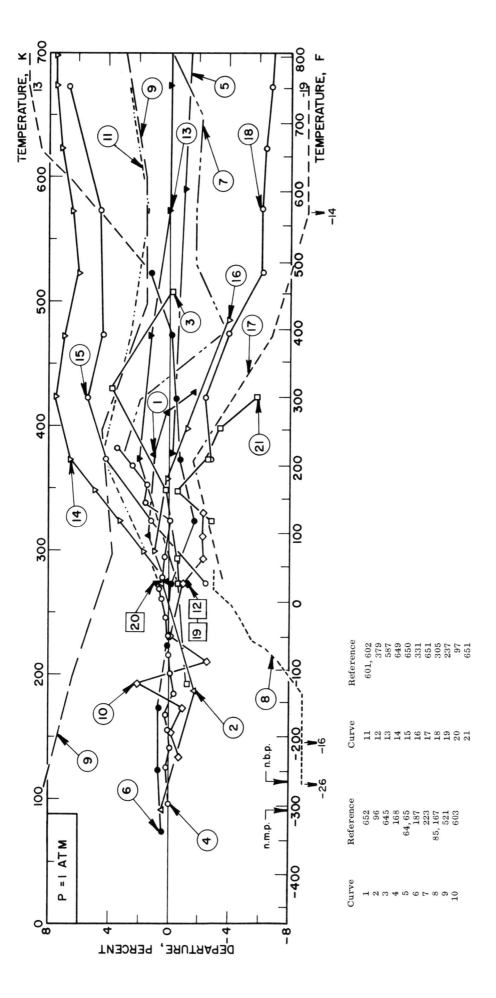

FIGURE 50   DEPARTURE PLOT FOR THERMAL CONDUCTIVITY OF GASEOUS METHANE

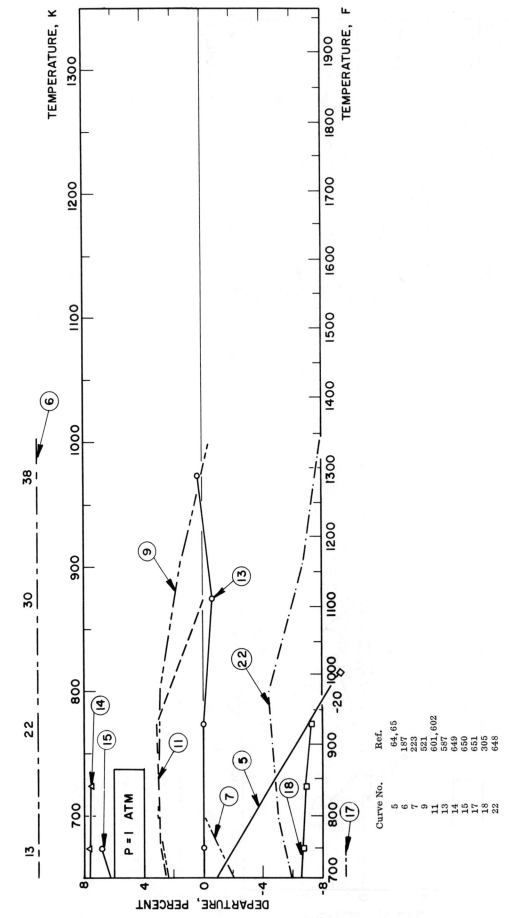

FIGURE 50 DEPARTURE PLOT FOR THERMAL CONDUCTIVITY OF GASEOUS METHANE (Continued)

TABLE 51 THERMAL CONDUCTIVITY OF METHYL ALCOHOL

[Temperature, T, K; Thermal Conductivity, k, mW cm^{-1}k^{-1}]

DISCUSSION

SATURATED LIQUID

There exist 18 experimental works on the thermal conductivity of liquid methyl alcohol. The discrepancy between the reported values of different investigators is rather small, especially near room temperature. The careful measurements of Filippov (442), Mason (475) and Riedel (487) are considered to be most reliable from the standpoint of their experimental methods and procedures; therefore, heavy weight is given to their data in this analysis. The single point values of Riedel (484), Sakiadis-Coates (489), Scheffy-Johnson (490, 491), and Van der Held-Van Drunen (507) are also included in the estimation of the most probable values.

The correlation formula obtained is

10^6k (cgsu) = 687.314 - 0.680519 T (T in K).

This equation fits the above-enumerated measurements with a mean deviation of 1.3 percent and a maximum of 3.6 percent, and is used to generate the recommended values. The values in the temperature range from 230 to 390 K should be substantially correct. Outside this range the uncertainty increases.

Two sets of extensive results up to the critical point were reported by Abas-Zade (3, 5); however, his values near room temperature are less reliable. Hence, no correlation is attempted for vapor pressures higher than one atm.

RECOMMENDED VALUES

SATURATED LIQUID

T	k
150	(2.45)†
160	(2.42)†
170	(2.39)†
180	2.36
190	2.34
200	2.31
210	2.28
220	2.25
230	2.221
240	2.192
250	2.164
260	2.135
270	2.107
280	2.078
290	2.050
300	2.022
310	1.993
320	1.965
330	1.936
340	1.908‡
350	1.879‡
360	1.851‡
370	1.822‡
380	1.794‡
390	1.765‡
400	1.737‡
410	1.71‡
420	1.68‡
430	1.65‡

† Extrapolated for the supercooled liquid. (n.m.p. = 175 K)
‡ Under saturation vapor pressures. (n.b.p. = 338 K)

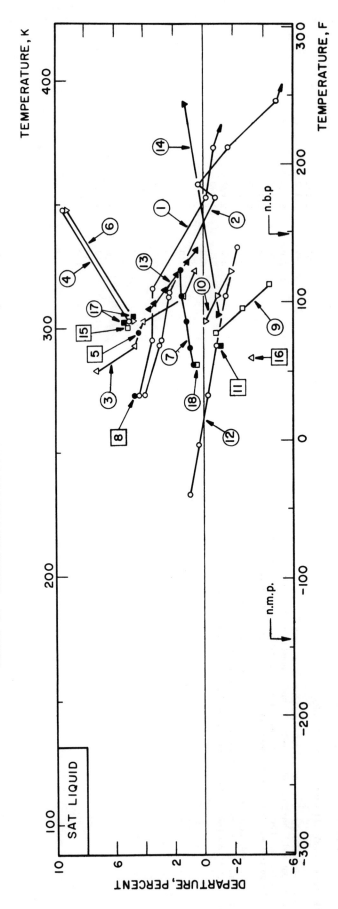

FIGURE 51   DEPARTURE PLOT FOR THERMAL CONDUCTIVITY OF LIQUID METHYL ALCOHOL

TABLE 51   THERMAL CONDUCTIVITY OF METHYL ALCOHOL

RECOMMENDED VALUES

[Temperature, T, K; Thermal Conductivity, k, mW cm^{-1}K^{-1}]

DISCUSSION

GAS

Only two sets of measurements were found for methyl alcohol, twelve data points in the range 333-500 K by Shushpanov (319) and four data points from 355-383 K by Vines (367). The measurements of Vines were relative, the values of Kannuluik and Carman (172) for air being used as reference data. A tabulation of data with no source references is also given by Lenoir (223) covering the range 255-422 K. These values follow the trend of Vines data.

Quadratic temperature fits were found accurate to within 1.1 percent and 0.05 percent for the data of (319) and (367) respectively. The formulas deduced from these fittings were accordingly used to compute values of thermal con-

10^5k (cgsu) = -12.0915 + 6.92493.10^{-2}T - 6.12440.10^{-5}T^2 (T in K)   (367)

10^5k (cgsu) = 2.57489 - 1.21163.10^{-2}T + 5.22159.10^{-5}T^2 (T in K)   (319)

ductivity from 300 to 550 K.

Examination of these data showed athat the Shushpanov data were consistently higher than those of Vines and that the trend with temperature of the two data disagreed. No factor could be found in either work which would account for the disagreement so that it was decided to cite as 'most probable' the mean values with an error estimate based upon the deviation of either curve from this mean. It will be seen that in the range 310-420 K the accuracy is within five percent but that at higher temperatures the uncertainty increases, reaching 20 percent at the highest temperature tabulated.

GAS

T	k
300	(0.143)*
310	(0.154)*
320	(0.164)*
330	(0.175)*
340	0.185
350	0.196
360	0.207
370	0.217
380	0.228
390	0.238
400	0.249
410	0.259
420	0.270
430	0.280
440	0.290
450	0.300
460	0.311
470	0.321
480	0.331
490	0.341
500	0.351
510	0.36
520	0.37
530	0.38
540	0.39
550	0.40

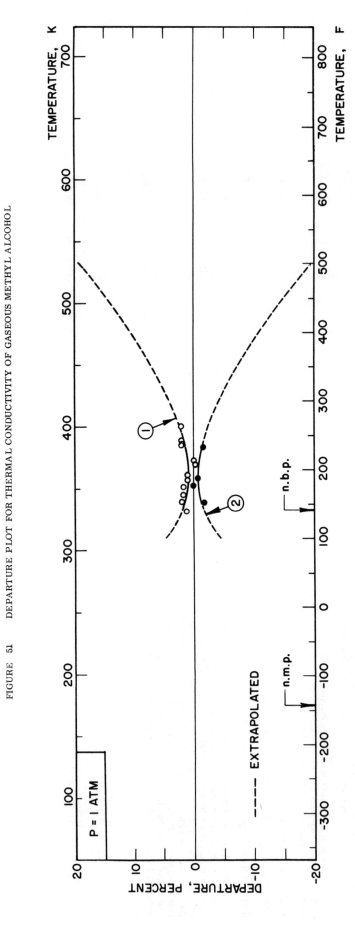

FIGURE 51 DEPARTURE PLOT FOR THERMAL CONDUCTIVITY OF GASEOUS METHYL ALCOHOL

TABLE 52 THERMAL CONDUCTIVITY OF METHYL CHLORIDE

RECOMMENDED VALUES

[Temperature, T, K; Thermal Conductivity, k mW cm^{-1}K^{-1}]

SATURATED LIQUID

T	k
200	2.41
210	2.32
220	2.24
230	2.15
240	2.07
250	1.984‡
260	1.899‡
270	1.815‡
280	1.730‡
290	1.645‡
300	1.561‡
310	1.476‡
320	1.39 ‡
330	1.31 ‡
340	1.22 ‡
350	1.14 ‡

DISCUSSION

SATURATED LIQUID

There exists only one set of experimental data for the thermal conductivity of liquid methyl chloride. The measurements were made by Kardos (462, 526) using a hot-wire apparatus, covering the temperature range from 260 to 298 K, with an accuracy of 2.5 percent. All of the original experimental values are given equal weight in the present analysis and are fitted with a linear correlation formula given by

$10^6 k$ (cgsu) = 979.41 - 2.0212 T (T in K).

This equation is found to reproduce all of the experimental points with a mean deviation of 0.65 percent and a maximum of 1.2 percent. The above formula is used to calculate the recommended values. The values between 250 and 310 K are considered to be substantially accurate within 2.5 percent. Outside this range the uncertainty increases.

Incidentally, Koch (527) calculated the thermal conductivity of saturated methyl chloride by means of an empirical correlation. His values are found to agree with the above formula within 2.7 percent. In the departure plot, Koch's data are also shown by a dotted line.

‡ Under saturated vapor pressures. (n.b.p. = 249 K)

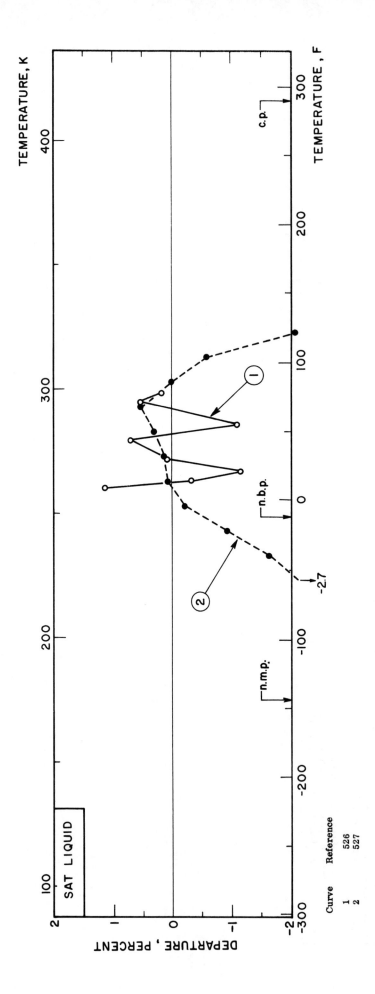

FIGURE 52 DEPARTURE PLOT FOR THERMAL CONDUCTIVITY OF LIQUID METHYL CHLORIDE

TABLE 52 THERMAL CONDUCTIVITY OF METHYL CHLORIDE

RECOMMENDED VALUES

[Temperature, T, K; Thermal Conductivity, k, mW cm^{-1}K^{-1}]

DISCUSSION

GAS

Values from 255 to 533 K are quoted without source reference by Lenoir (223). Vargaftik (601, 602, 653) quotes two sets of data, the first for 273(10) 373 K and the second for 273(100) 873 K. Graphical plotting of these values revealed that the first Vargaftik set were consistently higher than the Lenoir values by 2.5 percent while the second set were lower than the Lenoir values below 400 K and indistinguishable at higher temperatures.

The recommended values were obtained from a smooth curve drawn through the Lenoir values below 400 K and through all available information for higher temperatures. Due to the paucity of available experimental data no departure plot is given. The recommended values may be accurate to five percent below 400 K and ten percent for higher temperatures.

GAS

T	k	T	k
250	0.075	500	0.267
260	0.081	510	0.277
270	0.087	520	0.287
280	0.094	530	0.297
290	0.100	540	0.307
300	0.107	550	0.317
310	0.114	560	0.327
320	0.121	570	0.337
330	0.128	580	0.347
340	0.135	590	0.357
350	0.142	600	0.368
360	0.149	610	0.378
370	0.156	620	0.389
380	0.163	630	0.400
390	0.170	640	0.411
400	0.177	650	0.422
410	0.186	660	0.434
420	0.194	670	0.445
430	0.203	680	0.457
440	0.211	690	0.469
450	0.220	700	0.481
460	0.229	710	0.494
470	0.238	720	0.506
480	0.247	730	0.519
490	0.257	740	0.533
		750	0.546

TABLE 53   THERMAL CONDUCTIVITY OF n-NONANE

RECOMMENDED VALUES

[Temperature, T, K; Thermal Conductivity, k, mW cm^{-1} K^{-1}]

DISCUSSION

SATURATED LIQUID

Only three sets of experimental investigations are available on the thermal conductivity of liquid n-nonane. From the viewpoint of the results for other liquids, the single point value of Frontas'ev - Gusakov (447, 448) is considered to be the most reliable. The extensive measurements of Sakiadis - Coates (489) might be reasonable in magnitude but the temperature coefficient is thought to be less reliable. The data of Smith (501) are also very high because he measured the thermal conductivity using Bridgman's apparatus. The single point value calculated from an empirical equation by Vilim (540) is not considered to be accurate. Therefore, in order to estimate a correlation formula, the single experimental point of Frontas'ev - Gusakov and an empirical relation between the thermal conductivity and the molecular weight for the saturated normal hydrocarbons are used in this analysis.

The correlation formula obtained is given by

10^6 k (cgsu) = 542.27 - 0.72750 T   (T in K).

This equation is found to reproduce the results of Frontas'ev - Gusakov, Sakiadis - Coates and Vilim within the maximum deviation of 1.9 percent. The above formula is used to generate the recommended values. The data should be correct within two percent in the temperature range from 290 to 350 K. Outside this range the uncertainty increases.

SATURATED LIQUID

T	k
250	1.51
260	1.48
270	1.45
280	1.42
290	1.39
300	1.356
310	1.325
320	1.295
330	1.264
340	1.234
350	1.204
360	1.17
370	1.14
380	1.11
390	1.08
400	1.05
410	1.02
420	0.99
430	(0.96) ‡
440	(0.93) ‡

‡ Extrapolated for the liquid under vapor pressure, ignoring pressure dependence. (n.b.p. = 424 K)

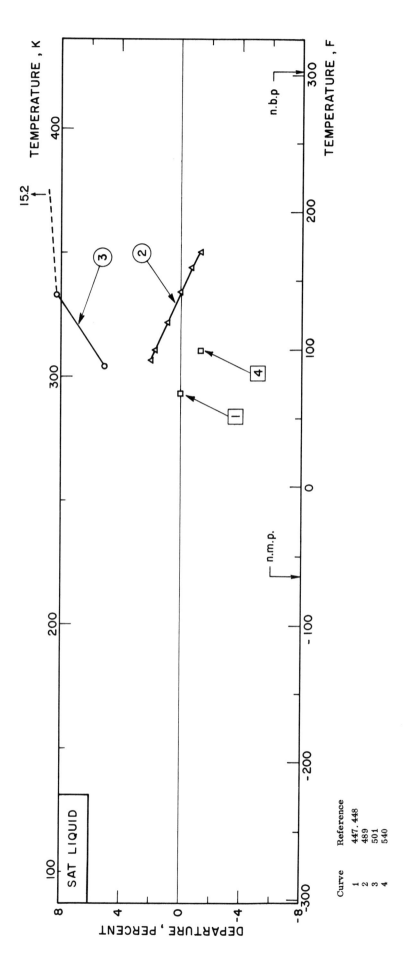

FIGURE 53   DEPARTURE PLOT FOR THERMAL CONDUCTIVITY OF LIQUID n-NONANE

TABLE 53   THERMAL CONDUCTIVITY OF n-NONANE

RECOMMENDED VALUES
[Temperature, T, K; Thermal Conductivity, k, mW cm^{-1}K^{-1}]

DISCUSSION

GAS

No experimental data for the thermal conductivity of nonane have been found. The data presented here are based partly on the values of Lenoir (223), who gives no information on the source of his values, and partly on values calculated by Bachman et. al. (577). It was found that to the accuracy of these values, the data could be represented by

$$10^3 k (B\,ft^{-1}hr^{-1}F^{-1}) = 0.518 + 2.482 \cdot 10^{-3} T \text{ (T in F)}$$

and this equation was used to generate the recommended values. These should not be considered as better than within 15 percent of the true values. Experimental measurements are urgently required for this substance.

GAS

T	k	T	k
250	0.086	550	0.317
260	0.093	560	0.325
270	0.101	570	0.333
280	0.109	580	0.341
290	0.116	590	0.348
300	0.124	600	0.356
310	0.132	610	0.364
320	0.140	620	0.372
330	0.147	630	0.379
340	0.155	640	0.387
350	0.163	650	0.395
360	0.171	660	0.403
370	0.178	670	0.410
380	0.186	680	0.418
390	0.194	690	0.426
400	0.202	700	0.433
410	0.209	710	0.441
420	0.217	720	0.449
430	0.225	730	0.457
440	0.232	740	0.464
450	0.240	750	0.472
460	0.248	800	0.511
470	0.256	850	0.550
480	0.263	900	0.588
490	0.271	950	0.627
500	0.279	1000	0.666
510	0.287		
520	0.294		
530	0.302		
540	0.310		

* n. b. p. = 424 K.

TABLE 54   THERMAL CONDUCTIVITY OF n-OCTANE

RECOMMENDED VALUES

[Temperature, T, K; Thermal Conductivity, k, mW cm^{-1}K^{-1}]

DISCUSSION

SATURATED LIQUID

Five experimental works are available in the literature on thermal conductivity of liquid n-octane. The single point value of Frontas'ev - Gusakov (447, 448) is considered to be the most reliable. The extensive results of Sakiadis-Coates (489) might be reasonable in magnitude, but their trend with temperature is thought to be less reliable considering the results for other liquids. Two sets of measurements by Smith (500, 501) give very high values because he used Bridgman's apparatus. The single point value calculated from an empirical equation by Vilim (540) is found to be unreliable. Therefore in the present analysis, only the results of Frontas'ev - Gusakov and an empirical relation between the thermal conductivity and the molecular weight for the saturated normal hydrocarbons is used for the estimation of the most probable correlation.

The correlation formula obtained is given by

10^6k (cgsu) = 519.81 - 0.68500 T   (T in K).

This equation is found to fit the data of Frontas'ev - Gusakov and Sakiadis-Coates within 1.5 percent, and is used to generate the recommended values. The tabulated values should be substantially correct in the temperature range from 290 to 350 K, and outside this range the uncertainty increases.

SATURATED LIQUID

T	k
250	1.46
260	1.43
270	1.40
280	1.37
290	1.34
300	1.32
310	1.29
320	1.26
330	1.23
340	1.20
350	1.17
360	1.14
370	1.11
380	1.09
390	1.06
400	(1.03) ‡

‡ Extrapolated for the liquid under vapor pressure, ignoring pressure dependence.   (n. b. p. = 399 K)

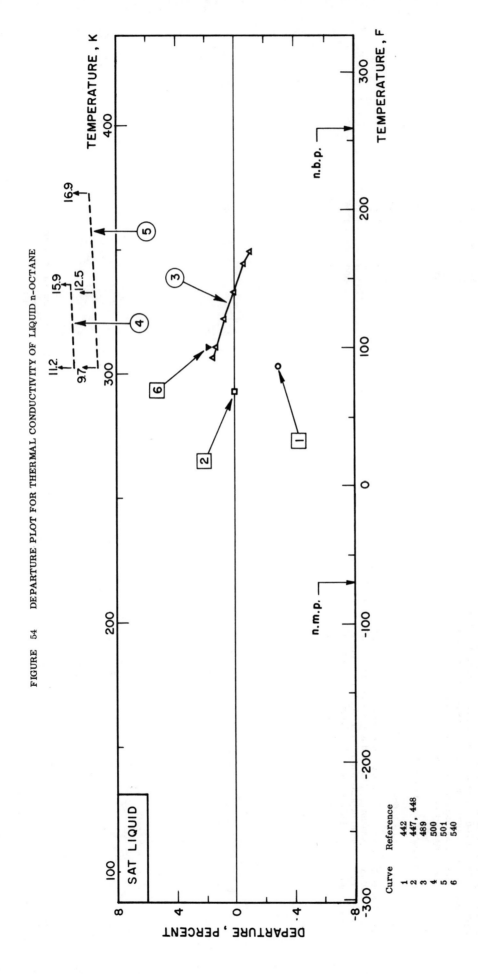

FIGURE 54   DEPARTURE PLOT FOR THERMAL CONDUCTIVITY OF LIQUID n-OCTANE

TABLE 54 THERMAL CONDUCTIVITY OF n-OCTANE

RECOMMENDED VALUES

[Temperature, T, K; Thermal Conductivity, k, mW cm^{-1} K^{-1}]

DISCUSSION

GAS

The only set of values available for octane are those of Lenoir (223), for the range 283 - 478 K. These are given without source references. A single experimental data point was also presented by Lambert, Cotton et. al. (211), at 355 K.

Analysis of the Lenoir data by least mean squares showed that, to ±0.4%, the values could be represented by the formula

$$10^5 k \text{ (cgsu)} = -0.980559 + 8.18249 \cdot 10^{-3} T + 1.950055 \cdot 10^{-5} T^2 \quad (T \text{ in K})$$

This formula was accordingly used to compute values at 250(10)500 K. The Lenoir value at 355 K is 8.5% higher than the only experimental value available. No recommendation concerning the accuracy of the Lenoir data can be made in view of the paucity of the data.

GAS

T	k
250	0.096*
260	0.103*
270	0.111*
280	0.119*
290	0.127*
300	0.135*
310	0.144*
320	0.152*
330	0.161*
340	0.170*
350	0.179*
360	0.188*
370	0.197*
380	0.207*
390	0.217*
400	0.226
410	0.236
420	0.247
430	0.257
440	0.268
450	0.278
460	0.289
470	0.300
480	0.311
490	0.323
500	0.334

*Ignoring pressure dependence. (n. b. p. = 399 K)

TABLE 55   THERMAL CONDUCTIVITY OF n-PENTANE

DISCUSSION

SATURATED LIQUID

Although there exist five experimental works reported on the thermal conductivity of liquid n-pentane, most of the data fall in a relatively narrow temperature range. The measurements of Bridgman (431) and Smith (501) give extremely high values, similar to their results for other liquids. A set of the old measurements by Goldschmidt (450), covering temperatures from 287 K down to the melting point, is not considered to be too reliable. However, his value at 287 K falls within several percent of the two single data points of both Riedel (484) and Vilim (540). In the present analysis, in addition to these three points, an empirical relation between the thermal conductivity and the molecular weight for the saturated normal hydrocarbons is used to estimate the most probably values.

The correlation formula obtained is given by

$10^6 k$ (cgsu) = 507.77 - 0.75604 T   (T in K).

This equation is used to generate the recommended values. The tabulated values below the boiling point are thought to be accurate within five percent. The accuracy above the boiling point cannot be estimated because of the lack of experimental data.

RECOMMENDED VALUES

[Temperature, T, K; Thermal Conductivity, k, mW cm^{-1}K^{-1}]

SATURATED LIQUID

T	k
200	1.49
210	1.46
220	1.43
230	1.40
240	1.37
250	1.33
260	1.30
270	1.27
280	1.24
290	1.21
300	1.18
310	(1.14) ‡
320	(1.11) ‡
330	(1.08) ‡
340	(1.05) ‡
350	(1.02) ‡
360	(0.99) ‡
370	(0.95) ‡
380	(0.92) ‡
390	(0.89) ‡
400	(0.86) ‡

‡ Extrapolated for the liquid under vapor pressure, ignoring pressure dependence. (n.b.p. = 309 K)

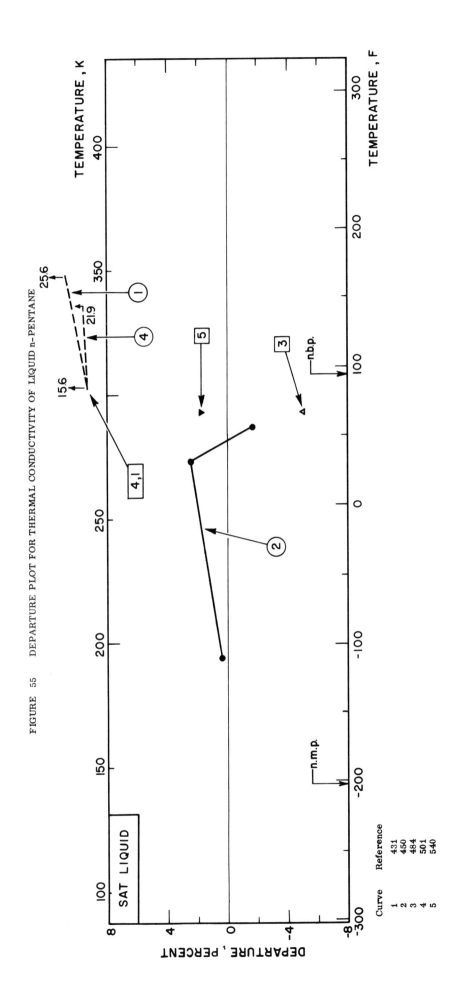

FIGURE 55 DEPARTURE PLOT FOR THERMAL CONDUCTIVITY OF LIQUID n-PENTANE

TABLE 55   THERMAL CONDUCTIVITY OF n-PENTANE

[Temperature, T, K; Thermal Conductivity, k, mWcm^{-1}K^{-1}]

DISCUSSION

Experimental measurements of the thermal conductivity of normal pentane are fragmentary. The most extensive source of values is the report by Lenoir (223) which tabulates values from 266 to 505 K without source references. Experimental values at mostly single temperatures are reported by (211, 237, 254) some of which are requoted by (51, 416).

In determining the most probable values the Lenoir tabulation was fitted to a polynomial temperature function and the equation

$10^5 k$ (cgsu) = $-1.52770 + 1.24036 \cdot 10^{-2} T + 1.59125 \cdot 10^{-5} T^2$ (T in K)

fitted the tabular data to within one percent. This equation was used to generate the recommended values.

Examination of the departure plot shows that the average uncertainty in the measurements is about four percent. Measurements of the thermal conductivity of this substance would certainly be helpful in determining the values at higher temperatures. The uncertainty in the recommended values may be assessed as about five percent below 375 K increasing to as much as ten percent about 500 K.

RECOMMENDED VALUES

GAS

T	k
250*	0.107
260*	0.116
270*	0.125
280*	0.134
290*	0.143
300*	0.152
310	0.161
320	0.170
330	0.180
340	0.189
350	0.199
360	0.209
370	0.219
380	0.229
390	0.240
400	0.250
410	0.261
420	0.271
430	0.282
440	0.293
450	0.304
460	0.316
470	0.327
480	0.339
490	0.350
500	0.362

*Ignoring pressure dependence.
(n.b.p. = 309 K)

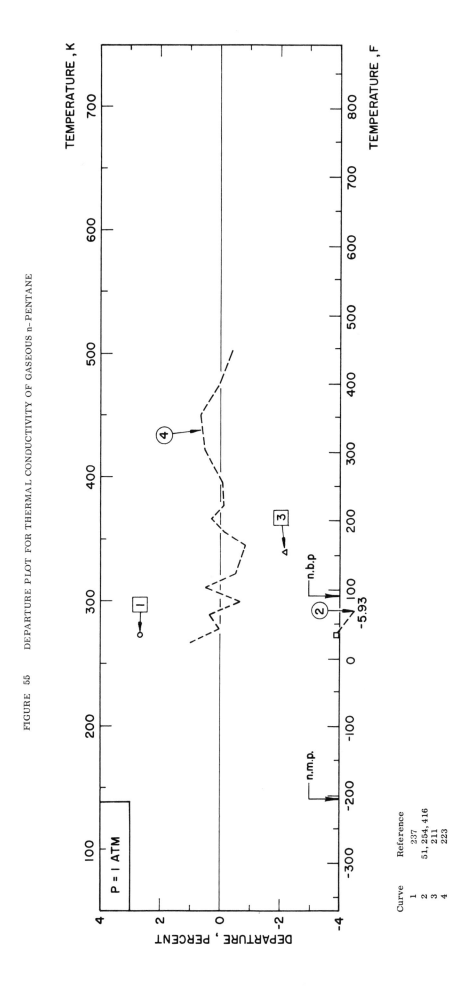

FIGURE 55   DEPARTURE PLOT FOR THERMAL CONDUCTIVITY OF GASEOUS n-PENTANE

TABLE 56   THERMAL CONDUCTIVITY OF PROPANE

DISCUSSION

GAS

Until the publication of calculated values at 100 (100) 5000 K by Svehla (521), the most extensive tabulation was by Lenoir (223) for the range 233 - 477 K, given without source references. Relatively few experimental data values are available for this substance (211, 221, 233, 237, 343, 368, 370, 401). The Lenoir tabulation was analyzed and it was found that the equation

$$10^5 k \text{ (cgsu)} = -2.57254 + 1.99967 \cdot 10^{-2} T + 1.01639 \cdot 10^{-5} T^2 \quad (T \text{ in K})$$

fitted the values to within five percent at 233 K and 0.2 percent at 477 K. This equation was used to generate the recommended values.

The departure plot shows that the experimental values agree with those recommended to within a few percent. The accuracy of the latter may be assessed at five percent below 270 K, three percent between 270 and 400 K, about five percent above 400 K.

RECOMMENDED VALUES

[Temperature, T, K; Thermal Conductivity, k, mW cm^{-1} K^{-1}]

GAS

T	k
200	(0.077)*
210	(0.087)*
220	(0.097)*
230	(0.107)*
240	0.118
250	0.129
260	0.139
270	0.149
280	0.160
290	0.171
300	0.183
310	0.193
320	0.204
330	0.215
340	0.226
350	0.237
360	0.249
370	0.260
380	0.272
390	0.284
400	0.295
410	0.307
420	0.319
430	0.331
440	0.343
450	0.355
460	0.367
470	0.380
480	0.392
490	0.405
500	0.417

*Extrapolated for the gas phase ignoring pressure dependence. (n. b. p. = 231 K)

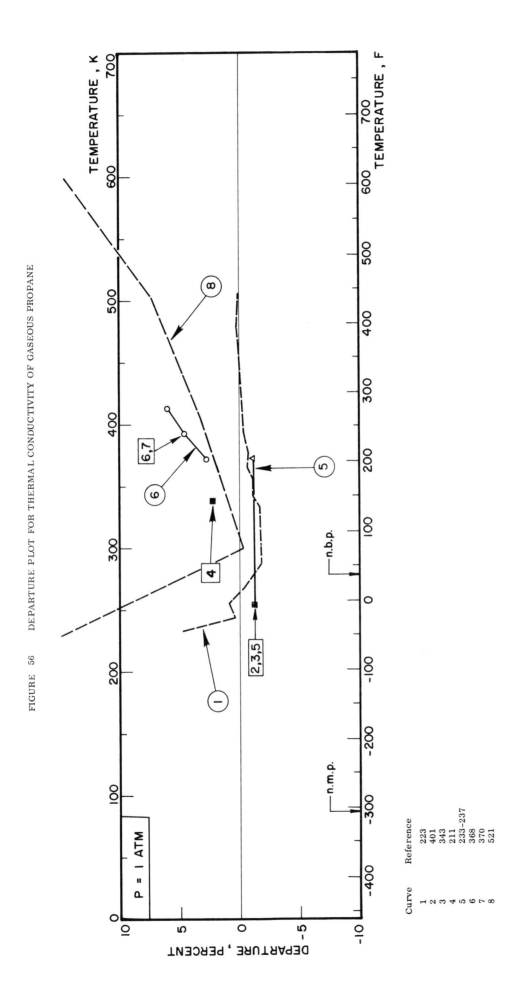

FIGURE 56 DEPARTURE PLOT FOR THERMAL CONDUCTIVITY OF GASEOUS PROPANE

Curve	Reference
1	223
2	401
3	343
4	211
5	233-237
6	368
7	370
8	521

TABLE 57   THERMAL CONDUCTIVITY OF TOLUENE

RECOMMENDED VALUES

[Temperature, T, K; Thermal Conductivity, k, mW cm^{-1}K^{-1}]

SATURATED LIQUID

T	k
170	(1.67)†
180	1.65
190	1.62
200	1.594
210	1.569
220	1.544
230	1.518
240	1.493
250	1.467
260	1.442
270	1.417
280	1.391
290	1.366
300	1.340
310	1.315
320	1.290
330	1.264
340	1.239
350	1.209
360	1.188
370	1.163
380	1.137
390	(1.112)‡
400	(1.09)‡
410	(1.06)‡
420	(1.04)‡
430	(1.01)‡
440	(0.99)‡
450	(0.96)‡
460	(0.93)‡
470	(0.91)‡
480	(0.88)‡
490	(0.86)‡
500	(0.83)‡

DISCUSSION

SATURATED LIQUID

Twenty-four experimental works are available on the thermal conductivity of liquid toluene. The discrepancy between the reported values of different investigators is extremely large. The results of several extensive measurements fall into two groups, one group being about 12 percent to 18 percent higher than the other. The results of Abas-Zade (1, 2), who used the hot-wire method, and those of Bridgman (431), Markwood - Benning (238), and Smith (500), who used the coaxial-cylinder method, all fall in the higher set. On the other hand, recent results of Challoner - Powell (434), Filippov (100, 441) McCready (471), Os'minin (478), Riedel (486), Schmidt - Leidenfrost (492), Vargaftik (508) and Ziebland (519) fall within the lower group. From the standpoint of the experimental method and procedure, the latter set of data are felt to be more reliable. Therefore, the eight sets of extensive data mentioned above are given equal weight in this analysis, and the single point values of Frontas' ev - Gusakov (447, 448) and Riedel (483, 484) are also included in the estimation of the most probable values.

The correlation formula obtained is

$10^6 k$ (cgsu) = 502.540 − 0.607275 T   (T in K).

This equation is found to fit the experimental values of the above-enumerated investigators with a mean deviation of 1.2 percent and a maximum of 3.9 percent.

The above equation is used for the calculation of the recommended values. The tabulated values should be correct in the temperature range between 190 and 390 K. Outside this range the uncertainty increases.

Although Abas-Zade (2) made measurements up to the critical point and Filippov (100, 441) also measured up to 511 K under vapor pressures, no correlation is attempted in the region where the vapor pressure is higher than one atm.

---

†Extrapolated for the supercooled liquid. (n.m.p. = 178 K).

‡Extrapolated for the liquid under vapor pressures, ignoring pressure dependence. (n.b.p. = 384 K).

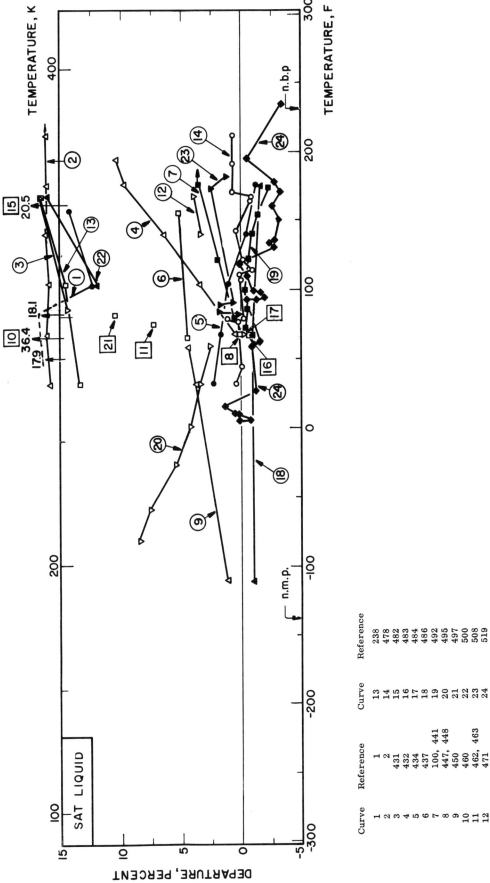

FIGURE 57 DEPARTURE PLOT FOR THERMAL CONDUCTIVITY OF LIQUID TOLUENE

244

FIGURE 57 DEPARTURE PLOT FOR THERMAL CONDUCTIVITY OF LIQUID TOLUENE (continued)

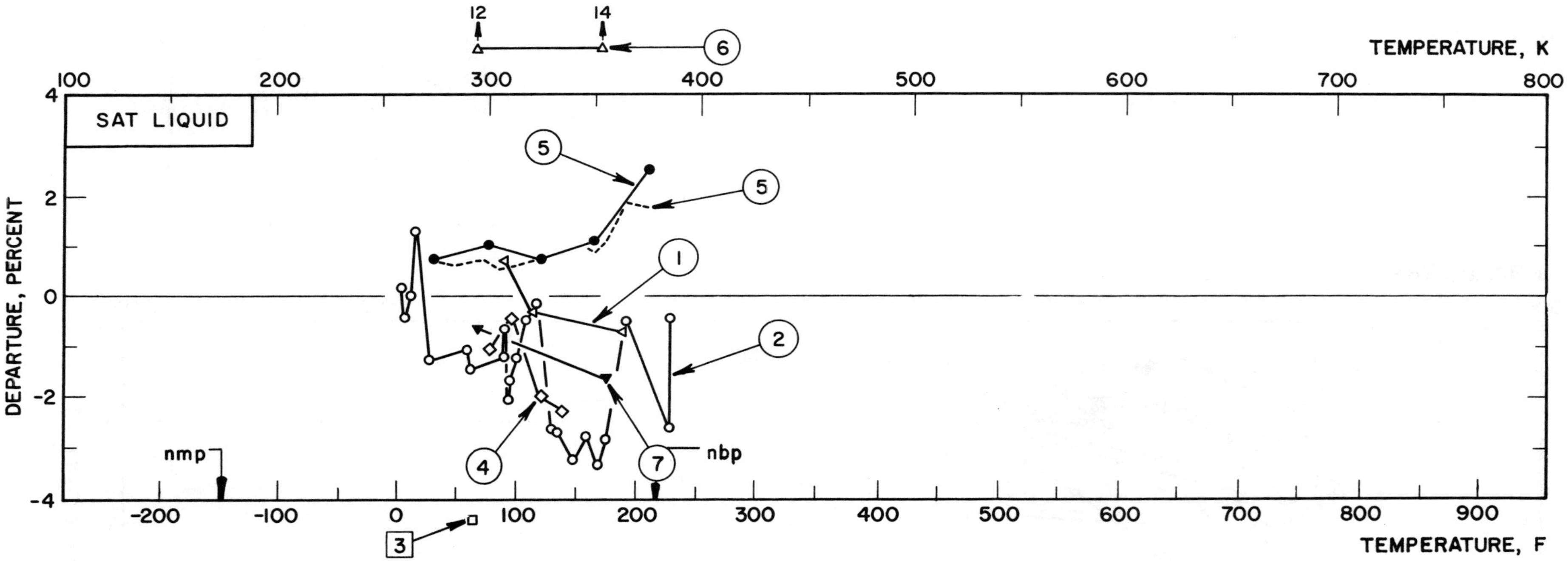

Curve No.	Reference No.
1	638
2	641
3	643
4	637
5	632
6	633
7	642

TABLE 57   THERMAL CONDUCTIVITY OF TOLUENE

RECOMMENDED VALUES

[Temperature, T, K; Thermal Conductivity, k, mW cm^{-1}K^{-1}]

GAS

DISCUSSION

Data on the thermal conductivity of gaseous toluene have been reported by Abas-Zade (2) for temperatures between 273 and 594 K. Examination of these data showed that between 373 and 573 K a linear variation of thermal conductivity with temperatures apparently occurs. The value quoted at the highest temperature appears anomalously high unless decomposition of the vapor occurred. In the preparation of the table of recommended values a smooth curve was drawn through the experimental points except for the value at 594 K. The recommended values were obtained from this curve which was assumed to be a linear above 373 K. The trend of the data with temperature is in need of rechecking by new experimental measurements. Provisionally, the accuracy can be assessed at two percent below 530 K and ten percent for the higher temperatures.

GAS

T	k
250	0.116
260	0.121
270	0.126
280	0.133
290	0.139
300	0.146
310	0.154
320	0.162
330	0.170
340	0.180
350	0.189
360	0.198
370	0.208
380	0.219
390	0.230
400	0.240
410	0.251
420	0.262
430	0.273
440	0.284
450	0.295
460	0.305
470	0.316
380	0.327
490	0.338
500	0.349
510	0.360
520	0.371
530	0.382
540	0.393
550	0.405
560	0.416
570	0.427
580	0.439
590	0.450
600	0.461

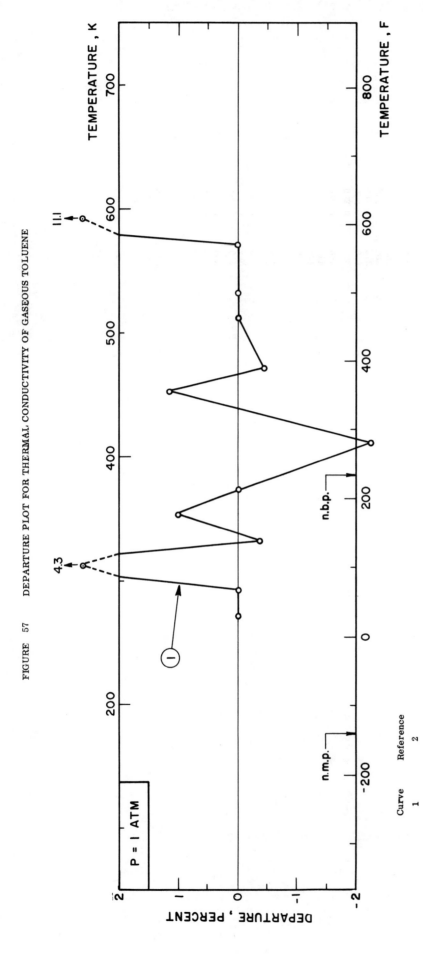

FIGURE 57 DEPARTURE PLOT FOR THERMAL CONDUCTIVITY OF GASEOUS TOLUENE

BINARY SYSTEMS

The raw data (expressed in $mW\,cm^{-1}K^{-1}$) and graphically smoothed values are reported in the Tables 58 through 139 for binary systems. In each table the a part of it gives the raw data and the b part the smoothed values. The thermal conductivity data tabulated in the table are shown plotted against the mole fraction of the heavier component in the mixture in a figure numbered identically. If all the data are not easily accommodated in one figure these are distributed in a set of figures numbered successively by the same numeral but distinguished by the addition of the lower case letters a, b, and c.

The raw data (expressed in $mW\,cm^{-1}K^{-1}$) for ternary and quaternary systems are recorded in the Tables 140 through 169. These data are not further processed like binary systems, but are grouped in different heads. The underlying idea being that the semitheoretical methods developed for computing the thermal conductivity of multicomponent mixtures may be assessed subsequently for their adequacy to reproduce the data of mixtures composed of gases of varying molecular complexity. This is of particular importance for multicomponent mixtures as it is impossible to determine thermal conductivity of all probable permutations. Once the most probable values of thermal conductivity of binary systems have been generated and the schemes of thermal conductivity predictions tested for them, it will be desirable to extend them for the multicomponent systems.

For binary systems, the basis of the graphical smoothing was to regard the two end points, referring to the two pure gases, as correct and then, consistent with the accuracy of the data, to pass a smooth curve through the experimental points. This approach, which has been used with a few exceptions, has many implications. First and foremost, the greatest reliability has been assumed in the end points of each worker. The accuracy of thermal conductivity data for the pure gases is likely to be better than that for a gas mixture because no errors (such as thermal diffusion) can add in the former which are dependent for their existance on the presence of two different gases, and also because of uncertainty in the mixture composition. In many cases uncertainties due to such factors are small and the end points are coincidant with the mixture points. In all relative measurements, the calibration is done at the end points and these are most reliable of all the reported points. Any systematic descrepancy present in the data which does not depend upon the magnitude of thermal conductivity (like cell constant), will not alter the shape of the thermal conductivity curve with composition, the magnitude will indeed be altered by a constant factor. A critical assessment of the pure conductivity values of a particular author in relation to others then can provide a basis for reconsideration of the entire set of data including mixtures. Also, invariably the uncertainty of the thermal conductivity measurement varies with the magnitude of thermal conductivity, provided that the thermal conductivity is reasonably large, and consequently will have extreme values at the two end points. To ensure that the shape of the curve giving the composition dependence of thermal conductivity is least vitiated by the experimental uncertainty of its determination, the above approach is preferable.

The entire data on 82 binary systems have been arranged in five different categories. This division is artificial but convenient and is based on the complexity of the molecular structure. We give here a brief mention of any special feature encountered for a certain gas pair or general comments applicable to a group.

(1) Experimental data exist for all ten binary systems of the five stable rare gases. Most data are for temperatures below 363.2 K and the maximum temperature of measurement is 793.2 K. Even in this temperature range, only for the argon-helium system are there enough intermediate temperatures at which measurements exist so that a second interpolation with reference to temperature could be made.

The precision of each set of data for all binaries is quite good in general and graphical smoothing is almost uniquely determined. The reproducibility of the measurements of Lindsay and Bromley [690] is poor -- not surprising in view of the nature and details of their unsteady state method. The precision of the data of Thornton and Baker [686] and Davidson and Music [83] is also relatively poor (about 5%) compared to the rest of the workers (less than 2%).

The thermal conductivity data for all the ten gas pairs are very well behaved. The thermal conductivity of a binary mixture decreases monotonically with concavity upward as the concentration of the heavier component increases in the mixture as long as the temperature is kept constant. For a given mixture the conductivity increases as the temperature increases. Further, thermal conductivity of these rare gases continuously decreases as the molecular weight of the gas increases. These conclusions are easily explained on the fact that all the rare gases form one isoelectronic sequence and are spherically symmetric in shape, Saxena [721].

(2) Eight binary systems in which argon occurs as a common component with polyatomic gases are considered here. In four of them the other constituent is one of the four nonpolar diatomic gases H_2, D_2, N_2 and O_2. For Ar-H_2 and Ar-D_2 systems the conductivity varies in the same fashion as for rare gas mixtures, while Ar-N_2 and Ar-O_2 behave with the difference that the upward concave nature of the curve is much less pronounced and for a few temperatures the variation is almost linear. A more precise statement is not possible due to the scatter in the data. Of the remaining four systems Ar-CH_4 and Ar-C_6H_6 again exhibit a decreasing thermal conductivity trend as the mole fraction of the heavier component increases in the mixture while opposite is the case with Ar-CO_2 and Ar-C_3H_8 systems. This is the first interesting case where one finds that even the qualitative variation of thermal conductivity changes as in a binary system keeping one component fixed, the other is replaced by different polyatomic gases of differing molecular complexity.

In Figure 75, referring to Ar-C_3H_8, the curve I has two alternatives corresponding to the two values of thermal conductivity of pure argon. The original value of Cheung, Bromley and Wilke [65, 688] refers to 594.2 K and dashed curve is drawn on this basis. If this value is reduced to 591.2 K, the temperature to which the rest of the measurements refer, on the basis of the data of Saxena and Saxena [316, 700] we get the continuous curve. The values given in Table 75b are on the basis of the continuous curve.

(3) In Tables 76 through 88 data are presented for thirteen gas pairs, in each of which helium is a common component. Of these, in five, the other component is a diatomic gas (H_2, D_2, N_2, O_2 or air). The three systems of He with N_2, O_2 and air show the normal variation of thermal conductivity viz., monotonic decrease, concave up, as the concentration of helium decreases in the mixture. In some of the curves for He-H_2 and He-D_2 there appears to be a minimum in the value of the thermal conductivity. The minimum is not well exhibited in He-D_2, the uncertainty of the data, purity of D_2, and the number of points actually measured come in the way. In He-H_2 system also, the dip in some curves is completely missing. Tondon, Gandhi and Saxena [722] made a careful study to investigate this anomaly and found that within the limit of the precision of their measurement (about 1%) the composition variation is normal. They [722] also found that none of the available theories of thermal conductivity predict any anomalous behavior. However, Minter [723] has recently again indicated the possibility of a very shallow (less than a percent) dip in the thermal conductivity plot of He-H_2 system at about 8 percent H_2 in the mixture. In the opinion of the present authors, Minter's work as well as those of Hansen, Frost and Murphy [724] and Neal, Greenway and Coutts [725] suffer from one common major weakness that their methods are not absolute in nature and probable uncertainties must be eliminated and experimental accuracy be improved to establish this small effect with any confidence. In fact, the existance of a dip is conspicuous only in the measurements of Mukhopadhyay et al. [704] while in almost every other case we find a tendency for the thermal conductivity curve to straighten out as the proportion of He increases in the mixture beyond 80 percent or so. It seems there is still a need to experimentally examine the He-H_2 system provided values can be obtained with an accuracy of better than a percent.

The remaining eight gas pairs of helium with carbon dioxide, cyclopropane, ethane, ethylene, methane, propane, propene and n-butane are normal. The shape of the curve of thermal conductivity versus mole fraction of the heavier component in each case is monotone decreasing, concave up.

(4) The thermal conductivity of krypton with four other diatomic gases (hydrogen, deuterium, nitrogen, and oxygen) is analyzed here and in each case the conductivity monotonically decreases as the amount of krypton in the mixture is increased. This is quite understandable for krypton is a spherically symmetric molecule while all these four are linear, nonpolar and diatomic. It is important to note that Kr-H_2 and Kr-D_2 systems do not show any strange feature

(5) Of the five binary systems reported here with neon as a common component Ne-H_2, Ne-D_2, and Ne-O_2 behave perfectly normal. For Ne-CO_2 and Ne-N_2 (only at one temperature) there is a peculiar variation in conductivity. With increasing proportion of the heavier component in the mixture the conductivity first decreases, then starts to increase and then decreases again continuously, probably only due to experimental uncertainties. For Ne-N_2 experimental data exist at seven other temperatures and none indicate such an irregular variation. Ne-O_2 is also widely studied and found to indicate no such trend. The uncertainty in the data of Ne-CO_2 is large enough and this is in all probability responsible for this strange behavior, in our opinion.

(6) The four combinations of xenon with hydrogen, deuterium, nitrogen, and oxygen all behave normally, the thermal conductivity monotonically decreasing with concavity upwards.

(7) Acetylene-Air system is peculiar as its thermal conductivity, when plotted against the mole fraction of acetylene, though monotonically decreasing, has a concavity downwards instead of upwards as in most cases. Further, the curve referring to the higher temperature has a maxima. Air-carbon monoxide and air-methane also have somewhat similar variation with the mole fraction of the heavier component. Neither system has a maximum and though the curve shape for air-CO is concave downward, the air-methane plot is more nearly linear.

(8) Benzene-hexane shows a different type of thermal conductivity dependence. Its thermal conductivity monotonically increases as the mole fraction of the heavier component in the mixture is increased, the variation is almost linear.

(9) Carbon dioxide-ethylene system shows the general trend of the thermal conductivity decreasing as the mole fraction of CO_2 is increased in the mixture but the change is almost linear with a possibility that it is concave downward. The two systems carbon dioxide-oxygen and carbon dioxide-hydrogen again show the general type of monotonically decreasing thermal conductivity with concavity upwards as the concentration of carbon-dioxide is increased in the mixture. The thermal conductivity of carbon dioxide-nitrogen has been studied at many temperatures and the nature of the qualitative variation undergoes a systematic change as the temperature is increased. The shape of the curves 1 and 2 (Figure 108a) is a normal one at low temperatures but as the temperature increases the thermal conductivity, though still monotonically decreasing, has a concavity changing from upwards to downwards (curves 4 and 6, Figure 108a) and of course in between the variation becomes linear (curve 3, Figure 108a). As the temperature further increases the conductivity curve shows a maxima. With increasing proportion of carbon dioxide in the mixture the conductivity does decrease (curves 7 through 9, Figures 108a and 108b) and eventually starts increasing with the content of carbon dioxide in the mixture though still exhibiting a maxima (curve 10 through 14, Figure 108b). Any precise quantitative statement is not possible because of the disagreement between the results of the two workers [189, 65] so that curves 5 and 6 appear in the wrong order. The former refers to a lower temperature than the latter, but is higher. The carbon dioxide-propane system has a still more peculiar thermal conductivity variation. The thermal conductivity monotonically increases (instead of a decrease) with concavity upwards as the amount of propane in the mixture is increased. This is one example where one can see how even the qualitative shape of the curve of thermal conductiviy variation with composition can drastically change as one of the two components of the binary systems is changed keeping the other the same.

(10) The thermal conductivity of carbon monoxide-hydrogen, deuterium-hydrogen, deuterium-nitrogen, ethylene-hydrogen, and ethylene-methane all show the normal variation of thermal conductivity with composition. However, ethylene-nitrogen system differs to the extent that the concavity of the curve is downward.

(11) The thermal conductivity of hydrogen-nitrogen, hydrogen-nitrous oxide, hydrogen-oxygen and methane-propane systems is normal. The thermal conductivity decreases monotonically as the amount of the heavier component in the mixture increases, curves having concavity upwards.

(12) The thermal conductivity of nitrogen-oxygen and nitrogen-propane systems is not normal because in both cases the thermal conductivity increases monotonically as the proportion of the heavier component in the mixture is increased. In the case of nitrogen-oxygen the curves assume all sorts of shapes, concavity downward with a maxima, as well as concavity upwards, while nitrogen-propane curves have concavity downwards.

(13) The thermal conductivity of polar-nonpolar gas mixtures in general has a different thermal conductivity dependence than mixtures of nonpolar gases. In many cases of the former category thermal conductivity will be found to increase with the increasing proportion of the heavier component in the mixture, with concavity downwards and in several cases a maxima in the curve.

(14) All the three thermal conductivity curves for the acetone-benzene system at three temperatures exhibit a maxima with the increasing concentration of benzene in the mixture.

(15) The variation for ammonia-carbon monoxide and ammonia-air systems is such that the thermal conductivity increases with the increasing concentration of the heavier component in the mixture but the curve exhibits a maxima

also. Ammonia-nitrogen thermal conductivity curves exhibit maxima except for one case (curve 1) the thermal conductivity decreases with the increasing proportion of nitrogen in the mixture. The thermal conductivity of ammonia-ethylene system decreases with increasing concentration of ethylene in the mixture with concavity downwards. On the other hand ammonia hydrogen system behaves like most of the mixtures of nonpolar gases, thermal conductivity decreasing with increasing ammonia, the curves having concavity upwards.

(16) Dimethyl ether-argon and Dimethyl-ether-propane systems behave very much alike. The thermal conductivity of both the systems increases with increasing proportion of the heavier component in the mixture with concavity upwards.

(17) The thermal conductivity of methanol-argon and methanol-hexane systems have a maxima when plotted against the increasing amount of the heavier component in the mixture.

(18) The thermal conductivity of methyl formate-propane system increases almost linearly with the increasing amount of propane in the mixture.

(19) The thermal conductivity of the binary systems of steam with air, carbon dioxide and nitrogen when plotted against the mole fraction of the heavier component in each case exhibit a maxima.

(20) The thermal conductivity of the three binary systems in which both the components are polars is analyzed here. In the two cases, chloroform-ether and dimethyl-ether-methyl formate, the thermal conductivity monotonically decreases as the heavier component in the mixture increases. The variation is approximately linear. More precisely the former system has a concavity downwards while the latter upwards. The thermal conductivity of diethylamine-ethyl ether system shows a gradual increase with increasing amount of ether in the mixture.

TABLE 58a. EXPERIMENTAL THERMAL CONDUCTIVITY DATA FOR ARGON-HELIUM SYSTEM

Curve No.	Fig. No.	Ref. No.	Author(s)	Temp. (K)	Mole Fraction of Ar	Thermal Cond. (mW cm^{-1} K^{-1})	Remarks
1	58a	684, 685	Mason, E.A. and von Ubisch, H.	302.2	0.000 0.106 0.276 0.541 0.710 1.000	1.537 1.223 0.8457 0.5108 0.3529 0.1817	He: Matheson Co., N.J., Ar: welding grade of AGA Gasaccumulator, Stockholm-Lidingo; thin hot wire method with constant resistance; precision ±2%.
2	58a	684, 685	Mason, E.A. and von Ubisch, H.	793.2	0.000 0.106 0.276 0.541 0.710 1.000	3.082 2.470 1.767 1.089 0.7494 0.3827	Same as above.
3	58b	686	Thornton, E. and Baker, W.A.D.	291.2	0.000 0.061 0.208 0.299 0.438 0.520 0.574 0.645 0.720 0.782 0.844 0.914 1.000	1.491 1.382 1.009 0.8541 0.6322 0.5317 0.4647 0.4019 0.3475 0.3056 0.2638 0.2261 0.1742	He: spectroscopically pure, Ar: impurities less than 0.2%; Katharometer method; maximum error in mixture composition ±0.3%, accuracy of these relative measurements decreased with the increase in conductivity and varied between ±2.2 to ±4.0%.
4	58b	289	Saxena, S.C.	311.2	0.0000 0.1412 0.2302 0.4164 0.6084 0.8398 1.0000	1.571 1.128 0.9408 0.6439 0.4342 0.2751 0.1834	He and Ar: spectroscopically pure; thick hot wire method; precision ±2.0%.
5	58c	687	Gambhir, R.S. and Saxena, S.C.	308.2	0.000 0.228 0.416 0.748 1.000	1.503 0.8960 0.6322 0.3266 0.1834	He and Ar: spectroscopically pure; thick hot wire method; accuracy ±1.0 to ±2.0%, precision ±1.0%.
6	58c	687	Gambhir, R.S. and Saxena, S.C.	323.2	0.000 0.228 0.416 0.748 1.000	1.549 0.9337 0.6573 0.3404 0.1905	Same as above.
7	58c	687	Gambhir, R.S. and Saxena, S.C.	343.2	0.000 0.228 0.416 0.748 1.000	1.612 0.9755 0.6866 0.3588 0.2001	Same as above.
8	58c	687	Gambhir, R.S. and Saxena, S.C.	363.2	0.000 0.228 0.416 0.748 1.000	1.671 1.013 0.7076 0.3772 0.2098	Same as above.

TABLE 58a. EXPERIMENTAL THERMAL CONDUCTIVITY DATA FOR ARGON-HELIUM SYSTEM (continued)

Curve No.	Fig. No.	Ref. No.	Author(s)	Temp. (K)	Mole Fraction of Ar	Thermal Cond. (mW cm^{-1} K^{-1})	Remarks
--	--	65, 688	Cheung, H., Bromley, L.A., and Wilke, C.R.	373.2 372.2 373.7 375.7 379.2	0.000 0.220 0.475 0.724 1.000	1.756 1.054 0.6657 0.3981 0.2146	He: U.S. Navy research grade, specified purity 99.99%, chief impurities H$_2$ and H$_2$O, Ar: Linde Air Products Co., standard grade, specified purity 99.97%, chief impurity N$_2$; coaxial cylinder method; total maximum error 5.7%, average error 1.2% and maximum uncertainty in mixture composition 0.25%.
--	--	65, 688	Cheung, H., Bromley, L.A., and Wilke, C.R.	588.2 589.2 589.2 591.2 594.2	0.000 0.226 0.427 0.694 1.000	2.459 1.391 0.9630 0.5594 0.3056	
9	58b	372	Wachsmuth, J.	273.2	0.0000 0.0539 0.1532 0.5463 0.7296 1.0000	1.418 1.231 0.9713 0.4509 0.3109 0.1630	Helium slightly impure and argon very pure as judged from density determination; hot wire potential lead method.
--	--	690	Lindsay, A.L., and Bromley, L.A.	295.0 294.3 294.5 294.7 294.7 294.6 294.6 294.5 295.1 296.5	0.000 0.202 0.202 0.412 0.412 0.603 0.603 0.793 0.793 1.000	1.501 0.9340 0.9547 0.5950 0.5172 0.4428 0.4013 0.2785 0.2767 0.1640	He: Matheson Co., 99.6% pure, Ar: Linde Air Products Co., 99.8% pure; unsteady state method; precision ±15% in the worst case.
10	58a	65	Cheung, H., Bromley, L.A., and Wilke, C.R.	373.2	0.000 0.220 0.475 0.724 1.000	1.756 1.057 0.6649 0.3952 0.2118	These authors have smoothed their original data reproduced above.
11	58a	65	Cheung, H., Bromley, L.A. and Wilke, C.R.	589.2	0.000 0.226 0.427 0.694 1.000	2.328 1.391 0.9630 0.5573 0.3038	These authors have smoothed their original data reproduced above.
12	58b	690	Lindsay, A.L. and Bromley, L.A.	295.0	0.000 0.202 0.412 0.603 0.793 1.000	1.501 0.944 0.555 0.422 0.277 0.164	We have generated these data from the original reproduced above by averaging the multiple values referring to the same composition of the mixture.
--	58b	686	Thornton, E. and Baker, W.A.D.	291.2	0.000 0.103 0.204 0.296 0.408 0.503 0.582 0.714 0.797 0.901 1.000	1.491 1.256 0.9839 0.8164 0.6280 0.5275 0.4396 0.3349 0.2805 0.2219 0.1742	Same as above. This is a repeated set and agrees with the previous one within a maximum deviation of 4.5%.

TABLE 58a. EXPERIMENTAL THERMAL CONDUCTIVITY DATA FOR ARGON-HELIUM SYSTEM (continued)

Curve No.	Fig. No.	Ref. No.	Author(s)	Temp. (K)	Mole Fraction of Ar	Thermal Cond. (mW cm^{-1} K^{-1})	Remarks
13	58b	691	Burge, H. L, and Robinson, L. B.	297	0.00 0.25 0.50 0.75 1.00	1.472 0.8411 0.5296 0.2780 0.1742	Line-source transient-heat-transfer technique; precision better than ±1.0%.

TABLE 58b. THERMAL CONDUCTIVITY (mW cm^{-1}K^{-1}) OF ARGON-HELIUM SYSTEM AS A FUNCTION OF COMPOSITION AT THE TEMPERATURE OF MEASUREMENT AS DERIVED BY GRAPHICAL SMOOTHING

Mole Fraction of Ar	273.2 K (Ref. 372)	291.2 K (Ref. 686)	295 K (Ref. 690)	297 K (Ref. 691)	302.2 K (Ref. 684)	308.2 K (Ref. 687)	311.2 K (Ref. 289)
0.00	1.42	1.49	1.50	1.47	1.54	1.50	1.57
0.05	1.24	1.36	1.33	1.32	1.48	1.33	1.40
0.10	1.10	1.24	1.18	1.18	1.24	1.18	1.24
0.15	0.978	1.13	1.05	1.05	1.13	1.06	1.11
0.20	0.880	1.03	0.937	0.942	0.993	0.948	1.00
0.25	0.795	0.932	0.842	0.841	0.892	0.860	0.906
0.30	0.721	0.842	0.755	0.758	0.809	0.785	0.820
0.35	0.656	0.760	0.679	0.682	0.734	0.716	0.740
0.40	0.596	0.684	0.612	0.614	0.667	0.654	0.667
0.45	0.542	0.617	0.554	0.554	0.607	0.594	0.601
0.50	0.494	0.554	0.501	0.499	0.551	0.539	0.542
0.55	0.448	0.498	0.454	0.450	0.500	0.489	0.489
0.60	0.406	0.447	0.411	0.405	0.451	0.444	0.442
0.65	0.367	0.402	0.372	0.366	0.406	0.400	0.402
0.70	0.331	0.362	0.336	0.331	0.364	0.362	0.366
0.75	0.298	0.325	0.302	0.300	0.326	0.326	0.331
0.80	0.267	0.291	0.271	0.272	0.294	0.294	0.299
0.85	0.239	0.250	0.242	0.246	0.263	0.264	0.269
0.90	0.211	0.230	0.214	0.223	0.234	0.236	0.240
0.95	0.186	0.202	0.189	0.199	0.207	0.209	0.212
1.00	0.163	0.174	0.164	0.174	0.182	0.183	0.183

Mole Fraction of Ar	323.2 K (Ref. 687)	343.2 K (Ref. 687)	363.2 K (Ref. 687)	373.2 K (Ref. 65)	589.2 K (Ref. 65)	793.2 K (Ref. 684)
0.00	1.55	1.61	1.67	1.76	2.33	3.08
0.05	1.37	1.43	1.50	1.56	2.07	2.79
0.10	1.22	1.28	1.34	1.39	1.83	2.50
0.15	1.09	1.15	1.20	1.24	1.64	2.26
0.20	0.987	1.03	1.08	1.11	1.47	1.95
0.25	0.896	0.935	0.970	0.997	1.32	1.86
0.30	0.816	0.801	0.880	0.905	1.20	1.69
0.35	0.744	0.775	0.800	0.823	1.09	1.54
0.40	0.678	0.707	0.729	0.748	0.982	1.40
0.45	0.617	0.646	0.666	0.680	0.891	1.28
0.50	0.560	0.588	0.608	0.618	0.810	1.16
0.55	0.508	0.534	0.555	0.562	0.738	1.06
0.60	0.461	0.485	0.506	0.510	0.672	0.957
0.65	0.417	0.440	0.461	0.461	0.611	0.860
0.70	0.377	0.398	0.418	0.417	0.557	0.772
0.75	0.341	0.358	0.378	0.376	0.507	0.693
0.80	0.308	0.324	0.341	0.340	0.463	0.622
0.85	0.275	0.290	0.306	0.305	0.421	0.555
0.90	0.246	0.260	0.273	0.272	0.381	0.494
0.95	0.218	0.230	0.241	0.240	0.342	0.436
1.00	0.191	0.200	0.210	0.212	0.304	0.383

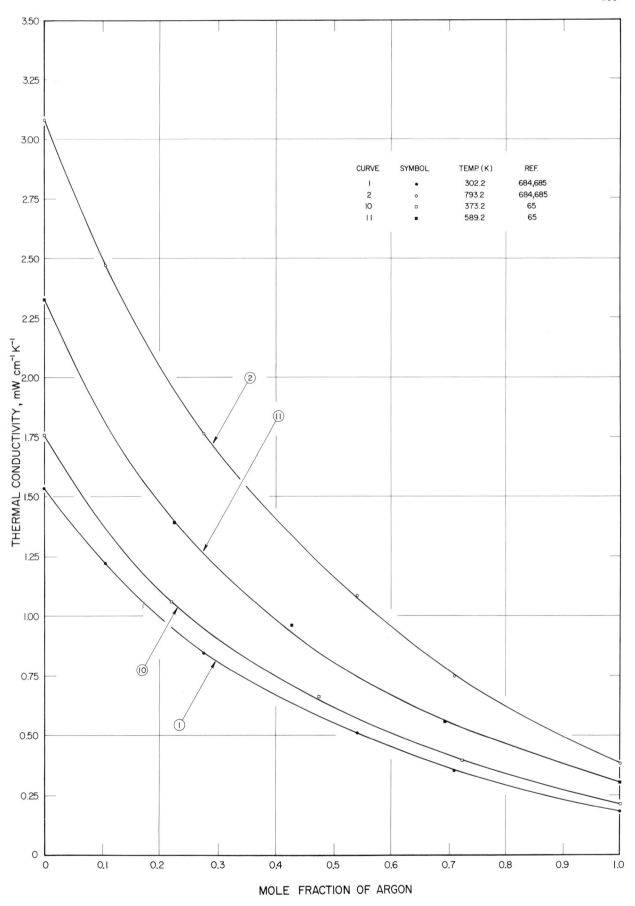

FIGURE 58a. THERMAL CONDUCTIVITY OF ARGON-HELIUM SYSTEM

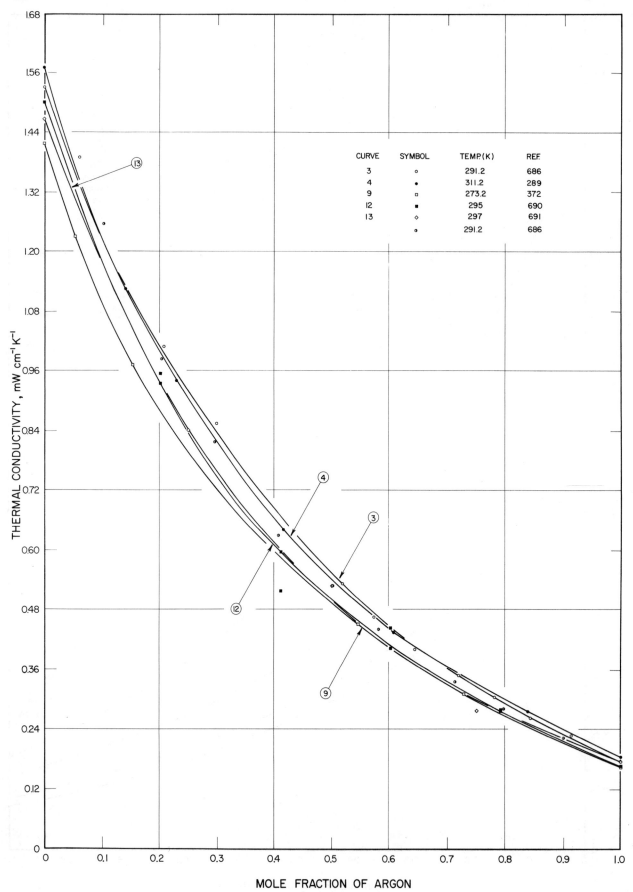

FIGURE 58b. THERMAL CONDUCTIVITY OF ARGON-HELIUM SYSTEM

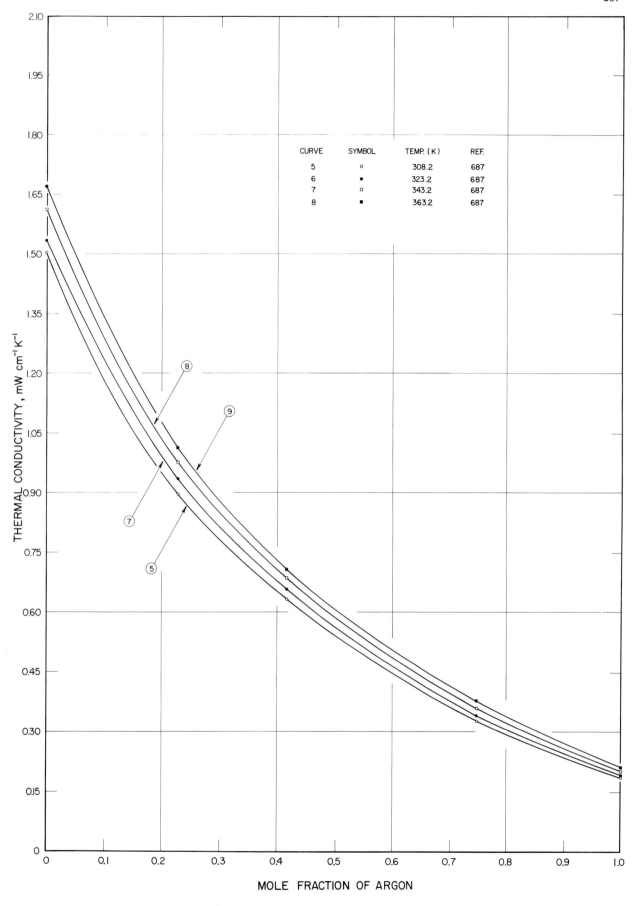

FIGURE 58c. THERMAL CONDUCTIVITY OF ARGON-HELIUM SYSTEM

TABLE 59a. EXPERIMENTAL THERMAL CONDUCTIVITY DATA FOR ARGON-NEON SYSTEM

Curve No.	Fig. No.	Ref. No.	Author(s)	Temp. (K)	Mole Fraction of Ar	Thermal Cond. (mW cm^{-1} K^{-1})	Remarks
1	59a	326	Srivastava, B.N. and Saxena, S.C.	311.2	0.0000 0.1183 0.1370 0.3124 0.3472 0.4215 0.6688 0.8286 0.8381 0.8660 1.0000	0.4953 0.4384 0.4291 0.3513 0.3475 0.3215 0.2541 0.2156 0.2160 0.2106 0.1834	Ne and Ar: spectroscopically pure; thick hot wire method; precision ± 2%.
2	59a	684, 685	Mason, E.A. and von Ubisch, H.	302.2	0.000 0.237 0.423 0.642 0.842 1.000	0.5179 0.3982 0.3341 0.2633 0.2173 0.1817	Ne: spectroscopically pure, Ar: welding grade of AGA Gasaccumulator, Stockholm-Lidingo; thin hot wire method with constant resistance; precision ± 2%.
3	59b	684, 685	Mason, E.A. and von Ubisch, H.	793.2	0.000 0.237 0.423 0.642 0.842 1.000	0.9881 0.7787 0.6615 0.5317 0.4312 0.3827	Same as above.
4	59a	686	Thornton, E. and Baker, W.A.D.	291.2	0.000 0.157 0.221 0.328 0.436 0.541 0.638 0.726 0.803 0.900 1.000	0.4857 0.4145 0.3894 0.3517 0.3182 0.2847 0.2596 0.2386 0.2177 0.1968 0.1742	Ne: spectroscopically pure; Ar: impurities less than 0.2%; katharometer method; maximum error in mixture composition ± 0.3%, accuracy of these relative measurements decreased with the increase in conductivity and varied between ± 2.2 to ± 4.0%.
5	59a	692	Mathur, S., Tondon, P.K., and Saxena, S.C.	313.2	0.00 0.20 0.40 0.60 0.80 1.00	0.5141 0.4132 0.3450 0.2860 0.2332 0.1851	Ne and Ar: spectroscopically pure; thick hot wire method; accuracy ± 2.0%, precision ± 1.0%.
6	59a	692	Mathur, S., Tondon, P.K., and Saxena, S.C.	338.2	0.00 0.20 0.40 0.60 0.80 1.00	0.5338 0.4342 0.3567 0.2948 0.2428 0.2001	Same as above.
7	59a	692	Mathur, S., Tondon, P.K., and Saxena, S.C.	363.2	0.00 0.20 0.40 0.60 0.80 1.00	0.5589 0.4568 0.3764 0.3132 0.2587 0.2110	Same as above.

TABLE 59a. EXPERIMENTAL THERMAL CONDUCTIVITY DATA FOR ARGON-NEON SYSTEM (cont.)

Curve No.	Fig. No.	Ref. No.	Author(s)	Temp. (K)	Mole Fraction of Ar	Thermal Cond. (mW cm^{-1} K^{-1})	Remarks
8	59a	325	Srivastava, B.N. and Madan, M.P.	273.2	0.0000 0.2406 0.5740 0.7900 1.0000	0.4551 0.3651 0.2554 0.2035 0.1796	Ne and Ar: spectroscopically pure; thin hot wire method with constant resistance; precision of these relative measurements is about ±2%.

TABLE 59b. THERMAL CONDUCTIVITY (mW cm^{-1}K^{-1}) OF ARGON-NEON SYSTEM AS A FUNCTION OF COMPOSITION AT THE TEMPERATURE OF MEASUREMENT AS DERIVED BY GRAPHICAL SMOOTHING

Mole Fraction of Ar	273.2 K (Ref. 325)	291.2 K (Ref. 686)	302.2 K (Ref. 684)	311.2 K (Ref. 326)
0.00	0.455	0.486	0.518	0.495
0.05	0.436	0.463	0.491	0.470
0.10	0.417	0.440	0.465	0.446
0.15	0.398	0.418	0.439	0.423
0.20	0.380	0.398	0.415	0.400
0.25	0.362	0.379	0.393	0.380
0.30	0.344	0.361	0.373	0.361
0.35	0.328	0.345	0.354	0.343
0.40	0.310	0.328	0.337	0.327
0.45	0.293	0.313	0.320	0.312
0.50	0.277	0.298	0.304	0.298
0.55	0.262	0.283	0.289	0.284
0.60	0.248	0.270	0.275	0.271
0.65	0.234	0.256	0.264	0.258
0.70	0.222	0.243	0.249	0.246
0.75	0.211	0.230	0.237	0.235
0.80	0.202	0.218	0.226	0.224
0.85	0.194	0.207	0.215	0.212
0.90	0.188	0.196	0.204	0.202
0.95	0.184	0.185	0.193	0.192
1.00	0.180	0.174	0.182	0.183

Mole Fraction of Ar	313.2 K (Ref. 692)	338.2 K (Ref. 692)	363.2 K (Ref. 692)	793.2 K (Ref. 684)
0.00	0.514	0.534	0.559	0.988
0.05	0.487	0.508	0.532	0.942
0.10	0.460	0.483	0.507	0.895
0.15	0.435	0.458	0.481	0.851
0.20	0.413	0.434	0.457	0.810
0.25	0.395	0.413	0.435	0.772
0.30	0.377	0.393	0.415	0.737
0.35	0.361	0.375	0.395	0.706
0.40	0.345	0.357	0.376	0.675
0.45	0.330	0.341	0.360	0.645
0.50	0.315	0.325	0.344	0.616
0.55	0.300	0.310	0.328	0.586
0.60	0.286	0.295	0.313	0.557
0.65	0.273	0.281	0.299	0.528
0.70	0.259	0.268	0.285	0.500
0.75	0.241	0.255	0.272	0.474
0.80	0.233	0.243	0.259	0.449
0.85	0.221	0.232	0.246	0.429
0.90	0.209	0.221	0.234	0.412
0.95	0.197	0.211	0.222	0.397
1.00	0.185	0.200	0.211	0.383

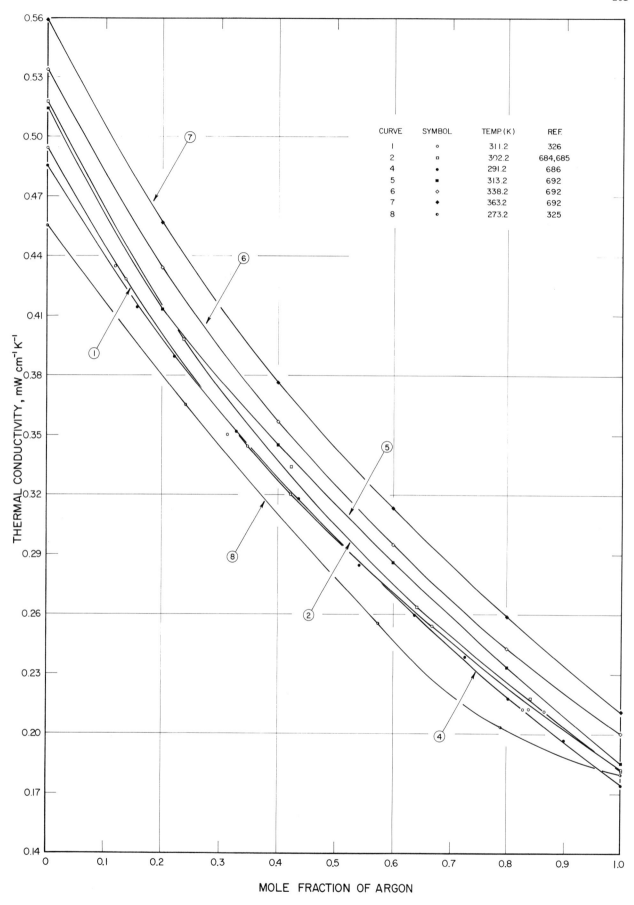

FIGURE 59a. THERMAL CONDUCTIVITY OF ARGON-NEON SYSTEM

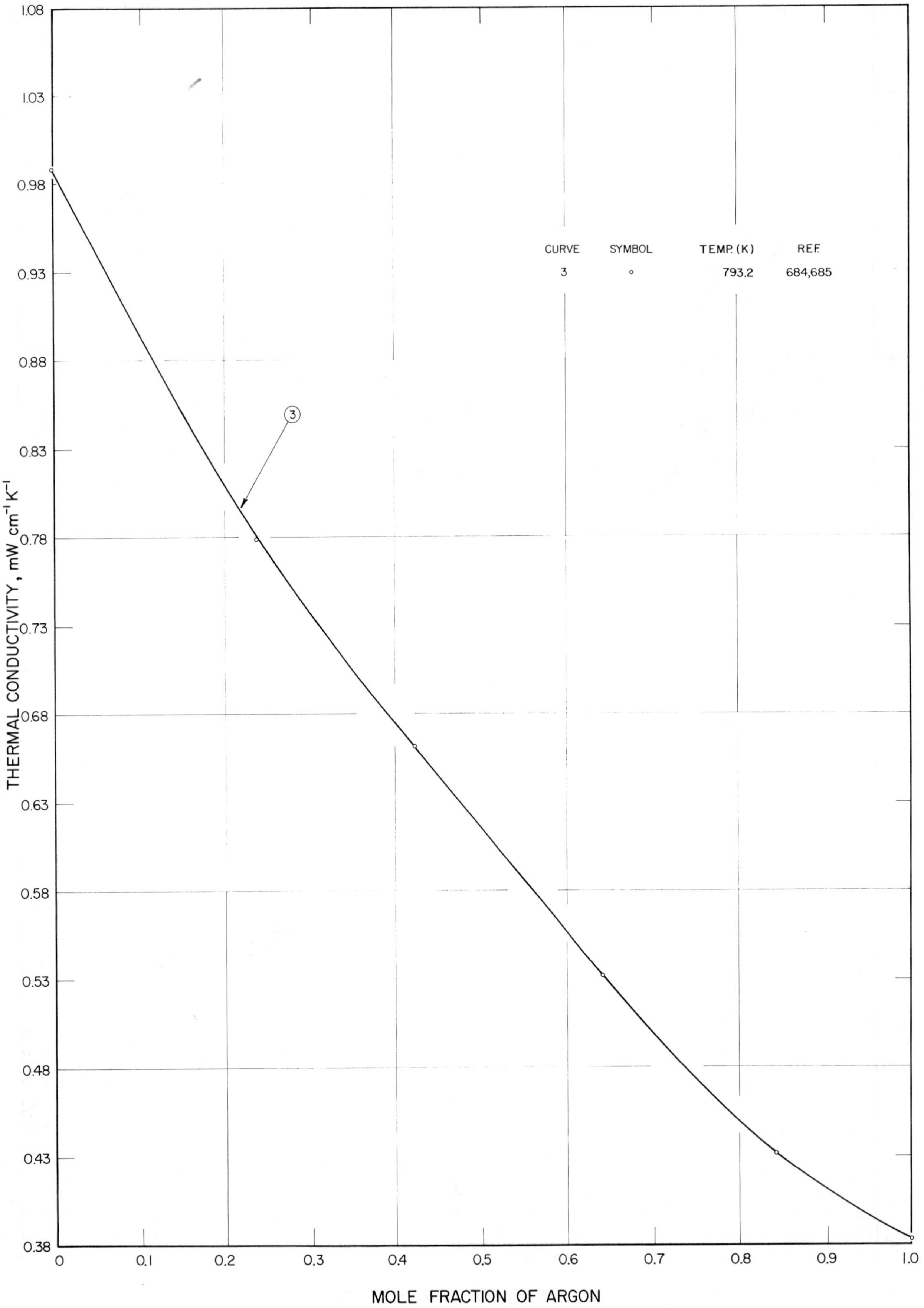

FIGURE 59b. THERMAL CONDUCTIVITY OF ARGON-NEON SYSTEM

TABLE 60a. EXPERIMENTAL THERMAL CONDUCTIVITY DATA FOR ARGON-KRYPTON SYSTEM

Curve No.	Fig. No.	Ref. No.	Author(s)	Temp. (K)	Mole Fraction of Kr	Thermal Cond. (mW cm^{-1}K^{-1})	Remarks
1	60a	326	Srivastava, B.N. and Saxena, S.C.	311.2	0.0000 0.0866 0.2338 0.3795 0.4840 0.6683 0.8115 1.0000	0.1834 0.1666 0.1532 0.1398 0.1327 0.1160 0.1072 0.0980	Ar: spectroscopically pure, Kr: traces of xenon; thick hot wire method; precision ± 2%.
2	60a	684, 685	Mason, E.A. and von Ubisch, H.	302.2	0.000 0.298 0.536 0.764 1.000	0.1817 0.1516 0.1290 0.1139 0.0971	Ar: welding grade of AGA Gas-accumulator, Stockholm-Lidingo, Kr: spectroscopically pure; thin hot wire method with constant resistance; precision ± 2%.
3	60b	684, 685	Mason, E.A. and von Ubisch, H.	793.2	0.000 0.298 0.536 0.764 1.000	0.3827 0.3241 0.2793 0.2495 0.2236	Same as above.
4	60a	693	Thornton, E.	291.2	0.000 0.109 0.228 0.333 0.443 0.546 0.673 0.777 0.865 1.000	0.1742 0.1620 0.1457 0.1365 0.1264 0.1160 0.1072 0.1013 0.0976 0.0921	Ar: impurities less than 0.2%, Kr: 99-100% pure, balance xenon; katharometer method; maximum error in mixture composition ± 0.3%, accuracy of these relative conductivity values is better than ± 2.2%.
5	60a	687	Gambhir, R.S. and Saxena, S.C.	308.2	0.000 0.256 0.558 0.742 1.000	0.1834 0.1541 0.1243 0.1101 0.0959	Ar: spectroscopically pure, Kr: 99-100% pure, balance xenon; thick hot wire meothd; accuracy ± 2.0%, precision ± 1.0%.
6	60a	687	Gambhir, R.S. and Saxena, S.C.	323.2	0.000 0.256 0.558 0.742 1.000	0.1905 0.1608 0.1294 0.1168 0.1017	Same as above.
7	60a	687	Gambhir, R.S. and Saxena, S.C.	343.2	0.000 0.256 0.558 0.742 1.000	0.2001 0.1683 0.1357 0.1239 0.1072	Same as above.
8	60a	687	Gambhir, R.S. and Saxena, S.C.	363.2	0.000 0.256 0.558 0.742 1.000	0.2098 0.1750 0.1424 0.1306 0.1114	Same as above.

TABLE 60b. THERMAL CONDUCTIVITY (mW cm^{-1} K^{-1}) OF ARGON-KRYPTON SYSTEM AS A FUNCTION OF COMPOSITION AT THE TEMPERATURE OF MEASUREMENT AS DERIVED BY GRAPHICAL SMOOTHING

Mole Fraction of Kr	291.2 K (Ref. 693)	302.2 K (Ref. 684)	308.2 K (Ref. 687)	311.2 K (Ref. 326)
0.00	0.174	0.182	0.183	0.183
0.05	0.168	0.177	0.178	0.176
0.10	0.162	0.172	0.172	0.170
0.15	0.156	0.167	0.166	0.163
0.20	0.150	0.162	0.160	0.157
0.25	0.144	0.157	0.155	0.151
0.30	0.139	0.151	0.149	0.145
0.35	0.134	0.147	0.144	0.141
0.40	0.129	0.142	0.139	0.136
0.45	0.124	0.138	0.134	0.132
0.50	0.120	0.133	0.130	0.128
0.55	0.116	0.129	0.125	0.125
0.60	0.112	0.125	0.121	0.121
0.65	0.109	0.122	0.117	0.118
0.70	0.106	0.118	0.113	0.115
0.75	0.103	0.114	0.110	0.112
0.80	0.100	0.111	0.106	0.109
0.85	0.0978	0.107	0.104	0.106
0.90	0.0958	0.104	0.101	0.103
0.95	0.0938	0.100	0.0982	0.100
1.00	0.0921	0.0971	0.0959	0.0980

Mole Fraction of Kr	323.2 K (Ref. 687)	343.2 K (Ref. 687)	363.2 K (Ref. 687)	793.2 K (Ref. 684)
0.00	0.191	0.200	0.210	0.383
0.05	0.185	0.194	0.203	0.372
0.10	0.179	0.187	0.196	0.362
0.15	0.173	0.181	0.189	0.352
0.20	0.167	0.175	0.182	0.341
0.25	0.162	0.169	0.176	0.331
0.30	0.156	0.163	0.170	0.321
0.35	0.150	0.158	0.164	0.312
0.40	0.145	0.152	0.159	0.303
0.45	0.140	0.147	0.154	0.294
0.50	0.135	0.142	0.149	0.286
0.55	0.131	0.137	0.144	0.278
0.60	0.127	0.133	0.140	0.271
0.65	0.123	0.129	0.136	0.264
0.70	0.119	0.125	0.132	0.257
0.75	0.116	0.122	0.128	0.251
0.80	0.113	0.119	0.125	0.246
0.85	0.110	0.116	0.121	0.240
0.90	0.107	0.113	0.118	0.234
0.95	0.104	0.110	0.114	0.229
1.00	0.102	0.107	0.111	0.224

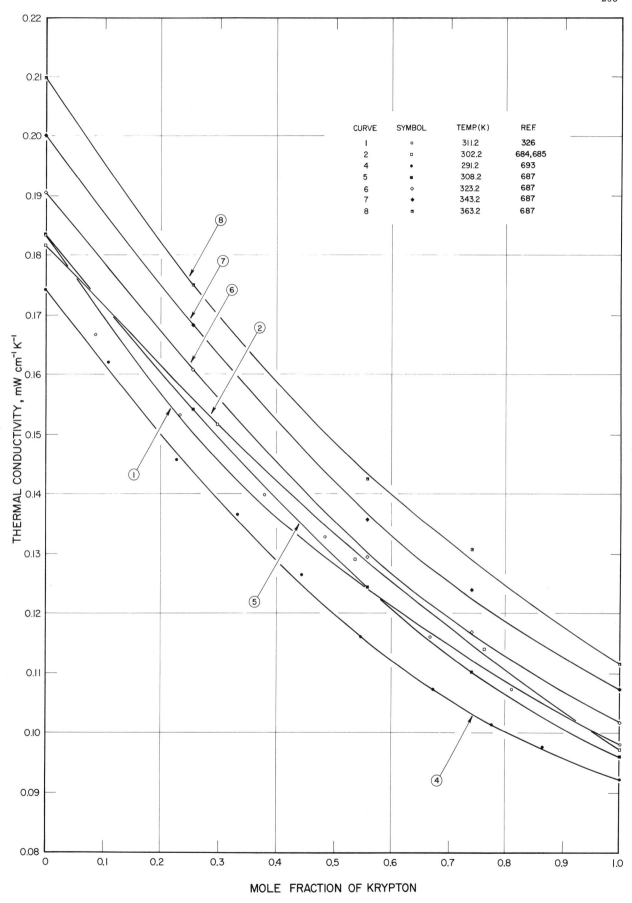

FIGURE 60a. THERMAL CONDUCTIVITY OF ARGON-KRYPTON SYSTEM

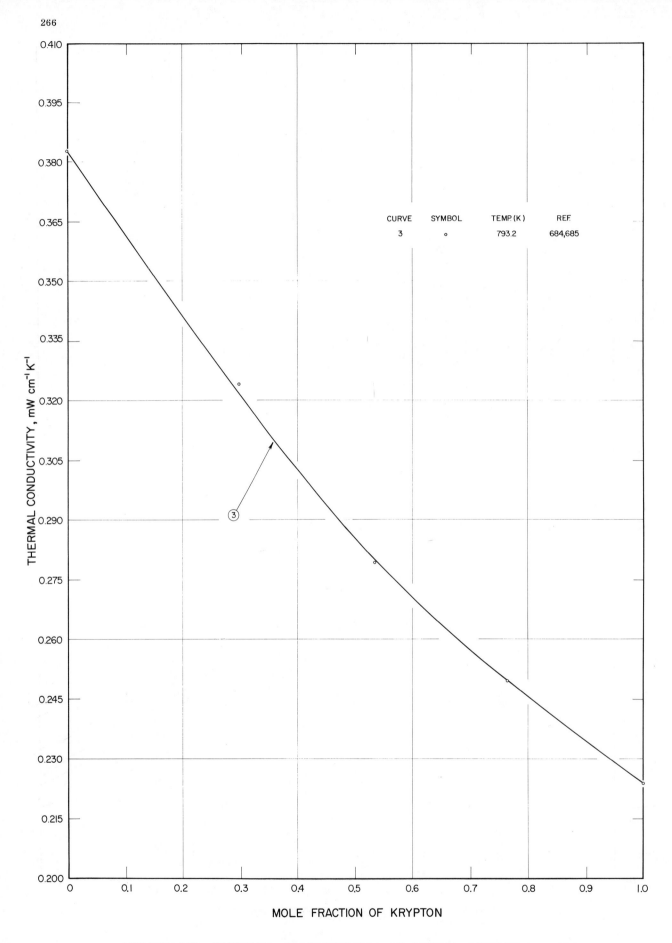

FIGURE 60b. THERMAL CONDUCTIVITY OF ARGON-KRYPTON SYSTEM

TABLE 61a.   EXPERIMENTAL THERMAL CONDUCTIVITY DATA FOR ARGON-XENON SYSTEM

Curve No.	Fig. No.	Ref. No.	Author(s)	Temp. (K)	Mole Fraction of Xe	Thermal Cond. (mW cm^{-1} K^{-1})	Remarks
1	61a	684, 685	Mason, E.A. and von Ubisch, H.	302.2	0.000 0.271 0.504 0.750 1.000	0.1817 0.1323 0.1001 0.0758 0.0597	Ar: welding grade of AGA Gasaccumulator, Stockholm-Lidingo, Xe: spectroscopically pure; thin hot wire method with constant resistance; precision ± 2%.
2	61b	684, 685	Mason, E.A. and von Ubisch, H.	793.2	0.000 0.271 0.504 0.750 1.000	0.3827 0.2851 0.2248 0.1742 0.1398	Same as above.
3	61a	694	Thornton, E.	291.2	0.000 0.109 0.213 0.300 0.405 0.498 0.598 0.701 0.792 0.905 1.000	0.1742 0.1499 0.1302 0.1172 0.1051 0.0946 0.0846 0.0754 0.0682 0.0607 0.0553	Ar: impurities less than 0.2%, Xe: 99-100% pure, balance krypton; katharometer method; maximum error in mixture composition ± 0.3%, and estimated maximum error of these relative measurements ± 2.2%.
4	61a	289	Saxena, S.C.	311.2	0.0000 0.1757 0.3231 0.5023 0.6727 0.7517 0.8339 1.0000	0.1834 0.1453 0.1223 0.0980 0.0804 0.0733 0.0682 0.0565	Ar: spectroscopically pure, Xe: traces of krypton; thick hot wire method; precision ± 2.0%.
5	61a	692	Mathur, S., Tondon, P.K., and Saxena, S.C.	311.2	0.00 0.20 0.40 0.60 0.80 1.00	0.1851 0.1583 0.1319 0.1055 0.0816 0.0620	Ar: spectroscopically pure, Xe: 99-100% pure, balance krypton; thick hot wire method; accuracy ± 2.0%, precision ± 1.0%.
6	61b	692	Mathur, S. Tondon, P.K., and Saxena, S.C.	366.2	0.00 0.20 0.40 0.60 0.80 1.00	0.2131 0.1784 0.1474 0.1193 0.0942 0.0745	Same as above.
7	61a	707, 708	Tondon, P.K. and Saxena, S.C.	313.2	0.000 0.241 0.758 1.000	0.1853 0.1407 0.0804 0.0593	Ar: spectroscopically pure, Xe: 99-100% pure, balance krypton; thick hot wire method; accuracy ± 2.0%, precision ± 1.0%.
8	61a	707, 708	Tondon, P.K. and Saxena, S.C.	338.2	0.000 0.241 0.758 1.000	0.1930 0.1499 0.0871 0.0620	Same as above.
9	61b	707, 708	Tondon, P.K. and Saxena, S.C.	366.2	0.000 0.241 0.758 1.000	0.2104 0.1637 0.0980 0.0706	Same as above.

TABLE 61b. THERMAL CONDUCTIVITY (mW cm^{-1} K^{-1}) OF ARGON-XENON SYSTEM AS A FUNCTION OF COMPOSITION AT THE TEMPERATURE OF MEASUREMENT AS DERIVED BY GRAPHICAL SMOOTHING

Mole Fraction of Xe	291.2 K (Ref. 694)	302.2 K (Ref. 684)	311.2 K (Ref. 289)	311.2 K (Ref. 692)	313.2 K (Ref. 707)
0.00	0.174	0.182	0.183	0.185	0.185
0.05	0.163	0.172	0.172	0.178	0.175
0.10	0.152	0.162	0.160	0.172	0.166
0.15	0.142	0.153	0.150	0.165	0.156
0.20	0.132	0.144	0.141	0.158	0.147
0.25	0.124	0.136	0.132	0.152	0.139
0.30	0.117	0.128	0.125	0.145	0.132
0.35	0.111	0.121	0.118	0.138	0.125
0.40	0.105	0.113	0.111	0.132	0.119
0.45	0.100	0.107	0.105	0.125	0.113
0.50	0.0945	0.101	0.0985	0.119	0.107
0.55	0.0894	0.0949	0.0928	0.112	0.102
0.60	0.0845	0.0895	0.0875	0.106	0.0963
0.65	0.0799	0.0845	0.0827	0.0993	0.0912
0.70	0.0755	0.0800	0.0781	0.0932	0.0862
0.75	0.0715	0.0758	0.0739	0.0873	0.0814
0.80	0.0678	0.0720	0.0701	0.0816	0.0767
0.85	0.0644	0.0685	0.0665	0.0765	0.0723
0.90	0.0613	0.0654	0.0631	0.0716	0.0679
0.95	0.0582	0.0625	0.0598	0.0667	0.0635
1.00	0.0553	0.0597	0.0565	0.0620	0.0593

Mole Fraction of Xe	338.2 K (Ref. 707)	366.2 K (Ref. 707)	366.2 K (Ref. 692)	793.2 K (Ref. 684)
0.00	0.193	0.210	0.213	0.383
0.05	0.184	0.200	0.204	0.363
0.10	0.175	0.190	0.196	0.344
0.15	0.165	0.180	0.187	0.326
0.20	0.157	0.171	0.178	0.309
0.25	0.149	0.162	0.170	0.292
0.30	0.141	0.155	0.163	0.277
0.35	0.135	0.148	0.155	0.262
0.40	0.128	0.141	0.147	0.249
0.45	0.122	0.135	0.140	0.236
0.50	0.116	0.128	0.133	0.224
0.55	0.110	0.123	0.126	0.213
0.60	0.105	0.117	0.119	0.202
0.65	0.0991	0.111	0.113	0.192
0.70	0.0935	0.105	0.106	0.183
0.75	0.0880	0.0990	0.100	0.175
0.80	0.0826	0.0933	0.0944	0.167
0.85	0.0773	0.0877	0.0890	0.160
0.90	0.0722	0.0821	0.0841	0.153
0.95	0.0671	0.0764	0.0793	0.146
1.00	0.0620	0.0706	0.0745	0.140

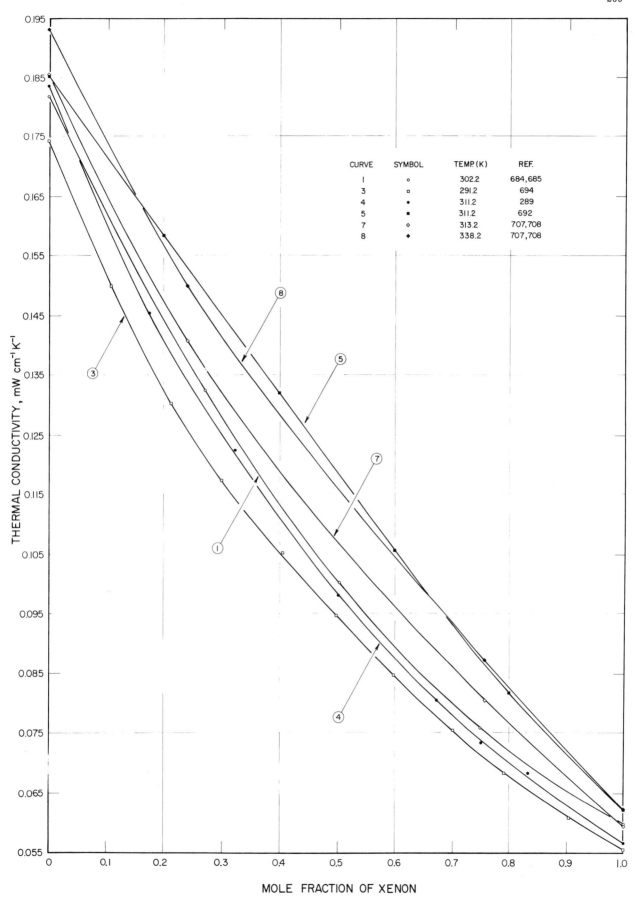

FIGURE 61a. THERMAL CONDUCTIVITY OF ARGON-XENON SYSTEM

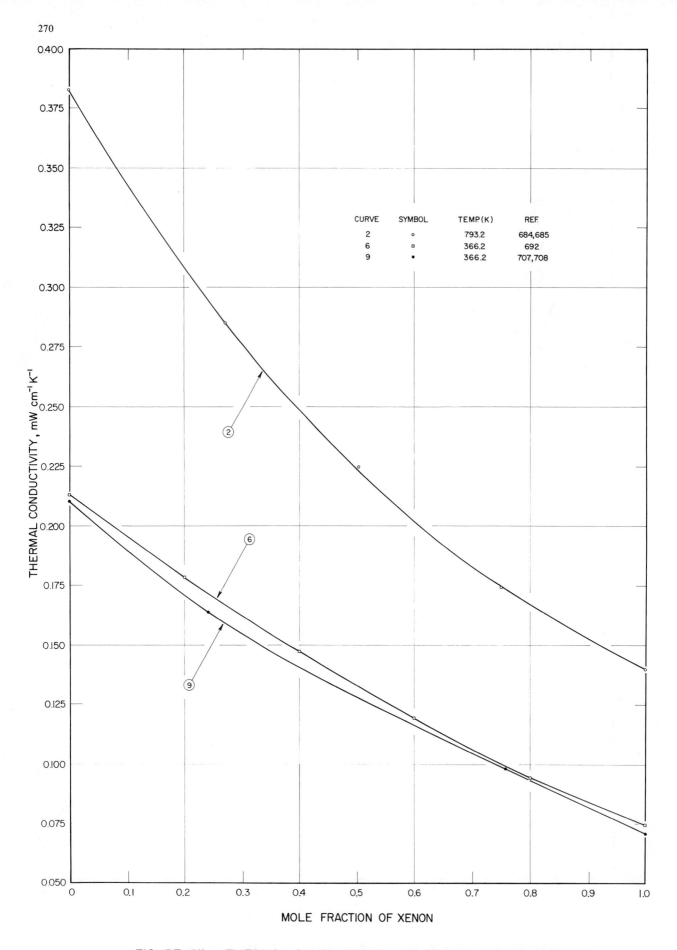

FIGURE 61b. THERMAL CONDUCTIVITY OF ARGON-XENON SYSTEM

TABLE 62a. EXPERIMENTAL THERMAL CONDUCTIVITY DATA FOR HELIUM-NEON SYSTEM

Curve No.	Fig. No.	Ref. No.	Author(s)	Temp. (K)	Mole Fraction of Ne	Thermal Cond. (mW cm^{-1} K^{-1})	Remarks
1	62a	686	Thornton, E. and Baker, W.A.D.	291.2	0.000 0.158 0.250 0.393 0.565 0.655 0.783 0.894 1.000	1.491 1.298 1.189 0.9923 0.8332 0.7494 0.6406 0.5568 0.4857	He and Ne: spectroscopically pure; katharometer method; maximum error in mixture composition ±0.3%, maximum estimated error of these relative measurements ±4.0%.
2	62a	684, 685	Mason, E.A. and von Ubisch, H.	302.2	0.000 0.119 0.130 0.382 0.755 1.000	1.537 1.348 1.306 0.9797 0.6573 0.5179	He: Matheson Co., N.J., Ne: spectroscopically pure; thin hot wire method with constant resistance; precision ±2%.
3	62b	684, 685	Mason, E.A. and von Ubisch, H.	793.2	0.000 0.119 0.130 0.382 0.755 1.000	3.082 2.755 2.755 1.985 1.327 0.9881	Same as above.
4	62a	83	Davidson, J.M. and Music, J.F.	273.2	0.00 0.25 0.51 0.75 1.00	1.390 1.015 0.7725 0.5899 0.4530	Unsteady state method, the rate of cooling of a solid inner cylinder of copper was determined; accuracy of these relative measurements ±5%.
5	62a	697	Gandhi, J.M. and Saxena, S.C.	303.2	0.0000 0.2566 0.4560 0.7552 1.0000	1.495 1.115 0.8826 0.6263 0.4940	He and Ne: spectroscopically pure, thick hot wire method; accuracy ±2.0%, precision ±1.0%.
6	62a	697	Gandhi, J.M. and Saxena, S.C.	323.2	0.0000 0.2566 0.4560 0.7552 1.0000	1.549 1.161 0.9110 0.6569 0.5150	Same as above
7	62a	697	Gandhi, J.M. and Saxena, S.C.	343.2	0.0000 0.2566 0.4560 0.7552 1.0000	1.612 1.207 0.9399 0.6883 0.5317	Same as above.
8	62a	697	Gandhi, J.M. and Saxena, S.C.	363.2	0.0000 0.2566 0.4560 0.7552 1.0000	1.671 1.254 0.9688 0.7189 0.5527	Same as above.

TABLE 62a. EXPERIMENTAL THERMAL CONDUCTIVITY DATA FOR HELIUM-NEON SYSTEM (continued)

Curve No.	Fig. No.	Ref. No.	Author(s)	Temp. (K)	Mole Fraction of Ne	Thermal Cond. (mW cm^{-1} K^{-1})	Remarks
9	62b	691	Burge, H. L. and Robinson, L. B.	297	0.00 0.25 0.50 0.75 1.00	1.472 1.005 0.6824 0.5200 0.4698	Line-source transient-heat-transfer technique; precision better than ±1.0%.

TABLE 62b. THERMAL CONDUCTIVITY (mW cm^{-1}K^{-1}) OF HELIUM-NEON SYSTEM AS A FUNCTION OF COMPOSITION AT THE TEMPERATURE OF MEASUREMENT AS DERIVED BY GRAPHICAL SMOOTHING

Mole Fraction of Ne	273.2 K (Ref. 83)	291.2 K (Ref. 686)	297 K (Ref. 691)	302.2 K (Ref. 684)	303.2 K (Ref. 697)
0.00	1.39	1.49	1.47	1.54	1.50
0.05	1.31	1.43	1.37	1.45	1.41
0.10	1.23	1.36	1.27	1.37	1.34
0.15	1.16	1.30	1.18	1.29	1.26
0.20	1.09	1.24	1.09	1.21	1.19
0.25	1.03	1.18	1.01	1.14	1.12
0.30	0.968	1.12	0.929	1.08	1.06
0.35	0.914	1.06	0.857	1.02	0.999
0.40	0.865	1.01	0.792	0.959	0.943
0.45	0.819	0.954	0.734	0.906	0.888
0.50	0.774	0.902	0.682	0.856	0.838
0.55	0.734	0.851	0.640	0.812	0.791
0.60	0.696	0.801	0.601	0.771	0.746
0.65	0.659	0.754	0.569	0.732	0.705
0.70	0.623	0.710	0.542	0.695	0.666
0.75	0.590	0.667	0.520	0.660	0.631
0.80	0.559	0.626	0.504	0.628	0.599
0.85	0.531	0.589	0.492	0.598	0.570
0.90	0.503	0.552	0.484	0.570	0.544
0.95	0.477	0.518	0.476	0.542	0.518
1.00	0.453	0.486	0.470	0.518	0.494

Mole Fraction of Ne	323.2 K (Ref. 697)	343.2 K (Ref. 697)	363.2 K (Ref. 697)	793.2 K (Ref. 684)
0.00	1.55	1.61	1.67	3.08
0.05	1.47	1.53	1.58	2.93
0.10	1.39	1.44	1.50	2.77
0.15	1.32	1.36	1.41	2.62
0.20	1.24	1.28	1.33	2.47
0.25	1.17	1.20	1.25	2.33
0.30	1.10	1.13	1.18	2.20
0.35	1.04	1.07	1.10	2.08
0.40	0.974	1.00	1.04	1.97
0.45	0.918	0.946	0.980	1.86
0.50	0.865	0.894	0.928	1.77
0.55	0.817	0.849	0.882	1.67
0.60	0.774	0.805	0.839	1.58
0.65	0.734	0.766	0.799	1.50
0.70	0.696	0.728	0.760	1.42
0.75	0.660	0.692	0.723	1.34
0.80	0.628	0.657	0.686	1.26
0.85	0.598	0.624	0.652	1.19
0.90	0.569	0.592	0.618	1.12
0.95	0.542	0.562	0.585	1.05
1.00	0.515	0.532	0.553	0.988

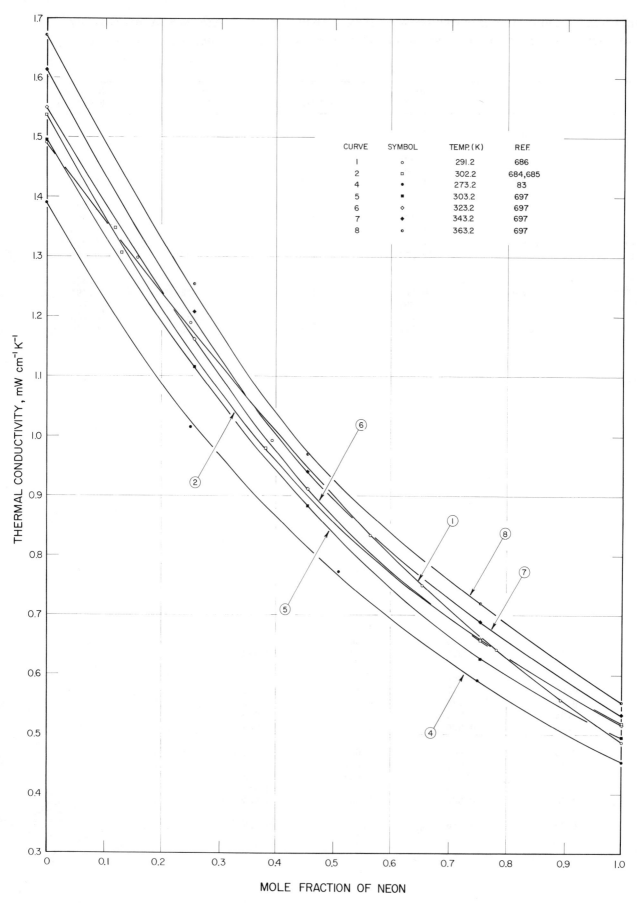

FIGURE 62a. THERMAL CONDUCTIVITY OF HELIUM-NEON SYSTEM

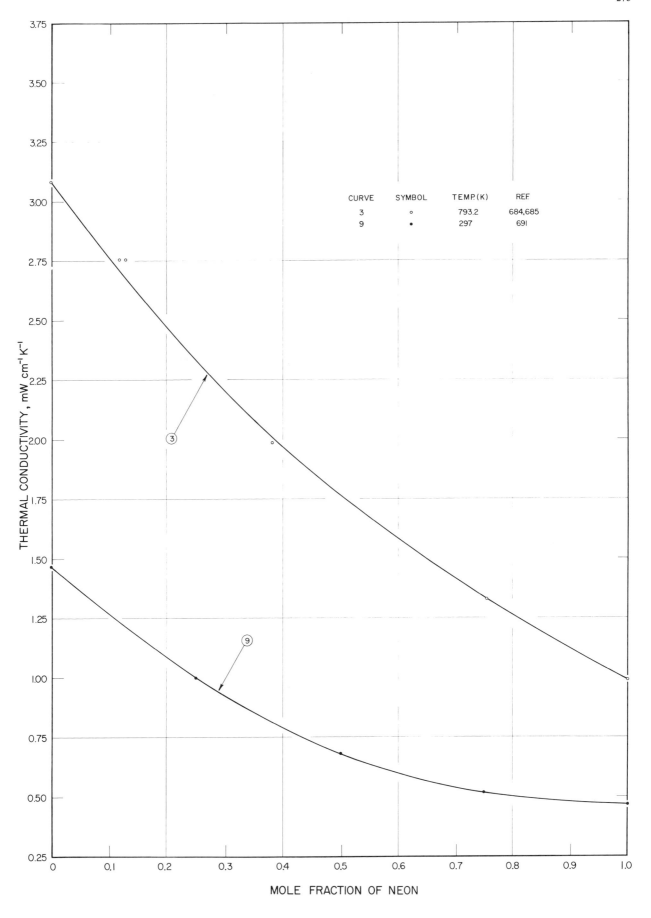

FIGURE 62b. THERMAL CONDUCTIVITY OF HELIUM−NEON SYSTEM

TABLE 63a. EXPERIMENTAL THERMAL CONDUCTIVITY DATA FOR HELIUM-KRYPTON SYSTEM

Curve No.	Fig. No.	Ref. No.	Author(s)	Temp. (K)	Mole Fraction of Kr	Thermal Cond. (mW cm^{-1} K^{-1})	Remarks
1	63a	693	Thornton, E.	291.2	0.000 0.069 0.151 0.272 0.353 0.439 0.600 0.698 0.797 0.891 1.000	1.491 1.277 1.047 0.7746 0.6406 0.5066 0.3454 0.2688 0.2010 0.1436 0.0921	He: spectroscopically pure; Kr: 99-100% pure, balance xenon; katharometer method; maximum error in mixture composition ±0.3%, accuracy of these relative measurements decreased with the increase in conductivity and varied between ±2.2 to ±4.0%.
2	63a	684, 685	Mason, E.A. and von Ubisch, H.	302.2	0.000 0.120 0.250 0.423 0.510 0.578 0.760 1.000	1.537 1.093 0.8081 0.5359 0.4312 0.3718 0.2265 0.0971	He: Matheson Co., N.J. Kr: spectroscopically pure; thin hot wire method with constant resistance; precision ±2%.
3	63b	684, 685	Mason, E.A. and von Ubisch, H.	793.2	0.000 0.120 0.250 0.423 0.510 0.578 0.760 1.000	3.082 2.307 1.633 1.118 0.9420 0.8164 0.5066 0.2236	Same as above.
4	63a	687	Gambhir, R.S. and Saxena, S.C.	308.2	0.000 0.079 0.247 0.541 0.898 1.000	1.503 1.210 0.8039 0.4049 0.1486 0.0959	He: spectroscopically pure, Kr: 99-100% pure, balance xenon; thick hot wire method; accuracy ±2.0%, precision ±1.0%.
5	63a	687	Gambhir, R.S. and Saxena, S.C.	323.2	0.000 0.079 0.247 0.541 0.898 1.000	1.549 1.244 0.8290 0.4187 0.1557 0.1017	Same as above.
6	63a	687	Gambhir, R.S. and Saxena, S.C.	343.2	0.000 0.079 0.247 0.541 0.898 1.000	1.612 1.298 0.8583 0.4396 0.1654 0.1072	Same as above.
7	63a	687	Gambhir, R.S. and Saxena, S.C.	363.2	0.000 0.079 0.247 0.541 0.898 1.000	1.671 1.348 0.8960 0.4564 0.1738 0.1114	Same as above.

TABLE 63b. THERMAL CONDUCTIVITY (mW cm^{-1} K^{-1}) OF HELIUM-KRYPTON SYSTEM AS A FUNCTION OF COMPOSITION AT THE TEMPERATURE OF MEASUREMENT AS DERIVED BY GRAPHICAL SMOOTHING

Mole Fraction of Kr	291.2 K (Ref. 693)	302.2 K (Ref. 684)	308.2 K (Ref. 687)	323.2 K (Ref. 687)	343.2 K (Ref. 687)	363.2 K (Ref. 687)	793.2 K (Ref. 684)
0.00	1.49	1.54	1.50	1.55	1.61	1.67	3.08
0.05	1.34	1.34	1.32	1.35	1.41	1.46	2.76
0.10	1.19	1.15	1.15	1.18	1.23	1.28	2.44
0.15	1.05	1.02	1.02	1.04	1.08	1.13	2.13
0.20	0.931	0.905	0.899	0.923	0.960	1.00	1.86
0.25	0.820	0.808	0.798	0.824	0.855	0.890	1.63
0.30	0.725	0.720	0.710	0.736	0.764	0.791	1.46
0.35	0.639	0.639	0.632	0.657	0.682	0.704	1.32
0.40	0.560	0.565	0.563	0.588	0.608	0.627	1.18
0.45	0.495	0.502	0.501	0.520	0.543	0.561	1.06
0.50	0.440	0.447	0.447	0.462	0.484	0.501	0.953
0.55	0.392	0.394	0.398	0.410	0.430	0.448	0.852
0.60	0.345	0.349	0.354	0.364	0.382	0.399	0.760
0.65	0.306	0.306	0.314	0.322	0.339	0.354	0.674
0.70	0.267	0.267	0.276	0.283	0.299	0.313	0.594
0.75	0.232	0.233	0.240	0.248	0.262	0.274	0.520
0.80	0.200	0.201	0.207	0.214	0.226	0.238	0.454
0.85	0.169	0.173	0.177	0.183	0.195	0.204	0.392
0.90	0.142	0.146	0.148	0.154	0.164	0.172	0.335
0.95	0.117	0.121	0.122	0.127	0.136	0.142	0.278
1.00	0.0921	0.0971	0.0959	0.102	0.1072	0.111	0.224

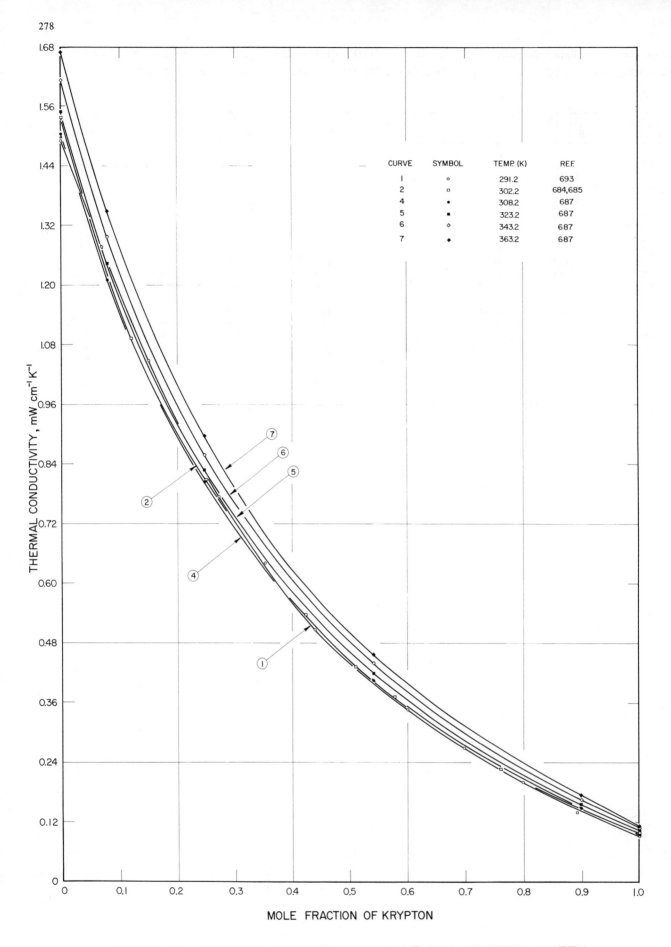

FIGURE 63a. THERMAL CONDUCTIVITY OF HELIUM-KRYPTON SYSTEM

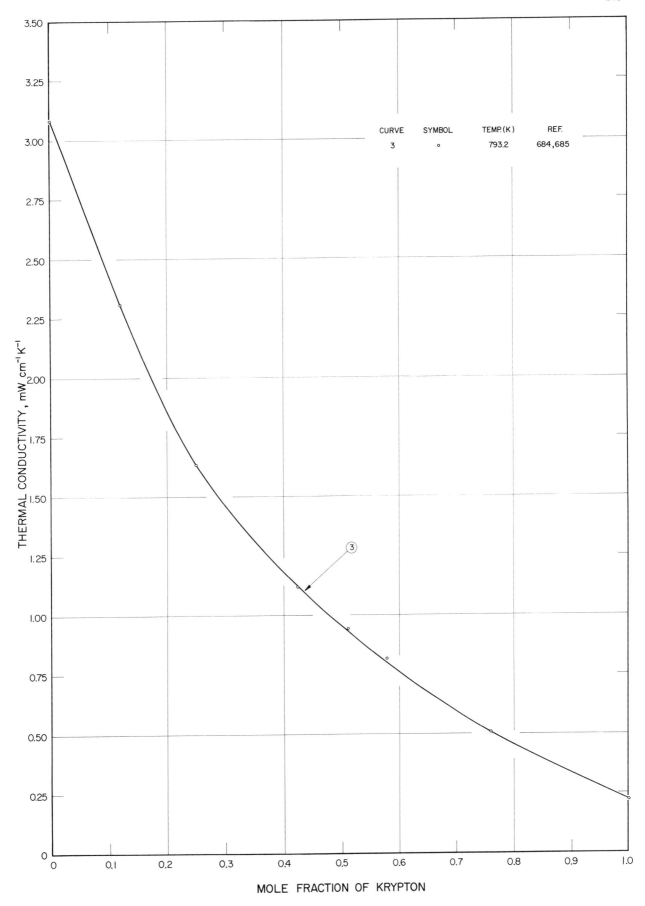

FIGURE 63b. THERMAL CONDUCTIVITY OF HELIUM-KRYPTON SYSTEM

TABLE 64a. EXPERIMENTAL THERMAL CONDUCTIVITY DATA FOR HELIUM-XENON SYSTEM

Curve No.	Fig. No.	Ref. No.	Author(s)	Temp. (K)	Mole Fraction of Xe	Thermal Cond. (mW cm^{-1} K^{-1})	Remarks
1	64a	289	Saxena, S.C.	311.2	0.0000 0.1139 0.2603 0.3460 0.4963 0.6333 0.8991 1.0000	1.571 1.081 0.7092 0.5527 0.3760 0.2579 0.1038 0.0565	He: spectroscopically pure, Xe: traces of krypton, thick hot wire method; precision ± 2%.
2	64b	684, 685	Mason, E.A. and von Ubisch, H.	793.2	0.000 0.213 0.283 0.582 0.798 1.000	3.082 1.658 1.369 0.6238 0.3224 0.1398	He: Matheson Co., N.J., Xe: spectroscopically pure; thin hot wire method with constant resistance; precision ± 2%.
3	64a	684, 685	Mason, E.A. and von Ubisch, H.	302.2	0.000 0.213 0.283 0.582 0.798 1.000	1.537 0.7880 0.6406 0.3002 0.1495 0.0597	Same as above.
4	64a	694	Thornton, E.	291.3	0.000 0.063 0.139 0.201 0.304 0.401 0.494 0.594 0.687 0.792 0.898 1.000	1.491 1.134 0.9127 0.7536 0.6029 0.4647 0.3789 0.2847 0.2169 0.1491 0.0971 0.0553	He: spectroscopically pure, Xe: 99-100% pure, balance krypton; katharometer method; maximum error in mixture composition ± 0.3%, accuracy of these relative measurements decreased with the increase in conductivity and varied between ± 2.2 to ± 4.0%.
5	64a	697	Gandhi, J.M. and Saxena, S.C.	303.2	0.0000 0.1203 0.3011 0.4810 0.7837 1.0000	1.495 1.029 0.6510 0.4028 0.1712 0.0574	He: spectroscopically pure, Xe: 99-100% pure, balance krypton; thick hot wire method; accuracy ± 2.0%, precision ± 1.0%.
6	64a	697	Gandhi, J.M. and Saxena, S.C.	323.2	0.0000 0.1203 0.3011 0.4810 0.7837 1.0000	1.549 1.074 0.6808 0.4191 0.1817 0.0611	Same as above.
7	64a	697	Gandhi, J.M. and Saxena, S.C.	343.2	0.0000 0.1203 0.3011 0.4810 0.7837 1.0000	1.612 1.120 0.7105 0.4384 0.1909 0.0649	Same as above.
8	64a	697	Gandhi, J.M. and Saxena, S.C.	363.2	0.0000 0.1203 0.3011 0.4810 0.7837 1.0000	1.671 1.166 0.7406 0.4576 0.1997 0.0682	Same as above.

TABLE 64b. THERMAL CONDUCTIVITY (mW cm^{-1} K^{-1}) OF HELIUM-XENON SYSTEM AS A FUNCTION OF COMPOSITION AT THE TEMPERATURE OF MEASUREMENT AS DERIVED BY GRAPHICAL SMOOTHING

Mole Fraction of Ar	291.3 K (Ref. 694)	302.2 K (Ref. 684)	303.2 K (Ref. 697)	311.2 K (Ref. 289)
0.00	1.49	1.54	1.50	1.57
0.05	1.20	1.33	1.28	1.33
0.10	1.00	1.13	1.09	1.12
0.15	0.872	0.962	0.957	0.971
0.20	0.760	0.821	0.847	0.844
0.25	0.673	0.704	0.745	0.730
0.30	0.599	0.612	0.654	0.629
0.35	0.532	0.538	0.573	0.545
0.40	0.471	0.475	0.501	0.480
0.45	0.417	0.421	0.438	0.424
0.50	0.367	0.372	0.384	0.373
0.55	0.320	0.327	0.347	0.326
0.60	0.278	0.286	0.295	0.283
0.65	0.240	0.247	0.258	0.245
0.70	0.205	0.210	0.224	0.211
0.75	0.174	0.177	0.191	0.181
0.80	0.146	0.147	0.162	0.153
0.85	0.121	0.122	0.134	0.127
0.90	0.097	0.099	0.108	0.103
0.95	0.076	0.078	0.082	0.079
1.00	0.0553	0.0597	0.0574	0.0565

Mole Fraction of Xe	323.2 K (Ref. 697)	343.2 K (Ref. 697)	363.2 K (Ref. 697)	793.2 K (Ref. 684)
0.00	1.55	1.61	1.67	3.08
0.05	1.34	1.38	1.44	2.67
0.10	1.14	1.18	1.24	2.31
0.15	0.998	1.04	1.08	1.98
0.20	0.882	0.916	0.955	1.71
0.25	0.778	0.810	0.845	1.50
0.30	0.683	0.714	0.744	1.31
0.35	0.597	0.621	0.652	1.15
0.40	0.521	0.543	0.570	1.01
0.45	0.454	0.476	0.497	0.890
0.50	0.398	0.419	0.436	0.780
0.55	0.349	0.370	0.383	0.681
0.60	0.307	0.325	0.337	0.594
0.65	0.269	0.285	0.296	0.516
0.70	0.234	0.248	0.258	0.446
0.75	0.202	0.213	0.223	0.381
0.80	0.171	0.181	0.189	0.323
0.85	0.143	0.151	0.158	0.270
0.90	0.115	0.122	0.128	0.222
0.95	0.088	0.093	0.098	0.178
1.00	0.0611	0.0649	0.0682	0.140

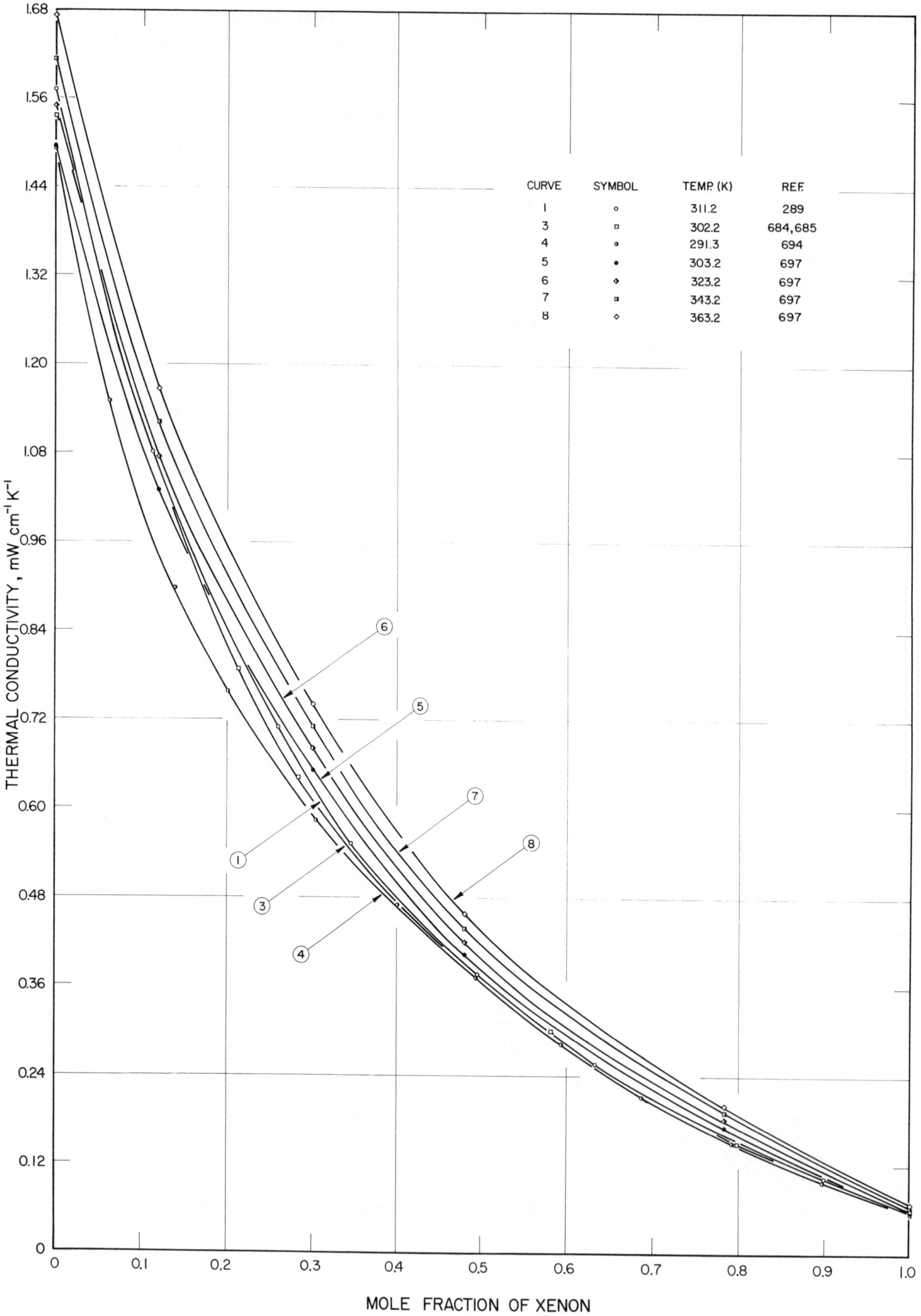
CURVE SYMBOL TEMP. (K) REF.
1 ○ 311.2 289
3 □ 302.2 684,685
4 ○ 291.3 694
5 ● 303.2 697
6 ◈ 323.2 697
7 ◘ 343.2 697
8 ◇ 363.2 697
THERMAL CONDUCTIVITY, mW cm⁻¹ K⁻¹
MOLE FRACTION OF XENON
1.68
1.56
1.44
1.32
1.20
1.08
0.96
0.84
0.72
0.60
0.48
0.36
0.24
0.12
0
0.1
0.2
0.3
0.4
0.5
0.6
0.7
0.8
0.9
1.0
6
5
7
8
1
3
4

FIGURE 64a. THERMAL CONDUCTIVITY OF HELIUM-XENON SYSTEM

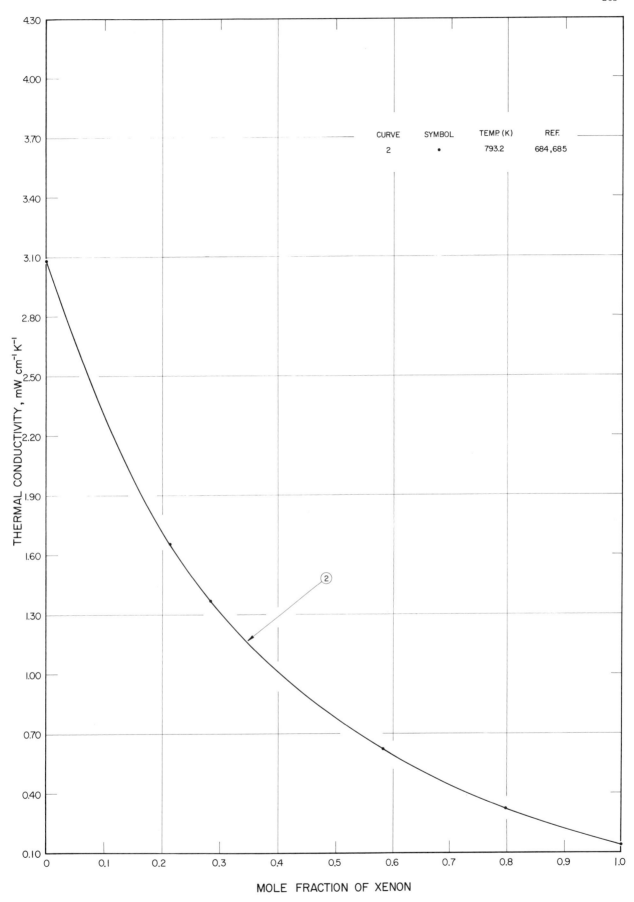

FIGURE 64b. THERMAL CONDUCTIVITY OF HELIUM-XENON SYSTEM

TABLE 65a. EXPERIMENTAL THERMAL CONDUCTIVITY DATA FOR KRYPTON-NEON SYSTEM

Curve No.	Fig. No.	Ref. No.	Author(s)	Temp. (K)	Mole Fraction of Kr	Thermal Cond. (mW cm^{-1} K^{-1})	Remarks
1	65a	326	Srivastava, B.N. and Saxena, S.C.	311.2	0.0000 0.0712 0.2076 0.3092 0.4277 0.5070 0.6707 0.8556 1.0000	0.4953 0.4333 0.3479 0.2914 0.2474 0.2160 0.1666 0.1285 0.0980	Ne: spectroscopically pure, Kr: traces of xenon; thick hot wire method; precision ± 2%.
2	65a	693	Thornton, E.	291.2	0.000 0.065 0.111 0.229 0.339 0.438 0.533 0.647 0.797 0.889 1.000	0.4857 0.4312 0.4019 0.3266 0.2721 0.2378 0.2018 0.1675 0.1277 0.1076 0.0921	Ne: spectroscopically pure, Kr: 99-100% pure, balance xenon; katharometer method; maximum error in mixture composition ± 0.3%, accuracy of these relative measurements decreased with the increase in conductivity and varied between ± 2.2 to ± 4.0%.
3	65a	684, 685	Mason, E.A. and von Ubisch, H.	302.2	0.000 0.308 0.460 0.750 1.000	0.5179 0.2998 0.2378 0.1495 0.0971	Ne and Kr: spectroscopically pure; thin hot wire method with constant resistance; precision ± 2%.
4	65b	684, 685	Mason, E.A. and von Ubisch, H.	793.2	0.000 0.308 0.460 0.750 1.000	0.9881 0.6155 0.4982 0.3266 0.2236	Same as above.
5	65a	692	Mathur, S., Tondon, P.K., and Saxena, S.C.	313.2	0.00 0.20 0.40 0.60 0.80 1.00	0.5141 0.3718 0.2680 0.1972 0.1440 0.1034	Ne: spectroscopically pure, Kr: 99-100% pure, balance xenon; thick hot wire method; accuracy ± 2.0%, precision ± 1.0%.
6	65a	692	Mathur, S., Tondon, P.K., and Saxena, S.C.	338.2	0.00 0.20 0.40 0.60 0.80 1.00	0.5426 0.3814 0.2814 0.2077 0.1520 0.1101	Same as above.
7	65a	692	Mathur, S., Tondon, P.K., and Saxena, S.C.	363.2	0.00 0.20 0.40 0.60 0.80 1.00	0.5589 0.3973 0.2960 0.2202 0.1620 0.1172	Same as above.

TABLE 65b. THERMAL CONDUCTIVITY (mW cm^{-1}K^{-1}) OF KRYPTON-NEON SYSTEM AS A FUNCTION OF COMPOSITION AT THE TEMPERATURE OF MEASUREMENT AS DERIVED BY GRAPHICAL SMOOTHING

Mole Fraction of Kr	291.2 K (Ref. 693)	302.2 K (Ref. 684)	311.2 K (Ref. 326)	313.2 K (Ref. 692)	338.2 K (Ref. 692)	363.2 K (Ref. 692)	793.2 K (Ref. 684)
0.00	0.486	0.518	0.495	0.514	0.543	0.559	0.988
0.05	0.446	0.476	0.453	0.477	0.497	0.513	0.933
0.10	0.408	0.436	0.416	0.440	0.455	0.470	0.868
0.15	0.375	0.398	0.382	0.405	0.416	0.432	0.807
0.20	0.344	0.364	0.350	0.372	0.381	0.397	0.749
0.25	0.316	0.333	0.322	0.343	0.353	0.368	0.693
0.30	0.292	0.306	0.296	0.316	0.327	0.342	0.642
0.35	0.269	0.281	0.274	0.291	0.303	0.318	0.594
0.40	0.248	0.259	0.253	0.268	0.281	0.296	0.549
0.45	0.230	0.239	0.234	0.248	0.261	0.276	0.507
0.50	0.213	0.220	0.217	0.230	0.242	0.256	0.470
0.55	0.196	0.203	0.201	0.213	0.224	0.238	0.437
0.60	0.181	0.188	0.186	0.197	0.208	0.220	0.406
0.65	0.166	0.174	0.172	0.183	0.192	0.205	0.378
0.70	0.152	0.162	0.160	0.169	0.178	0.189	0.351
0.75	0.139	0.150	0.148	0.156	0.165	0.175	0.327
0.80	0.127	0.138	0.137	0.144	0.152	0.162	0.304
0.85	0.115	0.128	0.127	0.134	0.141	0.150	0.282
0.90	0.105	0.117	0.117	0.123	0.130	0.139	0.261
0.95	0.0981	0.107	0.107	0.114	0.120	0.128	0.242
1.00	0.0921	0.0971	0.0980	0.103	0.110	0.117	0.224

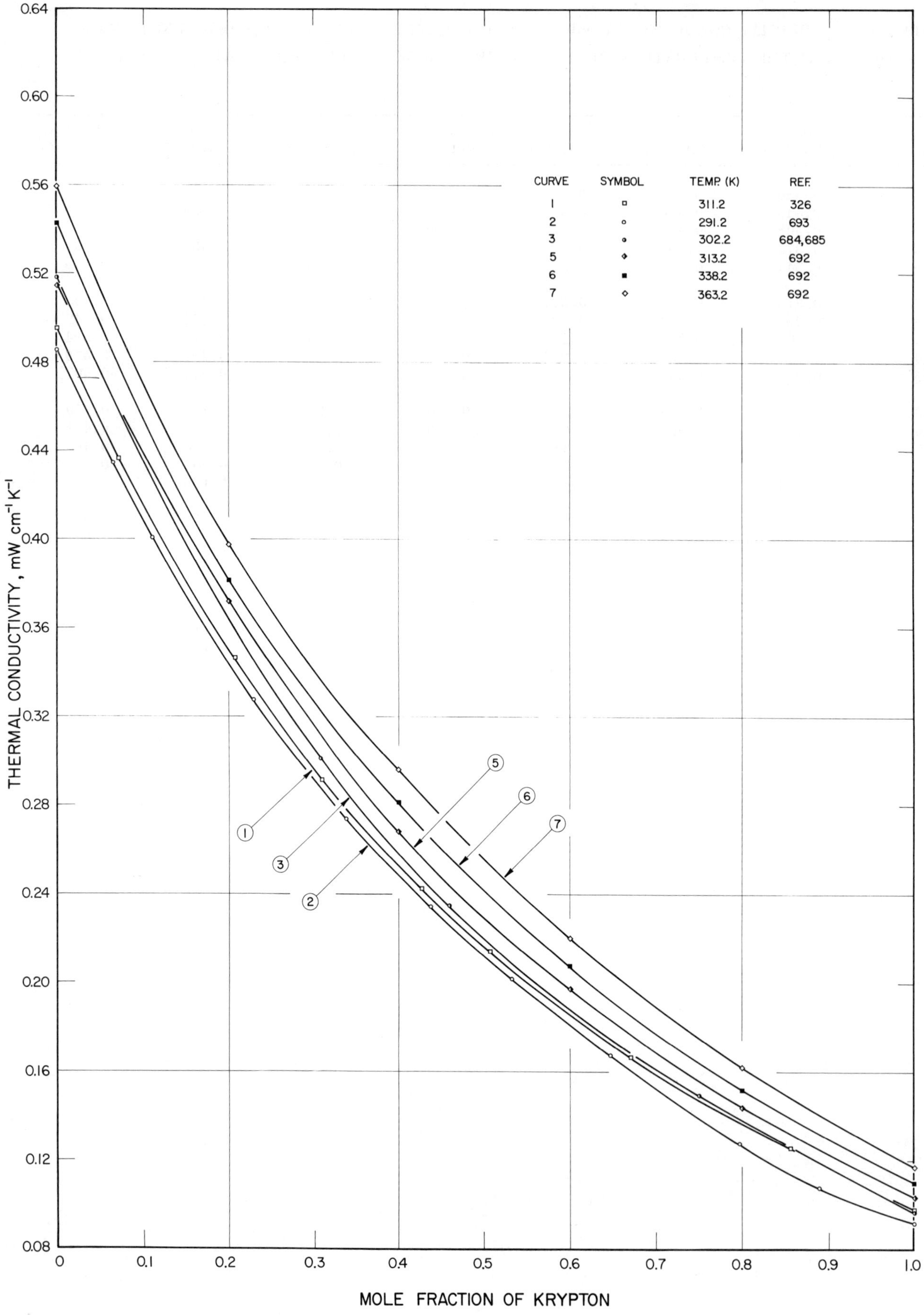

FIGURE 65a. THERMAL CONDUCTIVITY OF KRYPTON–NEON SYSTEM

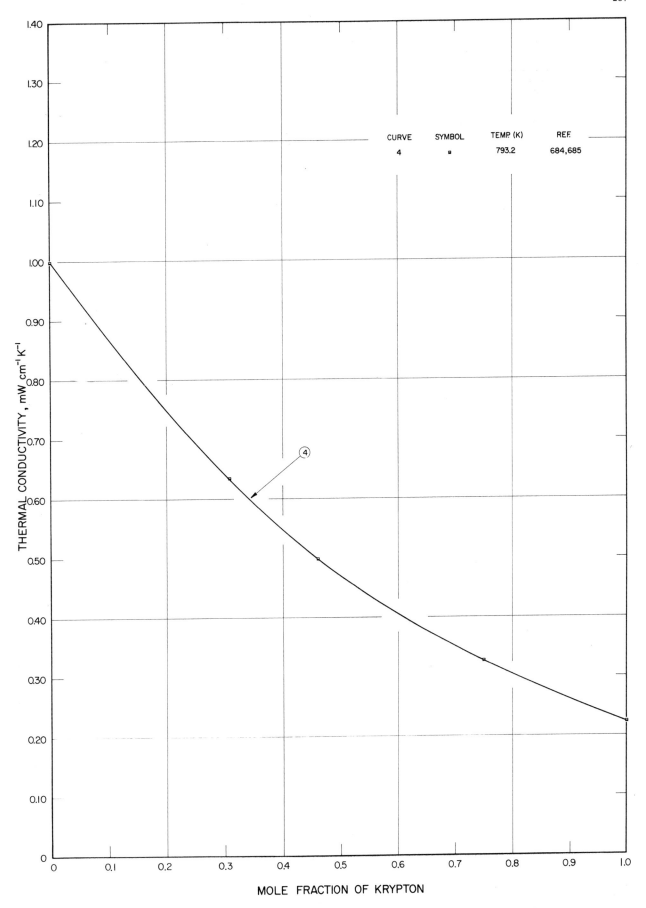

FIGURE 65b. THERMAL CONDUCTIVITY OF KRYPTON—NEON SYSTEM

TABLE 66a. EXPERIMENTAL THERMAL CONDUCTIVITY DATA FOR KRYPTON-XENON SYSTEM

Curve No.	Fig. No.	Ref. No.	Author(s)	Temp. (K)	Mole Fraction of Xe	Thermal Cond. (mW cm^{-1} K^{-1})	Remarks
1	66	684, 685	Mason, E.A. and von Ubisch, H.	302.2	0.000 0.215 0.490 0.724 0.842 0.890 1.000	0.0971 0.0862 0.0779 0.0662 0.0624 0.0607 0.0599	Kr and Xe: spectroscopically pure; thin hot wire method with constant resistance; precision ±2%.
2	66	684, 685	Mason, E.A. and von Ubisch, H.	793.2	0.000 0.215 0.490 0.724 0.842 0.890 1.000	0.2236 0.2005 0.1758 0.1499 0.1428 0.1415 0.1398	Same as above.
3	66	692	Mathur, S., Tondon, P.K., and Saxena, S.C.	313.2	0.00 0.20 0.40 0.60 0.80 1.00	0.1034 0.0929 0.0842 0.0762 0.0687 0.0620	Kr: 99-100% pure, balance xenon, Xe: 99-100% pure, balance krypton, thick hot wire method; accuracy ±2.0%, precision ±1.0%.
4	66	692	Mathur, S., Tondon, P.K., and Saxena, S.C.	338.2	0.00 0.20 0.40 0.60 0.80 1.00	0.1101 0.0988 0.0896 0.0816 0.0737 0.0670	Same as above.
5	66	692	Mathur, S., Tondon, P.K., and Saxena, S.C.	363.2	0.00 0.20 0.40 0.60 0.80 1.00	0.1172 0.1072 0.0976 0.0888 0.0804 0.0729	Same as above.
6	66	694	Thornton, E.	291.2	0.000 0.115 0.201 0.296 0.397 0.491 0.595 0.693 0.786 0.896 1.000	0.0921 0.0867 0.0825 0.0783 0.0745 0.0703 0.0670 0.0636 0.0607 0.0578 0.0553	Kr: 99-100% pure, balance xenon, Xe: 99-100% pure, balance krypton; katharometer method; maximum error in mixture composition ±0.3%, maximum estimated error of these relative measurements ±2.2%.

TABLE 66b. THERMAL CONDUCTIVITY (mW cm^{-1} K^{-1}) OF KRYPTON-XENON SYSTEM AS A FUNCTION OF COMPOSITION AT THE TMPERATURE OF MEASUREMENT AS DERIVED BY GRAPHICAL SMOOTHING

Mole Fraction of Xe	291.2 K (Ref. 694)	302.2 K (Ref. 684)	313.2 K (Ref. 692)	338.2 K (Ref. 692)	363.2 K (Ref. 692)	793.2 K (Ref. 684)
0.00	0.0921	0.0971	0.103	0.110	0.117	0.224
0.05	0.0897	0.0946	0.101	0.107	0.114	0.218
0.10	0.0873	0.0922	0.0981	0.104	0.112	0.213
0.15	0.0849	0.0897	0.0956	0.102	0.110	0.208
0.20	0.0826	0.0873	0.0932	0.0989	0.107	0.202
0.25	0.0804	0.0849	0.0908	0.0965	0.105	0.197
0.30	0.0782	0.0826	0.0886	0.0941	0.102	0.191
0.35	0.0762	0.0804	0.0864	0.0918	0.100	0.186
0.40	0.0742	0.0783	0.0842	0.0896	0.0976	0.181
0.45	0.0722	0.0763	0.0822	0.0875	0.0954	0.176
0.50	0.0703	0.0743	0.0802	0.0854	0.0931	0.170
0.55	0.0685	0.0724	0.0782	0.0832	0.0909	0.165
0.60	0.0667	0.0706	0.0762	0.0812	0.0887	0.161
0.65	0.0650	0.0688	0.0743	0.0797	0.0870	0.156
0.70	0.0633	0.0670	0.0724	0.0774	0.0845	0.152
0.75	0.0617	0.0653	0.0705	0.0755	0.0824	0.148
0.80	0.0603	0.0637	0.0687	0.0737	0.0804	0.145
0.85	0.0589	0.0621	0.0670	0.0720	0.0785	0.143
0.90	0.0577	0.0605	0.0654	0.0703	0.0766	0.141
0.95	0.0564	0.0600	0.0637	0.0686	0.0748	0.140
1.00	0.0553	0.0599	0.0620	0.0670	0.0729	0.139

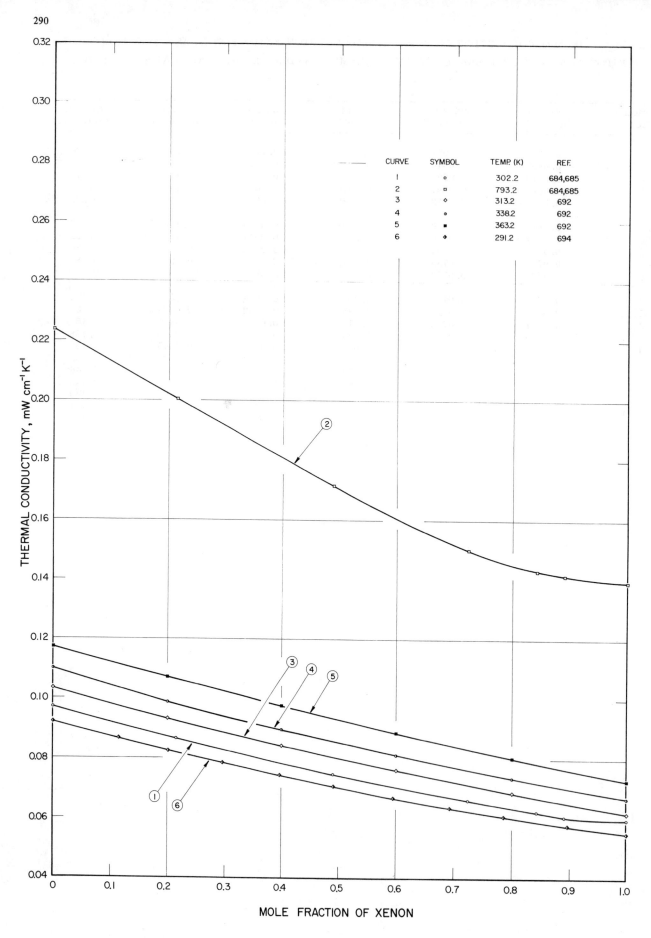

FIGURE 66. THERMAL CONDUCTIVITY OF KRYPTON-XENON SYSTEM

TABLE 67a.   EXPERIMENTAL THERMAL CONDUCTIVITY DATA FOR NEON-XENON SYSTEM

Curve No.	Fig. No.	Ref. No.	Author(s)	Temp. (K)	Mole Fraction of Xe	Thermal Cond. (mW cm^{-1} K^{-1})	Remarks
1	67a	694	Thornton, E.	291.2	0.000 0.103 0.199 0.285 0.393 0.504 0.594 0.673 0.794 0.903 1.000	0.4857 0.3873 0.3148 0.2613 0.2093 0.1666 0.1369 0.1160 0.0892 0.0695 0.0553	Ne: spectroscopically pure, Xe: 99-100% pure, balance krypton; katharometer method; maximum error in mixture composition ±0.3%, accuracy of these relative measurements decreased with the increase in conductivity and varied between ±2.2 to ±4.0%.
2	67a	684, 685	Mason, E.A. and von Ubisch, H.	302.2	0.000 0.330 0.430 0.704 1.000	0.5179 0.2495 0.2035 0.1168 0.0599	Ne and Xe: spectroscopically pure; thin hot wire method with constant resistance; precision ±2%.
3	67b	684, 685	Mason, E.A. and von Ubisch, H.	793.2	0.000 0.330 0.430 0.704 1.000	0.9881 0.5192 0.4354 0.2625 0.1398	Same as above.
4	67a	697	Gandhi, J.M. and Saxena, S.C.	303.2	0.0000 0.1537 0.5586 0.7715 1.0000	0.4940 0.3563 0.1641 0.1034 0.0574	Ne: spectroscopically pure; Xe: 99-100% pure, balance krypton; thick hot wire method; accuracy ±2.0%, precision ±1.0%.
5	67a	697	Gandhi, J.M. and Saxena, S.C.	323.2	0.0000 0.1537 0.5586 0.7715 1.0000	0.5150 0.3726 0.1704 0.1097 0.0611	Same as above.
6	67a	697	Gandhi, J.M. and Saxena, S.C.	343.2	0.0000 0.1537 0.5586 0.7715 1.0000	0.5317 0.3885 0.1763 0.1164 0.0649	Same as above.
7	67a	697	Gandhi, J.M. and Saxena, S.C.	363.2	0.0000 0.1537 0.5586 0.7715 1.0000	0.5527 0.4044 0.1825 0.1227 0.0682	Same as above.

TABLE 67b. THERMAL CONDUCTIVITY (mW cm^{-1}K^{-1}) OF NEON-XENON SYSTEM AS A FUNCTION OF COMPOSITION AT THE TEMPERATURE OF MEASUREMENT AS DERIVED BY GRAPHICAL SMOOTHING

Mole Fraction of Xe	291.2 K (Ref. 694)	302.2 K (Ref. 684)	303.2 K (Ref. 697)	323.2 K (Ref. 697)	343.2 K (Ref. 697)	363.2 K (Ref. 697)	793.2 K (Ref. 684)
0.00	0.486	0.518	0.494	0.515	0.532	0.553	0.988
0.05	0.437	0.464	0.446	0.466	0.483	0.502	0.903
0.10	0.391	0.414	0.400	0.417	0.435	0.453	0.822
0.15	0.350	0.370	0.359	0.374	0.391	0.407	0.743
0.20	0.314	0.331	0.326	0.338	0.353	0.368	0.673
0.25	0.282	0.296	0.298	0.307	0.320	0.333	0.608
0.30	0.254	0.266	0.271	0.278	0.290	0.302	0.552
0.35	0.229	0.239	0.247	0.253	0.263	0.274	0.502
0.40	0.207	0.216	0.224	0.231	0.239	0.248	0.459
0.45	0.187	0.195	0.204	0.210	0.217	0.225	0.420
0.50	0.169	0.177	0.184	0.191	0.197	0.204	0.384
0.55	0.152	0.160	0.167	0.173	0.179	0.185	0.351
0.60	0.136	0.145	0.150	0.157	0.163	0.169	0.320
0.65	0.122	0.131	0.136	0.142	0.148	0.155	0.292
0.70	0.110	0.118	0.122	0.128	0.135	0.141	0.265
0.75	0.0981	0.107	0.108	0.115	0.122	0.128	0.240
0.80	0.0879	0.0964	0.0966	0.103	0.109	0.115	0.217
0.85	0.0786	0.0869	0.0863	0.091	0.0972	0.103	0.196
0.90	0.0702	0.0775	0.0765	0.0805	0.0860	0.0912	0.176
0.95	0.0622	0.0682	0.0671	0.0706	0.0750	0.0798	0.157
1.00	0.0553	0.0599	0.0574	0.0611	0.0649	0.0682	0.140

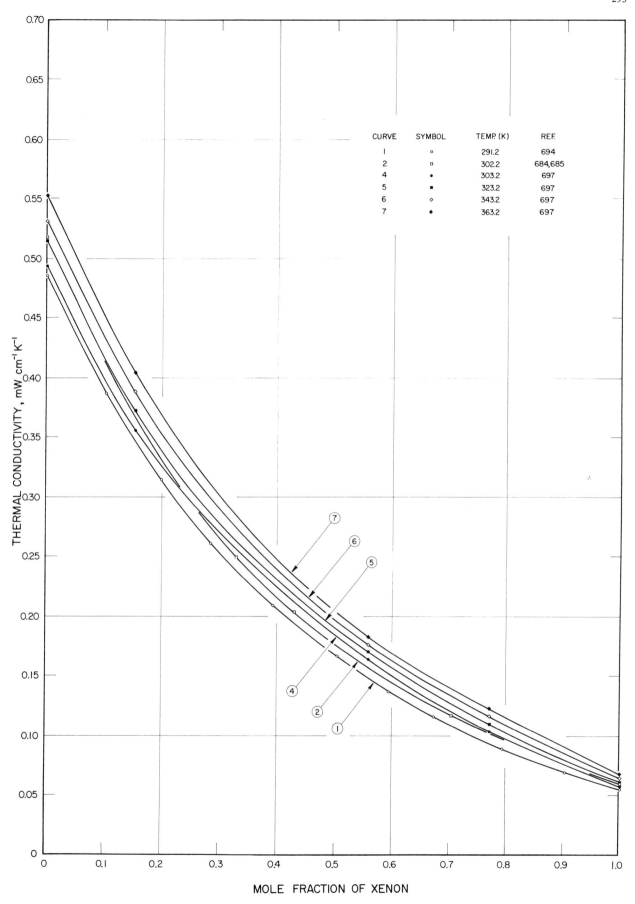

FIGURE 67a. THERMAL CONDUCTIVITY OF NEON-XENON SYSTEM

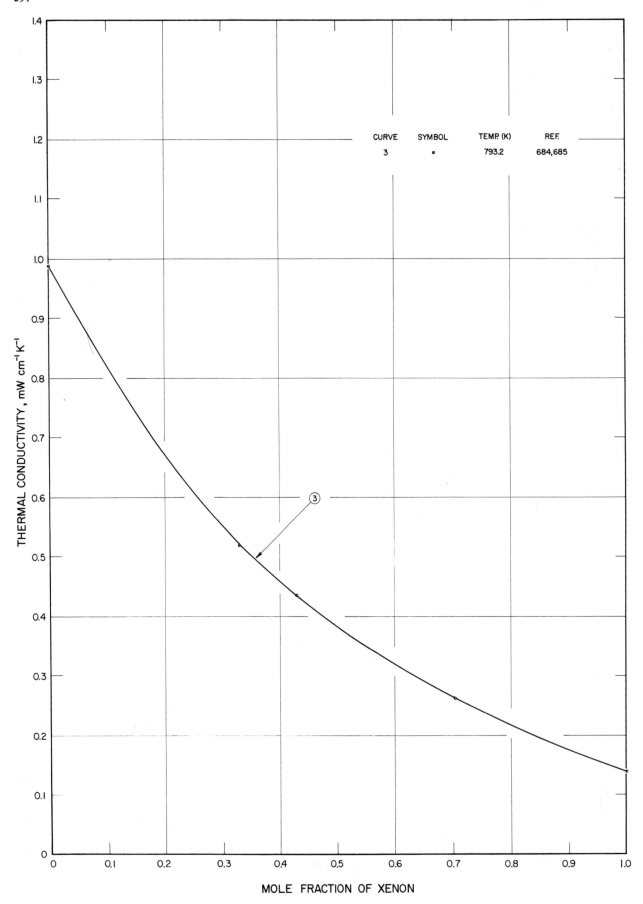

FIGURE 67b. THERMAL CONDUCTIVITY OF NEON-XENON SYSTEM

TABLE 68a. EXPERIMENTAL THERMAL CONDUCTIVITY DATA FOR ARGON-BENZENE SYSTEM

Curve No.	Fig. No.	Ref. No.	Author(s)	Temp. (K)	Mole Fraction of C_6H_6	Thermal Cond. (mW cm^{-1} K^{-1})	Remarks
1	68	32	Bennett, L.A. and Vines, R.G.	351.2	0.00 0.25 0.50 0.75 1.00	0.2031 0.1750 0.1591 0.1503 0.1449	Ar: 98% pure, C_6H_6: shaken with concentrated sulphuric acid, washed with water, dried over calcium chloride, and distilled; compensated hot wire method; estimated accuracy of these relative measurements ± 1%.
2	68	32	Bennett, L.A. and Vines, R.G.	373.8	0.00 0.25 0.50 0.75 1.00	0.2135 0.1909 0.1771 0.1708 0.1662	
3	68	32	Bennett, L.A. and Vines, R.G.	398.2	0.00 0.25 0.50 0.75 1.00	0.2269 0.2081 0.1972 0.1930 0.1905	Same as above.

TABLE 68b. THERMAL CONDUCTIVITY (mW cm^{-1} K^{-1}) OF ARGON-BENZENE SYSTEM AS A FUNCTION OF COMPOSITION AT THE TEMPERATURE OF MEASUREMENT AS DERIVED BY GRAPHICAL SMOOTHING

Mole Fraction of C_6H_6	351.2 K (Ref. 32)	373.8 K (Ref. 32)	398.2 K (Ref. 32)
0.00	0.203	0.214	0.227
0.05	0.197	0.209	0.223
0.10	0.191	0.204	0.219
0.15	0.186	0.199	0.215
0.20	0.180	0.195	0.211
0.25	0.175	0.191	0.208
0.30	0.171	0.187	0.205
0.35	0.168	0.184	0.203
0.40	0.164	0.182	0.201
0.45	0.162	0.179	0.199
0.50	0.159	0.177	0.197
0.55	0.157	0.175	0.196
0.60	0.155	0.174	0.195
0.65	0.153	0.173	0.194
0.70	0.152	0.172	0.194
0.75	0.150	0.171	0.193
0.80	0.149	0.170	0.193
0.85	0.148	0.169	0.192
0.90	0.147	0.168	0.192
0.95	0.146	0.167	0.192
1.00	0.145	0.166	0.191

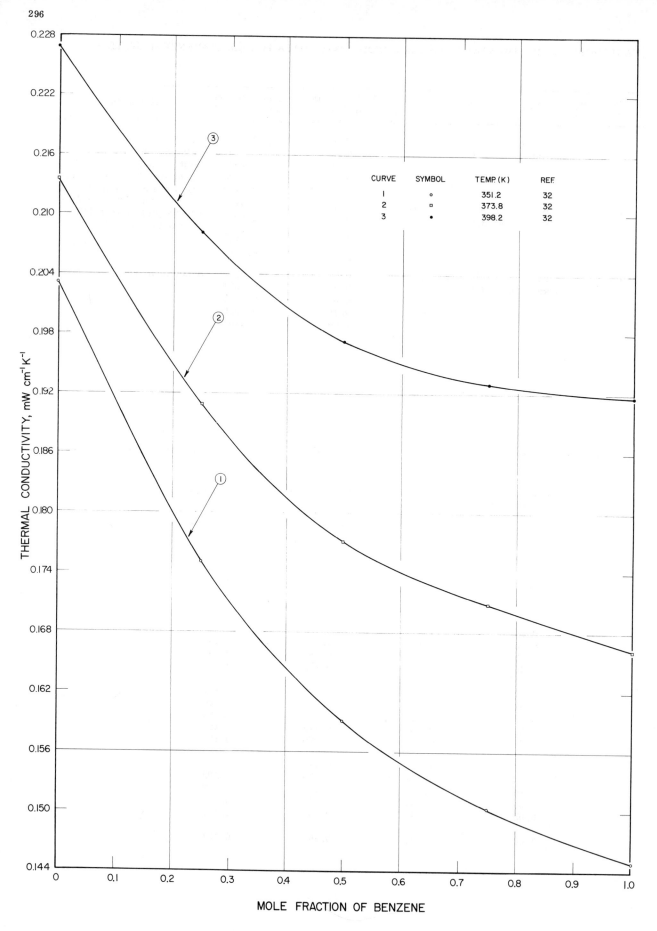

FIGURE 68. THERMAL CONDUCTIVITY OF ARGON–BENZENE SYSTEM

TABLE 69a. EXPERIMENTAL THERMAL CONDUCTIVITY FOR ARGON-CARBON DIOXIDE SYSTEM

Curve No.	Fig. No.	Ref. No.	Author(s)	Temp. (K)	Mole Fraction of CO_2	Thermal Cond. (mW cm^{-1} K^{-1})	Remarks
1	69	65	Cheung, H., Bromley, L.A., and Wilke, C.R.	594.2 593.2 593.2	0.0000 0.5065 1.0000	0.3056 0.3630 0.4042	Ar: Linde Air Prod. Co., standard grade, specified purity 99.97%, chief impurity N_2, CO_2: Pure Carbonic, Inc., specified purity 99.5%, chief impurity air; coaxial cylinder method; avg error 1.2%, max error 2% and max uncertainty in mixture composition 0.25%.
1	69	65, 316	Cheung, H., Bromley, L.A., and Wilke, C.R.	593.2	0.000	0.3053	Value obtained on the basis of data of Saxena and Saxena reported below.
--	--	316, 700	Saxena, V.K. and Saxena, S.C.	593.2	0.000	0.3001	Ar: spectroscopically pure; conductivity column method; estimated precision ±2%.

TABLE 69b. THERMAL CONDUCTIVITY (mW cm^{-1}K^{-1}) OF ARGON-CARBON DIOXIDE SYSTEM AS A FUNCTION OF COMPOSITION AT THE TEMPERATURE OF MEASUREMENT AS DERIVED BY GRAPHICAL SMOOTHING

Mole Fraction CO_2	593.2 K (Ref. 65)
0.00	0.305
0.05	0.312
0.10	0.317
0.15	0.323
0.20	0.329
0.25	0.335
0.30	0.341
0.35	0.347
0.40	0.352
0.45	0.357
0.50	0.362
0.55	0.367
0.60	0.372
0.65	0.377
0.70	0.381
0.75	0.385
0.80	0.389
0.85	0.393
0.90	0.397
0.95	0.401
1.00	0.404

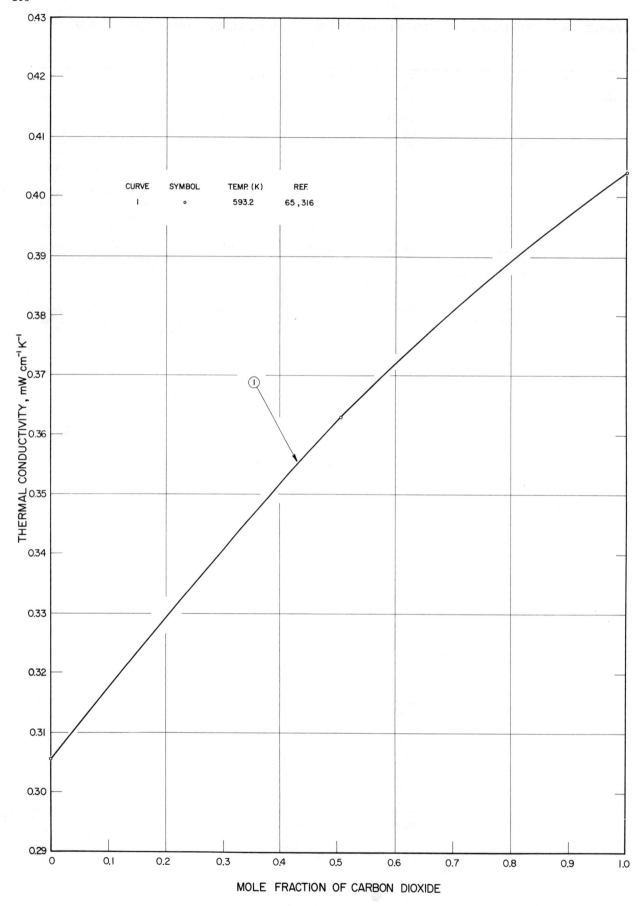

FIGURE 69. THERMAL CONDUCTIVITY OF ARGON-CARBON DIOXIDE SYSTEM

TABLE 70a. EXPERIMENTAL THERMAL CONDUCTIVITY DATA FOR ARGON-DEUTERIUM SYSTEM

Curve No.	Fig. No.	Ref. No.	Author(s)	Temp. (K)	Mole Fraction of Ar	Thermal Cond. (mW cm^{-1} K^{-1})	Remarks
1	70	698	Gambhir, R.S. and Saxena, S.C.	308.2	0.000 0.101 0.242 0.437 0.796 1.000	1.357 1.080 0.8625 0.5903 0.3123 0.1834	Ar: spectroscopically pure, D_2: 98.6% pure; impurities 0.8% H_2 and water vapor 0.6%; thick hot wire method; precision ±1%, accuracy ±1 to ±2%.
2	70	698	Gambhir, R.S. and Saxena, S.C.	323.2	0.000 0.101 0.242 0.437 0.796 1.000	1.398 1.114 0.8918 0.6238 0.3257 0.1905	Same as above.
3	70	698	Gambhir, R.S. and Saxena, S.C.	343.2	0.000 0.101 0.242 0.437 0.796 1.000	1.457 1.168 0.9295 0.6615 0.3437 0.2001	Same as above.
4	70	698	Gambhir, R.S. and Saxena, S.C.	363.2	0.000 0.101 0.242 0.437 0.796 1.000	1.511 1.235 0.9672 0.6908 0.3617 0.2098	Same as above.

TABLE 70b. THERMAL CONDUCTIVITY (mW cm^{-1}K^{-1}) OF ARGON-DEUTERIUM SYSTEM AS A FUNCTION OF COMPOSITION AT THE TEMPERATURE OF MEASUREMENT AS DERIVED BY GRAPHICAL SMOOTHING

Mole Fraction of Ar	308.2 K (Ref. 698)	323.2 K (Ref. 698)	343.2 K (Ref. 698)	363.2 K (Ref. 698)
0.00	1.36	1.40	1.46	1.51
0.05	1.20	1.30	1.30	1.37
0.10	1.09	1.17	1.17	1.24
0.15	0.998	1.08	1.08	1.13
0.20	0.921	0.993	0.994	1.04
0.25	0.846	0.916	0.916	0.955
0.30	0.772	0.842	0.842	0.878
0.35	0.700	0.771	0.771	0.805
0.40	0.635	0.706	0.706	0.738
0.45	0.577	0.646	0.646	0.677
0.50	0.530	0.594	0.594	0.622
0.55	0.488	0.546	0.546	0.572
0.60	0.448	0.500	0.500	0.525
0.65	0.411	0.458	0.458	0.480
0.70	0.376	0.417	0.417	0.438
0.75	0.342	0.378	0.378	0.398
0.80	0.308	0.341	0.341	0.357
0.85	0.277	0.305	0.305	0.320
0.90	0.245	0.270	0.270	0.282
0.95	0.214	0.234	0.235	0.245
1.00	0.183	0.191	0.200	0.210

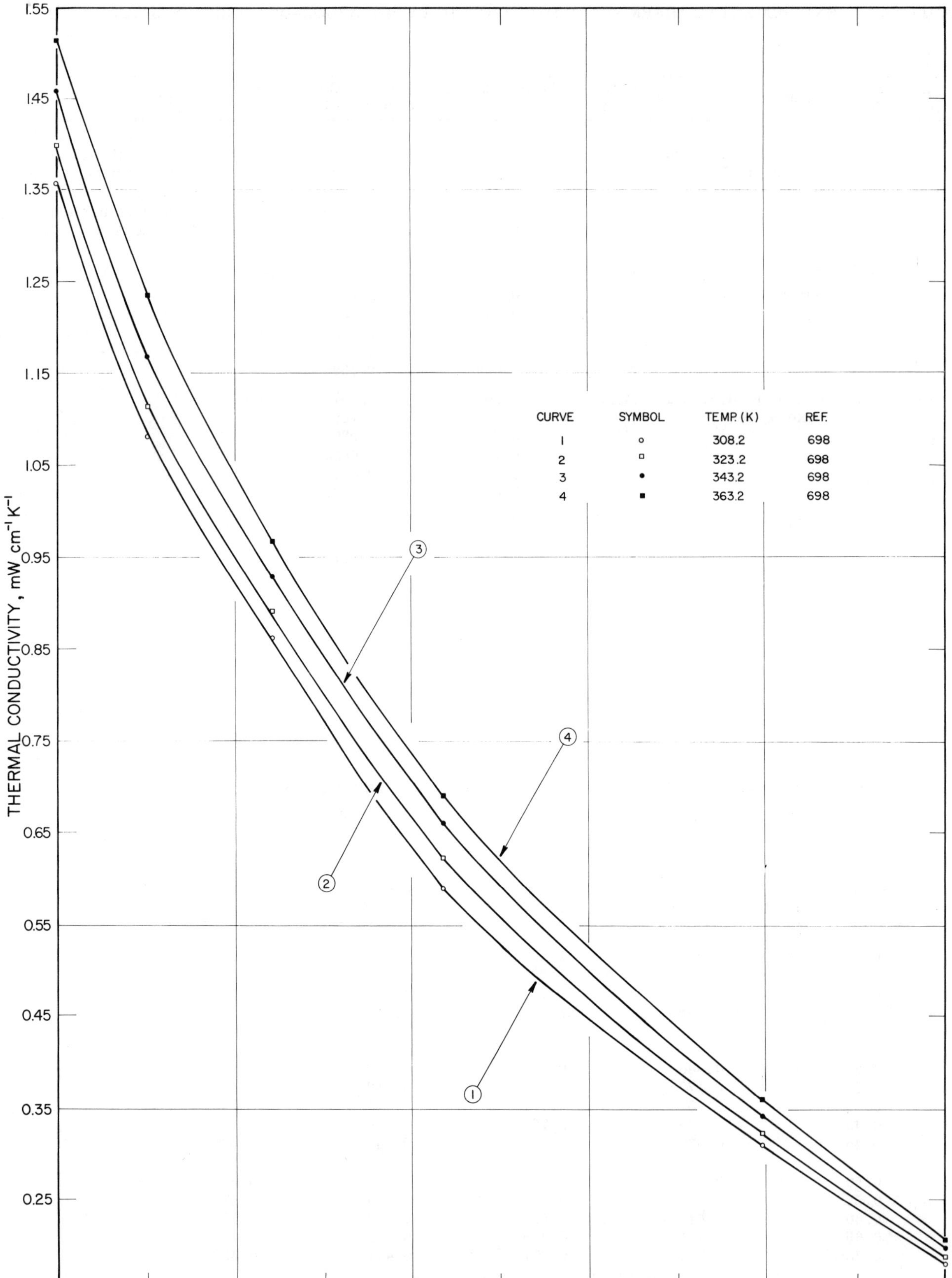

CURVE	SYMBOL	TEMP. (K)	REF.
1	○	308.2	698
2	□	323.2	698
3	●	343.2	698
4	■	363.2	698

FIGURE 70. THERMAL CONDUCTIVITY OF ARGON-DEUTERIUM SYSTEM

TABLE 71a. EXPERIMENTAL THERMAL CONDUCTIVITY DATA FOR ARGON-HYDROGEN SYSTEM

Curve No.	Fig. No.	Ref. No.	Author(s)	Temp. (K)	Mole Fraction of Ar	Thermal Cond. (mW cm^{-1} K^{-1})	Remarks
1	71	599	Srivastava, B.N. and Srivastava, R.C.	311.2	0.000 0.077 0.276 0.494 0.671 1.000	1.853 1.514 1.065 0.6322 0.5016 0.1821	Ar and H$_2$: spectroscopically pure; thick hot wire method; precision ±2%.
2	71	699, 696	Saxena, S.C. and Gupta, G.P.	313.2	0.000 0.202 0.388 0.614 1.000	1.820 1.245 0.891 0.557 0.185	Ar: spectroscopically pure, H$_2$: 99.9% pure; thick hot wire method; precision ±1%, accuracy ±1 to ±2%.
3	71	699, 696	Saxena, S.C. and Gupta, G.P.	338.2	0.000 0.202 0.388 0.614 1.000	1.940 1.325 0.955 0.589 0.193	Same as above.
4	71	699, 696	Saxena, S.C. and Gupta, G.P.	366.2	0.000 0.202 0.388 0.614 1.000	2.040 1.395 1.026 0.640 0.210	Same as above.
--	--	690	Lindsay, A.L. and Bromley, L.A.	297.0 296.9 296.7 295.9 295.7 295.5 295.3 295.3 295.0 297.0 297.0 296.5 296.6 296.4	0.000 0.000 0.000 0.209 0.209 0.396 0.396 0.618 0.618 0.803 0.803 1.000 1.000 1.000	1.809 1.816 1.835 1.124 1.212 0.8786 0.8510 0.5154 0.4895 0.3338 0.3321 0.1603 0.1743 0.1572	Ar: Linde Air Prod. Co., 99.8% pure, H$_2$: 99.9% pure; unsteady state method; precision better than ±10%.
5	71	156	Ibbs, T.L. and Hirst, A.A.	273.2	0.000 0.198 0.400 0.600 0.820 0.910 1.000	1.692 1.130 0.7829 0.5317 0.3056 0.2303 0.1633	Purity of gases as supplied in cylinders; katharometer method; these are relative measurements and for calibration thermal conductivity values for argon-helium system were used.
6	71	690	Lindsay, A.L. and Bromley, L.A.	296.5 297.0 295.2 295.4 295.8 296.9	0.000 0.209 0.396 0.618 0.803 1.000	1.820 1.168 0.8648 0.5016 0.3321 0.1640	We have generated these data from the original reproduced above by averaging the multiple values referring to the same composition of the mixture.

TABLE 71b. THERMAL CONDUCTIVITY (mW cm^{-1}K^{-1}) OF ARGON-HYDROGEN SYSTEM AS A FUNCTION OF COMPOSITION AT THE TEMPERATURE OF MEASUREMENT AS DERIVED BY GRAPHICAL SMOOTHING

Mole Fraction of Ar	273.2 K (Ref. 156)	296 K (Ref. 690)	311.2 K (Ref. 599)	313.2 K (Ref. 699)	338.2 K (Ref. 699)	366.2 K (Ref. 699)
0.00	1.69	1.82	1.85	1.82	1.94	2.04
0.05	1.51	1.62	1.63	1.67	1.77	1.85
0.10	1.36	1.46	1.46	1.51	1.62	1.70
0.15	1.23	1.32	1.31	1.40	1.47	1.55
0.20	1.13	1.20	1.19	1.25	1.33	1.41
0.25	1.03	1.09	1.07	1.14	1.21	1.29
0.30	0.939	0.987	0.974	1.03	1.10	1.18
0.35	0.856	0.897	0.882	0.937	1.00	1.08
0.40	0.784	0.815	0.798	0.850	0.909	0.985
0.45	0.716	0.741	0.720	0.772	0.824	0.898
0.50	0.652	0.671	0.650	0.701	0.745	0.828
0.55	0.590	0.606	0.587	0.635	0.673	0.742
0.60	0.532	0.545	0.530	0.573	0.606	0.670
0.65	0.477	0.488	0.478	0.516	0.544	0.604
0.70	0.424	0.434	0.430	0.463	0.486	0.542
0.75	0.373	0.383	0.384	0.414	0.433	0.484
0.80	0.324	0.336	0.341	0.365	0.383	0.426
0.85	0.279	0.291	0.300	0.318	0.333	0.371
0.90	0.238	0.248	0.260	0.272	0.285	0.316
0.95	0.198	0.205	0.220	0.228	0.238	0.262
1.00	0.163	0.164	0.182	0.185	0.193	0.210

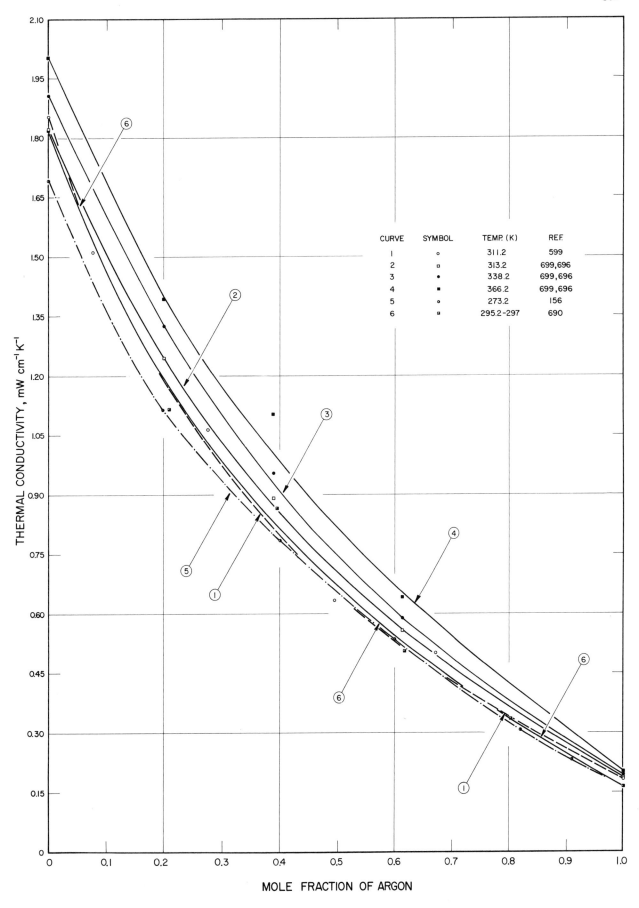

FIGURE 71. THERMAL CONDUCTIVITY OF ARGON-HYDROGEN SYSTEM

TABLE 72a. EXPERIMENTAL THERMAL CONDUCTIVITY DATA FOR ARGON-METHANE SYSTEM

Curve No.	Fig. No.	Ref. No.	Author(s)	Temp. (K)	Mole Fraction of Ar	Thermal Cond. (mW cm^{-1} K^{-1})	Remarks
1	72	65, 688	Cheung, H., Bromley, L.A., and Wilke, C.R.	811.2 810.2	0.0000 0.5233	1.057 0.7130	Ar: Linde Air Prod. Co., standard grade, specified purity 99.97%, chief impurity N$_2$; coaxial cylinder method; avg and max errors 1.2 and 2% respectively.
1	72	316, 700	Saxena, V.K. and Saxena, S.C.	811.2	1.000	0.3756	Ar: spectroscopically pure; conductivity column method; estimated precision ±2%.

TABLE 72b. THERMAL CONDUCTIVITY (mW cm^{-1}K^{-1}) OF ARGON-METHANE SYSTEM AS A FUNCTION OF COMPOSITION AT THE TEMPERATURE OF MEASUREMENT AS DERIVED BY GRAPHICAL SMOOTHING

Mole Fraction of Ar	811 K (Ref. 65)
0.00	1.06
0.05	1.02
0.10	0.992
0.15	0.959
0.20	0.926
0.25	0.894
0.30	0.862
0.35	0.828
0.40	0.795
0.45	0.762
0.50	0.729
0.55	0.695
0.60	0.661
0.65	0.627
0.70	0.593
0.75	0.559
0.80	0.524
0.85	0.488
0.90	0.454
0.95	0.414
1.00	0.376

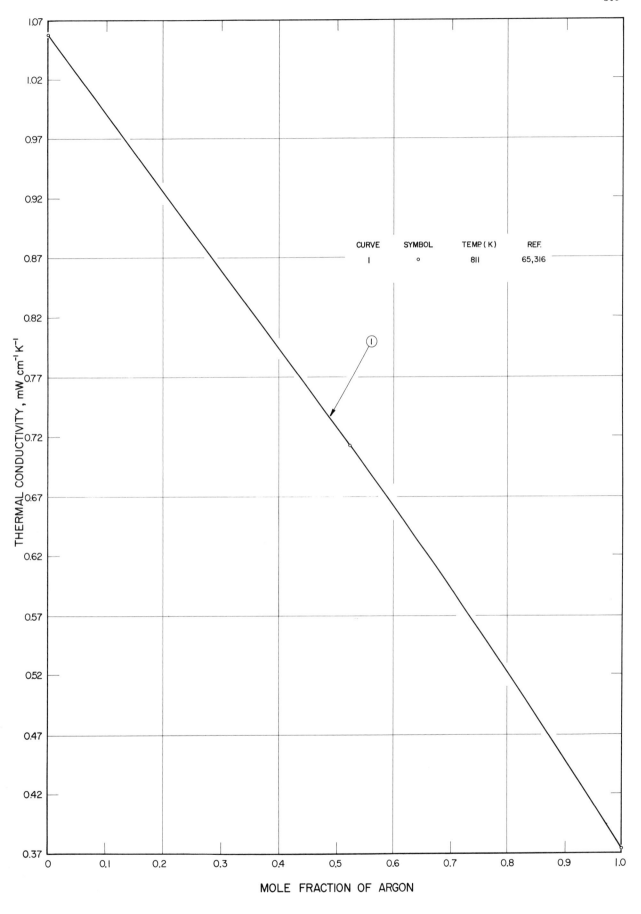

FIGURE 72. THERMAL CONDUCTIVITY OF ARGON-METHANE SYSTEM

TABLE 73a. EXPERIMENTAL THERMAL CONDUCTIVITY DATA FOR ARGON-NITROGEN SYSTEM

Curve No.	Fig. No.	Ref. No.	Author(s)	Temp. (K)	Mole Fraction of Ar	Thermal Cond. (mW cm^{-1} K^{-1})	Remarks
1	73a	326	Srivastava, B.N. and Srivastava, R.C.	311.2	0.000 0.161 0.238 0.540 0.659 0.848 1.000	0.2538 0.2342 0.2283 0.2104 0.2044 0.1911 0.1821	Ar and N_2: spectroscopically pure; thick hot wire method; precision ±2%.
2	73a	65, 688	Cheung, H., Bromley, L.A., and Wilke, C.R.	590.2 593.2 594.2	0.0000 0.5034 1.0000	0.4467 0.3679 0.3056	Ar: Linde Air Prod. Co., standard grade, specified purity 99.97%, chief impurity N_2, N_2: Linde Air Prod. Co., "water pumped" grade, specified purity 99.9%, chief impurities Ar and Ne; coaxial cylinder method; avg and max errors as 1.2 and 2.0% respectively.
3	73a	699, 696	Saxena, S.C. and Gupta, G.P.	313.2	0.000 0.188 0.492 0.767 1.000	0.268 0.255 0.227 0.207 0.185	Ar: spectroscopically pure, N_2: 99.9% pure; thick hot wire method; precision ±1%, accuracy ±1 to ±2%.
4	73a	699, 696	Saxena, S.C. and Gupta, G.P.	338.2	0.000 0.188 0.492 0.767 1.000	0.290 0.269 0.241 0.220 0.193	Same as above.
5	73a	699, 696	Saxena, S.C. and Gupta, G.P.	366.2	0.000 0.188 0.492 0.767 1.000	0.312 0.292 0.254 0.234 0.210	Same as above.
6	73a	380	Weber, S.	273.2	0.0000 0.2196 0.3892 0.6413 0.7962 1.0000	0.2370 0.2192 0.2050 0.1857 0.1747 0.1611	Ar and N_2: pure gases; thin hot wire potential lead method, vertical arrangement.
7	73b	689	Mukhopadhyay, P. and Barua, A.K.	90.2	0.000 0.073 0.129 0.394 0.650 0.806 0.888 0.926 0.932 0.950 0.984 1.000	0.0967 0.0938 0.0917 0.0783 0.0678 0.0628 0.0611 0.0595 0.0586 0.0603 0.0595 0.0586	Ar: spectroscopically pure, N_2: 99.95% pure; thick hot wire method; accuracy of these relative measurements ±1%.

TABLE 73a. EXPERIMENTAL THERMAL CONDUCTIVITY DATA FOR ARGON-NITROGEN SYSTEM (continued)

Curve No.	Fig. No.	Ref. No.	Author(s)	Temp. (K)	Mole Fraction of Ar	Thermal Cond. (mW cm^{-1} K^{-1})	Remarks
8	73b	689	Mukhopadhyay, P. and Barua, A.K.	258.3	0.000	0.2290	Same as above.
					0.163	0.2186	
					0.394	0.2043	
					0.600	0.1800	
					0.812	0.1687	
					0.920	0.1595	
					0.934	0.1578	
					0.951	0.1570	
					0.973	0.1595	
					1.000	0.1549	
9	73b	689	Mukhopadhyay, P. and Barua, A.K.	293.3	0.000	0.2567	Same as above.
					0.139	0.2211	
					0.383	0.2043	
					0.605	0.1905	
					0.790	0.1813	
					0.901	0.1754	
					0.946	0.1720	
					0.973	0.1712	
					0.990	0.1696	
					1.000	0.1738	
10	73b	689	Mukhopadhyay, P. and Barua, A.K.	393.2	0.000	0.3098	Same as above.
					0.190	0.2675	
					0.406	0.2567	
					0.599	0.2412	
					0.811	0.2286	
					0.957	0.2236	
					0.973	0.2215	
					0.982	0.2186	
					0.988	0.2206	
					1.000	0.2219	
11	73b	689	Mukhopadhyay, P. and Barua, A.K.	473.2	0.000	0.3529	Same as above.
					0.193	0.3182	
					0.403	0.3035	
					0.600	0.2692	
					0.804	0.2671	
					0.957	0.2395	
					0.973	0.2374	
					0.995	0.2361	
					1.000	0.2512	

TABLE 73b. THERMAL CONDUCTIVITY (mW cm^{-1}K^{-1}) OF ARGON-NITROGEN SYSTEM AS A FUNCTION OF COMPOSITION AT THE TEMPERATURE OF MEASUREMENT AS DERIVED BY GRAPHICAL SMOOTHING

Mole Fraction of Ar	273.2 K (Ref. 380)	311.2 K (Ref. 326)	313.2 K (Ref. 699)	338.2 K (Ref. 699)	366.2 K (Ref. 699)	593 K (Ref. 65)
0.00	0.237	0.254	0.270	0.290	0.318	0.447
0.05	0.233	0.248	0.265	0.284	0.307	0.438
0.10	0.229	0.242	0.261	0.279	0.300	0.430
0.15	0.225	0.236	0.257	0.273	0.294	0.421
0.20	0.220	0.231	0.253	0.268	0.288	0.413
0.25	0.216	0.227	0.248	0.263	0.282	0.405
0.30	0.212	0.223	0.244	0.258	0.276	0.397
0.35	0.208	0.220	0.240	0.253	0.270	0.389
0.40	0.204	0.216	0.235	0.248	0.264	0.382
0.45	0.200	0.213	0.231	0.244	0.259	0.375
0.50	0.196	0.210	0.227	0.239	0.254	0.368
0.55	0.193	0.207	0.223	0.235	0.249	0.362
0.60	0.189	0.204	0.219	0.230	0.244	0.356
0.65	0.185	0.201	0.215	0.226	0.240	0.350
0.70	0.181	0.198	0.211	0.221	0.235	0.343
0.75	0.178	0.196	0.206	0.217	0.231	0.338
0.80	0.175	0.193	0.202	0.213	0.228	0.332
0.85	0.171	0.190	0.198	0.209	0.224	0.326
0.90	0.168	0.188	0.194	0.204	0.220	0.321
0.95	0.164	0.185	0.190	0.200	0.216	0.315
1.00	0.161	0.182	0.186	0.196	0.213	0.306

Mole Fraction of Ar	90.2 K (Ref. 689)	258.3 K (Ref. 689)	293.3 K (Ref. 689)	393.2 K (Ref. 689)	473.2 K (Ref. 689)
0.00	0.0967	0.229	0.257	0.310	0.353
0.05	0.0948	0.226	0.236	0.291	0.345
0.10	0.0929	0.223	0.227	0.281	0.337
0.15	0.0908	0.220	0.221	0.274	0.329
0.20	0.0885	0.216	0.216	0.269	0.321
0.25	0.0860	0.212	0.212	0.264	0.314
0.30	0.0833	0.209	0.209	0.260	0.307
0.35	0.0806	0.204	0.206	0.256	0.301
0.40	0.0781	0.200	0.202	0.253	0.295
0.45	0.0758	0.195	0.199	0.249	0.290
0.50	0.0737	0.191	0.196	0.246	0.285
0.55	0.0716	0.186	0.194	0.243	0.281
0.60	0.0696	0.182	0.191	0.240	0.277
0.65	0.0678	0.178	0.188	0.237	0.273
0.70	0.0662	0.174	0.186	0.235	0.269
0.75	0.0646	0.170	0.183	0.232	0.266
0.80	0.0631	0.167	0.181	0.230	0.263
0.85	0.0617	0.164	0.179	0.228	0.260
0.90	0.0605	0.161	0.177	0.226	0.257
0.95	0.0595	0.168	0.175	0.224	0.254
1.00	0.0586	0.155	0.174	0.222	0.251

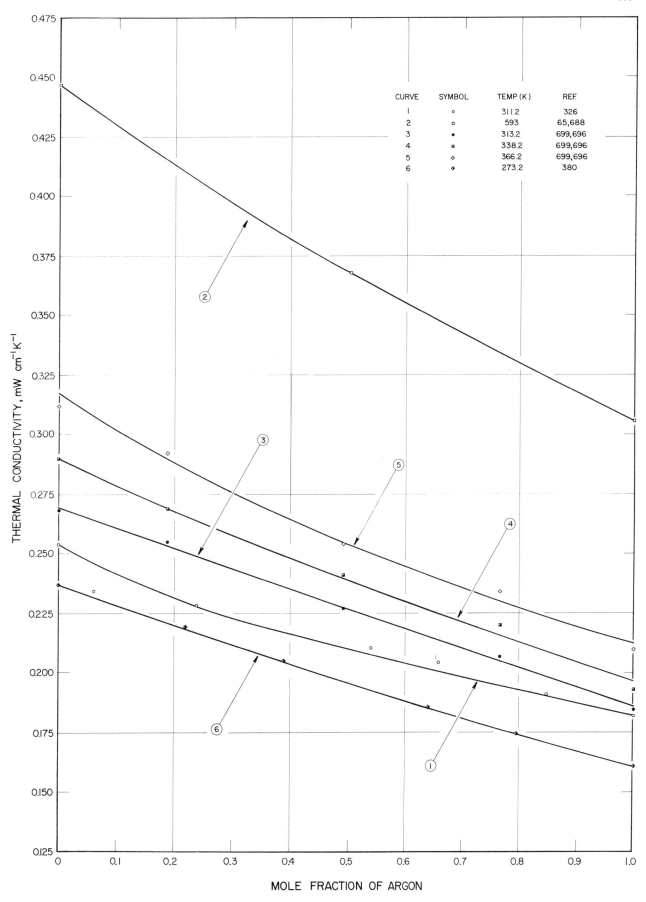

FIGURE 73a. THERMAL CONDUCTIVITY OF ARGON–NITROGEN SYSTEM

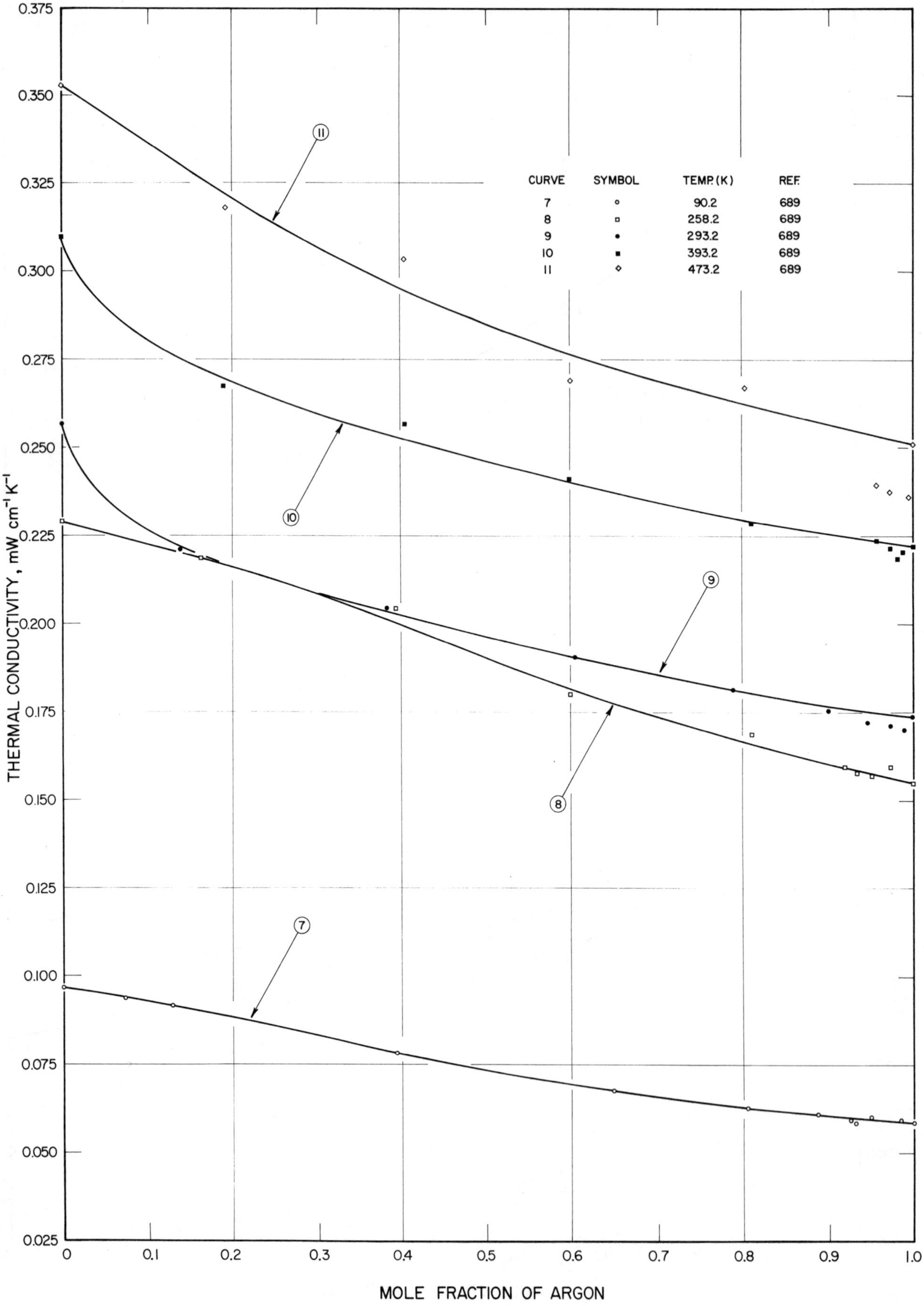

FIGURE 73b. THERMAL CONDUCTIVITY OF ARGON–NITROGEN SYSTEM

TABLE 74a. EXPERIMENTAL THERMAL CONDUCTIVITY DATA FOR ARGON-OXYGEN SYSTEM

Curve No.	Fig. No.	Ref. No.	Author(s)	Temp. (K)	Mole Fraction of Ar	Thermal Cond. (mW cm^{-1} K^{-1})	Remarks
1	74a	599	Srivastava, B.N. and Srivastava, R.C.	311.2	0.000 0.178 0.350 0.464 0.822 0.890 1.000	0.2705 0.2397 0.2219 0.2140 0.1922 0.1884 0.1821	Ar and O$_2$: spectroscopically pure; thick hot wire method; precision ±2%.
2	74a	699, 696	Saxena, S.C. and Gupta, G.P.	313.2	0.000 0.249 0.484 0.753 1.000	0.281 0.252 0.228 0.208 0.185	Ar: spectroscopically pure, O$_2$: 99.9% pure; thick hot wire method; precision ±1%, accuracy ±1 to ±2%.
3	74a	699, 696	Saxena, S.C. and Gupta, G.P.	338.2	0.000 0.249 0.484 0.753 1.000	0.291 0.272 0.245 0.221 0.193	Same as above.
4	74a	699, 696	Saxena, S.C. and Gupta, G.P.	366.2	0.000 0.249 0.484 0.753 1.000	0.313 0.292 0.262 0.234 0.210	Same as above.
5	74b	689	Mukhopadhyay, P. and Barua, A.K.	90.2	0.000 0.104 0.388 0.702 0.901 0.931 0.946 0.960 0.970 1.000	0.08918 0.08583 0.07620 0.06280 0.05987 0.05862 0.05778 0.05694 0.05820 0.05862	Ar: spectroscopically pure, O$_2$: 99.95% pure; thick hot wire method; accuracy of these relative measurements ±1%.
6	74b	689	Mukhopadhyay, P. and Barua, A.K.	258.3	0.000 0.108 0.391 0.709 0.973 0.986 0.988 0.993 1.000	0.2290 0.2265 0.2068 0.1742 0.1570 0.1566 0.1549 0.1545 0.1549	Same as above.
7	74b	689	Mukhopadhyay, P. and Barua, A.K.	293.3	0.000 0.087 0.399 0.717 0.964 0.975 0.987 0.992 0.994 1.000	0.2529 0.2374 0.2165 0.1938 0.1767 0.1763 0.1742 0.1729 0.1721 0.1738	Same as above.

TABLE 74a. EXPERIMENTAL THERMAL CONDUCTIVITY DATA FOR ARGON-OXYGEN SYSTEM (continued)

Curve No.	Fig. No.	Ref. No.	Author(s)	Temp. (K)	Mole Fraction of Ar	Thermal Cond. (mW cm^{-1} K^{-1})	Remarks
8	74b	689	Mukhopadhyay, P. and Barua, A.K.	393.2	0.000 0.084 0.404 0.692 0.957 0.973 0.985 0.993 1.000	0.3324 0.2914 0.2680 0.2453 0.2206 0.2198 0.2190 0.2169 0.2219	Same as above.
9	74b	689	Mukhopadhyay, P. and Barua, A.K.	473.2	0.000 0.192 0.390 0.596 0.801 0.956 0.977 0.987 1.000	0.3890 0.3349 0.2985 0.2713 0.2604 0.2474 0.2474 0.2512 0.2512	Same as above.

TABLE 74b. THERMAL CONDUCTIVITY (mW cm^{-1} K^{-1}) OF ARGON-OXYGEN SYSTEM AS A FUNCTION OF COMPOSITION AT THE TEMPERATURE OF MEASUREMENT AS DERIVED BY GRAPHICAL SMOOTHING

Mole Fraction of Ar	311.2 K (Ref. 599)	313.2 K (Ref. 699)	338.2 K (Ref. 699)	366.2 K (Ref. 699)	90.2 K (Ref. 689)
0.00	0.271	0.281	0.291	0.313	0.0892
0.05	0.262	0.274	0.287	0.308	0.0876
0.10	0.255	0.267	0.283	0.303	0.0860
0.15	0.247	0.261	0.279	0.298	0.0844
0.20	0.241	0.255	0.274	0.293	0.0829
0.25	0.235	0.249	0.270	0.287	0.0813
0.30	0.229	0.243	0.265	0.282	0.0796
0.35	0.224	0.238	0.260	0.277	0.0766
0.40	0.220	0.232	0.256	0.272	0.0756
0.45	0.215	0.227	0.251	0.267	0.0732
0.50	0.211	0.222	0.246	0.261	0.0708
0.55	0.207	0.218	0.241	0.256	0.0686
0.60	0.204	0.213	0.236	0.251	0.0664
0.65	0.200	0.209	0.231	0.246	0.0645
0.70	0.197	0.205	0.226	0.240	0.0630
0.75	0.195	0.201	0.220	0.235	0.0616
0.80	0.192	0.197	0.215	0.230	0.0606
0.85	0.189	0.194	0.210	0.225	0.0597
0.90	0.187	0.191	0.204	0.220	0.0591
0.95	0.184	0.188	0.199	0.215	0.0588
1.00	0.182	0.185	0.193	0.210	0.0586

Mole Fraction of Ar	258.3 K (Ref. 689)	293.3 K (Ref. 689)	393.2 K (Ref. 689)	473.2 K (Ref. 689)
0.00	0.229	0.253	0.332	0.389
0.05	0.228	0.246	0.302	0.371
0.10	0.227	0.239	0.292	0.354
0.15	0.225	0.233	0.286	0.341
0.20	0.222	0.228	0.282	0.331
0.25	0.219	0.224	0.278	0.321
0.30	0.215	0.220	0.274	0.313
0.35	0.211	0.217	0.270	0.305
0.40	0.206	0.213	0.266	0.298
0.45	0.201	0.210	0.261	0.291
0.50	0.196	0.206	0.258	0.285
0.55	0.191	0.203	0.254	0.279
0.60	0.185	0.200	0.250	0.274
0.65	0.180	0.197	0.246	0.270
0.70	0.175	0.193	0.242	0.265
0.75	0.171	0.190	0.238	0.262
0.80	0.167	0.187	0.235	0.258
0.85	0.163	0.184	0.232	0.256
0.90	0.160	0.180	0.228	0.253
0.95	0.157	0.177	0.225	0.250
1.00	0.155	0.174	0.222	0.248

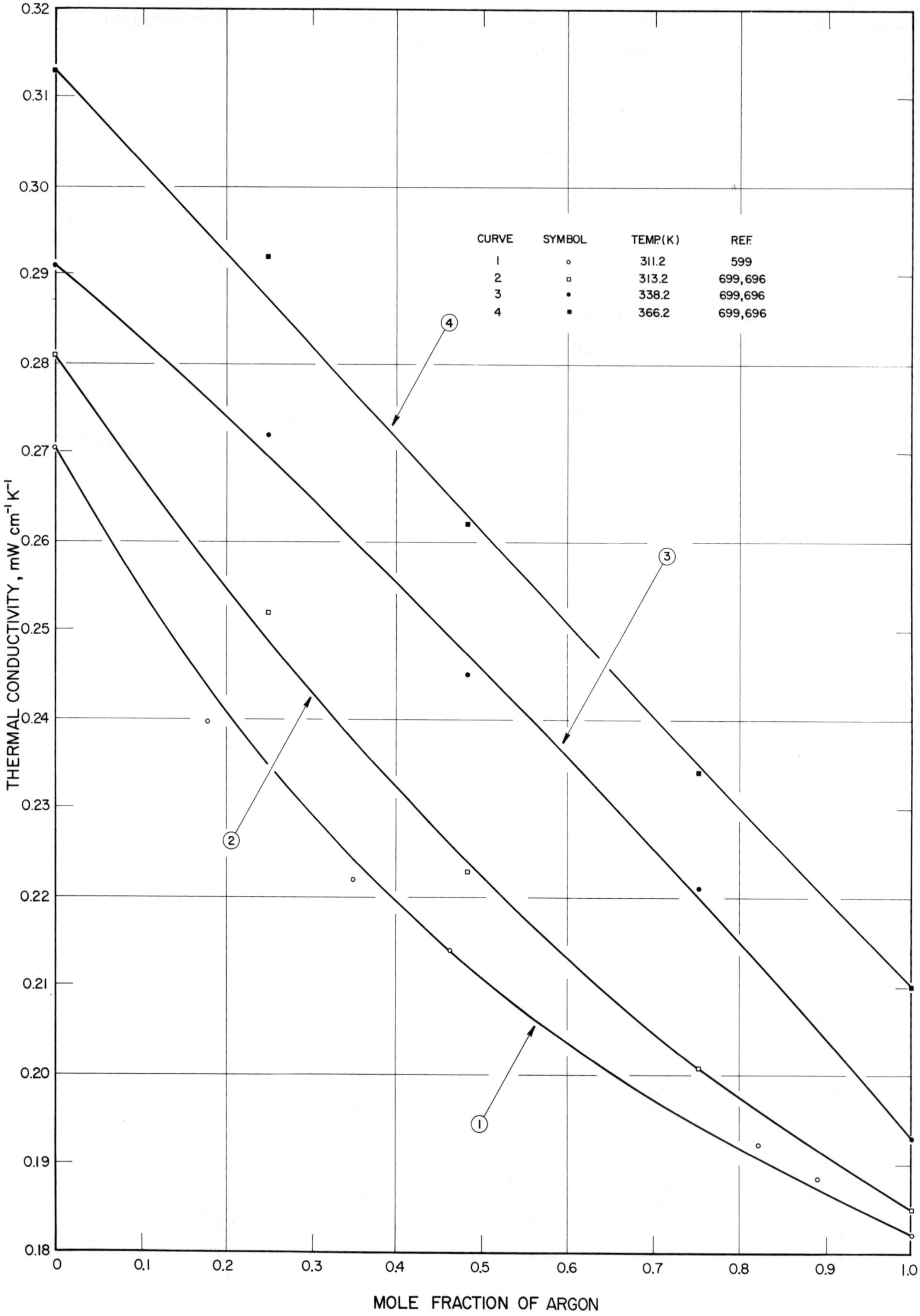

FIGURE 74a. THERMAL CONDUCTIVITY OF ARGON-OXYGEN SYSTEM

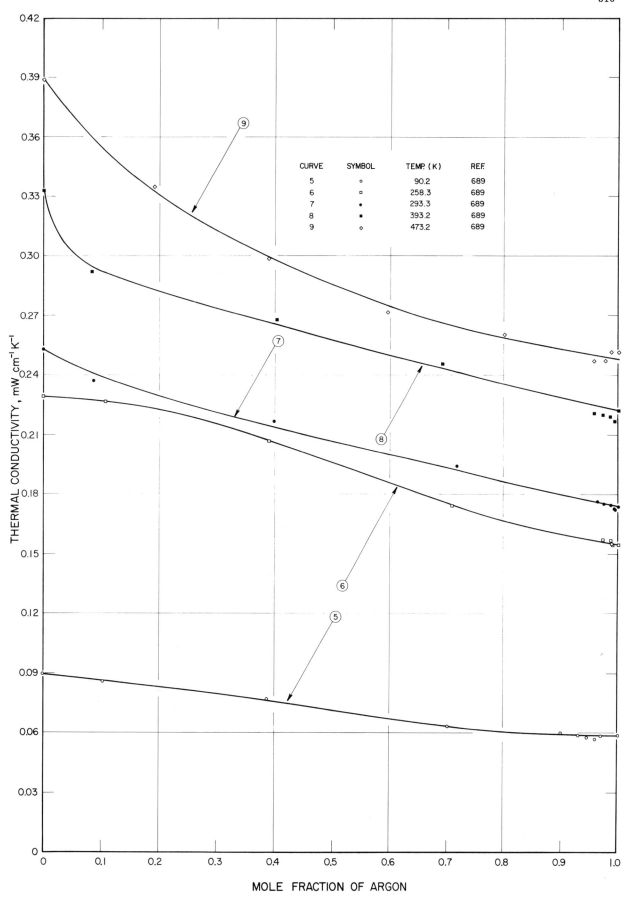

FIGURE 74b. THERMAL CONDUCTIVITY OF ARGON—OXYGEN SYSTEM

TABLE 75a. EXPERIMENTAL THERMAL CONDUCTIVITY DATA FOR ARGON-PROPANE SYSTEM

Curve No.	Fig. No.	Ref. No.	Author(s)	Temp. (K)	Mole Fraction of C_3H_8	Thermal Cond. (mW cm^{-1} K^{-1})	Remarks
1	75	65, 688	Cheung, H., Bromley, L.A., and Wilke, C.R.	594.2 591.2 591.2	0.0000 0.5288 1.0000	0.3056 0.4966 0.6134	Ar: Linde Air Prod. Co., standard grade, specified purity 99.97%, chief impurity N_2; propane: Matheson Co., Inc., instrument grade, specified purity 99.9% pure; coaxial cylinder method; avg and max error as 1.2% and 2% respectively.
2	75	65, 688	Cheung, H., Bromley, L.A., and Wilke, C.R.	811.2 810.2	0.5282 1.0000	0.829 1.050	
1	75	316, 700	Saxena, V.K. and Saxena, S.C.	591.2	0.00	0.2993	Ar: spectroscopically pure; conductivity column method; estimated precision ±2%.
2	75	316, 700	Saxena, V.K. and Saxena, S.C.	811.2	0.00	0.3756	Same as above.

TABLE 75b. THERMAL CONDUCTIVITY (mW cm^{-1}K^{-1}) OF ARGON-PROPANE SYSTEM AS A FUNCTION OF COMPOSITION AT THE TEMPERATURE OF MEASUREMENT AS DERIVED BY GRAPHICAL SMOOTHING

Mole Fraction of C_3H_8	591.2 K (Ref. 65)	811 K (Ref. 65)
0.00	0.299	0.376
0.05	0.322	0.429
0.10	0.346	0.484
0.15	0.368	0.536
0.20	0.388	0.582
0.25	0.407	0.627
0.30	0.426	0.669
0.35	0.446	0.708
0.40	0.458	0.744
0.45	0.474	0.779
0.50	0.488	0.812
0.55	0.503	0.842
0.60	0.516	0.869
0.65	0.529	0.895
0.70	0.542	0.919
0.75	0.554	0.944
0.80	0.567	0.964
0.85	0.579	0.986
0.90	0.590	1.01
0.95	0.602	1.03
1.00	0.613	1.05

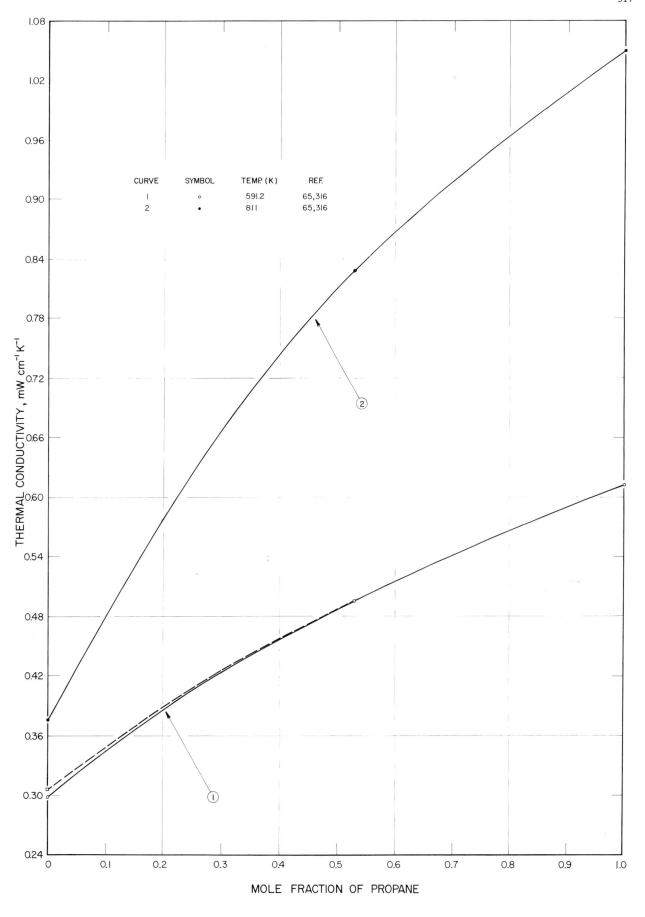

FIGURE 75. THERMAL CONDUCTIVITY OF ARGON-PROPANE SYSTEM

TABLE 76a.   EXPERIMENTAL THERMAL CONDUCTIVITY DATA FOR HELIUM-AIR SYSTEM

Curve No.	Fig. No.	Ref. No.	Author(s)	Temp. (K)	Mole Fraction of Air	Thermal Cond. (mW cm^{-1} K^{-1})	Remarks
1	76	701	Cotton, J.E.	328.3	0.0000 0.0295 0.0970 0.1814 0.3345 0.7147 1.0000	1.638 1.530 1.334 1.137 0.8729 0.4484 0.2813	Thin hot wire potential lead method; accuracy better than 0.2%.

TABLE 76b.   THERMAL CONDUCTIVITY (mW cm^{-1}K^{-1}) OF HELIUM-AIR SYSTEM AS A FUNCTION OF COMPOSITION AT THE TEMPERATURE OF MEASUREMENT AS DERIVED BY GRAPHICAL SMOOTHING

Mole Fraction of Air	328.3 K (Ref. 701)
0.00	1.64
0.05	1.47
0.10	1.33
0.15	1.21
0.20	1.10
0.25	1.01
0.30	0.926
0.35	0.850
0.40	0.780
0.45	0.716
0.50	0.656
0.55	0.600
0.60	0.548
0.65	0.500
0.70	0.459
0.75	0.422
0.80	0.390
0.85	0.360
0.90	0.332
0.95	0.306
1.00	0.281

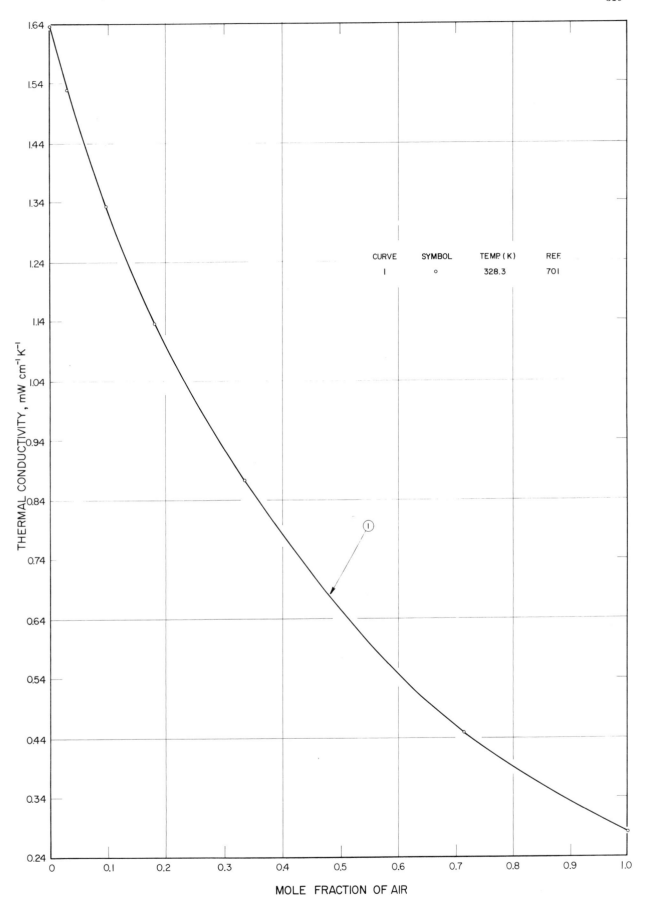

FIGURE 76. THERMAL CONDUCTIVITY OF HELIUM-AIR SYSTEM

TABLE 77a.   EXPERIMENTAL THERMAL CONDUCTIVITY DATA FOR HELIUM-n-BUTANE SYSTEM

Curve No.	Fig. No.	Ref. No.	Author(s)	Temp. (K)	Mole Fraction of n-Butane	Thermal Cond. (mW cm^{-1} K^{-1})	Remarks
1	77	701	Cotton, J.E.	328.4	0.0000 0.0903 0.1941 0.3560 0.7220 1.0000	1.638 1.139 0.8621 0.5506 0.3394 0.1895	Thin hot wire potential lead method; accuracy better than 0.2%.

TABLE 77b.   THERMAL CONDUCTIVITY (mW cm^{-1}K^{-1}) OF HELIUM-n-BUTANE SYSTEM AS A FUNCTION OF COMPOSITION AT THE TEMPERATURE OF MEASUREMENT AS DERIVED BY GRAPHICAL SMOOTHING

Mole Fraction of n-Butane	328.4 K (Ref. 701)
0.00	1.64
0.05	1.33
0.10	1.11
0.15	0.964
0.20	0.852
0.25	0.756
0.30	0.675
0.35	0.604
0.40	0.546
0.45	0.494
0.50	0.449
0.55	0.409
0.60	0.373
0.65	0.341
0.70	0.313
0.75	0.288
0.80	0.266
0.85	0.246
0.90	0.226
0.95	0.207
1.00	0.190

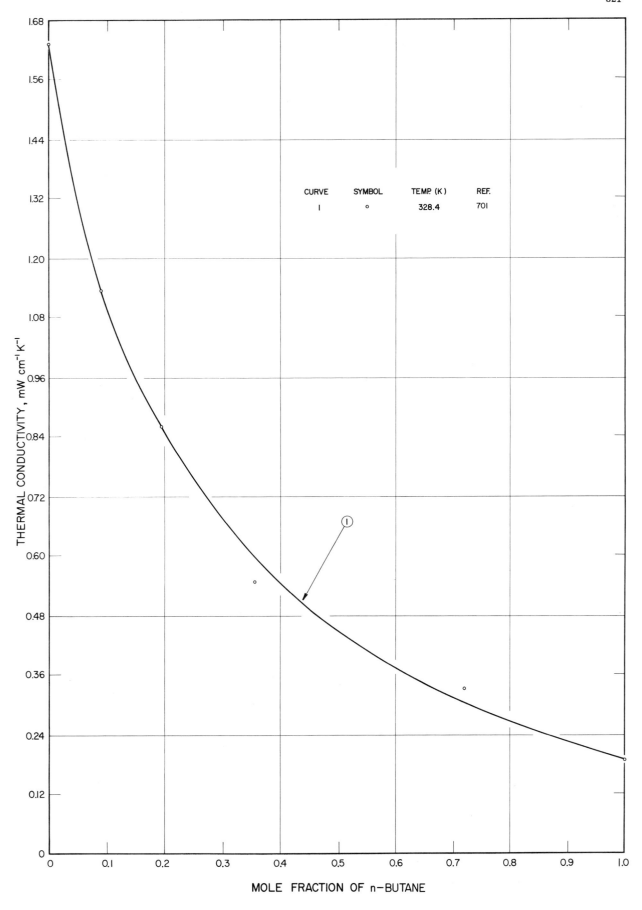

FIGURE 77. THERMAL CONDUCTIVITY OF HELIUM–n-BUTANE SYSTEM

TABLE 78a. EXPERIMENTAL THERMAL CONDUCTIVITY DATA FOR HELIUM-CARBON DIOXIDE SYSTEM

Curve No.	Fig. No.	Ref. No.	Author(s)	Temp. (K)	Mole Fraction of CO_2	Thermal Cond. (mW cm^{-1} K^{-1})	Remarks
1	78a	83	Davidson, J.M. and Music, J.F.	273.2	0.00 0.26 0.48 0.75 1.00	1.390 0.7465 0.4689 0.2659 0.1419	Unsteady state method, the rate of cooling of a solid inner cylinder of copper was determined; accuracy of these relative measurements ±5%.
2	78b	65, 688	Cheung, H., Bromley, L.A., and Wilke, C.R.	588.2 590.2 593.2	0.00 0.39 1.00	2.459 0.9902 0.4042	He: U.S. Navy research grade, specified purity 99.99%, chief impurities H_2 and H_2O; coaxial cylinder method; total max error 5.7%, avg error 1.2%, and max uncertainty in mixture composition 0.25%.

TABLE 78b. THERMAL CONDUCTIVITY (mW cm^{-1}K^{-1}) OF HELIUM-CARBON DIOXIDE SYSTEM AS A FUNCTION OF COMPOSITION AT THE TEMPERATURE OF MEASUREMENT AS DERIVED BY GRAPHICAL SMOOTHING

Mole Fraction of CO_2	273.2 K (Ref. 83)	590 K (Ref. 65)
0.00	1.39	2.46
0.05	1.21	2.18
0.10	1.06	1.92
0.15	0.943	1.70
0.20	0.845	1.51
0.25	0.760	1.35
0.30	0.684	1.20
0.35	0.616	1.08
0.40	0.555	0.968
0.45	0.500	0.872
0.50	0.451	0.788
0.55	0.406	0.714
0.60	0.366	0.648
0.65	0.329	0.592
0.70	0.296	0.543
0.75	0.266	0.502
0.80	0.238	0.468
0.85	0.212	0.442
0.90	0.187	0.422
0.95	0.165	0.408
1.00	0.142	0.404

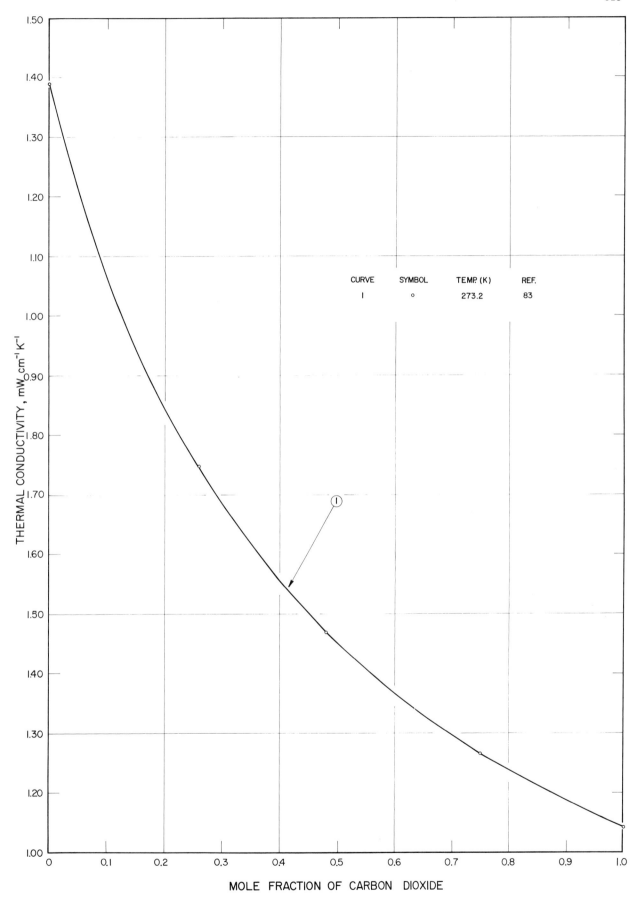

FIGURE 78a. THERMAL CONDUCTIVITY OF HELIUM-CARBON DIOXIDE SYSTEM

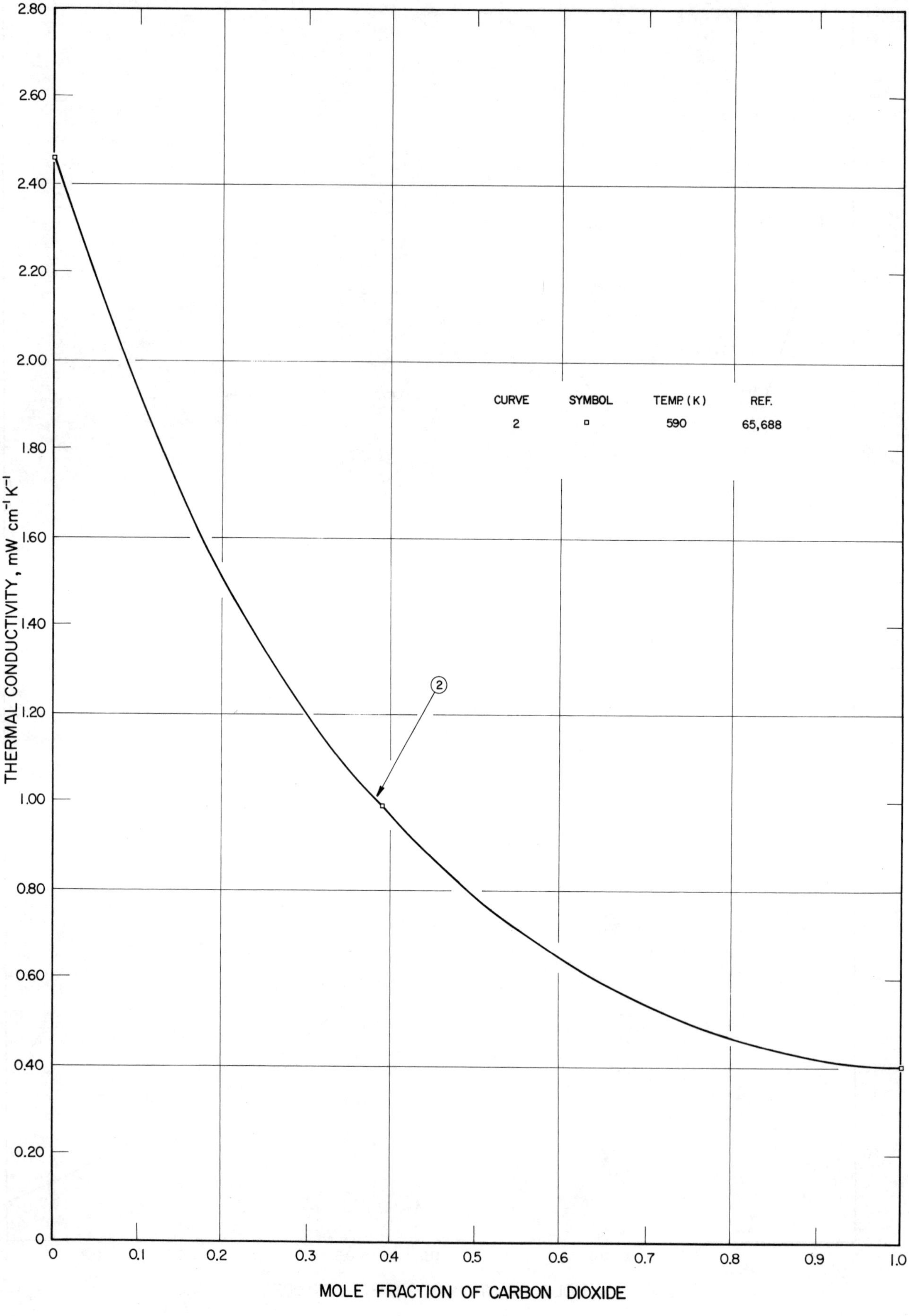

FIGURE 78b. THERMAL CONDUCTIVITY OF HELIUM-CARBON DIOXIDE SYSTEM

TABLE 79a. EXPERIMENTAL THERMAL CONDUCTIVITY DATA FOR HELIUM-CYCLOPROPANE SYSTEM

Curve No.	Fig. No.	Ref. No.	Author(s)	Temp. (K)	Mole Fraction of Cyclopropane	Thermal Cond. (mW cm^{-1} K^{-1})	Remarks
1	79	701	Cotton, J.E.	328.4	0.0000	1.638	Thin hot wire potential lead method; accuracy better than 0.2%.
					0.1145	1.130	
					0.1963	0.9152	
					0.3890	0.5933	
					0.7161	0.3000	
					0.8072	0.2633	
					0.8957	0.2220	
					1.0000	0.1842	

TABLE 79b. THERMAL CONDUCTIVITY (mW cm^{-1}K^{-1}) OF HELIUM-CYCLOPROPANE SYSTEM AS A FUNCTION OF COMPOSITION AT THE TEMPERATURE OF MEASUREMENT AS DERIVED BY GRAPHICAL SMOOTHING

Mole Fraction of Cyclopropane	328.4 K (Ref. 701)
0.00	1.64
0.05	1.39
0.10	1.18
0.15	1.03
0.20	0.911
0.25	0.809
0.30	0.722
0.35	0.645
0.40	0.580
0.45	0.523
0.50	0.473
0.55	0.428
0.60	0.388
0.65	0.352
0.70	0.320
0.75	0.292
0.80	0.267
0.85	0.245
0.90	0.224
0.95	0.204
1.00	0.184

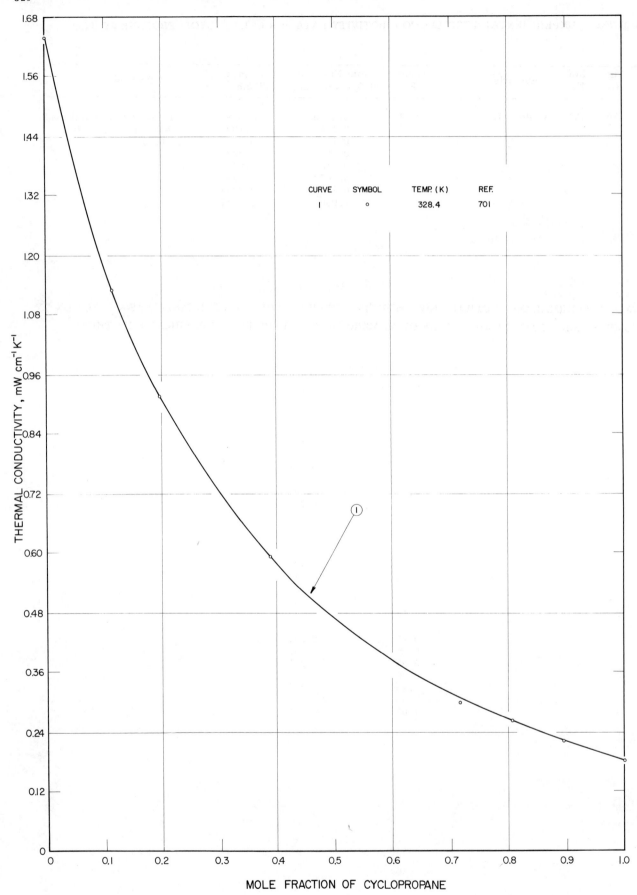

FIGURE 79. THERMAL CONDUCTIVITY OF HELIUM-CYCLOPROPANE SYSTEM

TABLE 80a. EXPERIMENTAL THERMAL CONDUCTIVITY DATA FOR HELIUM-DETERIUM SYSTEM

Curve No.	Fig. No.	Ref. No.	Author(s)	Temp. (K)	Mole Fraction of D_2	Thermal Cond. (mW cm^{-1} K^{-1})	Remarks
1	80	702	Gandhi, J.M. and Saxena, S.C.	303.2	0.0000 0.2965 0.4777 0.7205 1.0000	1.495 1.409 1.388 1.285 1.344	He: spectroscopically pure, D_2: 98.6% pure, impurities 0.8% H_2 and 0.6% H_2O; thick hot wire method; precision ±1%, accuracy ±1 to ±2%.
2	80	702	Gandhi, J.M. and Saxena, S.C.	323.2	0.0000 0.2965 0.4777 0.7205 1.0000	1.549 1.474 1.447 1.366 1.398	Same as above.
3	80	702	Gandhi, J.M. and Saxena, S.C.	343.2	0.0000 0.2965 0.4777 0.7205 1.0000	1.612 1.537 1.504 1.447 1.457	Same as above.
4	80	702	Gandhi, J.M. and Saxena, S.C.	363.2	0.0000 0.2965 0.4777 0.7205 1.0000	1.671 1.598 1.563 1.491 1.511	Same as above.

TABLE 80b. THERMAL CONDUCTIVITY (mW cm^{-1}K^{-1}) OF HELIUM-DEUTERIUM SYSTEM AS A FUNCTION OF COMPOSITION AT THE TEMPERATURE OF MEASUREMENT AS DERIVED BY GRAPHICAL SMOOTHING

Mole Fraction of D_2	303.2 K (Ref. 702)	323.2 K (Ref. 702)	343.2 K (Ref. 702)	363.2 K (Ref. 702)
0.00	1.50	1.55	1.61	1.67
0.05	1.49	1.54	1.60	1.66
0.10	1.48	1.53	1.59	1.65
0.15	1.46	1.52	1.58	1.64
0.20	1.45	1.51	1.57	1.63
0.25	1.44	1.50	1.56	1.62
0.30	1.43	1.49	1.55	1.60
0.35	1.41	1.47	1.53	1.59
0.40	1.39	1.45	1.52	1.58
0.45	1.37	1.44	1.50	1.56
0.50	1.35	1.42	1.49	1.54
0.55	1.33	1.40	1.48	1.53
0.60	1.31	1.38	1.47	1.51
0.65	1.30	1.37	1.46	1.50
0.70	1.29	1.36	1.45	1.49
0.75	1.28	1.36	1.45	1.49
0.80	1.29	1.36	1.44	1.49
0.85	1.29	1.37	1.44	1.49
0.90	1.31	1.38	1.45	1.50
0.95	1.32	1.39	1.45	1.50
1.00	1.34	1.40	1.46	1.51

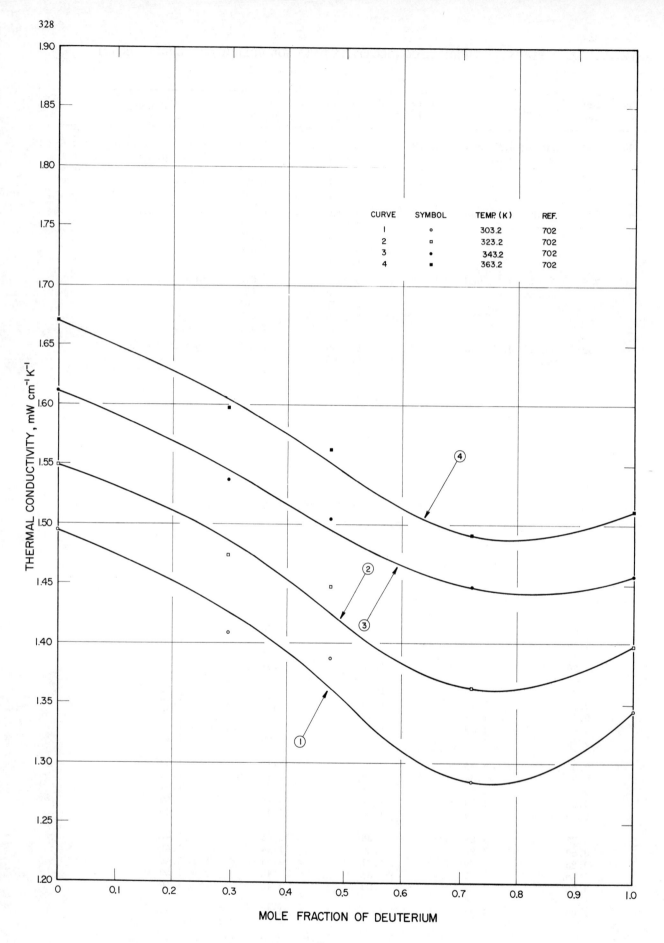

FIGURE 80. THERMAL CONDUCTIVITY OF HELIUM-DEUTERIUM SYSTEM

TABLE 81a. EXPERIMENTAL THERMAL CONDUCTIVITY DATA FOR HELIUM-ETHANE SYSTEM

Curve No.	Fig. No.	Ref. No.	Author(s)	Temp. (K)	Mole Fraction of Ethane	Thermal Cond. (mW cm^{-1} K^{-1})	Remarks
1	81	701	Cotton, J.E.	328.4	0.0000 0.0957 0.1964 0.3465 0.7184 1.0000	1.638 1.246 0.9726 0.7314 0.3860 0.2512	Thin hot wire potential lead method; accuracy better than 0.2%.

TABLE 81b. THERMAL CONDUCTIVITY (mW cm^{-1}K^{-1}) OF HELIUM-ETHANE SYSTEM AS A FUNCTION OF COMPOSITION AT THE TEMPERATURE OF MEASUREMENT AS DERIVED BY GRAPHICAL SMOOTHING

Mole Fraction of Ethane	328.4 K (Ref. 701)
0.00	1.64
0.05	1.42
0.10	1.22
0.15	1.08
0.20	0.966
0.25	0.874
0.30	0.796
0.35	0.726
0.40	0.664
0.45	0.606
0.50	0.556
0.55	0.510
0.60	0.468
0.65	0.431
0.70	0.398
0.75	0.368
0.80	0.340
0.85	0.316
0.90	0.294
0.95	0.271
1.00	0.251

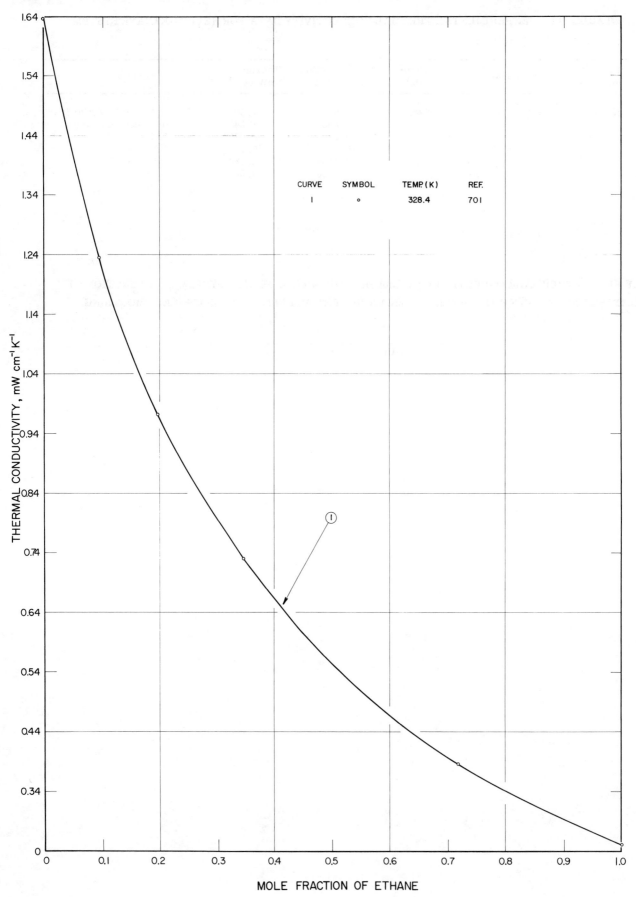

FIGURE 81. THERMAL CONDUCTIVITY OF HELIUM-ETHANE SYSTEM

TABLE 82a. EXPERIMENTAL THERMAL CONDUCTIVITY DATA FOR HELIUM-ETHYLENE SYSTEM

Curve No.	Fig. No.	Ref. No.	Author(s)	Temp. (K)	Mole Fraction of Ethylene	Thermal Cond. (mW cm^{-1} K^{-1})	Remarks
1	82	701	Cotton, J.E.	328.4	0.0000 0.0942 0.1857 0.3487 0.7106 1.0000	1.638 1.279 1.027 0.7331 0.3903 0.2402	Thin hot wire potential lead method; accuracy better than 0.2%.

TABLE 82b. THERMAL CONDUCTIVITY (mW cm^{-1}K^{-1}) OF HELIUM-ETHYLENE SYSTEM AS A FUNCTION OF COMPOSITION AT THE TEMPERATURE OF MEASUREMENT AS DERIVED BY GRAPHICAL SMOOTHING

Mole Fraction of Ethylene	328.4 K (Ref. 701)
0.00	1.64
0.05	1.44
0.10	1.26
0.15	1.12
0.20	0.993
0.25	0.891
0.30	0.806
0.35	0.731
0.40	0.668
0.45	0.711
0.50	0.560
0.55	0.513
0.60	0.471
0.65	0.432
0.70	0.397
0.75	0.366
0.80	0.338
0.85	0.312
0.90	0.288
0.95	0.264
1.00	0.240

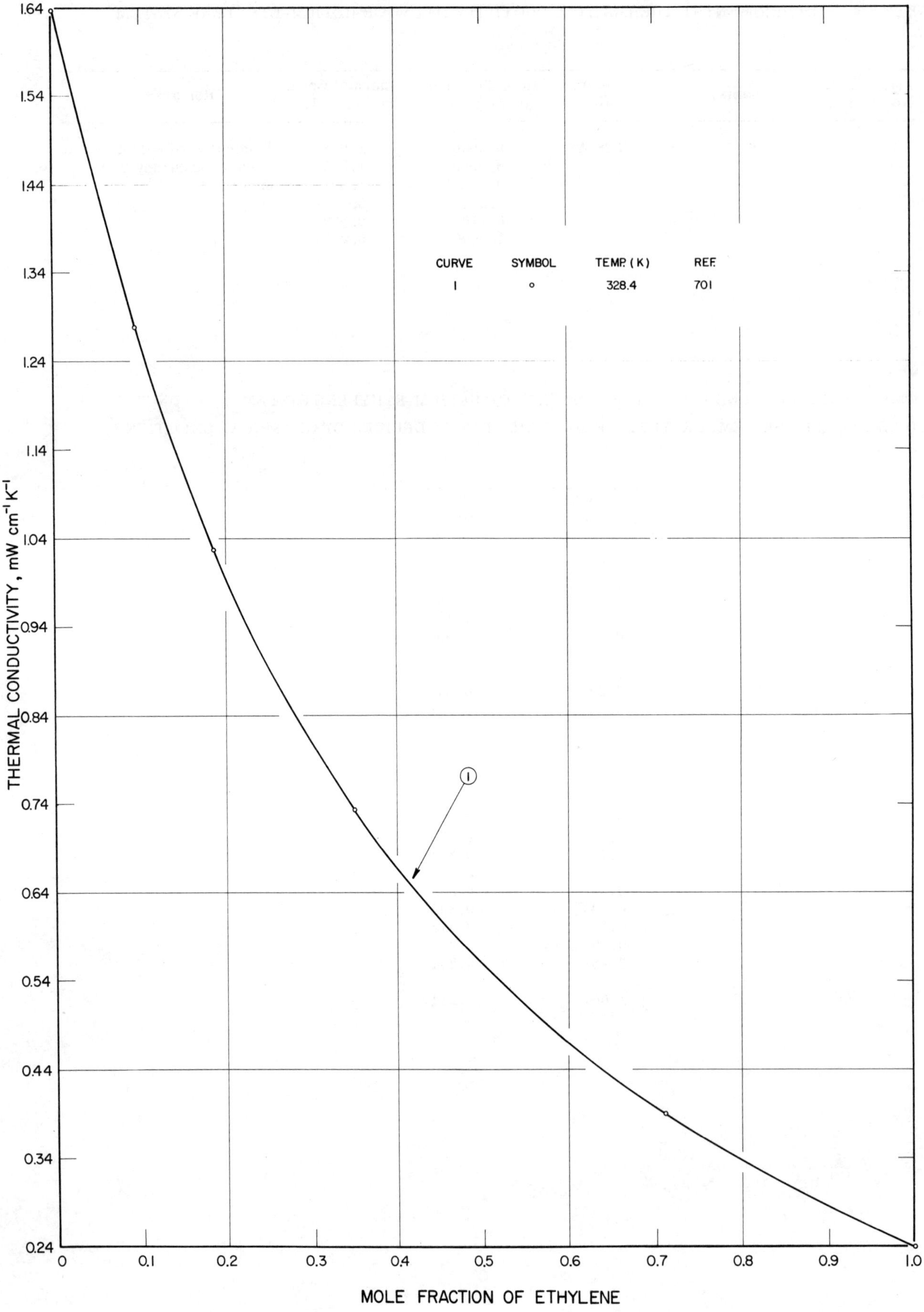

FIGURE 82. THERMAL CONDUCTIVITY OF HELIUM-ETHYLENE SYSTEM

TABLE 83a. EXPERIMENTAL THERMAL CONDUCTIVITY DATA FOR HELIUM-HYDROGEN SYSTEM

Curve No.	Fig. No.	Ref. No.	Author(s)	Temp. (K)	Mole Fraction of He	Thermal Cond. (mW cm^{-1} K^{-1})	Remarks
1	83a	703	Barua, A.K.	303.2	0.0000	1.809	He and H$_2$: spectroscopically pure; thick hot wire method; precision ±2%.
					0.1464	1.778	
					0.2891	1.697	
					0.4562	1.631	
					0.5328	1.606	
					0.7136	1.556	
					0.8713	1.534	
					1.0000	1.521	
2	83a	703	Barua, A.K.	318.2	0.0000	1.874	Same as above.
					0.1469	1.843	
					0.2786	1.773	
					0.4136	1.692	
					0.5462	1.643	
					0.7321	1.604	
					0.8675	1.577	
					1.0000	1.562	
3	83a	704	Mukhopadhyay, P. and Barua, A.K.	90.2	0.000	0.5966	He: spectroscopically pure, H$_2$: 99.95% pure; thick hot wire method, accuracy of these relative measurements ±1%.
					0.114	0.5702	
					0.193	0.5543	
					0.389	0.5916	
					0.492	0.6142	
					0.627	0.6347	
					0.810	0.6619	
					0.862	0.6661	
					0.909	0.6757	
					0.952	0.6866	
					1.000	0.6929	
4	83a	704	Mukhopadhyay, P. and Barua, A.K.	258.3	0.000	1.615	Same as above.
					0.153	1.558	
					0.303	1.499	
					0.606	1.426	
					0.701	1.405	
					0.807	1.364	
					0.855	1.301	
					0.910	1.356	
					0.948	1.388	
					1.000	1.410	
5	83a	704	Mukhopadhyay, P. and Barua, A.K.	273.3	0.000	1.692	Same as above.
					0.153	1.618	
					0.303	1.554	
					0.606	1.473	
					0.701	1.456	
					0.807	1.419	
					0.855	1.364	
					0.910	1.406	
					0.948	1.435	
					1.000	1.458	
6	83a	704	Mukhopadhyay, P. and Barua, A.K.	293.3	0.000	1.761	Same as above.
					0.153	1.704	
					0.303	1.639	
					0.606	1.553	
					0.701	1.537	
					0.807	1.477	
					0.855	1.414	
					0.910	1.480	
					0.948	1.511	
					1.000	1.528	

TABLE 83a. EXPERIMENTAL THERMAL CONDUCTIVITY DATA FOR HELIUM-HYDROGEN SYSTEM (cont.)

Curve No.	Fig. No.	Ref. No.	Author(s)	Temp. (K)	Mole Fraction of He	Thermal Cond. (mW cm^{-1} K^{-1})	Remarks
7	83b	704	Mukhopadhyay, P. and Barua, A.K.	393.3	0.000 0.130 0.396 0.804 0.875 0.950 1.000	2.186 2.106 1.976 1.815 1.756 1.840 1.870	Same as above.
8	83b	704	Mukhopadhyay, P. and Barua, A.K.	473.3	0.000 0.130 0.396 0.804 0.875 0.950 1.000	2.462 2.391 2.227 2.089 2.065 2.098 2.116	Same as above.
9	83b	701	Cotton, J.E.	303.3	0.0000 0.8014 0.8276 0.8779 0.9131 0.9615 1.0000	1.844 1.538 1.537 1.536 1.532 1.540 1.537	Thin hot wire potential lead method; accuracy better than 0.2%.
10	83b	701	Cotton, J.E.	328.2	0.0000 0.0702 0.2856 0.4961 0.7150 0.7999 0.8907 0.9375 1.0000	1.966 1.926 1.808 1.720 1.653 1.636 1.631 1.631 1.637	Same as above.
11	83b	701	Cotton, J.E.	378.3	0.0000 0.2867 0.5092 0.7192 0.8393 0.8988 0.9515 1.0000	2.185 2.015 1.908 1.832 1.805 1.799 1.806 1.808	Same as above.
12	83b	701	Cotton, J.E.	353.4	0.0000 0.2837 0.4956 0.7160 0.8630 0.9077 0.9509 1.0000	2.087 1.915 1.811 1.741 1.722 1.721 1.720 1.724	Same as above.
13	83b	701	Cotton, J.E.	398.2	0.0000 0.7116 0.8184 0.8568 0.8920 0.9215 0.9603 1.0000	2.282 1.920 1.882 1.880 1.877 1.881 1.882 1.884	Same as above.

TABLE 83b. THERMAL CONDUCTIVITY (mW cm^{-1}K^{-1}) OF HELIUM-HYDROGEN SYSTEM AS A FUNCTION OF COMPOSITION AT THE TEMPERATURE OF MEASUREMENT AS DERIVED BY GRAPHICAL SMOOTHING

Mole Fraction of He	90.2 K (Ref. 704)	258.3 K (Ref. 704)	273.3 K (Ref. 704)	293.3 K (Ref. 704)	303.2 K (Ref. 703)	303.3 K (Ref. 701)	318.2 K (Ref. 703)
0.00	0.597	1.62	1.69	1.76	1.81	1.84	1.87
0.05	0.587	1.60	1.67	1.74	1.81	1.81	1.87
0.10	0.574	1.57	1.64	1.72	1.79	1.79	1.86
0.15	0.558	1.56	1.62	1.70	1.78	1.76	1.84
0.20	0.555	1.54	1.60	1.68	1.75	1.73	1.82
0.25	0.560	1.52	1.57	1.66	1.73	1.71	1.79
0.30	0.569	1.50	1.55	1.64	1.70	1.68	1.76
0.35	0.581	1.49	1.54	1.62	1.67	1.66	1.73
0.40	0.594	1.47	1.52	1.61	1.65	1.64	1.70
0.45	0.606	1.46	1.51	1.59	1.63	1.62	1.68
0.50	0.615	1.45	1.49	1.58	1.61	1.60	1.66
0.55	0.623	1.44	1.48	1.57	1.59	1.59	1.64
0.60	0.631	1.43	1.47	1.56	1.58	1.57	1.63
0.65	0.638	1.42	1.47	1.55	1.57	1.56	1.62
0.70	0.644	1.41	1.46	1.54	1.56	1.55	1.61
0.75	0.651	1.39	1.44	1.52	1.55	1.54	1.60
0.80	0.658	1.37	1.44	1.49	1.54	1.54	1.59
0.85	0.665	1.30	1.36	1.41	1.54	1.54	1.58
0.90	0.674	1.35	1.40	1.47	1.53	1.53	1.57
0.95	0.684	1.39	1.44	1.51	1.53	1.54	1.57
1.00	0.693	1.41	1.46	1.53	1.52	1.54	1.56

Mole Fraction of He	328.2 K (Ref. 701)	353.4 K (Ref. 701)	378.3 K (Ref. 701)	393.3 K (Ref. 704)	398.2 K (Ref. 701)	473.3 K (Ref. 704)
0.00	1.97	2.09	2.19	2.19	2.28	2.46
0.05	1.94	2.05	2.16	2.16	2.25	2.43
0.10	1.91	2.02	2.12	2.12	2.22	2.40
0.15	1.88	1.99	2.10	2.10	2.19	2.37
0.20	1.85	1.96	2.07	2.07	2.16	2.34
0.25	1.83	1.93	2.04	2.04	2.13	2.31
0.30	1.80	1.91	2.01	2.02	2.10	2.28
0.35	1.78	1.88	1.99	2.00	2.07	2.25
0.40	1.76	1.86	1.96	1.97	2.04	2.23
0.45	1.74	1.83	1.93	1.96	2.02	2.20
0.50	1.72	1.81	1.92	1.94	1.99	2.18
0.55	1.70	1.79	1.89	1.92	1.97	2.17
0.60	1.68	1.77	1.87	1.90	1.95	2.15
0.65	1.67	1.76	1.85	1.89	1.93	2.14
0.70	1.66	1.74	1.84	1.87	1.91	2.12
0.75	1.64	1.73	1.82	1.85	1.90	2.11
0.80	1.64	1.73	1.81	1.82	1.89	2.09
0.85	1.63	1.72	1.81	1.77	1.88	2.07
0.90	1.63	1.72	1.80	1.76	1.88	2.07
0.95	1.63	1.72	1.80	1.84	1.88	2.10
1.00	1.64	1.72	1.81	1.87	1.88	2.12

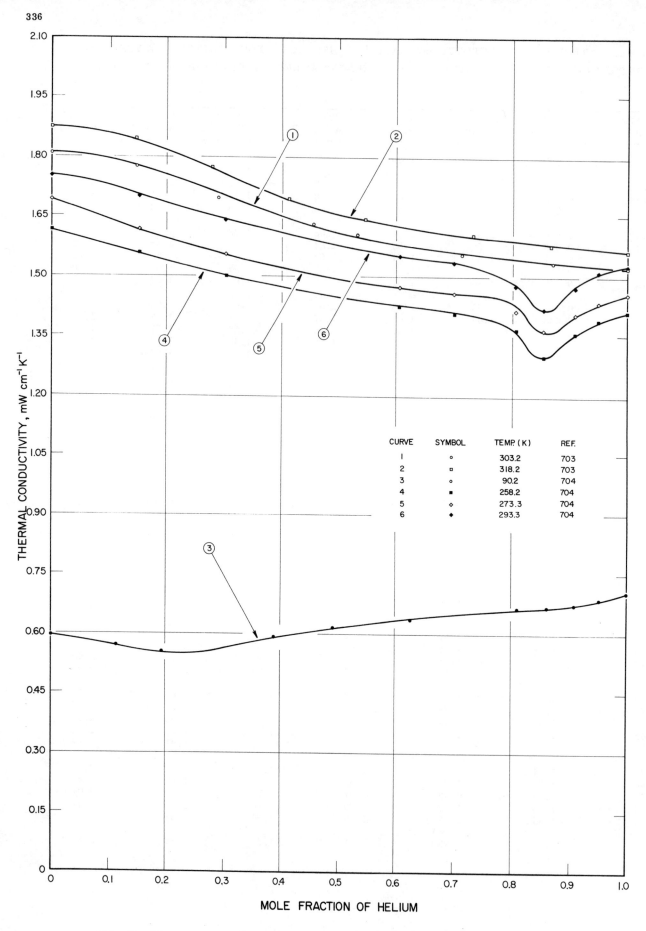

FIGURE 83a. THERMAL CONDUCTIVITY OF HELIUM-HYDROGEN SYSTEM

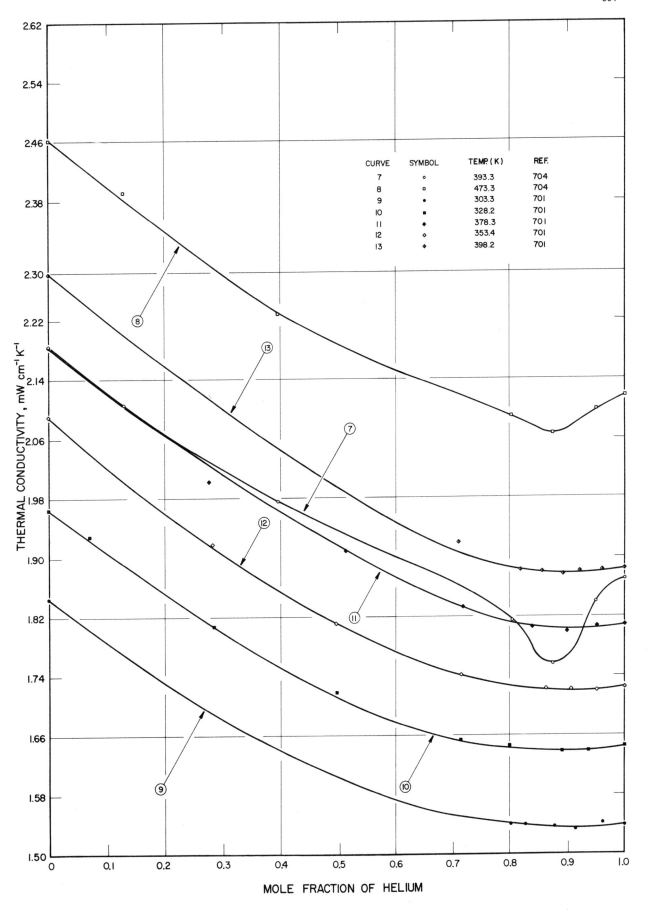

FIGURE 83b. THERMAL CONDUCTIVITY OF HELIUM-HYDROGEN SYSTEM

TABLE 84a. EXPERIMENTAL THERMAL CONDUCTIVITY DATA FOR HELIUM-METHANE SYSTEM

Curve No.	Fig. No.	Ref. No.	Author(s)	Temp. (K)	Mole Fraction of CH_4	Thermal Cond. (mW cm^{-1} K^{-1})	Remarks
--	--	688	Cheung, H., Bromley, L.A., and Wilke, C.R.	588.2 589.2 589.2 590.2 590.2	0.000 0.254 0.450 0.701 1.000	2.459 1.701 1.279 1.029 0.8516	He: U.S. Navy research grade, specified purity 99.99%, chief impurities H_2 and H_2O, CH_4: Phillips Petroleum Co., specified purity 99%, chief impurity N_2; coaxial cylinder method; max error 5.7%, avg error 1.2% and max uncertainty in mixture composition 0.25%.
1	84	65	Cheung, H., Bromley, L.A., and Wilke, C.R.	589.2	0.000 0.254 0.450 0.701 1.000	2.328 1.701 1.279 1.027 0.8495	These authors have reduced their data, reproduced above, to refer to a common temp.

TABLE 84b. THERMAL CONDUCTIVITY (mW cm^{-1}K^{-1}) OF HELIUM-METHANE SYSTEM AS A FUNCTION OF COMPOSITION AT THE TEMPERATURE OF MEASUREMENT AS DERIVED BY GRAPHICAL SMOOTHING

Mole Fraction of CH_4	589.2 K (Ref. 65)
0.00	2.33
0.05	2.20
0.10	2.08
0.15	1.95
0.20	1.82
0.25	1.69
0.30	1.57
0.35	1.46
0.40	1.36
0.45	1.28
0.50	1.24
0.55	1.16
0.60	1.11
0.65	1.07
0.70	1.03
0.75	0.992
0.80	0.962
0.85	0.931
0.90	0.902
0.95	0.874
1.00	0.850

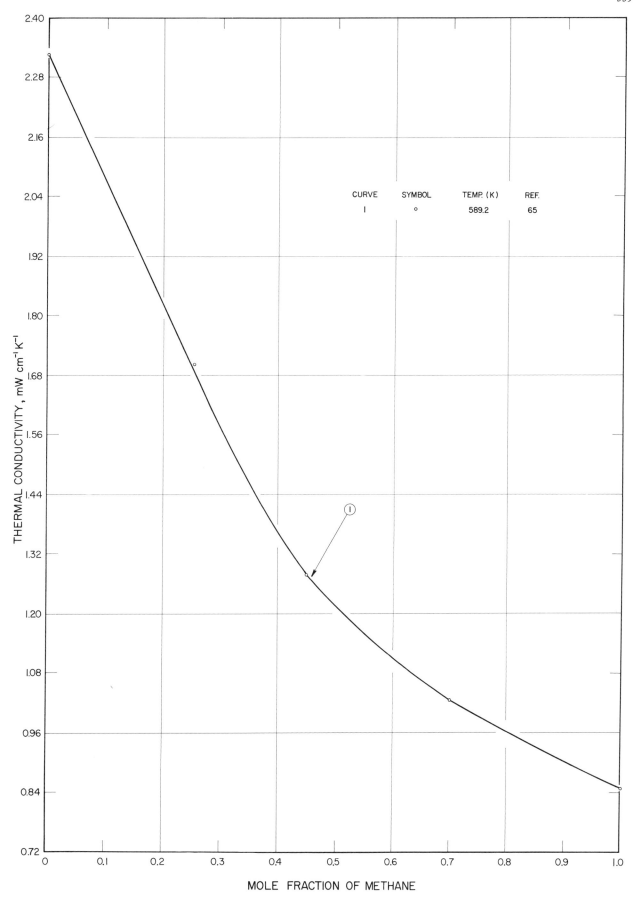

FIGURE 84. THERMAL CONDUCTIVITY OF HELIUM-METHANE SYSTEM

TABLE 85a. EXPERIMENTAL THERMAL CONDUCTIVITY DATA FOR HELIUM-NITROGEN SYSTEM

Curve No.	Fig. No.	Ref. No.	Author(s)	Temp. (K)	Mole Fraction of N_2	Thermal Cond. (mW cm^{-1} K^{-1})	Remarks
--	--	688	Cheung, H., Bromley, L.A., and Wilke, C.R.	373.2 373.2 380.7 377.2 377.7	0.000 0.163 0.591 0.781 1.000	1.756 1.270 0.6109 0.4484 0.3123	He: U.S. Navy Research grade, specified purity 99.99%, chief impurities H_2 and H_2O, N_2: Linde Air Prod. Co., water pumped, specified purity 99.9%, chief impurities Ar and Ne; co-axial cylinder method; max error 5.7%, avg error 1.2% and max uncertainty in mixture composition 0.25%.
--	--	688	Cheung, H., Bromley, L.A., and Wilke, C.R.	589.2 590.2 591.2 590.2	0.261 0.363 0.695 1.000	1.401 1.184 0.6812 0.4467	
1	85	83	Davidson, J.M., and Music, J.F.	273.2	0.000 0.240 0.450 0.740 1.000	1.390 0.8277 0.5958 0.3617 0.2416	Unsteady state method, the rate of cooling of a solid inner cylinder of copper was determined, accuracy of these relative measurements ±5%.
2	85	706	Barua, A.K.	303.2	0.0000 0.1136 0.2568 0.3959 0.5319 0.7107 0.8472 1.0000	1.522 1.190 0.9010 0.7118 0.5451 0.4147 0.3368 0.2561	He and N_2: spectroscopically pure; thick hot wire method; precision ±2%.
3	85	706	Barua, A.K.	318.2	0.0000 0.1349 0.2638 0.3759 0.5019 0.7038 0.8438 1.0000	1.576 1.182 0.9236 0.7570 0.6021 0.4245 0.3491 0.2672	Same as above.
4	85	65	Cheung, H., Bromley, L.A., and Wilke, C.R.	377.2	0.000 0.163 0.591 0.781 1.000	1.768 1.280 0.6050 0.4484 0.3119	These authors have reduced their data, reproduced above, to refer to a common temp.
5	85	65	Cheung, H., Bromley, L.A., and Wilke, C.R.	589.2	0.000 0.261 0.363 0.695 1.000	2.328 1.401 1.182 0.6787 0.4455	These authors have reduced their data, reproduced above, to refer to a common temp.

TABLE 85b. THERMAL CONDUCTIVITY (mW cm^{-1}K^{-1}) OF HELIUM-NITROGEN SYSTEM AS A FUNCTION OF COMPOSITION AT THE TEMPERATURE OF MEASUREMENT AS DERIVED BY GRAPHICAL SMOOTHING

Mole Fraction of N_2	273.2 K (Ref. 83)	303.2 K (Ref. 706)	318.2 K (Ref. 706)	377.2 K (Ref. 65)	589.2 K (Ref. 65)
0.00	1.39	1.52	1.58	1.77	2.33
0.05	1.25	1.37	1.43	1.61	2.05
0.10	1.12	1.23	1.29	1.46	1.85
0.15	0.999	1.11	1.16	1.32	1.69
0.20	0.896	1.00	1.05	1.19	1.55
0.25	0.813	0.906	0.947	1.08	1.43
0.30	0.741	0.826	0.863	0.990	1.32
0.35	0.679	0.754	0.788	0.911	1.21
0.40	0.621	0.690	0.721	0.840	1.11
0.45	0.571	0.632	0.659	0.774	1.02
0.50	0.526	0.580	0.603	0.710	0.934
0.55	0.484	0.484	0.552	0.653	0.857
0.60	0.447	0.490	0.507	0.596	0.788
0.65	0.412	0.450	0.465	0.545	0.726
0.70	0.379	0.414	0.426	0.498	0.672
0.75	0.350	0.380	0.392	0.455	0.626
0.80	0.317	0.350	0.360	0.417	0.584
0.85	0.284	0.321	0.332	0.384	0.546
0.90	0.275	0.296	0.311	0.355	0.511
0.95	0.231	0.274	0.285	0.331	0.477
1.00	0.242	0.256	0.267	0.312	0.446

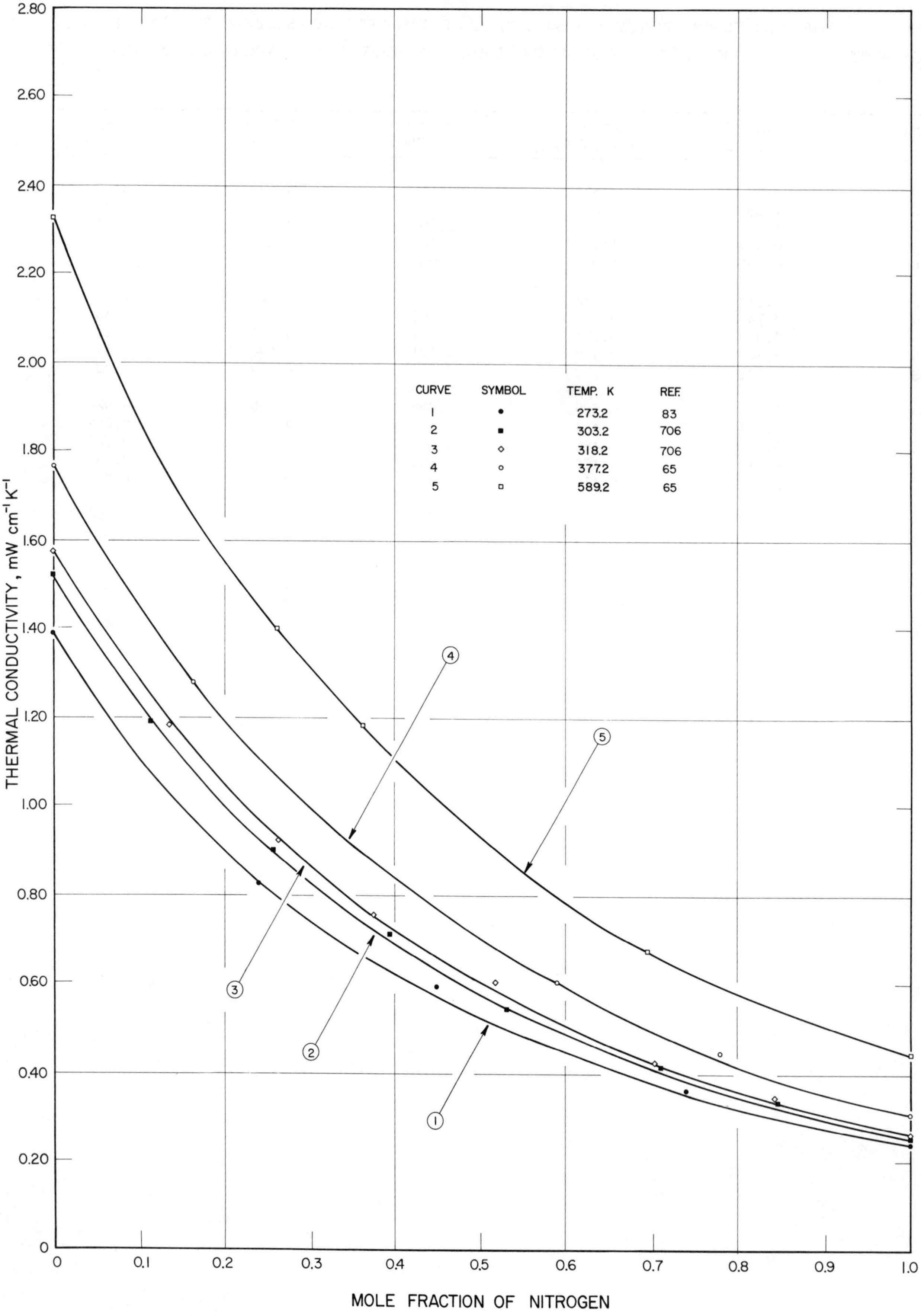

CURVE	SYMBOL	TEMP. K	REF.
1	●	273.2	83
2	■	303.2	706
3	◇	318.2	706
4	○	377.2	65
5	□	589.2	65

FIGURE 85. THERMAL CONDUCTIVITY OF HELIUM-NITROGEN SYSTEM

TABLE 86a. EXPERIMENTAL THERMAL CONDUCTIVITY DATA FOR HELIUM-OXYGEN SYSTEM

Curve No.	Fig. No.	Ref. No.	Author(s)	Temp. (K)	Mole Fraction of O_2	Thermal Cond. (mW cm^{-1} K^{-1})	Remarks
1	86	705	Srivastava, B.N. and Barua, A.K.	303.2	0.0000 0.1238 0.3159 0.4319 0.6619 0.7941 0.8616 1.0000	1.523 1.189 0.8131 0.6737 0.4693 0.3842 0.3267 0.2697	He and O_2: spectroscopically pure; thick hot wire method; precision ±2%.
2	86	705	Srivastava, B.N. and Barua, A.K.	318.2	0.0000 0.1636 0.2879 0.4537 0.6016 0.7416 0.8539 1.0000	1.578 1.147 0.8662 0.6829 0.5305 0.4179 0.3549 0.2800	Same as above.

TABLE 86b. THERMAL CONDUCTIVITY (mW cm^{-1}K^{-1}) OF HELIUM-OXYGEN SYSTEM AS A FUNCTION OF COMPOSITION AT THE TEMPERATURE OF MEASUREMENT AS DERIVED BY GRAPHICAL SMOOTHING

Mole Fraction of O_2	303.2 K (Ref. 705)	318.2 K (Ref. 705)
0.00	1.52	1.58
0.05	1.39	1.44
0.10	1.25	1.30
0.15	1.13	1.18
0.20	1.02	1.06
0.25	0.921	0.958
0.30	0.836	0.868
0.35	0.764	0.784
0.40	0.701	0.730
0.45	0.645	0.672
0.50	0.595	0.619
0.55	0.549	0.572
0.60	0.507	0.528
0.65	0.469	0.488
0.70	0.434	0.449
0.75	0.402	0.415
0.80	0.372	0.384
0.85	0.344	0.355
0.90	0.318	0.328
0.95	0.293	0.304
1.00	0.270	0.280

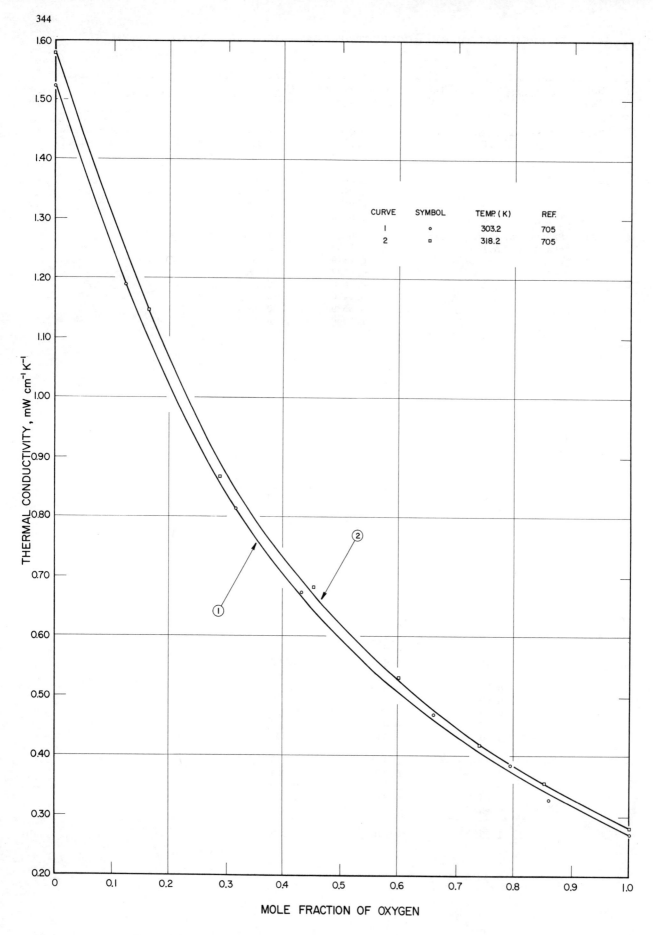

FIGURE 86. THERMAL CONDUCTIVITY OF HELIUM–OXYGEN SYSTEM

TABLE 87a. EXPERIMENTAL THERMAL CONDUCTIVITY DATA FOR HELIUM-PROPANE SYSTEM

Curve No.	Fig. No.	Ref. No.	Author(s)	Temp. (K)	Mole Fraction of Propane	Thermal Cond. (mW cm^{-1} K^{-1})	Remarks
1	87	701	Cotton, J. E.	328.4	0.0000	1.638	Thin hot wire potential lead method; accuracy better than 0.2%.
					0.1093	1.131	
					0.2283	0.8227	
					0.3618	0.6125	
					0.7165	0.3291	
					1.0000	0.2123	

TABLE 87b. THERMAL CONDUCTIVITY (mW cm^{-1}K^{-1}) OF HELIUM-PROPANE SYSTEM AS A FUNCTION OF COMPOSITION AT THE TEMPERATURE OF MEASUREMENT AS DERIVED BY GRAPHICAL SMOOTHING

Mole Fraction of Propane	328.4 K (Ref. 701)
0.00	1.64
0.05	1.38
0.10	1.16
0.15	1.01
0.20	0.882
0.25	0.781
0.30	0.699
0.35	0.628
0.40	0.568
0.45	0.518
0.50	0.472
0.55	0.432
0.60	0.397
0.65	0.365
0.70	0.338
0.75	0.313
0.80	0.290
0.85	0.269
0.90	0.249
0.95	0.230
1.00	0.212

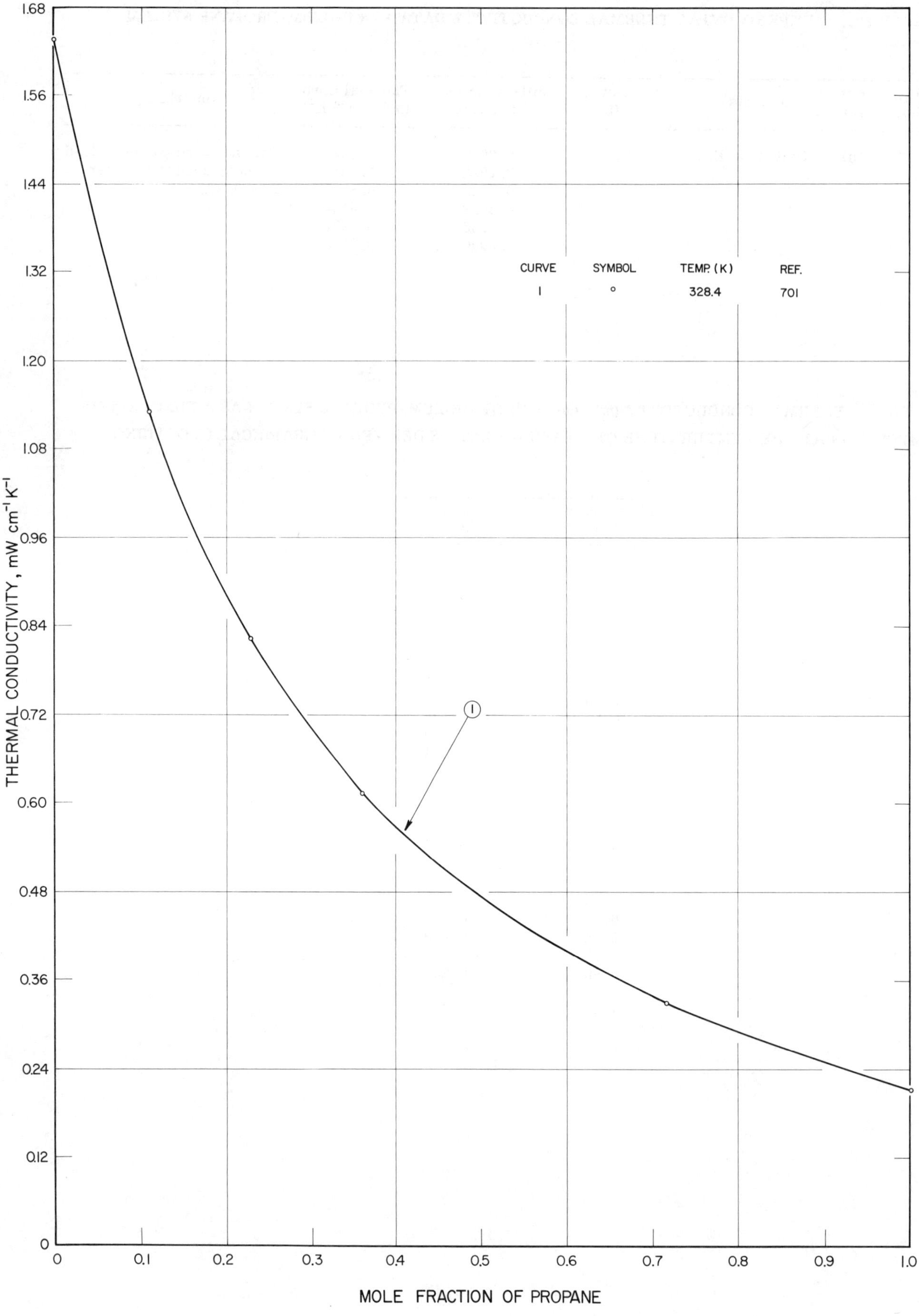

FIGURE 87. THERMAL CONDUCTIVITY OF HELIUM-PROPANE SYSTEM

TABLE 88a. EXPERIMENTAL THERMAL CONDUCTIVITY DATA FOR HELIUM-PROPYLENE SYSTEM

Curve No.	Fig. No.	Ref. No.	Author(s)	Temp. (K)	Mole Fraction of Propylene	Thermal Cond. (mW cm^{-1} K^{-1})	Remarks
1	88	701	Cotton, J.E.	328.4	0.0000 0.0881 0.1904 0.3243 0.7131 1.0000	1.638 1.220 0.9311 0.6837 0.3291 0.2055	Thin hot wire potential lead method, accuracy better than 0.2%.

TABLE 88b. THERMAL CONDUCTIVITY (mW cm^{-1}K^{-1}) OF HELIUM-PROPYLENE SYSTEM AS A FUNCTION OF COMPOSITION AT THE TEMPERATURE OF MEASUREMENT AS DERIVED BY GRAPHICAL SMOOTHING

Mole Fraction of Propylene	328.4 K (Ref. 701)
0.00	1.64
0.05	1.38
0.10	1.18
0.15	1.03
0.20	0.909
0.25	0.808
0.30	0.722
0.35	0.648
0.40	0.586
0.45	0.531
0.50	0.483
0.55	0.440
0.60	0.401
0.65	0.366
0.70	0.336
0.75	0.308
0.80	0.284
0.85	0.262
0.90	0.242
0.95	0.222
1.00	0.206

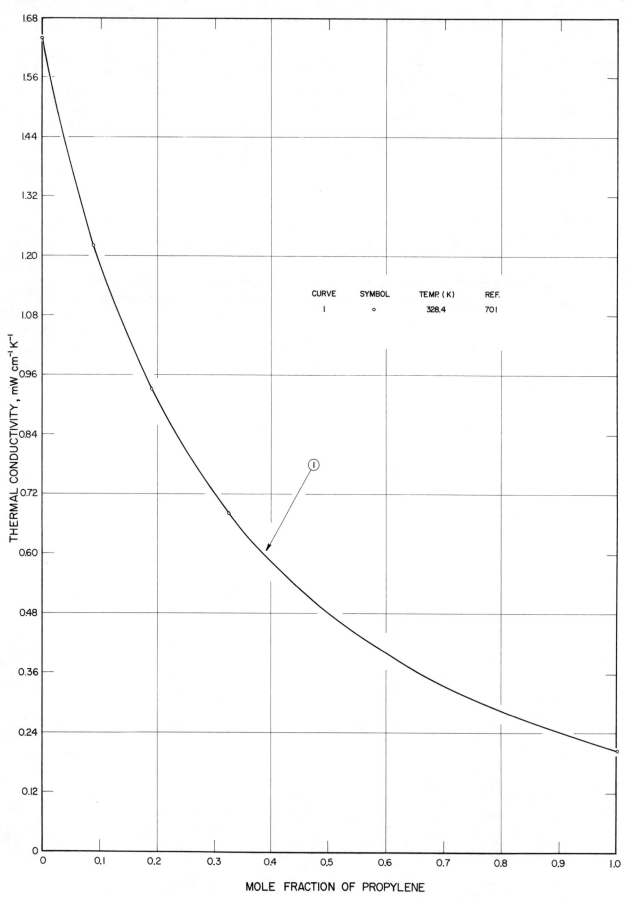

FIGURE 88. THERMAL CONDUCTIVITY OF HELIUM-PROPYLENE SYSTEM

TABLE 89a. EXPERIMENTAL THERMAL CONDUCTIVITY DATA FOR KRYPTON-DEUTERIUM SYSTEM

Curve No.	Fig. No.	Ref. No.	Author(s)	Temp. (K)	Mole Fraction of Kr	Thermal Cond. (mW cm^{-1} K^{-1})	Remarks
1	89	698	Gambhir, R.S. and Saxena, S.C.	308.2	0.000 0.084 0.222 0.446 0.822 1.000	1.357 1.089 0.8122 0.5066 0.2047 0.0959	D_2: 98.6% pure, 0.8% H_2 and 0.6% H_2O, Kr: 99-100% pure, balance Xe; thick hot wire method; accuracy ±1 to ±2%, precision ±1%.
2	89	698	Gambhir, R.S. and Saxena, S.C.	323.2	0.000 0.084 0.222 0.446 0.822 1.000	1.398 1.130 0.8457 0.5359 0.2139 0.1017	Same as above.
3	89	698	Gambhir, R.S. and Saxena, S.C.	343.2	0.000 0.084 0.222 0.446 0.822 1.000	1.457 1.181 0.8918 0.5694 0.2257 0.1072	Same as above.
4	89	698	Gambhir, R.S. and Saxena, S.C.	363.2	0.000 0.084 0.222 0.446 0.822 1.000	1.511 1.227 0.9253 0.5987 0.2378 0.1114	Same as above.

TABLE 89b. THERMAL CONDUCTIVITY (mW cm^{-1}K^{-1}) OF KRYPTON-DEUTERIUM SYSTEM AS A FUNCTION OF COMPOSITION AT THE TEMPERATURE OF MEASUREMENT AS DERIVED BY GRAPHICAL SMOOTHING

Mole Fraction of Kr	308.2 K (Ref. 698)	323.2 K (Ref. 698)	343.2 K (Ref. 698)	363.2 K (Ref. 698)
0.00	1.36	1.40	1.46	1.51
0.05	1.19	1.23	1.28	1.33
0.10	1.05	1.09	1.14	1.18
0.15	0.942	0.981	1.03	1.07
0.20	0.848	0.885	0.930	0.967
0.25	0.768	0.800	0.845	0.878
0.30	0.691	0.724	0.766	0.798
0.35	0.621	0.654	0.693	0.724
0.40	0.558	0.589	0.626	0.656
0.45	0.503	0.531	0.565	0.594
0.50	0.454	0.480	0.510	0.537
0.55	0.409	0.434	0.458	0.483
0.60	0.366	0.389	0.410	0.431
0.65	0.326	0.346	0.365	0.383
0.70	0.288	0.305	0.321	0.337
0.75	0.252	0.266	0.280	0.294
0.80	0.218	0.230	0.242	0.254
0.85	0.186	0.197	0.207	0.217
0.90	0.156	0.164	0.173	0.182
0.95	0.126	0.133	0.140	0.146
1.00	0.0959	0.102	0.107	0.111

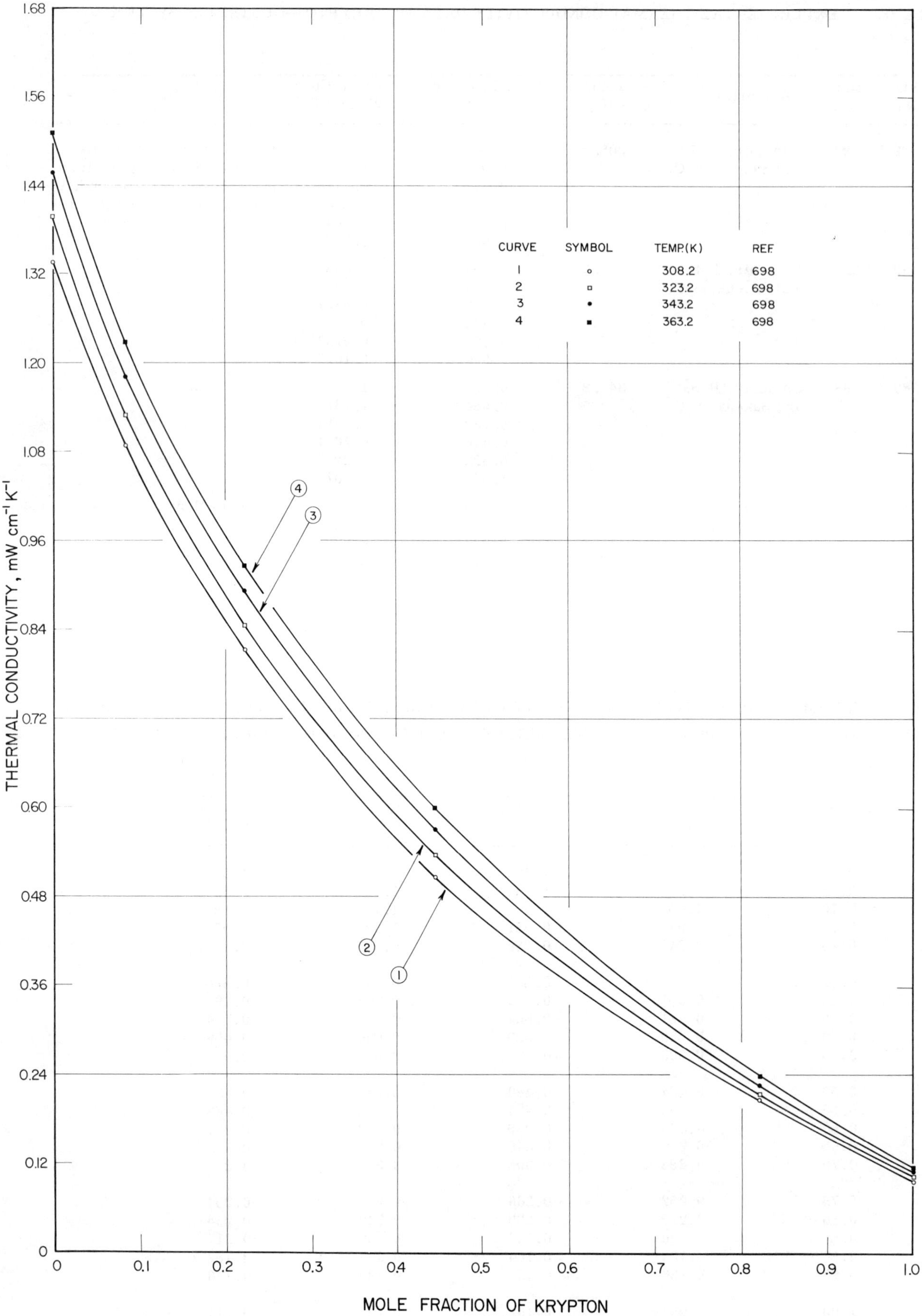

FIGURE 89. THERMAL CONDUCTIVITY OF KRYPTON-DEUTERIUM SYSTEM

TABLE 90a.   EXPERIMENTAL THERMAL CONDUCTIVITY DATA FOR KRYPTON-HYDROGEN SYSTEM

Curve No.	Fig. No.	Ref. No.	Author(s)	Temp. (K)	Mole Fraction of Kr	Thermal Cond. (mW cm^{-1} K^{-1})	Remarks
1	90	707, 708	Tondon, P.K. and Saxena, S.C.	313.2	0.000 0.253 0.469 0.653 1.000	1.822 1.005 0.6615 0.4229 0.0984	Kr: 99-100% pure, balance Xe, H$_2$: 99.95% pure; thick hot wire method; accuracy ±1 to ±2%, precision ±1%.
2	90	707, 708	Tondon, P.K. and Saxena, S.C.	338.2	0.000 0.253 0.469 0.653 1.000	1.939 1.101 0.6950 0.4438 0.1078	Same as above.
3	90	707, 708	Tondon, P.K. and Saxena, S.C.	366.2	0.000 0.253 0.469 0.653 1.000	2.037 1.181 0.7620 0.4773 0.1143	Same as above.
4	90	703	Barua, A.K.	303.2	0.0000 0.1363 0.2584 0.4462 0.5139 0.7326 0.8862 1.0000	1.838 1.351 1.021 0.6624 0.5573 0.3037 0.1261 0.0945	Kr: traces of impurities, H$_2$: spectroscopically pure; thick hot wire method; precision ±2%.
5	90	703	Barua, A.K.	318.2	0.0000 0.1542 0.2431 0.3864 0.5639 0.7241 0.8562 1.0000	1.875 1.382 1.091 0.6716 0.5263 0.3437 0.2124 0.0983	Same as above.

TABLE 90b. THERMAL CONDUCTIVITY (mW cm^{-1}K^{-1}) OF KRYPTON-HYDROGEN SYSTEM AS A FUNCTION OF COMPOSITION AT THE TEMPERATURE OF MEASUREMENT AS DERIVED BY GRAPHICAL SMOOTHING

Mole Fraction of Kr	303.2 K (Ref. 703)	313.2 K (Ref. 707)	318.2 K (Ref. 703)	338.2 K (Ref. 707)	366.2 K (Ref. 707)
0.00	1.84	1.82	1.88	1.94	2.04
0.05	1.66	1.59	1.69	1.72	1.84
0.10	1.47	1.40	1.52	1.52	1.64
0.15	1.30	1.25	1.35	1.36	1.47
0.20	1.17	1.12	1.21	1.23	1.31
0.25	1.04	1.01	1.08	1.11	1.19
0.30	0.932	0.920	0.962	1.00	1.08
0.35	0.834	0.838	0.864	0.905	0.979
0.40	0.742	0.761	0.773	0.813	0.886
0.45	0.656	0.688	0.690	0.728	0.797
0.50	0.578	0.618	0.614	0.649	0.710
0.55	0.508	0.552	0.544	0.577	0.628
0.60	0.442	0.488	0.481	0.510	0.552
0.65	0.386	0.430	0.422	0.448	0.483
0.70	0.334	0.376	0.368	0.392	0.420
0.75	0.289	0.325	0.319	0.340	0.364
0.80	0.246	0.275	0.271	0.289	0.310
0.85	0.207	0.228	0.226	0.242	0.259
0.90	0.168	0.184	0.183	0.195	0.209
0.95	0.131	0.140	0.141	0.152	0.162
1.00	0.0945	0.0984	0.0983	0.108	0.114

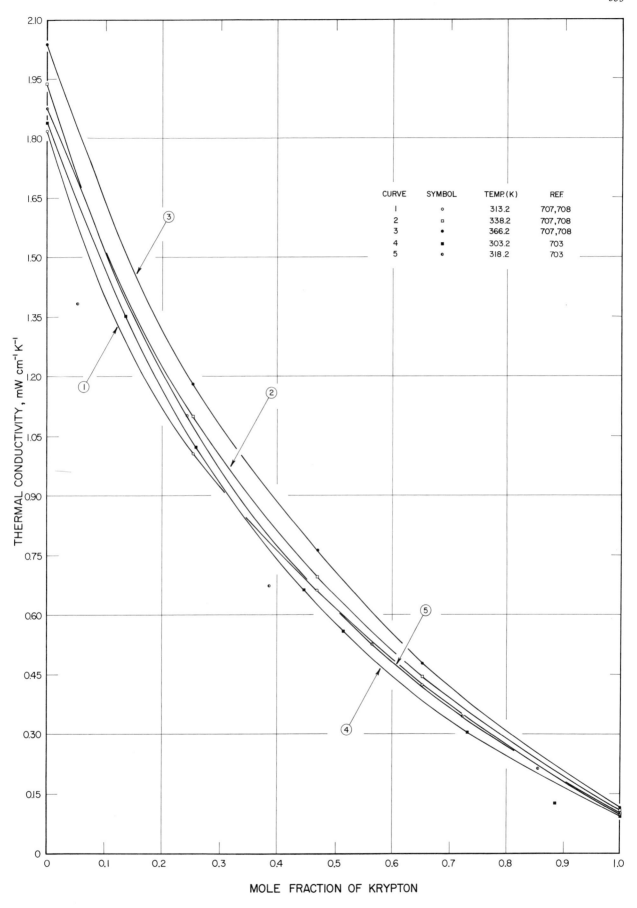

FIGURE 90. THERMAL CONDUCTIVITY OF KRYPTON-HYDROGEN SYSTEM

TABLE 91a. EXPERIMENTAL THERMAL CONDUCTIVITY DATA FOR KRYPTON-NITROGEN SYSTEM

Curve No.	Fig. No.	Ref. No.	Author(s)	Temp. (K)	Mole Fraction of Kr	Thermal Cond. (mW cm^{-1} K^{-1})	Remarks
1	91	706	Barua, A.K.	303.2	0.0000	0.2559	Kr: traces of Xe, N$_2$: spectroscopically pure; thick hot wire method; precision ±2%.
					0.1371	0.2212	
					0.2756	0.1916	
					0.3728	0.1764	
					0.5364	0.1476	
					0.7139	0.1234	
					0.8084	0.1139	
					0.8914	0.1021	
					1.0000	0.0962	
2	91	706	Barua, A.K.	318.2	0.0000	0.2672	Same as above.
					0.1545	0.2313	
					0.1872	0.2259	
					0.3641	0.1859	
					0.6089	0.1618	
					0.6451	0.1395	
					0.8882	0.1066	
					1.0000	0.0980	

TABLE 91b. THERMAL CONDUCTIVITY (mW cm^{-1}K^{-1}) OF KRYPTON-NITROGEN SYSTEM AS A FUNCTION OF COMPOSITION AT THE TEMPERATURE OF MEASUREMENT AS DERIVED BY GRAPHICAL SMOOTHING

Mole Fraction of Kr	303.2 K (Ref. 706)	318.2 K (Ref. 706)
0.00	0.256	0.267
0.05	0.244	0.256
0.10	0.232	0.244
0.15	0.220	0.233
0.20	0.209	0.221
0.25	0.199	0.199
0.30	0.188	0.200
0.35	0.179	0.190
0.40	0.170	0.180
0.45	0.161	0.171
0.50	0.153	0.163
0.55	0.146	0.154
0.60	0.138	0.146
0.65	0.132	0.132
0.70	0.125	0.132
0.75	0.119	0.119
0.80	0.114	0.119
0.85	0.109	0.109
0.90	0.104	0.108
0.95	0.100	0.103
1.00	0.0962	0.0980

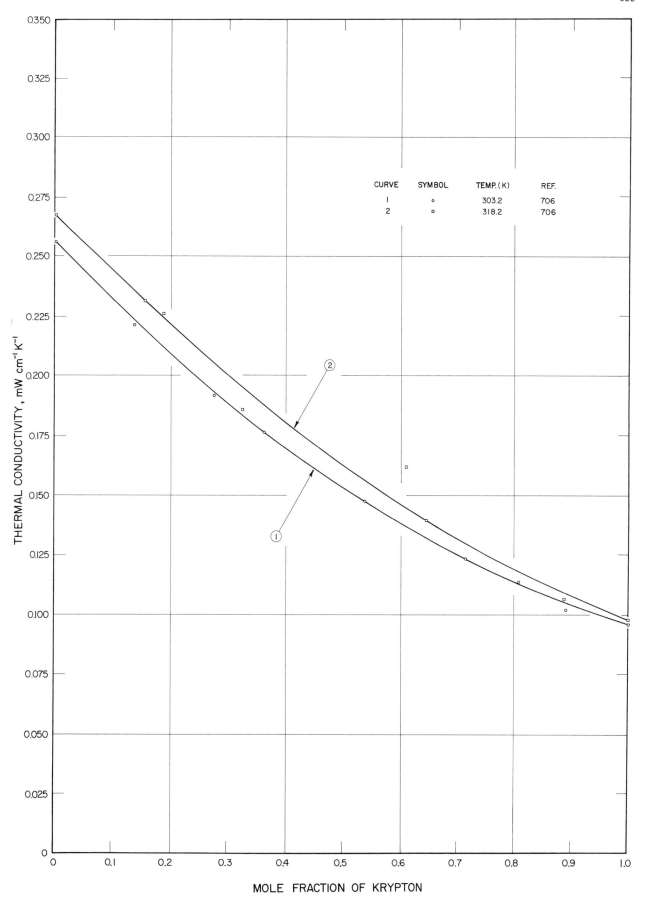

FIGURE 91. THERMAL CONDUCTIVITY OF KRYPTON—NITROGEN SYSTEM

TABLE 92a. EXPERIMENTAL THERMAL CONDUCTIVITY DATA FOR KRYPTON-OXYGEN SYSTEM

Curve No.	Fig. No.	Ref. No.	Author(s)	Temp. (K)	Mole Fraction of Kr	Thermal Cond. (mW cm^{-1} K^{-1})	Remarks
1	92	705	Srivastava, B.N. and Barua, A.K.	303.2	0.0000 0.1021 0.2631 0.3631 0.4978 0.6215 0.7410 1.0000	0.2697 0.2500 0.2031 0.1841 0.1642 0.1363 0.1226 0.0940	Kr: traces of impurities, O$_2$: spectroscopically pure; thick hot wire method; precision ±2%.
2	92	705	Srivastava, B.N. and Barua, A.K.	318.2	0.0000 0.1545 0.4751 0.6059 0.7384 0.8914 1.0000	0.2800 0.2395 0.1438 0.1433 0.1308 0.1112 0.0973	

TABLE 92b. THERMAL CONDUCTIVITY (mW cm^{-1}K^{-1}) OF KRYPTON-OXYGEN SYSTEM AS A FUNCTION OF COMPOSITION AT THE TEMPERATURE OF MEASUREMENT AS DERIVED BY GRAPHICAL SMOOTHING

Mole Fraction of Kr	303.2 K (Ref. 705)	318.2 K (Ref. 705)
0.00	0.270	0.280
0.05	0.256	0.267
0.10	0.243	0.253
0.15	0.230	0.241
0.20	0.218	0.218
0.25	0.206	0.216
0.30	0.195	0.204
0.35	0.184	0.193
0.40	0.174	0.183
0.45	0.165	0.171
0.50	0.156	0.166
0.55	0.147	0.155
0.60	0.140	0.147
0.65	0.132	0.139
0.70	0.126	0.132
0.75	0.120	0.125
0.80	0.114	0.119
0.85	0.109	0.113
0.90	0.104	0.108
0.95	0.0986	0.102
1.00	0.0940	0.0973

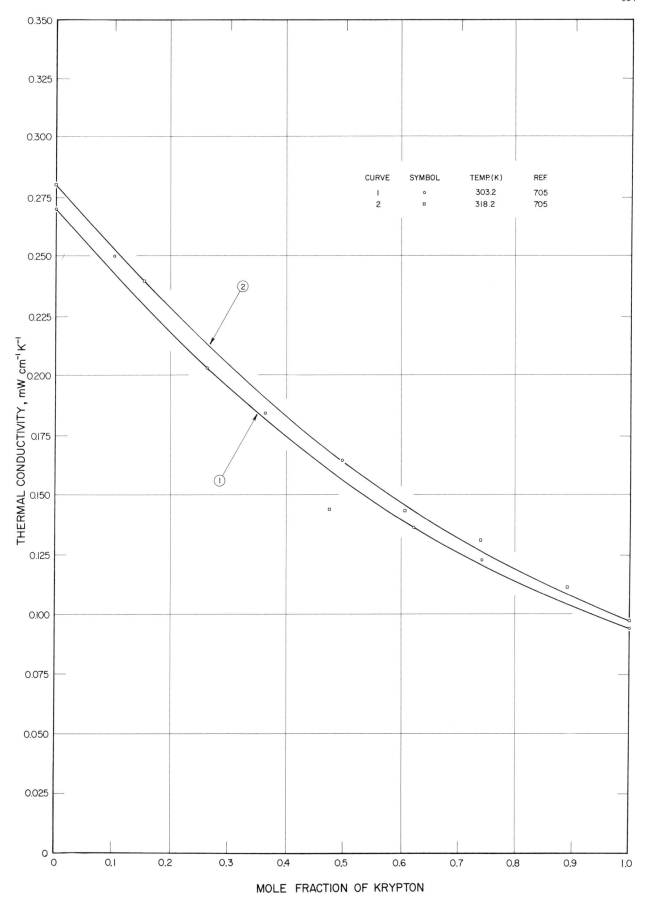

FIGURE 92. THERMAL CONDUCTIVITY OF KRYPTON–OXYGEN SYSTEM

TABLE 93a. EXPERIMENTAL THERMAL CONDUCTIVITY DATA FOR NEON-CARBON DIOXIDE SYSTEM

Curve No.	Fig. No.	Ref. No.	Author(s)	Temp. (K)	Mole Fraction of CO_2	Thermal Cond. (mW cm^{-1} K^{-1})	Remarks
1	93	83	Davidson, J.M. and Music, J.F.	273.2	0.00 0.31 0.40 0.53 0.74 1.00	0.4530 0.3061 0.3262 0.2462 0.1871 0.1419	Unsteady state method, the rate of cooling of a solid inner cylinder of copper was determined; accuracy of these relative measurements ±5%.

TABLE 93b. THERMAL CONDUCTIVITY (mW cm^{-1} K^{-1}) OF NEON-CARBON DIOXIDE SYSTEM AS A FUNCTION OF COMPOSITION AT THE TEMPERATURE OF MEASUREMENT AS DERIVED BY GRAPHICAL SMOOTHING

Mole Fraction of CO_2	273.2 K (Ref. 83)	273.2 K (Ref. 83)
0.00	0.453	0.453
0.05	0.424	0.424
0.10	0.396	0.386
0.15	0.371	0.371
0.20	0.346	0.349
0.25	0.323	0.329
0.30	0.307	0.310
0.35	0.314	0.293
0.40	0.326	0.276
0.45	0.299	0.260
0.50	0.262	0.245
0.55	0.238	0.234
0.60	0.221	0.218
0.65	0.207	0.206
0.70	0.195	0.195
0.75	0.185	0.185
0.80	0.176	0.176
0.85	0.167	0.167
0.90	0.158	0.158
0.95	0.150	0.150
1.00	0.142	0.142

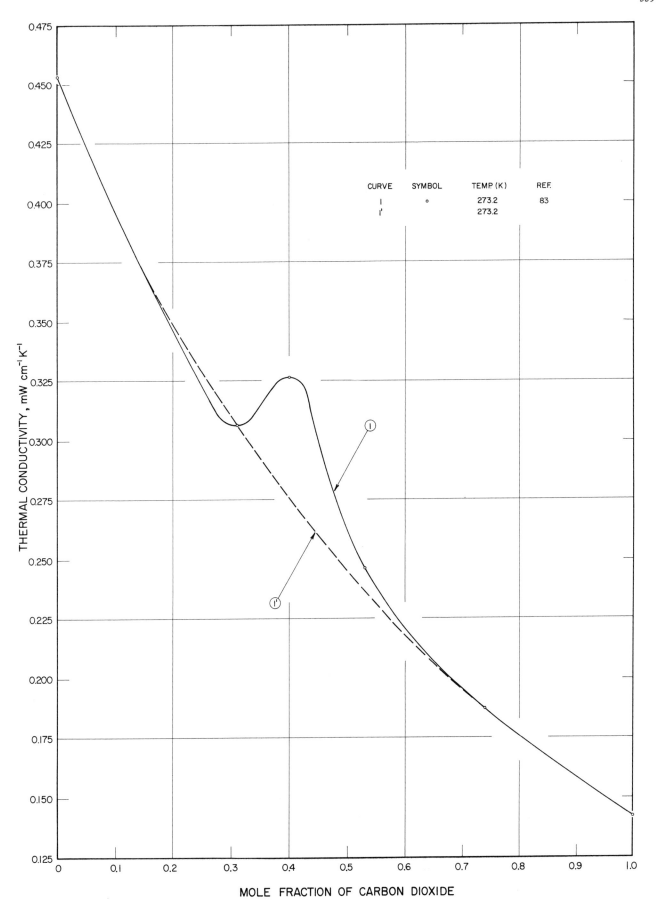

FIGURE 93. THERMAL CONDUCTIVITY OF NEON-CARBON DIOXIDE SYSTEM

TABLE 94a. EXPERIMENTAL THERMAL CONDUCTIVITY DATA FOR NEON-DEUTERIUM SYSTEM

Curve No.	Fig. No.	Ref. No.	Author(s)	Temp. (K)	Mole Fraction of Ne	Thermal Cond. (mW cm^{-1} K^{-1})	Remarks
1	94	702	Gandhi, J.M. and Saxena, S.C.	303.2	0.0000 0.0949 0.2558 0.4547 0.6534 1.0000	1.344 1.199 1.040 0.8596 0.6996 0.4940	Ne: spectroscopically pure, D_2: 98.6% pure, 0.8% H_2 and 0.6 H_2O; thick hot wire method; precision ±1%, accuracy ±1 to ±2%.
2	94	702	Gandhi, J.M. and Saxena, S.C.	323.2	0.0000 0.0949 0.2558 0.4547 0.6534 1.0000	1.398 1.243 1.081 0.8893 0.7281 0.5150	Same as above.
3	94	702	Gandhi, J.M. and Saxena, S.C.	343.2	0.0000 0.0949 0.2558 0.4547 0.6534 1.0000	1.457 1.286 1.123 0.9177 0.7566 0.5317	Same as above.
4	94	702	Gandhi, J.M. and Saxena, S.C.	363.2	0.0000 0.0949 0.2558 0.4547 0.6534 1.0000	1.511 1.330 1.164 0.9458 0.7817 0.5527	Same as above.

TABLE 94b. THERMAL CONDUCTIVITY (mW cm^{-1} K^{-1}) OF NEON-DEUTERIUM SYSTEM AS A FUNCTION OF COMPOSITION AT THE TEMPERATURE OF MEASUREMENT AS DERIVED BY GRAPHICAL SMOOTHING

Mole Fraction of Ne	303.2 K (Ref. 702)	323.2 K (Ref. 702)	343.2 K (Ref. 702)	363.2 K (Ref. 702)
0.00	1.34	1.40	1.46	1.51
0.05	1.26	1.32	1.37	1.42
0.10	1.19	1.25	1.29	1.34
0.15	1.13	1.18	1.23	1.27
0.20	1.08	1.12	1.17	1.21
0.25	1.04	1.07	1.11	1.15
0.30	0.988	1.02	1.06	1.10
0.35	0.941	0.976	1.01	1.04
0.40	0.896	0.931	0.962	0.996
0.45	0.854	0.888	0.918	0.950
0.50	0.814	0.847	0.876	0.907
0.55	0.775	0.806	0.836	0.864
0.60	0.738	0.767	0.797	0.823
0.65	0.703	0.731	0.759	0.785
0.70	0.670	0.696	0.724	0.748
0.75	0.638	0.663	0.689	0.712
0.80	0.606	0.631	0.656	0.677
0.85	0.577	0.600	0.623	0.644
0.90	0.547	0.571	0.593	0.611
0.95	0.520	0.542	0.562	0.581
1.00	0.494	0.515	0.532	0.553

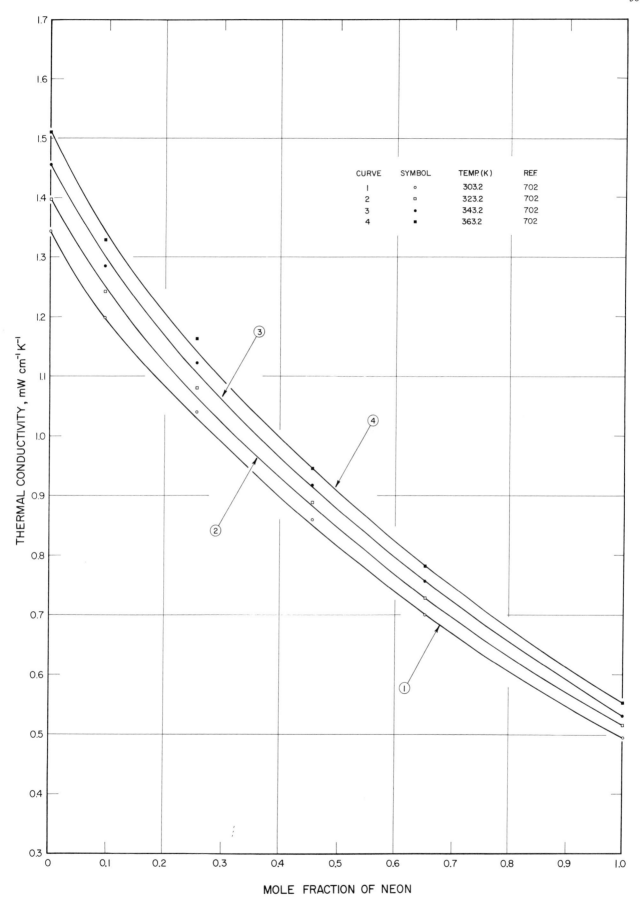

FIGURE 94. THERMAL CONDUCTIVITY OF NEON–DEUTERIUM SYSTEM

TABLE 95a. EXPERIMENTAL THERMAL CONDUCTIVITY DATA FOR NEON-HYDROGEN SYSTEM

Curve No.	Fig. No.	Ref. No.	Author(s)	Temp. (K)	Mole Fraction of Ne	Thermal Cond. (mW cm^{-1} K^{-1})	Remarks
1	95	703	Barua, A.K.	303.2	0.0000 0.1482 0.2893 0.4893 0.6131 0.7283 0.8561 1.0000	1.807 1.535 1.324 0.9923 0.8378 0.7134 0.5661 0.4878	Ne and H$_2$: spectroscopically pure; thick hot wire method; precision ±2%.
2	95	703	Barua, A.K.	318.2	0.0000 0.1462 0.2983 0.4654 0.6041 0.7486 0.8671 1.0000	1.877 1.468 1.320 1.070 0.8859 0.7247 0.6117 0.5024	Same as above.
3	95	707, 709	Tondon, P.K. and Saxena, S.C.	313.2	0.000 0.150 0.405 0.663 1.000	1.822 1.532 1.156 0.8541 0.4961	Ne: spectroscopically pure, H$_2$: 99.95% pure; thick hot wire method; precision ±1%, accuracy ±1 to ±2%.
4	95	707, 709	Tondon, P.K. and Saxena, S.C.	338.2	0.000 0.150 0.405 0.663 1.000	1.939 1.625 1.244 0.9002 0.5342	Same as above.
5	95	707, 709	Tondon, P.K. and Saxena, S.C.	366.2	0.000 0.150 0.405 0.663 1.000	2.037 1.717 1.311 0.9588 0.5627	Same as above.
6	95	707, 710	Tondon, P.K. and Saxena, S.C.	368.2	0.000 0.272 0.293 1.000	2.114 1.499 1.422 0.5711	Same as above.
7	95	707, 710	Tondon, P.K. and Saxena, S.C.	408.2	0.000 0.272 0.293 1.000	2.269 1.708 1.519 0.6029	Same as above.
8	95	707, 710	Tondon, P.K. and Saxena, S.C.	448.2	0.000 0.272 0.293 1.000	2.370 1.829 1.748 0.6414	Same as above.

TABLE 95b. THERMAL CONDUCTIVITY (mW cm^{-1}K^{-1}) OF NEON-HYDROGEN SYSTEM AS A FUNCTION OF COMPOSITION AT THE TEMPERATURE OF MEASUREMENT AS DERIVED BY GRAPHICAL SMOOTHING

Mole Fraction of Ne	303.2 K (Ref. 703)	313.2 K (Ref. 707)	318.2 K (Ref. 703)	338.2 K (Ref. 707)
0.00	1.81	1.82	1.88	1.94
0.05	1.71	1.72	1.78	1.83
0.10	1.62	1.62	1.68	1.72
0.15	1.53	1.53	1.58	1.63
0.20	1.44	1.45	1.49	1.54
0.25	1.36	1.37	1.40	1.46
0.30	1.28	1.30	1.32	1.39
0.35	1.20	1.23	1.24	1.32
0.40	1.12	1.16	1.16	1.25
0.45	1.05	1.10	1.09	1.18
0.50	0.976	1.04	1.02	1.11
0.55	0.913	0.982	0.955	1.05
0.60	0.852	0.924	0.890	0.980
0.65	0.796	0.868	0.832	0.918
0.70	0.742	0.813	0.776	0.858
0.75	0.694	0.758	0.723	0.801
0.80	0.648	0.704	0.674	0.744
0.85	0.605	0.650	0.628	0.690
0.90	0.564	0.598	0.584	0.637
0.95	0.525	0.546	0.543	0.585
1.00	0.488	0.496	0.502	0.534

Mole Fraction of Ne	366.2 K (Ref. 707)	368.2 K (Ref. 707)	408.2 K (Ref. 707)	448.2 K (Ref. 707)
0.00	2.04	2.11	2.27	2.37
0.05	1.91	1.96	2.10	2.24
0.10	1.81	1.82	1.97	2.13
0.15	1.72	1.71	1.86	2.03
0.20	1.63	1.61	1.76	1.93
0.25	1.55	1.51	1.66	1.84
0.30	1.47	1.43	1.57	1.75
0.35	1.39	1.35	1.48	1.66
0.40	1.32	1.28	1.40	1.58
0.45	1.25	1.21	1.33	1.49
0.50	1.18	1.14	1.25	1.41
0.55	1.11	1.08	1.18	1.32
0.60	1.04	1.02	1.11	1.24
0.65	0.976	0.959	1.04	1.16
0.70	0.914	0.900	0.975	1.08
0.75	0.852	0.842	0.910	1.01
0.80	0.792	0.786	0.846	0.931
0.85	0.734	0.732	0.784	0.858
0.90	0.676	0.678	0.722	0.786
0.95	0.619	0.624	0.661	0.713
1.00	0.563	0.571	0.603	0.641

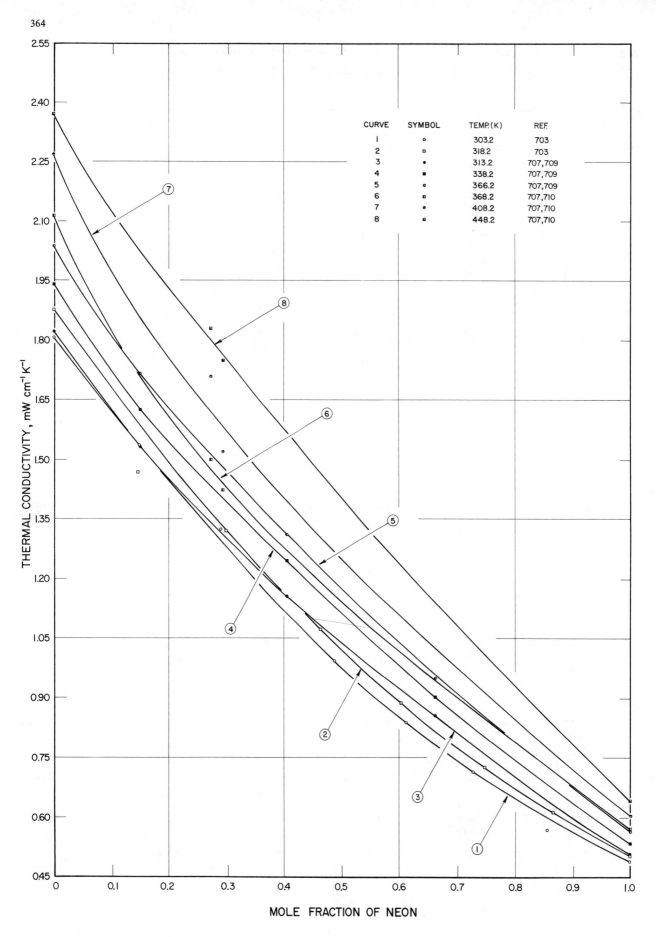

FIGURE 95. THERMAL CONDUCTIVITY OF NEON–HYDROGEN SYSTEM

TABLE 96a. EXPERIMENTAL THERMAL CONDUCTIVITY DATA FOR NEON-NITROGEN SYSTEM

Curve No.	Fig. No.	Ref. No.	Author(s)	Temp. (K)	Mole Fraction of N_2	Thermal Cond. (mW cm^{-1} K^{-1})	Remarks
1, 1'	96	706	Barua, A. K.	303.2	0.0000 0.0974 0.2246 0.3046 0.5504 0.6723 0.8714 1.0000	0.4865 0.4463 0.4068 0.4151 0.3323 0.3053 0.2731 0.2554	Ne and N_2: spectroscopically pure; thick hot wire method; precision ±2%.
2	96	706	Barua, A. K.	318.2	0.0000 0.1000 0.2388 0.3496 0.5063 0.6962 0.8520 1.0000	0.5016 0.4555 0.4136 0.3921 0.3509 0.3158 0.2944 0.2673	Same as above.
3	96	707, 709	Tondon, P. K. and Saxena, S. C.	313.2	0.000 0.203 0.511 0.805 1.000	0.4961 0.4354 0.3529 0.3031 0.2684	Ne: spectroscopically pure, N_2: 99.95% pure; thick hot wire method; precision ±1%, accuracy ±1 to ±2%.
4	96	707, 709	Tondon, P. K. and Saxena, S. C.	338.2	0.000 0.203 0.511 0.805 1.000	0.5342 0.4605 0.3739 0.3203 0.2901	Same as above.
5	96	707, 709	Tondon, P. K. and Saxena, S. C.	366.2	0.000 0.203 0.511 0.805 1.000	0.5627 0.4857 0.4024 0.3421 0.3125	Same as above.
6	96	707, 710	Tondon, P. K. and Saxena, S. C.	368.2	0.000 0.256 0.735 1.000	0.5711 0.4212 0.3303 0.3052	Same as above.
7	96	707, 710	Tondon, P. K. and Saxena, S. C.	408.2	0.000 0.256 0.735 1.000	0.6029 0.4861 0.3668 0.3329	Same as above.
8	96	707, 710	Tondon, P. K. and Saxena, S. C.	448.2	0.000 0.256 0.735 1.000	0.6414 0.5233 0.3915 0.3622	Same as above.

TABLE 96b. THERMAL CONDUCTIVITY (mW cm^{-1}K^{-1}) OF NEON-NITROGEN SYSTEM AS A FUNCTION OF COMPOSITION AT THE TEMPERATURE OF MEASUREMENT AS DERIVED BY GRAPHICAL SMOOTHING

Mole Fraction of N$_2$	303.2 K (Ref. 706)	303.2 K (Ref. 706)	313.2 K (Ref. 707)	318.2 K (Ref. 706)	338.2 K (Ref. 707)
	Curve 1	Curve 1'			
0.00	0.487	0.487	0.496	0.502	0.534
0.05	0.469	0.469	0.481	0.478	0.514
0.10	0.445	0.445	0.466	0.456	0.496
0.15	0.425	0.427	0.451	0.438	0.478
0.20	0.403	0.411	0.436	0.422	0.461
0.25	0.405	0.397	0.422	0.408	0.445
0.30	0.415	0.383	0.408	0.395	0.430
0.35	0.407	0.370	0.394	0.383	0.415
0.40	0.387	0.358	0.381	0.372	0.401
0.45	0.367	0.348	0.368	0.363	0.388
0.50	0.349	0.338	0.357	0.352	0.377
0.55	0.344	0.328	0.346	0.343	0.366
0.60	0.321	0.318	0.336	0.333	0.356
0.65	0.310	0.310	0.327	0.325	0.347
0.70	0.300	0.300	0.318	0.317	0.338
0.75	0.292	0.292	0.310	0.308	0.329
0.80	0.284	0.284	0.302	0.299	0.321
0.85	0.276	0.276	0.294	0.292	0.313
0.90	0.269	0.269	0.285	0.283	0.305
0.95	0.262	0.262	0.277	0.276	0.298
1.00	0.255	0.255	0.268	0.267	0.290

Mole Fraction of N$_2$	366.2 K (Ref. 707)	368.2 K (Ref. 707)	408.2 K (Ref. 707)	448.2 K (Ref. 707)
0.00	0.563	0.571	0.603	0.641
0.05	0.542	0.519	0.574	0.615
0.10	0.522	0.486	0.549	0.591
0.15	0.504	0.461	0.527	0.568
0.20	0.487	0.440	0.507	0.546
0.25	0.471	0.423	0.488	0.525
0.30	0.457	0.408	0.471	0.507
0.35	0.443	0.395	0.454	0.489
0.40	0.430	0.383	0.438	0.472
0.45	0.41	0.373	0.424	0.457
0.50	0.405	0.364	0.412	0.443
0.55	0.394	0.356	0.400	0.430
0.60	0.383	0.348	0.390	0.418
0.65	0.372	0.341	0.381	0.407
0.70	0.362	0.335	0.372	0.397
0.75	0.352	0.329	0.365	0.390
0.80	0.343	0.324	0.358	0.383
0.85	0.335	0.319	0.351	0.377
0.90	0.327	0.314	0.345	0.372
0.95	0.320	0.310	0.339	0.367
1.00	0.313	0.305	0.333	0.362

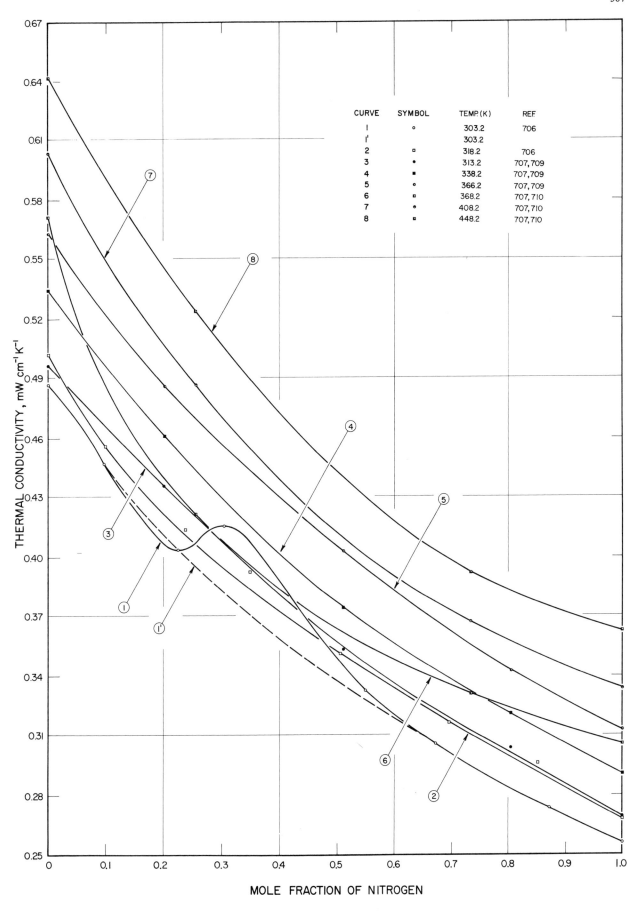

FIGURE 96. THERMAL CONDUCTIVITY OF NEON−NITROGEN SYSTEM

TABLE 97a. EXPERIMENTAL THERMAL CONDUCTIVITY DATA FOR NEON-OXYGEN SYSTEM

Curve No.	Fig. No.	Ref. No.	Author(s)	Temp. (K)	Mole Fraction of O_2	Thermal Cond. (mW cm^{-1} K^{-1})	Remarks
1	97	705	Srivastava, B.N. and Barua, A.K.	303.2	0.0000 0.2251 0.4236 0.5904 0.7634 0.8602 1.0000	0.4861 0.4271 0.3754 0.3412 0.3115 0.2970 0.2697	Ne and O_2: spectroscopically pure; thick hot wire method; precision ±2%.
2	97	705	Srivastava, B.N. and Barua, A.K.	318.2	0.0000 0.1597 0.2580 0.3485 0.5156 0.7007 0.8848 1.0000	0.5045 0.4513 0.4212 0.3896 0.3534 0.3251 0.3089 0.2800	Same as above.
3	97	707, 709	Tondon, P.K. and Saxena, S.C.	313.2	0.000 0.340 0.496 0.739 1.000	0.4961 0.4137 0.3756 0.3211 0.2811	Ne: spectroscopically pure, O_2: 99.95% pure; thick hot wire method; precision ±1%, accuracy ±1 to ±2%.
4	97	707, 709	Tondon, P.K. and Saxena, S.C.	338.2	0.000 0.340 0.496 0.739 1.000	0.5342 0.4271 0.3948 0.3412 0.2906	Same as above.
5	97	707, 709	Tondon, P.K. and Saxena, S.C.	366.2	0.000 0.340 0.496 0.739 1.000	0.5627 0.4522 0.4229 0.3626 0.3134	Same as above.
6	97	707, 710	Tondon, P.K. and Saxena, S.C.	368.2	0.000 0.229 0.492 0.744 1.000	0.5711 0.4547 0.3806 0.3680 0.3270	Same as above.
7	97	707, 710	Tondon, P.K. and Saxena, S.C.	408.2	0.000 0.229 0.492 0.744 1.000	0.6029 0.5108 0.4379 0.3965 0.3450	Same as above.
8	97	707, 710	Tondon, P.K. and Saxena, S.C.	448.2	0.000 0.229 0.492 0.744 1.000	0.6414 0.5460 0.5041 0.4116 0.3776	Same as above.

TABLE 97b. THERMAL CONDUCTIVITY (mW cm^{-1}K^{-1}) OF NEON-OXYGEN SYSTEM AS A FUNCTION OF COMPOSITION AT THE TEMPERATURE OF MEASUREMENT AS DERIVED BY GRAPHICAL SMOOTHING

Mole Fraction of O$_2$	303.2 K (Ref. 705)	313.2 K (Ref. 707)	318.2 K (Ref. 705)	338.2 K (Ref. 707)
0.00	0.486	0.496	0.505	0.534
0.05	0.473	0.484	0.488	0.517
0.10	0.458	0.472	0.472	0.500
0.15	0.444	0.460	0.455	0.484
0.20	0.430	0.447	0.439	0.469
0.25	0.417	0.435	0.424	0.454
0.30	0.404	0.422	0.411	0.440
0.35	0.392	0.411	0.398	0.427
0.40	0.380	0.400	0.387	0.415
0.45	0.369	0.388	0.375	0.403
0.50	0.358	0.377	0.365	0.391
0.55	0.348	0.365	0.354	0.380
0.60	0.338	0.355	0.344	0.369
0.65	0.328	0.344	0.335	0.359
0.70	0.320	0.335	0.326	0.349
0.75	0.311	0.325	0.317	0.339
0.80	0.302	0.316	0.309	0.329
0.85	0.294	0.307	0.302	0.320
0.90	0.286	0.298	0.294	0.310
0.95	0.277	0.290	0.287	0.300
1.00	0.270	0.281	0.280	0.291

Mole Fraction of O$_2$	366.2 K (Ref. 707)	368.2 K (Ref. 707)	408.2 K (Ref. 707)	448.2 K (Ref. 707)
0.00	0.563	0.571	0.603	0.641
0.05	0.545	0.554	0.580	0.620
0.10	0.528	0.538	0.559	0.599
0.15	0.512	0.522	0.539	0.577
0.20	0.496	0.506	0.521	0.558
0.25	0.481	0.491	0.504	0.539
0.30	0.467	0.476	0.489	0.523
0.35	0.453	0.461	0.474	0.507
0.40	0.441	0.448	0.461	0.493
0.45	0.428	0.434	0.449	0.480
0.50	0.417	0.422	0.438	0.468
0.55	0.406	0.409	0.427	0.457
0.60	0.395	0.398	0.418	0.446
0.65	0.384	0.387	0.408	0.437
0.70	0.373	0.376	0.399	0.427
0.75	0.363	0.367	0.390	0.418
0.80	0.353	0.358	0.380	0.410
0.85	0.343	0.350	0.371	0.402
0.90	0.333	0.343	0.362	0.394
0.95	0.323	0.335	0.353	0.386
1.00	0.313	0.327	0.345	0.378

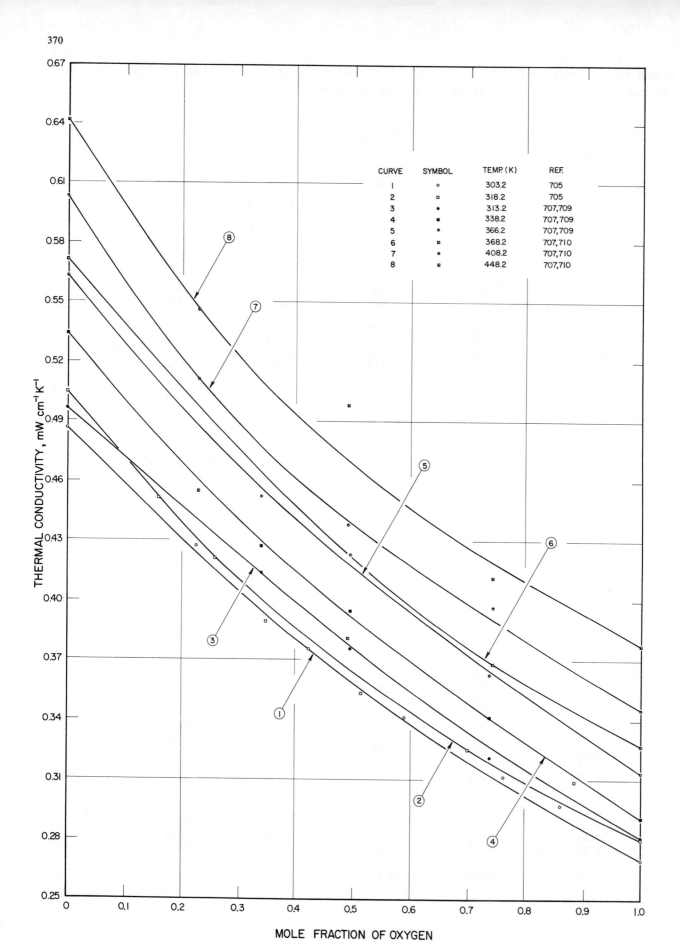

FIGURE 97. THERMAL CONDUCTIVITY OF NEON−OXYGEN SYSTEM

TABLE 98a. EXPERIMENTAL THERMAL CONDUCTIVITY DATA FOR XENON-DEUTERIUM SYSTEM

Curve No.	Fig. No.	Ref. No.	Author(s)	Temp. (K)	Mole Fraction of Xe	Thermal Cond. (mW cm^{-1} K^{-1})	Remarks
1	98	711, 707	Mathur, S., Tondon, P.K., and Saxena, S.C.	311.2	0.00 0.20 0.40 0.60 0.80 1.00	1.327 0.8478 0.5405 0.3349 0.1771 0.0615	Xe: 99-100% pure, balance Kr, D$_2$: 98.6% pure, 0.8% H$_2$ and 0.6% H$_2$O; thick hot wire method; precision ±1%, accuracy ±1 to ±2%.
2	98	711, 707	Mathur, S., Tondon, P.K., and Saxena, S.C.	366.2	0.00 0.20 0.40 0.60 0.80 1.00	1.478 0.9483 0.6347 0.3860 0.2010 0.0741	Same as above.
3	98	707, 708	Tondon, P.K. and Saxena, S.C.	313.2	0.000 0.255 0.496 0.759 1.000	1.350 0.7327 0.4271 0.2072 0.0593	Xe: 99-100% pure, balance Kr, D$_2$: 98.6% pure, 0.8% H$_2$ and 0.6% H$_2$O; thick hot wire method; precision ±1%, accuracy ±1 to ±2%.
4	98	707, 708	Tondon, P.K. and Saxena, S.C.	338.2	0.000 0.255 0.496 0.759 1.000	1.426 0.7955 0.4522 0.2206 0.0620	Same as above.
5	98	707, 708	Tondon, P.K. and Saxena, S.C.	366.2	0.000 0.255 0.496 0.759 1.000	1.495 0.8374 0.4940 0.2340 0.0706	Same as above.

TABLE 98b. THERMAL CONDUCTIVITY (mW cm^{-1}K^{-1}) OF XENON-DEUTERIUM SYSTEM AS A FUNCTION OF COMPOSITION AT THE TEMPERATURE OF MEASUREMENT AS DERIVED BY GRAPHICAL SMOOTHING

Mole Fraction of Xe	311.2 K (Ref. 711)	313.2 K (Ref. 707)	338.2 K (Ref. 707)	366.2 K (Ref. 711)	366.2 K (Ref. 707)
0.00	1.33	1.35	1.43	1.48	1.50
0.05	1.18	1.19	1.27	1.33	1.33
0.10	1.05	1.05	1.13	1.19	1.18
0.15	0.943	0.934	1.01	1.06	1.06
0.20	0.848	0.833	0.902	0.948	0.945
0.25	0.759	0.742	0.804	0.859	0.847
0.30	0.679	0.664	0.718	0.780	0.760
0.35	0.606	0.597	0.640	0.705	0.682
0.40	0.541	0.534	0.570	0.635	0.612
0.45	0.482	0.476	0.506	0.567	0.548
0.50	0.429	0.425	0.449	0.502	0.490
0.55	0.380	0.376	0.397	0.441	0.434
0.60	0.334	0.332	0.351	0.386	0.382
0.65	0.291	0.290	0.307	0.333	0.332
0.70	0.251	0.250	0.266	0.286	0.286
0.75	0.212	0.214	0.228	0.241	0.242
0.80	0.177	0.180	0.192	0.201	0.204
0.85	0.145	0.148	0.158	0.165	0.167
0.90	0.116	0.117	0.124	0.132	0.134
0.95	0.088	0.088	0.092	0.102	0.101
1.00	0.0615	0.0593	0.0620	0.0741	0.0706

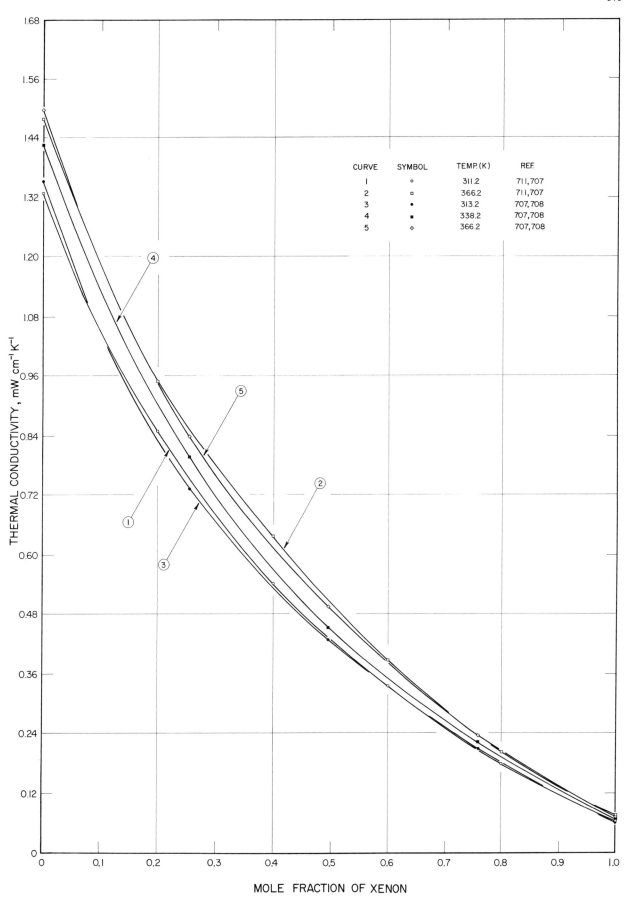

FIGURE 98. THERMAL CONDUCTIVITY OF XENON-DEUTERIUM SYSTEM

TABLE 99a. EXPERIMENTAL THERMAL CONDUCTIVITY DATA FOR XENON-HYDROGEN SYSTEM

Curve No.	Fig. No.	Ref. No.	Author(s)	Temp. (K)	Mole Fraction of Xe	Thermal Cond. (mW cm^{-1} K^{-1})	Remarks
1	99	703	Barua, A.K.	303.2	0.0000 0.1431 0.2568 0.4379 0.5462 0.7431 0.8624 1.0000	1.807 1.282 0.9119 0.6029 0.4668 0.2598 0.1413 0.0528	Xe: traces of impurity, H$_2$: spectroscopically pure; thick hot wire method; precision ±2%.
2	99	703	Barua, A.K.	318.2	0.0000 0.1286 0.2261 0.4039 0.5762 0.7231 0.8754 1.0000	1.874 1.414 1.080 0.7159 0.4806 0.2908 0.1375 0.0553	Same as above.
3	99	707, 708	Tondon, P.K. and Saxena, S.C.	313.2	0.000 0.160 0.434 0.608 1.000	1.939 1.306 0.6908 0.4438 0.0620	Xe: 99-100% pure, impurity Kr, H$_2$: 99.95% pure; thick hot wire method; precision ±1%, accuracy ±1 to ±2%.
4	99	707, 708	Tondon, P.K. and Saxena, S.C.	338.2	0.000 0.160 0.434 0.608 1.000	1.822 1.206 0.6448 0.3915 0.0593	Same as above.
5	99	707, 708	Tondon, P.K. and Saxena, S.C.	366.2	0.000 0.160 0.434 0.608 1.000	2.037 1.407 0.7494 0.4605 0.0706	Same as above.

TABLE 99b. THERMAL CONDUCTIVITY (mW cm^{-1}K^{-1}) OF XENON-HYDROGEN SYSTEM AS A FUNCTION OF COMPOSITION AT THE TEMPERATURE OF MEASUREMENT AS DERIVED BY GRAPHICAL SMOOTHING

Mole Fraction of Xe	303.2 K (Ref. 703)	313.2 K (Ref. 707)	318.2 K (Ref. 703)	338.2 K (Ref. 707)	366.2 K (Ref. 707)
0.00	1.81	1.94	1.87	1.82	2.04
0.05	1.62	1.73	1.69	1.61	1.84
0.10	1.44	1.51	1.51	1.40	1.63
0.15	1.26	1.34	1.34	1.23	1.44
0.20	1.09	1.20	1.17	1.09	1.27
0.25	0.939	1.07	1.01	0.970	1.12
0.30	0.822	0.958	0.898	0.863	0.994
0.35	0.726	0.849	0.797	0.766	0.886
0.40	0.642	0.752	0.708	0.680	0.789
0.45	0.568	0.665	0.624	0.601	0.702
0.50	0.499	0.588	0.550	0.534	0.624
0.55	0.437	0.516	0.484	0.474	0.554
0.60	0.380	0.452	0.422	0.418	0.490
0.65	0.328	0.392	0.365	0.366	0.429
0.70	0.281	0.340	0.312	0.316	0.370
0.75	0.238	0.289	0.263	0.270	0.316
0.80	0.198	0.241	0.217	0.226	0.266
0.85	0.160	0.195	0.174	0.183	0.216
0.90	0.124	0.150	0.132	0.142	0.167
0.95	0.0880	0.106	0.0920	0.100	0.118
1.00	0.0528	0.0620	0.0553	0.0953	0.0706

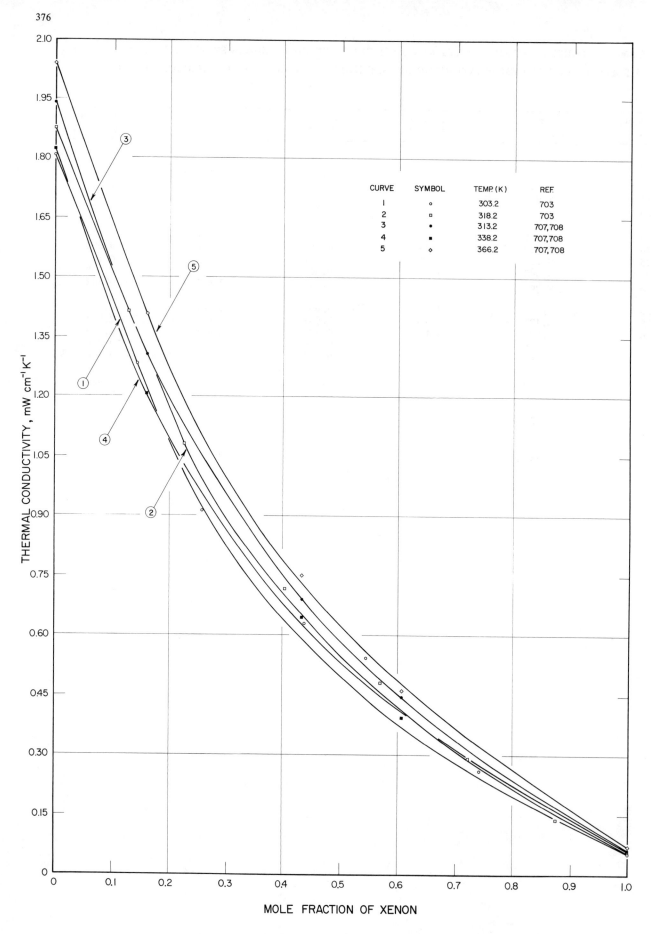

FIGURE 99. THERMAL CONDUCTIVITY OF XENON–HYDROGEN SYSTEM

TABLE 100a. EXPERIMENTAL THERMAL CONDUCTIVITY DATA FOR XENON-NITROGEN SYSTEM

Curve No.	Fig. No.	Ref. No.	Author(s)	Temp. (K)	Mole Fraction of Xe	Thermal Cond. (mW cm^{-1} K^{-1})	Remarks
1	100	706	Barua, A.K.	303.2	0.0000	0.2558	Xe: traces of impurity, N$_2$: spectroscopically pure, thick hot wire method; precision ±2%.
					0.1313	0.2091	
					0.2683	0.1677	
					0.4274	0.1303	
					0.5889	0.1071	
					0.7265	0.0842	
					0.8821	0.0652	
					1.0000	0.0525	
2	100	706	Barua, A.K.	318.2	0.0000	0.2657	Same as above.
					0.1424	0.2134	
					0.2752	0.1776	
					0.4475	0.1382	
					0.5863	0.1141	
					0.7021	0.0945	
					0.8345	0.0774	
					1.0000	0.0554	

TABLE 100b. THERMAL CONDUCTIVITY (mW cm^{-1}K^{-1}) OF XENON-NITROGEN SYSTEM AS A FUNCTION OF COMPOSITION AT THE TEMPERATURE OF MEASUREMENT AS DERIVED BY GRAPHICAL SMOOTHING

Mole Fraction of Xe	303.2 K (Ref. 706)	318.2 K (Ref. 706)
0.00	0.256	0.266
0.05	0.237	0.246
0.10	0.220	0.227
0.15	0.203	0.211
0.20	0.188	0.197
0.25	0.173	0.184
0.30	0.160	0.172
0.35	0.148	0.160
0.40	0.137	0.148
0.45	0.128	0.138
0.50	0.119	0.128
0.55	0.110	0.119
0.60	0.103	0.110
0.65	0.0954	0.103
0.70	0.0882	0.0950
0.75	0.0816	0.0880
0.80	0.0752	0.0812
0.85	0.0692	0.0744
0.90	0.0634	0.0678
0.95	0.0574	0.0616
1.00	0.0525	0.0554

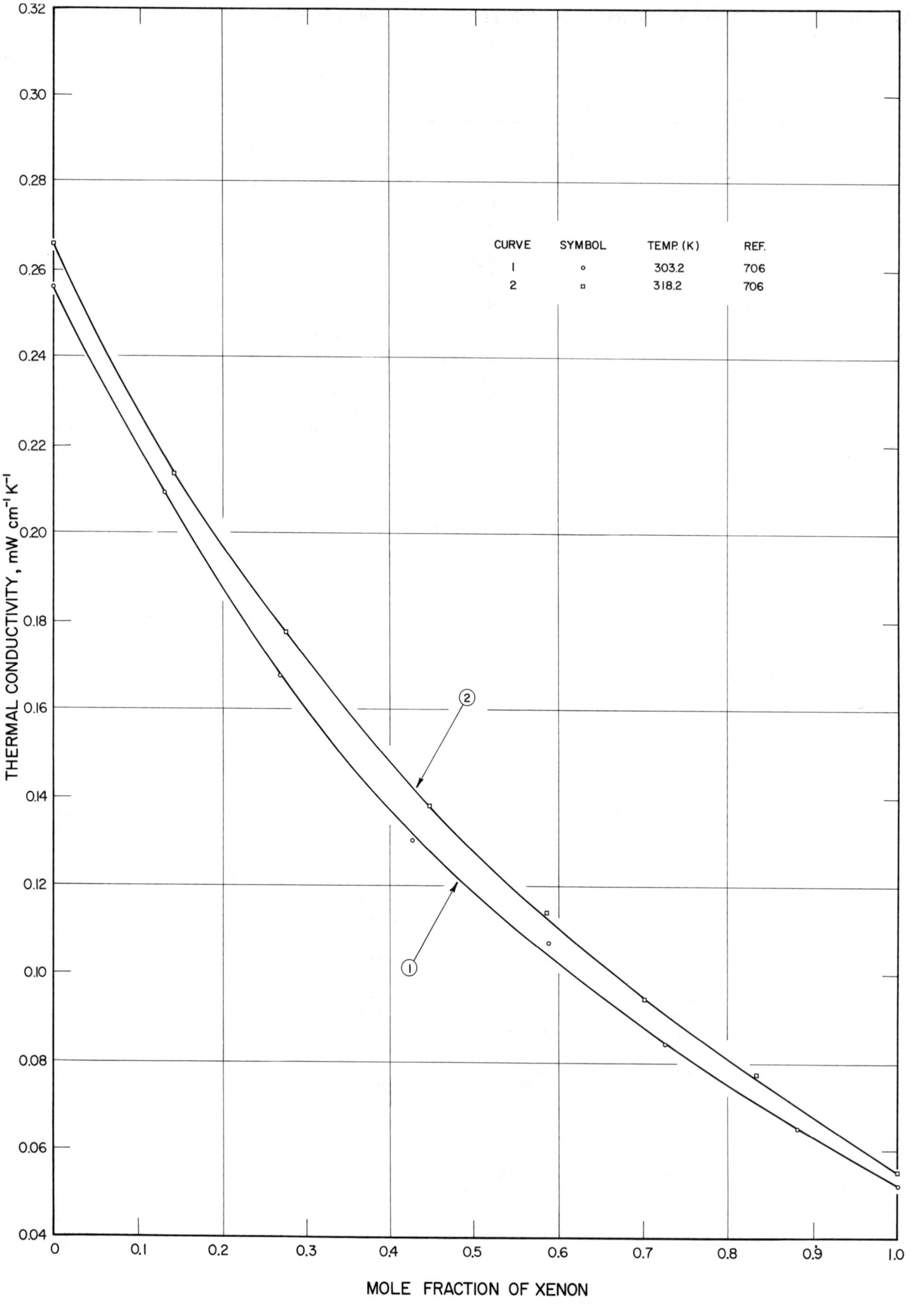

FIGURE 100. THERMAL CONDUCTIVITY OF XENON−NITROGEN SYSTEM

TABLE 101a. EXPERIMENTAL THERMAL CONDUCTIVITY DATA FOR XENON-OXYGEN SYSTEM

Curve No.	Fig. No.	Ref. No.	Author(s)	Temp. (K)	Mole Fraction of Xe	Thermal Cond. (mW cm^{-1} K^{-1})	Remarks
1	101	705	Srivastava, B.N. and Barua, A.K.	303.2	0.0000 0.1217 0.2638 0.4241 0.5879 0.7265 0.8713 1.0000	0.2697 0.2206 0.1805 0.1448 0.1165 0.0902 0.0667 0.0542	Xe: traces of impurities, O$_2$: spectroscopically pure; thick hot wire method; precision ±2%.
2	101	705	Srivastava, B.N. and Barua, A.K.	318.2	0.0000 0.1281 0.2564 0.4138 0.5772 0.7238 0.8543 1.0000	0.2800 0.2315 0.1909 0.1450 0.1246 0.0960 0.0750 0.0566	Same as above.

TABLE 101b. THERMAL CONDUCTIVITY (mW cm^{-1}K^{-1}) OF XENON-OXYGEN SYSTEM AS A FUNCTION OF COMPOSITION AT THE TEMPERATURE OF MEASUREMENT AS DERIVED BY GRAPHICAL SMOOTHING

Mole Fraction of Xe	303.2 K (Ref. 705)	318.2 K (Ref. 705)
0.00	0.270	0.280
0.05	0.248	0.260
0.10	0.228	0.241
0.15	0.211	0.224
0.20	0.197	0.208
0.25	0.184	0.193
0.30	0.171	0.180
0.35	0.160	0.168
0.40	0.149	0.156
0.45	0.139	0.146
0.50	0.129	0.135
0.55	0.120	0.126
0.60	0.111	0.117
0.65	0.102	0.108
0.70	0.0940	0.100
0.75	0.0867	0.0920
0.80	0.0796	0.0846
0.85	0.0730	0.0774
0.90	0.0666	0.0704
0.95	0.0604	0.0636
1.00	0.0542	0.0566

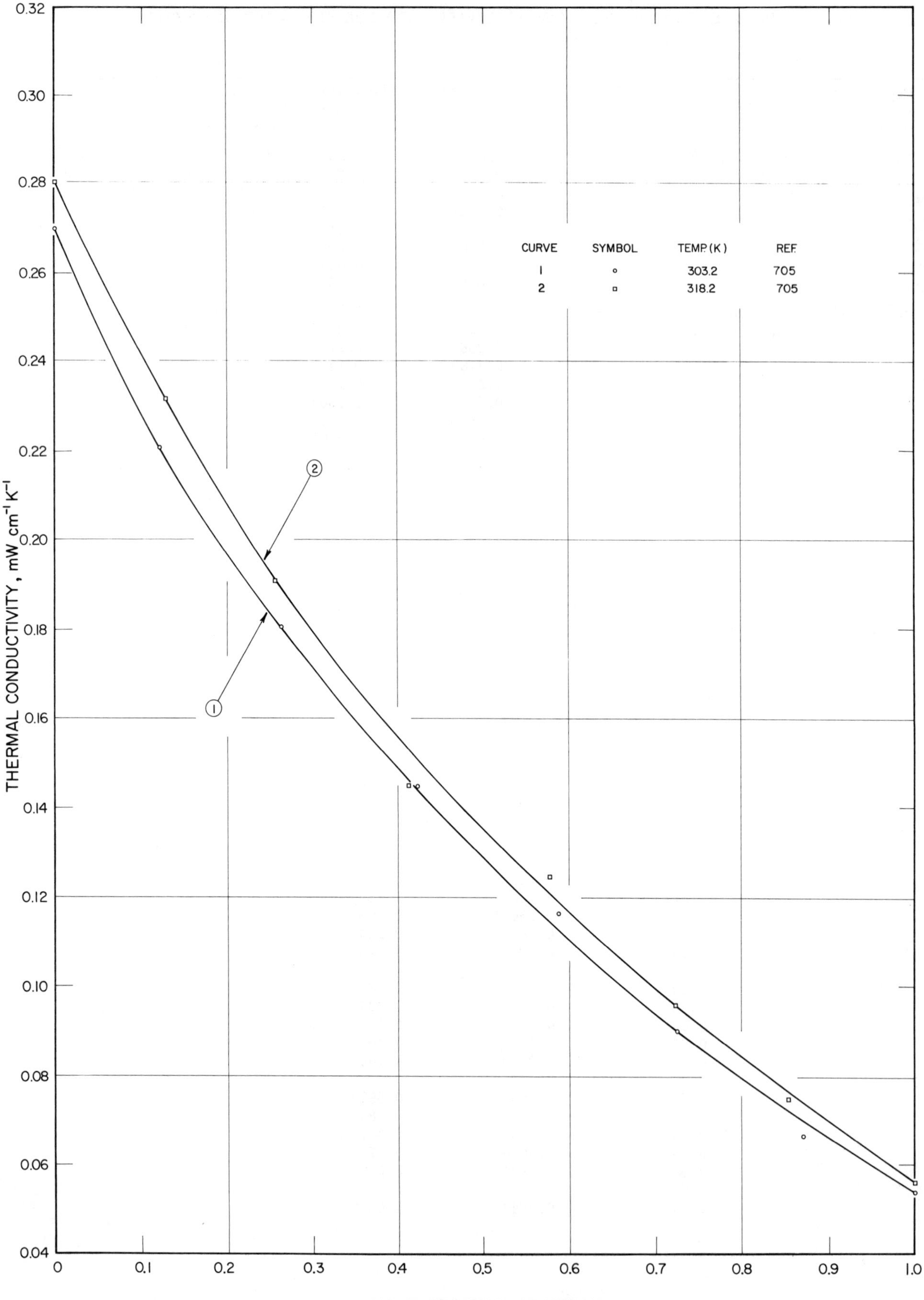

FIGURE 101. THERMAL CONDUCTIVITY OF XENON-OXYGEN SYSTEM

TABLE 102a. EXPERIMENTAL THERMAL CONDUCTIVITY DATA FOR ACETYLENE-AIR SYSTEM

Curve No.	Fig. No.	Ref. No.	Author(s)	Temp. (K)	Mole Fraction of C_2H_2	Thermal Cond. (mW cm^{-1} K^{-1})	Remarks
1	102	135	Gruss, H. and Schmick, H.	293.2	0.000 0.141 0.320 0.536 0.630 0.900 1.000	0.2511 0.2494 0.2442 0.2370 0.2323 0.2217 0.2184	Air: dry and CO_2 free; vertical compensated hot wire method.
2	102	135	Gruss, H. and Schmick, H.	338.2	0.000 0.211 0.464 0.646 0.821 1.000	0.2804 0.2830 0.2785 0.2752 0.2686 0.2629	Same as above.

TABLE 102b. THERMAL CONDUCTIVITY (mW cm^{-1}K^{-1}) OF ACETYLENE-AIR SYSTEM AS A FUNCTION OF COMPOSITION AT THE TEMPERATURE OF MEASUREMENT AS DERIVED BY GRAPHICAL SMOOTHING

Mole Fraction of C_2H_2	293.2 K (Ref. 135)	338.2 K (Ref. 135)
0.00	0.251	0.280
0.05	0.251	0.282
0.10	0.250	0.282
0.15	0.259	0.283
0.20	0.248	0.283
0.25	0.247	0.283
0.30	0.245	0.282
0.35	0.244	0.282
0.40	0.242	0.281
0.45	0.239	0.280
0.50	0.238	0.278
0.55	0.236	0.272
0.60	0.234	0.276
0.65	0.233	0.274
0.70	0.230	0.273
0.75	0.224	0.271
0.80	0.226	0.263
0.85	0.224	0.268
0.90	0.222	0.266
0.95	0.222	0.265
1.00	0.218	0.263

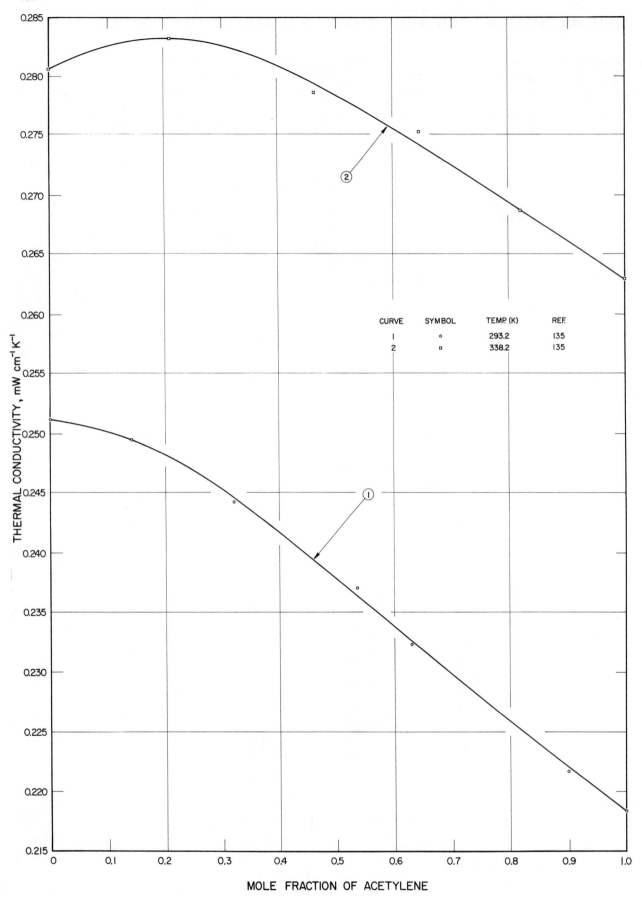

FIGURE 102. THERMAL CONDUCTIVITY OF ACETYLENE–AIR SYSTEM

TABLE 103a. EXPERIMENTAL THERMAL CONDUCTIVITY DATA FOR AIR-CARBON MONOXIDE SYSTEM

Curve No.	Fig. No.	Ref. No.	Author(s)	Temp. (K)	Mole Fraction of CO	Thermal Cond. (mW cm^{-1} K^{-1})	Remarks
1	103	135	Gruss, H. and Schmick, H.	291.2	0.000 0.108 0.321 0.562 0.978 1.000	0.2499 0.2491 0.2468 0.2439 0.2380 0.2376	Air: dry and CO_2 free; vertical compensated hot wire method.

TABLE 103b. THERMAL CONDUCTIVITY (mW cm^{-1}K^{-1}) OF AIR-CARBON MONOXIDE SYSTEM AS A FUNCTION OF COMPOSITION AT THE TEMPERATURE OF MEASUREMENT AS DERIVED BY GRAPHICAL SMOOTHING

Mole Fraction of CO	291.2 K (Ref. 135)
0.00	0.250
0.05	0.250
0.10	0.249
0.15	0.249
0.20	0.248
0.25	0.248
0.30	0.247
0.35	0.247
0.40	0.246
0.45	0.245
0.50	0.245
0.55	0.244
0.60	0.244
0.65	0.243
0.70	0.242
0.75	0.242
0.80	0.241
0.85	0.240
0.90	0.239
0.95	0.239
1.00	0.238

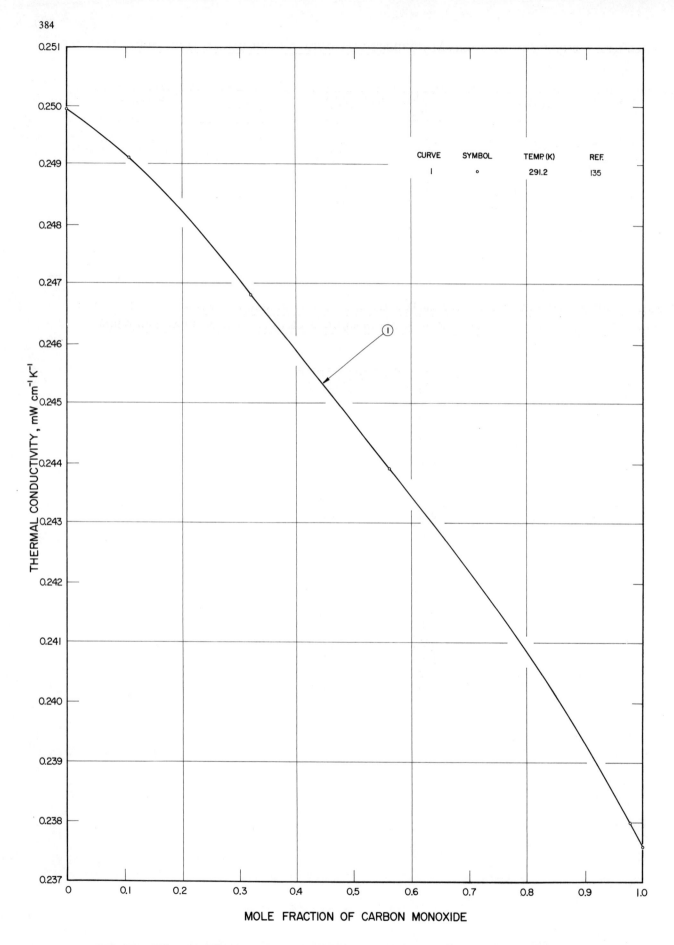

FIGURE 103. THERMAL CONDUCTIVITY OF AIR–CARBON MONOXIDE SYSTEM

TABLE 104a.   EXPERIMENTAL THERMAL CONDUCTIVITY DATA FOR AIR-METHANE SYSTEM

Curve No.	Fig. No.	Ref. No.	Author(s)	Temp. (K)	Mole Fraction of Air	Thermal Cond. (mW cm^{-1} K^{-1})	Remarks
1	104	135	Gruss, H. and Schmick, H.	295.2	0.000 0.120 0.300 0.610 0.924 1.000	0.3022 0.2961 0.2875 0.2717 0.2563 0.2525	Air: dry and CO$_2$ free, CH$_4$: impurities N$_2$, H$_2$ and air; vertical compensated hot wire method.

TABLE 104b.   THERMAL CONDUCTIVITY (mW cm^{-1} K^{-1}) OF AIR-METHANE SYSTEM AS A FUNCTION OF COMPOSITION AT THE TEMPERATURE OF MEASUREMENT AS DERIVED BY GRAPHICAL SMOOTHING

Mole Fraction of Air	295.2 K (Ref. 135)
0.00	0.302
0.05	0.300
0.10	0.297
0.15	0.295
0.20	0.292
0.25	0.290
0.30	0.288
0.35	0.285
0.40	0.282
0.45	0.280
0.50	0.277
0.55	0.275
0.60	0.272
0.65	0.270
0.70	0.267
0.75	0.265
0.80	0.262
0.85	0.260
0.90	0.258
0.95	0.255
1.00	0.253

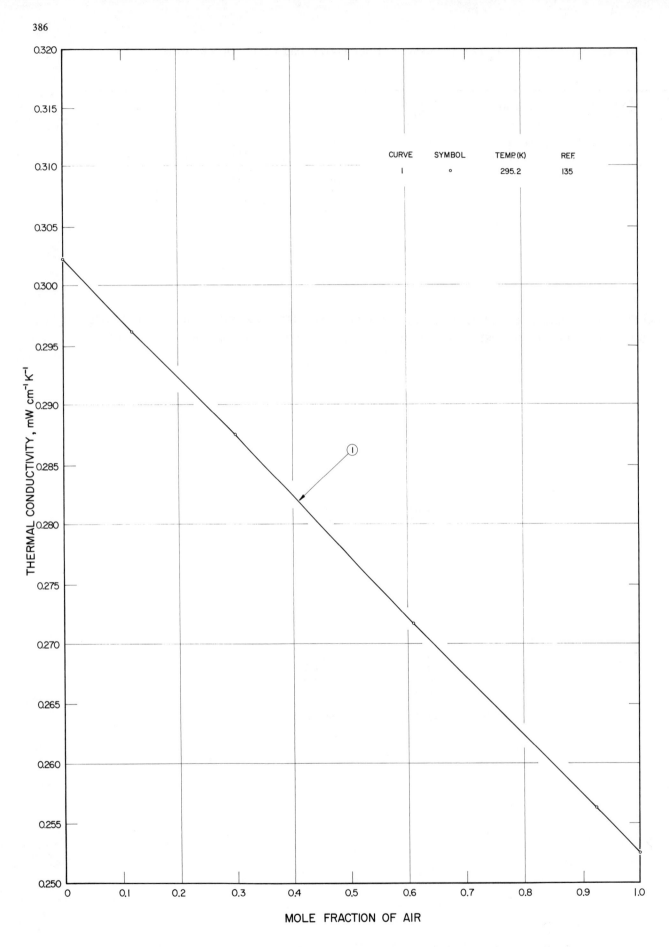

FIGURE 104. THERMAL CONDUCTIVITY OF AIR—METHANE SYSTEM

TABLE 105a. EXPERIMENTAL THERMAL CONDUCTIVITY DATA FOR BENZENE-HEXANE SYSTEM

Curve No.	Fig. No.	Ref. No.	Author(s)	Temp. (K)	Mole Fraction of C_6H_{14}	Thermal Cond. (mW cm^{-1} K^{-1})	Remarks
1	105	32	Bennett, L.A. and Vines, R.G.	360.9	0.00 0.25 0.50 0.75 1.00	0.1528 0.1624 0.1704 0.1809 0.1884	C_6H_6: shaken with concentrated sulphuric acid, washed with water, dried over calcium chloride, and distilled, C_6H_{14}: impurities less than a percent; Phillips, Oklahoma, pure grade; compensated hot wire method; estimated accuracy of these relative measurements ±1%.
2	105	32	Bennett, L.A. and Vines, R.G.	398.2	0.00 0.25 0.50 0.75 1.00	0.1901 0.2010 0.2081 0.2202 0.2290	

TABLE 105b. THERMAL CONDUCTIVITY (mW cm^{-1} K^{-1}) OF BENZENE-HEXANE SYSTEM AS A FUNCTION OF COMPOSITION AT THE TEMPERATURE OF MEASUREMENT AS DERIVED BY GRAPHICAL SMOOTHING

Mole Fraction of C_6H_{14}	360.9 K (Ref. 32)	398.2 K (Ref. 32)
0.00	0.153	0.190
0.05	0.155	0.192
0.10	0.157	0.194
0.15	0.159	0.196
0.20	0.160	0.198
0.25	0.162	0.200
0.30	0.164	0.202
0.35	0.166	0.204
0.40	0.168	0.206
0.45	0.170	0.208
0.50	0.171	0.210
0.55	0.173	0.212
0.60	0.175	0.214
0.65	0.177	0.216
0.70	0.178	0.218
0.75	0.180	0.220
0.80	0.182	0.222
0.85	0.183	0.224
0.90	0.185	0.226
0.95	0.187	0.227
1.00	0.188	0.229

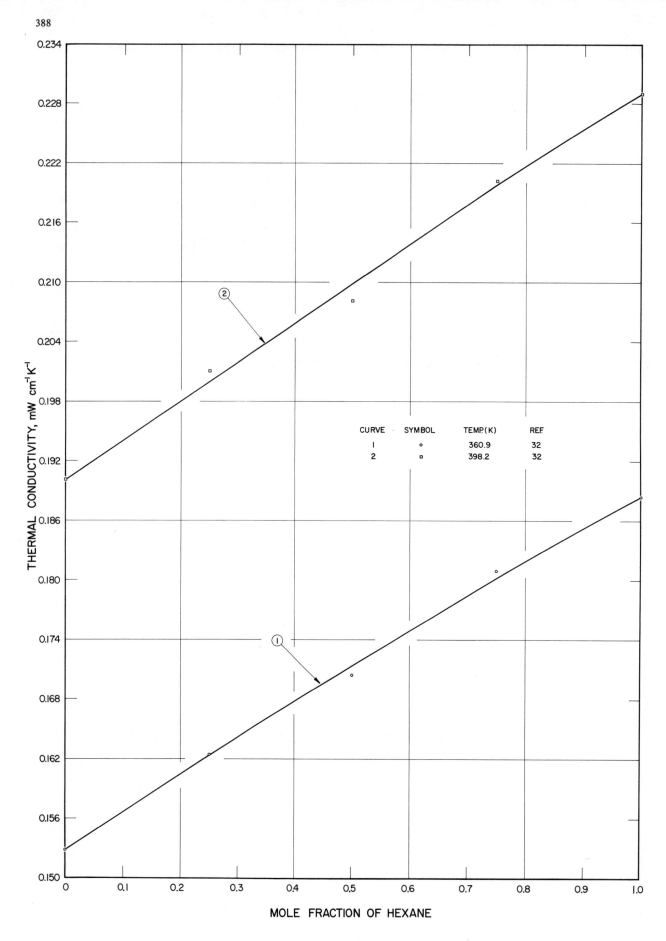

FIGURE 105. THERMAL CONDUCTIVITY OF BENZENE-HEXANE SYSTEM

TABLE 106a. EXPERIMENTAL THERMAL CONDUCTIVITY DATA FOR CARBON DIOXIDE-ETHYLENE SYSTEM

Curve No.	Fig. No.	Ref. No.	Author(s)	Temp. (K)	Mole Fraction of CO_2	Thermal Cond. (mW cm^{-1} K^{-1})	Remarks
1	106	65, 688	Cheung, H., Bromley, L.A., and Wilke, C.R.	591.2 591.2 593.2	0.0000 0.4992 1.0000	0.6406 0.5313 0.4042	CO_2: Pure Carbonic Inc., specified purity 99.5%, chief impurity air, C_2H_4: Matheson Co., Inc., C.P. grade, specified purity 99.5%; coaxial cylinder method; avg error 1.2% and max error 2%.

TABLE 106b. THERMAL CONDUCTIVITY (mW cm^{-1}K^{-1}) OF CARBON DIOXIDE-ETHYLENE SYSTEM AS A FUNCTION OF COMPOSITION AT THE TEMPERATURE OF MEASUREMENT AS DERIVED BY GRAPHICAL SMOOTHING

Mole Fraction of CO_2	591 K (Ref. 65)
0.00	0.641
0.05	0.631
0.10	0.621
0.15	0.611
0.20	0.601
0.25	0.590
0.30	0.579
0.35	0.568
0.40	0.557
0.45	0.545
0.50	0.533
0.55	0.521
0.60	0.508
0.65	0.496
0.70	0.482
0.75	0.469
0.80	0.456
0.85	0.443
0.90	0.430
0.95	0.417
1.00	0.404

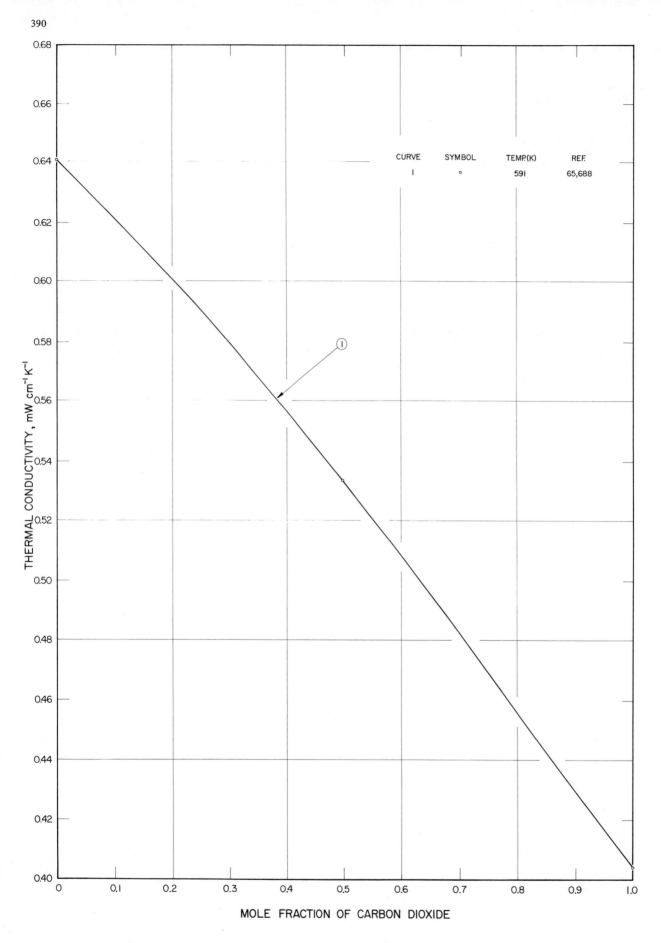

FIGURE 106. THERMAL CONDUCTIVITY OF CARBON DIOXIDE–ETHYLENE SYSTEM

TABLE 107a. EXPERIMENTAL THERMAL CONDUCTIVITY DATA FOR CARBON DIOXIDE-HYDROGEN SYSTEM

Curve No.	Fig. No.	Ref. No.	Author(s)	Temp. (K)	Mole Fraction of CO_2	Thermal Cond. (mW cm^{-1} K^{-1})	Remarks
--	--	690	Lindsay, A. L. and Bromley, L. A.	296.7	0.000	1.835	H_2: 99.9% pure, CO_2: 99.5%; unsteady state method; precision about 11%.
				296.9	0.000	1.818	
				297.0	0.000	1.809	
				296.9	0.216	1.065	
				296.7	0.216	1.193	
				296.0	0.415	0.7351	
				296.0	0.415	0.7731	
				296.0	0.595	0.4791	
				295.5	0.595	0.4808	
				295.0	0.811	0.2646	
				295.3	0.811	0.2819	
				295.4	1.000	0.1673	
				295.7	1.000	0.1735	
1	107a	380	Weber, S.	273.2	0.0000	1.743	CO_2: pure, H_2: pure electrolytic; thin hot wire potential lead method, vertical arrangement.
					0.0570	1.503	
					0.1654	1.172	
					0.3932	0.7218	
					0.6302	0.4329	
					0.8299	0.2541	
					0.9060	0.1997	
					0.9247	0.1876	
					1.0000	0.1419	
2	107a	156	Ibbs, T. L. and Hirst, A. A.	273.2	0.000	1.691	Purity of gases as supplied in cylinders; katharometer method; these are relative measurements and for calibration thermal conductivity values for argon-helium system were used.
					0.050	1.486	
					0.099	1.319	
					0.250	0.9504	
					0.500	0.5652	
					0.645	0.4187	
					0.750	0.3224	
					0.858	0.2386	
					0.900	0.2135	
					0.951	0.1842	
					1.000	0.1507	
3	107a	204	Kornfeld, G. and Hilferding, K.	298.0	0.0000	1.830	Compensated hot wire method.
					0.0362	1.683	
					0.0941	1.465	
					0.5040	0.6330	
					0.8070	0.3170	
					0.9530	0.1854	
					1.0000	0.1707	
4	107a	690	Lindsay, A. L. and Bromley, L. A.	296	0.000	1.820	We have generated these data from the original reproduced above by averaging the multiple values referring to the same composition of the mixture.
					0.216	1.129	
					0.415	0.7541	
					0.595	0.4791	
					0.811	0.2733	
					1.000	0.1704	

TABLE 107a. EXPERIMENTAL THERMAL CONDUCTIVITY DATA FOR CARBON DIOXIDE-HYDROGEN SYSTEM (continued)

Curve No.	Fig. No.	Ref. No.	Author(s)	Temp. (K)	Mole Fraction of Xe	Thermal Cond. (mW cm^{-1} K^{-1})	Remarks
5	107b	712	Mukhopadhyay, P. and Barua, A. K.	258.3	0.000 0.193 0.359 0.580 0.811 1.000	1.615 1.030 0.7113 0.4342 0.2537 0.1386	H_2: 99.95% pure, CO_2: obtained pure by the thermal decomposition of $BaCO_3$; thick hot wire method; accuracy of these relative measurements ± 1%.
6	107b	712	Mukhopadhyay, P. and Barua, A. K.	273.3	0.000 0.193 0.359 0.580 0.811 1.000	1.692 1.067 0.7369 0.4534 0.2667 0.1478	Same as above.
7	107b	712	Mukhopadhyay, P. and Barua, A. K.	293.3	0.000 0.193 0.359 0.580 0.811 1.000	1.761 1.130 0.7779 0.4790 0.2776 0.1557	Same as above.
8	107b	712	Mukhopadhyay, P. and Barua, A. K.	353.3	0.000 0.210 0.423 0.626 0.748 1.000	2.046 1.256 0.7959 0.5171 0.3931 0.2001	Same as above.
9	107b	712	Mukhopadhyay, P. and Barua, A. K.	393.3	0.000 0.215 0.374 0.672 0.792 1.000	2.186 1.342 0.9521 0.5112 0.3948 0.2437	Same as above.
10	107b	712	Mukhopadhyay, P. and Barua, A. K.	433.3	0.000 0.215 0.406 0.586 0.821 1.000	2.312 1.442 0.9605 0.6670 0.4015 0.2738	Same as above.
11	107b	712	Mukhopadhyay, P. and Barua, A. K.	473.3	0.000 0.251 0.382 0.639 0.846 1.000	2.462 1.398 1.072 0.6272 0.4149 0.3094	Same as above.

TABLE 107b. THERMAL CONDUCTIVITY (mW cm^{-1} K^{-1}) OF CARBON DIOXIDE-HYDROGEN SYSTEM AS A FUNCTION OF COMPOSITION AT THE TEMPERATURE OF MEASUREMENT AS DERIVED BY GRAPHICAL SMOOTHING

Mole Fraction of CO_2	296 K (Ref. 690)	273.2 K (Ref. 156)	273.2 K (Ref. 380)	298.0 K (Ref. 204)
0.00	1.82	1.69	1.74	1.83
0.05	1.63	1.49	1.53	1.63
0.10	1.45	1.32	1.36	1.44
0.15	1.30	1.18	1.21	1.30
0.20	1.17	1.06	1.09	1.17
0.25	1.05	0.951	0.977	1.06
0.30	0.944	0.860	0.878	0.952
0.35	0.849	0.775	0.790	0.860
0.40	0.762	0.698	0.711	0.772
0.45	0.682	0.628	0.642	0.696
0.50	0.608	0.565	0.577	0.628
0.55	0.542	0.508	0.519	0.567
0.60	0.484	0.455	0.464	0.511
0.65	0.430	0.408	0.412	0.460
0.70	0.382	0.364	0.364	0.412
0.75	0.340	0.322	0.319	0.367
0.80	0.302	0.284	0.277	0.323
0.85	0.266	0.247	0.239	0.284
0.90	0.234	0.214	0.204	0.245
0.95	0.202	0.182	0.172	0.208
1.00	0.170	0.151	0.142	0.171

Mole Fraction of CO_2	258.3 K (Ref. 712)	273.3 K (Ref. 712)	293.3 K (Ref. 712)	353.3 K (Ref. 712)	393.3 K (Ref. 712)	433.3 K (Ref. 712)	473.3 K (Ref. 712)
0.00	1.62	1.69	1.76	2.05	2.19	2.31	2.46
0.05	1.45	1.52	1.59	1.85	1.99	2.11	2.21
0.10	1.28	1.34	1.41	1.64	1.78	1.89	1.97
0.15	1.14	1.18	1.25	1.44	1.57	1.68	1.75
0.20	1.01	1.05	1.11	1.28	1.39	1.49	1.56
0.25	0.905	0.936	0.992	1.15	1.24	1.33	1.40
0.30	0.810	0.837	0.884	1.03	1.11	1.19	1.27
0.35	0.725	0.751	0.792	0.927	1.00	1.07	1.14
0.40	0.649	0.674	0.711	0.835	0.903	0.972	1.03
0.45	0.580	0.605	0.638	0.752	0.814	0.880	0.931
0.50	0.518	0.542	0.571	0.676	0.734	0.797	0.839
0.55	0.463	0.485	0.511	0.609	0.661	0.720	0.756
0.60	0.414	0.433	0.458	0.547	0.594	0.647	0.680
0.65	0.371	0.387	0.410	0.490	0.534	0.581	0.612
0.70	0.332	0.346	0.366	0.437	0.479	0.522	0.550
0.75	0.295	0.308	0.325	0.390	0.430	0.468	0.497
0.80	0.261	0.274	0.286	0.350	0.386	0.420	0.451
0.85	0.228	0.240	0.250	0.312	0.347	0.378	0.412
0.90	0.196	0.208	0.216	0.274	0.310	0.340	0.375
0.95	0.167	0.178	0.186	0.238	0.276	0.306	0.341
1.00	0.139	0.148	0.156	0.200	0.244	0.274	0.309

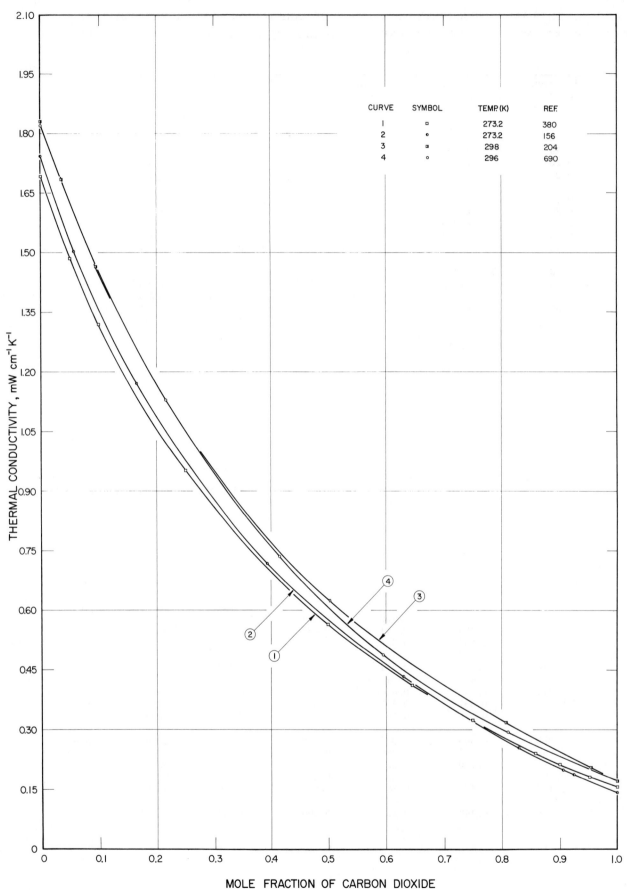

FIGURE 107a. THERMAL CONDUCTIVITY OF CARBON DIOXIDE–HYDROGEN SYSTEM

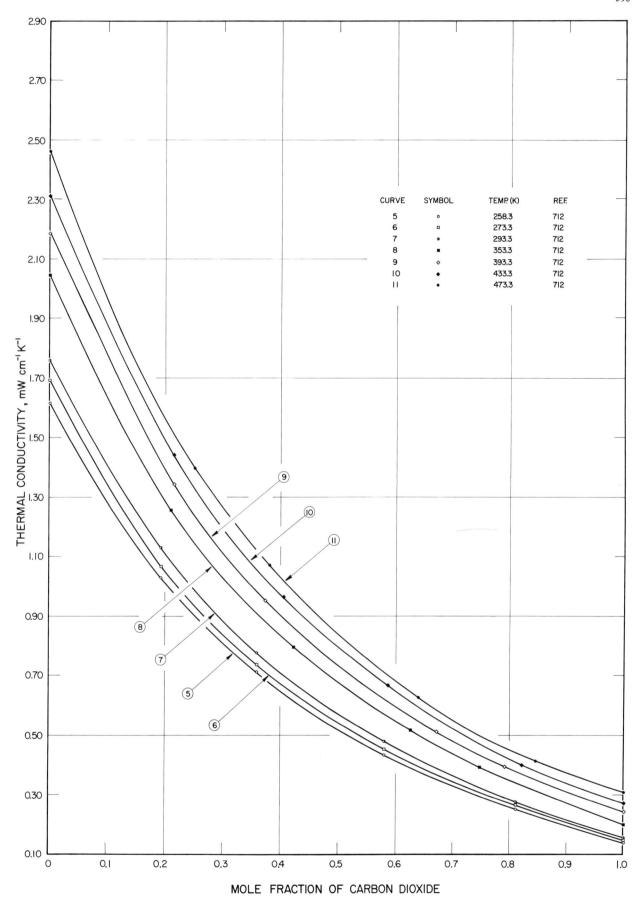

FIGURE 107b. THERMAL CONDUCTIVITY OF CARBON DIOXIDE-HYDROGEN SYSTEM

TABLE 108a. EXPERIMENTAL THERMAL CONDUCTIVITY DATA FOR CARBON DIOXIDE-NITROGEN SYSTEM

Curve No.	Fig. No.	Ref. No.	Author(s)	Temp. (K)	Mole Fraction of CO_2	Thermal Cond. (mW cm^{-1} K^{-1})	Remarks
1	108a	83	Davidson, J.M. and Music, J.F.	273.2	0.00 0.25 0.52 0.66 1.00	0.2416 0.2144 0.1905 0.1641 0.1419	Unsteady state method, the rate of cooling of a solid inner cylinder of copper was determined; accuracy of these relative measurements ±5%.
2	108a	189	Keyes, F.G.	323.2	0.0000 0.3350 0.4712 0.6594 1.0000	0.2780 0.2374 0.2248 0.2089 0.1817	Coaxial cylinder method.
3	108a	189	Keyes, F.G.	423.2	0.0000 0.3350 0.4712 0.6594 1.0000	0.3479 0.3199 0.3090 0.2939 0.2625	Same as above.
4	108a	189	Keyes, F.G.	523.2	0.0000 0.3350 0.4712 0.6594 1.0000	0.4116 0.3919 0.3856 0.3730 0.3500	Same as above.
5	108a	189	Keyes, F.G.	623.2	0.0000 0.3350 0.4712 0.6594 1.0000	0.4698 0.4731 0.4719 0.4622 0.4430	Same as above.
6	108a	65, 281, 280	Rothman, A.J. and Bromley, L.A.,	642.2	0.00 0.25 0.47 0.50 1.00	0.4668 0.4614 0.4568 0.4375 0.4354	Coaxial cylinder method; estimated error increases with temp from about 1% to 3%.
7	108a	65, 281, 280	Rothman, A.J. and Bromley, L.A.,	645.2	0.00 0.17 0.33 0.50 0.67 1.00	0.4970 0.4986 0.4949 0.4861 0.4756 0.4631	Same as above.
8	108a	65, 281, 280	Rothman, A.J. and Bromley, L.A.,	648.2	0.00 0.17 0.33 0.50 0.67 1.00	0.4970 0.5058 0.5024 0.4932 0.4827 0.4685	Same as above.
9	108b	65, 281, 280	Rothman, A.J. and Bromley, L.A.,	745.2	0.00 0.44 1.00	0.5158 0.5259 0.5087	Same as above.
10	108b	65, 281, 280	Rothman, A.J. and Bromley, L.A.	842.2	0.00 0.33 0.50 0.67 1.00	0.5493 0.5836 0.5811 0.5769 0.5702	Same as above.
11	108b	65, 281, 280	Rothman, A.J. and Bromley, L.A.,	846.2	0.00 0.33 0.50 0.67 1.00	0.5606 0.5941 0.5933 0.5874 0.5807	Same as above.

TABLE 108a. EXPERIMENTAL THERMAL CONDUCTIVITY DATA FOR CARBON DIOXIDE-NITROGEN SYSTEM (cont.)

Curve No.	Fig. No.	Ref. No.	Author(s)	Temp. (K)	Mole Fraction of CO_2	Thermal Cond. (mW cm^{-1} K^{-1})	Remarks
12	108b	65, 281, 280	Rothman, A.J. and Bromley, L.A.	950.2	0.00 0.50 1.00	0.6330 0.6812 0.6741	Same as above.
13	108b	65, 281, 280	Rothman, A.J. and Bromley, L.A.	961.2	0.00 0.25 0.50 0.75 1.00	0.6423 0.6820 0.6912 0.6883 0.6787	Same as above.
14	108b	65, 281, 280	Rothman, A.J. and Bromley, L.A.	1047.2	0.00 0.50 1.00	0.7113 0.7653 0.7658	Same as above.

TABLE 108b. THERMAL CONDUCTIVITY (mW cm^{-1}K^{-1}) OF CARBON DIOXIDE-NITROGEN SYSTEM AS A FUNCTION OF COMPOSITION AT THE TEMPERATURE OF MEASUREMENT AS DERIVED BY GRAPHICAL SMOOTHING

Mole Fraction of CO_2	273.2 K (Ref. 83)	323.2 K (Ref. 189)	423.2 K (Ref. 189)	523.2 K (Ref. 189)	623.2 K (Ref. 189)	642.2 K (Ref. 65)	645.2 K (Ref. 65)
0.00	0.242	0.278	0.348	0.412	0.470	0.467	0.499
0.05	0.236	0.271	0.343	0.409	0.471	0.466	0.498
0.10	0.230	0.264	0.339	0.407	0.472	0.465	0.498
0.15	0.224	0.258	0.335	0.403	0.472	0.464	0.499
0.20	0.219	0.252	0.331	0.401	0.472	0.463	0.498
0.25	0.213	0.246	0.327	0.398	0.472	0.461	0.498
0.30	0.207	0.241	0.323	0.395	0.472	0.460	0.496
0 35	0.202	0.236	0.318	0.392	0.472	0.459	0.494
0.40	0.196	0.231	0.314	0.389	0.472	0.458	0.492
0.45	0.191	0.227	0.310	0.386	0.471	0.457	0.489
0.50	0.185	0.222	0.306	0.383	0.470	0.456	0.486
0.55	0.180	0.218	0.301	0.380	0.469	0.455	0.483
0.60	0.175	0.214	0.297	0.377	0.467	0.453	0.480
0.65	0.169	0.210	0.293	0.373	0.464	0.452	0.477
0.70	0.164	0.206	0.288	0.370	0.462	0.450	0.474
0.75	0.159	0.201	0.283	0.367	0.459	0.449	0.472
0.80	0.155	0.197	0.279	0.364	0.456	0.447	0.470
0.85	0.151	0.193	0.275	0.360	0.453	0.444	0.468
0.90	0.147	0.189	0.271	0.357	0.450	0.442	0.467
0.95	0.144	0.186	0.267	0.354	0.447	0.439	0.465
1.00	0.142	0.182	0.263	0.350	0.443	0.435	0.463

Mole Fraction of CO_2	648.2 K (Ref. 65)	745.2 K (Ref. 65)	842.2 K (Ref. 65)	846.2 K (Ref. 65)	950.2 K (Ref. 65)	961.2 K (Ref. 65)	1047.2 K (Ref. 65)
0.00	0.497	0.516	0.549	0.561	0.633	0.642	0.711
0.05	0.502	0.518	0.571	0.578	0.652	0.657	0.724
0.10	0.504	0.520	0.576	0.585	0.661	0.666	0.733
0.15	0.506	0.522	0.579	0.589	0.667	0.673	0.740
0.20	0.506	0.523	0.581	0.591	0.671	0.678	0.745
0.25	0.505	0.524	0.583	0.593	0.674	0.682	0.750
0.30	0.504	0.525	0.583	0.594	0.676	0.685	0.755
0.35	0.501	0.525	0.583	0.594	0.678	0.688	0.759
0.40	0.499	0.526	0.583	0.594	0.680	0.689	0.762
0.45	0.496	0.526	0.582	0.593	0.681	0.691	0.764
0.50	0.493	0.526	0.581	0.593	0.681	0.691	0.765
0.55	0.490	0.526	0.580	0.592	0.682	0.691	0.766
0.60	0.487	0.526	0.579	0.591	0.682	0.691	0.767
0.65	0.484	0.525	0.578	0.590	0.681	0.691	0.768
0.70	0.481	0.525	0.577	0.589	0.681	0.690	0.768
0.75	0.479	0.524	0.576	0.588	0.680	0.688	0.768
0.80	0.477	0.523	0.575	0.586	0.679	0.687	0.768
0.85	0.475	0.522	0.574	0.585	0.678	0.685	0.767
0.90	0.473	0.520	0.572	0.584	0.677	0.683	0.767
0.95	0.470	0.516	0.571	0.582	0.676	0.681	0.766
1.00	0.469	0.509	0.570	0.581	0.674	0.679	0.766

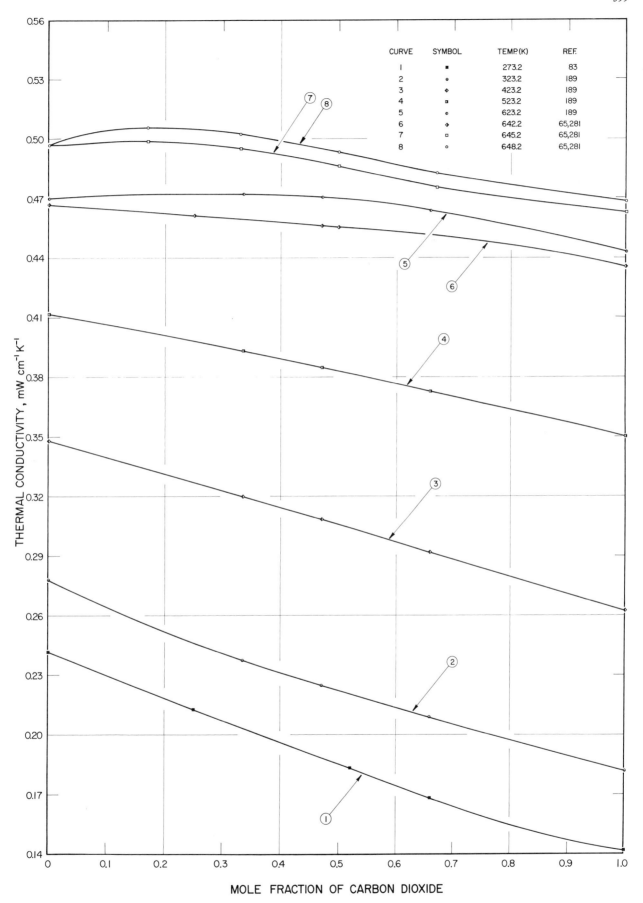

FIGURE 108a. THERMAL CONDUCTIVITY OF CARBON DIOXIDE–NITROGEN SYSTEM

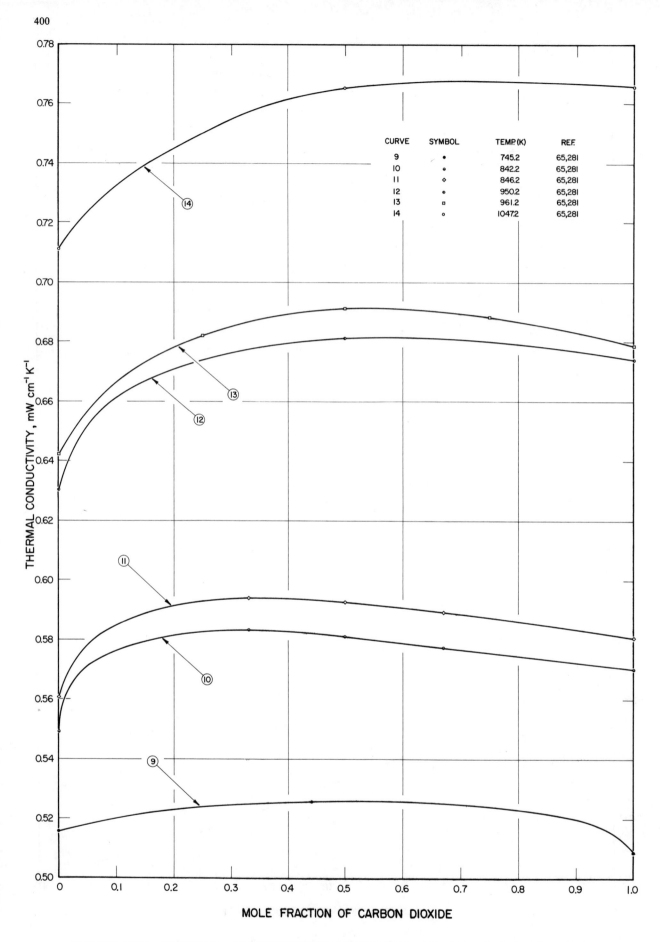

FIGURE 108b. THERMAL CONDUCTIVITY OF CARBON DIOXIDE-NITROGEN SYSTEM

TABLE 109a. EXPERIMENTAL THERMAL CONDUCTIVITY DATA FOR CARBON DIOXIDE-OXYGEN SYSTEM

Curve No.	Fig. No.	Ref. No.	Author(s)	Temp. (K)	Mole Fraction of CO_2	Thermal Cond. (mW cm^{-1} K^{-1})	Remarks
1	109	688	Cheung, H., Bromley, L.A., and Wilke, C.R.	374.2 370.2 369.2 369.2 370.2 376.2	0.0000 0.2240 0.4644 0.6847 0.7301 1.0000	0.3237 0.2942 0.2665 0.2456 0.2419 0.2232	O_2: Liquid Carbonic Co., commercial grade, specified purity 99.5%, chief impurity Ar, CO_2: Pure Carbonic, Inc., specified purity 99.5%, chief impurity air; coaxial cylinder method; average error 1.2%, max error 2%.

TABLE 109b. THERMAL CONDUCTIVITY (mW cm^{-1}K^{-1}) OF CARBON DIOXIDE-OXYGEN SYSTEM AS A FUNCTION OF COMPOSITION AT THE TEMPERATURE OF MEASUREMENT AS DERIVED BY GRAPHICAL SMOOTHING

Mole Fraction of CO_2	370 K (Ref. 688)
0.00	0.324
0.05	0.317
0.10	0.310
0.15	0.304
0.20	0.297
0.25	0.291
0.30	0.285
0.35	0.279
0.40	0.274
0.45	0.268
0.50	0.263
0.55	0.258
0.60	0.253
0.65	0.249
0.70	0.244
0.75	0.240
0.80	0.237
0.85	0.233
0.90	0.230
0.95	0.226
1.00	0.223

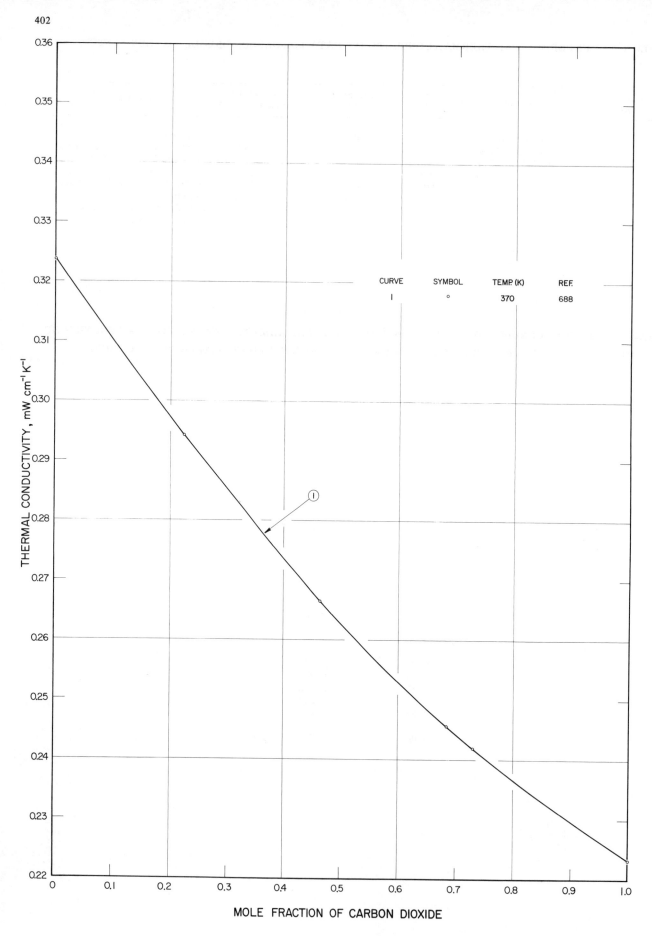

FIGURE 109. THERMAL CONDUCTIVITY OF CARBON DIOXIDE—OXYGEN SYSTEM

TABLE 110a. EXPERIMENTAL THERMAL CONDUCTIVITY DATA FOR CARBON DIOXIDE-PROPANE SYSTEM

Curve No.	Fig. No.	Ref. No.	Author(s)	Temp. (K)	Mole Fraction of C_3H_8	Thermal Cond. (mW cm^{-1} K^{-1})	Remarks
1	110	65, 688	Cheung, H., Bromley, L.A., and Wilke, C.R.	376.2 369.2 368.2 368.2 373.2	0.0000 0.3646 0.5510 0.7088 1.0000	0.2232 0.2402 0.2464 0.2547 0.2721	CO_2: Pure Carbonic, Inc., specified purity 99.5%, chief impurity air, C_3H_8: Matheson Co., Inc., instrument grade, specified purity 99.9%; coaxial cylinder method; avg error 1.2% and max error 2%.

TABLE 110b. THERMAL CONDUCTIVITY (mW cm^{-1}K^{-1}) OF CARBON DIOXIDE-PROPANE SYSTEM AS A FUNCTION OF COMPOSITION AT THE TEMPERATURE OF MEASUREMENT AS DERIVED BY GRAPHICAL SMOOTHING

Mole Fraction of C_3H_8	369 K (Ref. 65)
0.00	0.223
0.05	0.225
0.10	0.227
0.15	0.230
0.20	0.232
0.25	0.234
0.30	0.236
0.35	0.238
0.40	0.240
0.45	0.243
0.50	0.245
0.55	0.247
0.60	0.249
0.65	0.252
0.70	0.254
0.75	0.257
0.80	0.260
0.85	0.263
0.90	0.266
0.95	0.269
1.00	0.272

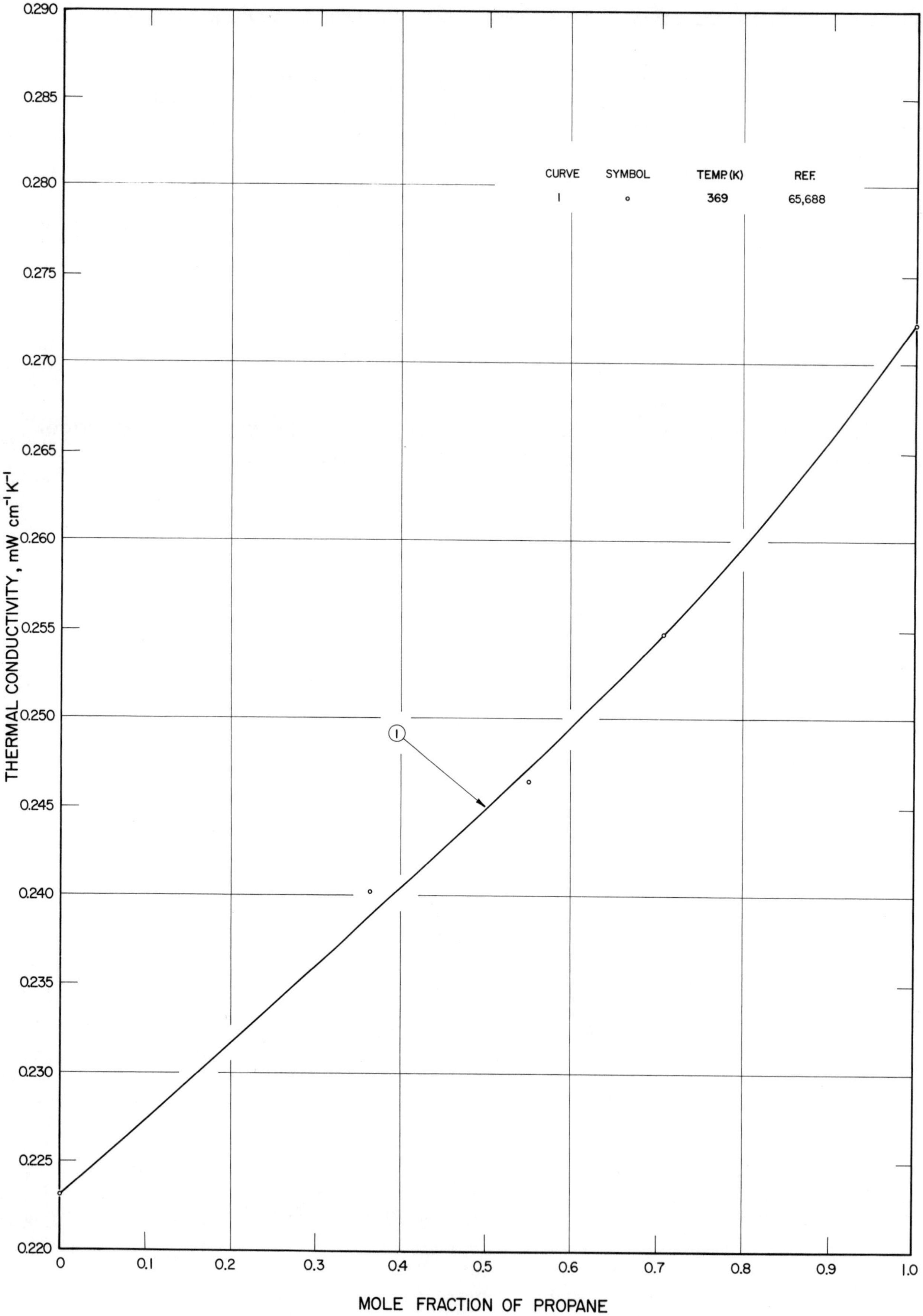

FIGURE 110. THERMAL CONDUCTIVITY OF CARBON DIOXIDE—PROPANE SYSTEM

TABLE 111a. EXPERIMENTAL THERMAL CONDUCTIVITY DATA FOR CARBON MONOXIDE-HYDROGEN SYSTEM

Curve No.	Fig. No.	Ref. No.	Author(s)	Temp. (K)	Mole Fraction of CO	Thermal Cond. (mW cm^{-1} K^{-1})	Remarks
1	111	156	Ibbs, T. L. and Hirst, A. A.	273.2	0.000	1.692	Purity of gases as supplied in cylinders; katharometer method, these are relative measurements and for calibration thermal conductivity values for argon-helium system were used.
					0.206	1.130	
					0.366	0.8750	
					0.434	0.7536	
					0.728	0.4271	
					0.837	0.3349	
					1.000	0.2219	

TABLE 111b. THERMAL CONDUCTIVITY (mW cm^{-1}K^{-1}) OF CARBON MONOXIDE-HYDROGEN SYSTEM AS A FUNCTION OF COMPOSITION AT THE TEMPERATURE OF MEASUREMENT AS DERIVED BY GRAPHICAL SMOOTHING

Mole Fraction of CO	273.2 K (Ref. 156)
0.00	1.69
0.05	1.53
0.10	1.39
0.15	1.26
0.20	1.15
0.25	1.05
0.30	0.963
0.35	0.880
0.40	0.804
0.45	0.734
0.50	0.669
0.55	0.609
0.60	0.554
0.65	0.502
0.70	0.453
0.75	0.408
0.80	0.364
0.85	0.324
0.90	0.287
0.95	0.254
1.00	0.222

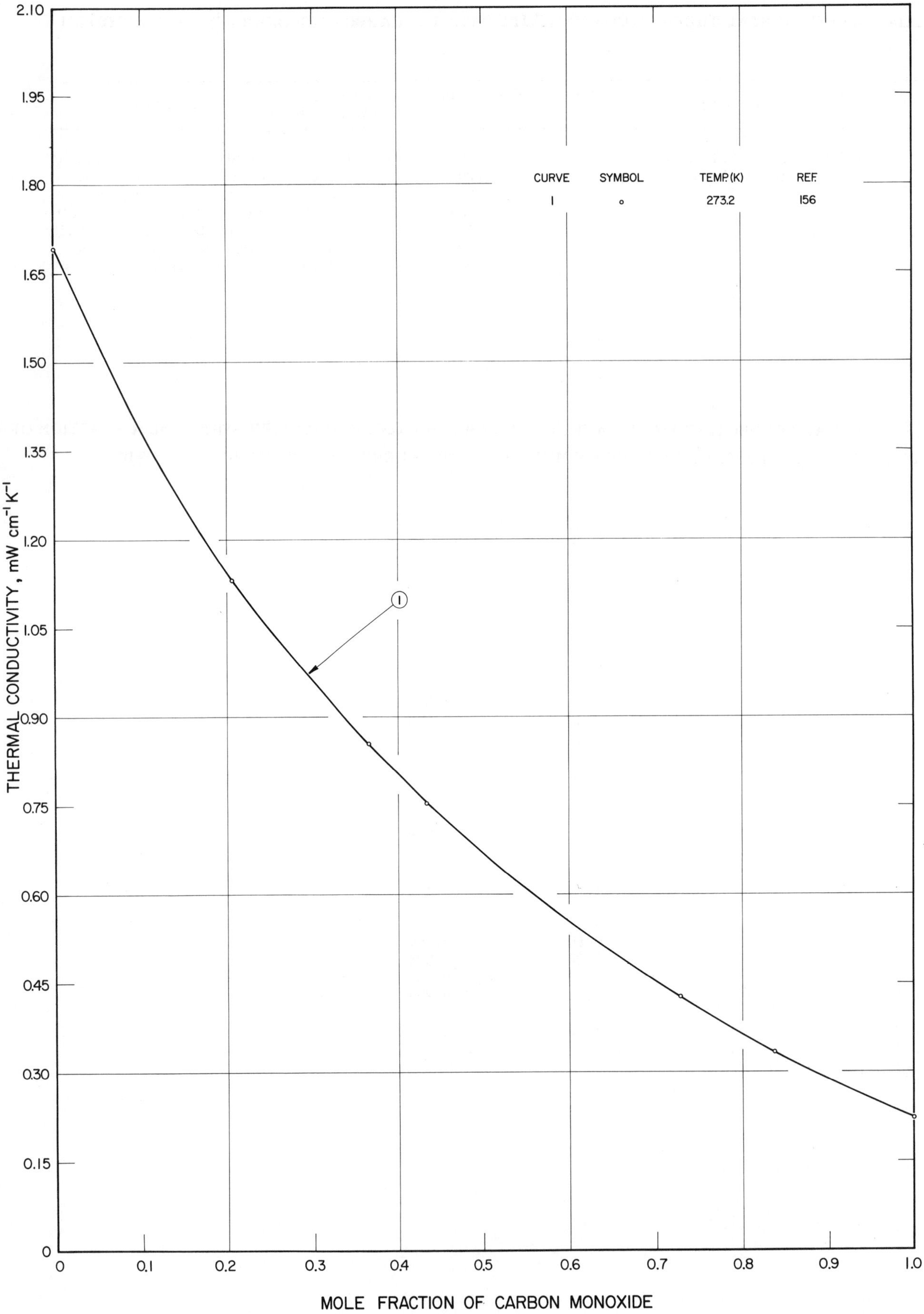

FIGURE III. THERMAL CONDUCTIVITY OF CARBON MONOXIDE–HYDROGEN SYSTEM

TABLE 112a. EXPERIMENTAL THERMAL CONDUCTIVITY DATA FOR DEUTERIUM-HYDROGEN SYSTEM

Curve No.	Fig. No.	Ref. No.	Author(s)	Temp. (K)	Mole Fraction of D_2	Thermal Cond. (mW cm^{-1} K^{-1})	Remarks
1	112	21	Archer, C.T.	273.2	0.000 0.198 0.345 0.504 0.605 0.813 1.000	1.750 1.600 0.527 1.467 1.427 1.353 1.289	D_2: prepared from deuterium oxide of 99.95% concentration, H_2: from distilled water; vertical compensated hot wire method; accuracy 0.25%.
2	112	707, 708	Tondon, P.K. and Saxena, S.C.	313.2	0.000 0.253 0.497 0.762 1.000	1.822 1.616 1.524 1.357 1.350	H_2: 99.95% pure, D_2: 98.6% pure, 0.8% H_2, 0.6% H_2O; thick hot wire method; precision ±1%, accuracy ±1 to ±2%.
3	112	707, 708	Tondon, P.K. and Saxena, S.C.	338.2	0.000 0.253 0.497 0.762 1.000	1.939 1.700 1.578 1.440 1.426	Same as above.
4	112	707, 708	Tondon, P.K. and Saxena, S.C.	366.2	0.000 0.253 0.497 0.762 1.000	2.037 1.867 1.683 1.545 1.495	Same as above.
5	112	707, 710	Tondon, P.K. and Saxena, S.C.	368.2	0.000 0.243 0.488 0.762 0.936 1.000	2.114 1.922 1.717 1.586 1.581 1.540	Same as above.
6	112	707, 710	Tondon, P.K. and Saxena, S.C.	408.2	0.000 0.243 0.488 0.762 0.936 1.000	2.269 1.978 1.871 1.653 1.655 1.625	Same as above.
7	112	707, 710	Tondon, P.K. and Saxena, S.C.	448.2	0.000 0.243 0.488 0.762 0.936 1.000	2.370 2.146 1.986 1.782 1.769 1.746	Same as above.

TABLE 112b. THERMAL CONDUCTIVITY (mW cm^{-1}K^{-1}) OF DEUTERIUM-HYDROGEN SYSTEM AS A FUNCTION OF COMPOSITION AT THE TEMPERATURE OF MEASUREMENT AS DERIVED BY GRAPHICAL SMOOTHING

Mole Fraction of D$_2$	273.2 K (Ref. 21)	313.2 K (Ref. 707)	338.2 K (Ref. 707)	366.2 K (Ref. 707)	368.2 K (Ref. 707)	408.2 K (Ref. 707)	448.2 K (Ref. 707)
0.00	1.75	1.82	1.94	2.04	2.11	2.27	2.37
0.05	1.72	1.78	1.88	2.01	2.07	2.21	2.32
0.10	1.69	1.74	1.83	1.97	2.03	2.15	2.28
0.15	1.66	1.69	1.79	1.94	1.99	2.09	2.23
0.20	1.63	1.66	1.74	1.90	1.94	2.04	2.19
0.25	1.60	1.62	1.70	1.87	1.90	1.99	2.14
0.30	1.57	1.59	1.66	1.83	1.86	1.95	2.10
0.35	1.54	1.56	1.63	1.79	1.82	1.91	2.06
0.40	1.52	1.53	1.60	1.76	1.79	1.87	2.02
0.45	1.49	1.51	1.58	1.72	1.76	1.84	1.98
0.50	1.47	1.49	1.55	1.68	1.73	1.81	1.95
0.55	1.45	1.47	1.53	1.65	1.70	1.78	1.91
0.60	1.43	1.45	1.51	1.62	1.67	1.75	1.89
0.65	1.41	1.43	1.49	1.59	1.65	1.73	1.86
0.70	1.39	1.42	1.47	1.57	1.63	1.70	1.84
0.75	1.37	1.41	1.46	1.55	1.61	1.69	1.82
0.80	1.35	1.39	1.45	1.53	1.59	1.67	1.80
0.85	1.34	1.38	1.44	1.52	1.58	1.66	1.78
0.90	1.32	1.37	1.44	1.51	1.56	1.64	1.77
0.95	1.30	1.36	1.43	1.50	1.55	1.63	1.76
1.00	1.29	1.35	1.43	1.50	1.54	1.63	1.75

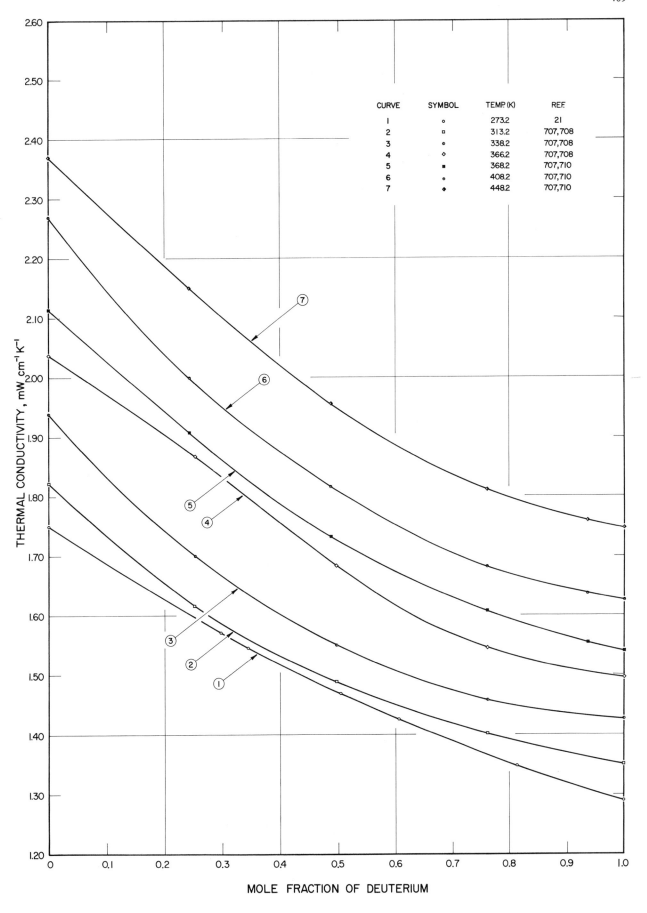

FIGURE 112. THERMAL CONDUCTIVITY OF DEUTERIUM—HYDROGEN SYSTEM

TABLE 113a. EXPERIMENTAL THERMAL CONDUCTIVITY DATA FOR DEUTERIUM-NITROGEN SYSTEM

Curve No.	Fig. No.	Ref. No.	Author(s)	Temp. (K)	Mole Fraction of N_2	Thermal Cond. (mW cm^{-1} K^{-1})	Remarks
1	113	696, 695	Gupta, G.P. and Saxena, S.C.	313.2	0.000 0.222 0.601 1.000	1.350 0.938 0.550 0.268	D_2: 98.6% pure, 0.8% H_2 and 0.6% H_2O; N_2: 99.95% pure; thick hot wire method; precision ±1%, accuracy ±1 to ±2%.
2	113	696, 695	Gupta, G.P. and Saxena, S.C.	338.2	0.000 0.222 0.601 1.000	1.430 1.004 0.578 0.290	Same as above.
3	113	696, 695	Gupta, G.P. and Saxena, S.C.	366.2	0.000 0.222 0.601 1.000	1.490 1.063 0.621 0.312	Same as above.
4	113	707, 710	Tondon, P.K. and Saxena, S.C.	368.2	0.00 0.332 0.496 1.000	1.540 0.9337 0.7310 0.3052	D_2: 98.6% pure, 0.8% H_2 and 0.6% H_2O; N_2: 99.95% pure; thick hot wire method; precision ±1%, accuracy ±1 to ±2%.
5	113	707, 710	Tondon, P.K. and Saxena, S.C.	408.2	0.00 0.332 0.496 1.000	1.624 1.007 0.7808 0.3329	Same as above.
6	113	707, 710	Tondon, P.K. and Saxena, S.C.	448.2	0.00 0.332 0.496 1.000	1.746 1.071 0.8570 0.3622	Same as above.

TABLE 113b.   THERMAL CONDUCTIVITY (mW cm^{-1}K^{-1}) OF DEUTERIUM-NITROGEN SYSTEM AS A FUNCTION OF COMPOSITION AT THE TEMPERATURE OF MEASUREMENT AS DERIVED BY GRAPHICAL SMOOTHING

Mole Fraction of N$_2$	313.2 K (Ref. 696)	338.2 K (Ref. 696)	366.2 K (Ref. 696)	368.2 K (Ref. 707)	408.2 K (Ref. 707)	448.2 K (Ref. 707)
0.00	1.35	1.43	1.49	1.54	1.62	1.75
0.05	1.24	1.32	1.39	1.43	1.51	1.61
0.10	1.14	1.21	1.28	1.32	1.41	1.49
0.15	1.05	1.12	1.19	1.23	1.31	1.39
0.20	0.971	1.04	1.10	1.14	1.22	1.29
0.25	0.900	0.963	1.02	1.06	1.14	1.20
0.30	0.837	0.894	0.952	0.978	1.06	1.12
0.35	0.780	0.830	0.889	0.908	0.980	1.05
0.40	0.728	0.772	0.830	0.843	0.908	0.977
0.45	0.680	0.718	0.774	0.783	0.840	0.913
0.50	0.635	0.669	0.720	0.727	0.778	0.853
0.55	0.592	0.623	0.670	0.678	0.721	0.795
0.60	0.552	0.579	0.622	0.626	0.668	0.740
0.65	0.512	0.539	0.578	0.580	0.620	0.688
0.70	0.474	0.500	0.536	0.536	0.572	0.636
0.75	0.438	0.463	0.496	0.494	0.528	0.588
0.80	0.402	0.426	0.458	0.455	0.487	0.541
0.85	0.366	0.392	0.422	0.417	0.448	0.494
0.90	0.332	0.358	0.384	0.380	0.408	0.449
0.95	0.300	0.324	0.348	0.342	0.370	0.406
1.00	0.268	0.290	0.312	0.305	0.333	0.362

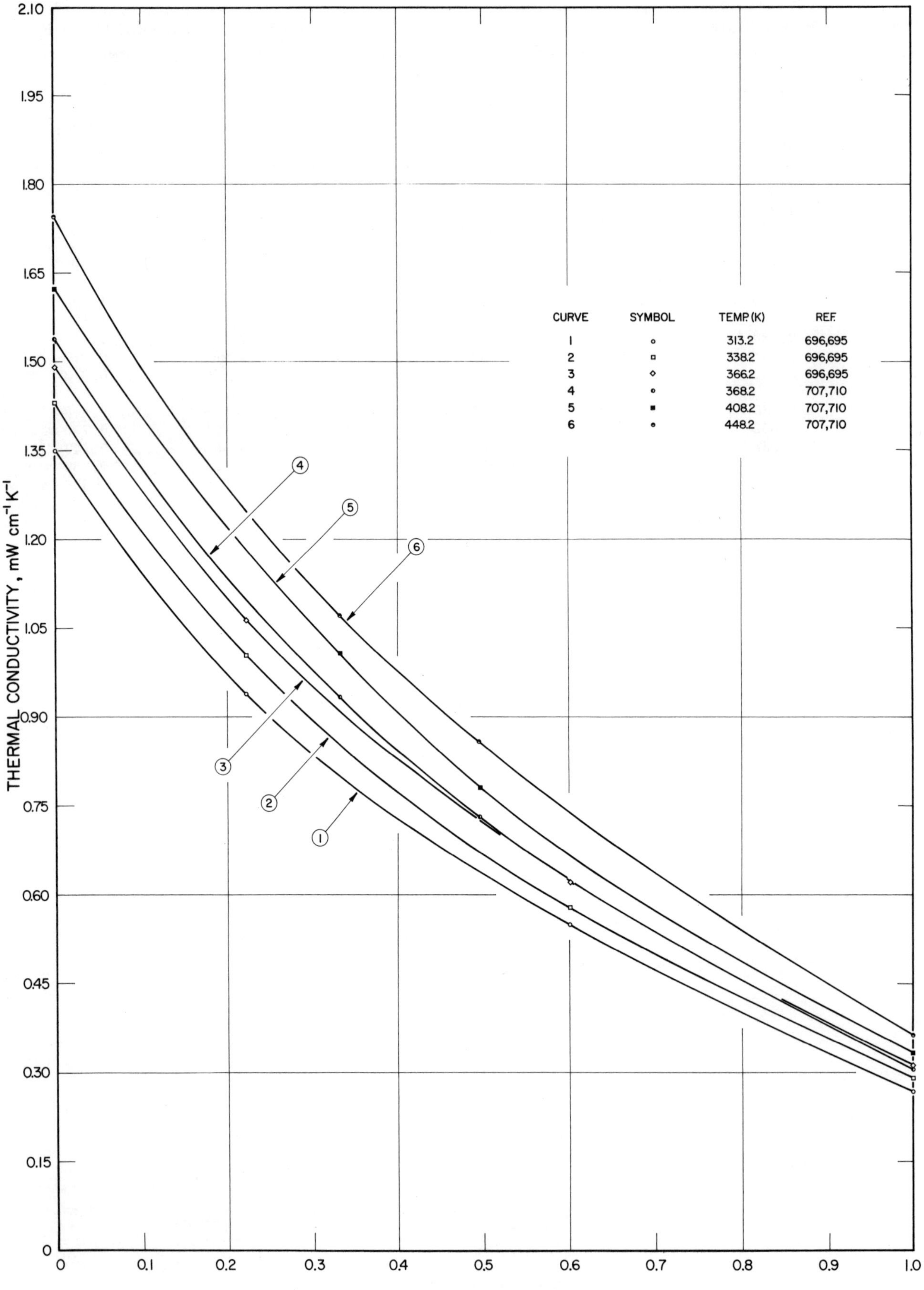
2.10
1.95
1.80
1.65
1.50
1.35
1.20
1.05
0.90
0.75
0.60
0.45
0.30
0.15
0
THERMAL CONDUCTIVITY, mW cm⁻¹ K⁻¹
0
0.1
0.2
0.3
0.4
0.5
0.6
0.7
0.8
0.9
1.0
MOLE FRACTION OF NITROGEN
CURVE SYMBOL TEMP. (K) REF.
1 ○ 313.2 696,695
2 □ 338.2 696,695
3 ◇ 366.2 696,695
4 ○ 368.2 707,710
5 ■ 408.2 707,710
6 ○ 448.2 707,710
1
2
3
4
5
6

FIGURE 113. THERMAL CONDUCTIVITY OF DEUTERIUM—NITROGEN SYSTEM

TABLE 114a. EXPERIMENTAL THERMAL CONDUCTIVITY DATA FOR ETHYLENE-HYDROGEN SYSTEM

Curve No.	Fig. No.	Ref. No.	Author(s)	Temp. (K)	Mole Fraction of C_2H_4	Thermal Cond. (mW cm^{-1} K^{-1})	Remarks
1	114	204	Kornfeld, G. and Hilferding, K.	298.2	0.0000 0.1351 0.3890 0.4863 0.6860 0.8302 1.0000	1.830 1.377 0.8625 0.7076 0.4806 0.3605 0.2206	Compensated hot wire method.

TABLE 114b. THERMAL CONDUCTIVITY (mW cm^{-1} K^{-1}) OF ETHYLENE-HYDROGEN SYSTEM AS A FUNCTION OF COMPOSITION AT THE TEMPERATURE OF MEASUREMENT AS DERIVED BY GRAPHICAL SMOOTHING

Mole Fraction of C_2H_4	298.2 K (Ref. 204)
0.00	1.83
0.05	1.66
0.10	1.49
0.15	1.34
0.20	1.21
0.25	1.10
0.30	1.00
0.35	0.914
0.40	0.834
0.45	0.762
0.50	0.694
0.55	0.634
0.60	0.579
0.65	0.524
0.70	0.474
0.75	0.426
0.80	0.382
0.85	0.340
0.90	0.300
0.95	0.260
1.00	0.221

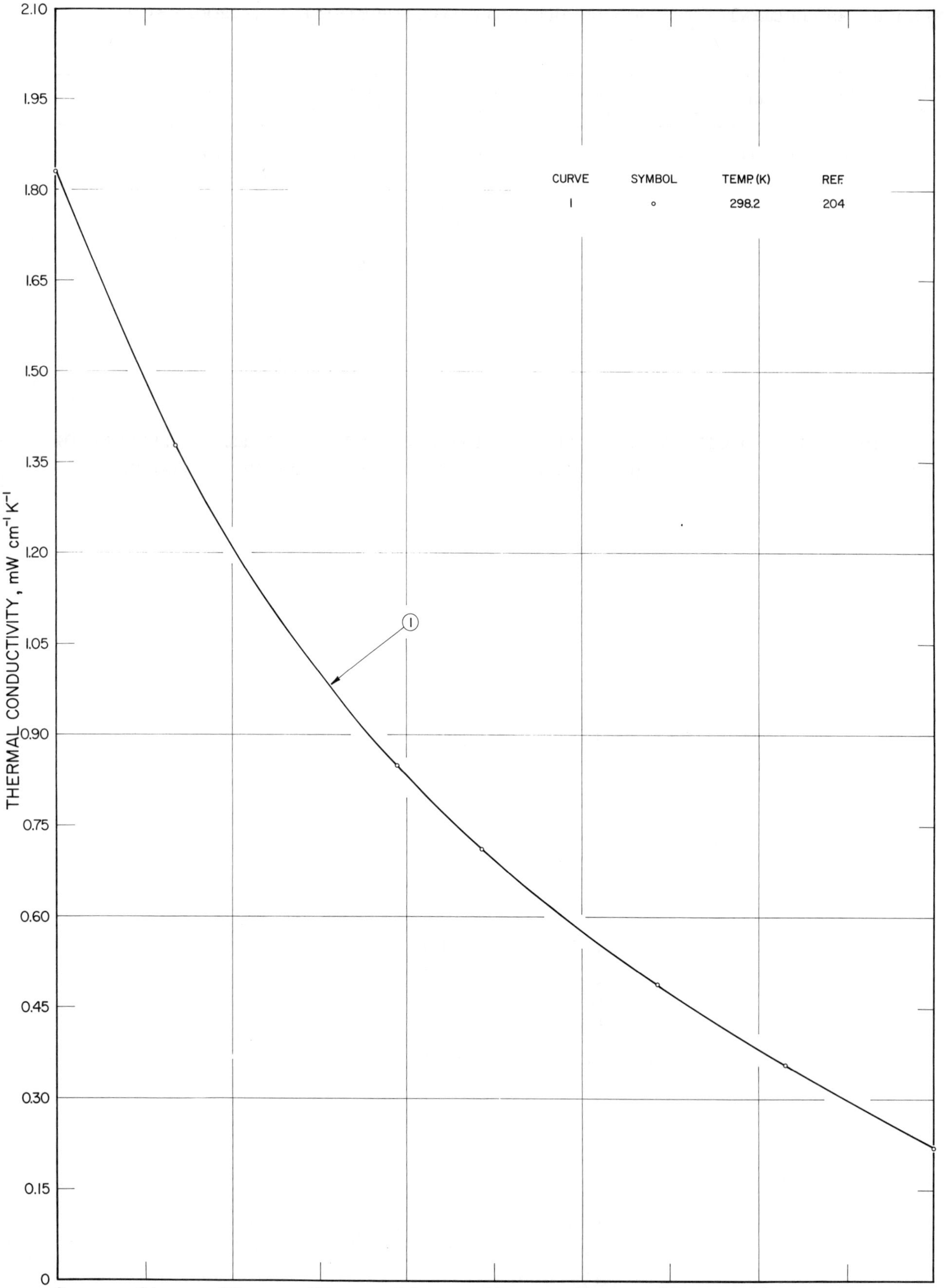

FIGURE 114. THERMAL CONDUCTIVITY OF ETHYLENE−HYDROGEN SYSTEM

TABLE 115a. EXPERIMENTAL THERMAL CONDUCTIVITY DATA FOR ETHYLENE-METHANE SYSTEM

Curve No.	Fig. No.	Ref. No.	Author(s)	Temp. (K)	Mole Fraction of C_2H_4	Thermal Cond. (mW cm^{-1} K^{-1})	Remarks
1	115	65, 688	Cheung, H., Bromley, L.A., and Wilke, C.R.	590.2 590.2 591.2	0.0000 0.5106 1.0000	0.8516 0.7005 0.6406	CH_4: Phillips Petroleum Co., specified purity 99%, chief impurity C_2H_6, C_2H_4: Matheson Co., C.P. grade, specified purity 99.5%; coaxial cylinder method; average error 1.2%, max error 2%.

TABLE 115b. THERMAL CONDUCTIVITY (mW cm^{-1}K^{-1}) OF ETHYLENE-METHANE SYSTEM AS A FUNCTION OF COMPOSITION AT THE TEMPERATURE OF MEASUREMENT AS DERIVED BY GRAPHICAL SMOOTHING

Mole Fraction of C_2H_4	590 K (Ref. 65)
0.00	0.852
0.05	0.831
0.10	0.815
0.15	0.797
0.20	0.780
0.25	0.770
0.30	0.749
0.35	0.736
0.40	0.724
0.45	0.713
0.50	0.703
0.55	0.694
0.60	0.685
0.65	0.678
0.70	0.671
0.75	0.665
0.80	0.660
0.85	0.655
0.90	0.650
0.95	0.645
1.00	0.641

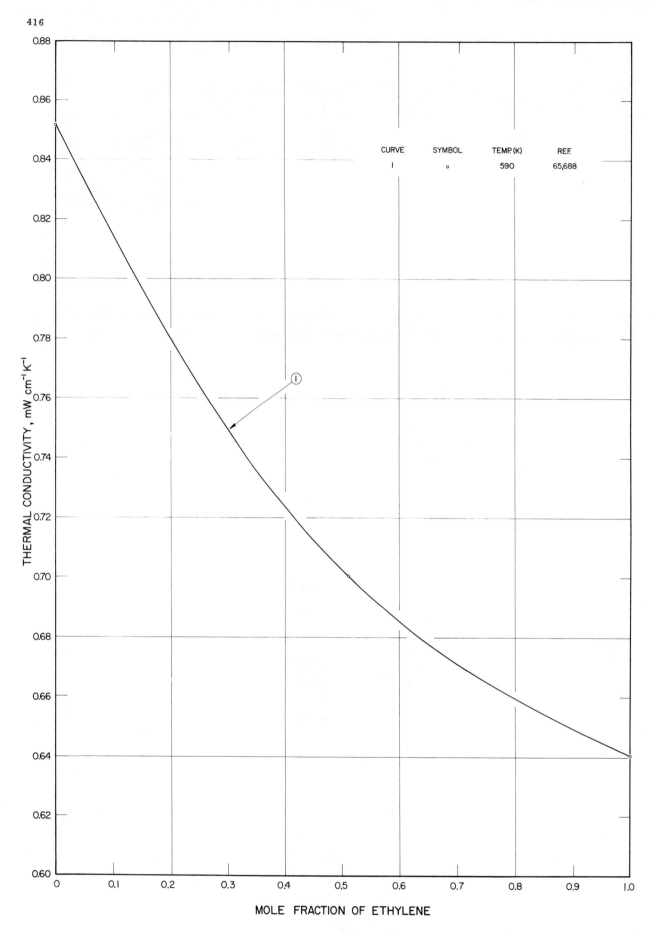

FIGURE 115. THERMAL CONDUCTIVITY OF ETHYLENE—METHANE SYSTEM

TABLE 116a. EXPERIMENTAL THERMAL CONDUCTIVITY DATA FOR ETHYLENE-NITROGEN SYSTEM

Curve No.	Fig. No.	Ref. No.	Author(s)	Temp. (K)	Mole Fraction of N_2	Thermal Cond. (mW cm^{-1} K^{-1})	Remarks
1	116	65, 688	Cheung, H., Bromley, L.A., and Wilke, C.R.	591.2 591.2 592.2 590.2	0.0000 0.4980 0.7558 1.0000	0.6406 0.5535 0.5045 0.4467	N_2: Linde Air Products Co., water pumped, specified purity 99.9%, chief impurities Ar and Ne, C_2H_4: Matheson Co, C.P. grade, specified purity 99.5% pure; coaxial cylinder method; avg error 1.2%, max error 2%.

TABLE 116b. THERMAL CONDUCTIVITY (mW cm^{-1}K^{-1}) OF ETHYLENE-NITROGEN SYSTEM AS A FUNCTION OF COMPOSITION AT THE TEMPERATURE OF MEASUREMENT AS DERIVED BY GRAPHICAL SMOOTHING

Mole Fraction of N_2	591 K (Ref. 65)
0.00	0.641
0.05	0.632
0.10	0.623
0.15	0.614
0.20	0.606
0.25	0.597
0.30	0.588
0.35	0.580
0.40	0.571
0.45	0.562
0.50	0.553
0.55	0.544
0.60	0.535
0.65	0.525
0.70	0.516
0.75	0.506
0.80	0.495
0.85	0.484
0.90	0.460
0.95	0.458
1.00	0.447

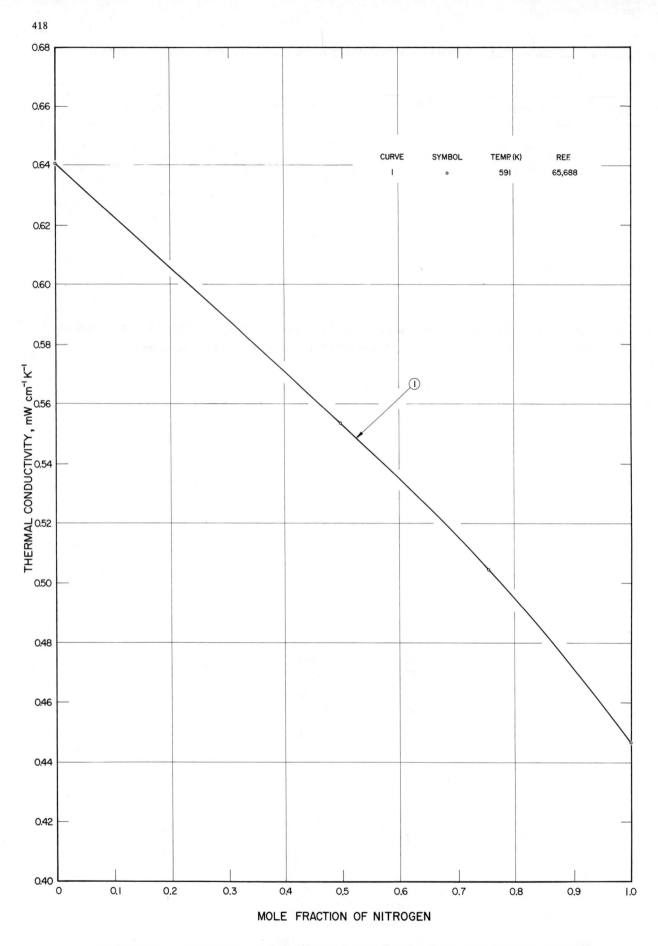

FIGURE 116. THERMAL CONDUCTIVITY OF ETHYLENE—NITROGEN SYSTEM

TABLE 117a. EXPERIMENTAL THERMAL CONDUCTIVITY DATA FOR HYDROGEN-NITROGEN SYSTEM

Curve No.	Fig. No.	Ref. No.	Author(s)	Temp. (K)	Mole Fraction of N_2	Thermal Cond. (mW cm^{-1} K^{-1})	Remarks
1	117a	696, 695	Gupta, G. P. and Saxena, S. C.	313.2	0.000 0.147 0.338 0.592 1.000	1.820 1.414 1.027 0.677 0.268	H_2 and N_2: 99.95% pure; thick hot wire method; accuracy ±1 to ±2%, precision ±1%.
2	117a	696, 695	Gupta, G. P. and Saxena, S. C.	338.2	0.000 0.147 0.338 0.592 1.000	1.940 1.496 1.078 0.715 0.290	Same as above.
3	117a	696, 695	Gupta, G. P. and Saxena, S. C.	366.2	0.000 0.147 0.338 0.592 1.000	2.040 1.590 1.161 0.766 0.312	Same as above.
4	117a	707, 710	Tondon, P. K. and Saxena, S. C.	368.2	0.00 0.260 0.513 0.880 1.000	2.114 1.340 0.8851 0.4358 0.3052	H_2 and N_2: 99.95% pure; thick hot wire method; accuracy ±1 to ±2%, precision ±1%.
5	117a	707, 710	Tondon, P. K. and Saxena, S. C.	408.2	0.00 0.260 0.513 0.880 1.000	2.269 1.418 0.9663 0.4551 0.3329	Same as above.
6	117b	707, 710	Tondon, P. K. and Saxena, S. C.	448.2	0.000 0.260 0.513 0.880 1.000	2.370 1.525 1.009 0.4928 0.3622	Same as above.
--	--	690	Lindsay, A. L. and Bromley, L. A.	297.0 296.9 296.7 298.0 298.0 298.0 298.0 297.8 297.8 296.0 296.0 296.3 296.0 300.2 300.2 300.0 298.0 298.0 298.0	0.000 0.000 0.000 0.214 0.214 0.214 0.410 0.410 0.410 0.680 0.680 0.824 0.824 1.00 1.00 1.00 1.00 1.00 1.00	1.809 1.818 1.835 1.517 1.437 1.465 0.9288 0.9392 0.7904 0.5050 0.5016 0.3131 0.3113 0.2698 0.2612 0.2560 0.2664 0.2594 0.2525	H_2: 99.9% pure, N_2: 99.99% pure; unsteady state method; precision ±16%.
7	117b	156	Ibbs, T. L. and Hirst, A. A.	273.2	0.000 0.197 0.205 0.348 0.610 0.841 1.000	1.692 1.076 1.055 0.8122 0.5317 0.3349 0.2303	Purity of gases as supplied in cylinders; katharometer method; these are relative measurements and for calibration thermal conductivity values for argon-helium system were used.

419

TABLE 117a. EXPERIMENTAL THERMAL CONDUCTIVITY DATA FOR HYDROGEN-NITROGEN SYSTEM (cont.)

Curve No.	Fig. No.	Ref. No.	Author(s)	Temp. (K)	Mole Fraction of N_2	Thermal Cond. (mW cm^{-1} K^{-1})	Remarks
8	117b	588	Gray, P. and Wright, P.G.	298.5	0.000 0.100 0.199 0.2985 0.335 0.497 0.594 0.690 0.770 0.890 1.000	1.763 1.470 1.239 1.047 1.022 0.7704 0.6280 0.5338 0.4291 0.3446 0.2596	H_2: purified by diffusion through palladium, N_2: obtained from sodium azide; two-wire type conductivity cell; accuracy of these relative measurements about 1%.
9	117b	588	Gray, P. and Wright, P.G.	348.0	0.000 0.1415 0.3145 0.504 0.711 0.853 1.000	2.077 1.566 1.181 0.8583 0.5694 0.4187 0.2927	Same as above.
10	117b	588	Gray, P. and Wright, P.G.	372.3	0.000 0.082 0.1875 0.352 0.6045 0.814 1.000	2.081 1.817 1.528 1.183 0.7327 0.4731 0.3098	Same as above.
11	117b	588	Gray, P. and Wright, P.G.	422.5	0.000 0.082 0.1415 0.3105 0.504 0.6045 0.711 0.814 0.853 1.000	2.261 1.978 1.750 1.344 0.9630 0.7997 0.6573 0.5317 0.5066 0.3408	Same as above.
12	117a	690	Lindsay, A.L. and Bromley, L.A.	298	0.000 0.214 0.410 0.680 0.824 1.000	1.820 1.472 0.8856 0.5033 0.3113 0.2594	We have generated these data from the original reproduced above by averaging the multiple values referring to the same composition of the mixture.
13	117c	712	Mukhopadhyay, P., Das Gupta, A., and Barua, A.K.	258.3	0.000 0.222 0.425 0.599 0.809 1.000	1.615 1.076 0.7520 0.5543 0.3609 0.2290	H_2 and N_2: 99.95% pure; thick hot wire method; accuracy of these relative measurements ±1%.
14	117c	712	Mukhopadhyay, P., Das Gupta, A., and Barua, A.K.	273.3	0.000 0.222 0.425 0.599 0.809 1.000	1.692 1.130 0.7779 0.5673 0.3802 0.2428	Same as above.

TABLE 117a. EXPERIMENTAL THERMAL CONDUCTIVITY DATA FOR HYDROGEN-NITROGEN SYSTEM (cont.)

Curve No.	Fig. No.	Ref. No.	Author(s)	Temp. (K)	Mole Fraction of N_2	Thermal Cond. (mW cm^{-1} K^{-1})	Remarks
15	117c	712	Mukhopadhyay, P., Das Gupta, A., and Barua, A. K.	293.3	0.000 0.222 0.425 0.599 0.809 1.000	1.761 1.178 0.8420 0.5983 0.3923 0.2567	Same as above.
16	117c	712	Mukhopadhyay, P., Das Gupta, A., and Barua, A. K.	353.3	0.000 0.208 0.351 0.512 0.694 1.000	2.046 1.403 1.088 0.8156 0.5895 0.2801	Same as above.
17	117c	712	Mukhopadhyay, P., Das Gupta, A., and Barua, A. K.	393.3	0.000 0.208 0.351 0.512 0.694 1.000	2.186 1.528 1.185 0.8746 0.6356 0.3098	Same as above.
18	117c	712	Mukhopadhyay, P., Das Gupta, A., and Barua, A. K.	433.3	0.000 0.210 0.395 0.472 0.750 1.000	2.312 1.574 1.174 0.9977 0.6054 0.3303	Same as above.
19	117c	712	Mukhopadhyay, P., Das Gupta, A., and Barua, A. K.	473.3	0.000 0.210 0.395 0.472 0.750 1.000	2.462 1.689 1.221 1.069 0.6272 0.3529	Same as above.

TABLE 117b. THERMAL CONDUCTIVITY (mW cm^{-1} K^{-1}) OF HYDROGEN-NITROGEN SYSTEM AS A FUNCTION OF COMPOSITION AT THE TEMPERATURE OF MEASUREMENT AS DERIVED BY GRAPHICAL SMOOTHING

Mole Fraction of N$_2$	313.2 K (Ref. 696)	338.2 K (Ref. 696)	366.2 K (Ref. 696)	368.2 K (Ref. 707)	408.2 K (Ref. 707)	448.2 K (Ref. 707)	273.2 K (Ref. 156)
0.00	1.82	1.94	2.04	2.11	2.27	2.37	1.69
0.05	1.67	1.78	1.87	1.93	2.07	2.17	1.50
0.10	1.53	1.63	1.72	1.77	1.88	2.00	1.33
0.15	1.41	1.49	1.58	1.62	1.71	1.83	1.19
0.20	1.29	1.36	1.46	1.49	1.57	1.69	1.07
0.25	1.19	1.25	1.34	1.36	1.44	1.55	0.966
0.30	1.09	1.15	1.24	1.26	1.34	1.43	0.881
0.35	1.01	1.06	1.14	1.16	1.24	1.32	0.811
0.40	0.931	0.975	1.05	1.06	1.15	1.22	0.749
0.45	0.859	0.900	0.973	0.981	1.07	1.12	0.693
0.50	0.792	0.830	0.896	0.904	0.987	1.03	0.641
0.55	0.728	0.765	0.825	0.833	0.909	0.950	0.591
0.60	0.667	0.704	0.757	0.766	0.834	0.872	0.542
0.65	0.612	0.647	0.692	0.702	0.760	0.798	0.495
0.70	0.559	0.591	0.630	0.640	0.688	0.727	0.450
0.75	0.508	0.538	0.570	0.582	0.618	0.658	0.407
0.80	0.458	0.487	0.513	0.525	0.553	0.592	0.366
0.85	0.410	0.436	0.460	0.468	0.492	0.530	0.330
0.90	0.360	0.386	0.410	0.413	0.435	0.472	0.295
0.95	0.313	0.337	0.360	0.359	0.384	0.416	0.262
1.00	0.268	0.290	0.312	0.305	0.333	0.362	0.230

Mole Fraction of N$_2$	298.5 K (Ref. 588)	348.0 K (Ref. 588)	372.3 K (Ref. 588)	422.5 K (Ref. 588)	298 K (Ref. 690)	258.3 K (Ref. 712)	273.3 K (Ref. 712)
0.00	1.76	2.08	2.08	2.26	1.82	1.62	1.69
0.05	1.61	1.89	1.92	2.08	1.71	1.48	1.55
0.10	1.47	1.70	1.76	1.90	1.60	1.35	1.42
0.15	1.35	1.54	1.62	1.73	1.50	1.23	1.29
0.20	1.24	1.42	1.50	1.60	1.40	1.12	1.18
0.25	1.14	1.31	1.39	1.49	1.30	1.02	1.07
0.30	1.06	1.21	1.29	1.38	1.20	0.938	0.981
0.35	0.975	1.12	1.19	1.27	1.10	0.861	0.899
0.40	0.897	1.03	1.09	1.16	0.996	0.788	0.824
0.45	0.822	0.945	0.993	1.06	0.896	0.719	0.752
0.50	0.751	0.865	0.904	0.972	0.802	0.655	0.685
0.55	0.684	0.788	0.818	0.885	0.713	0.597	0.622
0.60	0.620	0.716	0.740	0.808	0.628	0.543	0.566
0.65	0.560	0.647	0.669	0.738	0.552	0.495	0.516
0.70	0.504	0.584	0.604	0.674	0.484	0.450	0.470
0.75	0.454	0.526	0.544	0.612	0.426	0.408	0.428
0.80	0.410	0.472	0.489	0.553	0.378	0.368	0.388
0.85	0.370	0.424	0.440	0.497	0.340	0.332	0.350
0.90	0.332	0.378	0.394	0.444	0.310	0.296	0.313
0.95	0.295	0.335	0.350	0.391	0.284	0.262	0.277
1.00	0.260	0.293	0.310	0.341	0.259	0.229	0.243

TABLE 117b. THERMAL CONDUCTIVITY (mW cm^{-1} K^{-1}) OF HYDROGEN-NITROGEN SYSTEM AS A FUNCTION OF COMPOSITION AT THE TEMPERATURE OF MEASUREMENT AS DERIVED BY GRAPHICAL SMOOTHING (cont.)

Mole Fraction of N_2	293.3 K (Ref. 712)	353.3 K (Ref. 712)	393.3 K (Ref. 712)	433.3 K (Ref. 712)	473.3 K (Ref. 712)
0.00	1.76	2.05	2.19	2.31	2.46
0.05	1.61	1.89	2.04	2.12	2.26
0.10	1.47	1.71	1.88	1.93	2.06
0.15	1.34	1.56	1.70	1.76	1.89
0.20	1.23	1.42	1.55	1.60	1.72
0.25	1.12	1.31	1.42	1.47	1.57
0.30	1.03	1.19	1.30	1.35	1.44
0.35	0.951	1.09	1.19	1.24	1.32
0.40	0.873	0.998	1.08	1.14	1.21
0.45	0.800	0.911	0.981	1.04	1.11
0.50	0.728	0.833	0.892	0.948	1.02
0.55	0.661	0.765	0.816	0.865	0.925
0.60	0.597	0.701	0.751	0.791	0.840
0.65	0.541	0.642	0.689	0.723	0.763
0.70	0.491	0.583	0.628	0.662	0.692
0.75	0.444	0.532	0.575	0.605	0.627
0.80	0.402	0.480	0.521	0.550	0.568
0.85	0.362	0.429	0.466	0.495	0.512
0.90	0.326	0.378	0.413	0.439	0.457
0.95	0.291	0.329	0.361	0.384	0.405
1.00	0.257	0.280	0.310	0.330	0.353

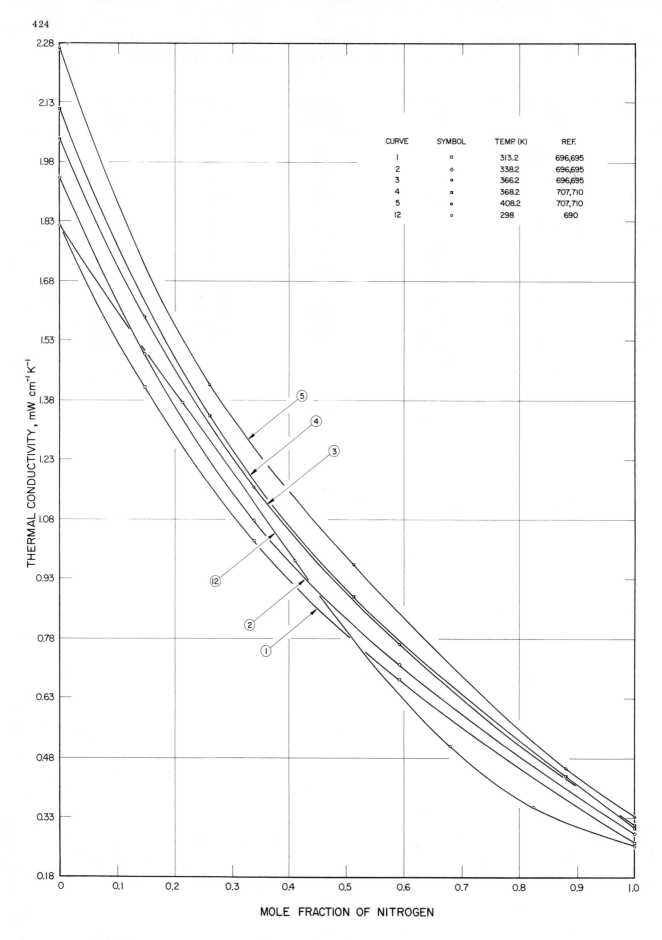

FIGURE 117a. THERMAL CONDUCTIVITY OF HYDROGEN-NITROGEN SYSTEM

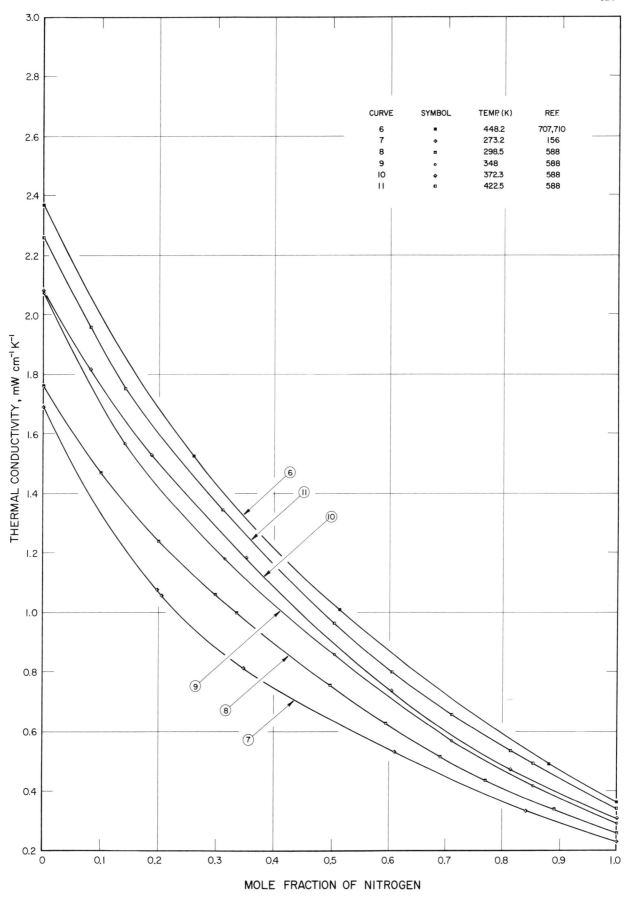

FIGURE 117b. THERMAL CONDUCTIVITY OF HYDROGEN—NITROGEN SYSTEM

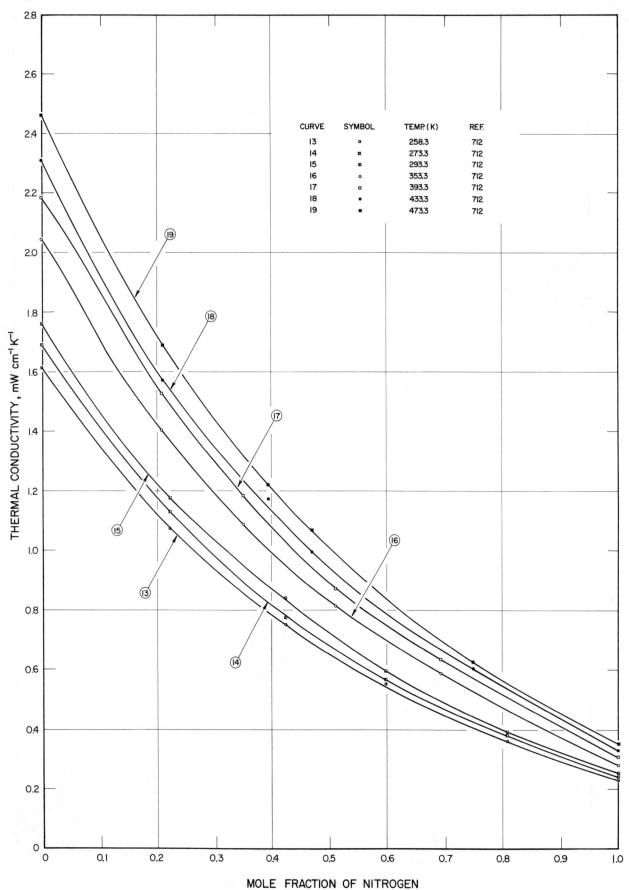

FIGURE 117c. THERMAL CONDUCTIVITY OF HYDROGEN−NITROGEN SYSTEM

TABLE 118a. EXPERIMENTAL THERMAL CONDUCTIVITY DATA FOR HYDROGEN-NITROUS OXIDE SYSTEM

Curve No.	Fig. No.	Ref. No.	Author(s)	Temp. (K)	Mole Fraction of N_2O	Thermal Cond. (mW cm^{-1} K^{-1})	Remarks
1	118	156	Ibbs, T. L. and Hirst, A. A.	273.2	0.000 0.188 0.401 0.614 0.791 0.925 1.000	1.691 1.139 0.7118 0.4480 0.2973 0.2010 0.1591	Purity of gases as supplied in cylinders; katharometer method, these are relative measurements and for calibration thermal conductivity values of argon-helium system were used.

TABLE 118b. THERMAL CONDUCTIVITY (mW cm^{-1}K^{-1}) OF HYDROGEN-NITROUS OXIDE SYSTEM AS A FUNCTION OF COMPOSITION AT THE TEMPERATURE OF MEASUREMENT AS DERIVED BY GRAPHICAL SMOOTHING

Mole Fraction of N_2O	273.2 K (Ref. 156)
0.00	1.69
0.05	1.54
0.10	1.38
0.15	1.24
0.20	1.11
0.25	0.996
0.30	0.892
0.35	0.797
0.40	0.712
0.45	0.637
0.50	0.571
0.55	0.512
0.60	0.458
0.65	0.408
0.70	0.363
0.75	0.322
0.80	0.284
0.85	0.248
0.90	0.216
0.95	0.186
1.00	0.159

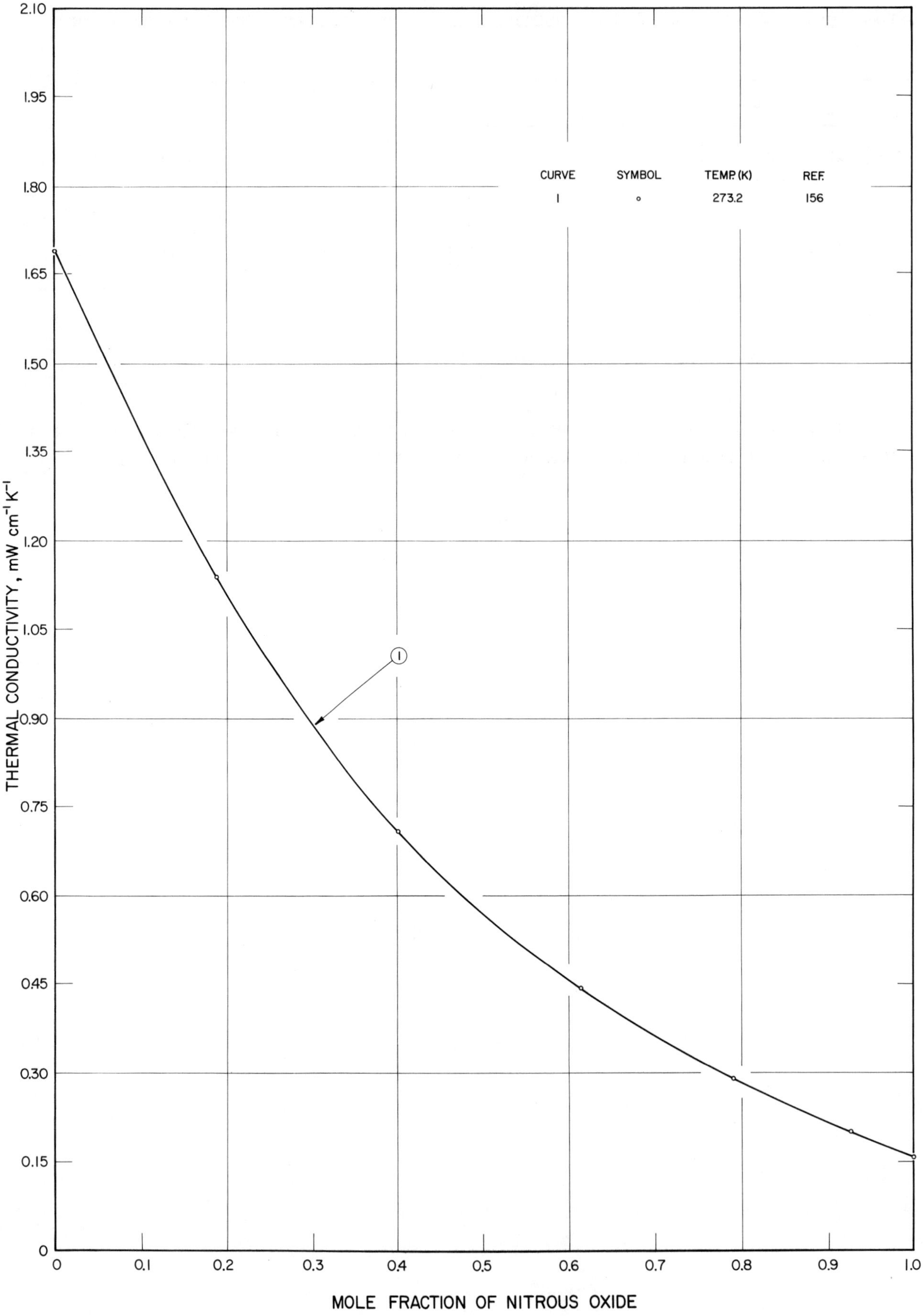

FIGURE 118. THERMAL CONDUCTIVITY OF HYDROGEN–NITROUS OXIDE SYSTEM

TABLE 119a. EXPERIMENTAL THERMAL CONDUCTIVITY DATA FOR HYDROGEN-OXYGEN SYSTEM

Curve No.	Fig. No.	Ref. No.	Author(s)	Temp. (K)	Mole Fraction of O_2	Thermal Cond. (mW cm^{-1} K^{-1})	Remarks
1	119	696, 695	Gupta, G. P. and Saxena, S. C.	313.2	0.000 0.209 0.491 0.796 1.000	1.820 1.280 0.812 0.451 0.281	H_2 and O_2: 99.95% pure; thick hot wire method; precision ±1%, accuracy ±1 to ±2%.
2	119	696, 695	Gupta, G. P. and Saxena, S. C.	338.2	0.000 0.209 0.491 0.796 1.000	1.940 1.377 0.868 0.489 0.291	Same as above.
3	119	696, 695	Gupta, G. P. and Saxena, S. C.	366.2	0.000 0.209 0.491 0.791 1.000	2.040 1.456 0.928 0.527 0.313	Same as above.
4	119	377	Wassiljewa, A.	295.2	0.000 0.0526 0.1429 0.2500 0.3333 0.5000 0.7500 0.8000 0.8462 0.875 0.9394 0.9664 1.0000	1.665 1.568 1.347 1.151 0.9935 0.7649 0.4656 0.4141 0.3848 0.3492 0.2989 0.2726 0.2650	Cooling thermometer method. These are relative measurements and the k values of the two pure gases were used for calibration.

TABLE 119b. THERMAL CONDUCTIVITY (mW cm^{-1} K^{-1}) OF HYDROGEN-OXYGEN SYSTEM AS A FUNCTION OF COMPOSITION AT THE TEMPERATURE OF MEASUREMENT AS DERIVED BY GRAPHICAL SMOOTHING

Mole Fraction of O$_2$	295.2 K (Ref. 377)	313.2 K (Ref. 696)	338.2 K (Ref. 696)	366.2 K (Ref. 696)
0.00	1.67	1.82	1.94	2.04
0.05	1.56	1.68	1.80	1.89
0.10	1.45	1.54	1.66	1.74
0.15	1.34	1.41	1.52	1.60
0.20	1.25	1.30	1.40	1.48
0.25	1.15	1.20	1.29	1.37
0.30	1.07	1.06	1.19	1.26
0.35	0.985	1.02	1.10	1.17
0.40	0.906	0.943	1.01	1.08
0.45	0.832	0.869	0.932	0.994
0.50	0.763	0.800	0.857	0.917
0.55	0.696	0.736	0.787	0.842
0.60	0.634	0.672	0.722	0.770
0.65	0.574	0.620	0.659	0.704
0.70	0.519	0.566	0.599	0.639
0.75	0.466	0.514	0.540	0.577
0.80	0.418	0.465	0.486	0.518
0.85	0.372	0.418	0.434	0.462
0.90	0.332	0.372	0.385	0.411
0.95	0.296	0.326	0.339	0.362
1.00	0.265	0.281	0.291	0.313

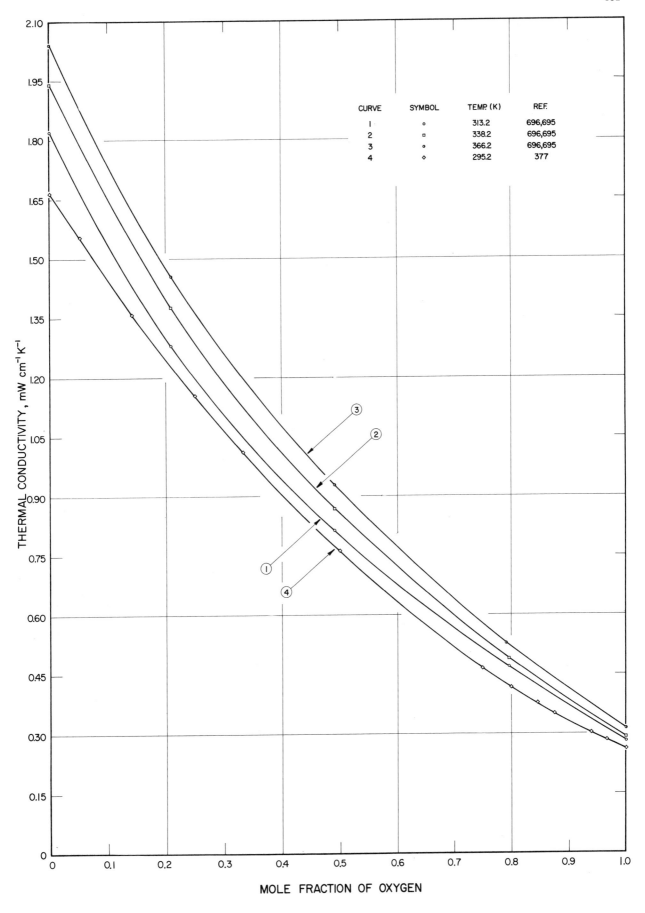

FIGURE 119. THERMAL CONDUCTIVITY OF HYDROGEN-OXYGEN SYSTEM

TABLE 120a. EXPERIMENTAL THERMAL CONDUCTIVITY DATA FOR METHANE-PROPANE SYSTEM

Curve No.	Fig. No.	Ref. No.	Author(s)	Temp. (K)	Mole Fraction C_3H_8	Thermal Cond. (mW cm^{-1} K^{-1})	Remarks
1	120	65, 688	Cheung, H., Bromley, L.A., and Wilke, C.R.	371.2 366.2 368.2 367.2 373.2	0.0000 0.3208 0.5145 0.6870 1.0000	0.4438 0.3702 0.3197 0.2952 0.2721	CH_4: Phillips Petroleum Co., specified purity 99%, chief impurity C_2H_6; C_3H_8: Matheson Co., instrument grade, specified purity 99.9%; coaxial cylinder method; avg error 1.2%, maximum error 2%.

TABLE 120b. THERMAL CONDUCTIVITY (mW cm^{-1}K^{-1}) OF METHANE-PROPANE SYSTEM AS A FUNCTION OF COMPOSITION AT THE TEMPERATURE OF MEASUREMENT AS DERIVED BY GRAPHICAL SMOOTHING

Mole Fraction of C_3H_8	368 K (Ref. 65)
0.00	0.444
0.05	0.431
0.10	0.419
0.15	0.407
0.20	0.395
0.25	0.383
0.30	0.372
0.35	0.360
0.40	0.349
0.45	0.338
0.50	0.327
0.55	0.317
0.60	0.308
0.65	0.300
0.70	0.293
0.75	0.288
0.80	0.284
0.85	0.280
0.90	0.277
0.95	0.274
1.00	0.272

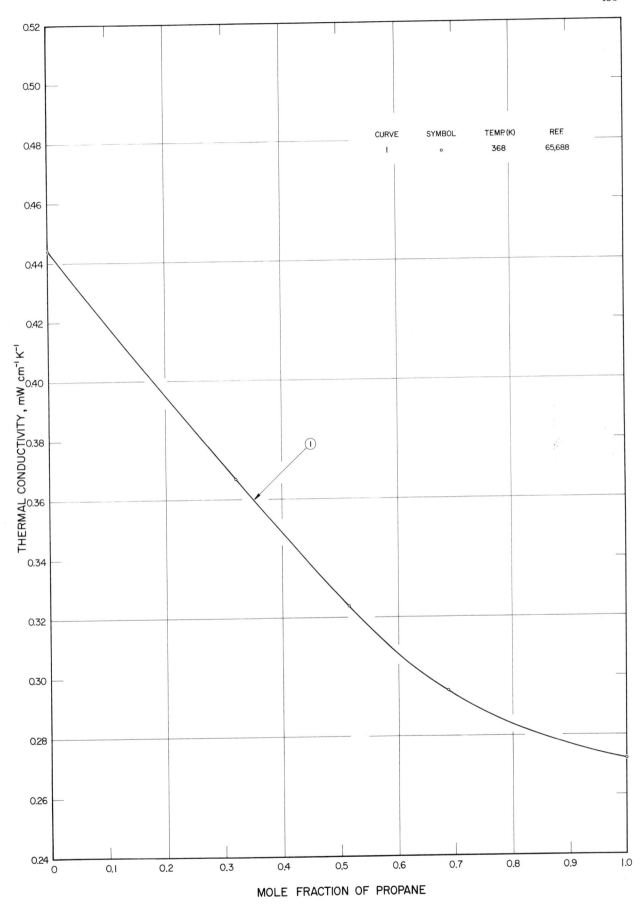

FIGURE 120. THERMAL CONDUCTIVITY OF METHANE-PROPANE SYSTEM

TABLE 121a. EXPERIMENTAL THERMAL CONDUCTIVITY DATA FOR NITROGEN-OXYGEN SYSTEM

Curve No.	Fig. No.	Ref. No.	Author(s)	Temp. (K)	Mole Fraction of O_2	Thermal Cond. (mW cm^{-1} K^{-1})	Remarks
1, 1'	121a	696, 695	Gupta, G. P. and Saxena, S. C.	313.2	0.000 0.249 0.529 0.762 1.000	0.268 0.272 0.271 0.272 0.281	N_2 and O_2: 99.95% pure; thick hot wire method; precision ±1%, accuracy ±1 to ±2%.
2	121a	696, 695	Gupta, G. P. and Saxena, S. C.	338.2	0.000 0.249 0.529 0.762 1.000	0.290 0.291 0.291 0.295 0.291	Same as above.
3	121a	696, 695	Gupta, G. P. and Saxena, S. C.	366.2	0.000 0.249 0.529 0.762 1.000	0.312 0.313 0.313 0.310 0.313	Same as above.
4	121a	707, 710	Tondon, P. K. and Saxena, S. C.	368.2	0.000 0.227 0.514 0.782 1.000	0.3052 0.3140 0.3308 0.3249 0.3270	N_2 and O_2: 99.95% pure; thick hot wire method; precision ±1%, accuracy ±1 to ±2%.
5	121a	707, 710	Tondon, P. K. and Saxena, S. C.	408.2	0.000 0.227 0.514 0.782 1.000	0.3329 0.3400 0.3554 0.3538 0.3450	Same as above.
6	121a	707, 710	Tondon, P. K. and Saxena, S. C.	448.2	0.000 0.227 0.514 0.782 1.000	0.3622 0.3588 0.3743 0.3806 0.3776	Same as above.
7	121b	65, 688	Cheung, H., Bromley, L. A., and Wilke, C. R.	590.2 592.2 592.2	0.000 0.6098 1.000	0.4467 0.4685 0.4865	N_2: Linde Air Products Co., water pumped, 99.9% pure, chief impurities Ar and Ne; O_2: Liquid Carbonic Co., commercial grade, 99.5% pure, chief impurity Ar; coaxial cylinder method; avg error 1.2%, max error 2%.
7	121b	688, 713	Cheung, H., Bromley, L. A., and Wilke, C. R.	592.2	0.000	0.4478	Their value at 590.2 K given above is reduced to refer to 592.2 K on the basis of data of Gupta et al. given below.
--	--	713	Gupta, G. P., Saxena, V. K., and Saxena, S. C.	592.2	0.000	0.4340	N_2: 99.95% pure; conductivity column method; precision ±2%.

TABLE 121b. THERMAL CONDUCTIVITY (mW cm^{-1}K^{-1}) OF NITROGEN-OXYGEN SYSTEM AS A FUNCTION OF COMPOSITION AT THE TEMPERATURE OF MEASUREMENT AS DERIVED BY GRAPHICAL SMOOTHING

Mole Fraction of O$_2$	313.2 K (Ref. 696)	313.2 K (Ref. 696)	338.2 K (Ref. 696)	366.2 K (Ref. 696)
	Curve 1	Curve 1'		
0.00	0.268	0.268	0.290	0.312
0.05	0.269	0.269	0.290	0.312
0.10	0.270	0.269	0.291	0.312
0.15	0.270	0.270	0.291	0.313
0.20	0.271	0.270	0.292	0.313
0.25	0.272	0.271	0.292	0.313
0.30	0.273	0.271	0.292	0.313
0.35	0.274	0.272	0.292	0.313
0.40	0.274	0.272	0.292	0.313
0.45	0.275	0.273	0.293	0.313
0.50	0.276	0.273	0.293	0.3131
0.55	0.276	0.273	0.293	0.313
0.60	0.277	0.274	0.293	0.314
0.65	0.278	0.274	0.293	0.314
0.70	0.278	0.275	0.293	0.314
0.75	0.279	0.275	0.292	0.314
0.80	0.279	0.275	0.292	0.314
0.85	0.280	0.276	0.292	0.313
0.90	0.280	0.276	0.292	0.313
0.95	0.281	0.276	0.291	0.313
1.00	0.281	0.277	0.291	0.313

Mole Fraction of O$_2$	368.2 K (Ref. 707)	408.2 K (Ref. 707)	448.2 K (Ref. 707)	592.2 K (Ref. 65)
0.00	0.305	0.333	0.362	0.448
0.05	0.308	0.336	0.364	0.449
0.10	0.311	0.338	0.365	0.451
0.15	0.314	0.341	0.366	0.452
0.20	0.317	0.343	0.368	0.453
0.25	0.319	0.346	0.369	0.455
0.30	0.321	0.348	0.371	0.457
0.35	0.322	0.349	0.372	0.458
0.40	0.324	0.351	0.373	0.460
0.45	0.325	0.352	0.375	0.462
0.50	0.326	0.353	0.376	0.464
0.55	0.327	0.354	0.377	0.466
0.60	0.328	0.354	0.378	0.468
0.65	0.329	0.355	0.379	0.470
0.70	0.329	0.354	0.379	0.473
0.75	0.329	0.354	0.379	0.475
0.80	0.329	0.353	0.379	0.477
0.85	0.329	0.351	0.379	0.480
0.90	0.329	0.350	0.379	0.482
0.95	0.328	0.344	0.378	0.484
1.00	0.327	0.345	0.378	0.487

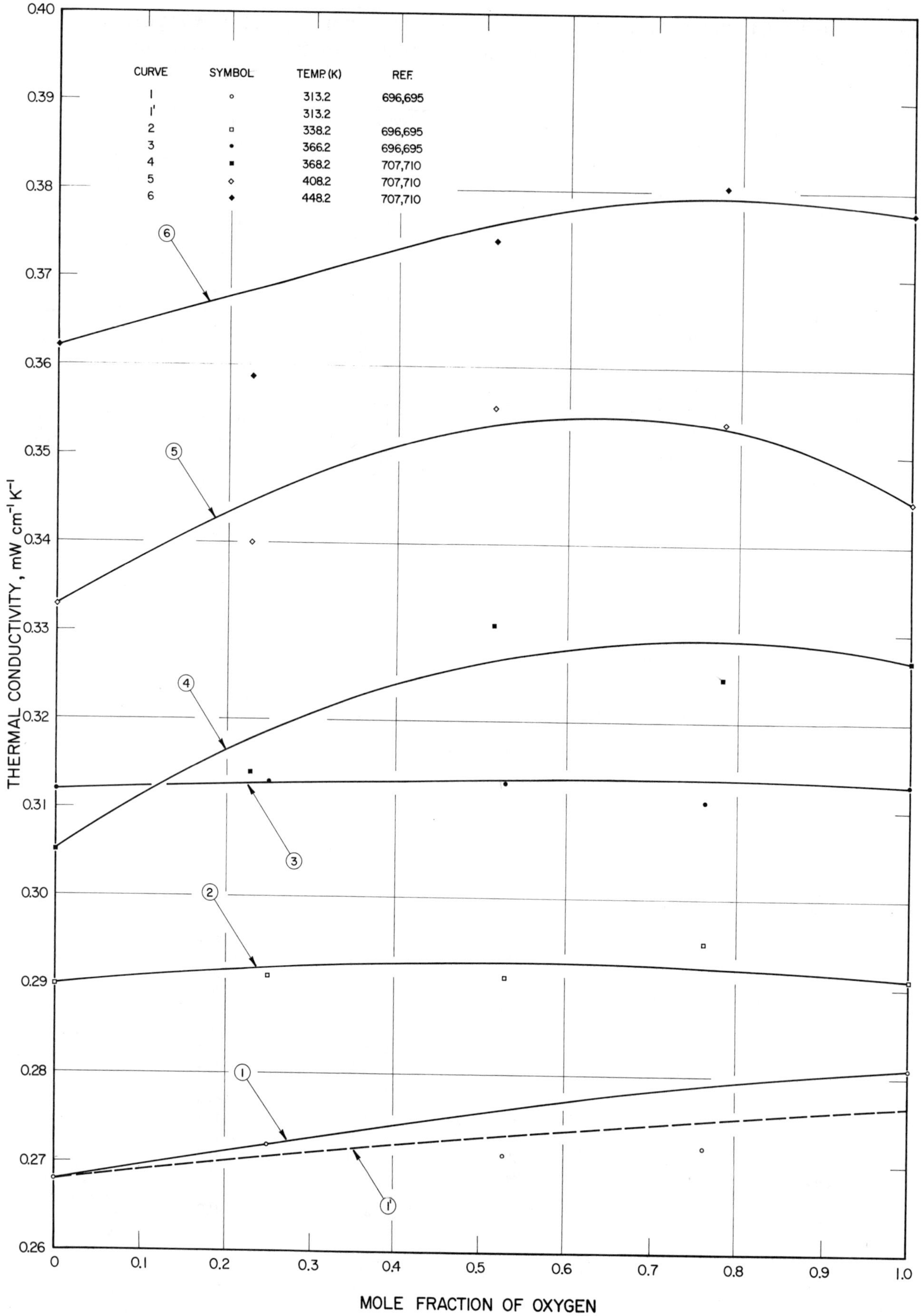

FIGURE 121a. THERMAL CONDUCTIVITY OF NITROGEN-OXYGEN SYSTEM

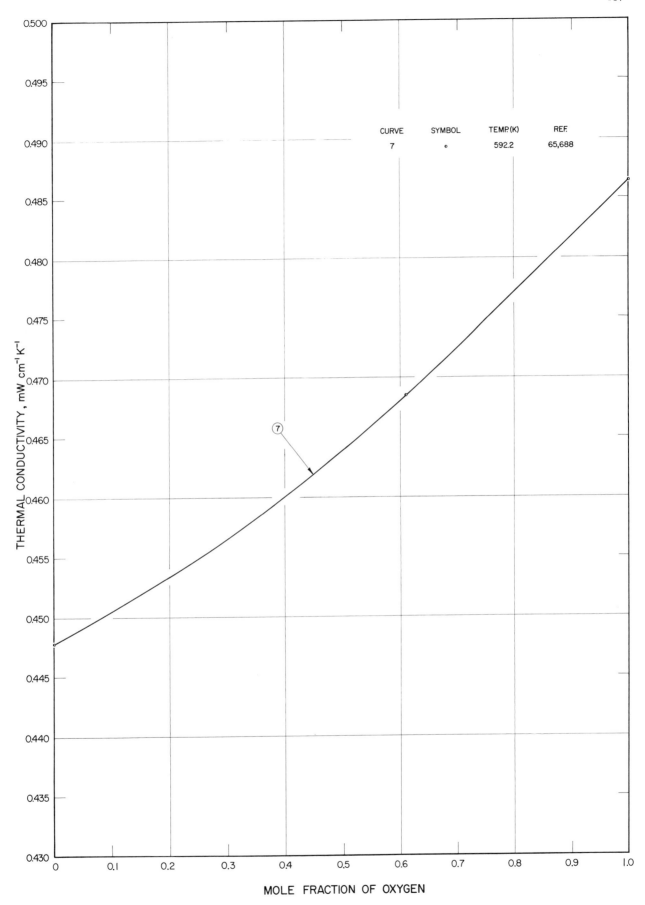

FIGURE 121b. THERMAL CONDUCTIVITY OF NITROGEN–OXYGEN SYSTEM

TABLE 122a. EXPERIMENTAL THERMAL CONDUCTIVITY DATA FOR NITROGEN-PROPANE SYSTEM

Curve No.	Fig. No.	Ref. No.	Author(s)	Temp. (K)	Mole Fraction of C_3H_8	Thermal Cond. (mW cm^{-1} K^{-1})	Remarks
1	122	65, 688	Cheung, H., Bromley, L.A., and Wilke, C.R.	590.2 591.2 591.2	0.0000 0.4747 1.0000	0.4467 0.5468 0.6134	N_2: Linde Air Products Co., water pumped, 99.9% pure, chief impurity Ar and Ne, C_3H_8: Matheson Co., instrument grade, 99.9% pure, coaxial cylinder method; average error 1.2%, max error 2%.
2	122	65, 688	Cheung, H., Bromley, L.A., and Wilke, C.R.	811.2 810.2	0.5239 1.0000	0.8788 1.050	
2	122	713, 700	Gupta, G.P., Saxena, V.K., and Saxena, S.C.	591.2 811.2	0.00 0.00	0.4333 0.5487	N_2: 99.95% pure; conductivity column method; precision ±2%.
1	122	688, 713	Cheung, H., Bromley, L.A., and Wilke, C.R.	591.2	0.00	0.4471	Their value at 590.2 K given above is reduced to refer to 592.2 K on the basis of data of Gupta et al. given above.

TABLE 122b. THERMAL CONDUCTIVITY (mW cm^{-1}K^{-1}) OF NITROGEN-PROPANE SYSTEM AS A FUNCTION OF COMPOSITION AT THE TEMPERATURE OF MEASUREMENT AS DERIVED BY GRAPHICAL SMOOTHING

Mole Fraction of C_3H_8	591.2 K (Ref. 688)	811 K (Ref. 65)
0.00	0.447	0.549
0.05	0.460	0.588
0.10	0.473	0.626
0.15	0.484	0.664
0.20	0.496	0.700
0.25	0.506	0.734
0.30	0.516	0.766
0.35	0.526	0.795
0.40	0.534	0.822
0.45	0.543	0.846
0.50	0.550	0.869
0.55	0.559	0.894
0.60	0.565	0.900
0.65	0.572	0.929
0.70	0.578	0.948
0.75	0.585	0.965
0.80	0.591	0.982
0.85	0.597	0.999
0.90	0.603	1.02
0.95	0.608	1.03
1.00	0.613	1.05

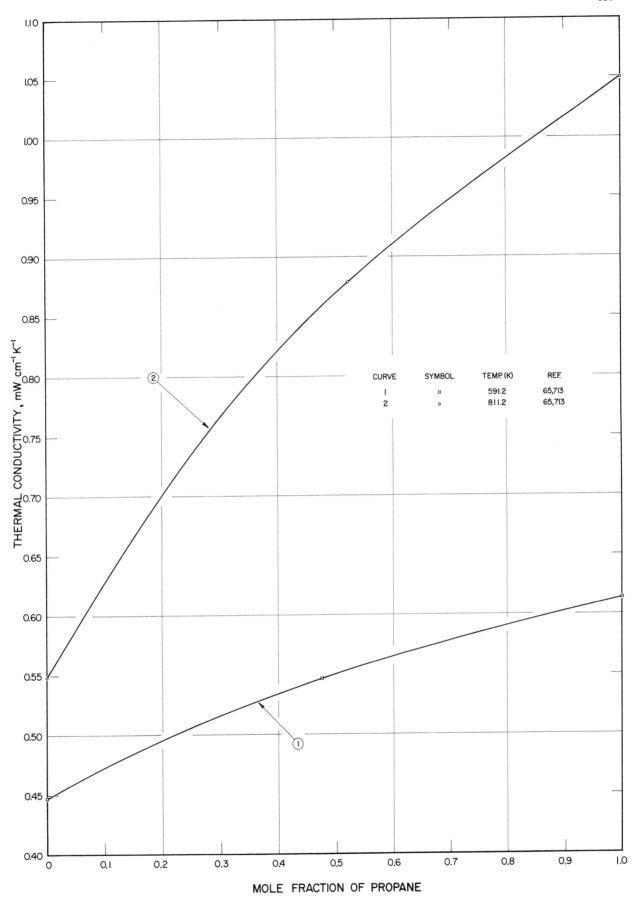

FIGURE 122. THERMAL CONDUCTIVITY OF NITROGEN—PROPANE SYSTEM

TABLE 123a. EXPERIMENTAL THERMAL CONDUCTIVITY DATA FOR ACETONE-BENZENE SYSTEM

Curve No.	Fig. No.	Ref. No.	Author(s)	Temp. (K)	Mole Fraction of Benzene	Thermal Cond. (mW cm^{-1} K^{-1})	Remarks
1	123	32	Bennett, L.A. and Vines, R.G.	349.9	0.00 0.25 0.50 0.75 1.00	0.1541 0.1561 0.1543 0.1486 0.1427	Acetone: analar grade, dried over potassium carbonate and distilled, C_6H_6: shaken with concentrated sulphuric acid, washed with water, dried over calcium chloride, and distilled; compensated hot wire method, estimated accuracy of these relative measurements ± 1%.
2	123	32	Bennett, L.A. and Vines, R.G.	376.0	0.00 0.25 0.50 0.75 1.00	0.1768 0.1799 0.1783 0.1743 0.1678	
3	123	32	Bennett, L.A. and Vines, R.G.	398.3	0.00 0.25 0.50 0.75 1.00	0.1991 0.2028 0.2004 0.1965 0.1897	Same as above.

TABLE 123b. THERMAL CONDUCTIVITY (mW cm^{-1} K^{-1}) OF ACETONE-BENZENE SYSTEM AS A FUNCTION OF COMPOSITION AT THE TEMPERATURE OF MEASUREMENT AS DERIVED BY GRAPHICAL SMOOTHING

Mole Fraction of Benzene	349.9 K (Ref. 32)	376.0 K (Ref. 32)	398.3 K (Ref. 32)
0.00	0.154	0.177	0.199
0.05	0.155	0.178	0.200
0.10	0.155	0.178	0.201
0.15	0.156	0.179	0.202
0.20	0.156	0.180	0.203
0.25	0.156	0.180	0.203
0.30	0.156	0.180	0.203
0.35	0.156	0.180	0.202
0.40	0.155	0.179	0.202
0.45	0.155	0.179	0.201
0.50	0.154	0.178	0.201
0.55	0.153	0.178	0.200
0.60	0.152	0.177	0.199
0.65	0.151	0.176	0.198
0.70	0.150	0.175	0.197
0.75	0.149	0.174	0.196
0.80	0.148	0.173	0.195
0.85	0.146	0.172	0.194
0.90	0.145	0.171	0.193
0.95	0.144	0.169	0.191
1.00	0.143	0.168	0.190

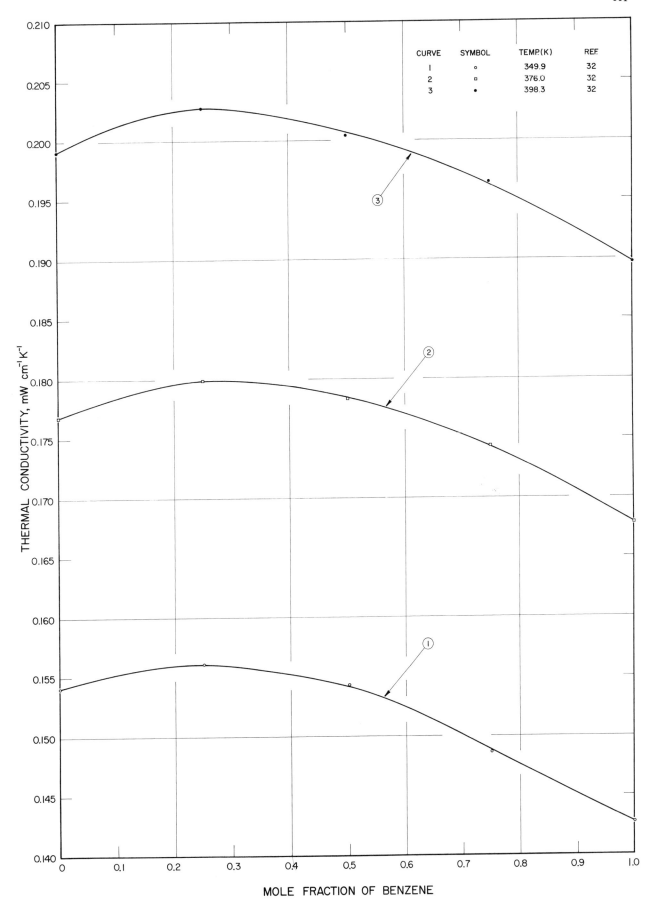

FIGURE 123. THERMAL CONDUCTIVITY OF ACETONE-BENZENE SYSTEM

TABLE 124a. EXPERIMENTAL THERMAL CONDUCTIVITY DATA FOR AMMONIA-AIR SYSTEM

Curve No.	Fig. No.	Ref. No.	Author(s)	Temp. (K)	Mole Fraction of Air	Thermal Cond. (mW cm^{-1} K^{-1})	Remarks
1	124	135	Gruss, H. and Schmick, H.	293.2	0.000 0.195 0.392 0.634 0.754 1.000	0.2300 0.2408 0.2549 0.2627 0.2641 0.2511	NH$_3$: prepared from commercial liquid ammonia and dried; air: dry and CO$_2$ free; vertical compensated hot wire method.
2	124	135	Gruss, H. and Schmick, H.	353.2	0.000 0.285 0.424 0.590 0.784 1.000	0.3013 0.3113 0.3151 0.3155 0.3084 0.2869	Same as above.

TABLE 124b. THERMAL CONDUCTIVITY (mW cm^{-1}K^{-1}) OF AMMONIA-AIR SYSTEM AS A FUNCTION OF COMPOSITION AT THE TEMPERATURE OF MEASUREMENT AS DERIVED BY GRAPHICAL SMOOTHING

Mole Fraction of Air	293.2 K (Ref. 135)	353.2 K (Ref. 135)
0.00	0.230	0.301
0.05	0.233	0.303
0.10	0.236	0.305
0.15	0.239	0.307
0.20	0.242	0.309
0.25	0.245	0.310
0.30	0.248	0.312
0.35	0.251	0.313
0.40	0.255	0.315
0.45	0.258	0.316
0.50	0.260	0.316
0.55	0.262	0.316
0.60	0.263	0.315
0.65	0.263	0.314
0.70	0.264	0.312
0.75	0.264	0.310
0.80	0.263	0.308
0.85	0.263	0.305
0.90	0.262	0.301
0.95	0.259	0.295
1.00	0.251	0.287

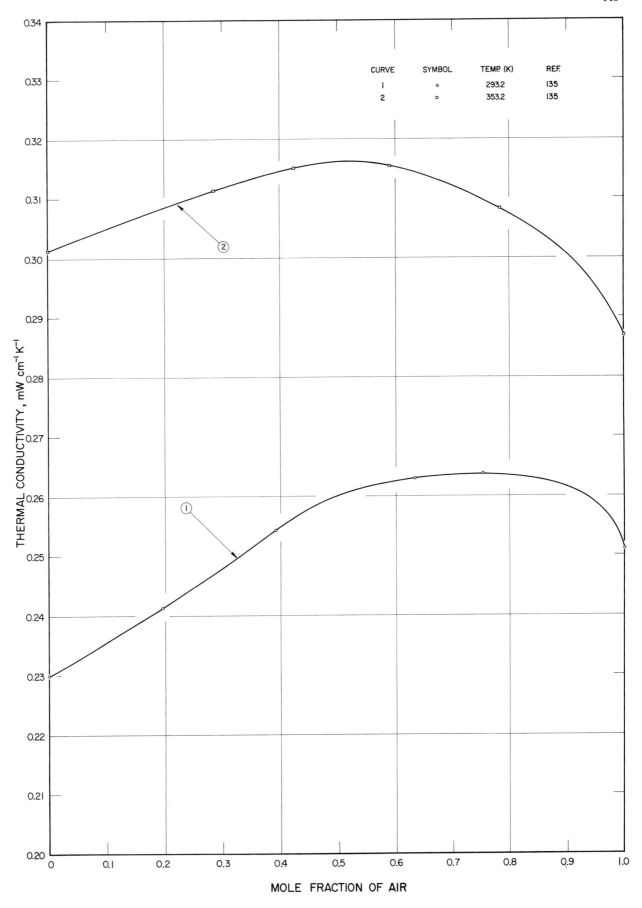

FIGURE 124. THERMAL CONDUCTIVITY OF AMMONIA–AIR SYSTEM

TABLE 125a. EXPERIMENTAL THERMAL CONDUCTIVITY DATA FOR AMMONIA-CARBON MONOXIDE SYSTEM

Curve No.	Fig. No.	Ref. No.	Author(s)	Temp. (K)	Mole Fraction of CO	Thermal Cond. (mW cm^{-1} K^{-1})	Remarks
1	125	135	Gruss, H. and Schmick, H.	295.2	0.000 0.210 0.380 0.662 0.780 1.000	0.2325 0.2439 0.2487 0.2523 0.2499 0.2404	NH$_3$: prepared from commercial liquid ammonia and dried; vertical compensated hot wire method.

TABLE 125b. THERMAL CONDUCTIVITY (mW cm^{-1}K^{-1}) OF AMMONIA-CARBON MONOXIDE SYSTEM AS A FUNCTION OF COMPOSITION AT THE TEMPERATURE OF MEASUREMENT AS DERIVED BY GRAPHICAL SMOOTHING

Mole Fraction of CO	295.2 K (Ref. 135)
0.00	0.233
0.05	0.235
0.10	0.238
0.15	0.241
0.20	0.243
0.25	0.245
0.30	0.247
0.35	0.248
0.40	0.249
0.45	0.250
0.50	0.250
0.55	0.251
0.60	0.251
0.65	0.251
0.70	0.251
0.75	0.251
0.80	0.251
0.85	0.250
0.90	0.249
0.95	0.247
1.00	0.240

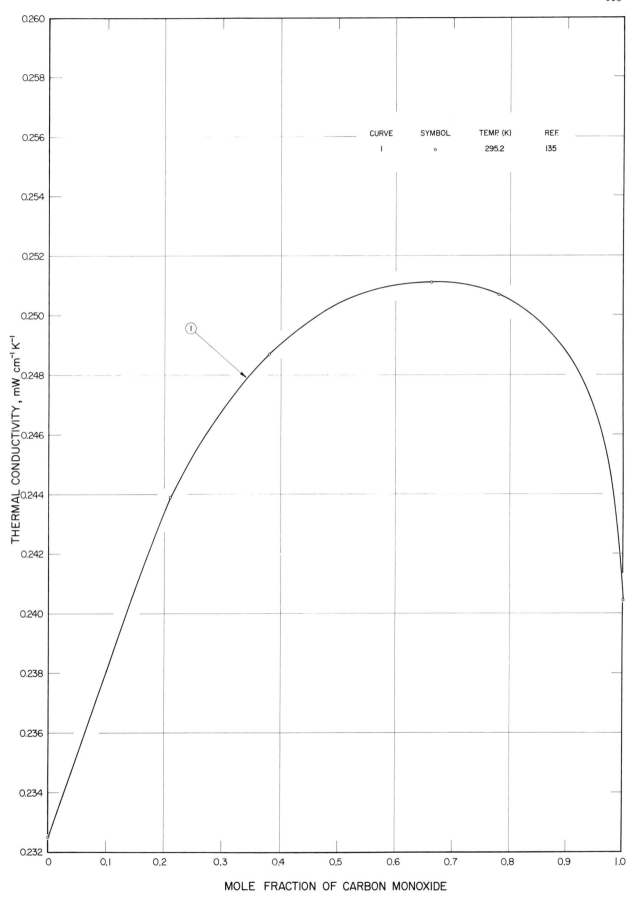

FIGURE 125. THERMAL CONDUCTIVITY OF AMMONIA-CARBON MONOXIDE SYSTEM

TABLE 126a. EXPERIMENTAL THERMAL CONDUCTIVITY DATA FOR AMMONIA-ETHYLENE SYSTEM

Curve No.	Fig. No.	Ref. No.	Author(s)	Temp. (K)	Mole Fraction of C_2H_4	Thermal Cond. (mW cm^{-1} K^{-1})	Remarks
1	126	204	Kornfeld, G. and Hilferding, K.	298.2	0.0000 0.2268 0.4121 0.7360 1.0000	0.2638 0.2620 0.2563 0.2409 0.2209	Compensated hot wire method.

TABLE 126b. THERMAL CONDUCTIVITY (mW cm^{-1}K^{-1}) OF AMMONIA-ETHYLENE SYSTEM AS A FUNCTION OF COMPOSITION AT THE TEMPERATURE OF MEASUREMENT AS DERIVED BY GRAPHICAL SMOOTHING

Mole Fraction of C_2H_4	298.2 K (Ref. 204)
0.00	0.264
0.05	0.264
0.10	0.263
0.15	0.262
0.20	0.262
0.25	0.261
0.30	0.259
0.35	0.258
0.40	0.252
0.45	0.255
0.50	0.253
0.55	0.251
0.60	0.244
0.65	0.246
0.70	0.243
0.75	0.240
0.80	0.237
0.85	0.233
0.90	0.230
0.95	0.225
1.00	0.221

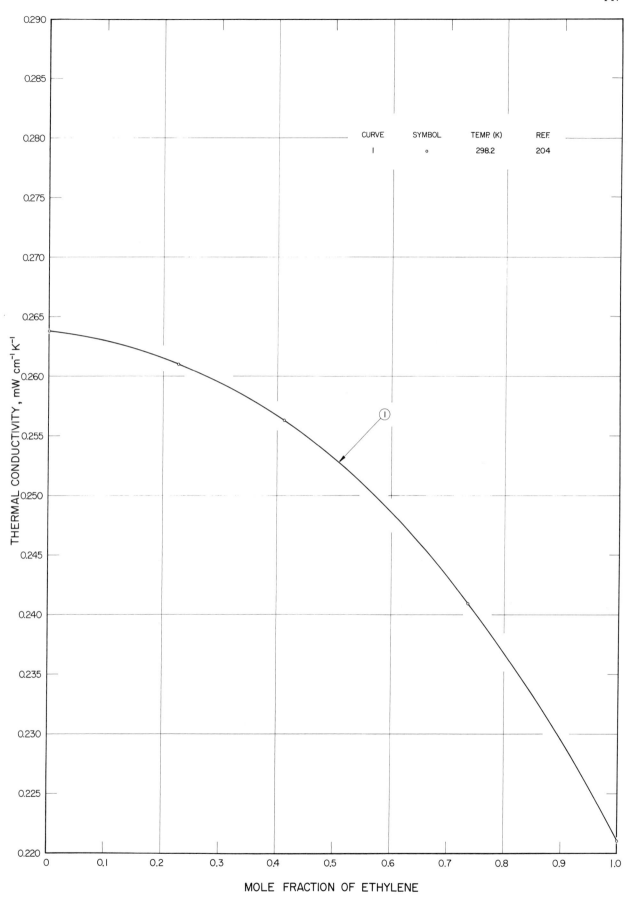

FIGURE 126. THERMAL CONDUCTIVITY OF AMMONIA-ETHYLENE SYSTEM

TABLE 127a. EXPERIMENTAL THERMAL CONDUCTIVITY DATA FOR AMMONIA-HYDROGEN SYSTEM

Curve No.	Fig. No.	Ref. No.	Author(s)	Temp. (K)	Mole Fraction of NH$_3$	Thermal Cond. (mW cm^{-1} K^{-1})	Remarks
--	--	690	Lindsay, A. L. and Bromley, L. A.	297.0	0.000	1.809	NH$_3$: 99.98% pure, H$_2$: 99.9% pure; unsteady state method; precision about 13% in the worst case.
				296.9	0.000	1.818	
				296.7	0.000	1.835	
				299.7	0.206	1.474	
				299.7	0.206	1.519	
				299.6	0.422	1.024	
				299.9	0.422	1.052	
				299.9	0.422	1.027	
				299.9	0.640	0.7524	
				299.8	0.640	0.6503	
				299.6	0.798	0.4981	
				299.5	0.798	0.4912	
				299.2	1.000	0.2715	
				298.9	1.000	0.2577	
1	127	588	Gray, P. and Wright, W. G.	298.5	0.000	1.763	NH$_3$: generated from aqueous solution and dried first by distillation and then over sodium, finally being distilled between liquid nitrogen traps, H$_2$: purified by diffusion through palladium; two wire type conductivity cell; accuracy of these relative measurements about 1%.
					0.090	1.545	
					0.145	1.424	
					0.208	1.256	
					0.324	1.068	
					0.416	0.9127	
					0.523	0.7662	
					0.599	0.6866	
					0.666	0.5987	
					0.769	0.4564	
					0.887	0.3529	
					1.000	0.2437	
2	127	588	Gray, P. and Wright, W. G.	348.0	0.0000	1.964	Same as above.
					0.1750	1.545	
					0.3875	1.141	
					0.4880	1.009	
					0.7510	0.6071	
					0.8315	0.4857	
					1.0000	0.2962	
3	127	588	Gray, P. and Wright, W. G.	372.3	0.0000	2.081	Same as above.
					0.0945	1.813	
					0.1650	1.679	
					0.4230	1.151	
					0.5980	0.8143	
					0.8200	0.5108	
					1.0000	0.3182	
4	127	588	Gray, P. and Wright, W. G.	422.5	0.0000	2.261	Same as above.
					0.0780	2.014	
					0.2760	1.503	
					0.4620	1.164	
					0.6850	0.7787	
					0.8675	0.5485	
					1.0000	0.3726	
5	127	690	Lindsay, A. L. and Bromley, L. A.	299	0.000	1.820	We have generated these data from the original reproduced above by averaging multiple values referring to the same composition of the mixture.
					0.206	1.496	
					0.422	1.034	
					0.640	0.7005	
					0.798	0.4947	
					1.000	0.2646	

TABLE 127b. THERMAL CONDUCTIVITY (mW cm^{-1}K^{-1}) OF AMMONIA-HYDROGEN SYSTEM AS A FUNCTION OF COMPOSITION AT THE TEMPERATURE OF MEASUREMENT AS DERIVED BY GRAPHICAL SMOOTHING

Mole Fraction of NH$_3$	298.5 K (Ref. 588)	299 K (Ref. 690)	348.0 K (Ref. 588)	372.3 K (Ref. 588)	422.5 K (Ref. 588)
0.00	1.76	1.82	1.96	2.08	2.26
0.05	1.64	1.72	1.84	1.95	2.10
0.10	1.52	1.62	1.72	1.81	1.95
0.15	1.41	1.53	1.60	1.69	1.81
0.20	1.29	1.43	1.48	1.59	1.68
0.25	1.20	1.34	1.39	1.48	1.56
0.30	1.10	1.25	1.30	1.39	1.46
0.35	1.02	1.16	1.21	1.29	1.36
0.40	0.940	1.07	1.13	1.20	1.28
0.45	0.866	0.988	1.05	1.10	1.19
0.50	0.796	0.908	0.966	0.995	1.09
0.55	0.730	0.831	0.890	0.900	1.00
0.60	0.668	0.758	0.816	0.812	0.916
0.65	0.608	0.688	0.744	0.735	0.833
0.70	0.550	0.620	0.674	0.665	0.756
0.75	0.494	0.554	0.607	0.597	0.684
0.80	0.439	0.492	0.540	0.534	0.617
0.85	0.388	0.432	0.476	0.476	0.551
0.90	0.337	0.376	0.414	0.422	0.490
0.95	0.289	0.321	0.354	0.368	0.431
1.00	0.244	0.265	0.296	0.318	0.373

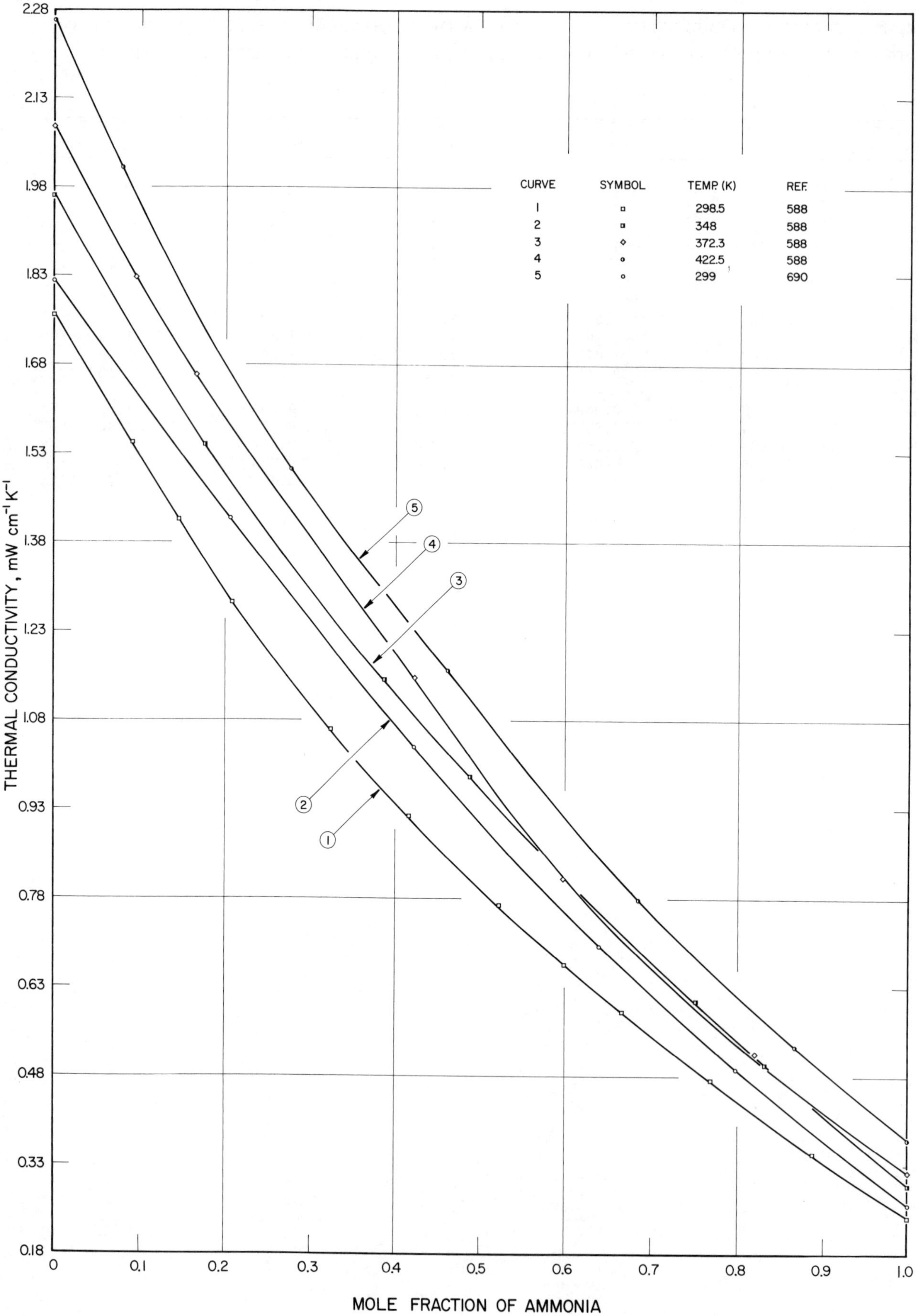

FIGURE 127. THERMAL CONDUCTIVITY OF AMMONIA—HYDROGEN SYSTEM

CURVE	SYMBOL	TEMP. (K)	REF.
1	□	298.5	588
2	◘	348	588
3	◇	372.3	588
4	○	422.5	588
5	○	299	690

TABLE 128a. EXPERIMENTAL THERMAL CONDUCTIVITY DATA FOR AMMONIA-NITROGEN SYSTEM

Curve No.	Fig. No.	Ref. No.	Author(s)	Temp. (K)	Mole Fraction of N_2	Thermal Cond. (mW cm^{-1} K^{-1})	Remarks
--	--	690	Lindsay, A. L. and Bromley, L. A.	299.2	0.000	0.2715	NH_3: 99.98% pure, N_2: 99.99% pure; unsteady state method; precision ± 15% in the worst case.
				298.9	0.000	0.2577	
				298.7	0.177	0.3667	
				298.6	0.177	0.3442	
				298.5	0.395	0.3684	
				298.5	0.395	0.3580	
				300.6	0.597	0.3304	
				300.6	0.597	0.2819	
				300.2	0.789	0.3649	
				300.2	0.789	0.3425	
				300.2	1.000	0.2698	
				300.2	1.000	0.2612	
				300.0	1.000	0.2560	
				298.0	1.000	0.2664	
				298.0	1.000	0.2594	
				298.0	1.000	0.2525	
1	128	588	Gray, P. and Wright, P. G.	298.5	0.0000	0.2437	NH_3: generated from aqueous solution and dried first by distillation and then over sodium, finally being distilled between liquid nitrogen traps, N_2: prepared from sodium azide; two wire type conductivity cell; accuracy of these relative measurements about 1%.
					0.0975	0.2541	
					0.1830	0.2567	
					0.2450	0.2583	
					0.3730	0.2675	
					0.5040	0.2663	
					0.5810	0.2709	
					0.6565	0.2709	
					0.7260	0.2726	
					0.9080	0.2663	
					1.0000	0.2596	
2	128	588	Gray, P. and Wright, P. G.	348.0	0.0000	0.2962	Same as above.
					0.1500	0.3174	
					0.3410	0.3316	
					0.5000	0.3358	
					0.5125	0.3333	
					0.6670	0.3312	
					0.8555	0.3249	
					1.0000	0.2927	
3	128	588	Gray, P. and Wright, P. G.	372.3	0.0000	0.3182	Same as above.
					0.0975	0.3224	
					0.3480	0.3506	
					0.6565	0.3345	
					0.8240	0.3148	
					1.0000	0.3098	
4	128	588	Gray, P. and Wright, P. G.	422.5	0.000	0.3726	Same as above.
					0.112	0.3944	
					0.709	0.3617	
					0.891	0.3617	
					1.000	0.3408	
5	128	690	Lindsay, A. L. and Bromley, L. A.	300	0.000	0.2646	We have generated these data from the original reproduced above by averaging the multiple values referring to the same composition of the mixture.
					0.177	0.3546	
					0.395	0.3632	
					0.597	0.3061	
					0.789	0.3528	
					1.000	0.2594	

TABLE 128b. THERMAL CONDUCTIVITY (mW cm^{-1}K^{-1}) OF AMMONIA-NITROGEN SYSTEM AS A FUNCTION OF COMPOSITION AT THE TEMPERATURE OF MEASUREMENT AS DERIVED BY GRAPHICAL SMOOTHING

Mole Fraction of N$_2$	298.5 K (Ref. 588)	300 K (Ref. 690)	348.0 K (Ref. 588)	372.3 K (Ref. 588)	422.5 K (Ref. 588)
0.00	0.244	0.265	0.296	0.318	0.373
0.05	0.248	0.330	0.304	0.324	0.384
0.10	0.251	0.344	0.311	0.330	0.391
0.15	0.255	0.351	0.317	0.334	0.395
0.20	0.258	0.356	0.322	0.338	0.397
0.25	0.261	0.360	0.326	0.341	0.398
0.30	0.263	0.362	0.329	0.344	0.398
0.35	0.265	0.363	0.332	0.346	0.399
0.40	0.267	0.363	0.334	0.347	0.397
0.45	0.269	0.362	0.335	0.347	0.395
0.50	0.270	0.361	0.335	0.347	0.392
0.55	0.271	0.358	0.336	0.346	0.389
0.60	0.271	0.354	0.335	0.345	0.385
0.65	0.271	0.350	0.335	0.342	0.380
0.70	0.271	0.344	0.333	0.339	0.375
0.75	0.270	0.337	0.331	0.335	0.370
0.80	0.270	0.328	0.328	0.331	0.364
0.85	0.268	0.316	0.324	0.326	0.359
0.90	0.267	0.301	0.318	0.321	0.353
0.95	0.264	0.283	0.308	0.316	0.347
1.00	0.259	0.259	0.293	0.310	0.341

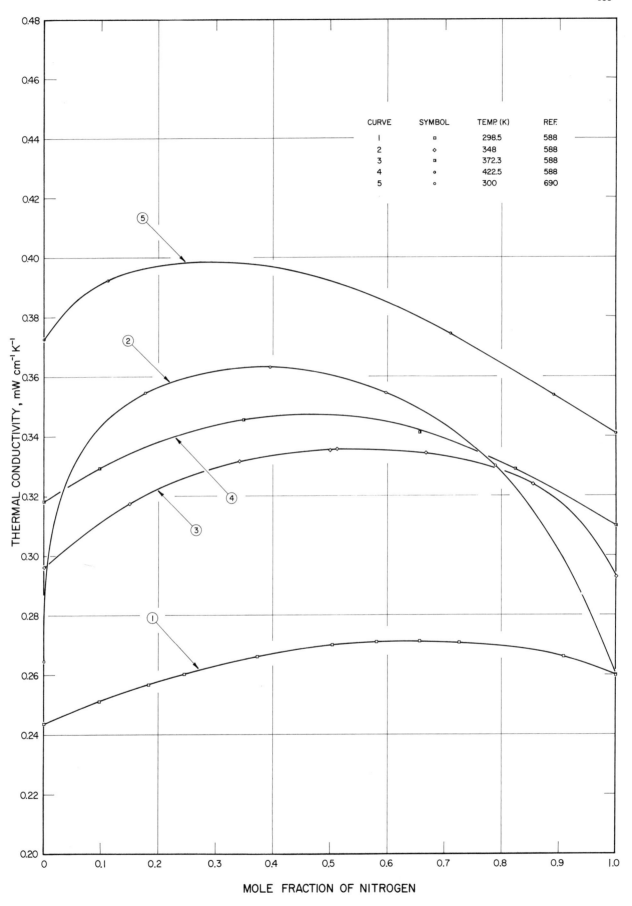

FIGURE 128. THERMAL CONDUCTIVITY OF AMMONIA-NITROGEN SYSTEM

TABLE 129a. EXPERIMENTAL THERMAL CONDUCTIVITY DATA FOR DIMETHYL-ETHER-ARGON SYSTEM

Curve No.	Fig. No.	Ref. No.	Author(s)	Temp. (K)	Mole Fraction of $(CH_3)_2O$	Thermal Cond. ($mW\ cm^{-1}\ K^{-1}$)	Remarks
1	129	65	Cheung, H., Bromley, L.A., and Wilke, C.R.	379.2 369.2 369.2 368.2 374.2	0.0000 0.3295 0.5123 0.6834 1.0000	0.2146 0.2170 0.2249 0.2284 0.2518	Ar: Linde Air Prod. Co., standard grade, specified purity 99.97%, chief impurity N_2 dimethyl ether: Matheson Co., specified purity 99.9%; coaxial cylinder method; total max error 2%, avg error 1.2% and max uncertainty in mixture composition 0.25%.

TABLE 129b. THERMAL CONDUCTIVITY ($mW\ cm^{-1}K^{-1}$) OF DIMETHYL-ETHER-ARGON SYSTEM AS A FUNCTION OF COMPOSITION AT THE TEMPERATURE OF MEASUREMENT AS DERIVED BY GRAPHICAL SMOOTHING

Mole Fraction of $(CH_3)_2O$	369 K (Ref. 65)
0.00	0.211
0.05	0.212
0.10	0.213
0.15	0.214
0.20	0.215
0.25	0.216
0.30	0.217
0.35	0.218
0.40	0.220
0.45	0.221
0.50	0.223
0.55	0.225
0.60	0.227
0.65	0.229
0.70	0.231
0.75	0.234
0.80	0.237
0.85	0.240
0.90	0.243
0.95	0.246
1.00	0.249

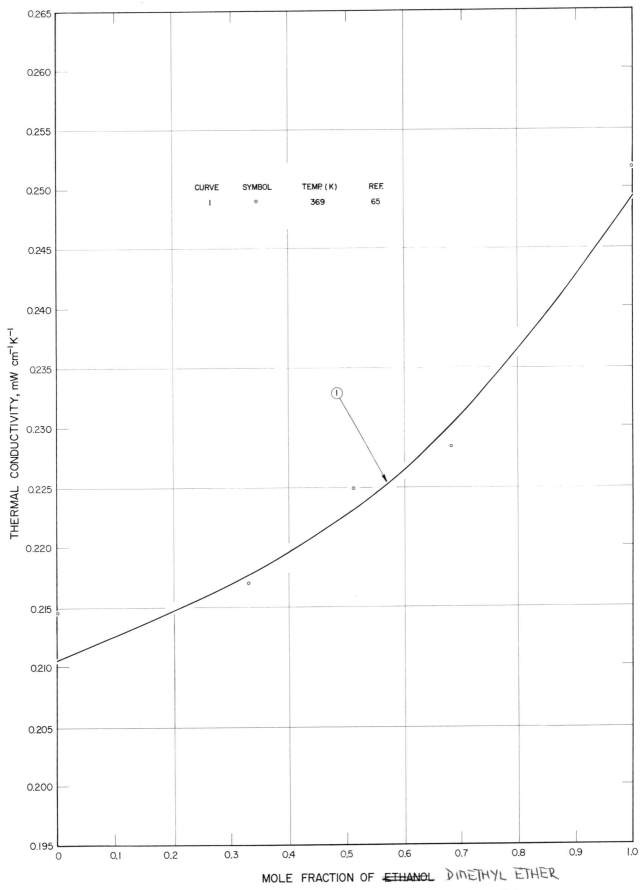

FIGURE 129. THERMAL CONDUCTIVITY OF DIMETHYL ETHER-ARGON SYSTEM

TABLE 130a. EXPERIMENTAL THERMAL CONDUCTIVITY DATA FOR DIMETHYL-ETHER-PROPANE SYSTEM

Curve No.	Fig. No.	Ref. No.	Author(s)	Temp. (K)	Mole Fraction of C_3H_8	Thermal Cond. (mW cm^{-1} K^{-1})	Remarks
1	130	65, 688	Cheung, H., Bromley, L.A., and Wilke, C.R.	374.2 368.2 368.2 373.2	0.0000 0.5017 0.6851 1.0000	0.2518 0.2587 0.2564 0.2721	$(CH_3)_2O$: Matheson Co., purity 99.9%, C_3H_8: Matheson Co., instrument grade, purity 99.9%; coaxial cylinder method; avg error 1.2%, max error 2%.
2	130	65, 688	Cheung, H., Bromley, L.A., and Wilke, C.R.	591.2 590.2 591.2	0.0000 0.4966 1.0000	0.5426 0.5769 0.6134	

TABLE 130b. THERMAL CONDUCTIVITY (mW cm^{-1}K^{-1}) OF DIMETHYL-ETHER-PROPANE SYSTEM AS A FUNCTION OF COMPOSITION AT THE TEMPERATURE OF MEASUREMENT AS DERIVED BY GRAPHICAL SMOOTHING

Mole Fraction of C_3H_8	368 K (Ref. 65)	591 K (Ref. 65)
0.00	0.252	0.543
0.05	0.252	0.546
0.10	0.253	0.549
0.15	0.253	0.553
0.20	0.253	0.556
0.25	0.254	0.559
0.30	0.254	0.563
0.35	0.255	0.566
0.40	0.255	0.570
0.45	0.256	0.574
0.50	0.256	0.577
0.55	0.257	0.581
0.60	0.258	0.584
0.65	0.258	0.588
0.70	0.259	0.592
0.75	0.260	0.595
0.80	0.262	0.599
0.85	0.264	0.603
0.90	0.266	0.606
0.95	0.269	0.610
1.00	0.272	0.613

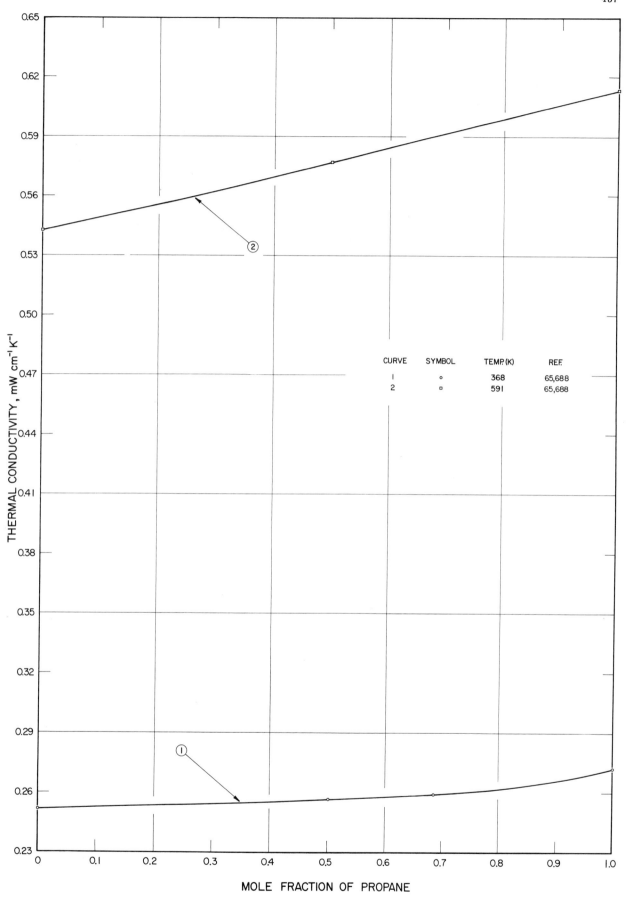

FIGURE 130. THERMAL CONDUCTIVITY OF DIMETHYL-ETHER–PROPANE SYSTEM

TABLE 131a. EXPERIMENTAL THERMAL CONDUCTIVITY DATA FOR METHANOL-ARGON SYSTEM

Curve No.	Fig. No.	Ref. No.	Author(s)	Temp. (K)	Mole Fraction of Ar	Thermal Cond. (mW cm^{-1} K^{-1})	Remarks
1	131	32	Bennett, L.A. and Vines, R.G.	351.2	0.00 0.25 0.50 0.75 1.00	0.1959 0.2024 0.2073 0.2094 0.2032	Methanol: refluxed with magnesium and a trace of iodine and distilled, Ar: 98% pure; compensated hot wire method; estimated accuracy of these relative measurements ±1%.
2	131	32	Bennett, L.A. and Vines, R.G.	373.2	0.00 0.25 0.50 0.75 1.00	0.2191 0.2225 0.2248 0.2237 0.2138	

TABLE 131b. THERMAL CONDUCTIVITY (mW cm^{-1} K^{-1}) OF METHANOL-ARGON SYSTEM AS A FUNCTION OF COMPOSITION AT THE TEMPERATURE OF MEASUREMENT AS DERIVED BY GRAPHICAL SMOOTHING

Mole Fraction of Ar	351.2 K (Ref. 32)	373.2 K (Ref. 32)
0.00	0.196	0.219
0.05	0.197	0.220
0.10	0.199	0.221
0.15	0.200	0.221
0.20	0.201	0.222
0.25	0.202	0.223
0.30	0.204	0.223
0.35	0.205	0.224
0.40	0.206	0.224
0.45	0.207	0.225
0.50	0.207	0.225
0.55	0.208	0.225
0.60	0.209	0.225
0.65	0.209	0.225
0.70	0.209	0.224
0.75	0.209	0.224
0.80	0.209	0.223
0.85	0.208	0.211
0.90	0.206	0.219
0.95	0.205	0.216
1.00	0.203	0.214

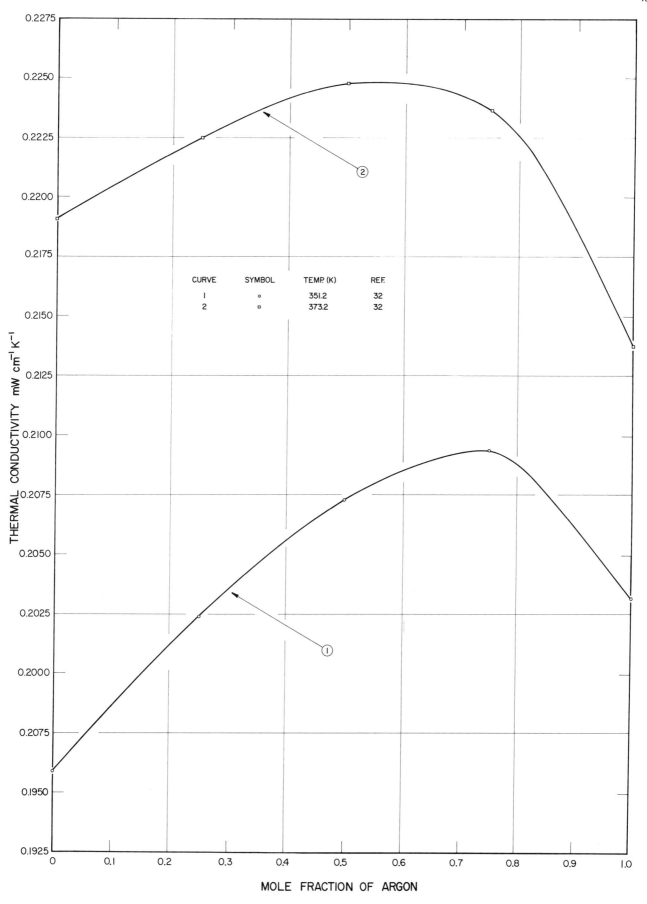

FIGURE 131. THERMAL CONDUCTIVITY OF METHANOL–ARGON SYSTEM

TABLE 132a. EXPERIMENTAL THERMAL CONDUCTIVITY DATA FOR METHANOL-HEXANE SYSTEM

Curve No.	Fig. No.	Ref. No.	Author(s)	Temp. (K)	Mole Fraction of Hexane	Thermal Cond. (mW cm^{-1} K^{-1})	Remarks
1	132	32	Bennett, L.A. and Vines, R.G.	351.2	0.00 0.25 0.50 0.75 1.00	0.1958 0.2030 0.2010 0.1916 0.1777	Methanol: refluxed with magnesium and a trace of iodine and distilled, C_6H_{14}: impurities less than a percent, Phillips, Oklahoma, pure grade; compensated hot wire method; estimated accuracy of these relative measurements ± 1%.
2	132	32	Bennett, L.A. and Vines, R.G.	371.6	0.00 0.25 0.50 0.75 1.00	0.2181 0.2257 0.2218 0.2127 0.1992	
3	132	32	Bennett, L.A. and Vines, R.G.	394.6	0.00 0.50 0.75 1.00	0.2416 0.2446 0.2398 0.2250	Same as above.

TABLE 132b. THERMAL CONDUCTIVITY (mW cm^{-1} K^{-1}) OF METHANOL-HEXANE SYSTEM AS A FUNCTION OF COMPOSITION AT THE TEMPERATURE OF MEASUREMENT AS DERIVED BY GRAPHICAL SMOOTHING

Mole Fraction of Hexane	351.2 K (Ref. 32)	371.6 K (Ref. 32)	394.6 K (Ref. 32)
0.00	0.196	0.218	0.242
0.05	0.198	0.220	0.242
0.10	0.200	0.222	0.243
0.15	0.201	0.224	0.243
0.20	0.202	0.225	0.243
0.25	0.203	0.226	0.244
0.30	0.203	0.226	0.244
0.35	0.203	0.225	0.244
0.40	0.203	0.224	0.244
0.45	0.202	0.223	0.245
0.50	0.201	0.222	0.245
0.55	0.200	0.220	0.244
0.60	0.198	0.219	0.244
0.65	0.196	0.217	0.243
0.70	0.193	0.215	0.242
0.75	0.192	0.213	0.240
0.80	0.189	0.210	0.237
0.85	0.186	0.208	0.235
0.90	0.184	0.205	0.232
0.95	0.181	0.201	0.228
1.00	0.178	0.199	0.225

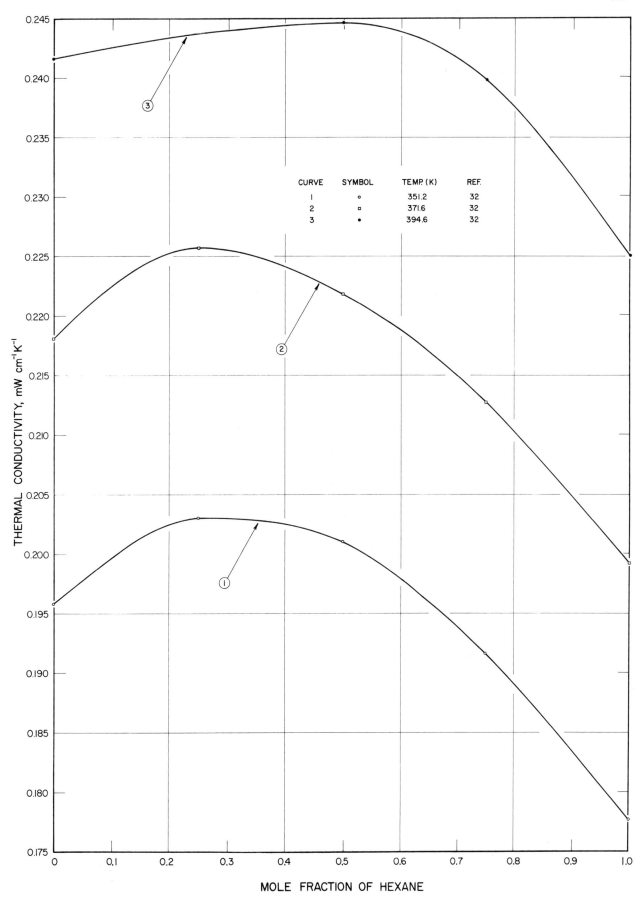

FIGURE 132. THERMAL CONDUCTIVITY OF METHANOL–HEXANE SYSTEM

TABLE 133a. EXPERIMENTAL THERMAL CONDUCTIVITY DATA FOR METHYL FORMATE-PROPANE SYSTEM

Curve No.	Fig. No.	Ref. No.	Author(s)	Temp. (K)	Mole Fraction of C_3H_8	Thermal Cond. (mW cm^{-1} K^{-1})	Remarks
1	133	688, 65	Cheung, H., Bromley, L.A., and Wilke, C.R.	372.2 369.2 373.2	0.00 0.53 1.00	0.1766 0.2276 0.2721	C_3H_8: Matheson Co., instrument grade, specified purity 99.9%, $C_2H_4O_2$: Eastman Kodak Co., S1227 spectro grade; coaxial cylinder method; average error 1.2%, max error 2%.

TABLE 133b. THERMAL CONDUCTIVITY (mW cm^{-1}K^{-1}) OF METHYL FORMATE-PROPANE SYSTEM AS A FUNCTION OF COMPOSITION AT THE TEMPERATURE OF MEASUREMENT AS DERIVED BY GRAPHICAL SMOOTHING

Mole Fraction of C_3H_8	372 K (Ref. 688)
0.00	0.177
0.05	0.181
0.10	0.186
0.15	0.191
0.20	0.196
0.25	0.201
0.30	0.205
0.35	0.210
0.40	0.215
0.45	0.220
0.50	0.225
0.55	0.230
0.60	0.234
0.65	0.239
0.70	0.244
0.75	0.249
0.80	0.253
0.85	0.258
0.90	0.263
0.95	0.267
1.00	0.272

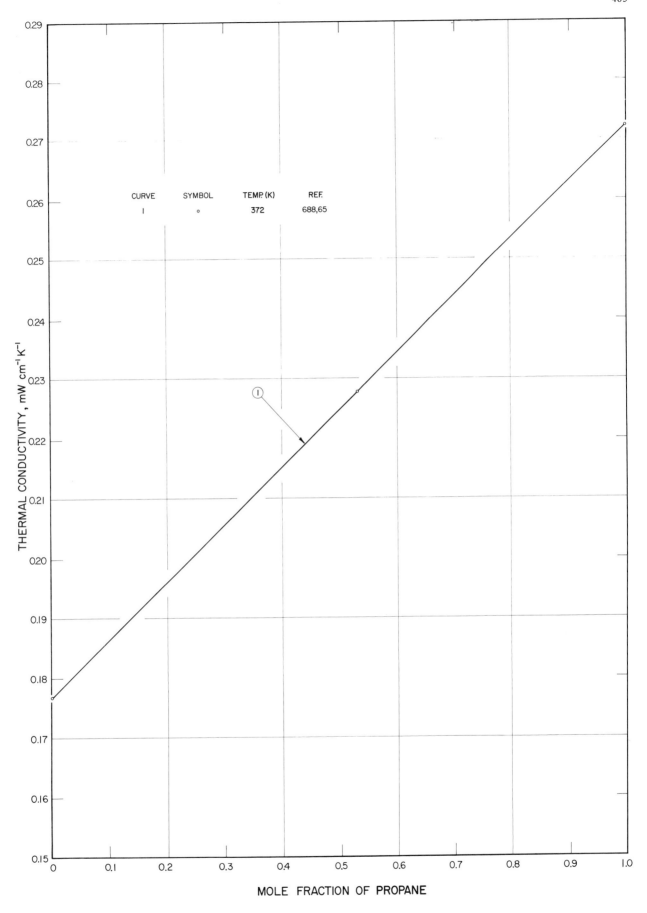

FIGURE 133. THERMAL CONDUCTIVITY OF METHYL FOMATE—PROPANE SYSTEM

TABLE 134a. EXPERIMENTAL THERMAL CONDUCTIVITY DATA FOR STEAM-AIR SYSTEM

Curve No.	Fig. No.	Ref. No.	Author(s)	Temp. (K)	Mole Fraction of Air	Thermal Cond. (mW cm^{-1} K^{-1})	Remarks
1	134	135	Gruss, H. and Schmick, H.	353.2	0.000 0.481 0.556 0.694 0.803 1.000	0.2190 0.2814 0.2885 0.2961 0.2992 0.2869	Air: dry and CO_2 free; vertical compensated hot wire method.

TABLE 134b. THERMAL CONDUCTIVITY (mW cm^{-1}K^{-1}) OF STEAM-AIR SYSTEM AS A FUNCTION OF COMPOSITION AT THE TEMPERATURE OF MEASUREMENT AS DERIVED BY GRAPHICAL SMOOTHING

Mole Fraction of Air	353.2 K (Ref. 134)
0.00	0.219
0.05	0.227
0.10	0.235
0.15	0.242
0.20	0.249
0.25	0.255
0.30	0.262
0.35	0.268
0.40	0.273
0.45	0.278
0.50	0.283
0.55	0.288
0.60	0.292
0.65	0.294
0.70	0.296
0.75	0.298
0.80	0.298
0.85	0.299
0.90	0.296
0.95	0.292
1.00	0.287

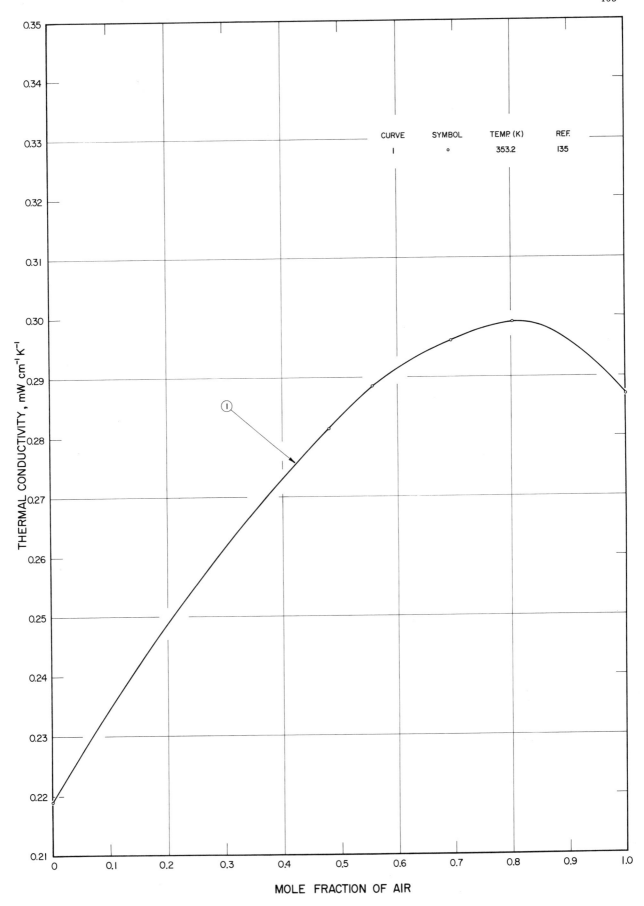

FIGURE 134. THERMAL CONDUCTIVITY OF STEAM—AIR SYSTEM

TABLE 135a. EXPERIMENTAL THERMAL CONDUCTIVITY DATA FOR STEAM-CARBON DIOXIDE SYSTEM

Curve No.	Fig. No.	Ref. No.	Author(s)	Temp. (K)	Mole Fraction of CO_2	Thermal Cond. (mW cm^{-1} K^{-1})	Remarks
1	135	358	Timrot, D. L. and Vargaftik, N. B.	338.2	0.000 0.614 0.805 0.900 1.000	0.2104 0.2220 0.2173 0.2092 0.1999	Thin hot wire method with constant current; estimated accuracy of the measurements 1-2%.
2	135	358	Timrot, D. L. and Vargaftik, N. B.	603.2	0.000 0.58 0.799 0.900 1.000	0.4719 0.4800 0.4602 0.4440 0.4161	Same as above.

TABLE 135b. THERMAL CONDUCTIVITY (mW cm^{-1} K^{-1}) OF STEAM-CARBON DIOXIDE SYSTEM AS A FUNCTION OF COMPOSITION AT THE TEMPERATURE OF MEASUREMENT AS DERIVED BY GRAPHICAL SMOOTHING

Mole Fraction of CO_2	338.2 K (Ref. 358)	603.2 K (Ref. 358)
0.00	0.210	0.472
0.05	0.212	0.473
0.10	0.214	0.473
0.15	0.215	0.474
0.20	0.216	0.475
0.25	0.218	0.476
0.30	0.219	0.476
0.35	0.220	0.477
0.40	0.221	0.478
0.45	0.221	0.478
0.50	0.222	0.479
0.55	0.222	0.480
0.60	0.222	0.479
0.65	0.222	0.476
0.70	0.221	0.472
0.75	0.220	0.467
0.80	0.218	0.460
0.85	0.214	0.453
0.90	0.210	0.443
0.95	0.205	0.431
1.00	0.200	0.416

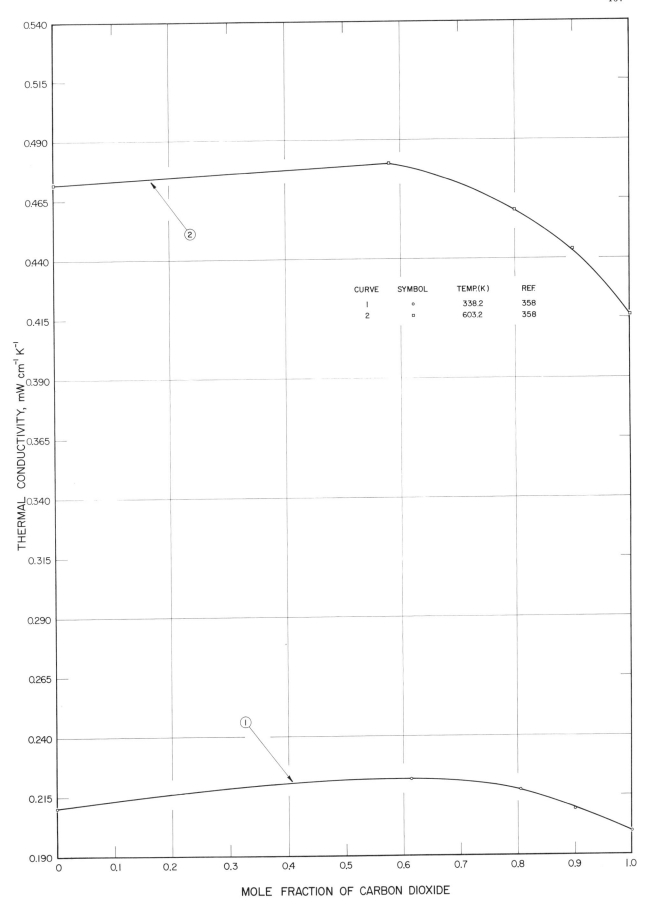

FIGURE 135. THERMAL CONDUCTIVITY OF STEAM—CARBON DIOXIDE SYSTEM

TABLE 136a. EXPERIMENTAL THERMAL CONDUCTIVITY DATA FOR STEAM-NITROGEN SYSTEM

Curve. No.	Fig. No.	Ref. No.	Author(s)	Temp. (K)	Mole Fraction of N_2	Thermal Cond. (mW cm^{-1} K^{-1})	Remarks
1	136	358	Timrot, D. L. and Vargaftik, N. B.	338.2	0.000 0.598 0.801 0.911 1.000	0.2150 0.2859 0.2940 0.2964 0.2917	Thin hot wire method with constant current; estimated accuracy of the measurement 1-2%.
2	136	358	Timrot, D. L. and Vargaftik, N. B.	603.2	0.000 0.602 0.805 0.901 1.000	0.4695 0.5242 0.5195 0.5009 0.4672	Same as above.

TABLE 136b. THERMAL CONDUCTIVITY (mW cm^{-1} K^{-1}) OF STEAM-NITROGEN SYSTEM AS A FUNCTION OF COMPOSITION AT THE TEMPERATURE OF MEASUREMENT AS DERIVED BY GRAPHICAL SMOOTHING

Mole Fraction of N_2	338.2 K (Ref. 358)	603.2 K (Ref. 358)
0.00	0.215	0.470
0.05	0.222	0.475
0.10	0.229	0.481
0.15	0.236	0.486
0.20	0.242	0.491
0.25	0.249	0.496
0.30	0.255	0.501
0.35	0.261	0.506
0.40	0.267	0.510
0.45	0.272	0.514
0.50	0.278	0.518
0.55	0.282	0.521
0.60	0.286	0.524
0.65	0.290	0.526
0.70	0.292	0.526
0.75	0.295	0.524
0.80	0.296	0.520
0.85	0.296	0.512
0.90	0.296	0.501
0.95	0.294	0.486
1.00	0.292	0.467

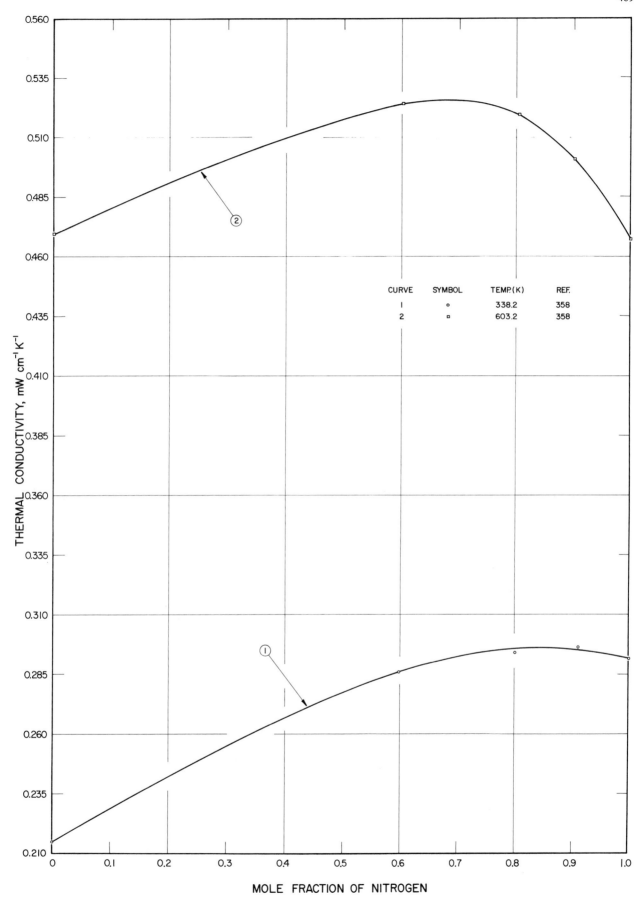

FIGURE 136. THERMAL CONDUCTIVITY OF STEAM−NITROGEN SYSTEM

TABLE 137a. EXPERIMENTAL THERMAL CONDUCTIVITY DATA FOR CHLOROFORM-ETHER SYSTEM

Curve No.	Fig. No.	Ref. No.	Author(s)	Temp. (K)	Mole Fraction of Chloroform	Thermal Cond. (mW cm^{-1} K^{-1})	Remarks
1	137	32	Bennett, L.A. and Vines, R.G.	332.5	0.00 0.25 0.50 0.75 1.00	0.1796 0.1583 0.1365 0.1072 0.08415	Chloroform: shaken with concentrated sulphuric acid, washed, and dried over potassium carbonate before distilling, ether: analar
2	137	32	Bennett, L.A. and Vines, R.G.	347.2	0.00 0.25 0.50 0.75 1.00	0.1943 0.1691 0.1440 0.1181 0.08918	grade, shaken with ferrous sulphate and distilled, dried over calcium chloride and stored over sodium; compensated hot wire method; estimated accuracy of these
3	137	32	Bennett, L.A. and Vines, R.G.	377.8	0.00 0.25 0.50 0.75 1.00	0.2278 0.1959 0.1687 0.1352 0.1017	relative measurements ± 1%.

TABLE 137b. THERMAL CONDUCTIVITY (mW cm^{-1} K^{-1}) OF CHLOROFORM-ETHER SYSTEM AS A FUNCTION OF COMPOSITION AT THE TEMPERATURE OF MEASUREMENT AS DERIVED BY GRAPHICAL SMOOTHING

Mole Fraction of Chloroform	332.5 K (Ref. 32)	347.2 K (Ref. 32)	377.8 K (Ref. 32)
0.00	0.180	0.194	0.228
0.05	0.176	0.181	0.222
0.10	0.171	0.178	0.216
0.15	0.167	0.173	0.210
0.20	0.163	0.172	0.204
0.25	0.158	0.168	0.199
0.30	0.154	0.164	0.193
0.35	0.149	0.160	0.187
0.40	0.145	0.155	0.181
0.45	0.140	0.150	0.174
0.50	0.135	0.145	0.168
0.55	0.130	0.139	0.164
0.60	0.125	0.134	0.155
0.65	0.120	0.129	0.148
0.70	0.115	0.124	0.142
0.75	0.110	0.118	0.135
0.80	0.105	0.113	0.129
0.85	0.0997	0.107	0.122
0.90	0.0945	0.101	0.115
0.95	0.0894	0.0952	0.109
1.00	0.0842	0.0892	0.102

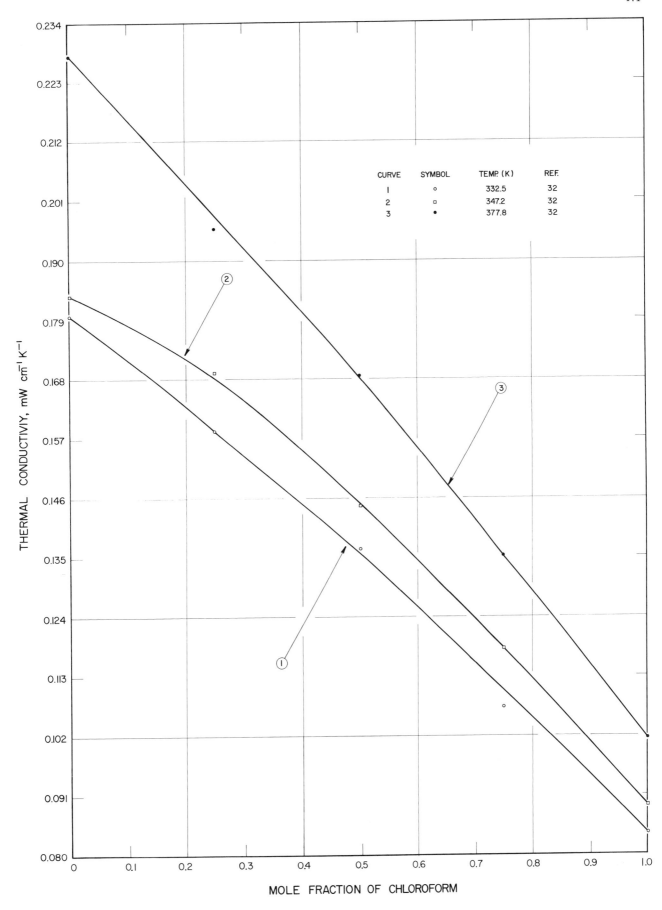

FIGURE 137. THERMAL CONDUCTIVITY OF CHLOROFORM—ETHER SYSTEM

TABLE 138a. EXPERIMENTAL THERMAL CONDUCTIVITY DATA FOR DIETHYLAMINE-ETHER SYSTEM

Curve No.	Fig. No.	Ref. No.	Author(s)	Temp. (K)	Mole Fraction of Ether	Thermal Cond. (mW cm^{-1} K^{-1})	Remarks
1	138	32	Bennett, L.A. and Vines, R.G.	335.6	0.00 0.25 0.50 0.75 1.00	0.1788 0.1798 0.1802 0.1806 0.1820	Diethylamine: dried with solid potassium hydroxide and carefully fractionated on distillation, ether: analar grade, shaken with ferrous sulphate and distilled, dried over calcium chloride and stored over sodium compensated hot wire method; estimated accuracy of these relative measurements ± 1%.
2	138	32	Bennett, L.A. and Vines, R.G.	369.4	0.00 0.25 0.50 0.75 1.00	0.2139 0.2142 0.2160 0.2160 0.2173	
3	138	32	Bennett, L.A. and Vines, R.G.	398.3	0.00 0.25 0.50 0.75 1.00	0.2496 0.2496 0.2507 0.2508 0.2508	Same as above.

TABLE 138b. THERMAL CONDUCTIVITY (mW cm^{-1} K^{-1}) OF DIETHYLAMINE-ETHER SYSTEM AS A FUNCTION OF COMPOSITION AT THE TEMPERATURE OF MEASUREMENT AS DERIVED BY GRAPHICAL SMOOTHING

Mole Fraction of Ether	335.6 K (Ref. 32)	369.4 K (Ref. 32)	398.3 K (Ref. 32)
0.00	0.179	0.214	0.250
0.05	0.179	0.214	0.250
0.10	0.179	0.214	0.250
0.15	0.179	0.214	0.250
0.20	0.179	0.215	0.250
0.25	0.180	0.215	0.250
0.30	0.180	0.215	0.250
0.35	0.180	0.215	0.250
0.40	0.180	0.215	0.251
0.45	0.180	0.215	0.251
0.50	0.180	0.216	0.251
0.55	0.181	0.216	0.251
0.60	0.181	0.216	0.251
0.65	0.181	0.216	0.251
0.70	0.181	0.216	0.251
0.75	0.181	0.216	0.251
0.80	0.181	0.217	0.251
0.85	0.182	0.217	0.251
0.90	0.182	0.217	0.251
0.95	0.182	0.217	0.251
1.00	0.182	0.217	0.251

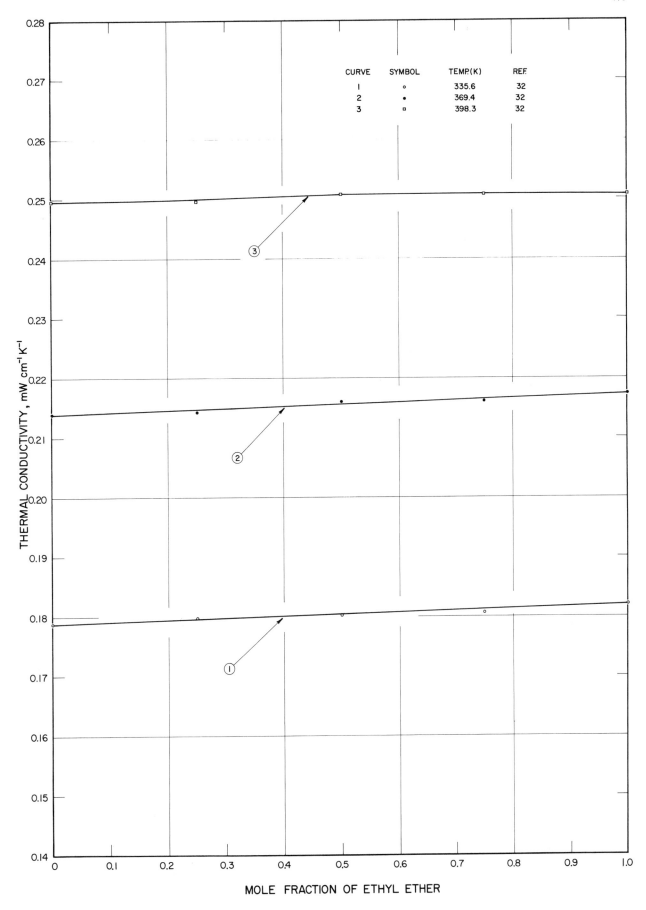

FIGURE 138. THERMAL CONDUCTIVITY OF DIETHYLAMINE—ETHYL ETHER SYSTEM

TABLE 139a. EXPERIMENTAL THERMAL CONDUCTIVITY DATA FOR DIMETHYL-ETHER-METHYL FORMATE SYSTEM

Curve No.	Fig. No.	Ref. No.	Author(s)	Temp. (K)	Mole Fraction of $C_2H_4O_2$	Thermal Cond. (mW cm^{-1} K^{-1})	Remarks
1	139	65, 688	Cheung, H., Bromley, L.A., and Wilke, C.R.	374.2 373.2 370.2 374.2 374.2	0.000 0.2752 0.5159 0.6941 1.0000	0.2518 0.2245 0.2063 0.1971 0.1766	$(CH_3)_2O$: Matheson Co., Inc., specified purity 99.9%, $C_2H_4O_2$: Eastman Kodak Co., S1227 spectro grade; coaxial cylinder method; avg error 1.2%, max error 2%.

TABLE 139b. THERMAL CONDUCTIVITY (mW cm^{-1} K^{-1}) OF DIMETHYL-ETHER-METHYL FORMATE SYSTEM AS A FUNCTION OF COMPOSITION AT THE TEMPERATURE OF MEASUREMENT AS DERIVED BY GRAPHICAL SMOOTHING

Mole Fraction of $C_2H_4O_2$	374 K (Ref. 65)
0.00	0.252
0.05	0.247
0.10	0.241
0.15	0.236
0.20	0.231
0.25	0.227
0.30	0.223
0.35	0.219
0.40	0.215
0.45	0.211
0.50	0.207
0.55	0.204
0.60	0.201
0.65	0.197
0.70	0.194
0.75	0.191
0.80	0.188
0.85	0.185
0.90	0.182
0.95	0.180
1.00	0.177

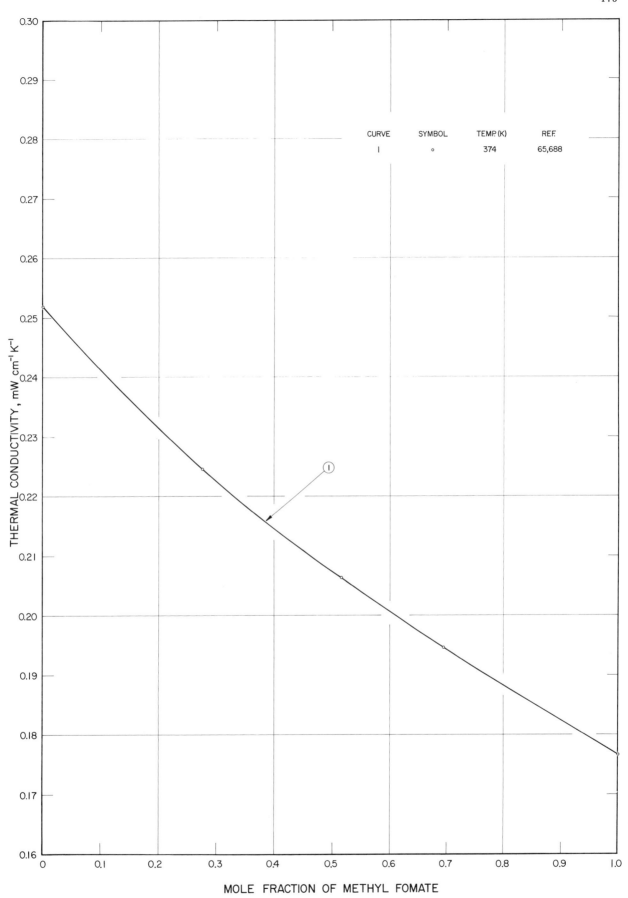

FIGURE 139. THERMAL CONDUCTIVITY OF ~~ETHANOL~~ DIMETHYL-ETHER—METHYL FOMATE SYSTEM

TERNARY SYSTEMS

The experimental thermal conductivity data for mixtures containing three different gases are reported in Tables 140 through 162. These data are, however, presented in four different categories depending upon the nature of the molecules in the mixture and their molecular structure. Group A deals with six systems of rare gases, while Group B deals with thirteen systems which involve both rare gases and nonpolar polyatomic gases. Two ternary systems which consist of only nonpolar polyatomic gases are dealt with in Group C. In category D, is presented data on two such systems which involve both nonpolar as well as polar gases.

TABLE 140.   EXPERIMENTAL THERMAL CONDUCTIVITY DATA FOR NEON-ARGON-KRYPTON SYSTEM

Author(s)	Ref. No.	Temp. (K)	Mole Fractions of Ne	Ar	Kr	Thermal Cond. (mW cm^{-1} K^{-1})	Remarks
Srivastava, B.N. and Saxena, S.C.	714	311.2	0.1387	0.7172	0.1441	0.188	Ne and Ar: spectroscopically pure, Kr: traces of Xe; thick hot wire method; precision ± 2%.
			0.1861	0.1449	0.6690	0.145	
			0.3019	0.3848	0.3133	0.201	
			0.4537	0.1567	0.3896	0.228	
			0.5984	0.1301	0.2715	0.276	
			0.1279	0.1569	0.7152	0.136	
			0.7919	0.1417	0.0664	0.0354	
			0	0	1	0.0976	
			0	1	0	0.183	
			1	0	0	0.495	
Mathur, S., Tondon, P.K., and Saxena, S.C.	692	313.2	0.099	0.599	0.302	0.283	Ne and Ar: spectroscopically pure, Kr: 99-100% pure, balance Xe; thick hot wire method; accuracy ± 1 to ± 2%, precision ± 1%.
			0.310	0.170	0.520	0.168	
			0.518	0.341	0.141	0.246	
			0.330	0.329	0.341	0.213	
			0	0	1	0.103	
			0	1	0	0.185	
			1	0	0	0.514	
Mathur, S., Tondon, P.K., and Saxena, S.C.	692	338.2	0.099	0.599	0.302	0.297	Same as above.
			0.310	0.170	0.520	0.178	
			0.518	0.341	0.141	0.257	
			0.330	0.329	0.341	0.225	
			0	0	1	0.110	
			0	1	0	0.200	
			1	0	0	0.534	
Mathur, S., Tondon, P.K., and Saxena, S.C.	692	363.2	0.099	0.599	0.302	0.309	Same as above.
			0.310	0.170	0.520	0.188	
			0.518	0.341	0.141	0.268	
			0.330	0.329	0.341	0.236	
			0	0	1	0.117	
			0	1	0	0.211	
			1	0	0	0.559	

TABLE 141. EXPERIMENTAL THERMAL CONDUCTIVITY DATA FOR HELIUM-ARGON-XENON SYSTEM

Author(s)	Ref. No.	Temp. (K)	Mole Fraction of He	Ar	Xe	Thermal Cond. (mW cm^{-1} K^{-1})	Remarks
Saxena, S.C.	715	311.2	0.1138	0.1495	0.7367	0.119	He and Ar: spectroscopically pure, Xe: traces of Kr; thick hot wire method; precision ± 2%.
			0.1967	0.1800	0.6233	0.168	
			0.3901	0.3675	0.2424	0.354	
			0.3202	0.6065	0.0733	0.353	
			0.6801	0.1880	0.1319	0.696	
			0	0	1	0.0565	
			0	1	0	0.183	
			1	0	0	1.57	

TABLE 142. EXPERIMENTAL THERMAL CONDUCTIVITY DATA FOR HELIUM-KRYPTON-XENON SYSTEM

Author(s)	Ref. No.	Temp. (K)	Mole Fractions of He	Kr	Xe	Thermal Cond. (mW cm^{-1} K^{-1})	Remarks
Mason, E.A. and von Ubisch, H.	685	302.2	0.219	0.086	0.695	0.170	He: Matheson Co., N.J., Kr and Xe: spectroscopically pure; thin hot wire method with constant resistance; precision ±2%.
			0.486	0.057	0.457	0.384	
			0.709	0.032	0.259	0.662	
			0.864	0.016	0.121	1.03	
			0.248	0.119	0.633	0.191	
			0.480	0.082	0.438	0.374	
			0.742	0.041	0.217	0.716	
			0.865	0.021	0.114	1.00	
			0.215	0.217	0.568	0.173	
			0.507	0.136	0.357	0.404	
			0.706	0.081	0.213	0.636	
			0.859	0.039	0.102	0.997	
			0.227	0.394	0.379	0.188	
			0.479	0.266	0.255	0.389	
			0.729	0.138	0.133	0.729	
			0.856	0.073	0.071	0.963	
			0.245	0.593	0.162	0.223	
			0.519	0.378	0.103	0.444	
			0.743	0.202	0.055	0.770	
			0.865	0.106	0.029	1.05	
			0	0	1	0.0598	
			0	1	0	0.0971	
			1	0	0	1.54	
Mason, E.A. and von Ubisch, H.	685	793.2	0.219	0.086	0.695	0.360	
			0.486	0.057	0.457	0.833	
			0.709	0.032	0.259	1.44	
			0.864	0.016	0.121	2.13	
			0.248	0.119	0.633	0.409	
			0.480	0.082	0.438	0.787	
			0.742	0.041	0.217	1.52	
			0.865	0.021	0.114	2.04	
			0.215	0.217	0.568	0.401	
			0.507	0.136	0.357	0.850	
			0.706	0.081	0.213	1.42	
			0.859	0.039	0.102	2.02	
			0.227	0.394	0.379	0.417	
			0.479	0.266	0.255	0.816	
			0.729	0.138	0.133	1.58	
			0.856	0.073	0.071	2.24	
			0.245	0.593	0.162	0.477	
			0.519	0.378	0.103	1.942	
			0.743	0.202	0.055	1.61	
			0.865	0.106	0.029	2.14	
			0	0	1	0.140	
			0	1	0	0.224	
			1	0	0	3.08	

TABLE 143. EXPERIMENTAL THERMAL CONDUCTIVITY DATA FOR HELIUM-ARGON-KRYPTON SYSTEM

Author(s)	Ref. No.	Temp. (K)	Mole Fractions of			Thermal Cond. (mW cm^{-1} K^{-1})	Remarks
			He	Ar	Kr		
Gambhir, R.S. and Saxena, S.C.	687	308.2	0.871	0.068	0.061	1.06	He and Ar: spectroscopically pure, Kr: 99-100% pure, balance Xe; thick hot wire method; accuracy ± 1 to ± 2%; precision ± 1%.
			0.775	0.077	0.148	0.842	
			0.689	0.109	0.202	0.729	
			0.143	0.584	0.274	0.219	
			0	0	1	0.0959	
			0	1	0	0.183	
			1	0	0	1.50	
Gambhir, R.S. and Saxena, S.C.	687	323.2	0.871	0.068	0.061	1.13	Same as above.
			0.775	0.077	0.148	0.888	
			0.689	0.109	0.202	0.749	
			0.143	0.584	0.273	0.233	
			0	0	1	0.102	
			0	1	0	0.191	
			1	0	0	1.55	
Gambhir, R.S. and Saxena, S.C.	687	343.2	0.871	0.068	0.061	1.19	Same as above.
			0.775	0.077	0.148	0.934	
			0.689	0.109	0.202	0.775	
			0.143	0.584	0.273	0.247	
			0	0	1	0.107	
			0	1	0	0.200	
			1	0	0	1.61	
Gambhir, R.S. and Saxena, S.C.	687	363.2	0.871	0.068	0.061	1.23	Same as above.
			0.775	0.077	0.148	0.967	
			0.689	0.109	0.202	0.800	
			0.143	0.584	0.273	0.260	
			0	0	1	0.114	
			0	1	0	0.210	
			1	0	0	1.67	

TABLE 144. EXPERIMENTAL THERMAL CONDUCTIVITY DATA FOR HELIUM-NEON-XENON SYSTEM

Author(s)	Ref. No.	Temp. (K)	Mole Fractions of He	Ne	Xe	Thermal Cond. (mW cm^{-1} K^{-1})	Remarks
Gandhi, J.M. and Saxena, S.C.	697	303.2	0.7486	0.1291	0.1223	0.899	He and Ne: spectroscopically pure, Xe: 99-100% pure; thick hot wire method; accuracy ±1 to ±2%, precision ±1%.
			0.5431	0.3667	0.0902	0.745	
			0.3899	0.3693	0.2408	0.473	
			0.1833	0.3291	0.4887	0.250	
			0.0855	0.3297	0.5848	0.179	
			0	0	1	0.0574	
			0	1	0	0.494	
			1	0	0	1.50	
Gandhi, J.M. and Saxena, S.C.	697	323.2	0.7486	0.1291	0.1223	0.935	Same as above.
			0.5431	0.3667	0.0902	0.780	
			0.3899	0.3693	0.2408	0.497	
			0.1822	0.3291	0.4887	0.263	
			0.0855	0.3297	0.5848	0.188	
			0	0	1	0.0611	
			0	1	0	0.515	
			1	0	0	1.55	
Gandhi, J.M. and Saxena, S.C.	697	343.2	0.7486	0.1291	0.1223	0.971	Same as above.
			0.5431	0.3667	0.0902	0.814	
			0.3899	0.3693	0.2408	0.523	
			0.1822	0.3291	0.4887	0.276	
			0.0855	0.3297	0.5848	0.197	
			0	0	1	0.0649	
			0	1	0	0.532	
			1	0	0	1.61	
Gandhi, J.M. and Saxena, S.C.	697	363.2	0.7486	0.1291	0.1223	1.01	Same as above.
			0.5431	0.3667	0.0902	0.847	
			0.3899	0.3693	0.2408	0.547	
			0.1822	0.3291	0.4887	0.289	
			0.0855	0.3297	0.5848	0.206	
			0	0	1	0.0682	
			0	1	0	0.553	
			1	0	0	1.67	

TABLE 145. EXPERIMENTAL THERMAL CONDUCTIVITY DATA FOR ARGON-KRYPTON-XENON SYSTEM

Author(s)	Ref. No.	Temp. (K)	Mole Fractions of			Thermal Cond. (mW cm^{-1} K^{-1})	Remarks
			Ar	Kr	Xe		
Mathur, S., Tondon, P.K., and Saxena, S.C.	692	311.3	0.249	0.252	0.499	0.0946	Ar: spectroscopically pure, Kr: 99-100% pure, balance Xe. Xe: 99-100% pure, balance Kr; thick hot wire method; accuracy ± 1 to ± 2%, precision ± 1%.
			0.742	0.136	0.122	0.159	
			0	0	1	0.0616	
			0	1	0	0.103	
			1	0	0	0.184	
Mathur, S., Tondon, P.K. and Saxena, S.C.	692	366.8	0.249	0.252	0.499	0.110	
			0.742	0.136	0.122	0.183	
			0	0	1	0.0741	
			0	1	0	0.118	
			1	0	0	0.212	

TABLE 146. EXPERIMENTAL THERMAL CONDUCTIVITY DATA FOR HELIUM-OXYGEN-METHANE SYSTEM

Author(s)	Ref. No.	Temp. (K)	Mole Fractions of			Thermal Cond. (mW cm^{-1} K^{-1})	Remarks
			He	O_2	CH_4		
Clingman, W.H., Brokaw, R.S., and Pease, R.N.	716, 718	273.2	0.743	0.197	0.060	0.837	O_2 and He: American Oxygen Co., c.p. methane: Matheson Co., A model RCT Gow-Mac thermal conductivity cell and bridge arrangement; relative measurements and calibration based on Ar-He data of Wachsmuth; the pure conductivity values are from the literature.
			0.712	0.188	0.100	0.816	
			0.684	0.181	0.135	0.779	
			0	0	1	0.301	
			0	1	0	0.242	
			1	0	0	1.42	

TABLE 147. EXPERIMENTAL THERMAL CONDUCTIVITY DATA FOR ARGON-OXYGEN-METHANE SYSTEM

Author(s)	Ref. No.	Temp. (K)	Mole Fractions of Ar	O_2	CH_4	Thermal Cond. (mW cm^{-1} K^{-1})	Remarks
Clingman, W.H., Brokaw, R.S., and Pease, R.N.	716, 718	273.2	0.751	0.199	0.050	0.192	Ar and c.p. CH_4: Matheson Co., O_2: American Oxygen Co.; A model RCT Gow-Mac thermal conductivity cell and bridge arrangement; relative measurements and calibration based on Ar-He data of Wachsmuth; the pure conductivity values are from the literature.
			0.712	0.188	0.100	0.197	
			0.677	0.179	0.144	0.206	
			0	0	1	0.301	
			0	1	0	0.242	
			1	0	0	0.163	

TABLE 148. EXPERIMENTAL THERMAL CONDUCTIVITY DATA FOR HELIUM-ARGON-NITROGEN SYSTEM

Author(s)	Ref. No.	Temp. (K)	Mole Fractions of			Thermal Cond. ($mW\ cm^{-1}\ K^{-1}$)	Remarks
			He	Ar	N_2		
Cheung, H., Bromley, L.A., and Wilke, C.R.	65, 688	372.7	0.415	0.117	0.468	0.584	He: U.S. Navy Research grade, specified purity 99.99%, chief impurities H_2 and H_2O. Ar: Linde Air Products Co., standard grade, specified purity 99.97%, chief impurity N_2. N_2: Linde Air Products Co., water pumped, specified purity 99.9%, chief impurities Ar and Ne; coaxial cylinder method; max error 5.7% and avg error 1.2%.
			0	0	1	0.309	
			0	1	0	0.212	
			1	0	0	1.75	

TABLE 149. EXPERIMENTAL THERMAL CONDUCTIVITY DATA FOR HELIUM-NITROGEN-METHANE SYSTEM

Author(s)	Ref. No.	Temp. (K)	Mole Fractions of			Thermal Cond. (mW cm^{-1} K^{-1})	Remarks
			He	N_2	CH_4		
Cheung, H., Bromley, L.A., and Wilke, C.R.	65, 688	590.2	0.159	0.476	0.365	0.701	He: U.S. Navy Research grade, specified purity 99.99%, chief impurities H_2 and H_2O, N_2: Linde Air Products Co., water pumped, specified purity 99.9%, chief impurities Ar and Ne, CH_4: Phillips Petroleum Co., specified purity 99%, chief impurity C_2H_6; coaxial cylinder method; max error 5.7% and avg error 1.2%.
			0	0	1	0.851	
			0	1	0	0.447	
			1	0	0	2.33	

TABLE 150. EXPERIMENTAL THERMAL CONDUCTIVITY DATA FOR ARGON-KRYPTON-DEUTERIUM SYSTEM

Author(s)	Ref. No.	Temp. (K)	Mole Fractions of			Thermal Cond. (mW cm^{-1} K^{-1})	Remarks
			Ar	Kr	D_2		
Gambhir, R.S. and Saxena, S.C.	698	308.2	0.173	0.218	0.609	0.620	Ar: spectroscopically pure, Kr: 99-100% pure, balance Xe. D_2: 98.6% pure, 0.8% H_2 and 0.6% H_2O; thick hot wire method; accuracy ± 1 to ± 2%, precision ± 1%.
			0.602	0.298	0.100	0.198	
			0	0	1	1.36	
			0	1	0	0.0959	
			1	0	0	0.183	
Gambhir, R.S. and Saxena, S.C.	698	323.2	0.173	0.218	0.609	0.649	
			0.602	0.298	0.100	0.209	
			0	0	1	1.40	
			0	1	0	0.102	
			1	0	0	0.191	
Gambhir, R.S. and Saxena, S.C.	698	343.2	0.173	0.218	0.609	0.682	Same as above.
			0.602	0.298	0.100	0.222	
			0	0	1	1.46	
			0	1	0	0.107	
			1	0	0	0.200	
Gambhir, R.S. and Saxena, S.C.	698	363.2	0.173	0.218	0.609	0.0255	Same as above.
			0.602	0.298	0.100	0.235	
			0	0	1	1.51	
			0	1	0	0.111	
			1	0	0	0.210	

TABLE 151. EXPERIMENTAL THERMAL CONDUCTIVITY DATA FOR HELIUM-NEON-DEUTERIUM SYSTEM

Author(s)	Ref. No.	Temp. (K)	Mole Fractions of He	Ne	D_2	Thermal Cond. (mW cm^{-1} K^{-1})	Remarks
Gandhi, J.M. and Saxena, S.C.	702	303.2	0.3059	0.2155	0.4786	1.11	He and Ne: spectroscopically pure, D_2: 98.6% pure, 0.8% H_2 and 0.6% H_2O; thick hot wire method; accuracy ±1 to ±2%, precision ±1%.
			0.2814	0.5636	0.1550	0.793	
			0.5484	0.2519	0.1997	1.11	
			0	0	1	1.34	
			0	1	0	0.494	
			1	0	0	1.50	
Gandhi, J.M. and Saxena, S.C.	702	323.2	0.3059	0.2155	0.4786	1.15	Same as above.
			0.2814	0.5636	0.1550	0.827	
			0.5484	0.2519	0.1997	1.15	
			0	0	1	1.40	
			0	1	0	0.515	
			1	0	0	1.55	
Gandhi, J.M. and Saxena, S.C.	702	343.2	0.3059	0.2155	0.4786	1.20	Same as above.
			0.2814	0.5636	0.1550	0.858	
			0.5484	0.2519	0.1997	1.19	
			0	0	1	1.46	
			0	1	0	0.532	
			1	0	0	1.61	
Gandhi, J.M. and Saxena, S.C.	702	363.2	0.3059	0.2155	0.4786	1.25	Same as above.
			0.2814	0.5636	0.1550	0.891	
			0.5484	0.2519	0.1997	1.23	
			0	0	1	1.51	
			0	1	0	0.553	
			1	0	0	1.67	

TABLE 152. EXPERIMENTAL THERMAL CONDUCTIVITY DATA FOR NEON-ARGON-DEUTERIUM SYSTEM

Authors(s)	Ref. No.	Temp. (K)	Mole Fractions of Ne	Ar	D_2	Thermal Cond. (mW cm^{-1} K^{-1})	Remarks
Mathur, S., Tondon, P.K., and Saxena, S.C.	711, 717	313.2	0.474	0.248	0.278	0.531	Ne and Ar: spectroscopically pure, D_2: 98.6% pure, 0.8% H_2 and 0.6% H_2O; thick hot wire method; accuracy ±1 to ±2%, precision ±1%.
			0.295	0.112	0.593	0.861	
			0.088	0.602	0.310	0.443	
			0	0	1	1.34	
			0	1	0	0.185	
			1	0	0	0.514	
Mathur, S., Tondon, P.K., and Saxena, S.C.	711, 717	338.2	0.474	0.248	0.278	0.577	Same as above.
			0.295	0.112	0.593	0.895	
			0.088	0.602	0.310	0.452	
			0	0	1	1.41	
			0	1	0	0.200	
			1	0	0	0.534	
Mathur, S., Tondon, P.K., and Saxena, S.C.	711, 717	363.2	0.474	0.248	0.278	0.606	Same as above.
			0.295	0.112	0.593	0.935	
			0.088	0.602	0.310	0.482	
			0	0	1	1.47	
			0	1	0	0.211	
			1	0	0	0.559	

TABLE 153. EXPERIMENTAL THERMAL CONDUCTIVITY DATA FOR NEON-KRYPTON-DEUTERIUM SYSTEM

Author(s)	Ref. No.	Temp. (K)	Mole Fractions of Ne	Kr	D_2	Thermal Cond. (mW cm^{-1} K^{-1})	Remarks
Mathur, S., Tondon, P.K., and Saxena, S.C.	711, 717	313.2	0.192	0.298	0.510	0.601	Ne: spectroscopically pure, Kr: 99-100% pure, balance Xe, D_2: 98.6% pure, 0.8% H_2 and 0.6% H_2O; thick hot wire method; accuracy ±1 to ±2%, precision ±1%.
			0.591	0.096	0.313	0.611	
			0.336	0.328	0.336	0.474	
			0.290	0.614	0.096	0.236	
			0	0	1	1.34	
			0	1	0	0.103	
			1	0	0	0.514	
Mathur, S., Tondon, P.K., and Saxena, S.C.	711, 717	338.2	0.192	0.298	0.510	0.635	Same as above.
			0.591	0.096	0.313	0.646	
			0.336	0.328	0.336	0.499	
			0.290	0.614	0.096	0.248	
			0	0	1	1.41	
			0	1	0	0.110	
			1	0	0	0.534	
Mathur, S., Tondon, P.K., and Saxena, S.C.	711, 717	363.2	0.192	0.298	0.510	0.672	Same as above.
			0.591	0.096	0.313	0.678	
			0.336	0.328	0.336	0.533	
			0.290	0.614	0.096	0.261	
			0	0	1	1.47	
			0	1	0	0.117	
			1	0	0	0.559	

TABLE 154. EXPERIMENTAL THERMAL CONDUCTIVITY DATA FOR NEON-HYDROGEN-OXYGEN SYSTEM

Author(s)	Ref. No.	Temp. (K)	Mole Fractions of			Thermal Cond. ($mW\ cm^{-1}\ K^{-1}$)	Remarks
			Ne	H_2	O_2		
Gupta, G.P. and Saxena, S.C.	695, 696	313.2	0.135	0.234	0.631	1.04	Ne: spectroscopically pure, H_2 and O_2: 99.95% pure; thick hot wire method; accuracy ±1 to ±2%, precision ±1%.
			0.215	0.655	0.130	0.437	
			0	0	1	0.281	
			0	1	0	1.82	
			1	0	0	0.496	
Gupta, G.P. and Saxena, S.C.	695, 696	338.2	0.135	0.234	0.631	1.18	Same as above.
			0.215	0.655	0.130	0.447	
			0	0	1	0.291	
			0	1	0	1.94	
			1	0	0	0.534	
Gupta, G.P. and Saxena, S.C.	695, 696	366.2	0.135	0.234	0.631	1.18	Same as above.
			0.215	0.655	0.130	0.490	
			0	0	1	0.313	
			0	1	0	2.04	
			1	0	0	0.563	

TABLE 155. EXPERIMENTAL THERMAL CONDUCTIVITY DATA FOR ARGON-HYDROGEN-NITROGEN SYSTEM

Author(s)	Ref. No.	Temp. (K)	Mole Fractions of Ar	H$_2$	N$_2$	Thermal Cond. (mW cm^{-1} K^{-1})	Remarks
Saxena, S.C. and Gupta, G.P.	699, 696	313.2	0.298	0.201	0.501	0.417	Ar: spectroscopically pure, H$_2$ and N$_2$: 99.95% pure; thick hot wire method; accuracy ±1 to ±2%, precision ±1%.
			0.103	0.608	0.289	0.918	
			0	0	1	0.268	
			0	1	0	1.82	
			1	0	0	0.185	
Saxena, S.C. and Gupta, G.P.	699, 696	338.2	0.298	0.201	0.501	0.439	Same as above.
			0.103	0.608	0.289	1.02	
			0	0	1	0.290	
			0	1	0	1.94	
			1	0	0	0.193	
Saxena, S.C. and Gupta, G.P.	699, 696	366.2	0.298	0.201	0.501	0.471	Same as above.
			0.103	0.608	0.289	1.07	
			0	0	1	0.312	
			0	1	0	2.04	
			1	0	0	0.210	

TABLE 156. EXPERIMENTAL THERMAL CONDUCTIVITY DATA FOR NEON-HYDROGEN-NITROGEN SYSTEM

Author(s)	Ref. No.	Temp. (K)	Mole Fractions of Ne	H_2	N_2	Thermal Cond. (mW cm^{-1} K^{-1})	Remarks
Tondon, P.K. and Saxena, S.C.	709, 707	313.2	0.242	0.512	0.246	0.900	Ne: spectroscopically pure, H_2 and N_2: 99.95% pure; thick hot wire method; accuracy ±1 to ±2%, precision ±1%.
			0.394	0.093	0.513	0.427	
			0	0	1	0.268	
			0	1	0	1.82	
			1	0	0	0.496	
Tondon, P.K. and Saxena, S.C.	709, 707	338.2	0.242	0.512	0.246	0.938	Same as above.
			0.394	0.093	0.513	0.440	
			0	0	1	0.290	
			0	1	0	1.94	
			1	0	0	0.534	
Tondon, P.K. and Saxena, S.C.	709, 707	366.2	0.242	0.512	0.246	1.06	Same as above.
			0.394	0.093	0.513	0.477	
			0	0	1	0.313	
			0	1	0	2.04	
			1	0	0	0.563	
Tondon, P.K. and Saxena, S.C.	709, 707	368.2	0.160	0.192	0.648	0.525	Same as above.
			0	0	1	0.305	
			0	1	0	2.11	
			1	0	0	0.571	
Tondon, P.K. and Saxena, S.C.	709, 707	408.2	0.160	0.192	0.648	0.598	Same as above.
			0	0	1	0.333	
			0	1	0	2.27	
			1	0	0	0.603	
Tondon, P.K. and Saxena, S.C.	709, 707	448.2	0.160	0.192	0.648	0.616	Same as above.
			0	0	1	0.362	
			0	1	0	2.37	
			1	0	0	0.641	

TABLE 157. EXPERIMENTAL THERMAL CONDUCTIVITY DATA FOR NEON-NITROGEN-OXYGEN SYSTEM

Author(s)	Ref. No.	Temp. (K)	Mole Fractions of Ne	N_2	O_2	Thermal Cond. (mW cm^{-1} K^{-1})	Remarks
Tondon, P.K. and Saxena, S.C.	709, 707	313.2	0.258	0.495	0.247	0.306	Ne: spectroscopically pure, N_2 and O_2: 99.95% pure; thick hot wire method; accuracy ± 1 to ± 2%, precision ± 1%.
			0.330	0.269	0.401	0.334	
			0	0	1	0.281	
			0	1	0	0.268	
			1	0	0	0.496	
Tondon, P.K. and Saxena, S.C.	709, 707	338.2	0.258	0.495	0.247	0.329	Same as above.
			0.330	0.269	0.401	0.343	
			0	0	1	0.291	
			0	1	0	0.290	
			1	0	0	0.534	
Tondon, P.K. and Saxena, S.C.	709, 707	366.2	0.258	0.495	0.247	0.344	Same as above.
			0.330	0.269	0.401	0.371	
			0	0	1	0.313	
			0	1	0	0.313	
			1	0	0	0.563	
Tondon, P.K. and Saxena, S.C.	709, 707	368.2	0.103	0.416	0.481	0.315	Same as above.
			0	0	1	0.327	
			0	1	0	0.305	
			1	0	0	0.571	
Tondon, P.K. and Saxena, S.C.	709, 707	408.2	0.103	0.416	0.481	0.351	Same as above.
			0	0	1	0.345	
			0	1	0	0.333	
			1	0	0	0.603	
Tondon, P.K. and Saxena, S.C.	709, 707	448.2	0.103	0.416	0.481	0.362	Same as above.
			0	0	1	0.378	
			0	1	0	0.362	
			1	0	0	0.641	

TABLE 158. EXPERIMENTAL THERMAL CONDUCTIVITY DATA FOR ARGON-KRYPTON-HYDROGEN SYSTEM

Author(s)	Ref. No.	Temp. (K)	Mole Fractions of			Thermal Cond. ($mW\ cm^{-1}\ K^{-1}$)	Remarks
			Ar	Kr	H_2		
Tondon, P.K. and Saxena, S.C.	708, 707	313.2	0.236	0.257	0.480	0.611	Ar: spectroscopically pure, Kr: 99-100% pure, balance Xe, H_2: 99.95% pure; thick hot wire method; accuracy ±1 to ±2%, precision ±1%.
			0.496	0.373	0.131	0.237	
			0	0	1	1.82	
			0	1	0	0.0984	
			1	0	0	0.185	
Tondon, P.K. and Saxena, S.C.	708, 707	338.2	0.236	0.257	0.480	0.682	Same as above.
			0.496	0.373	0.131	0.272	
			0	0	1	1.94	
			0	1	0	0.108	
			1	0	0	0.193	
Tondon, P.K. and Saxena, S.C.	708, 707	366.2	0.236	0.257	0.480	0.737	Same as above.
			0.496	0.373	0.131	0.291	
			0	0	1	2.04	
			0	1	0	0.114	
			1	0	0	0.210	

TABLE 159. EXPERIMENTAL THERMAL CONDUCTIVITY DATA FOR NITROGEN-OXYGEN-CARBON DIOXIDE SYSTEM

Author(s)	Ref. No.	Temp. (K)	Mole Fractions of			Thermal Cond. (mW cm^{-1} K^{-1})	Remarks
			N_2	O_2	CO_2		
Cheung, H., Bromley, L.A. and Wilke, C.R.	65, 688	370.2	0.3231	0.3729	0.3040	0.282	N_2: Linde Air Products Co., water pumped, specified purity 99.9%, chief impurities Ar and Ne, O_2: Liquid Carbonic Co., commercial grade, specified purity 99.99%, chief impurities H_2 and H_2O, CO_2: Pure Carbonic, Inc., specified purity 99.5%, chief impurity air; coaxial cylinder method; max error 2% and avg error 1.2%.
			0	0	1	0.219	
			0	1	0	0.321	
			1	0	0	0.307	

TABLE 160. EXPERIMENTAL THERMAL CONDUCTIVITY DATA FOR HYDROGEN-NITROGEN-OXYGEN SYSTEM

Author(s)	Ref. No.	Temp. (K)	Mole Fractions of H_2	N_2	O_2	Thermal Cond. (mW cm^{-1} K^{-1})	Remarks
Gupta, G.P. and Saxena, S.C.	695, 696	313.2	0.607	0.097	0.296	0.946	H_2, N_2 and O_2: 99.95% pure; thick hot wire method; accuracy ±1 to ±2%, precision ±1%.
			0.199	0.535	0.266	0.454	
			0	0	1	0.281	
			0	1	0	0.268	
			1	0	0	1.82	
Gupta, G.P. and Saxena, S.C.	695, 696	338.2	0.607	0.097	0.296	1.01	Same as above.
			0.199	0.535	0.266	0.475	
			0	0	1	0.291	
			0	1	0	0.290	
			1	0	0	1.94	
Gupta, G.P. and Saxena, S.C.	695, 696	366.2	0.607	0.097	0.296	1.08	Same as above.
			0.199	0.535	0.266	0.509	
			0	0	1	0.313	
			0	1	0	0.312	
			1	0	0	2.04	

TABLE 161. EXPERIMENTAL THERMAL CONDUCTIVITY DATA FOR ARGON-PROPANE-DIMETHYL-ETHER SYSTEM

Author(s)	Ref. No.	Temp. (K)	Mole Fractions of			Thermal Cond. $(mW\ cm^{-1}\ K^{-1})$	Remarks
			Ar	C_3H_8	$(CH_3)_2O$		
Cheung, H., Bromley, L.A., and Wilke, C.R.	65, 688	371.2 374.2 373.2 379.2	0.3660 0 0 1	0.3080 0 1 0	0.3260 1 0 0	0.242 0.252 0.272 0.215	Ar: Linde Air Products Co., standard grade, specified purity 99.97%, chief impurity N_2, C_3H_8: Matheson Co., instrument grade, specified purity 99.9%, $(CH_3)_2O$: Matheson Co., specified purity 99.9%; coaxial cylinder method; max error 2% and avg error 1.2%.
Cheung, H., Bromley, L.A., and Wilke, C.R.	65, 688	591.2 591.2 591.2 594.2	0.5348 0 0 1	0.2310 0 1 0	 1 0 0	0.461 0.543 0.613 0.306	

TABLE 162. EXPERIMENTAL THERMAL CONDUCTIVITY DATA FOR HYDROGEN-NITROGEN-AMMONIA SYSTEM

Author(s)	Ref. No.	Temp. (K)	Mole Fractions of H_2	N_2	NH_3	Thermal Cond. (mW cm^{-1} K^{-1})	Remarks
Gray, P. and Wright, P.G.	588	298.5	0.232	0.572	0.195	0.435	H_2: purified by diffusion through palladium, N_2: obtained from sodium azide, NH_3: generated from aqueous solution and dried first by distillation and then over sodium, finally being distilled between liquid nitrogen traps; two wire type conductivity cell; accuracy of these relative measurements about 1%.
			0.232	0.572	0.195	0.463	
			0.448	0.376	0.176	0.703	
			0.2775	0.086	0.6365	0.523	
			0.558	0.151	0.291	0.873	
			0.077	0.389	0.5335	0.366	
			0.1725	0.230	0.5975	0.417	
			0.135	0.6305	0.234	0.384	
			0.2615	0.6185	0.120	0.502	
			0.427	0.102	0.471	0.699	
			0.348	0.493	0.159	0.586	
			0.5455	0.2955	0.159	0.835	
			0.372	0.244	0.384	0.620	
			0.5845	0.282	0.133	0.925	
			0.556	0.091	0.353	0.886	
			0.632	0.238	0.130	0.888	
			0.6725	0.238	0.0895	1.04	
			0.771	0.117	0.112	1.31	
			0.6525	0.1215	0.226	1.03	
			0	0	1	0.244	
			0	1	0	0.260	
			1	0	0	1.76	
Gray, P. and Wright, P.G.	588	348.0	0.389	0.390	0.221	0.749	Same as above.
			0.276	0.568	0.156	0.624	
			0.468	0.4265	0.105	0.858	
			0.244	0.3145	0.442	0.569	
			0.550	0.326	0.124	0.992	
			0.278	0.152	0.570	0.632	
			0.490	0.177	0.333	0.873	
			0.411	0.135	0.454	0.783	
			0.564	0.1675	0.2685	1.03	
			0.138	0.149	0.713	0.452	
			0.146	0.687	0.167	0.473	
			0.5605	0.271	0.1685	1.01	
			0.142	0.558	0.300	0.473	
			0.1405	0.419	0.4405	0.471	
			0.150	0.2805	0.569	0.471	
			0	0	1	0.296	
			0	1	0	0.293	
			1	0	0	1.96	
Gray, P. and Wright, P.G.	588	372.3	0.324	0.181	0.495	0.743	Same as above.
			0.705	0.168	0.127	1.44	
			0.705	0.168	0.127	1.37	
			0.189	0.174	0.637	0.536	
			0.554	0.165	0.281	1.04	
			0.4825	0.175	0.3425	0.938	
			0.280	0.654	0.066	0.616	
			0.171	0.784	0.045	0.486	
			0.180	0.578	0.243	0.515	
			0.159	0.3645	0.4765	0.511	
			0.402	0.392	0.206	0.796	
			0.456	0.319	0.2255	0.896	
			0.5735	0.319	0.108	1.10	
			0.363	0.323	0.3135	0.758	
			0.156	0.317	0.527	0.507	
			0.2295	0.332	0.439	0.561	
			0.2295	0.332	0.439	0.580	
			0	0	1	0.318	
			0	1	0	0.310	
			1	0	0	2.08	

TABLE 162. EXPERIMENTAL THERMAL CONDUCTIVITY DATA FOR HYDROGEN-NITROGEN-AMMONIA SYSTEM (cont.)

Author(s)	Ref. No.	Temp. (K)	Mole Fractions of			Thermal Cond. ($mW\ cm^{-1}\ K^{-1}$)	Remarks
			H_2	N_2	NH_3		
Gray, P. and Wright, P.G.	588	422.5	0.484	0.195	0.3215	1.08	Same as above.
			0.643	0.2245	0.133	1.29	
			0.342	0.192	0.466	0.846	
			0.279	0.192	0.529	0.699	
			0.114	0.1915	0.694	0.519	
			0.199	0.654	0.147	0.626	
			0.199	0.654	0.147	0.611	
			0.5945	0.138	0.2675	1.29	
			0.139	0.375	0.486	0.553	
			0.179	0.408	0.413	0.561	
			0.1385	0.269	0.5925	0.540	
			0.389	0.390	0.221	0.896	
			0.276	0.568	0.156	0.699	
			0.468	0.4265	0.105	1.01	
			0.244	0.3145	0.442	0.695	
			0.550	0.326	0.124	1.14	
			0	0	1	0.373	
			0	1	0	0.341	
			1	0	0	2.26	

QUATERNARY SYSTEMS

Not many systems involving four differents gases have been investigated. In Tables 163 through 169 we report data on seven different quaternary systems. In each case experiments have been conducted at either two or three different temperatures and for several compositions of the mixture. There is only one combination of rare gases which has been investigated experimentally and these conductivity values are reported in Table 163, category A. Data on six different systems which permute from different monatomic and nonpolar polyatomic gases are reported in category B.

TABLE 163. EXPERIMENTAL THERMAL CONDUCTIVITY DATA FOR NEON-ARGON-KRYPTON-XENON SYSTEM

Author(s)	Ref. No.	Temp. (K)	Mole Fractions of				Thermal Cond. ($mW\ cm^{-1}\ K^{-1}$)	Remarks
			Ne	Ar	Kr	Xe		
Mathur, S., Tondon, P.K., and Saxena, S.C.	692, 717	311.3	0.239	0.258	0.267	0.236	0.155	Ne and Ar: spectroscopically pure, Kr: 99-100% pure, balance Xe, Xe: 99-100% pure, balance Kr; thick hot wire method; accuracy ±1 to ±2%, precision ±1%.
			0.117	0.252	0.159	0.472	0.118	
			0.504	0.094	0.274	0.128	0.214	
			0	0	0	1	0.0616	
			0	0	1	0	0.103	
			0	1	0	0	0.184	
			1	0	0	0	0.513	
Mathur, S., Tondon, P.K., and Saxena, S.C.	692, 717	366.8	0.239	0.258	0.267	0.236	0.180	Same as above.
			0.117	0.252	0.159	0.472	0.133	
			0.504	0.094	0.274	0.128	0.240	
			0	0	0	1	0.0741	
			0	0	1	0	0.118	
			0	1	0	0	0.212	
			1	0	0	0	0.563	

TABLE 164. EXPERIMENTAL THERMAL CONDUCTIVITY DATA FOR ARGON-KRYPTON-XENON-HYDROGEN SYSTEM

| Author(s) | Ref. No. | Temp. (K) | Mole Fractions of | | | | Thermal Cond. (mW cm^{-1} K^{-1}) | Remarks |
			Ar	Kr	Xe	H_2		
Gupta, G. P. and Saxena, S. C.	695, 696	313.2	0.200	0.102	0.306	0.392	0.469	Ar: spectroscopically pure, Kr: 99-100% pure, balance Xe, Xe: 99-100% pure, balance Kr, H_2: 99.95% pure; thick hot wire method; accuracy ±1 to ±2%, precision ±1%.
			0.453	0.165	0.240	0.142	0.230	
			0	0	0	1	1.82	
			0	0	1	0	0.0593	
			0	1	0	0	0.0984	
			1	0	0	0	0.185	
Gupta, G. P. and Saxena, S. C.	695, 696	338.2	0.200	0.102	0.306	0.392	0.499	Same as above.
			0.453	0.165	0.240	0.142	0.256	
			0	0	0	1	1.94	
			0	0	1	0	0.0619	
			0	1	0	0	0.108	
			1	0	0	0	0.193	
Gupta, G. P. and Saxena, S. C.	695, 696	366.2	0.200	0.102	0.306	0.392	0.538	Same as above.
			0.453	0.165	0.240	0.142	0.280	
			0	0	0	1	2.04	
			0	0	1	0	0.0706	
			0	1	0	0	0.114	
			1	0	0	0	0.210	

TABLE 165. EXPERIMENTAL THERMAL CONDUCTIVITY DATA FOR ARGON-KRYPTON-XENON-DEUTERIUM SYSTEM

Author(s)	Ref. No.	Temp. (K)	Mole Fractions of				Thermal Cond. ($mW\ cm^{-1}\ K^{-1}$)	Remarks
			Ar	Kr	Xe	D_2		
Mathur, S., Tondon, P.K., and Saxena, S.C.	711, 717	311.3	0.237	0.182	0.104	0.477	0.488	Ar: spectroscopically pure, Kr: 99-100% pure, balance Xe, Xe: 99-100% pure, balance Kr, D_2: 98.6% pure, 0.8% H_2 and 0.6% H_2O; thick hot wire method; accuracy ±1 to ±2%, precision ±1%.
			0.256	0.273	0.222	0.249	0.273	
			0.121	0.239	0.516	0.124	0.157	
			0	0	0	1	1.33	
			0	0	1	0	0.0616	
			0	1	0	0	0.103	
			1	0	0	0	0.184	
Mathur, S., Tondon, P.K., and Saxena, S.C.	711, 717	366.8	0.237	0.182	0.104	0.477	0.556	Same as above.
			0.256	0.273	0.222	0.249	0.311	
			0.121	0.239	0.516	0.124	0.180	
			0	0	0	1	1.48	
			0	0	1	0	0.0741	
			0	1	0	0	0.118	
			1	0	0	0	0.212	

TABLE 166. EXPERIMENTAL THERMAL CONDUCTIVITY DATA FOR ARGON-HYDROGEN-DEUTERIUM-NITROGEN SYSTEM

Author(s)	Ref. No.	Temp. (K)	Mole Fractions of				Thermal Cond. $(mW\ cm^{-1}\ K^{-1})$	Remarks
			Ar	H_2	D_2	N_2		
Gupta, G.P. and Saxena, S.C.	699, 696	313.2	0.152	0.358	0.346	0.144	0.942	Ar: spectroscopically pure, H_2 and N_2: 99.95% pure, D_2: 98.6% pure, 0.8% H_2 and 0.6% H_2O; thick hot wire method; accuracy ±1 to ±2%, precision ±1%.
			0.132	0.206	0.158	0.504	0.552	
			0	0	0	1	0.268	
			0	0	1	0	1.35	
			0	1	0	0	1.82	
			1	0	0	0	0.185	
Gupta, G.P. and Saxena, S.C.	699, 696	338.2	0.152	0.358	0.346	0.144	1.01	Same as above.
			0.132	0.206	0.158	0.504	0.589	
			0	0	0	1	0.290	
			0	0	1	0	1.43	
			0	1	0	0	1.94	
			1	0	0	0	1.93	
Gupta, G.P. and Saxena, S.C.	699, 696	366.2	0.152	0.358	0.346	0.144	1.07	Same as above.
			0.132	0.206	0.158	0.504	0.624	
			0	0	0	1	0.312	
			0	0	1	0	1.49	
			0	1	0	0	2.04	
			1	0	0	0	2.10	

TABLE 167. EXPERIMENTAL THERMAL CONDUCTIVITY DATA FOR ARGON-HYDROGEN-NITROGEN-OXYGEN SYSTEM

Author(s)	Ref. No.	Temp. (K)	Ar	Mole Fractions of H_2	N_2	O_2	Thermal Cond. (mW cm^{-1} K^{-1})	Remarks
Saxena, S.C. and Gupta, G.P.	699, 696	313.2	0.132	0.379	0.125	0.364	0.615	Ar: spectroscopically pure, H_2, N_2 and O_2: 99.95% pure; thick hot wire method; accuracy ±1 to ±2%, precision ±1%.
			0.250	0.132	0.374	0.244	0.360	
			0	0	0	1	0.281	
			0	0	1	0	0.268	
			0	1	0	0	1.82	
			1	0	0	0	1.85	
Saxena, S.C. and Gupta, G.P.	699, 696	338.2	0.132	0.379	0.125	0.364	0.671	Same as above.
			0.250	0.132	0.374	0.244	0.383	
			0	0	0	1	0.291	
			0	0	1	0	0.290	
			0	1	0	0	1.94	
			1	0	0	0	1.93	
Saxena, S.C. and Gupta, G.P.	699, 696	366.2	0.132	0.379	0.125	0.364	0.717	Same as above.
			0.250	0.132	0.374	0.244	0.423	
			0	0	0	1	0.313	
			0	0	1	0	0.312	
			0	1	0	0	2.04	
			1	0	0	0	2.10	

TABLE 168. EXPERIMENTAL THERMAL CONDUCTIVITY DATA FOR NEON-ARGON-HYDROGEN-NITROGEN SYSTEM

Author(s)	Ref. No.	Temp. (K)	Ne	Mole Fractions of Ar	H$_2$	N$_2$	Thermal Cond. (mW cm^{-1} K^{-1})	Remarks
Tondon, P.K. and Saxena, S.C.	710, 707	313.2	0.302	0.107	0.396	0.195	0.754	Ne and Ar: spectroscopically pure, H$_2$ and N$_2$: 99.95% pure; thick hot wire method; accuracy ±1 to ±2%, precision ±1%.
			0.254	0.250	0.255	0.241	0.523	
			0	0	0	1	0.268	
			0	0	1	0	1.82	
			0	1	0	0	0.185	
			1	0	0	0	0.496	
Tondon, P.K. and Saxena, S.C.	710, 707	338.2	0.302	0.107	0.396	0.195	0.804	Same as above.
			0.254	0.250	0.255	0.241	0.557	
			0	0	0	1	0.290	
			0	0	1	0	1.94	
			0	1	0	0	0.193	
			1	0	0	0	0.534	
Tondon, P.K. and Saxena, S.C.	710, 707	366.2	0.302	0.107	0.396	0.195	0.846	Same as above.
			0.254	0.250	0.255	0.241	0.603	
			0	0	0	1	0.313	
			0	0	1	0	2.04	
			0	1	0	0	0.210	
			1	0	0	0	0.563	

TABLE 169. EXPERIMENTAL THERMAL CONDUCTIVITY DATA FOR ARGON-XENON-HYDROGEN-DEUTERIUM SYSTEM

Author(s)	Ref. No.	Temp. (K)	Mole Fractions of				Thermal Cond. (mW cm^{-1} K^{-1})	Remarks
			Ar	Xe	H_2	D_2		
Tondon, P.K. and Saxena, S.C.	708, 707	313.2	0.152	0.158	0.348	0.342	0.842	Ar: spectroscopically pure, Xe: 99-100% pure, balance Kr, H_2: 99.95% pure, D_2: 98.6% pure, 0.8% H_2 and 0.6% H_2O; thick hot wire method; accuracy ±1 to ±2%, precision ±1%.
			0.375	0.360	0.128	0.137	0.311	
			0	0	0	1	1.35	
			0	0	1	0	1.82	
			0	1	0	0	0.0593	
			1	0	0	0	0.185	
Tondon, P.K. and Saxena, S.C.	708, 707	338.2	0.152	0.158	0.348	0.342	0.875	Same as above.
			0.375	0.360	0.128	0.137	0.327	
			0	0	0	1	1.43	
			0	0	1	0	1.94	
			0	1	0	0	0.0620	
			1	0	0	0	0.193	
Tondon, P.K. and Saxena, S.C.	708, 707	366.2	0.152	0.158	0.348	0.342	0.917	Same as above.
			0.375	0.360	0.128	0.137	0.359	
			0	0	0	1	1.50	
			0	0	1	0	2.04	
			0	1	0	0	0.0706	
			1	0	0	0	0.210	

MULTICOMPONENT SYSTEMS

The only system considered to date is air. Here we depart from the format adopted for mixtures which consisted of a direct tabulation of all the experimental data for various mixture compositions and instead, in the present work, we have considered air to be effectively a single compound of unique composition. The reasons for this procedure include the following (a) many authors have not specified the composition of the air that they measured and this has therefore been assumed as the usual atmospheric substance, (b) only in a few cases was information given as to whether the air was dried and/or carbon dioxide removed. Not considering these cases as a separate group yielded agreement to well within the individual experimental errors with the overall correlation, (c) possible variation in the composition of the atmosphere with time is small and such as can be determined should have no effect on the values to well within the probable errors.

A more precise analysis should consider the above factors and simultaneously possible pressure effects for measurements which we have considered as exactly atmospheric but which, in reality, cover pressures from close to zero to a few atmospheres. Involved in such an analysis would be the theoretical correlation of air treated as a multicomponent system with the pure component data for the individual fluids tabulated elsewhere in this volume. As even ternary system computations have resulted in difficulty, the extension to air is thought technologically important and scientifically difficult. In future work it is hoped that at least a partial consideration of such factors may be possible. The present tables must be considered as subject to the above limitations. For this reason the present work does not tabulate composition and experimental data but merely reports deviations in departure plot from the recommended values which consider air as a single compound.

Many experimental investigations of the thermal conductivity of air have appeared, yet, surprisingly, a fewer number have covered extensive temperature ranges than might be expected. Measurements at the ice point have received extensive consideration, both for the information in itself and also for use in calibration of relative apparatus. Over thirty different values have been located in over sixty different sources. The value here considered most probable, 0.241 mW cm^{-1}K^{-1}, is within one percent of forty of the sources and two percent of fifty. However, an estimate better than one percent is not presently considered possible.

Graphical plotting of the many data and estimates proves instructive. Several sets (114, 146, 147, 187, 357, 369, 591) show a similar trend with temperature above about 400 K. As later stated (618), the original Vines (369) data appear too high and curve 10 shows the result of reducing the original data (curve 2) by the factor suggested. Presumably, a similar reduction should be made in the (187) data. Such a reduction results in values in much better agreement with the Geier and Schafer (587) data for higher temperatures.

The recommended values were obtained from large scale graphs of the different values and were checked by differencing. Below about 400 K all available data were considered while above 873 K the Geier and Schafer data were selected as most probable. Values from 400 to 873 K were obtained by extrapolation of the two sets.

The choice of the upper limit of 1500 K for these tables is felt to require explanation in view of the many tables which have appeared to even 100,000 K. While numerous high temperature estimates have been noted only one evaluation of the thermal conductivity from shock tube studies to 6000 K has been found. The estimates have been made for various pressures and/or density ratios while the shock tube studies were obtained for nominally a density ratio of 0.1. Intercomparison of all the different estimates and the shock tube data reveals poor to moderate agreement between any sources except in some limited ranges of temperature for temperatures to 15,000 K. For higher temperatures the agreement is even worse. If significant recommended values are required for temperatures above 1500 K a detailed analysis of the various contributions to the total thermal conductivity will be necessary. A further complication is that such calculations will require accurate thermodynamic data. A presently unknown uncertainty in the available calculated data is produced by errors in the compositions for the components of air assumed by the different authors.

The departure plots present only a selection of the various experimental and estimated values which were felt to be important from the standpoint of experimental accuracy or temperature range. None of the higher temperature calculations which extend to temperatures below 1500 K were included for reasons stated in the preceding paragraph. Below 400 K the recommended values should be accurate to within about one percent, the uncertainty then increasing to about five percent at 1500 K.

512

TABLE 170   THERMAL CONDUCTIVITY OF AIR

RECOMMENDED VALUES

[Temperature, T, K; Thermal Conductivity, k, mW cm^{-1}K^{-1}]

GAS

T	k	T	k	T	k
50	(0.046)*	450	0.3633	850	0.597
60	(0.055)*	460	0.3697	860	0.603
70	(0.065)*	470	0.3761	870	0.608
80	(0.0738)*	480	0.3825	880	0.614
90	0.0830	490	0.3888	890	0.619
100	0.0922	500	0.3951	900	0.625
110	0.1015	510	0.402	910	0.630
120	0.1106	520	0.408	920	0.635
130	0.1197	530	0.414	930	0.639
140	0.1287	540	0.420	940	0.644
150	0.1375	550	0.426	950	0.649
160	0.1463	560	0.432	960	0.654
170	0.1550	570	0.438	970	0.658
180	0.1637	580	0.444	980	0.663
190	0.1723	590	0.450	990	0.668
200	0.1810	600	0.456	1000	0.672
210	0.1895	610	0.462	1010	0.677
220	0.1980	620	0.468	1020	0.682
230	0.2063	630	0.473	1030	0.686
240	0.2145	640	0.479	1040	0.691
250	0.2226	650	0.484	1050	0.695
260	0.2305	660	0.490	1060	0.699
270	0.2384	670	0.496	1070	0.704
280	0.2461	680	0.501	1080	0.708
290	0.2538	690	0.507	1090	0.713
300	0.2614	706	0.513	1100	0.717
310	0.2687	710	0.518	1110	0.721
320	0.2759	720	0.524	1120	0.726
330	0.2830	730	0.530	1130	0.730
340	0.2900	740	0.535	1140	0.734
350	0.2970	750	0.541	1150	0.738
360	0.3039	760	0.546	1160	0.743
370	0.3107	770	0.552	1170	0.747
380	0.3173	780	0.558	1180	0.751
390	0.3239	790	0.563	1190	0.755
400	0.3305	800	0.569	1200	0.759
410	0.3371	810	0.575	1210	0.763
420	0.3437	820	0.580	1220	0.767
430	0.3503	830	0.586	1230	0.771
440	0.3568	840	0.592	1240	0.775
				1250	0.779
				1260	0.782
				1270	0.786
				1280	0.790
				1290	0.794
				1300	0.797
				1310	0.801
				1320	0.805
				1330	0.809
				1340	0.813
				1350	0.816
				1360	0.820
				1370	0.824
				1380	0.827
				1390	0.831
				1400	0.835
				1410	0.838
				1420	0.842
				1430	0.846
				1440	0.849
				1450	0.853
				1460	0.856
				1470	0.860
				1480	0.863
				1490	0.867
				1500	0.870

*Extrapolated (Bubble pt. = 79 K, Dew pt. = 82 K).

DISCUSSION

GAS

Many experimental investigations of the thermal conductivity of air have appeared, yet, surprisingly, a fewer number have covered extensive temperature ranges than might be expected. Measurements at the ice point have received extensive consideration, both for the information in itself and also for use in calibration of relative apparatus. Over thirty different values have been located in over sixty different sources. The value here considered most probable, 0.241 mw. cm^{-1}K^{-1}, is within one percent of forty of the sources and two percent of fifty. However, an estimate better than one percent is not presently considered possible.

Graphical plotting of the many data and estimates proves instructive. Several sets (114, 146, 147, 187, 357, 369, 591) show a similar trend with temperature above about 400 K. As later stated (618), the original Vines (369) data appear too high and curve 10 shows the result of reducing the original data (curve 2) by the factor suggested. Presumably, a similar reduction should be made in the (187) data. Such a reduction results in values in much better agreement with the Geier and Schafer (587) data for higher temperatures.

The recommended values were obtained from large scale graphs of the different values and were checked by differencing. Below about 400 K all available data were considered while above 873 K the Geier and Schafer data were selected as most probable. Values from 400 to 873 K were obtained by extrapolation of the two sets.

The choice of the upper limit of 1500 K for these tables is felt to require explanation in view of the many tables which have appeared to even 100,000 K. While numerous high temperature estimates have been noted only one evaluation of the thermal conductivity from shock tube studies to 6000 K has been found. The estimates have been made for various pressures and/or density ratios while the shock tube studies were obtained for nominally, a density ratio of 0.1. Inter comparison of all the different estimates and the shock tube data reveals poor to moderate agreement between any sources except in some limited ranges of temperature for temperatures to 15,000 K. For higher temperatures the agreement is even worse. If significant recommended values are required for temperatures above 1500 K a detailed analysis of the various contributions to the total thermal conductivity will be necessary. A further complication is that such calculations will require accurate thermodynamic data. A presently unknown uncertainty in the available calculated data is produced by errors in the compositions for the components of air assumed by the different authors.

The departure plots present only a selection of the various experimental and estimated values which were felt to be important from the standpoint of experimental accuracy or temperature range. None of the higher temperature calculations which extend to temperatures below 1500 K were included for reasons stated in the preceding paragraph. Below 400 K the recommended values should be accurate to within about one percent, the uncertainty then increasing to about five percent at 1500 K.

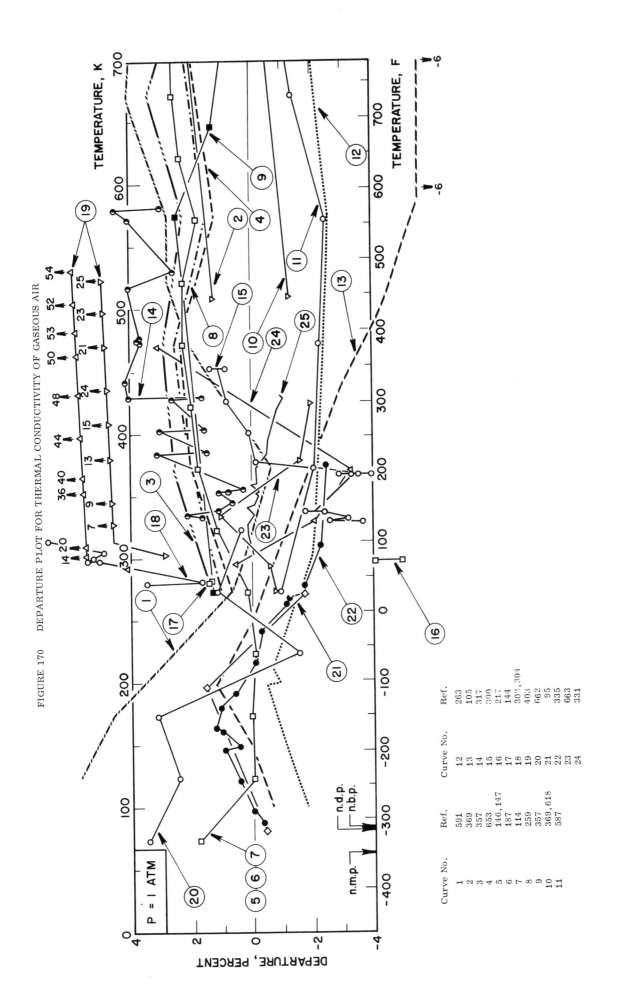

FIGURE 170 DEPARTURE PLOT FOR THERMAL CONDUCTIVITY OF GASEOUS AIR

FIGURE 170 DEPARTURE PLOT FOR THERMAL CONDUCTIVITY OF GASEOUS AIR (continued)

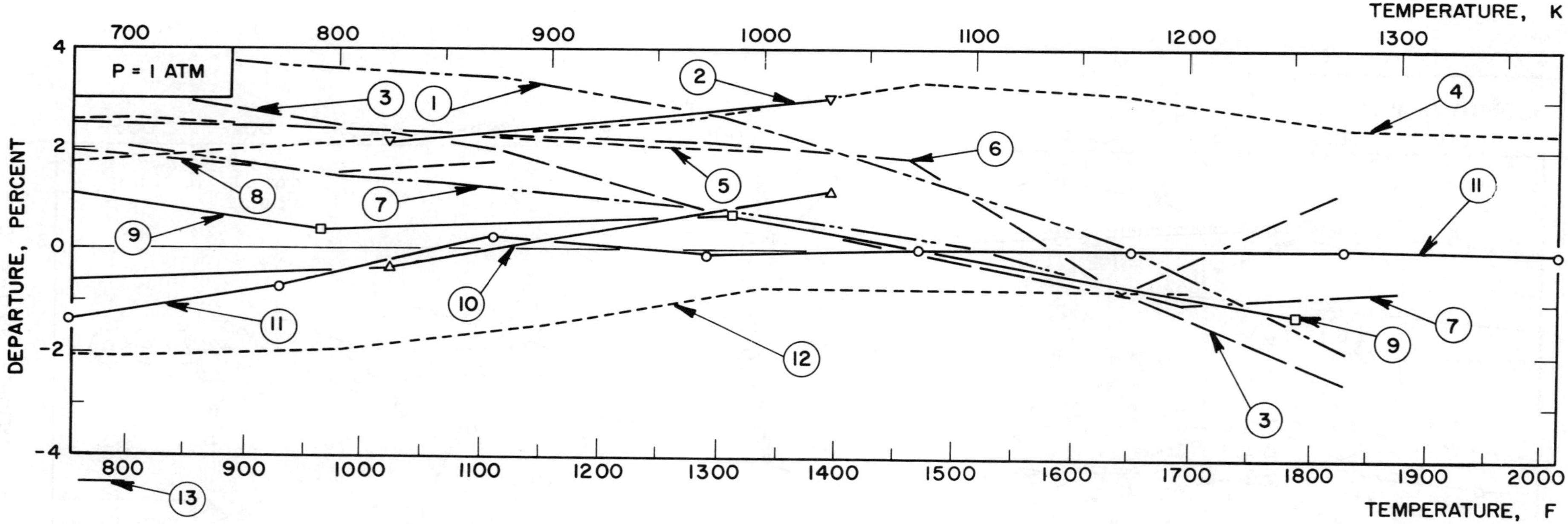

Curve	Reference
1	591
2	369
3	357
4	653
5	146, 147
6	187
7	114
8	259
9	357
10	369, 618

Curve	Reference
11	587
12	263
13	105

REFERENCES TO DATA SOURCES

Ref. No.	TPRC No.	
1	79	Abas-Zade, A.K., Doklady Akad. Nauk Azerbaidzhan SSR, 3(1), 3-7, 1947.
2	1946	Abas-Zade, A.K., Doklady Akad. Nauk (SSSR), 68, 665-8, 1949.
3	823	Abas-Zade, A.K., Doklady Akad. Nauk (SSSR), 99, 227-30, 1954.
4	842	Abas-Zade, A.K., Zhur. Eksptl. i Teoret. Fiz., 23, 60-7, 1952.
5	8899	Abas-Zade, A.K. and Amiraslanov, A.M., Zhur. Fiz. Khim., 31, 1459-67, 1957.
6	6950	Aerodynamics Handbook Staff, Handbook of Supersonic Aerodynamics, NAVORD Rept. 1488, 5, 1953.
7	3079	Akin, S.W., Trans. ASME, 72, 751-7, 1950.
8	6929	Allcut, E.A., J. Inst. Heating Ventilating Engrs., 17, 151-195, 1949.
9	385	Amdur, I., J. Chem. Phys., 14, 339-42, 1946.
10	231	Amdur, I., J. Chem. Phys., 15, 482-7, 1947.
11	5339	Amdur, I., J. Chem. Phys., 16, 190-4, 1948.
12	9302	Amdur, I. and Mason, E.A., Phys. Fluids, 1, 370-83, 1958.
13	67	Andrussow, L., J. Chim. Phys., 49, 599-604, 1952.
14	5313	Andrussow, L., Z. Elektrochem., 56, 54-8, 1952.
15	238	Andrussow, L., Z. Elektrochem., 57, 124-30, 1953.
16	242	Andrussow, L., Z. Elektrochem., 57, 374-82, 1953.
17	822	Andrussow, L., J. Chim. Phys., 52, 295, 1955.
18	14	Andrussow, L., Z. Elektrochem., 60, 412-20, 1956.
19	30	Andrussow, L., Z. Elektrochem., 61, 253-65, 1957.
20	15816	Archer, C.T., Nature (London), 138, 286-7, 1936.
21	13407	Archer, C.T., Proc. Roy. Soc. (London), A165, 474-85, 1938.
22	1118	Awano, M., Busseiron Kenkyu, (43), 34-42, 1951.
23	26048	Bain, R.W., Heat Div. Rept. 153, Mech. Eng. Res. Lab. (U.K.), 1-12, 1958.
24	1932	Band, W. and Meyer, L., Phys. Rev., 73, 226-9, 1948.
25	9813	Bateman, J.S., Proc. Conf. "Thermodynamic and Transport Properties of Fluids", Inst. Mech. Eng. (London), 169-181, 1958.
26	9813	Bateman, J.S., Private Communication from F.G. Keyes, 1956. [See(25)]
27	9079	Bauer, E. and Zlotnick, M., AVCO Res. Develop. Div. Rept. RAD-TR-58-12, 1-67, 1958.
28	11350	Baulknight, C.W., "Transport Properties in Gases", Proc. 2nd. Gas Dynamics Symposium (Evanston), 89-95, 1958.
29	854	Baxter, S., Vodden, H.A., and Davies, S., J. Appl. Chem. (London), 3, 477-80, 1953.
30	7073	Beckett, C.W. and Fano, L., NBS Rept. 2562, 1-51, 1953 [AD 20002]
31	4288	Benedicks, C., Arkiv Mat. Astron, Fysik, 35A, 1-21, 1948.
32	763	Bennett, L.A. and Vines, R.G., J. Chem. Phys., 23, 1587-91, 1955.
33	25399	Bannawitz, E., Ann. Physik, 48, 577-92, 1915.
34	9954	Bentley, R., Brown, G., and Schlegel, R., USAEC Rept. CP 3061, 1-48, 1945.
35	1161	Berman, R., Foster, E.L., and Ziman, J.M., Proc. Roy. Soc. (London), A237, 344-54, 1956.
36	6968	Boegli, J.S. and Deissler, R.G., NACA RM E54L10, 1-20, 1955.
37	7684	Boelter, L.M.K. and Sharp, W.H., NACA TN 1912, 1-39, 1949.
38	5469	deBoer, J., Physica, 10, 348-56, 1943.
39	3853	deBoer, J. and Cohen, E.G.D., Physica, 17, 993-1000, 1951.
40	335	deBoer, J. and van Kranendonk, J., Physica, 14, 442-52, 1948.
41	24607	Bonilla, C.F., Brooks, R.D., and Walker, P.L., Jr., Proc. Gen. Disc. Heat Transfer, Inst. Mech. Eng. (London), 167-173, 1951.

Ref. No.	TPRC No.	
42	769	Borovik, E., Zhur. Eksptl. i Teoret. Fiz., 17, 328-35, 1947.
43		Chu, Ph.D. Thesis, Univ. of London. [See (357)]
44	1968	Borovik, E., Zhur. Eksptl. i Teoret. Fiz., 19, 561-4, 1949.
45	9347	Borovik, E., Zhur. Eksptl. i Teoret. Fiz., 19, 561-4, (U.K.) TIB/T 4194, (Translation of No. 44), 1949.
46	2381	Borovik, E., Matveev, A. and Panina, E., J. Tech. Phys. (USSR), 10, 988-98, 1940.
47	6517	Botzen, A., Thesis, Amsterdam, 1-74, 1952. [AD 108 543]
48	1726	Bowers, R., Proc. Phys. Soc. (London), 65A, 511-18, 1952.
49	70	Bowers, R. and Mendelssohn, K., Nature (London), 167, 111, 1951.
50	6959	Brokaw, R.S., NACA RM E57K19a, 1-17, 1958.
51	729	Bromley, LeR.A., USAEC Rept. UCRL-1852, 1-31, 1952.
52	263	Rothman, A.J., USAEC Rept. UCRL-2339, 1-115, 1954.
53	24663	Brown, G.B. and Dean, R.W., Instr. Engr., 1, 145, 1955.
54	5465	Brunot, A.W., Trans. ASME, 62, 613-19, 1940.
55	826	Bryan, R.T. and Pere, J.G., Trans. ASME, 73, 601-2, 1951.
56	6584	Burton, J.T.A. and Ziebland, H., ERDE Rept. 2/R/56, 1-15, 1956. [AD 90 759]
57	6396	Burton, J.T.A. and Ziebland, H., ERDE Rept. 11/R/57, 1-13, 1957. [AD 145 956]
58	1105	Callear, A.B., Private Communication, 1958. (See No. 59)
59	1105	Callear, A.B., and Robb, J.C., Trans. Faraday Soc., 51, 630-8, 1955.
60	24664	Chapman, S. and Cowling, T.G., "The Mathematical Theory of Non-Uniform Gases", Cambridge Univ. Press, 1939, (Pub. 1952).
61	16895	Chelton, D.B. and Mann, D.B., USAEC Rept. UCRL-3421, 1956.
62	8364	Chen, L.H., "Thermodynamic and Transport Properties of Gases, Liquids and Solids", ASME, N.Y., 358-69, 1959.
63	9370	Cherneyeva, L., Kholodil'naya Tekh., 32, 45-7, 1955.
64	7158	Cheung, H., USAEC Rept. UCRL-8230, 1-146, 1958.
65	26049	Cheung, H., LeBromley, R.A., and Wilke, C.R., USAEC Rept. UCRL-8230 Rev., 1-64, 1959.
66	8295	Christiansen, C., Ann. Physik, 14, 23-33, 1881.
67	13904	van Cleave, A.B. and Maass, O., Can. J. Research, 12, 372-6, 1935.
68	13454	van Cleave, A.B. and Maass, O., Can. J. Research, B13, 140-8, 1935.
69	5804	Codegone, C., Atti. Accad. Sci. Torino, 86, 126-8, 1951-2.
70	5805	Codegone, C., Atti. Accad. Sci. Torino, 86, 288-90, 1951-2.
71	112	Codegone, C., Termotecnica (Milan), 6, 407-11, 1952.
72	3926	Codegone, C., Inst. Intern. Froid, Journees Mons, Belg., Communs., 61-6, 1953.
73	16067	Codegone, C., Allgem. Warmetech., 8, 49-53, 1957.
74	1608	Cohen, E.G.D., Offerhaus, M.J., and deBoer, J., Physica, 20, 501-15, 1954.
75	9595	Cohen, E.G.D., Offerhaus, M.J., van Leeuwen, J.M.J., Roos, B.W., and deBoer, J., Physica, 22, 791-815, 1956.
76	580	Comings, E.W. and Nathan, M.F., Ind. Eng. Chem., 39, 964-70, 1947.
77	19360	Comings, E.W., Lee, W.B., and Kramer, F.R., Proc. Conf. "Thermodynamic and Transport Properties of Fluids", Inst. Mech. Eng. (London), 188-92, 1958.
78	1659	Cooke, B.A. and Mackenzie, H.A.E., J.S. African Chem. Inst., 6, 8-13, 1953.
79	810	Cooper, A., Trans. Inst. Rubber Ind., 27, 84-100, 1951.
80	5455	Craven, P.M. and Lambert, J.D., Proc. Roy. Soc. (London), A205, 439-49, 1951.
81	8353	Curie, M. and Lepape, A., Compt. Rend., 193, 842-3, 1931.
82	9362	Curie, M. and Lepape, A., J. Phys. Radium, 2, 393-97, 1931.
83	168	Davidson, J.M. and Music, J.F., USAEC Rept. HW-29021, 7-30, 1953.
84		Davies, R.H. and Kendall, J.T., Nature (London), 165, 487, 1950.
85	5314	Davis, D.S., Ind. Eng. Chem., 33, 675-8, 1941.

Ref. No.	TPRC No.	
86	9915	Dickens, B.G., Proc. Roy. Soc. (London), A143, 517-40, 1934.
87	1086	Dingle, R.R., Proc. Phys. Soc. (London), 62A, 154-66, 1949.
88	1624	Dognin, A., Compt. Rend., 243, 840-2, 1956.
89	6464	Dommett, R.L., (U.K.) RAE-TN-GW-429, 1-39, 1956. [AD 115386]
90	6464	Dommett, R.L., Private Communication, 1959. (See No. 89)
91	25402	Eckerlein, P.A., Ann. Physik, 3, 120-54, 1900.
92	780	Eckert, E.R.G., and Irvine, T.F., Jr., J. Appl. Mechanics, 24, 25-8, 1957.
93	8335	Eckert, E.R.G., Ibele, W.E., and Irvine, T.F., Jr., "Thermodynamic and Transport Properties of Gases, Liquids and Solids", ASME, N.Y., 295-300, 1959.
94	7166	Elenbaas, W., Philips Research Repts., 3, 450-65, 1948.
95	9884	Eucken, A., Physik. Z., 12, 1101-7, 1911.
96	10407	Eucken, A., Physik. Z., 14, 324-32, 1913.
97	6801	Eucken, A., Forsch. Gebiete Ingenieurw., 11B, 6-20, 1940.
98	7072	Fano, L., Hubbell, J.H., and Beckett, C.W., NBS Rept. 2535, 1953. [AD 19904]
99	750	Fano, L., Hubbell, J.H., and Beckett, C.W., NACA-TN-3273, 1-61, 1956.
100	7041	Filippov, L.P., Vestnik Moskov. Univ., 8 (9), Ser. Fiz.-Mat. i Estestven. Nauk No. 6, 109-14, 1953.
101	1925	Foz Gazulla, O.R. and Perez, S., Anales Fis. y quim. (Madrid), 39, 399-409, 1943.
102	8937	Foz Gazulla, O.R. and Perez, S., Z. Physik. Chem. (Leipzig), 193, 162-7, 1944.
103	3629	Foz Gazulla, O.R., Colomina, M., and Garcia Banda, J.F., Anales real soc. espan. fis. y quim. (Madrid), 44B, 1055-82, 1948.
104	1921	Foz Gazulla, O.R., Colomina, M., and Garcia Banda, J.F., Anales real soc. espan. fis. y quim. (Madrid), 44B, 1083-1100, 1948.
105	1913	Franck, E.U., Z. Elektrochem., 55, 636-43, 1951.
106	9304	Franck, E.U., Z. Physik. Chem. (Leipzig), 201, 16-31, 1952.
107	1620	Franck, E.U., Chem.-Ingr.-Tech., 25, 238-44, 1953.
108	6811	Franck, E.U., Chem.-Ingr.-Tech., 25, 442-6, 1953.
109	1914	Franck, E.U. and Wicke, E., Z. Elektrochem., 55, 643-7, 1951.
110	153	Furber, B.N., Proc. Inst. Mech. Engrs. (London), 168, 847-60, 1954.
111	26047	Gallo, Rend dell' Istituto Sperimentale Aeranautico, 9 (2), No. 1, 37-41, 1921.
112	836	Gardiner, W.C. and Schafer, K., Z. Elektrochem., 60, 588-94, 1956.
113		Gazulla, O.R.F. (See Foz Gazulla, O.R.)
114	6733	Glassman, I. and Bonilla, C.F., Chem. Eng. Progr. Symposium Ser., 49, 153-62, 1953.
115	1667	Godridge, A.M., Bull. Brit. Coal Utilization Research Assoc., 18, 1-21, 1954.
116	134	Granet, I., Petrol. Refiner, 33 (5), 205-6, 1954.
117	1665	Granet, I. and Kass, P., Petrol. Refiner, 31 (10), 113-4, 1952.
118	1662	Granet, I. and Kass, P., Petrol. Refiner, 31 (11), 137-8, 1952.
119	1664	Granet, I. and Kass, P., Petrol. Refiner, 32 (3), 149-50, 1953.
120	25415	Graetz, Ann. Physik, 14, 232-60, 1881.
121	15669	Green, M.S., NBS Rept. 5237, 1-17, 1956.
122	15668	Green, M.S., NBS Rept. 5238, 1-7, 1957.
123	11145	Gregory, H.S., Proc. Soc. (London), A149, 35-56, 1935.
124	24661	Gregory, H.S. and Archer, C.T., Phil. Mag., 1, 593-606, 1926.
125	22422	Gregory, H.S. and Archer, C.T., Proc. Roy. Soc. (London), A110, 91-122, 1926.
126	22429	Gregory, H.S. and Archer, C.T., Proc. Roy. Soc. (London), A121, 285-93, 1928.
127	12909	Gregory, H.S. and Archer, C.T., Phil. Mag., 15, 301-9, 1933.
128	13526	Gregory, H.S. and Dock, E.H., Phil. Mag., 25, 129-47, 1938.
129	22422	See No. 125.
130	9912	Gregory, H.S. and Marshall, S., Proc. Roy. Soc. (London), A114, 354-66, 1927.

Ref. No.	TPRC No.	
131	9879	Gregory, H.S. and Marshall, S., Proc. Roy. Soc. (London), A118, 594-607, 1928.
132	6946	Greifinger, P.S., Rand Corp. RM-1794, 1-25, 1957. [AD 133 002]
133	1089	Grenier, C., Phys. Rev., 83, 598-603, 1951.
134	5316	Gribkova, S.I., J. Expt. Theoret. Phys. (USSR), 11, 364-71, 1941.
135	25405	Gruss, H. and Schmick, H., Wiss. Veroffentl. Siemens Konzern, 7, 202-24, 1928.
136	9262	Guildner, L.A., Proc. Natl. Acad. Sci., U.S., 44, 1149-53, 1958.
137	26046	Gunther, W., Ph.D. Diss., Halle Univ., Germany, 1906.
138	2465	Hall, T.A. and Tsao, P.H., Proc. Roy. Soc. (London), A191, 6-21, 1947.
139	8298	Hammann, G., Ann. Physik, 32, 593-607, 1938.
140	6961	Hansen, C.F., NACA-TN-4150, 1-43, 1958.
141	8264	Hansen, C.F. and Heims, S.P., NACA-TN-4359, 1-33, 1958.
142	6303	Hawkins, G.A., Trans. ASME, 70, 655-9, 1948.
143	25410	Hercus, E.O. and Laby, T.H., Proc. Roy. Soc. (London), A95, 190-210, 1919.
144	22300	Hercus, E.O. and Laby, T.H., Phil. Mag., 3, 1061-4, 1927.
145	25411	Hercus, E.O. and Sutherland, D.M., Proc. Roy. Soc. (London), A145, 599-611, 1934.
146	1558	Hilsenrath, J., Beckett, C.W., Benedict, W.S., Fano, L., Hoge, H.J., Masi, J.F., Nuttall, R.L., Touloukian, Y.S., and Woolley, H.W., NBS Circ. 564, 1-488, 1955.
147	1661	Hilsenrath, J. and Touloukian, Y.S., Trans. Am. Soc. Mech. Engrs., 76, 967-85, 1954.
148	8160	Hirschfelder, J.O., Univ. Wisconsin NRL Rept. WIS-ONR-18, 1-14, 1956.
149	6570	Hirschfelder, J.O., Univ. Wisconsin NRL Rept. WIS-ONR-22, 1-14, 1956. [AD 91 783]
150	6382	Hirschfelder, J.O., Univ. Wisconsin NRL Rept. WIS-ONR-25, 1-22, 1957. [AD 146 165]
151	3418	Hirschfelder, J.O., Bird, R.B. and Spotz, E.L., Chem. Revs., 44, 205-31, 1949.
152	2565	Hirschfelder, J.O., Bird, R.B. and Spotz, E.L., Trans. Am. Soc. Mech. Engrs., 71, 921-37, 1949.
153	24667	Hirschfelder, J.O., Curtiss, C.F., and Bird, R.B., "Molecular Theory of Gases and Liquids", Wiley, N.Y., 1219 pp., 1954.
154	7905	Holleran, E.M., J. Chem. Phys., 23, 847-53, 1955.
155	817	Hougen, J.O. and Piret, E.L., Chem. Eng. Progr. Symp. Ser., 47, 295-303, 1951.
156	22433	Ibbs, T.L., and Hirst, A.A., Proc. Roy. Soc. (London), A123, 134-42, 1929.
157	780	Irvine, T.F., Jr., Private Communication, 1956. (See No. 92)
158	623	Ishikawa, F. and Yagi, K., Bull. Inst. Phys. Chem. Research (Tokyo), 22, 12-7, 1943.
159	3021	Isikawa, H. and Hijikata, K., Bull. Inst. Phys. Chem. Research (Tokyo), 18, 401-15, 1939.
160	24666	Jeans, J., "Dynamical Theory of Gases", Dover, N.Y., 1954.
161	1465	Johannin, P., Compt. Rend., 244, 2700-3, 1957.
162	19361	Johannin, P., Prov. Conf. "Thermodynamic and Transport Properties of Fluids", Inst. Mech. Eng. (London), 193-4, 1958.
163	26041	Johannin, P., Thesis, Univ. Paris, 1-39, 1958.
164	8366	Johannin, P., J. recherches centre nat. recherche sci., Labor, Bellevue (Paris), No. 43, 116-55, 1958.
165	8648	Johannin, P. and Vodar, B., Ind. Eng. Chem., 49, 2040-1, 1957.
166		Johansen, E.S., "Heat, Thermodynamics and Molecular Kinetic Theory", Gjellerups-Forlag, Copenhagen, 34 pp., 1950.
167	16702	Johnson, V.J., WADD-TR-60-56, Pt. I, 1960. [AD 249 644]
168	5363	Johnston, H.L. and Grilly, E.R., J. Chem. Phys., 14, 233-8, 1946.
169	66	Junk, W.A. and Comings, E.W., Chem. Eng. Progr. Symp. Ser., 49, 263-6, 1953.
170	5411	Kampmeyer, P.M., J. Appl. Phys., 23, 99-102, 1952.
171	15815	Kannuluik, W.G., Nature (London), 137, 741, 1936.
172	6302	Kannuluik, W.G. and Carman, E.H., Australian J. Sci. Research Ser., A4, 305-14, 1951.
173	2126	Kannuluik, W.G. and Carman, E.H., Proc. Phys. Soc. (London), 65B, 701-9, 1952.
174	69	Kannuluik, W.G. and Donald, H.B., Australian J. Sci. Research Ser., A3, 417-27, 1950.

Ref. No.	TPRC No.	
175	7644	Kannuluik, W.G. and Martin, L.H., Proc. Roy. Soc. (London), $A141$, 144-58, 1933.
176		Deleted.
177	10424	Kannuluik, W.G. and Martin, L.H., Proc. Roy. Soc., $A144$, 496-513, 1934.
178	8696	Kapitsa, P.L., J. Expt. Theoret. Phys. (USSR), 11, 1-31, 1941.
179	1178	Kapitsa, P.L., J. Expt. Theoret. Phys. (USSR), 11, 581-91, 1941.
180	9738	Kayan, C.F. and Glaser, P.E., Congr. intern. Froid, 9e, (Paris), Compt. rend. Trav. comm. I, 2051-62, 1955.
181	24668	Kennard, E.H., "Kinetic Theory of Gases", McGraw-Hill, N.Y., 483 pp., 1938.
182	576	Kersten, J.A.H., Physica, 14, 567-8, 1948.
183	7370	Keyes, F.G., Trans. ASME, 71, 939, 1949.
184	1153	Keyes, F.G., J. Am. Chem. Soc., 72, 433-6, 1950.
185	827	Keyes, F.G., Trans. ASME, 73, 589-96, 1951.
186	826	Keyes, F.G., Trans. ASME, 73, 597-603, 1951.
187	10637	Keyes, F.G., Technical Rept. 37, Project Squid, 1-33, 1952. [ATI 16173]
188	10011	Keyes, F.G., Memo. MIT-1, Project Squid, 1-34, 1952. [AD 5117]
189	201	Keyes, F.G., Trans. Am. Soc. Mech. Engrs., 74, 1303-6, 1952.
190		Liley, P.E., Unpublished Calculations, Purdue Univ., 1960-62.
191	245	Keyes, F.G., Trans. ASME, 76, 809-16, 1954.
192	760	Keyes, F.G., Trans. ASME, 77, 1395-6, 1955.
193	24649	Keyes, F.G., "Transport Properties in Gases", Proc. 2nd Gas Dynamics Symp., Evanston, Ill., 1957.
194	5380	Keyes, F.G. and Humphrey, B.G., Private Communication to Mason and Rice, 1953.
195	1919	Keyes, F.G. and Sandell, D.J., Jr., Trans. ASME, 72, 767-8, 1950.
196	9267	Khalatnikov, I.M., Uspekhi Fiz. Nauk, 59, 673-753, 1956.
197	9704	Khalatnikov, I.M., Zhur. Eksptl. i Teoret. Fiz., 23, 8-20, 1952.
198	783	Khalatnikov, I.M., Zhur. Eksptl. i Teoret. Fiz., 23, 21-34, 1952.
199	6334	Kharbanda, Om. P., Chem. Eng., 62, 236, 1955.
200	1587	Kilpatrick, J.E., Keller, W.E. and Hammel, F., Jr., Phys. Rev., 97, 9-12, 1955.
201	5321	Koch, B. and Fritz, W., Warme-u. Kaltetech., 42, 113-7, 1940.
202	15910	Konowalow, D.D., Linder, B., and Hirschfelder, J.O., Univ. Wisconsin NRL Rept. WIS-AEC-17, 1958, J. Chem. Phys., 31, 1575-9, 1959.
203	26040	Konowalow, D.D., Hirschfelder, J.O., and Linder, B., Univ. Wisconsin TCL Rept. WIS-AEC-22, 1959.
204	25398	Kornfeld, G. and Helferding, K., Bodenstein-Festband (Erg. zu Z. Phys. Chem.), Suppl. No. 1, 792-800, 1931.
205	5399	Kotani, M., Proc. Phys.-Math. Soc. (Japan), 24, 76-95, 1942.
206	19934	Kramer, F.R., Ph.D. Thesis, Purdue Univ., 1959.
207	26046	Krey, Ph.D. Thesis, Halle Univ., 1912.
208	25417	Kundt, A. and Warburg, E., Ann. Physik, 156, 177-211, 1875.
209	2343	Kuss, E., Z. Angew. Phys., 4, 203-7, 1952.
210	25409	Laby, T.H., Proc. Roy. Soc. (London), $A144$, 494, 1934.
211	1117	Lambert, J.D., Cotton, K.J., Pailthorpe, M.W., Robinson, A.M., Scrivins, J., Vale, W.R.F., and Young, R.M., Proc. Roy. Soc. (London), $A231$, 280-90, 1955.
212	1943	Lambert, J.D., Staines, E.N., and Woods, S.D., Proc. Roy. Soc. (London), $A200$, 262-71, 1950.
213	7694	Lander, R., Univ. Minn. Inst. Technol., Eng. Exp. Sta. Tech. Paper No. 40, 1-13, 1942.
214	24671	Landolt Bornstein Tabellen, Tables 272-3, 1923-6.
215	5242	Lanneau, K.P., Ind. Eng. Chem., 45, 2381-6, 1953.
216	19357	Lazarre, F. and Vodar, B., Proc. Conf. "Thermodynamic and Transport Properties of Fluids", Inst. Mech. Eng. (London), 159-62, 1958.
217	9285	Lees, C.H., Phil. Trans. Roy. Soc. (London), $191A$, 399-440, 1898.

Ref. No.	TPRC No.	
218	5232	Lehmann, H., Chem. Tech. (Berlin), 9, 530-7, 1957.
219	7061	Lemmon, A.W., Jr., Daniels, D.J., Sparrow, D.E., Geankoplis, C.J., Ward, J.J., and Clegg, J.W., AEC Rept. BMI 858, 1-64, 1953. [AD 18337]
220	1163	Leng, D.E., Univ. Microfilms Publ. No. 21690, 1-150, 1957.
221	6692	Leng, D.E. and Comings, E.W., Ind. Eng. Chem., 49, 2042-5, 1957.
222	8268	Lenoir, J.M., Univ. Arkansas, Eng. Exp. Sta. Bull. 16, 1-26, 1952.
223	24658	Lenoir, J.M., Univ. Arkansas, Eng. Exp. Sta. Bull. 18, 1-48, 1953.
224	811	Lenoir, J.M. and Comings, E.W., Chem. Eng. Progr. Symp. Ser., 47, 223-31, 1951.
225	1621	Lenoir, J.M., Junk, W.A., and Comings, E.W., Chem. Eng. Progr. Symp. Ser., 49, 539-42, 1953.
226	8302	Liley, P.E., "Thermodynamic and Transport Properties of Gases, Liquids and Solids", ASME, N.Y., 40-69, 1959.
227	7676	Linde, W. and Rogers, L.B., Anal. Chem., 30, 1250-52, 1958.
228	1915	Lindsay, A.L. and Bromley, LeR.A., Ind. Eng. Chem., 42, 1508-11, 1950.
229	24672	Loeb, L.B., "Kinetic Theory of Gases", McGraw-Hill, N.Y., 687 pp., 1934.
230	6833	Lohrisch, F.W., J. Appl. Chem. (London), 2, 464-9, 1952.
231	847	Long, E. and Meyer, L., Phys. Rev., 87, 153, 1952.
232	24673	Lottes, P.A., AEC Reactor Handbook-Engineering, McGraw-Hill, N.Y., 1075 pp., 1955.
233	24674	McAdams, W.H., "Heat Transmission", McGraw-Hill, N.Y., 532 pp., 1954.
234	1659	Mackenzie, H.A.E. and Raw, C.J.G., J.S. African Chem. Inst., 6, 4-13, 1953.
235	6837	Madan, M.P., J. Chem. Phys., 27, 113-5, 1957.
236	8789	Madan, M.P., Nuovo cimento, 5, 1369-70, 1957.
237	9303	Mann, W.B. and Dickins, B.G., Proc. Roy. Soc. (London), A134, 77-96, 1931.
238	9924	Markwood, W.H., Jr. and Benning, A.F., Refrig. Eng., 45, 95-9, 1943.
239	1669	Mason, E.A., J. Chem. Phys., 23, 49-56, 1955.
240	5379	Mason, E.A. and Rice, W.E., J. Chem. Phys., 22, 522-35, 1954.
241	5380	Mason, E.A. and Rice, W.E., J. Chem. Phys., 22, 843-51, 1954.
242	15912	Mason, E.A. and Saxena, S.C., Inst. Mol. Phys., Univ. Maryland, Rept. IMP-AEC-6, 1-35, 1958.
243	8125	Mason, E.A. and Saxena, S.C., Phys. Fluids, 1, 361-69, 1958.
244	26046	Mehliss, Ph.D. Thesis, Halle Univ., 1902, International Critical Tables, McGraw-Hill, N.Y., 5, 212-5, 1929.
245	797	Meyer, L. and Band, W., Naturwissenschaften, 36, 5-16, 1949.
246	1686	Michels, A. and Botzen, A., Physica, 18, 605-12, 1952.
247	1622	Michels, A. and Botzen, A., Physica, 19, 585-98, 1953.
248	5376	Michels, A., Botzen, A., and Schuulman, W., Physica, 23, 95-102, 1957.
249	1396	Michels, A., Botzen, A., Friedman, A.S., and Sengers, J.V., Physica, 22, 121-8, 1956.
250	9341	Michels, A., Cox. J.A.M., Botzen, A., and Friedman, A.S., J. Appl. Phys., 26, 843-5, 1955.
251	12913	Milverton, S.W., Phil. Mag., 17, 397-422, 1934.
252	26025	Milverton, S.W., Proc. Roy. Soc. (London) A150, 287-308, 1935.
253	15546	Montgomery, R.B., J. Meterol., 4, 193-6, 1947.
254	26046	Moser, Ph.D. Thesis, Univ. (Berlin), 1913, International Critical Tables, McGraw-Hill, N.Y., 5, 212-15, 1929.
255	25403	Muller, E., Wied. Ann., 60, 82-118, 1897.
256	1150	Mund, W., Huyskens, P. and Lories, R., Bull. classe sci., Acad. Roy. Belg., 35, 995-1007, 1949.
257	26042	National Defense Research Committee, OSRD Rept. Division 11, Washington, D.C., 1, 343-94, 1946.
258	6663	Newman, B.O., GE Rept. GI-401, 1-29, 1947. [AD 64951]
259	434	Nicklin, A.W., UKAEA, Ind. Group Tech. Note No. 36, 1-20, 1956.
260	13556	Northdurft, Ann. Physik, 28, 137-56, 1937.
261	9016	Novikov, I.I., Atomnaya Energ., 2 (5), 468-9, 1957.

Ref. No.	TPRC No.	
262	9305	Novotny, J.L., M.S. Thesis, Univ. Minnesota, 1958.
263	25397	Nusselt, W., Z. Ver. Deut. Ing., 61, 685-9, 1917.
264	26043	Nuttall, R.L., NBS-NACA "Tables of Thermal Properties of Gases", Table 17.42, 4 pp., 1951.
265	1944	Nuttall, R.L. and Ginnings, D.C., J. Research NBS, 58, 271-8, 1957.
266	9893	Onnes, H.K. and Holst, G., Proc. Koninkl. Ned. Akad. Wetenschap. No. 142c, 760-7, 1914.
267	92	Orlicek, A.F., Mitt. Chem. Forschungs-Inst. Wirtsch. Osterr., 7, 124-5, 1953.
268	7312	Owens, E.J. and Thodos, G., AIChE Journal, 3, 454-61, 1957.
269	19358	Owens, E.J. and Thodos, G., Proc. Conf. "Thermodynamic and Transport Properties of Fluids", (London), 163-8, 1958.
270	5312	van Paemel, O., Verhandel. Koninkl. Vlaam. Acad. Wetenschap., Belgie, Kl. Wetenschap., 3, 3-59, 1941.
271	10624	Patterson, G.N., AGARD Rept. 134, 1-58, 1957.
272	25400	Pauli, E., Ann. Physik, 23, 907-31, 1907.
273	6374	Peck, R.E. and Lokay, J.D., 1-80, 1955. [AD 62 517]
274	24675	Perry, J.H., Chemical Engineers Handbook, McGraw-Hill, N.Y., 1942 pp., 1957.
275	745	Powell, R.W., Modern Refrigeration, 59, 434-8, 1956.
276	6239	Powers, R.W., Johnston, H.L., and Mattox, R.W., Proc. 8th Int. Congr. Refrig. (London), 186-94, 1951.
277	19354	Prigogine, I. and Waelbroeck, F., Proc. Conf. "Thermodynamic and Transport Properties of Fluids", (London), 128-32, 1958.
278	9974	Przybycien, W.M. and Linde, D.W., OTS KAPL-M-WMP-1, 1-24, 1957.
279	1083	Riedel, L., Chem.-Ingr.-Tech., 21, 340-1, 1949.
280	263	Rothman, A.J., USAEC Rept. UCRL-2339, 1-115, 1953. (See No. 52)
281	143	Rothman, A.J. and Bromley, LeR.A., Ind. Eng. Chem., 47, 899-906, 1955.
282	5900	Rudorff, D.W., Eng. and Boiler House Rev., 60, 100-5, 115, 1946.
283	1466	Ruhemann, M., Atti Accad. Sci. Torino, 90, 290-3, 1955-56.
284	829	Salceanu, C. and Bojin, S., Compt. Rend., 243, 237-9, 1956.
285	5324	Salvetti, C., Atti Reale Accad. Italia, Mem. Classe Sci. Fis. mat. e nat., 13, 651-75, 1942.
286	1147	Saurel, J., Bergeon, R., Johannin, P., Dapoigny, J., Kieffer, J., and Vodar, B., Discussions Faraday Soc. No. 22, 64-9, 1956.
287	4	Saxena, S.C., Indian J. Phys., 29, 587-602, 1955.
288	1134	Saxena, S.C., Indian J. Phys., 31, 146-55, 1957.
289	2088	Saxena, S.C., Indian J. Phys., 31, 597-606, 1957.
290	1754	Schafer, K., Fortschr. Chem., Forsch, 1, 61-118, 1949.
291	5326	Schafer, K. and Foz Gazulla, O.R., Anales Real Soc. Espan. Fis. y Quim. (Madrid), 38, 316-46, 1942.
292	9819	Schafer, K. and Reiter, F.W., Naturwissenschaften, 43, 296-7, 1956.
293	8893	Schafer, K. and Reiter, F.W., Z. Elektrochem., 61, 1230-5, 1957.
294	861	Schafer, K. and Riggert, K.H., Naturwissenschaften, 40, 219, 1953.
295	136	Schirmer, H., Appl. Sci. Research, 5B, 196-200, 1955.
296	22854	Schreiner, E., Z. Physik. Chem. (Leipzig), 112, 1-67, 1924.
297	25414	Schliermacher, Ann. Physik, 34, 623-46, 1888.
298	26030	Schliermacher, Ann. Physik, 36, 346-57, 1889.
299	260	Schlinger, W.G., Hsu, N.T., Cavers, S.D., and Sage, B.H., Ind. Eng. Chem., 45, 864-70, 1953.
300	1652	Schmidt, A.F. and Spurlock, B.H., Jr., Trans. ASME, 76, 823-30, 1954.
301	221	Schmidt, E. and Leidenfrost, W., Chem.-Ingr.-Tech., 26, 35-8, 1954.
302	26039	Schmidt, W.D., M.S. Thesis, Illinois Inst. Techn., 1-57, 1952.
303	21403	Schneider, E., Ann. Physik, 79, 177-203, 1926.
304	21403	Schneider, E., Ann. Physik, 80, 215-6, 1927.
305	1618	Schottky, W.F., Z. Elektrochem., 56, 889-92, 1952.

Ref. No.	TPRC No.	
306	26051	Schreck, H., Ph.D. Thesis, Heidelberg Univ., 1-63, 1953.
307	25401	Schwarze, W., Ann. Physik, 11, 303-30, 1903.
308	25401	Schwarze, W., Ann. Physik, 11, 1144, 1903.
309	25406	Schwarze, W., Physik. Z., 4, 229, 1903.
310	8173	Sekiguchi, T. and Herndon, R.C., Phys. Rev., 112, 1-10, 1958.
311	9973	Sellschopp, W., Z. Ver. Deut. Ing., 77, 1158, 1933.
312	8343	Sellschopp, W., Forsch. Gebiete Ingenieurw., 5, 162-72, 1934.
313	210	Senftleben, H., Z. Angew. Phys., 5, 33-9, 1953.
314	20	Senftleben, H. and Gladisch, H., Z. Physik, 125, 653-6, 1949.
315	5327	Sherif, I.I., Nuovo cimento, 3, 6-11, 1956.
316	42564	Saxena, V.K. and Saxena, S.C., Chem. Phys. Letters, 2, 44-46, 1968.
317	13528	Sherratt, G.G. and Griffiths, E., Phil. Mag., 27, 68, 1939.
318	25798	Owens, E.J. and Thodos, G., A.I.Ch.E. Journal, 6, 676-81, 1960.
319	5328	Shushpanov, P.I., J. Exptl. Theoret. Phys. (USSR), 9, 875-83, 1939.
320	24676	Simmons, J.T., W.R. Whittaker Co. Rept. D9027, 1957.
321	51	Singwi, K.S. and Kothari, L.S., Phys. Rev., 76, 305-6, 1949.
322	24634	Society of Automotive Engineers, Aeron. Info. Rept. 24, 1952.
323	9908	Soddy, F. and Berry, A.J., Proc. Roy. Soc. (London), A84, 576-85, 1911.
324	5289	Srivastava, B.N. and Madan, M.P., Proc. Natl. Acad. Sci., India, 21A, 254-60, 1952.
325	1623	Srivastava, B.N. and Madan, M.P., Proc. Natl. Inst. Sci., India, 20A, 587-97, 1954.
326	7352	Srivastava, B.N. and Saxena, S.C., Proc. Phys. Soc. (London), 70B, 369-78, 1957.
327	25404	Srivastava, K.P., Indian J. Phys., 31, 404-14, 1957.
328	26026	Stefan, J., Sitzber. Akad. Wiss. Wien. Abt. IIa, 65, 45-69, 1872.
329	25909	Stefan, J., Sitzber. Akad. Wiss. Wien. Abt. IIa, 72, 69-101, 1875.
330	1148	Stolyarov, E.A., Zhur. Fiz. Khim., 24, 279-91, 1950.
331	1171	Stolyarov, E.A., Ipatev, V.V., and Teodorovich, V.P., Zhur. Fiz. Khim., 24, 166-76, 1950.
332	9342	Stolyarov, E.A., Ipatev, V.V., and Teodorovich, V.P., MOS (U.K.) TIB/T4445, (Translation of 331). [AD 92320]
333	805	Stops, D.W., Nature (London), 164, 966-7, 1949.
334	19	Strickler, H.S., J. Chem. Phys., 17, 427-8, 1949.
335	5361	Taylor, W.J. and Johnston, H.L., J. Chem. Phys., 14, 219-33, 1946.
336	1616	Thomas, L.B. and Golike, R.C., J. Chem. Phys., 22, 300-5, 1954.
337	18387	Timrot, D.L., "Tables of Thermal Properties of Water and Water Vapor Based on Experimental Data", (Moscow), 104 pp., 1952, 1958.
338	25407	Timrot, D.L. and Vargaftik, N.B., J. Phys. (USSR), 2, 101-11, 1940.
339	5330	Timrot, D.L. and Vargaftik, N.B., J. Tech. Phys. (USSR), 10, 1063-73, 1940.
340	20967	Todd, G.W., Proc. Roy. Soc. (London), A83, 19-39, 1909.
341	7070	Touloukian, Y.S., NBS Rept. 2568, 1-23, 1953. [AD 19902]
342	6724	Traustel, S., Brennstoff-Warme-Kraft, 3, 120-2, 1951.
343	13864	Trautz, M. and Zundel, A., Ann. Physik, 17, 345-75, 1933.
344	8779	Tribus, M. and Boelter, L.M.K., NACA Rept. W-9, 1-14, 1946.
345	1136	Tsederberg, N.V., Teploenergetica, 4 (1), 45-8, 1957.
346	821	Tsederberg, N.V. and Timrot, D.L., Zhur. Tekh. Fiz., 26, 1849-56, 1956.
347	1926	Ubbink, J.B., Physica, 13, 629-34, 1947.
348	1927	Ubbink, J.B., Physica, 13, 659-68, 1947.
349	1928	Ubbink, J.B., Physica, 14, 165-74, 1948.
350	5331	Ubbink, J.B. and deHaas, W.J., Physica, 10, 451-64, 1943.
351	1718	Ubbink, J.B. and deHaas, W.J., Physica, 10, 465-70, 1943.
352	825	von Ubisch, H., Appl. Sci. Research, A2, 364-402, 1951.

Ref. No.	TPRC No.	
353	1414	Uhlir, A., Jr., J. Chem. Phys., $\underline{20}$, 463-72, 1952.
354	425	Vanderkooi, W.N., Univ. Microfilms Publ. No. 16498, 1-269, 1956.
355	375	Vanderkooi, W.N. and De Vries, T., J. Phys. Chem., $\underline{60}$, 636-9, 1956.
356	13341	Vargaftik, N.B., Tech. Phys. (USSR), $\underline{4}$, 341-60, 1937.
357	26068	Stops, D.W., Ph.D. Thesis, Univ. of London, 1948.
358	26037	Vargaftik, N.B., World Power Conference, 1950.
359	3138	Vargaftik, N.B., Izvest. Vsesoyuz. Teplotekh. Inst. Im. Feliksa Dzerzhinskogo, $\underline{21}$ (1), 13-7, 1952.
360	9397	Vargaftik, N.B., 5th International Conf. on Properties of Steam (London), 12 pp., 1955.
361	19356	Vargaftik, N.B., Proc. Cond. "Thermodynamic and Transport Properties of Fluids", Inst. Mech. Eng. (London), 142-9, 1958.
362	24659	Vargaftik, N.B. and Belyakova, P.E., Teploenergetika, $\underline{1}$ (5), 45-51, 1954.
363	1445	Vargaftik, N.B. and Oleshchuk, O.N., Izvest. VTI, $\underline{15}$ (5), 7-15, 1946.
364	13379	Vargaftik, N.B. and Parfenov, L.D., Zhur. Ekspt1. i Teoret. Fiz., $\underline{8}$, 189-97, 1938.
365	1398	Vargaftik, N.B. and Smirnova, E.V., Zhur. Tekh. Fiz., $\underline{26}$, 1251-61, 1956.
366	1464	Vargaftik, N.B. and Smirnova, E.V., Soviet Phys., Tech. Phys., $\underline{1}$, 1221-31, 1957.
367	71	Vines, R.G., Australian J. Chem., $\underline{6}$, 1-26, 1953.
368	19352	Vines, R.G., Proc. Conf. "Thermodynamic and Transport Properties of Fluids", Inst. Mech. Eng. (London), 120-3, 1958.
369	17439	Vines, R.G., Trans. ASME, $\underline{82}$, 48-52, 1960.
370	1619	Vines, R.G. and Bennett, L.A., J. Chem. Phys., $\underline{22}$, 360-6, 1954.
371	8360	Vogel, H., Ann. Physik, $\underline{43}$, 1235-72, 1914.
372	15575	Wachsmuth, J., Physik, Z., $\underline{9}$, 235, 1908.
373	47	Waelbroeck, F.G., Lafleur, S., and Prigogine, I., Physica, $\underline{21}$, 667-75, 1955.
374	6909	Waelbroeck, F.G. and Zuckerbrodt, P., J. Chem. Phys., $\underline{28}$, 523-4, 1958.
375	6908	Waelbroeck, F.G. and Zuckerbrodt, P., J. Chem. Phys., $\underline{28}$, 524-6, 1958.
376	26036	Waldmann, L., "Handbuch der Physik", Springer-Verlag, Berlin, XII, 1958.
377	26035	Wassiljewa, A., Physik. Z., $\underline{5}$, 737-42, 1904.
378	26029	Weber, S., Ann. Physik, $\underline{54}$, 325-56, 1917.
379	8362	Weber, S., Ann. Physik, $\underline{54}$, 437-62, 1917.
380	26027	Weber, S., Ann. Physik, $\underline{54}$, 481-502, 1917.
381	16326	Weber, S., Commun. Kamerlingh Onnes Lab. Univ. Leiden, Suppl. 42b, 1918.
382		Weber, S., Verslag Gewone Vergader. Afdeel. Natuurk., Ned. Akad. Wetenschap., $\underline{26}$, 1338-53, 1917.
383	15873	Weber, S., Proc. Koninkl. Ned. Akad. Wetenschap, $\underline{21}$, 342-56, 1918.
384	6712	Weber, S., Ann. Physik, $\underline{82}$, 479-503, 1927.
385		Weber, S., Ann. Physik, $\underline{9}$, 981, 1931.
386	5332	Weber, S., Kgl. Danske Videnskab. Selskab, Math-Fys. Medd., $\underline{19}$ (11), 39, 1942.
387	812	Weininger, J.L. and Schneider, W.G., Ind. Eng. Chem., $\underline{43}$, 1229-33, 1951.
388	20089	Wellman, E.J. and Sibbitt, W.L., Combustion, $\underline{26}$, 51-6, 1955.
389	775	West, J.R., J. Phys. and Colloid Chem., $\underline{55}$, 402-5, 1951.
390	1625	Westenberg, A.A., Combustion and Flame, $\underline{1}$, 346-59, 1957.
391	3384	Wheeler, H.P., Jr., US Bur. Mines Circ. 7344, 1-66, 1946.
392	214	White, D., Chou, C., and Johnston, H.L., J. Chem. Phys., $\underline{20}$, 1819-20, 1952.
393	1918	Wicke, E., Angew. Chem., $\underline{A60}$, 65, 1948.
394	1668	Wicke, E. and Franck, E.U., Angew. Chem., $\underline{66}$, 701-10, 1954.
395	21410	Wilner, T. and Borelius, G., Ann. Physik, $\underline{4}$, 316-22, 1930.
396	26028	Winkelmann, A., Ann. Physik, $\underline{156}$, 497-531, 1875.
397	8359	Winkelmann, A., Ann. Physik, $\underline{29}$, 68-113, 1886.

Ref. No.	TPRC No.	
398	26031	Winkelmann, A., Ann. Physik, 44, 177-205, 1891.
399	25413	Winkelmann, A., Wied. Ann., 44, 429-56, 1891.
400	25412	Winkelmann, A., Ann. Physik, 48, 180-7, 1892.
401	1615	Wirth, H. and Klemenc, A., Monatsh., 83, 879-82, 1952.
402	139	Wirth, J. F. C., Rev. quim. e ing. quim. (Monterrey), 1, 46-82, 1954.
403	26046	de Wit, Ph.D. Thesis, Zurich Univ., 1913, International Critical Tables, McGraw-Hill, N.Y., 5, 212-5, 1929.
404	996	Woolley, H.W., NACA-TN 3271, 1-114, 1956.
405	1771	Woolley, H.W., J. Research NBS, 40, 163-8, 1948.
406	7071	Woolley, H.W., NBS Rept. 2287, 1-39, 1953. [AD 19 903]
407	7069	Woolley, H.W., NBS Rept. 2611, 1953. [AD 19 900]
408	15661	Woolley, H.W., Scott, R.B. and Brickwedde, F.G., J. Research NBS, 41, 379-475, 1948.
409	16651	Wright, J.M., AECD Rept. 4197, 1-17, 1951.
410	25416	Wullner, Ann. Physik, 4, 321-41, 1878.
411		Zahn, C.T., Phys. Rev., 27, 455-9, 1927.
412	167	Ziebland, H. and Burton, J.T.A., Brit. J. Appl. Phys., 6, 416-20, 1955.
413	6725	Ziebland, H. and Burton, J.T.A., Brit. J. Appl. Phys., 9, 52-9, 1958.
414	26046	Ziegler, Dissertation, Halle Univ., 1904, International Critical Tables, McGraw-Hill, N.Y., 5, 212-5, 1929.
415	7022	Richter, G.N. and Sage, B.H., Proj. Squid Rept. CIT-3-P, 1-16, 1958. [AD 200 655]
416	26046	International Critical Tables, McGraw-Hill, N.Y., 1-7, 1910-26.
417	7021	Richter, G.N. and Sage, B.H., Proj. Squid Rept. CIT-2-P, 1-17, 1957. [AD 139 905]
418	10378	Cherneyeva, L., Kholodil'naya Tekh., 30 (3), 60-3, 1953.
419	26045	Isotron Refrigerant leaflet, Pennsalt Chemicals Corp., Philadelphia, Pa., 1-3, 1957.
420	16004	Brokaw, R.S., NASA-TR-R-81, 1-34, 1960.
421	8280	Coffin, K.P. and O'Neal, C., Jr., NACA TN-4209, 1-22, 1958.
422	10377	Cherneyeva, L., Kholodil'naya Tekh., 29 (3), 55-8, 1952.
423	15588	Abas-Zade, A.K., AEC-TR-3989, 1-12, 1960.
424	6905	Bates, O.K., Ind. Eng. Chem., 25, 431-7, 1933.
425	8351	Bates, O.K., Ind. Eng. Chem., 28, 494-8, 1936.
426	8348	Bates, O.K., Hazzard, G., and Palmer, G., Ind. Eng. Chem., Anal. Ed., 10, 314-8, 1938.
427	5481	Bates, O.K., Hazzard, G., and Palmer, G., Ind. Eng. Chem., 33, 375-6, 1941.
428	721	Bates, O.K. and Hazzard, G., Ind. Eng. Chem., 37, 193-5, 1945.
429	9904	Bottomley, J.T., Proc. Roy. Soc. (London), 28, 462-3, 1879.
430	9905	Bottomley, J.T., Proc. Roy. Soc. (London), 31, 300, 1881.
431	9890	Bridgman, P.W., Proc. Am. Acad. Arts. Sci., 59, 141-69, 1923.
432	739	Briggs, D.K.H., Ind. Eng. Chem., 49, 418-21, 1957.
433	138	Cecil, O.B. and Munch, R.H., Ind. Eng. Chem., 48, 437-40, 1956.
434	1164	Challoner, A.R. and Powell, R.W., Proc. Roy. Soc. (London), A238, 90-106, 1956.
435	9339	Daniloff, M., J. Am. Chem. Soc., 54, 1328-32, 1932.
436	10376	Danilova, G., Kholodil'naya Tekh., 28, 22-6, 1951.
437	9391	Davis, A.H., Phil. Mag., 47, 972-5, 1924.
438	26009	Davis, A.H., Phil. Mag., 47, 1057-92, 1924.
439	1675	Dick, M.F. and McCready, D.W., Trans. Am. Soc. Mech. Engrs., 76, 831-9, 1954.
440	18387	Dsrjinskogo, F.E., "Tables of thermodynamic properties of Water and Water vapor based on experimental data", 104 pp., 1952.
441	15563	Filippov, L.P., LC or SLA 61-10145, 1-7, 1960.
442	1122	Filippov, L.P., Vestnik Moskov Univ., 9 (12), Ser. Fiz.-Mat. i Estestven. Nauk No. 8, 45-8, 1954.

Ref. No.	TPRC No.	
443	3074	Frontas'ev, V.P., J. Phys. Chem. (USSR), 20, 91-104, 1946.
444	1462	Frontas'ev, V.P., Doklady Akad. Nauk, (SSSR), 111, 1014-6, 1956.
445	9940	Frontas'ev, V.P., Soviet Phys.-Tech. Phys., 3, 1696-1700, 1958.
446	19618	Frontas'ev, V.P., Zhur. Tekh. Fiz., 28, 1839-44, 1958.
447	18002	Frontas'ev, V.P. and Gusakov, M.Ya., Zhur. Tekh. Fiz., 29, 1277-84, 1959.
448	18004	Frontas'ev, V.P. and Gusakov, M.Ya., Soviet Phys.-Tech. Phys., 4, 1171-7, 1960.
449	789	Gillam, D.G. and Lamm, O., Acta Chem. Scand., 9, 657-60, 1955.
450	9882	Goldschmidt, R., Physik. Z., 12, 417-24, 1911.
451	8357	Graetz, L., Ann. Physik, 3, 18, 79-94, 1883.
452	8299	Graetz, L., Ann. Physik, 3, 25, 337-57, 1885.
453	10673	Graetz, L., AERE Lib/Trans 794, 1-12, 1958. [AD 203 923]
454	9284	Guthrie, F., Phil. Trans. Roy. Soc. (London), 159A, 637-60, 1869.
455	9388	Guthrie, F., Phil. Mag., 37, 468-70, 1869.
456	8300	Henneberg, H., Ann. Physik, 3, 36, 146-64, 1889.
457	1159	Hill, R.A.W., Proc. Roy. Soc. (London), A239, 476-86, 1957.
458	1081	Hutchinson, E., Trans. Faraday Soc., 41, 87-90, 1945.
459	8312	Jakob, M., Ann. Physik, 4, 63, 537-70, 1920.
460	16358	Jakob, M., Z. Ver. deut. Ingr., 66, 688-93, 1922.
461	9373	Kallam, F.L., Mech. Eng., 45, 479-82, 1923.
462	9959	Kardos, A., Z. Ver. deut. Ingr., 77, 1158, 1933.
463	13734	Kardos, A., Z. Tech. Physik, 15, 79-82, 1934.
464	9913	Kaye, G.W.C. and Higgins, W.F., Proc. Roy. Soc. (London), A117, 459-70, 1928.
465	10510	Kays, W.M. and London, A.L., Tech. Rept. No. 23, Stanford Univ., 1954. [AD 49 051]
466	9758	Kerzhentsev, V.V., Trudy Moskov. Aviatsionn. Inst., (51), 72-82, 1955.
467	19934	Kramer, F.R., Univ. Microfilms, Mic. 59-4159, 1-161, 1959.
468	6	Kraus, W., Z. Angew. Phys., 1, 173-9, 1948.
469	1738	Kurtener, A.V. and Malyshev, E.K., J. Tech. Phys. (USSR), 13, 641-4, 1943.
470	18276	Lawson, A.W., Lowell, R., and Jain, A.L., J. Chem. Phys., 30, 643-7, 1959.
471	7693	McCready, D.W., WADC TR 58-405, 1-51, 1959. [AD 211693]
472	26010	McNall, P.E., Ph.D. Thesis, Mechanical Engineering, Purdue Univ., Lafayette, Indiana, 1-44, 1951.
473	7300	Malhotra, B.R., Modern Refrig., 60, 497-502, 1957.
474	9896	Martin, L.H. and Lang, K.C., Proc. Phys. Soc. (London), 45, 523-9, 1933.
475	1676	Mason, H.L., Trans. Am. Soc. Mech. Engrs., 76, 817-21, 1954.
476	24596	Milner, S.R. and Chattock, A.P., Phil. Mag., 48, 46-64, 1899.
477	24599	Nukiyama, S. and Yoshizawa, Y., J. Soc. Mech. Engrs. (Japan), 37, 347-50, 1934.
478	2566	Os'minin, Yu.P., Vestnik Moskov. Univ., 12, Ser. Mat., Mekhan., Astron., Fiz. i Khim., 117-25, 1957.
479	13435	Powell, R.W., Advances in Phys., 7, 276-97, 1958.
480	15742	Powell, R.W. and Challoner, A.R., Phil. Mag., 4, 1183-6, 1959.
481	15740	Powell, R.W. and Challoner, A.R., Modern Refrig., 63, 3-7, 1960.
482	16839	Priest, H.F., MDDC-179, 1-11, 1944.
483	12995	Riedel, L., Forsch. Gebiete Ingenieurw., 11B, 340-8, 1940.
484	16071	Riedel, L., Mitt. Kaltetech. Insts. u Reichsforschung. Anstalt Lebensmittelfrischhalt. Tech. Hochschule Karlsruhe No. 2, 1-47, 1948.
485	16090	Riedel, L., Kaltetechnik, 2 (4), 99-101, 1950.
486	1085	Riedel, L., Chem.-Ingr.-Tech., 23, 321-4, 1951.
487	82	Riedel, L., Chem.-Ingr.-Tech., 23, 465-9, 1951.
488	860	Sakiadis, B.C. and Coates, J., Louisiana State Univ., Eng. Expt. Sta. Bull. 35, 1-53, 1953.

Ref. No.	TPRC No.	
489	24595	Sakiadis, B.C. and Coates, J., A.I.Ch.E. Journal, 1, 275-88, 1955.
490	10679	Scheffy, W.J., PR-85-R, 1-99, 1958. [AD 204 891]
491	10584	Scheffy, W.J. and Johnson, E.F., PR-71-M, 1-19, 1957. [AD 142 428]
492	16077	Schmidt E. and Leidenfrost, W., Forsch. Gebiete Ingenieurw., 19, 65-80, 1953.
493	16068	Schmidt, E. and Leidenfrost, W., Forsch. Gebiete Ingenieurw., 21, 176-80, 1955.
494	14368	Schmidt, E. and Sellschopp, W., Forsch. Gebiete Ingenieurw., 3, 277-86, 1932.
495	6539	Schrock, V.E., Gott, R.E., and Starkman, E.S., WADC TR 56-104, 1-144, 1956. [AD 103 097]
496	845	Schulz, A.K., J. Chem. Phys., 51, 530-3, 1954.
497	9335	Shiba, H., Sci. Papers Inst. Phys. Chem. Res., 16, 205-41, 1931.
498	16862	Siletz, V.S. and Fridkin, L.M., (Editors), Westinghouse Electric Corp. Transl., 1-104, 1952.
499	12	Slawecki, T.K. and Molstad, M.C., Ind. Eng. Chem., 48, 1100-3, 1956.
500	8349	Smith, J.F.D., Ind. Eng. Chem., 22, 1246-51, 1930.
501	9953	Smith, J.F.D., Trans. Am. Soc. Mech. Engrs., 58, 719-25, 1936.
502	1077	Soonawala, M.F., Indian J. Phys., 18, 71-3, 1944.
503	10642	Spells, K.E., FPRC Rept. 1033, 1-10, 1958. [AD 200 754]
504	10593	Spencer, A.N., IGR-R/CA-215, 1-9, 1957. [AD 138 556]
505	10948	Spencer, A.N. and Todd, M.C.J., Brit. J. Appl. Phys., 11, 60-4, 1960.
506	1100	Tsederberg, N.V. and Timrot, D.L., Zhur. Tekh. Fiz., 25, 2458-62, 1955.
507	9296	Van der Held, E.F.M. and van Drunen, F.G., Physica, 15, 865-81, 1949.
508	19356	Vargaftik, N.B., Proc. Conf. Thermodynamic and Transport Properties Fluids, (London), 142-9, 1957.
509	16975	Vargaftik, N.B. and Oleshchuk, O.N., Teploenergetika, 6 (10), 70-4, 1959.
510	16973	Vargaftik, N.B. and Oleshchuk, O.N., TIL/T. 5171 MDF-V-138 N-99589, 1-6, 1961. [AD 259 797]
511	8311	Wachsmuth, R., Ann. Physik., 3, 48, 158-79, 1892.
512	8291	Weber, H.F., Ann. Physik., 3, 10, 304-20, 1880.
513	8292	Weber, H.F., Ann. Physik, 3, 10, 472-500, 1880.
514	24598	Weber, H.F., Berliner Ber., 2, 809-15, 1885.
515	8293	Weber, R., Ann. Physik., 4, 11, 1047-70, 1903.
516	8315	Winkelmann, A., Ann. Physik, 2, 153, 481-98, 1874.
517	1111	Woolf, J.R. and Sibbitt, W.L., Ind. Eng. Chem., 46, 1947-52, 1954.
518	19632	Wright, R.M., UCRL-9744, 1-155, 1961.
519	18674	Ziebland, H., Intern. J. Heat Mass Transfer, 2, 273-9, 1961.
520	16711	Nowak, E.S. and Grosh, R.J., USAEC ANL-6064, 1-102, 1959.
521	24648	Svehla, R.A., NASA TR-R-132, 1-140, 1962.
522	6986	Challoner, A.R., Gundry, H.A., and Powell, R.W., Proc. Roy. Soc. (London), A245, 259-67, 1958.
523	16875	Chelton, D.B. and Mann, D.B., WADC TR 59-8, 1-144, 1959. [AD 208 155]
524	20934	Friedman, A.S. and Hilsenrath, J., NBS Rept. 3163, 1-12, 1954.
525	4673	Griffiths, E., Awbery, J.H., and Powell, R.W., Dept. Sci. Ind. Res. (Brit.), Rept. of Food Invest. Board for the year 1938, 265-74, 1939.
526	16396	Kardos, A., Z. ges. Kalte-Ind., 41, 1-6, 29-35, 1934.
527	28791	Koch, B., "Grundlagen des Warmeaustausches", Verlagsaustalt H. Beucke and Sohne, 1950.
528	16595	Liberto, R.R., AFFTC TR-60-62, 1-70, 1960. [AD 249 542]
529	17005	Little, A.D., AFFTC TR-60-19, 1-246, 1960. [AD 242 285]
530	16396	Plank, R., Z. ges. Kalte-Ind., 41, 214, 1934.
531	1687	Powers, R.W., Mattox, R.W., and Johnston, H.L., J. Am. Chem. Soc., 76, 5968-71, 1954.
532	5916	Powers, R.W., Mattox, R.W., and Johnston, H.L., J. Am. Chem. Soc., 76, 5972-3, 1954.
533	5943	Powers, R.W., Mattox, R.W., and Johnston, H.L., J. Am. Chem. Soc., 76, 5974, 1954.
534	1420	Prosad, S., Brit. J. Appl. Phys., 3, 58-9, 1952.

Ref. No.	TPRC No.	
535	1461	Prosad, S., Current Sci. (India), 20, 264, 1951.
536	10122	Richter, G.N. and Sage, B.H., Project Squid Rept. CIT-1-P, 1956. [AD 104 498]
537	8959	Richter, G.N. and Sage, B.H., Chem. and Eng. Data Series, 2, 61-6, 1957.
538	28790	Sellschopp, W., Z. Ver. deutsch Ing., 79, 69, 1935.
539	3080	Tsederberg, N.V. and Timrot, D.L., Soviet Phys., Tech. Phys., 1, 1791-7, 1957. (Translation of 346)
540	17707	Vilim, O., Collection of Czechoslovak Chemical Communications, 25 (4), 993-9, 1960.
541	8338	Waterman, T.E., "Thermodynamic and Transport Properties of Gases, Liquids, and Solids (ASME Symp. on Thermal Properties)", 301-3, 1959.
542	17749	Ziebland, H., Dechema Monograph, 32, 74-82, 1959.
543	18305	Waterman, T.E., Kirsh, D.P., and Brabets, R.I., J. Chem. Phys., 29, 905-8, 1958.
544	29195	Lee, D.M. and Fairbank, H.A., Proc. 5th International Conference of Low Temperature Physics (Madison, Wis.), 90-3, 1957.
545	6436	Grenier, C.G., "Low Temperature Physics", Low Temperature Lab., Rice Institute, 25-8, 1950. [AD 121 132]
546	176	Grenier, C.G., Low-temperature Physics, NBS Circ. 519, 193-4, 1952.
547	12350	Keesom, W.H. and Keesom, A.P., Physica, 3, 359-60, 1936.
548	29196	Keesom, W.H., "Helium", Elsevier Publ. Co., Amsterdam, 1942.
549	29197	Lifshits, E.M. and Andronikashvili, E.L., "A Supplement to Helium", Consultants Bureau, Inc., New York, 1959.
550	13415	Allen, J.F. and Ganz, E., Proc. Roy. Soc. (London), A171, 242-50, 1939.
551	6691	Atkins, K.R., Phys. Rev., 108, 911-3, 1957.
552	29198	Brewer, D.F. and Edwards, D.O., Proc. Phys. Soc., 71, 117-25, 1958.
553	19405	Brewer, D.F. and Edwards, D.O., Proc. Roy. Soc. (London), A251, 247-64, 1959.
554	20738	Chase, C.E., Phys. Rev., 120, 688-96, 1960.
555	1688	Fairbank, H.A. and Wilks, J., Phys. Rev., 95, 277-8, 1954.
556	25441	Fairbank, H.A. and Wilks, J., Proc. Roy. Soc. (London), A231, 545-55, 1955.
557	3492	Ganz, E., Helv. Phys. Acta, 12, 294-5, 1939.
558	22509	Kapitza, P.L., J. Phys. (USSR), 4, 181-210, 1941.
559	1640	Keesom, W.H. and Duyckaerts, G., Physica, 13, 153-79, 1947.
560	4313	Keesom, W.H. and Saris, B.F., Physica, 7, 241-52, 1940.
561	4734	Keesom, W.H., Saris, B.F., and Meyer, L., Physica, 7, 817-30, 1940.
562	5862	Mellink, J.H., Physica, 13, 180-96, 1947.
563	5868	Meyer, L. and Mellink, J.H., Physica, 13, 197-215, 1947.
564	10517	Reppy, J.D. and Reynolds, C.A., Tech. Rept. No. 4, Cryogenic Group, Univ. of Connecticut, 1-18, 1956. [AD 103 530]
565	4539	Strelkov, P.G., J. Exptl. Theor. Phys. (USSR), 10, 1225-8, 1940.
566	1689	White, D., Gonzales, O.D., and Johnston, H.L., Phys. Rev., 89, 593-4, 1953.
567	19061	Wilks, J., Nuovo Cimento Supplemento, 9, 84-94, 1958.
568	27702	Richter, G.N. and Sage, B.H., J. Chem. Eng. Data, 9, 75-8, 1964.
569	16674	Blais, N.C. and Mann, J.B., J. Chem. Phys., 32, 1459-65, 1960. See also Mann, J.B. and Blais, N.C., AEC R. and D. Rept. LA 2316, 1-39, 1959.
570	25408	Rienda, J.M.B., Ph.D. Thesis, Madrid Univ., Spain, 1963. [AD 254 599]
571	18218	Masia, A.P. and Sonet, M.D., PB Rept. 155596, 1-58, 1961.
572	23552	E.I. duPont de Nemours and Co. Inc. Tech. Bull. B-2, 1-11, 1957.
573	29989	Richter, G.N. and Sage, B.H., J. Chem. Eng. Data, 8, 221-5, 1963.
574	24839	Andrussow, L., Progr. Int. Res. Thermodynamic and Transport Properties, Academic Press, N.Y., 279-87, 1962.
575	10073	Schaefer, C.A. and Thodos, G., A.I.Ch.E. Journal, 5, 367-72, 1959.
576	26104	Carmichael, L.T., Berry, V., and Sage, B.H., J. Chem. Eng. Data, 8, 281-5, 1963.

Ref. No.	TPRC No.	
577	23064	Bachman, K.C., Matthews, E.K., and Zudkevitch, D., ASD-TDR-62-254, 1962. [AD 281 898]
578	27935	Masia, A.P. and Alvarez, M.D.S., Anales Real Soc. Espan fis Quim (Madrid), B58, 3-12, 1962.
579	33602	Baker, C.B. and de Haas, N., Phys. Fluids, 7, 1400-2, 1964.
580	21309	Srivastava, B.N. and Barua, A.K., J. Chem. Phys., 35, 329-34, 1961.
581	24872	Srivastava, B.N. and Barua, A.K., J. Chem. Phys., 35, 649-51, 1961.
582	14811	Barua, A.K. and Chakraborti, P.K., J. Chem. Phys., 36, 2817-20, 1962.
583	32640	Srivastava, B.N., Barua, A.K., and Chakraborti, P.K., Trans. Faraday Soc., 59, 2522-7, 1963.
584	25719	Kennedy, J.T. and Thodos, G., A.I.Ch.E. Journal, 7, 625-31, 1961.
585	16745	Chen, L.H., AEC Report, GA 1038, 1-157, 1959.
586	28351	Sullivan, K., U.K.A.E.A. TRG Rept., 438 (R), 1-42, 1963.
587	24007	Geier, H. and Schafer, K., Allgem. Warmetech., 10, 70-5, 1961.
588	25951	Gray, P. and Wright, R.G., Proc. Roy. Soc., A263, 161-88, 1961.
589	33604	Needham, D.P. and Ziebland, H., Ministry of Aviation, Rept. ERDE-14 R 63, 1963. [AD 445 673]
590	27532	Keyes, F.G. and Vines, R.G., Proj. Squid Rept. MIT-32-P, 1964. [AD 410 093]
591		Grober, H. and Erk, S., "Die Grundgesetze der Warmeubertragung", Berlin, 1933.
592	26382	Van Iterson, F.K.Th., De Ingenieur, 47 (10), W 15-W 22, 1932.
593	33683	Jakob, M., Engineering, 132, 744-6, 1931.
594	33601	Keenan, J.H. and Keyes, F.G., "Thermodynamic Properties of Steam", Wiley, N.Y., 1-89, 1936.
595	30398	Theiss, R.V. and Thodos, G., J. Chem. Eng. Data, 8 (3), 390, 1963.
596		Keyes, F.G., Vth Int. Conf. Properties of Steam, London, 1956.
597		Vargaftik, N.B., Zhur. Tek. Fiz., 7, 1199, 1937.
598	19679	Ristikivi, J., Tek. Tidskr., 89, 1325-6, 1959.
599	10055	Srivastava, B.N. and Srivastava, R.C., J. Chem. Phys., 30, 1200-5, 1959.
600	28701	Bradford, C.S., Phil. Mag., 34, 433-71, 1943.
601	24750	Vargaftik, N.B., "Teplofizicheskiye Svoystvo Veshchestv", Moscow and Leningrad, 368 pp., 1956.
602	27952	Vargaftik, N.B., "Thermophysical Properties of Substances", F-TS-9537/V, 485 pp., 1959.
603	37862	Tsederberg, N.V., "Thermal Conductivity of Gases and Liquids", M.I.T. Press, 246 pp., 1965.
604	28850	Grier, N.T., NASA TN D-1406, 1-64, 1962.
605	10463	King, C.R., NASA TN D-275, 1-71, 1960. [AD 235 647]
606	21315	Rogers, J.D., Zeigler, R.K., et al., LA 2719, 1-40, 1962.
607	18822	Rosner, D.E., Aerochem. Res. Lab. Rept. TP 27, 1-27, 1961. [AD 257 863]
608	25625	Vanderslice, J.T., Weissman, S., et al., Phys. Fluids, 5, 155-64, 1962.
609		Mausa, S.A., Thesis, Cairo Univ., Egypt, 1943.
610	17305	El Nadi, M. and Salam, E., Z. Phys. Chem., 215, 121-32, 1960.
611	33385	Sherman, M.P., Aeron. Engng. Lab. Rept. 673, Princeton Univ., 1-74, 1963. [AD 437 625], [AD 435 141]
612	37875	Ahtye, W.F., NASA TN D-2611, 1-110, 1965.
613	34717	deVoto, R.S., Aeron. and Astron. Dept. Rept. SUDAER 217, Stanford Univ., 1-53, 1956 = Inst. for Plasma Res. Rept. SU AA 217, Stanford Univ., 1965.
614	24957	Arave, R.J. and Huseby, O.A., Boeing Aircraft Co. Rept. D2-11238, 1-227, 1962. [AD 285 668]
615	31971	Pindroh, A.L., Boeing Aircraft Co. Rept. D2-11253, 1-63, Sept. 1961. [AD 434 829]
616	29016	Chirkin, V.S., Heat Conductivity of Gases, F-TS-9928/V, 1-21, March 1960.
617	31011	Vargaftik, N.B. and Zimina, N.H., Teplofiz. Vysokikh Temp., 2, 716-24, 1964: High Temp., 2 (5), 645-51, 1964.
618	36295	Keyes, F.G. and Vines, R.G., Trans. A.S.M.E. (J. Heat Transfer), 87C, 177-83, 1965.
619	25186	Michels, A., Sengers, J.V. et al., Physica, 29, 149-60, 1963.
620	26276, 26277	Kulik, P.P., Panevin, I.G. Et al., Teplofiz. Vysokikh Temp., 1, 56-63, 1963: High Temp., 1, 45-51, 1963.
621	25080	Yos, J.M., RAD-TM-63-7, March 1963.

Ref. No.	TPRC No.	
622	18937	Yun, K-S., Weisman, S., et al., Phys. Fluids, 5, 672-7, 1962.
623	28918	Ahtye, W. F. and Peng, T-C., NASA TN D-1303, 1-108, 1962.
624	23120	Ludwig, H.C., Welding, J., (N.Y.), 38, 296s-300s, 1959.
625	31566	Schmitz, G. and Patt, H.J., Z. Physik, 171, 449-62, 1963.
626	25626	Westenberg, A.A. and deHaas, N., Phys. Fluids, 5, 266-73, 1962.
627	25900	Pereira, A.N.G. and Raw, C.J.G., Phys. Fluids, 6, 1091-6, 1963.
628	38397	Petersen, J.R. and Bonilla, C.F., 3rd ASME Symp. on Thermophysical Properties, Purdue Univ., Lafayette, Indiana, 264-76, 1965.
629	31023	Vargaftik, N.B. and Zimina, N.Kh., Teplofiz. Vysokikh. Temp., 2, 869-78, 1964; High Temp., 2, 782-90, 1964.
630	16058	Eckert, E.R.G., Ibele, W.E., and Irvine, T.F., Jr., NASA TN D-533, 1-39, 1960. [AD 241 819]
631	33413	Ibele, W.E., Novotny, J.L., and Eckert, E.R.G., NASA CR 55273, 1-49, 1963.
632	22201	Usmanov, A.G. and Mukhamedzyanov, G.Kh., Intern. Chem. Eng., 3, 369-74, 1963.
633	24731	Riedel, L., Arch. tech. Messen, 227, 273-6, 1954.
634	23284	Frontas'ev, V.P., Ind. Lab. USSR, 26, 787-9, 1960.
635	27296	Vasilevaskaya, Y.D., Tr. Mosk. Aviats. Inst. No. 132, 144-60, 1961.
636	20103	Stone, J.P., Ewing, C.T., et al., NRL Rept. 5675, 1-21, 1961. [AD 268 341]
637	23077	Horrocks, J.K. and McLaughlin, E., Proc. Roy. Soc. (London), A273, 259-74, 1963.
638	24981	Ziebland, H. and Burton, J.T.A., J. Chem. Eng. Data, 6, 579-83, 1961.
639	26551	Reiter, F.W., Euratom EuR-582d; N64-19006, 1-32, 1964.
640	29257	Ellard, J.A., King, C.D., et al., Monsanto Res. Corp. Rept. IDO-1101, 1-15, 1962.
641	20878	Ziebland, H., E.R.D.E. Rept. 12/R/61, 1-6, 1961. [AD 257 007]
642	27291	Riedel, L., AEC-TR-1822, 1-18, 1962.
643	27926	El'darov, F.G., Russ. J. Phys. Chem., 34, 677-9, 1960.
644	34006	Baker, C.E. and Brokaw, R.S., J. Chem. Phys., 43, 3519-28, 1965.
645	38073, 38882	Golubev, I.F. and Sokolova, V.P., Teploenergetika, 11 (9), 64-7, 1964; Thermal Eng., 11 (9), 78-82, 1964.
646	34000	Needham, D.P. and Ziebland, H., Intern. J. Heat Mass Transfer, 8 (11), 1387-1414, 1965.
647	36412	Missenard, A., Thermal Conductivity of Solids, Liquids and Gases and Their Mixtures, Editions Eyrolles, Paris, 576 pp., 1965.
648	17164	Ibele, W.E. and Irvine, T.F., Jr., Trans. A.S.M.E. (J. Heat Transfer), 82C, 381-6, 1960.
649	20027	Senftleben, H., Arch. Eisenhuttenw., 31, 709-10, 1960.
650	32701	Senftleben, H., Z. Angew. Phys., 16 (2), 111-5, 1963.
651	21304	Smith, W.J.S., Durbin, L.D. and Kobayashi, R., J. Chem. Eng. Data, 5, 316-21, 1960.
652	34509	Carmichael, L.T., Reamer, H.H., and Sage, B.H., J. Chem. Eng. Data, 11 (1), 52-7, 1966.
653	30965	Vargaftik, N.B., "Handbook of Thermophysical Properties of Gases and Liquids" (in Russian), Gosudarst. Izdatel. Fiz. Mat. i Lit., Moscow, 708 pp., 1963.
654	34723	Svehla, R.A. and Brokaw, R.S., NASA TN D-3327, 1-57, 1966.
655	39455	Svehla, R.A. and Brokaw, R.S., J. Chem. Phys., 44, 4643-5, 1966.
656	21265, 17877	Zaitseva, L.S., Zhur. tekh. Fiz., 29, 497-505, 1959; Soviet Phys.-Tech. Phys., 4, 444-50, 1959.
657	34444	Collins, D.J. and Menard, W.A., J. Heat Transfer, 88C, 52-6, 1966.
658	38904, 34557	Naziev, Ya.M., Zhur. fiz. Khim., 39, 1359-64, 1965; Russian J. Phys. Chem., 39, 724-6, 1965.
659	31305	Golubev, I.F. and Naziev, Ya.M., Tr. Energ. Inst., Akad. Nauk. Azerb. SSR, 15, 84-102, 1962.
660	39527	Choy, P. and Raw, C.J.G., J. Chem. Phys., 45, 1413, 1966.
661	39300, 39301	Dresvyannikov, F.N., Teploenergetika, 13 (2), 86-7, 1966; Thermal Engng., 13 (2), 101-3, 1966.
662	28817	Cammerer, W.F., Kaltetechnik, 12 (4), 107-10, 1960.
663	36625, 30275	Golubev, I.F., Teploenergetika, 10 (12), 78-82, 1963; TT-64-19120, RTS 2536, 1-15, 1964.

Ref. No.	TPRC No.	
664	10074	Schaefer, C.A. and Thodos, G., Ind. Eng. Chem., 50, 1585-8, 1958.
665	39064	Jones, I.W., Min. of Aviation (U.K.) E.R.D.E. Rept. 20/R/65, 1-18, 1965.
666		Corrucini, R.J., Liquid Hydrogen Technology Paper, Grenoble, France, 1965.
667	28075	Kerrisk, J.F., Rogers, J.D., and Hammel, E.F., LADC 5791, 1-21, 1963.
668	33225	Dwyer, R.F., Cook, G.A., and Berwaldt, O.E., J. Chem. Eng. Data, 11, 351-3, 1966.
669	7478	Hill, R.W., Schneidmesser, B., Z. Physik. Chem., 16, 257-66, 1958.
670		Aerojet General Corp. Publn. "Properties of Principal Cryogenics", PRA-SA-DSR, 8/20/64, 3rd printing, 1966.
671	24843	Stiel, L.I. and Thodos, G., Progr. Int. Res. Thermodyn. and Transport Properties, A.S.M.E., N.Y., 352-65, 1962.
672	30426	Ikenberry, L.D. and Rice, S.A., J. Chem. Phys., 39 (6), 1561-71, 1963.
673	35216	Julian, C.L., Phys. Rev., A137, 128-37, 1965.
674	6581	White, G.K. and Woods, S.B., Phil. Mag., 3, 785-97, 1958.
675		Dobbs, E.R. and Jones, G.O., Rept. Progr. Phys., 20, 556- , 1957.
676	30336	Chao, M.S. and Stenger, V.A., Talanta, 11, 271-81, 1964.
677	24053	Roder, H.M., Cryogenics, 2, 302-4, 1962.
678	18289	Keyes, R.W., J. Chem. Phys., 31, 452-4, 1959.
679	21693	Brewer, J., ASD-TR-61-625, 1-151, 1961. [AD 275728]
680		Berne, A., Boato, G., and de Paz, M., Nuovo Cimento, 46, 182-209, 1966.
681	1391	White, G.K. and Woods, S.B., Nature, 177, 851-2, 1956.
682	30810	Westenberg, A.A. and deHaas, N., Phys. Fluids, 6, 617-20, 1963.
683	30193, 40220	Lochtermann, E., Cryogenics, 3, 44, 1963; Monatsber. Deut. Akad. Wiss. Berlin, 5, 329, 1963.
684	17432	Von Ubisch, H., Arkiv Fysik, 16, 93-100, 1959.
685	13562	Mason, E.A. and Von Ubisch, H., Phys. Fluids, 3, 355-61, 1960.
686	30962	Thornton, E. and Baker, W.A.D., Proc. Phys. Soc. (London), 80, 1171-5, 1962.
687	41713	Gambhir, R.S. and Saxena, S.C., Mol. Phys., 11, 233-41, 1966.
688	27871	Cheung, H., Bromley, L.A., and Wilke, C.R., A.I.Ch.E.J., 8, 221-8, 1962.
689	35085	Mukhopadhyay, P. and Barua, A.K., Brit. J. Appl. Phys., 18, 1307-10, 1967.
690	16858	Lindsay, A.L. and Bromley, R.A., UCRL-1128, 1-49, 1951.
691	47765	Burge, H.L. and Robinson, L.B., J. Appl. Phys., 39, 51-4, 1968.
692	39920	Mathur, S., Tondon, P.K., and Saxena, S.C., Mol. Phys., 12, 509-79, 1967.
693	24144	Thornton, E., Proc. Phys. Soc. (London), 77, 1166-9, 1961.
694	15622	Thornton, E., Proc. Phys. Soc. (London), 76, 104-12, 1960.
695	---	Gupta, G.P. and Saxena, S.C., "Thermal Conductivity of Multicomponent Mixtures of Monatomic and Diatomic Gases," to be published (see also Ref. 696).
696	42584	Gupta, G.P., "Studies on Thermal Conductivity and Other Properties of Gases," Ph.D. Thesis, Department of Physics, University of Rajasthan, Jaipur, India, 1968.
697	44737	Gandhi, J.M. and Saxena, S.C., Mol. Phys., 12, 57-68, 1967.
698	40169	Gambhir, R.S. and Saxena, S.C., Physica, 32, 2037-43, 1966.
699	42571	Saxena, S.C. and Gupta, G.P., "Thermal Conductivity of Binary, Ternary and Quaternary Mixtures of Polyatomic Gases," in Thermal Conductivity - Proc. of the Seventh Conference, NBS Spec. Publ. 302, 1968.
700	42651	Saxena, V.K., "Studies on the Transport Properties of Gases," Ph.D. Thesis, Department of Physics, University of Rajasthan, Jaipur, India, 1967.
701	29690	Cotton, J.E., "Thermal Conductivity of Binary Mixtures of Gases," Ph.D. Thesis, Department of Chemistry, University of Oregon, 1962.
702	40111	Gandhi, J.M. and Saxena, S.C., Brit. J. Appl. Phys., 18, 807-12, 1967.
703	13976	Barua, A.K., Indian J. Phys., 34, 169-83, 1960.
704	43068	Mukhopadhyay, P. and Barua, A.K., Brit. J. Appl. Phys., 18, 635-40, 1967.
705	15054	Srivastava, B.N. and Barua, A.K., J. Chem. Phys., 32, 427-35, 1960.

Ref. No.	TPRC No.	
706	17233	Barua, A.K., Physica, $\underline{25}$, 1275-86, 1959.
707	42577	Tondon, P.K., "Studies on Nonequilibrium Properties of Gases," Ph.D. Thesis, Department of Physics, University of Rajasthan, Jaipur, India, 1968.
708	---	Tondon, P.K. and Saxena, S.C., "Thermal Conductivity of Binary, Ternary and Quaternary Gas Systems," to be published (see also Ref. 707).
709	---	Tondon, P.K. and Saxena, S.C., "Thermal Conductivity of the Gas Systems: Neon-Hydrogen, Neon-Nitrogen, Neon-Oxygen, Neon-Hydrogen-Nitrogen and Neon-Oxygen-Nitrogen," to be published (see also Ref. 707).
710	---	Tondon, P.K. and Saxena, S.C., "Thermal Conductivity of Binary and Ternary Gas Systems," to be published (see also Ref. 707).
711	---	Mathur, S., Tondon, P.K., and Saxena, S.C., "Thermal Conductivity of the Gas Mixtures: D_2-Xe, D_2-N-Kr, D_2-Ne-Ar and D_2-Ar-Kr-Xe," J. Phys. Soc. (Japan), $\underline{25}$, 1968, in press.
712	35084	Mukhopadhyay, P., Das Gupta, A., and Barua, A.K., Brit. J. Appl. Phys., $\underline{18}$, 1301-6, 1967.
713	---	Saxena, S.C., Gupta, G.P., and Saxena, V.K., "Measurement of the Thermal Conductivity of Nitrogen (350 to 1500 K) by the Column Method," Proc. Eighth Conference on Thermal Conductivity, 1968, Plenum Publishing Corp., in press.
714	2066	Srivastava, B.N. and Saxena, S.C., J. Chem. Phys., $\underline{27}$, 583-4, 1957.
715	751	Saxena, S.C., J. Chem. Phys., $\underline{25}$, 360-1, 1956.
716	49238	Clingman, W.H., Brokaw, R.S., and Pease, R.N., Fourth Symposium (International) on Combustion, Williams and Wilkins Co., Baltimore, 1953, p. 310-3.
717	42666	Mathur, S., "A Few Properties of Gases and Solids," Ph.D. Thesis, Department of Physics, University of Rajasthan, Jaipur, India, 1966.
718	1129	Brokaw, R.S., Ind. Eng. Chem., $\underline{47}$, 2398-400, 1955.
719	---	Touloukian, Y.S. (Editor), Thermophysical Properties Research Literature Retrieval Guide, Books I, II and III, Second Edition, Plenum Press, New York, 1967.
720	---	Nelson, R.D., Lide, D.R., and Maryott, A.A., "Selected Values of Electric Dipole Moments for Molecules in the Gas Phase," National Standard Reference Data Series - National Bureau of Standards, NSRDS-NBS 10, 1967.
721	10051	Saxena, S.C., Physica, $\underline{22}$, 1242-6, 1956.
722	35300	Tondon, P.K., Gandhi, J.M., and Saxena, S.C., Proc. Phys. Soc. (London), $\underline{92}$, 253-5, 1967.
723	48946	Minter, C.C., J. Phys. Chem., $\underline{72}$, 1924-6, 1968.
724	37550	Hansen, R.S., Frost, R.R., and Murphy, J.A., J. Phys. Chem., $\underline{68}$, 2028-9, 1964.
725	34566	Neal, W.E.J., Greenway, J.E., and Coutts, P.W., Proc. Phys. Soc. (London), $\underline{87}$, 577-9, 1966.

Material Index

MATERIAL INDEX TO THERMAL CONDUCTIVITY COMPANION VOLUMES 1, 2, AND 3

Material Name	Vol.	Page	Material Name	Vol.	Page
"A" nickel	1	239, 241, 1029, 1039	AISI 304	1	1161, 1165, 1168
Acetone $[(CH_3)_2CO]$	3	129	AISI 310	1	1167, 1168
Acetone - benzene system	3	440	AISI 316	1	1165, 1166
Acetylene (CHCH)	3	133	AISI 347	1	1165, 1166, 1168
Acetylene - air system	3	381			
Acid potassium sulfate $(KHSO_4)$ (see potassium hydrogen sulfate)			AISI 403	1	1149
Acheson graphite	2	73	AISI 410	1	1150
Acrylate rubber	2	982	AISI 420	1	1162
Acrylic rubber	2	982	AISI 430	1	1154
Adiprene rubber	2	982	AISI 440 C	1	1154
ADP (see ammonium dihydrogen orthophosphate)			AISI 446	1	1155, 1156
Advance	1	970 564	AISI 1010	1	1185
African ivory	2	1076	AISI 1095 (see SAE 1095)		
AgCu	1	1338	AISI 2515	1	1198, 1199, 1200
$Ag_{0.25}Cu_{0.75}InTe_2$	1	1406			
Aggregate concrete (see under modifiers)			AISI 4130 (see SAE 4130)		
$Ag_6Sb_6PbSe_{13}$	1	1379	AISI 4140 (see SAE 4140)		
$AgSbTe_2$	1	1335	AISI 4340	1	1213, 1214
$AgSbTe_2 + SnTe$	1	1410	AISI C 1010 (see SAE 1010)		
$AgSbTe_2 \cdot SnTe$	1	1411	AISI C 1015	1	1186
Ag_2Se	1	1339	AISI C 1020 (see SAE 1020)		
$Ag_{2-x}Te$	1	1342	Alloy steel	1	1214
Ag_2Te	1	1342	Alloy steel, high	1	1214
Air	3	512	Alpax	1	481
Air - carbon monoxide system	3	383	Alpax gamma	1	918
Air - methane system	3	385	Alum	2	688
AISI 301	1	1165	Alumel	1	1015, 1039
AISI 302	1	1161			
AISI 303	1	1165, 1168	Alumina (see aluminum oxide)		

Material Name	Vol.	Page	Material Name	Vol.	Page
Alumina + Mullite	2	322	Aluminum alloys (specific types) (continued)		
Alumina fused brick	2	897	2014 (same as aluminum alloy 14S)	1	901
Alumina porcelain	2	937	2024 (same as aluminum alloy 24S)	1	898, 901
Aluminate silicate 723 glass	2	923			
Aluminum	1	1	2358	1	481
Aluminum + Antimony	1	469	3003 (same as aluminum alloy 3S)	1	912
Aluminum + Copper	1	470	3004 (same as aluminum alloy 4S)	1	912
Aluminum + Copper + ΣX_i	1	895	5052 (same as aluminum alloy 52S)	1	478, 909
Aluminum + Iron	1	474	5083 (same as aluminum alloy LK183)	1	909
Aluminum + Iron + ΣX_i	1	905	5086 (same as aluminum alloy K186)	1	909
Aluminum + Magnesium	1	477	5154 (same as aluminum alloy A54S)	1	478, 909
Aluminum + Magnesium + ΣX_i	1	908			
Aluminum + Manganese + ΣX_i	1	911	5456	1	909
Aluminum + Nickel + ΣX_i	1	914	6063 (same as aluminum alloy 63S)	1	909
Aluminum + Silicon	1	480	7075 (same as aluminum alloy 75S)	1	923
Aluminum + Silicon + ΣX_i	1	917	A54S (see aluminum alloy 5154)		
Aluminum + Tin	1	483	Alpax	1	481
Aluminum + Uranium	1	484	Alpax gamma	1	918
Aluminum + Zinc	1	487	Alusil	1	481
Aluminum + Zinc + ΣX_i	1	922	British 2L-11	1	900
Aluminum + ΣX_i	1	925	British L-5	1	923
Aluminum alloys (specific types)			British L-8	1	899
2S (see aluminum alloy 1100)			British Y-1	1	900
3S (see aluminum alloy 3003)			British Y-2	1	900
4S (see aluminum alloy 3004)			Cond-Al	1	906
12	1	897, 899, 900	D (zeppelin)	1	900
			DIN 712	1	475
14S (see aluminum alloy 2014)			Duralumin	1	896
24S (see aluminum alloy 2024)			German Y alloy	1	896, 898
52S (see aluminum alloy 5052)			J51	1	906
63S (see aluminum alloy 6063)			Japanese 2E-8	1	899
75S (see aluminum alloy 7075)			Japanese M-1	1	899
132 (see aluminum alloy Lo-Ex)			K186 (see aluminum alloy 5086)		
1100 (same as aluminum alloy 2S)	1	906, 920	K-S alloy 245	1	920
			K-S alloy 280	1	920

Material Name	Vol.	Page	Material Name	Vol.	Page
Aluminum alloys (specific types) (continued)			Aluminum oxide (Al_2O_3) (continued)		
K-S alloy special	1	902	E98	2	101
LK183 (see aluminum alloy 5083)			Gulton HS. B	2	103
Lo-Ex (same as aluminum alloy 132)	1	919	Hi alumina	2	99
Magnalium	1	478	Ignited alumina	2	106
Nelson-Kebbenleg 10	1	896	Linde synthetic sapphire	2	94
RAE 40 C	1	915	Lucalox	2	106
RAE 47 B	1	915	Norton 38-900	2	103, 104
RAE 47 D	1	915	Sapphire	2	93
RAE 55	1	915	Synthetic sapphire	2	94
RR 50	1	918, 919, 920	TC 352	2	107
			Wesgo Al-300	2	101, 107, 108
RR 53	1	901			
RR 53 C	1	918	Aluminum oxide + Aluminum silicate	2	321
RR 59	1	898	Aluminum oxide + (di)Chromium trioxide	2	324
RR 77	1	923	Aluminum oxide + (di)Manganese trioxide	2	327
RR 131 D	1	909	Aluminum oxide + Silicon dioxide	2	328
SA 1	1	918, 919	Aluminum oxide + Silicon dioxide + ΣX_i	2	453
			Aluminum oxide + Titanium dioxide + ΣX_i	2	456
SA 44	1	918, 919	Aluminum oxide + Zirconium dioxide	2	331
Silumin, sodium modified	1	920	Aluminum oxide - chromium cermets	2	707
γ-Silumin, modified	1	920	Aluminum silicate ($3Al_2O_3 \cdot 2SiO_2$)	2	254
Y-alloy	1	896, 898	Aluminum silicate + Aluminum oxide	2	334
			Alundum	2	456
Aluminum borosilicate complex, natural (see tourmaline)			Alusil	1	481
Aluminum bronze	1	531, 532, 953	Amalgam	1	216
			Amber glass	2	924
Aluminum fluosilicate ($2AlFO \cdot SiO_2$)	2	251	American white wood	2	1090
Brazil topaz	2	252	Ammonia (NH_3)	3	95
Aluminum nitride (AlN)	2	653	Ammonia - air system	3	442
Aluminum oxide (Al_2O_3)	2	98	Ammonia - carbon monoxide system	3	444
AP-30	2	99	Ammonia - ethylene system	3	446
AV-30	2	102	Ammonia - hydrogen system	3	448
B45F	2	101	Ammonia - nitrogen system	3	451
Corundum	2	94, 99	Ammonium acid phosphate [$NH_4H_2PO_4$] (see ammonium dihydrogen phosphate)		

Material Name	Vol.	Page	Material Name	Vol.	Page
Ammonium perchlorate (NH_4ClO_4), reagent grade	2	757	Argon - carbon dioxide system	3	297
Ammonium dihydrogen phosphate ($NH_4H_2PO_4$)	2	679	Argon - deuterium system	3	299
Ammonium dihydrogen orthophosphate [$NH_4H_2PO_4$] (see ammonium dihydrogen phosphate)			Argon - helium system	3	251
			Argon - hydrogen system	3	301
Ammonium hydrogen sulfate (NH_4HSO_4)	2	687	Argon - hydrogen - deuterium - nitrogen system	3	507
Ammonium phosphate, monobasic [$NH_4H_2PO_4$] (see ammonium dihydrogen phosphate)			Argon - hydrogen - nitrogen system	3	493
Ammonium biphosphate [$NH_4H_2PO_4$] (see ammonium dihydrogen phosphate)			Argon - hydrogen - nitrogen - oxygen system	3	508
Ammonium bisulfate [NH_4HSO_4] (see ammonium hydrogen sulfate)			Argon - krypton system	3	263
			Argon - krypton - deuterium system	3	488
AMS 4908 A (see Ti-8Mn)			Argon - krypton - hydrogen system	3	496
AMS 4925 A (see titanium alloy C-130 AM, or titanium alloy RC-1308)			Argon - krypton - xenon system	3	483
			Argon - krypton - xenon - deuterium system	3	506
AMS 4926 (see titanium alloy A-110AT)			Argon - krypton - xenon - hydrogen system	3	505
AMS 4928 (see Ti-6Al-4V)			Argon - methane system	3	304
AMS 4929 (see Ti-155A)			Argon - neon system	3	258
AMS 4969 (see Ti-155A)			Argon - nitrogen system	3	306
AMS 5385 C (see Haynes stellite alloy 21)			Argon - oxygen system	3	311
Angora wool	2	1092	Argon - oxygen - methane system	3	485
Angren brown coal	2	808	Argon - propane system	3	316
Anthracene [$C_6H_4(CH)_2C_6H_4$]	2	985	Argon - propane - dimethyl-ether ~~ethanol~~ system	3	499
Anthracin [$C_6H_4(CH)_2C_6H_4$] (see anthracene)			Argon - xenon system	3	267
Antimony	1	10	Argon - xenon - hydrogen - deuterium system	3	510
Antimony + Aluminum	1	488			
Antimony + Beryllium + ΣX_i	1	926	Armalon lamintes (nonmetallic)	2	1032
Antimony + Bismuth	1	489	Armco iron	1	157, 158, 159, 160, 161, 163
Antimony + Cadmium	1	492			
Antimony + Copper	1	495			
Antimony + Lead	1	496			
Antimony - tellurium intermetallic compound Sb_2Te_3	1	1241	Arsenic	1	15
			Arsenic - tellurium intermetallic compound As_2Te_3	1	1244
Antimony + Tin	1	497			
Antimony telluride [Sb_2Te_3] (see antimony - tellurium intermetallic compound)			Arsenic telluride [As_2Te_3] (see arsenic - tellurium intermetallic compound)		
Argentum (see silver)			Asbestos cement board	2	1107
Argon	3	1	Asbestos fiber	2	1135
Argon - benzene system	3	295	Ash	2	1059

Material Name	Vol.	Page
Ashkhabad clay	2	804, 805
Asphalt-glass wood pad	2	1108
Asphaltic bituminous concrete	2	863
As_2Te_3	1	1244
ASTM B 265-58T, grade 6 (see titanium alloy A-110AT)		
ASTM B 265-58T, grade 7 (see Ti-8Mn)		
Astrolite	2	1029, 1030, 1052
Aurum (see gold)		
Austenitic stainless steel	1	1165, 1183
Balsa	2	1060
Pseudo	2	1060
Waterproofed	2	1060
Ba_2Pb	1	1245
Barium-lead intermetallic compound Ba_2Pb	1	1245
Barium-tin intermetallic compound Ba_2Sn	1	1246
Barium difluoride (BaF_2)	2	627
Barium oxide (BaO)	2	120
Barium oxide + Silicon dioxide + ΣX_i	2	457
Barium oxide + Strontium oxide	2	337
Barium oxide + Strontium oxide + ΣX_i	2	460
Barium stannide [Ba_2Sn] (see barium - tin intermetallic compound)		
Barium titanates		
$BaTiO_3$	2	257
$BaO \cdot 2TiO_3$	2	260
Barium metatitanate ($BaTiO_3$)	2	257
Barium metatitanate + Calcium metatitanate	2	340
$Ca_{0.034}Ba_{0.966}TiO_3$	2	341
$Ca_{0.099}Ba_{0.901}TiO_3$	2	341
$Ca_{0.19}Ba_{0.81}TiO_3$	2	341
Barium metatitanate + Magnesium zirconate	2	343
Barium metatitanate + Manganese niobate	2	344
Barium dititanate ($BaTi_2O_5$)	2	260

Material Name	Vol.	Page
Barytes concrete	2	871
Basalt	2	797
NTS basalt	2	798
Olivine basalt	2	798
Ba_2Sn	1	1246
Bauxite brick	2	901, 902
Beef fat	2	1072
$Be_{12}Nb$	1	1248
$Be_{17}Nb_2$	1	1248
Benzene (C_6H_6)	3	135
Benzene, p-dibromo ($C_6H_4Br_2$)	2	986
Benzene, p-dichloro ($C_6H_4Cl_2$)	2	987
Benzene, p-diiodo ($C_6H_4I_2$)	2	988
Benzene - hexane system	3	387
Beryl	2	800
Brazil	2	801
India	2	801
Beryllia (see beryllium oxide)		
Beryllium	1	18
Beryllium + Aluminum	1	498
Beryllium + Beryllium oxide	1	1416
Beryllium + Fluorine + ΣX_i	1	929
Beryllium + Magnesium	1	499
Beryllium + Magnesium + ΣX_i	1	932
Beryllium - niobium intermetallic compounds		
Be_xNb_y	1	1247
$Be_{12}Nb$	1	1248
$Be_{17}Nb_2$	1	1248
Beryllium - tantalum intermetallic compounds		
Be_xTa_y	1	1250
$TaBe_{12}$	1	1251
Ta_2Be_{17}	1	1251
Beryllium - uranium intermetallic compounds		
Be_xU_y	1	1253

Material Name	Vol.	Page	Material Name	Vol.	Page
Beryllium - uranium intermetallic compounds			Biphenyl $[C_6H_5C_6H_5]$ (see diphenyl)		
UBe$_{13}$	1	1254	Biphenyl + o-, m-, p-Terphenyl + Higher phenyls (see santowax R)		
Beryllium - zirconium intermetallic compounds Be$_{13}$Zr	1	1256	BiSbTe$_{3.13}$	1	1390
Beryllium bronze	1	539	Bi$_{1.33}$Sb$_{0.67}$Te$_{3.13}$	1	1389
(di)Beryllium carbide (Be$_2$C)	2	571	Bi$_{1.5}$Sb$_{0.5}$Te$_{3.13}$	1	1389
Beryllium copper	1	539	Bi$_{1.75}$Sb$_{0.25}$Te$_{3.1}$	1	1390
Beryllium oxide (BeO)	2	123	Bi$_{1.75}$Sb$_{0.25}$Te$_{3.13}$	1	1389
3008 (refractory grade)	2	125	Bi$_{1.75}$Sb$_{0.25}$Te$_{3.19}$	1	1390
4811 BeO porcelain	2	124	Bi$_{1.75}$Sb$_{0.25}$Te$_{3.26}$	1	1390
AOX grade	2	127, 129	Bismuth	1	25
BD-98	2	125	Bismuth + Antimony	1	502
Brush SP grade	2	125	Bismuth + Cadmium	1	505
Clifton metal grade	2	127	Bismuth + Cadmium + ΣX_i	1	935
Grade I	2	128	Bismuth + Lead	1	508
Grade II	2	128	Bismuth + Lead + ΣX_i	1	938
Porcelain	2	124	Bismuth - lead eutectic	1	509
Triangle beryllia	2	126	Bismuth - tellurium intermetallic compound Bi$_2$Te$_3$	1	1257
UOX grade	2	124, 127, 128, 129	Bismuth + Tin	1	511
			Bismuth alloys (specific types)		
Beryllium oxide + Aluminum oxide + ΣX_i	2	461	Hutchin's alloy	1	512
Beryllium oxide + Magnesium oxide + ΣX_i	2	464	Lipowitz alloy	1	939
Beryllium oxide + Thorium dioxide + ΣX_i	2	467	Rose metal	1	939
Beryllium oxide + Uranium dioxide	2	347	Wood's metal	1	939
Beryllium oxide + Zirconium dioxide + ΣX_i	2	470	Bismuth stannate $[Bi_2(SnO_3)_3]$	2	261
Beryllium oxide - beryllium cermets	2	708	Bismuth tristannate $[Bi_2(SnO_3)_3]$ (see bismuth stannate)		
Beryllium oxide - beryllium - molybdenum cermets	2	711	Bismuth telluride $[Bi_2Te_3]$ (see bismuth - tellurium intermetallic compound)		
Beryllium oxide - beryllium - silicon cermets	2	714	Bi$_2$Te$_3$	1	1257
Beskhudnikov clay	2	804	Bi$_2$Te$_3$ + Bi$_2$Se$_3$	1	1393
Be$_{12}$Ta	1	1251	Bi$_2$Te$_3$ + Sb$_2$Te$_3$	1	1388
Be$_{17}$Ta$_2$	1	1251	Bi$_2$Te$_3$ + Sb$_2$Te$_3$ + Sb$_2$Se$_3$	1	1392
Be$_{13}$U	1	1254	Bi$_2$Te$_3$ + Te	1	1415
Be$_{13}$Zr	1	1256	Bi$_2$Te$_{3.19}$	1	1415
			Bi$_2$Te$_{3.26}$	1	1415

Material Name	Vol.	Page
Bitter spar (see dolomite)		
Bitumen	2	1155
Bitumin concrete	2	863
Bituminous concrete aggregate, blended	2	863
Black temper cast iron	1	1137
Bone char	2	1156
Bone fat	2	1072
Boralloy (see boron nitride)		
Boric anhydride [B_2O_3] (see boron oxide)		
Boric oxide [B_2O_3] (see boron oxide)		
Boron	1 2	41 1
Boron - silicon intermetallic compounds		
SiB_4	1	1262
SiB_6	1	1262
(tetra)Boron carbide (B_4C)	2	572
(tetra)Boron carbide + Sodium metasilicate	2	541
(tetra)Boron carbide - aluminum cermets	2	717
Boron trifluoride (BF_3)	3	99
Boron nitride (BN)	2	656
Boron oxide (B_2O_3)	2	138
Boron sesquioxide [B_2O_3] (see boron oxide)		
(di)Boron trioxide [B_2O_3] (see boron oxide)		
Boron silicides (see boron - silicon intermetallic compounds)		
Boronated graphite	2	61
Borosilicate glass	2	923, 924
Borosilicate 3235 glass	2	923
Borosilicate crown glass	2	923
Boxwood	2	1061
Brass	1	591, 592, 980, 981, 982
Brass (specific types)		
70/30	1	590
B.S. 249	1	981

Material Name	Vol.	Page
Brass (specific types) (continued)		
Cast	1	980
High (see yellow brass)		
High tensile	1	980
Leaded free cutting	1	981
MS 58	1	980
MS 76/22/2	1	980
Red	1	591
Red, German	1	981
Rolled	1	980
Yellow	1	981, 982
Brazil beryl	2	801
Brazil topaz	2	252
Brazil tourmaline	2	855
Bricks	2	889
Alumina fused	2	897
Aluminous fire clay	2	900
Bauxite	2	329, 901, 902
Carbofrax	2	897
Carbofrax carborundum	2	895
Carbon	2	890, 896
Carsiat carborundum	2	895
Cement porous	2	890
Ceramic	2	890
Chamotte	2	890
Chrome	2	454, 897, 898
Chrome fire brick	2	897
Chrome magnesite	2	890
Chromite	2	473, 899
Chromomagnesite	2	481
Common	2	492, 897

Material Name	Vol.	Page	Material Name	Vol.	Page
Bricks (continued)			Bricks (continued)		
Corundum	2	454, 905	Magnesia	2	485, 897, 898, 899
Dense	2	443, 904	Magnesite	2	478, 483, 892, 895, 905
Dense fireclay	2	403			
Diatomaceous	2	890, 891			
Diatomaceous insulating	2	906, 907	Magnesite fire	2	897
			Magnezit	2	899, 902
Dinas	2	891	Marksa	2	899
Egyptian fire clay	2	491, 901	Metallurgical	2	892, 893
Fire	2	491, 891, 895, 902, 903	Metallurgical porous	2	893
			Mica	2	892
			Missouri fire	2	492, 905
Fireclay	2	403, 404, 490, 491, 896, 901, 903	Normal	2	488, 489, 900, 901
			Ordzhonikidze	2	899
Fire clay, dense	2	903	Penn. fire	2	905
Fire clay, superduty	2	890	Porous	2	894
Georgia fire	2	896	Porous concrete	2	894
Hand-burned face	2	891	Porous fire (Italy)	2	895
High temp. insulating	2	891	Red	2	405, 492, 898
High temp. insulating blast furnace	2	488, 899			
			Red, hard burned	2	896
Hytex hydraulic pressed building	2	896	Red, soft burned	2	896
Insulating	2	443, 891, 904	Red shamotte	2	405
			Refractory insulating	2	892
Insulating fire	2	891	Refractory insulating common chamotte	2	892
Kaolin fire	2	404, 405, 904	Shamotte	2	492, 894, 898
Kaolin insulating refractory	2	895	Shamotte, white	2	405
Light weight	2	488, 489, 892, 899, 900	Silica	2	408, 489, 492, 502, 894, 896,
Lime sand	2	892			

Material Name	Vol.	Page
Bricks (continued)		
Silica (continued)	2	897, 898, 900, 902, 904, 906
Silica fire	2	894, 895, 905
Silica refractory	2	185
Silicon carbide	2	555, 586, 895
Silicon carbide, refrax	2	586, 906
Silicious	2	492, 902
Sillimanite	2	329, 902
Sillimanite refractory	2	329, 403, 902, 903
Sil-O-Cel	2	896
Sil-O-Cel, calcined	2	896
Sil-O-Cel, natural	2	896
Sil-O-Cel, special	2	896
Sil-O-Cel, super	2	896
Slag	2	898
Spinel fire	2	905
Star-brand	2	185
Tripolite	2	894
Vermiculite	2	894
Zirconia	2	535, 895, 905
Brimstone (see sulfur)		
British 2L-11	1	900
British C-32	1	948
British carbon steel	1	1186
British L-5	1	923
British L-8	1	899

Material Name	Vol.	Page
British Y-1	1	900
British Y-2	1	900
British steel	1	1114, 1118, 1187
Brom-graphite	2	768
Bromine	3	13
Bromyride (see silver bromide)		
Bronze	1	585, 586, 976, 980
Bronze, aluminum	1	531, 532, 953
Bronze, beryllium	1	539
Bronze, phosphor	1	585, 586, 976
Bronze, silicon	1	973
Bronze, silver	1	579, 980
B_4Si	1	1262
B_6Si	1	1262
Butane, i-(i-C_4H_{10})	3	139
Butane, n-(n-C_4H_{10})	3	141
Butaprene E rubber	2	982
Butter of zinc (see zinc dichloride)		
Cadmium	1	45
Cadmium + Antimony	1	514
Cadmium - antimony intermetallic compound CdSb	1	1264
Cadmium + Bismuth	1	517
Cadmium + Bismuth + ΣX_i	1	941
Cadmium - tellurium intermetallic compound CdTe	1	2167
Cadmium + Thallium	1	520
Cadmium + Tin	1	521
Cadmium + Zinc	1	524
Cadmium antimonide [CdSb] (see cadmium - antimony intermetallic compound)		

Material Name	Vol.	Page
Cadmium germanium phosphide (CdGeP$_2$)	2	758
Cadmium telluride [CdTe] (see cadmium - tellurium intermetallic compound)		
Calcia (see calcium oxide)		
Calcite	2	761
Calcium - lead intermetallic compounds		
$\quad$ Ca$_x$Pb$_y$	1	1270
$\quad$ Ca$_2$Pb	2	1271
$\quad$ Ca$_{2.02}$Pb	1	1271
$\quad$ Ca$_{2.10}$Pb	1	1271
$\quad$ Ca$_{2.19}$Pb	1	1271
Calcium - tin intermetallic compound		
$\quad$ Ca$_2$Sn	1	1273
Calcium carbonate (CaCO$_3$)	2	759
Calcium carbonate (CaCO$_3$)		
$\quad$ Black marble	2	761
$\quad$ Brown marble	2	761
$\quad$ Calcite	2	761
$\quad$ Marble	2	760, 761
$\quad$ Marble powder	2	760, 761
$\quad$ Natural (see limestone)		
$\quad$ White marble	2	761
$\quad$ White Alabama marble	2	761
Calcium difluoride (CaF$_2$)	2	630
Calcium oxide (CaO)	2	141
Calcium phosphate + Lithium carbonate + Magnesium carbonate	2	763
Calcium stannate (CaSnO$_3$)	2	264
Calcium stannide [Ca$_2$Sn] (see calcium - tin intermetallic compound)		
Calcium metatitanate (CaTiO$_3$)	2	267
Calcium tungstate (CaWO$_4$)	2	270
Calcium wolframate [CaWO$_4$] (see calcium tungstate)		
Canadian natural graphite	2	54
Ca$_2$Pb	1	1271

Material Name	Vol.	Page
Ca$_{2.02}$Pb	1	1271
Ca$_{2.10}$Pb	1	1271
Ca$_{2.19}$Pb	1	1271
Carbofrax brick	2	897
Carbofrax carborundum brick	2	895
Carbon	2	5
$\quad$ Diamond	2	9
$\quad$ Graphite (see each individual graphite)		
$\quad$ Lampblack	2	6
$\quad$ Petroleum coke	2	6
Carbon + Oxygen	2	764
Carbon + Volatile materials	2	765
Carbon brick	2	890, 896
Carbon tetrachloride (CCl$_4$)	3	156
Carbon monoxide (CO)	3	151
Carbon monoxide - hydrogen system	3	405
Carbon dioxide (CO$_2$)	3	145
Carbon dioxide and ethylene system	3	389
Carbon dioxide - hydrogen system	3	391
Carbon dioxide - nitrogen system	3	396
Carbon dioxide - oxygen system	3	401
Carbon dioxide - propane system	3	403
Carbon steel	1	1118, 1119, 1126, 1180, 1185
Carbon steel, British	1	1186
Carbon steel, Japanese	1	1185
Carborundum	2	553, 555, 596
Carboxy nitrile rubber	2	982
Cardboard	2	1109
Carsiat carborundum brick	2	895
Cartridge brass 70% (see brass 70/30)		
Ca$_2$Sn	1	1273

Material Name	Vol.	Page	Material Name	Vol.	Page
Cassiopeium (see lutetium)			Cellular glass	2	923
Cast iron	1	1129, 1130, 1133, 1134, 1136, 1137, 1205, 1222	Cellulose fiberboard	1	1110
			Celtium (see hafnium)		
			Cement		
			Hydraulic (see Portland cement)	2	861
			Portland	2	861
Cast irons (specific types)			Slag	2	861
Black temper	1	1137	Slag - Portland	2	861
Gray	1	1130, 1135	Cement porous brick	2	890
Heat resistant	1	1146	Ceramic brick	2	890
High duty	1	1133, 1135	Ceramics, miscellaneous	2	915
Hot mold, gray	1	1135	Cerium	1	50
			Cerium dioxide (CeO_2)	2	144
Nickel-resist	1	1204	Cerium dioxide + Magnesium oxide	2	350
Nr 1510, spherical	1	1222	Cerium dioxide + Uranium dioxide	2	353
Nr 1520, pearlitic matrix	1	1222	Cerium sulfides		
Soft, gray	1	1135	CeS	2	697
White	1	1130, 1135	Ce_2S_3	2	698
White temper	1	1137	Cermets (see each individual cermet)		
$Cd_3As_2 + Zn_3As_2$	1	1396	Cesium	1	54
$Cd_{0.04}Hg_{0.96}Te$	1	1408	Cesium bromide (CsBr)	2	565
$Cd_{0.07}Hg_{0.93}Te$	1	1408	Cesium iodide (CsI)	2	561
CdSb	1	1264	Chamotte brick	2	890
CdSb + ZnSb	1	1397	Chamotte clay	2	804
CdSb · ZnSb	1	1398	Channel carbon black	2	764
2CdSb · 3ZnSb	1	1413	Charcoal	2	1157
3CdSb · 2ZnSb	1	1398	Chlorine	3	17
3CdSb · 7ZnSb	1	1413	Chlorodifluoromethane [$ClCHF_2$] (see Freon 22)		
7CdSb · 3ZnSb	1	1398			
CdTe	1	1267	Chloroform ($CHCl_3$)	3	161
$Cd_{1.6}Zn_{1.4}As_2$	1	1396	Chloroform - ethyl ether system	3	470
Cd_2ZnAs_2	1	1396	Chloromethane [CH_3Cl] (see methyl chloride)		
$Cd_{2.5}Zn_{0.5}As_2$	1	1396	Chloroprene rubber	2	983
Cedar	2	1062	Chlorotrifluoromethane [$ClCF_3$] (see Freon 13)		
Ceiba (see kapok)					

Material Name	Vol.	Page
Chroman	1	1018
Chrome brick	2	454, 897, 898
Chrome fire brick	2	897
Chrome magnesite brick	2	890
Chromel 502	1	1210
Chromel A	1	698
Chromel C	1	1036
Chromel P	1	698
Chromite brick	2	473, 899
Chromium	1	60
Chromium + Aluminum oxide	1	1419
Chromium + Iron + ΣX_i	1	944
Chromium + Nickel	1	525
Chromium alloy, ferrochromium	1	945
(di)Chromium trioxide + Magnesium oxide + ΣX_i	2	473
Cinder aggregate concrete	2	869, 870
Clays	2	803
Ashkhabad	2	804, 805
Beskhudnikov	2	804
Chamotte	2	804
Fire clay	2	804
Kuchin	2	804
Sandy clay	2	805
Clay aggregate concrete, expanded burned	2	870
Climax	1	1198, 1213
Coal	2	807
Angren brown coal	2	808
Donets anthracite	2	808
Donets gas coal	2	808
Coal tar fractions	2	1158
Coatings, applied (nonmetallic)	2	1009

Material Name	Vol.	Page
Cobalt	1	64
Cobalt + Carbon	1	526
Cobalt + Chromium	1	527
Cobalt + Chromium + ΣX_i	1	947
Cobalt + Iron + ΣX_i	1	950
Cobalt + Nickel	1	528
Cobalt + Nickel + ΣX_i	1	951
Cobalt - silicon intermetallic compound		
CoSi	1	1274
Cobalt alloys (specific types)		
British C-32	1	948
Haynes stellite 21	1	948
Haynes stellite 23	1	948
S 816	1	948
WI 52	1	948
X-40	1	948
(tri)Cobalt strontium metatitanate (Co_3SrTiO_3)	2	271
Cobalt zinc ferrate [$Co(Zn)Fe_2O_4$]	2	272
Coke, petroleum	2	765
Colloidal aggregate polystyrene	2	965
Colorless glass	2	924
Columbium (see niobium)		
Columbium alloys (see niobium alloys)		
Commercial castable concrete	2	871, 875, 876, 877, 878
Common brick	2	492, 897
Concretes	2	862
Asphaltic bituminous	2	863
Barytes	2	871
Bitumin	2	863
Bituminous aggregate, blended	2	863
Cinder aggregate	2	869, 870
Clay aggregate, expanded burned	2	870

Material Name	Vol.	Page	Material Name	Vol.	Page
Concretes (continued)			Copper, electrolytic tough pitch	1	70, 72
Commercial castable	2	871, 875, 876, 877, 878	Copper, free-cutting	1	582
			Copper, oxygen-free high-conducting	1	69, 74
Diatomaceous aggregate	2	874	Copper, phosphorus deoxidized	1	72
Haydite aggregate	2	870	Copper-126, leaded	1	555
Leuna slag	2	864	Copper + Aluminum	1	530
Light weight	2	874	Copper + Aluminum + ΣX_i	1	952
Light weight, foamed	2	881	Copper + Antimony	1	534
Limestone aggregate	2	869	Copper - antimony - selenium intermetallic compound $CuSbSe_2$	1	1275
Limestone gravel	2	864, 865	Copper + Arsenic	1	535
Lummite cement	2	871	Copper + Beryllium	1	538
Metallurgical pumice	2	863, 864	Copper + Beryllium + ΣX_i	1	955
Paraffin	2	863	Copper + Cadmium	1	541
Portland cement	2	871	Copper + Cadmium + ΣX_i	1	956
Sand cement	2	874	Copper + Chromium	1	542
Sand and gravel aggregate	2	868, 869	Copper + Cobalt	1	545
Slag	2	864, 880, 881	Copper + Cobalt + ΣX_i	1	957
			Copper + Gold	1	548
Slag, direct process	2	864	Copper + Iron	1	551
Slag, expanded	2	878, 879	Copper + Iron + ΣX_i	1	960
			Copper + Lead	1	554
Slag aggregate, limestone treated	2	870	Copper + Lead + ΣX_i	1	961
Cond-Al	1	906	Copper + Manganese	1	557
Constantan	1	564	Copper + Manganese + ΣX_i	1	964
Contracid	1	1036	Copper + Nickel	1	561
Contracid B 7 M	1	1036	Copper + Nickel + ΣX_i	1	969
Copoly(chloroethylene-vinyl-acetate)	2	943	Copper + Palladium	1	568
Copoly-[1,1-difluoro-ethylene-hexafluoro propene], Viton A rubber (see Viton rubber)			Copper + Phosphorus	1	571
			Copper + Platinum	1	574
Copoly(formaldehyde - urea)	2	944	Copper - selenium intermetallic compound Cu_3Se_2	1	1276
Copper	1	68			
Copper, coalesced	1	69, 72	Copper + Silicon	1	575
Copper, electrolytic	1	72, 73	Copper + Silicon + ΣX_i	1	972
			Copper + Silver	1	578

Material Name	Vol.	Page	Material Name	Vol.	Page
Copper + Tellurium	1	581	Copper alloys (specific types) (continued)		
Copper + Tin	1	584	Cuppralloy type 5, Russian	1	543
Copper + Tin + ΣX_i	1	975	Cupro nickel	1	970
Copper + Zinc	1	588	Cupro nickel, NM-81, Russian	1	562
Copper + Zinc + ΣX_i	1	979	Eureka	1	563
Copper + Zirconium + ΣX_i	1	985	German silver	1	980, 981
Copper alloys (specific types)			Gun-metal, admiralty	1	976
Advance	1	970	Gun-metal, ordinary	1	976
ASTM B301-58T	1	582	Lohm	1	564
Beryllium copper	1	539	Manganin	1	965
Brass	1	591, 592, 980, 981, 982	Manganin NM Mts	1	965
			Navy M	1	977
			Nickel silver	1	981
Brass 70/30	1	570	SAE bearing alloy 40	1	976
Brass B.S. 249	1	981	SAE bearing alloy 62	1	976
Brass, cast	1	980	SAE bearing alloy 64	1	976
Brass, high tensile	1	980	SAE bearing alloy 66	1	962
Brass, leaded free cutting	1	981	Copper glance [see (di)copper sulfide]		
Brass MS 58	1	980	Copper iodide (CuI)	2	562
Brass MS 76/22/2	1	980	Copper hemioxide (Cu_2O) [see (di)copper oxide]		
Brass, red, German	1	591, 981	(di)Copper oxide (Cu_2O)	2	147
Brass, rolled	1	980	Copper protooxide (Cu_2O) [see (di)copper oxide]		
Brass, yellow	1	981, 982	Copper selenide [Cu_3Se_2] (see copper - selenium intermetallic compound)		
Bronze	1	585, 586, 976, 980	(di)Copper sulfide (Cu_2S)	2	699
Bronze, aluminum	1	531, 532, 953	(di)Copper sulfide + Iron sulfide + (tri)Nickel disulfide	2	700
			(di)Copper sulfide + (tri)Nickel disulfide	2	701
Bronze, beryllium	1	539	Copperous oxide (Cu_2O) [see (di)copper oxide]		
Bronze, phosphor	1	585, 586, 976	Copperous sulfide (Cu_2S) [see (di)copper sulfide]		
Bronze, silicon	1	973	Cordierite	2	918
Bronze, silver	1	579, 980	Cordierite 202	2	919
			Rutgers	2	919
Constantan	1	564	Steatite	2	919

Material Name	Vol.	Page
Cork	2	1063
Corning 0080 glass	2	511, 928
Corning 7740 glass	2	933
Cornstalk wallboard	2	1111
Corronil	1	1032
Corundum	2	94, 99
Corundum brick	2	454, 905
Cotton	2	1068
Waste	2	1070
Medical	2	1069, 1070
Cotton fabric	2	1093
Cotton silicate felt fabric	2	1094
Cotton wool	2	1096
Crucible HNM	1	1168
Crucible steel, Japanese	1	1204
Cu + BeCo	1	1420
CuAu	1	1281
Cu_3Au	1	1281
Cupralloy, Russian, type 5	1	543
Cupronickel	1	970
Cupronickel, Russian, NM-81	1	562
Cuprum (see copper)		
$CuSbSe_2$	1	1275
$CuSbSe_2 + Cu_3Se_2$	1	1400
$(CuSbSe_2)_{0.1}(Cu_3Se_2)_{0.9}$	1	1401
$(CuSbSe_2)_{0.2}(Cu_3Se_2)_{0.8}$	1	1401
$(CuSbSe_2)_{0.25}(Cu_3Se_2)_{0.75}$	1	1401
$(CuSbSe_2)_{0.3}(Cu_3Se_2)_{0.7}$	1	1401
$(CuSbSe_2)_{0.33}(Cu_3Se_2)_{0.67}$	1	1401
$(CuSbSe_2)_{0.4}(Cu_3Se_2)_{0.6}$	1	1401
$(CuSbSe_2)_{0.5}(Cu_3Se_2)_{0.5}$	1	1401
$(CuSbSe_2)_{0.6}(Cu_3Se_2)_{0.4}$	1	1400
$(CuSbSe_2)_{0.7}(Cu_3Se_2)_{0.3}$	1	1400

Material Name	Vol.	Page
$(CuSbSe_2)_{0.8}(Cu_3Se_2)_{0.2}$	1	1400
$(CuSbSe_2)_{0.9}(Cu_3Se_2)_{0.1}$	1	1400
Cu_3Se_2	1	1276
$Cu_3Se_2 + CuSbSe_2$	1	1401
"D" nickel	1	1039
Decane, n-$(C_{10}H_{22})$	3	164
Dense brick	2	443, 904
Deuterium	3	21
Deuterium - hydrogen system	3	407
Deuterium - nitrogen system	3	410
Diamond	2	9
Type I	2	10
Type II	2	10
Diatomaceous aggregate concrete	2	874
Diatomaceous brick	2	890, 891
Diatomaceous earth	2	814
Diatomaceous insulating brick	2	906, 907
Diatomite (see diatomaceous earth)		
Diatomite aggregate	2	1112
Sil-O-Cel coarse grade	2	1112
Dichlorodifluoromethane $[Cl_2CF_2]$ (see Freon 12)		
Dichlorofluoromethane $[Cl_2CHF]$ (see Freon 21)		
1,2-Dichloro-1,1,2,2-tetrafluoroethane $[CClF_2CClF_2]$ (see Freon 114)		
Diethylamine - ethyl ether system	3	472
Dimethyl ketone $[(CH_3)_2CO]$ (see acetone)		
Dimethyl methane $[C_3H_8]$ (see propane)		
Dinas brick	2	891
Diphenyl $(C_6H_5C_6H_5)$	2	989
Diphenylamine $[(C_6H_5)_2NH]$	2	991
Diphenylmethane + Naphthalene	2	994
Diphenyl oxide $[(C_6H_5)_2O]$	2	990
Dolomite	2	810

Material Name	Vol.	Page	Material Name	Vol.	Page
Dolomite (continued)			Enamel (continued)		
NTS dolomite	2	811	Silicon	2	921
Domestic graphite, Japan	2	56	Erbium	1	86
Donets anthracite coal	2	808	Ethane (C_2H_6)	3	167
Donets gas coal	2	808	Ethanol [C_2H_5OH] (see ethyl alcohol)		
Dow metal	1	999	Dimethyl ether ~~Ethanol~~ - argon system	3	454
Duralumin	1	896	Dimethyl ether ~~Ethanol~~ - methyl formate system	3	474
Duranickel	1	1015	Dimethyl ether ~~Ethanol~~ - propane system	3	456
Duranickel alloy 301 (see duranickel)			Ethyl alcohol (C_2H_5OH)	3	169
Duroid 5600	2	968	Ethyl ether [$(C_2H_5)_2O$]	3	179
Dyna quartz fiber	2	1144	Ethylene (CH_2CH_2)	3	173
Dysprosium	1	82	Ethylene - hydrogen system	3	413
Earth	2	813	Ethylene - methane system	3	415
Diatomaceous	2	814	Ethylene - nitrogen system	3	417
Kieselguhr	2	814	Ethylene glycol (CH_2OHCH_2OH)	3	177
Kieselguhr, ignited	2	814	Eureka	1	563
Kieselguhr, ordinary	2	814	Europium	1	90
Easy-Flo silver solder silver alloy	1	1059	Excelsior	2	1113
Ebonite rubber	2	971	Fat	2	1072
Egyptian fire clay brick	2	491, 901	Beef	2	1072
EI-257, Russian	1	1166, 1214	Bone	2	1073
			Pig	2	1073
EI-435, Russian	1	1022	Ferrocarbontitanium, Russian	1	1081
EI-572, Russian	1	1167	Ferrochromium, Russian	1	945
EI-606, Russian	1	1167	Ferromanganese, Russian	1	684, 1010
EI-607, Russian	1	1019, 1020, 1021	Ferromanganese, low carbon, Russian	1	1010
EI-802, Russian	1	1156, 1157	Ferromanganese, normal, Russian	1	1010
			Ferromolybdenum, Russian	1	690, 1013
EI-855, Russian	1	1214	Ferrosilicon, Russian	1	765
Elastomer rubber	2	974	Ferrosilicon 45%, Russian	1	1218
Elckton 2	1	999	Ferrotitanium, Russian	1	1225
Electrical porcelain	2	937	Ferrotungsten, Russian	1	1090
Electrolytic iron	1	157, 159	Ferrovanadium, Russian	1	875
Enamel	2	921	Ferrum (see iron)		

Material Name	Vol.	Page
Fiberglass	2	1115
Fiberite	2	1052
Fir	2	1073
Fir plywood	2	1114
Fire brick	2	491, 891, 895, 902, 903
Fire clay	2	804
Fire clay, Aluminous	2	489
Fire clay, light weight	2	403, 404
Fire clay, pressed	2	403
Fire clay brick	2	403, 404, 490, 491, 896, 901, 903
Fire clay brick, aluminous	2	900
Fire clay brick, dense	2	903
Fire clay brick, superduty	2	890
Fissium alloy	1	1095
Flowers of tin (see tin dioxide)		
Fluorine	3	26
Foam glass	2	924, 925
Forsterite (Mg_2SiO_4)	2	275
Freon 10 [CCl_4] (see carbon tetrachloride)		
Freon 11 (Cl_3CF)	3	183
Freon 12 (Cl_2CF_2)	3	187
Freon 13 ($ClCF_3$)	3	191
Freon 20 [$CHCl_3$] (see chloroform)		
Freon 21 (Cl_2CHF)	3	193
Freon 22 ($ClCHF_2$)	3	197
Freon 113 (CCl_2FCClF_2)	3	201
Freon 114 ($CClF_2CClF_2$)	3	205
Fuel-filled graphite	2	545, 548, 558

Material Name	Vol.	Page
Fused quartz [see silicon dioxide (fused)]		
GaAs	1	1277
GaAs + GaP	1	1423
$GaAs_{0.5}P_{0.5}$	1	1424
$GaAs_{0.65}P_{0.35}$	1	1424
$GaAs_{0.67}P_{0.33}$	1	1424
$GaAs_{0.8}P_{0.2}$	1	1424
$GaAs_{0.9}P_{0.1}$	1	1424
Gabbro	2	816
Gadolinium	1	93
Gadolinium oxide + Samarium oxide	2	356
Gallium	1	97
Gallium - arsenic intermetallic compound GaAs	1	1277
Gallium arsenide [GaAs] (see gallium - arsenic intermetallic compound)		
Garnet [$M_3^{II}M_2^{III}(SiO_4)_3$]	2	278
Genetron 11 [Cl_3CF] (see Freon 11)		
Genetron 12 [Cl_2CF_2] (see Freon 12)		
Genetron 13 [$ClCF_3$] (see Freon 13)		
Genetron 22 [$ClCHF_2$] (see Freon 22)		
Genetron 113 [CCl_2FCClF_2] (see Freon 113)		
Genetron 114 [$CClF_2CClF_2$] (see Freon 114)		
Georgia fire brick	2	896
German chromin	1	1018
German silver	1	980, 981
German steel	1	1118
German Y alloy	1	896, 898
Germanium	1	108
Germanium + Silicon	1	597
Germanium - tellurium intermetallic compound GeTe	1	1280
Germanium 74, enriched	1	112
Germanium telluride [GeTe] (see germanium - tellurium intermetallic compound)		
GeTe	1	1280

Material Name	Vol.	Page	Material Name	Vol.	Page
Glasses	2	922	Glasses (continued)		
Aluminate silicate 723	2	923	Soda-lime silica	2	511, 924, 927
Amber	2	924			
Borosilicate	2	923, 924	Soda-lime silica plate 9330	2	923
Borosilicate 3235	2	923	Soft	2	511
Borosilicate crown	2	923	Solex 2808 plate	2	923
Cellular	2	923	Solex 2808 X	2	925
Colorless	2	924	Solex "S"	2	925
Corning 0080	2	511, 928	Soldex "S" plate	2	923
Foam	2	924, 925	Thuringian	2	923, 924
Golden plate (see amber glass)			Vycor-brand	2	926
Green	2	923	White plate	2	923, 925
Jena Geräte	2	924	Window	2	923, 924
Lead	2	923	X-ray protection	2	924
Monax	2	924	Glass fiber blankets (same as fiberglass)	2	1115
Phoenix	2	924	Insulation	2	1117
Plate	2	923, 924, 925, 926	Superfine	2	1116
			Glass fiber board	2	1124
Pyrex	2	499, 923, 924, 926, 927	Glucinum (see beryllium)		
			Glycerol ($CH_2OHCHOHCH_2OH$)	3	209
			Gnome salt	2	832
Pyrex 7740	2	499, 923, 924, 925, 926	Gold	1	132
			Gold + Cadmium	1	600
			Gold + Chromium	1	603
			Gold + Cobalt	1	606
Quartz	2	923, 924	Gold + Copper	1	609
Silica	2	923, 925, 926	Gold - copper intermetallic compounds		
			Au_xCu_y	1	1281
Silica, fused	2	925	CuAu	1	1282
Silicate	2	511	Cu_3Au	1	1282
Soda	2	923	Gold + Palladium	1	614
Soda-lime	2	926	Gold + Platinum	1	617
Soda-lime plate	2	926	Gold + Silver	1	620

Material Name	Vol.	Page	Material Name	Vol.	Page
Gold + Zinc	1	623	Graphite (continued)		
Golden plate glass (see amber glass)			Grade CEQ	2	63, 65
Government rubber-styrene rubber	2	977	Grade CFW	2	67
Granite	2	817	Grade CFZ	2	67, 71, 72
NTS granite	2	818			
Graphite	2	53			
Acheson	2	73	Grade CS	2	54, 55, 56, 64
Boronated	2	61			
British reactor grade A	2	69			
British reactor grade carbon	2	69, 70	Grade CS-112	2	63
			Grade CS-312	2	63
Brom-graphite	2	768	Grade CSF	2	55
Brookhaven	2	26	Grade CSF-MTR	2	63
Canadian natural graphite	2	54	Grade EY 9	2	69, 70, 71
Carbon resistor	2	73			
Deposited carbon	2	32	Grade EY 9A	2	70
Domestic, Japan	2	56	Grade G-5	2	60, 61
Fuel-filled	2	545, 548, 558	Grade G-9	2	60, 61
Grade 875 S	2	45	Grade GBE	2	54, 55
Grade 890 S	2	49			
Grade AGA	2	64	Grade GBH	2	55
Grade AGHT	2	57	Grade H4LM	2	61
Grade AGOT	2	13	Grade JTA	2	70, 72
Grade AGOT-KC	2	17	Grade L-117	2	63
Grade AGOT-CSF-MTR	2	17	Grade MH4LM	2	70
Grade AGSR	2	57, 58, 63, 64	Grade P1	2	35
			Grade R-0008	2	60
Grade AGSX	2	64	Grade R0025	2	71
Grade ATJ	2	20	Grade RT-0003	2	54
Grade ATL	2	64	Grade RVA	2	66, 67
Grade ATL-82	2	71	Grade RVD	2	67
Grade AUC	2	63, 64, 65	Grade SA-25	2	42
			Grade TS-148	2	59
Grade AWG	2	24	Grade TS-160	2	59
Grade CDG	2	65			

Material Name	Vol.	Page	Material Name	Vol.	Page
Graphite (continued)			Greenheart	2	1074
Nuclear grade TSP	2	60	Gulton HS. B aluminum oxide	2	103
Grade ZT	2	60, 61, 71	Gun metal, admiralty	1	976
			Gun metal, ordinary	1	976
Grade ZTA	2	65, 66, 70	"H" Monel	1	1032
			Hafnia (see hafnium oxide)		
Grade ZTB	2	66	Hafnium	1	138
Grade ZTC	2	66	Hafnium - boron intermetallic compound HfB_2	1	1284
Grade ZTD	2	66			
Grade ZTE	2	66	Hafnium + Zirconium	1	624
Grade ZTF	2	66	Hafnium carbide (HfC)	2	575
			Hafnium nitride (HfN)	2	659
Graphitized carbon black	2	60	Hafnium oxide (HfO_2)	2	150
Karbate	2	59	Hair felt	2	1099
Korite	2	55	Hand-burned face brick	2	891
Moderator graphite	2	70	Hardwood	2	1075
Natural Ceylon block	2	55	Hastelloy A	1	1036
Ohmite	2	73	Hastelloy B	1	1042
Pencil lead graphite	2	65	Hastelloy C	1	1018
Porous-40	2	63	Hastelloy R-235	1	1019
Porous-60	2	63	Haydite aggregate concrete	2	870
Pyrolytic	2	32	Haynes alloy N-155	1	1177
Pyrolytic graphite filament	2	32	Haynes alloy Nb-752	1	1056
Reactor grade carbon stock	2	73	Haynes stellite alloy 21	1	948
Spektral Kohle 1	2	54	Haynes stellite alloy 23	1	948
Supertemp pyrolytic	2	72	Haynes stellite alloy 27	1	1029
U. B. carbon	2	62	Haynes stellite alloy 31 (same as cobalt alloy X40)	1	948
U. B. graphite	2	62			
Graphite + Bromine	2	767	Heavy hydrogen (see deuterium, or tritium)		
Graphite + Thorium dioxide	2	544	Helium	3	29
Graphite + Uranium dicarbide	2	770	Helium - air system	3	318
Graphite + Uranium dioxide	2	547	Helium - argon - krypton system	3	481
Gray cast iron	1	1130, 1135	Helium - argon - nitrogen system	3	486
			Helium - argon - xenon system	3	479
Gray cast iron, hot mold	1	1135	Helium - n-butane system	3	320
Green glass	2	923	Helium - carbon dioxide system	3	322

Material Name	Vol.	Page
Helium - cyclopropane system	3	325
Helium - deuterium system	3	327
Helium - ethane system	3	329
Helium - ethylene system	3	331
Helium - hydrogen system	3	333
Helium - krypton system	3	276
Helium - krypton - xenon system	3	480
Helium - methane system	3	338
Helium - neon system	3	271
Helium - neon - deuterium system	3	489
Helium - neon - xenon system	3	482
Helium - nitrogen system	3	340
Helium - nitrogen - methane system	3	487
Helium - oxygen system	3	343
Helium - oxygen - methane system	3	484
Helium - propane system	3	345
Helium - propylene system	3	347
Helium - xenon system	3	280
Heptane, n-(C_7H_{16})	3	211
Hevea rubber	2	983
Hexane, n-(C_6H_{14})	3	214
HfB_2	1	1284
HgSe	1	1320
HgTe	1	1321
HgTe + CdTe	1	1407
Hi alumina	2	99
High carbon steel, Japanese	1	1119
High-perm-49	1	1199
High temp. insulating brick	2	891
High temp. insulating blast furnace brick	2	899
High zircon porcelain	2	937
Holmium	1	142
Honeycomb structures (metallic - nonmetallic)	2	1015
Honeycomb structures (nonmetallic)	2	1010
Hutchins alloy	1	512

Material Name	Vol.	Page
Hydrargyrum (see mercury)		
Hydriodic acid [HI] (see hydrogen iodide)		
Hydrochloric acid [HCl] (see hydrogen chloride)		
Hydrogen	3	41
Hydrogen - oxygen system	3	429
Hydrogen - nitrogen system	3	419
Hydrogen - nitrogen - ammonia system	3	500
Hydrogen - nitrogen - oxygen system	3	498
Hydrogen - nitrous oxide system	3	427
Hydrogen chloride (HCl)	3	101
Hydrogen iodide (HI)	3	103
Hydrogen sulfide (H_2S)	3	104
Hypalon S2 rubber	2	983
Hypo (see sodium thiosulfate)		
Hytex hydraulic pressed building brick	2	896
Ignited alumina	2	106
Illinium (see promethium)		
InAs	1	1292
InAs + InP	1	1426
$InAs_{0.6}P_{0.4}$	1	1427
$InAs_{0.8}P_{0.2}$	1	1427
$InAs_{0.9}P_{0.1}$	1	1427
$InAs_{0.95}P_{0.05}$	1	1427
Inco "713 C"	1	1022
Inconel	1	1018, 1019, 1021
Inconel alloy 600 (see inconel)		
Inconel alloy 702	1	1022
Inconel alloy 713 (see Inco "713 C")		
Inconel alloy X-750 (see inconel X)		
Inconel X	1	1018
India beryl	2	801
Indiana limestone	2	821
Indium	1	146

Material Name	Vol.	Page	Material Name	Vol.	Page
Indium - antimony intermetallic compound InSb	1	1287	Iron	1	156
Indium - arsenic intermetallic compound InAs	1	1292	Iron + Aluminum + ΣX_i (I)	1	1142
			Iron + Aluminum + ΣX_i (II)	1	1145
Indium + Lead	1	627	Iron + Carbon + ΣX_i (I) (C $\leq$ 2.00%)	1	1113
Indium - selenium intermetallic compound In_2Se_3	1	1295	Iron + Carbon + ΣX_i (II) (C $\leq$ 2.00%)	1	1124
			Iron + Carbon + ΣX_i (I) (C > 2.00%)	1	1128
Indium - tellurium intermetallic compound In_2Te_3	1	1298	Iron + Carbon + ΣX_i (II) (C > 2.00%)	1	1132
Indium + Thallium	1	630	Iron + Chromium + ΣX_i (I)	1	1148
Indium + Tin	1	634	Iron + Chromium + ΣX_i (II)	1	1152
Indium antimonide [InSb] (see indium - antimony intermetallic compound)			Iron + Chromium + Nickel + ΣX_i (I)	1	1160
			Iron + Chromium + Nickel + ΣX_i (II)	1	1164
Indium arsenide [InAs] (see indium - arsenic intermetallic compound)			Iron + Cobalt + ΣX_i (II)	1	1176
Indium oxide (InO)	2	153	Iron + Copper + ΣX_i (I)	1	1179
Indium selenide [In_2Se_3] (see indium - selenium intermetallic compound)			Iron + Manganese + ΣX_i (I)	1	1182
			Iron + Manganese + ΣX_i (II)	1	1191
Indium telluride [In_2Te_3] (see indium - tellurium intermetallic compound)			Iron + Molybdenum + ΣX_i (II)	1	1194
Ingot iron	1	1134	Iron + Nickel + ΣX_i (I)	1	1197
InSb	1	1287	Iron + Nickel + ΣX_i (II)	1	1202
InSb + In_2Te_3	1	1403	Iron + Nickel + Chromium + ΣX_i (I)	1	1209
In_2Se_3	1	1295	Iron + Nickel + Chromium + ΣX_i (II)	1	1212
Insulating brick	2	443, 891, 904	Iron + Phosphor + ΣX_i (I)	1	1216
			Iron + Silicon + ΣX_i (I)	1	1217
Insulating fire brick	2	891	Iron + Silicon + ΣX_i (II)	1	1221
Insulation fiberglass	2	1117	Iron + Titanium + ΣX_i (I)	1	1225
Insurok	2	1023, 1024	Iron + Tungsten + ΣX_i (I)	1	1226
			Iron + Tungsten + ΣX_i (II)	1	1229
In_2Te_3	1	1298	Iron, Armco	1	157, 158, 159, 160, 161, 163
In_2Te_3 + Cu_2Te + Ag_2Te	1	1406			
Intermetallic compounds (see each individual intermetallic compound)					
Invar	1	1199	Iron, electrolytic	1	157, 159
Invar, free cut	1	1205			
Iodine	2	83	Iron, nodular	1	1222
Iodyride [AgI] (see silver iodide)			Iron, silal	1	1222, 1223
Ionium (see thorium)					
Iridium	1	152	Iron, Swedish	1	158

Material Name	Vol.	Page
Iron, wrought	1	1185, 1219
(tri)Iron carbide (Fe$_3$C)	2	578
(tri)Iron tetraoxide (Fe$_3$O$_4$)	2	154
Iron oxide, magnetic [Fe$_3$O$_4$] (see (tri)iron tetraoxide)		
Isotron 11 (see Freon 11)		
Isotron 12 (see Freon 12)		
Isotron 13 (see Freon 13)		
Isotron 22 (see Freon 22)		
Isotron 113 (see Freon 113)		
Isotron 114 (see Freon 114)		
Ivory	2	1076
African	2	1076
Japanese 2E-8	1	899
Japanese fish-plate	1	1119
Japanese M-1	1	899
Japanese steel	1	1195, 1210
Jena Geräte glass	2	924
Jodium (see iodine)		
"K" Monel	1	1032
K.S. alloy 245	1	920
K.S. alloy 280	1	920
K.S. alloy special	1	902
K.S. magnet steel	1	1177
Kalium (see potassium)		
Kaolin fire brick	2	404, 405, 904
Kaolin insulating refractory brick	2	895
Kapok	2	1077
Karbate graphite	2	59
Kel-F	2	970
Kel-F 3700	2	983
Kennametals K161B	2	728
Ketopropane [(CH$_3$)$_2$CO] (see acetone)		
Kh80 T, Russian	1	1019

Material Name	Vol.	Page
Kieselguhr earth	2	814
Kieselguhr earth, ignited	2	814
Kieselguhr earth, ordinary	2	814
Knapic	1	327
Koldboard	2	1125
Korite graphite	2	55
Kovar	1	1203
Krupp steel	1	1115, 1184
Krypton	3	50
Krypton - deuterium system	3	349
Krypton - hydrogen system	3	351
Krypton - neon system	3	284
Krypton - nitrogen system	3	354
Krypton - oxygen system	3	356
Krypton - xenon system	3	288
Kuchin clay	2	804
"L" nickel	1	238, 239
Lamicoid	2	1023, 1024
Laminates (metallic - nonmetallic)	2	1036
Laminates (nonmetallic)	2	1021
Armalon	2	1032
Astrolite	2	1029, 1030
Insurok	2	1023, 1024
Lamicoid	2	1023, 1024
Scotchply	2	1029
Laminate, epoxy resin (see scotch ply laminate)		
Lampblack	2	6
Lanthanum	1	171
Lanthanum + Neodymium + ΣX_i	1	988
Lanthanum - selenium intermetallic compound		
LaSe	1	1301

Material Name	Vol.	Page
Lanthanum – tellurium intermetallic compound		
LaTe	1	1304
Lanthanum trifluoride (LaF_3)	2	633
Lanthanum selenide [LaSe] (see lanthanum – selenium intermetallic compound)		
Lanthanum sulfide (LaS)	2	702
Lanthanum telluride [LaTe] (see lanthanum – tellurium intermetallic compound)		
LaSe	1	1301
LaTe	1	1304
Laughing gas (see nitrous oxide)		
Lead	1	175
Lead, pyrometric standard	1	183, 184
Lead + Antimony	1	637
Lead + Antimony + ΣX_i	1	991
Lead + Bismuth	1	640
Lead + Indium	1	643
Lead + Silver	1	646
Lead – tellurium intermetallic compound		
PbTe	1	1307
Lead + Thallium	1	649
Lead + Tin	1	652
Lead alloy, SAE bearing alloy 12	1	991
Lead glass	2	923
Lead oxide + Silicon dioxide	2	359
Lead oxide + Silicon dioxide + ΣX_i	2	474
Lead telluride [PbTe] (see lead – tellurium intermetallic compound)		
Lead metatitanate ($PbTiO_3$)	2	279
Lead zirconate ($PbZrO_3$)	2	282
Light weight brick	2	488, 489, 892, 899, 900
Light weight concrete	2	874
Light weight concrete, foamed	2	881

Material Name	Vol.	Page
Lignum Vitae	2	1079
Lime sand brick	2	892
Limestone	2	820
Indiana	2	821
Queenstone grey	2	821
Rama	2	821
Limestone aggregate concrete	2	869
Limestone gravel concrete	2	864, 865
Lipowitz alloy	1	939
Lithia (see lithium oxide)		
Lithium	1	192
Lithium + Boron + ΣX_i	1	992
Lithium + Sodium	1	655
Lithium + Sodium + ΣX_i	1	995
Lithium fluoride (LiF)	2	636
Lithium fluoride + Potassium fluoride + ΣX_i	2	641
Lithium hydride (LiH)	2	773
Lithium oxide (Li_2O)	2	157
Lohm	1	564
Low alloy steel	1	1213
Low-exp-42	1	1205
Lowell sand	2	834, 835
Lucalox	2	106
Lummite cement concrete	2	871
Lutetium	1	198
Macloy G steel	1	1213
Magnalium	1	478
Magnesia (see magnesium oxide)		
Magnesia brick	2	485, 897, 898, 899
Magnesite brick	2	478, 483, 892, 895, 905

Material Name	Vol.	Page
Magnesite fire brick	2	897
Magnesium	1	202
Magnesium + Aluminum	1	658
Magnesium + Aluminum + ΣX_i	1	998
Magnesium - antimony intermetallic compound		
Mg_3Sb_2	1	1310
Magnesium + Cadmium	1	661
Magnesium + Calcium	1	662
Magnesium + Cerium	1	663
Magnesium + Cerium + ΣX_i	1	1001
Magnesium + Cobalt + ΣX_i	1	1004
Magnesium + Copper	1	666
Magnesium + Copper + ΣX_i	1	1005
Magnesium - germanium intermetallic compound		
Mg_2Ge	1	1311
Magnesium + Manganese	1	669
Magnesium + Nickel	1	672
Magnesium + Nickel + ΣX_i	1	1008
Magnesium + Silicon	1	675
Magnesium - silicon intermetallic compound		
Mg_2Si	1	1314
Magnesium + Silver	1	678
Magnesium + Tin	1	679
Magnesium - tin intermetallic compound		
Mg_2Sn	1	1317
Magnesium + Zinc	1	680
Magnesium alloys (specific types)		
AN-M-29	1	999
AZ 31 A (see magnesium alloy, AN-M-29)		
Dow metal	1	999
Elckton 2	1	999
Magnesium aluminates		
$MgO \cdot Al_2O_3$	2	283
$MgO \cdot 3.5Al_2O_3$	2	286

Material Name	Vol.	Page
Magnesium aluminates (continued)		
Natural ruby spinel	2	284
Spinel	2	284
Synthetic spinel	2	287
Magnesium aluminate + Magnesium oxide	2	362
Magnesium aluminate + Silicon dioxide	2	365
Magnesium aluminate + (di)Sodium oxide	2	368
Magnesium metaaluminate [$MgAl_2O_4$] (see magnesium aluminate)		
Magnesium antimonide [Mg_3Sb_2] (see magnesium - antimony intermetallic compound)		
Magnesium carbonate ($MgCO_3$)	2	776
Magnesium oxide (MgO)	2	158
Magnesium oxide + Beryllium oxide	2	371
Magnesium oxide + Calcium oxide + ΣX_i	2	477
Magnesium oxide + (di)Chromium trioxide + ΣX_i	2	480
Magnesium oxide + Clay	2	374
Magnesium oxide + (di)Iron trioxide + ΣX_i	2	483
Magnesium oxide + Magnesium aluminate	2	375
Magnesium oxide + Magnesium orthosilicate	2	378
Magnesium oxide + Nickel oxide	2	381
Magnesium oxide + Silicon dioxide	2	384
Magnesium oxide + Silicon dioxide + ΣX_i	2	484
Magnesium oxide + Talc	2	550
Magnesium oxide + Tin dioxide	2	387
Magnesium oxide + Uranium dioxide	2	390
Magnesium oxide + Zinc oxide	2	391
Magnesium silicate (see Forsterite)		
Magnesium orthosilicate + Magnesium oxide	2	394
Magnesium silicide [Mg_2Si] (see magnesium - silicon intermetallic compound)		
Magnesium stannate ($MgSnO_3$)	2	289
Magnesium stannide [Mg_2Sn] (see magnesium - tin intermetallic compound)		
Magnesium titanate porcelain	2	937
Magnezit	2	385, 481

Material Name	Vol.	Page	Material Name	Vol.	Page
Magnezit brick	2	899, 902	Marsh gas (see methane)		
Mahogany	2	1080	Marksa brick	2	899
Manganese	1	208	Medical cotton	2	1069, 1070
Manganese + Copper	1	683	Mercury	1	212
Manganese + Iron	1	684	Mercury - selenium intermetallic compound		
Manganese + Iron + ΣX_i	1	1009	HgSe	1	1320
Manganese + Nickel	1	685	Mercury + Sodium	1	686
Manganese + Silicon + ΣX_i	1	1012	Mercury - tellurium intermetallic compound		
Manganese alloys (specific types)			HgTe	1	1321
Ferromanganese, Russian	1	684, 1010	Mercury selenide [HgSe] (see mercury - selenium intermetallic compound)		
Silicomanganese, Russian	1	1010, 1012	Mercury telluride [HeTe] (see mercury - tellurium intermetallic compound)		
Manganese ferrate ($MnFe_2O_4$)	2	292	Metallurgical brick	2	892, 893
Manganese oxides			Metallurgical porous brick	2	893
MnO	2	168	Metallurgical pumice concrete	2	863, 864
Mn_3O_4	2	170			
Manganese monoxide [MnO] (see manganese oxides)			Methacrylate rubber	2	983
(di)Manganese trioxide + Aluminum oxide	2	397	Methane (CH_4)	3	218
(di)Manganese trioxide + Magnesium oxide	2	398	Methane - propane system	3	432
(di)Manganese trioxide + Silicon dioxide	2	399	Methanol [CH_3OH] (see methyl alcohol)		
(tri)Manganese tetraoxide [Mn_3O_4] (see manganese oxides)	2	170	Methanol - argon system	3	458
			Methanol - hexane system	3	460
Manganese zinc ferrate [$Mn(Zn)Fe_2O_4$]	2	295	Methyl alcohol (CH_3OH)	3	223
Manganin	1	965	Methyl chloride (CH_3Cl)	3	227
Manganin NM Mts, Russian	1	965	Methyl formate - propane system	3	462
Manganomanganic oxide [Mn_3O_4] (see (tri)manganese tetraoxide)			Mg_2Ge	1	1311
			Mg_3Sb_2	1	1310
Maple	2	1081	Mg_2Si	1	1314
Marbles			Mg_2Sn	1	1317
Black	2	761	Mica	2	823, 892
Brown	2	761			
Powder	2	760, 761	Canadian phlogopites	2	824, 825
White	2	761	Granulated vermiculite	2	825
White Alabama	2	761	Madagascan phlogopites	2	824

Material Name	Vol.	Page
Mica (continued)		
Synthetic	2	825
Mica, bonded	2	825
Micanite	2	1138
Mild steel	1	1186
Mineral cotton (see mineral wool)		
Mineral fiber	2	1139
Mineral wool	2	1147
Mineral wood, processed	2	1140
Board	2	1141
Felt	2	1141
Mipora	2	944
Missouri firebrick	2	905
Moderator graphite	2	70
Molybdenum	1	222
Molybdenum + Iron	1	690
Molybdenum + Iron + ΣX_i	1	1013
Molybdenum - silicon intermetallic compound		
$MoSi_2$	1	1324
Molybdenum + Thorium dioxide	1	1429
Molybdenum + Titanium	1	691
Molybdenum + Tungsten	1	694
Molybdenum alloy, ferromolybdenum, Russian	1	690, 1013
(di)Molybdenum carbide (Mo_2C)	2	579
Molybdenum disilicide [$MoSi_2$] (see molybdenum - silicon intermetallic compound)		
Monax glass	2	924
Monel	1	1032
Monel, cast	1	1032
Monel, "H"	1	1032
Monel, "K"	1	1032
Monel, "R"	1	1032
Monel, "S"	1	1032
Monel alloy 400 (see monel)		

Material Name	Vol.	Page
Monel alloy 505 (see "S" monel)		
Monel alloy 506 (see "H" monel)		
Monel alloy K-500 (see "K" monel)		
Monel alloy R-405 (see "R" monel)		
Monolithic wall	2	1126
$MoSi_2$	1	1324
MSM-4Al-4Mn (see titanium alloy C-130 AM or titanium alloy RC-1308)		
MSM-6Al-4V (see Ti-6Al-4V)		
MST-6Al-4V (see Ti-6Al-4V)		
MST-8Mn (see Ti-8Mn)		
Mullite	2	254, 934
Mullite + Alumina	2	335
Multimet N-155	1	1165
Mystic slag	2	1150
N.S. nickel	1	708
Naphthalene ($C_{10}H_8$)	2	995
Naphthalin [$C_{10}H_8$] (see naphthalene)		
Naphthol ($C_{10}H_7OH$)	2	998
Natrium (see sodium)		
Natural Ceylon graphite	2	55
Navy M	1	977
Nelson - Kebbenleg 10	1	896
Neodymium	1	230
Neon	3	56
Neon - argon - deuterium system	3	490
Neon - argon - hydrogen - nitrogen system	3	509
Neon - argon - krypton system	3	478
Neon - argon - krypton - xenon system	3	504
Neon - carbon dioxide system	3	358
Neon - deuterium system	3	360
Neon - hydrogen system	3	362
Neon - hydrogen - nitrogen system	3	494
Neon - hydrogen - oxygen system	3	492
Neon - krypton - deuterium system	3	491

Material Name	Vol.	Page
Neon - nitrogen system	3	365
Neon - nitrogen - oxygen system	3	495
Neon - oxygen system	3	368
Neon - xenon system	3	291
Neptunium	1	234
80 Ni-20 Cr (see chromel A)		
Ni-Cr steel	1	1167, 1168, 1210, 1213
Nickrom (see chromel A)		
Nichrome	1	1018, 1019, 1021, 1036
Nichrome N	1	698
Nichrome V (see chromel A)		
Nickel	1	237
Nickel, "A"	1	239, 241, 1029, 1039
Nickel, "D"	1	1039
Nickel, electrolytic	1	238, 239, 240
Nickel, "L"	1	238, 239
Nickel, "O"	1	239
Nickel, "Z" (see duranickel)		
Nickel 200 (see nickel, A)		
Nickel 211 (see nickel, D)		
Nickel + Aluminum + ΣX_i	1	1014
Nickel - antimony intermetallic compound		
NiSb	1	1327
Nickel + Chromium	1	697
Nickel + Chromium + ΣX_i	1	1017
Nickel + Cobalt	1	700
Nickel + Cobalt + ΣX_i	1	1028
Nickel + Copper	1	703
Nickel + Copper + ΣX_i	1	1031

Material Name	Vol.	Page
Nickel + Iron	1	707
Nickel + Iron + ΣX_i	1	1035
Nickel + Manganese	1	710
Nickel + Manganese + ΣX_i	1	1038
Nickel + Molybdenum + ΣX_i	1	1041
Nickel + ΣX_i	1	1044
Nickel alloys (specific types)		
"A" nickel	1	711
Alumel	1	1015, 1039
Chroman	1	1018
Chromel A	1	698
Chromel C	1	1036
Chromel P	1	698
Contracid	1	1036
Contracid B7M	1	1036
Corronil	1	1032
"D" nickel	1	1039
Duranickel	1	1015
EI-435, Russian	1	1022
EI-607, Russian	1	1019, 1020, 1021
German chromin	1	1018
Grade A	1	711, 1044
H monel	1	1032
Hastelloy A	1	1036
Hastelloy B	1	1042
Hastelloy C	1	1018
Hastelloy R-235	1	1019
Haynes stellite 27	1	1029
HyMn-80	1	1036
INCO "713 C"	1	1022
Inconel	1	1018, 1019, 1021
Inconel 702	1	1022

Material Name	Vol.	Page	Material Name	Vol.	Page
Nickel alloys (specific types) (continued)			Nickel alloys (specific types) (continued)		
Inconel alloy 713 (see Inco "713C")			Refralloy 26	1	1029
Inconel X	1	1018	Rene 41	1	1022
Inconel X-750 (see inconel X)			"S" monel	1	1032
INOR-8	1	1042	Silicon monel	1	1032
K monel	1	1032	"Z" nickel (see duranickel)		
Kh80T, Russian	1	1019	Nickel antimonide [NiSb] (see nickel - antimony intermetallic compound)		
"L" nickel	1	238, 239	Nickel bronze	1	1032
M 252	1	1022	Nickel oxide (NiO)	2	171
Monel	1	1032	Nickel silver	1	981
Monel, cast	1	1032	Nickel silver 12% (see german silver)		
Monel alloy 400 (see monel)			(tri)Nickel disulfide (Ni_3S_2)	2	705
Monel alloy 505 (see "S" monel)			Nickel zinc ferrate [$Ni(Zn)Fe_2O_4$]	2	298
Monel alloy K-500 (see "K" monel)			Nicrosilal, British	1	1204
80Ni-20Cr	1	1019	Nigrine (see rutile)		
Nichrome	1	1018, 1019, 1021, 1036	Nil alba (see zinc oxide)		
			Nimocast 713 C	1	1022
			Nimonic 75	1	1019
Nichrome N	1	698	Nimonic 75, French	1	1019
Nickel bronze	1	1032	Nimonic 80	1	1018
Nimocast 713 C	1	1022	Nimonic 80/80 A, French	1	1019
Nimonic 75	1	1019	Nimonic 90	1	1019
Nimonic 75, French	1	1019	Nimonic 95	1	1019
Nimonic 80	1	1018, 1019	Nimonic 100	1	1029
Nimonic 80/80A, French	1	1019	Nimonic 105	1	1029
Nimonic 90	1	1019	Nimonic 115	1	1029
Nimonic 95	1	1019	Nimonic DS, French	1	1213
Nimonic 100	1	1029	Nimonic PE 7	1	1206
Nimonic 105	1	1029	Niobium	1	245
Nimonic 115	1	1029	Niobium + Molybdenum + ΣX_i	1	1046
N.S. nickel	1	708	Niobium + Tantalum + ΣX_i	1	1049
"O" nickel	1	239	Niobium + Titanium + ΣX_i	1	1052
OKh 20N 60B	1	1022	Niobium + Tungsten + ΣX_i	1	1055
"R" monel	1	1032	Niobium + Uranium	1	713

Material Name	Vol.	Page
Niobium + Zirconium	1	716
Niobium alloys (specific types)		
D-36 (see niobium alloy Nb-10W-5Zr)		
Haynes alloy Nb-752	1	1056
Nb-5Mo-5V-1Zr	1	1047
Nb-27Ta-12W-0.2Zr	1	1050
Nb-10Ti-5Zr	1	1053
Nb-15W-5Mo-1Zr-0.05C	1	1056
Nb-10W-1Zr-0.1C	1	1056
Nb-10W-5Zr	1	1056
Nb-0.5Zr	1	717
Niobium carbide (NbC)	2	582
NiSb	1	1327
Niton (see radon)		
Nitric oxide (NO)	3	106
Nitrile rubber	2	982
Nitrogen	3	64
Nitrogen - oxygen system	3	434
Nitrogen - oxygen - carbon dioxide system	3	497
Nitrogen - propane system	3	438
Nitrogen dioxide [NO_2] (see nitrogen peroxide)		
Nitrogen peroxide (NO_2)	3	108
Nitrogen monoxide [N_2O] (see nitrous oxide)		
Nitrophenol ($NO_2C_6H_4OH$)	2	1001
Nitrous oxide (N_2O)	3	114
Nivac	1	238
Nodular iron	1	1137, 1222
Nonane, n-(C_9H_{20})	3	230
Normal brick	2	488, 489, 900, 901
NTS basalt	2	798
NTS dolomite	2	811
NTS granite	2	818

Material Name	Vol.	Page
Nylon	2	945
Nylon 6 (see polyhexahydro-2H-azepin-2-one)		
"O" nickel	1	239
Oak	2	1082
White	2	1082
Octane, n-(C_8H_{18})	3	233
Ohmite graphite	2	73
OKh20 N60 B, Russian	1	1022
Olivine [521] (see forsterite)		
Olivine basalt	2	798
Ordzhonikidze brick	2	899
Osmium	1	254
Oxygen	3	76
Palladium	1	258
Palladium + Copper	1	720
Palladium + Gold	1	723
Palladium + Platinum	1	726
Palladium + Silver	1	727
Paper	2	1127
Paraffin concrete	2	863
PbTe	1	1307
Pearlitic matrix cast iron, Nr. 1520	1	1222
Pearlitic pig iron, Russian	1	1137
Pencil lead graphite	2	65
Penn. fire brick	2	905
Pentane, n-(C_5H_{12})	3	236
Periclase	2	160
Perlite	2	827
Petalite	2	935
Petroleum coke	2	765
Phenanthrene ($C_{14}H_{10}$)	2	1004
Phenanthrin [$C_{14}H_{10}$] (see phenanthrene)		
Phenyl ether [$(C_6H_5)_2O$] (see diphenyl oxide)		
Phoenix glass	2	924

Material Name	Vol.	Page
Phosphor bronze	1	585, 586, 976
Phosphorus	2	86
Pig fat	2	1073
Pines	2	1083
Pitch	2	1083
White	2	1083
Pitch pines	2	1083
Pladuram	1	416
Plaster	2	887
Plate glass	2	923, 924, 925, 926
Platinoid	1	981
Platinum	1	262
Platinum + Copper	1	730
Platinum + Gold	1	733
Platinum + Iridium	1	734
Platinum + Palladium	1	737
Platinum + Rhodium	1	738
Platinum + Ruthenium	1	743
Platinum + Silver	1	745
Plexiglas	2	960
Plexiglas AN-P-44A	2	961
Pliofoam	2	950
Pluton cloth	2	1100
Plutonium	1	270
Plutonium, α-	1	271
Plutonium + Aluminum	1	746
Plutonium + Iron	1	747
Plutonium alloy, delta-stabilized	1	746
Polychloroethylene (polyvinyl chloride)	2	953
Polychloroethylene (polyvinyl chloride), plasticized	2	954
Polychlorotrifluoroethylene (see polytrifluorochloroethylene)		
Polyethylene	2	956
Polyethylene, chlorosulfonated (see rubber, hypalon)		
Polyhexahydro-2H-azepin-2-one, silon	2	959
Poly(methyl methacrylate) [same as plexiglas]	2	960
AN-P-44A	2	961
Perspex	2	961
Polystyrene	2	963
Colloidal aggregate	2	965
Styrofoam	2	965
Polysulfide rubber (see rubber, Thiokol)		
Polytetrafluoroethylene (same as Teflon)	2	967
Polytrifluorochloroethylene	2	970
Polyurethane [881] (see rubber, Adiprene)		
Polyvinyl chloride	2	953
Porcelains	2	936
Alumina	2	937
Electrical	2	937
High zircon	2	937
$MgTiO_3$ porcelain	2	937
Porcelain 576	2	937
Wet process	2	937
Porous brick	2	894
Porous concrete brick	2	894
Porous fire brick (Italy)	2	895
Portland cement	2	861
Portland cement concrete	2	871
Potassium	1	274
Potassium + Sodium	1	748
Potassium acid phosphate [KH_2PO_4] (see potassium dihydrogen phosphate)		
Potassium bromide (KBr)	2	566
Potassium bromide + Potassium chloride	2	779
Potassium chloride (KCl)	2	613
Potassium chloride + Potassium bromide	2	782

Material Name	Vol.	Page
Potassium chrome alum salt	2	689
Potassium chromium sulfate [KCr(SO$_4$)$_2 \cdot$ 12H$_2$O]	2	688
Potassium dideuterium phosphate (KD$_2$PO$_4$)	2	680
Potassium dihydrogen arsenate (KH$_2$AsO$_4$)	2	785
Potassium dihydrogen phosphate (KH$_2$PO$_4$)	2	684
Potassium hydrogen sulfate (KHSO$_4$)	2	691
Potassium nitrate (KNO$_3$)	2	647
Potassium phosphate, monobasic [KH$_2$PO$_4$] (see potassium dihydrogen phosphate)		
Potassium biphosphate [KH$_2$PO$_4$] (see potassium dihydrogen phosphate)		
Potassium diphosphate [KH$_2$PO$_4$] (see potassium dihydrogen phosphate)		
Potassium rhodanide [KSCN] (see potassium thiocyanate)		
Potassium sulfocyanate [KSCN] (see potassium thiocyannate)		
Potassium sulfocyanide [KSCN] (see potassium thiocyanide)		
Potassium thiocyanate (KSCN)	2	788
Powders (nonmetallic)	2	1040
Praseodymium	1	281
Promethium	1	285
Propane (C$_3$H$_8$)	3	240
2-Propanone [(CH$_3$)$_2$CO] (see acetone)		
Pseudo balsa	2	1060
Pyrex	2	499, 923, 924, 926, 927
Pyrex 7740	2	499, 923, 924, 925, 926
Pyroacetic acid (see acetone)		
Pyroceram 9606	2	940
Pyroceram brand glass-ceramic	2	939
Pyrolytic graphite	2	30
Quartz [see silicon dioxide (crystalline)]		

Material Name	Vol.	Page
Quartz fiber	2	1143
Dyna	2	1144
Quartz glass	2	187, 188, 923, 924
Quartz sand	2	834, 835, 836, 837
Queenstone grey limestone	2	821
Quick silver (see mercury)		
"R" monel	1	1032
Radon	3	84
Rama limestone	2	821
RCA N91	1	701
RCA N97	1	701
Re$_3$As$_7$	1	1330
Red brass	1	591
Red brass, German	1	981
Red brick	2	405, 492, 898
Red brick, hard burned	2	896
Red brick, soft burned	2	896
Redwood	2	1084
Bark	2	1084
Red wood fiber	2	1091
Refractory insulating brick	2	892
Refractory insulating common chamotte brick	2	892
Refralloy 26	1	1029
Refrax	2	586
ReGe	1	1331
ReGe$_2$	1	1331
Rene 41	1	1022
Rene 41 cloth	2	1102
ReSe$_2$	1	1332
Rex 78	1	1213

Material Name	Vol.	Page
Rhenium	1	288
Rhenium - arsenic intermetallic compound		
Re$_3$As$_7$	1	1330
Rhenium - germanium intermetallic compounds		
ReGe	1	1331
ReGe$_2$	1	1331
Rhenium - selenium intermetallic compound		
ReSe$_2$	1	1332
Rhenium selenide [ReSe$_2$] (see rhenium selenium intermetallic compound)		
Rhodium	1	292
Rock	2	828
Rock cork	2	1146
Rock wool	2	1148
Rose metal	1	939
Rubatex rubber	2	981
Rubatex R203-H rubber	2	981
Rubbers	2	980
Acrylate	2	982
Acrylic	2	982
Adiprene	2	982
Buna-N foam (see rubber, Rubatex R203-H)		
Butaprene E	2	982
Carboxy nitrile	2	982
Chloroprene	2	983
Dibenzo GMF-cured butyl	2	983
Ebonite	2	971
Elastomer	2	974
Government rubber-styrene	2	977
Hard	2	972, 981
Hevea	2	983
Hypalon S2	2	983
Kel-F 3700	2	983
Methacrylate	2	983

Material Name	Vol.	Page
Rubbers (continued)		
Nitrile	2	982
Poly(ethyl acrylate)	2	983
Polysulfide (see rubber, Thiokol)		
Resin-cured butyl	2	983
Rubatex	2	981
Rubatex R203-H (same as Buna-N foam)	2	981
Silicone	2	983
Tellurace-cured butyl	2	983
Thiokel ST	2	982
Viton	2	983
X-ray protective	2	981
Rubidium	1	296
Rubidium + Cesium	1	751
Russian alloy	1	1192, 1218, 1222
Russian cupralloy, type 5	1	543
Russian cupro nickel, NM-81	1	562
Russian stainless steel (see stainless steel)		
Russian steel	1	1118
Rutgers cordierite	2	919
Ruthenium	1	300
Rutile	2	203
"S" monel	1	1032
SAE 1010	1	1183
SAE 1015 (see AISI C 1015)		
SAE 1020	1	1183
SAE 1095	1	1114
SAE 4130	1	1153
SAE 4140	1	1155
SAE 4340 (see AISI 4340)		
SAE bearing alloy 10	1	1070
SAE bearing alloy 11	1	1070
SAE bearing alloy 12	1	991
SAE bearing alloy 40	1	976

Material Name	Vol.	Page	Material Name	Vol.	Page
SAE bearing alloy 62	1	976	$Sb_{1.4}Bi_{0.6}Te_{3.13}$	1	1381
SAE bearing alloy 64	1	976	$Sb_{1.4}Bi_{0.6}Te_{3.19}$	1	1383
SAE bearing alloy 66	1	962	$Sb_{1.4}Bi_{0.6}Te_{3.26}$	1	1384
Salt, gnome	2	832	$Sb_{1.5}Bi_{0.5}Te_3$	1	1381
Samarium	1	305	$Sb_{1.5}Bi_{0.5}Te_{3.06}$	1	1384
Sand	2	833	$Sb_{1.5}Bi_{0.5}Te_{3.13}$	1	1382
Lowell	2	834, 835	$Sb_{1.5}Bi_{0.5}Te_{3.19}$	1	1384
Quartz	2	834, 835, 836, 837	$Sb_{1.5}Bi_{0.5}Te_{3.26}$	1	1384
			$Sb_{1.6}Bi_{0.4}Te_3$	1	1381
			$Sb_{1.6}Bi_{0.4}Te_{3.06}$	1	1384
Silica	2	441, 837	$Sb_{1.6}Bi_{0.4}Te_{3.13}$	1	1383
			$Sb_{1.6}Bi_{0.4}Te_{3.19}$	1	1384
Sand cement concrete	2	874	$Sb_{1.6}Bi_{0.4}Te_{3.26}$	1	1384
Sand and gravel aggregate concrete	2	868, 869	$Sb_{1.7}Bi_{0.3}Te_3$	1	1381
Sandstone	2	840	$Sb_{1.8}Bi_{0.2}Te_3$	1	1381
Berea	2	841, 842	$Sb_{1.8}Bi_{0.2}Te_{3.13}$	1	1383
Berkeley	2	841, 842	$Sb_2Se_3 + Ag_2Se + PbSe$	1	1379
			Sb_2Te_3	1	1241
St. Peters	2	841	$Sb_2Te_3 + Bi_2Te_3$	1	1380
Teapot	2	842	$Sb_2Te_3 + In_2Te_3$	1	1386
Tensleep	2	841, 842	Scandium	1	309
Tripolite	2	842	Scotchply laminate (nonmetallic)	2	1029
Sandwiches (nonmetallic)	2	1044	Sea-weed product	2	1128
Sandwiches (metallic - nonmetallic)	2	1047	Selenium	1	313
Sandy clay	2	805	Selenium + Bromine	1	754
Santowax R	2	1005	Selenium + Cadmium	1	755
Sapphire	2	93	Selenium + Chlorine	1	756
Sapphire, synthetic	2	95	Selenium + Iodine	1	757
Sapphire, Linde synthetic	2	94	Selenium + Thallium	1	758
Satin walnut	2	1089	Shamotte brick	2	894, 898
Sawdust	2	1085	Sheep wool	2	1092
$Sb_{1.2}Bi_{0.8}Ti_{3.13}$	1	1381	Silat iron	1	1222, 1223
$Sb_{1.33}Bi_{0.67}Te_{3.13}$	1	1381			
$Sb_{1.4}Bi_{0.6}Te_{3.06}$	1	1383	Silica (see silicon dioxide)		

Material Name	Vol.	Page	Material Name	Vol.	Page
Silica brick	2	408, 489, 492, 502, 894, 896, 897, 898, 900, 902, 904, 906	Silicon dioxide (SiO_2)		
			Crystalline	2	174
			Domestic (USA)	2	175
			Foamed fused silica	2	184
			Fused	2	183
			Linde silica	2	184
			Slip 10	2	189
Silica fire brick	2	894, 895, 905	Slip 18	2	188
			Quartz glass	2	187, 188
Silica glass	2	923, 925, 926	Silica gel	2	185
Silica glass, fused	2	925	Silica refractory brick	2	185
Silica sand	2	837	Slip cast fused silica	2	184
Silicate glass	2	511	Star-brand brick	2	185
Silicous brick	2	492, 902	Vitreous	2	184, 185, 187
Silicomanganese, Russian	1	1010, 1012	Silicon dioxide + Aluminum oxide	2	402
Silicon	1	326	Silicon dioxide + Aluminum oxide + ΣX_i	2	487
Silicon + Germanium	1	761	Silicon dioxide + Barium oxide + ΣX_i	2	495
Silicon + Iron	1	764	Silicon dioxide + Boron oxide + ΣX_i	2	498
Silicon alloy, ferrosilicon, Russian	1	765	Silicon dioxide + Calcium oxide	2	407
Silicon bronze	1	973	Silicon dioxide + Calcium oxide + ΣX_i	2	501
Silicon carbide (SiC)	2	585	Silicon dioxide + (di)Iron trioxide	2	410
Crystolon SiC	2	586	Silicon dioxide + Lead oxide + ΣX_i	2	504
SiC brick, refrax	2	586	Silicon dioxide + (di)Potassium oxide + ΣX_i	2	507
Silicon carbide, refractory (see refrax)			Silicon dioxide + (di)Sodium oxide + ΣX_i	2	510
Silicon carbide + Graphite	2	789	Silicone rubber	2	983
Silicon carbide - silicon cermets	2	718	Silk fabric	2	1105
Silicon carbide + Silicon dioxide	2	553	Sillimanite	2	454, 845
Silicon carbide + Silicon dioxide + ΣX_i	2	554			
Silicon carbide brick	2	895	Sillimanite brick	2	902
Silicon carbide brick, refrax	2	586, 906	Sillimanite refractory brick	2	902, 903
			Sil-O-Cel brick	2	896
Silicon enamel	2	921	Sil-O-Cel brick, calcined	2	896
Silicon monel	1	1032	Sil-O-Cel brick, natural	2	896
(tri)Silicon tetranitride (Si_3N_4)	2	662			

Material Name	Vol.	Page	Material Name	Vol.	Page
Sil-O-Cel brick, special	2	896	Silver chloride (AgCl)	2	620
Sil-O-Cel brick, super	2	896	Silver iodide (AgI)	2	563
Sil-O-Cel coarse grade diatomite aggregate	2	1112	Silver nitrate (AgNO$_3$)	2	650
Silon	2	959	Silver selenide [Ag$_2$Se] (see silver - selenium intermetallic compound)		
Silumin, sodium modified	2	920			
γ-Silumin, modified	1	920	Silver solder, Easy-Flo	1	1059
Silver	1	340	Silver steel	1	1114
Silver + Antimony	1	767	Silver telluride [Ag$_2$Te] (see silver - tellurium intermetallic compound)		
Silver - antimony - tellurium intermetallic compound			Slag aggregate concrete, limestone treated	2	870
AgSbTe$_2$	1	1335	Slag brick	2	898
Silver + Cadmium	1	770	Slag cement	2	861
Silver + Cadmium + ΣX_i	1	1058	Slag concrete	2	864, 880, 881
Silver + Copper	1	773			
Silver - copper intermetallic compound			Slag concrete, direct process	2	864
AgCu	1	1338	Slag concrete, expanded	2	878, 879
Silver + Gold	1	774	Slag concrete, Leuna	2	864
Silver + Indium	1	777	Slag-Portland cement	2	861
Silver + Lead	1	780	Slag wool (same as mineral wool)	2	1151
Silver + Manganese	1	783	Slate	2	846
Silver + Palladium	1	786	SnSe$_2$	1	1352
Silver + Platinum	1	790	SnTe	1	1355
Silver - selenium intermetallic compound			SnTe + AgSbTe$_2$	1	1411
Ag$_2$Se	1	1339	Soapstone	2	853
Silver - tellurium intermetallic compounds			Soda glass	2	923
Ag$_{2-x}$Te	1	1342	Soda-lime glass	2	926
Ag$_2$Te	1	1342	Soda-lime plate glass	2	926
Silver + Tin	1	791	Soda-lime silica glass	2	511, 924, 927
Silver + Zinc	1	792			
Silver + ΣX_i	1	1061	Soda-lime silica plate glass, 9330	2	923
Silver alloy, silver solder, Easy-Flo	1	1059	Sodium	1	349
Silver antimony telluride [AgSbTe$_2$] (see silver - antimony - tellurium intermetallic compound)			Sodium + Mercury	1	795
			Sodium + Potassium	1	798
			Sodium + (di)Sodium oxide	1	1432
Silver bromide (AgBr)	2	569			
Silver bronze	1	579, 980	Sodium acetate (NaC$_2$H$_3$O$_2 \cdot$ 3H$_2$O)	2	1006

Material Name	Vol.	Page
Sodium chloride (NaCl)	2	621
Sodium fluoride (NaF)	2	642
Sodium fluoride + Beryllium difluoride	2	645
Sodium fluoride + Zirconium tetrafluoride + ΣX_i	2	646
Sodium hydrate [NaOH] (see sodium hydroxide)		
Sodium hydrogen sulfate (NaHSO$_4$)	2	692
Sodium hydroxide (NaOH)	2	790
Sodium nitrate (NaNO$_3$)	2	651
(di)Sodium oxide - sodium cermets	2	721
Sodium hyposulfite [Na$_2$S$_2$O$_3 \cdot$ 5H$_2$O] (see sodium thiosulfate)		
Sodium thiosulfate (Na$_2$S$_2$O$_3 \cdot$ 5H$_2$O)	2	693
Sodium tungsten bronze (Na$_X$WO$_3$)	2	301
Sodium tungsten oxide [Na$_X$WO$_3$] (see sodium tungsten bronze)		
Soft cast iron, gray	1	1135
Soft glass	2	511
Soft steel	1	1126
Soil	2	847
Solder, soft	1	840
Solex 2808 plate glass	2	923
Solex 2808 X glass	2	925
Solex "S" glass	2	925
Solex "S" plate galss	2	923
Spektral Kohle 1	2	54
Spherical cast iron, Nr 1510	1	1222
Spinel	2	284, 369, 848
Spinel, natural ruby	2	284
Spinel firebrick	2	905
Spodumene	2	851
Spruce	2	1086
Sr$_2$Si	1	1343
Sr$_2$Sn	1	1344

Material Name	Vol.	Page
Stainless steels (specific types)		
1 Kh 18 N9T (Russian)	1	1168
15 Kh 12 VMF, Russian (see steel EI 802, Russian)		
17-4 PH	1	1168
17-7	1	1165
17-7 PH	1	1166
18-8	1	1161, 1162, 1167, 1168
416	1	1168
3754	1	1161
AISI 301	1	1165
AISI 302	1	1161
AISI 303	1	1165, 1168
AISI 304	1	1161, 1165, 1168
AISI 310	1	1168
AISI 316	1	1165, 1166, 1169, 1170
AISI 347	1	1165, 1166, 1168
AISI 403	1	1149
AISI 410	1	1150
AISI 420	1	1162
AISI 430	1	1150, 1154
AISI 440 C	1	1154
AISI 446	1	1149, 1150, 1155, 1156
AM 355 (Russian)	1	1168
AS 21	1	1161
Austenitic	1	1165, 1183
Crucible HNM	1	1168
EI 572, Russian (same as stainless steel 18-8)	1	1168

Material Name	Vol.	Page
Stainless steels (specific types) (continued)		
EY a 1 T (see stainless steel 1 Kh 18 N9 T)		
F. H. (British)	1	1161
Russian	1	1150, 1161
SF 11, British (see stainless steel AISI 403)		
Staybrite	1	1161
Stannic anhydride [SnO_2] (see tin dioxide)		
Stannic selenide [$SnSe_2$] (see tin - selenium intermetallic compound)		
Stannous telluride [$SnTe$] (see tin - tellurium intermetallic compound)		
Stannum (see tin)		
Staybrite steel, British	1	1161
Steam - air system	3	464
steam - carbon dioxide system	3	466
steam - nitrogen system	3	468
Steam bronze (see navy M)		
Steatite	2	852
10 B 2	2	853
12 C 2	2	853
228	2	853
Soapstone	2	853
Steatite cordierite	2	919
Steels (specific types)		
1 Kh 14 N 14 V2M (see steel EI 257)		
5 ZA 2, Russian	1	1213
12 MKH, Russian	1	1192
AISI 1010	1	1185
AISI 1095 (see steel SAE 1095)		
AISI 2515	1	1198, 1199, 1200
AISI 4130 (see steel SAE 4130)		
AISI 4140 (see steel SAE 4140)		
AISI 4340	1	1213, 1214

Material Name	Vol.	Page
Steels (specific types) (continued)		
AISI C 1010 (see steel SAE 1010)		
AISI C 1015 (same as steel SAE 1015)	1	1186
AISI C 1020 (same as steel SAE 1020)		
Alloy steel	1	1214
Alloy steel, high	1	1214
AMS 2713	1	1210
AMS 2714	1	1213
Haynes alloy N-155	1	1177
High carbon, Japanese	1	1119
High-perm-49	1	1199
High speed	1	1230, 1231, 1232, 1234
High speed, 18	1	1233
High speed, 18-4-1	1	1233
High speed, M1	1	1195
High speed, M2	1	1233
High speed, M10	1	1195
High speed, T1	1	1233
Invar	1	1199
Invar, free cut	1	1205
Japanese	1	1195, 1210
Jessop G 17, British	1	1213
Kh Zn (Russian)	1	1210
Kovar	1	1203
Krupp	1	1115, 1184
K. S. magnet	1	1177
Low alloy	1	1213
Low-exp-42	1	1205
Low Mn	1	1183
Macloy G	1	1213
Mild steel	1	1186
Ni-Cr steel	1	1167, 1168,

Material Name	Vol.	Page	Material Name	Vol.	Page
Steels (specific types) (continued)			Steels (specific types) (continued)		
Ni-Cr steel (continued)	1	1210, 1213	Crucible	1	1204, 1213
NI-Span-C	1	1214	EI-257 (Russian)	1	1166, 1214
Nichrome	1	1210, 1213	EI-606 (Russian)	1	1168
Nicrosilal, British	1	1204	EI-802 (Russian)	1	1156
Nimonic DS, French	1	1213	EI-855 (Russian)	1	1214
Nimonic PE7	1	1206	En8 (CMK), British	1	1184, 1186
Oil-hardening non-deforming	1	1125	En 19 (British)	1	1153
R7 (Russian)	1	1236	En 31 (British)	1	1153, 1154
R10 (Russian)	1	1236	En 32 A (BGKI), British	1	1192
R12 (Russian)	1	1236	Era ATV (British)	1	1213
R15 (Russian)	1	1235	EYA-2	1	1166
R18 (Russian)	1	1236	Ferrosilicon 45%, Russian	1	1218
R15 Kh 3 (Russian)	1	1235	Ferrotitanium, Russian	1	1225
R15 Kh 3 K 5 (Russian)	1	1235	Fish-plate, Japanese	1	1119
R15 Kh 3 K 10 (Russian)	1	1235	FNCT	1	1213
R15 Kh 3 K 12 (Russian)	1	1235, 1236	G 18B, British	1	1165, 1213
R15 Kh 4 (Russian)	1	1236	German	1	1118
R20, British	1	1165	H. 20, British	1	1154
Rex 78	1	1213	H. 27, British	1	1154
Russian	1	1118, 1166	H. 46, British	1	1154
Russian alloy	1	1192, 1218, 1222	SAE 1020	1	1183
SAE 1010	1	1183	SAE 1095	1	1114
SAE 1015 (see steel AISI C 1015)			SAE 4130	1	1153
British	1	1114, 1118, 1187	SAE 4140	1	1155
Carbon	1	1118, 1119, 1126, 1180, 1185	SAE 4340 (see steel AISI 4340)		
			Silver steel	1	1114
			Soft	1	1126
Carbon, British	1	1186	St 42.11 (German)	1	1186, 1218
Carbon, Japanese	1	1185	Stainless steels (see separate entries under stainless steels)		
Chromel 502	1	1210			
Climax	1	1198, 1213	Tool steel	1	1115

Material Name	Vol.	Page
Steels (specific types) (continued)		
Vacromin F	1	1213
WF 100 (Russian)	1	1166
Stibium (see antimony)		
Strontia (see strontium oxide)		
Strontium - silicon intermetallic compound		
Sr_2Si	1	1343
Strontium - tin intermetallic compound		
Sr_2Sn	1	1344
Strontium difluoride + ΣX_i	2	791
Strontium oxide (SrO)	2	194
Strontium oxide + Lithium aluminate + ΣX_i	2	513
Strontium oxide + Lithium zirconium silicate + ΣX_i	2	514
Strontium oxide + Titanium dioxide + ΣX_i	2	517
Strontium oxide + Zinc oxide + ΣX_i	2	520
Strontium silicide [Sr_2Si] (see strontium - silicon intermetallic compound)		
Strontium stannide [Sr_2Sn] (see strontium - tin intermetallic compound)		
Strontium metatitanate ($SrTiO_3$)	2	304
Strontium metatitanate - cobalt cermets	2	722
Strontium zirconate ($SrZrO_3$)	2	307
Styrofoam polystyrene	2	965
Sulfothiorine [$Na_2S_2O_3 \cdot 5H_2O$] (see sodium thiosulfate)		
Sulfur	2	89
Sulfur dioxide (SO_2)	3	116
Sulfurous acid anhydride [SO_2] (see sulfur dioxide)		
Supertemp pyrolytic graphite	2	72
Swedish iron	1	158
Systems, miscellaneous (metallic - nonmetallic)	2	1055
Systems, miscellaneous (nonmetallic)	2	1051
Ta-30Nb-7.5V	1	1063
Ta-8W-2Hf	1	1066
TaB_2	1	1345
$TaBe_{12}$ (see beryllium - tantalum interm. comp.)		
Ta_2Be_{17} (see beryllium - tantalum interm. comp.)		
$TaGe_2$	1	1348
Tantalum	1	355
Tantalum - boron intermetallic compound		
TaB_2	1	1345
Tantalum - germanium intermetallic compound		
$TaGe_2$	1	1348
Tantalum + Niobium	1	801
Tantalum + Niobium + ΣX_i	1	1062
Tantalum + Tungsten	1	802
Tantlum + Tungsten + ΣX_i	1	1065
Tantalum alloys (specific types)		
T 222	1	1066
Ta-30Nb-7.5V	1	1063
Ta-8W-2Hf	1	1066
Tantalum boride [TaB_2] (see tantalum - boron intermetallic compound)		
Tantalum carbide (TaC)	2	589
Tantalum nitride (TaN)	2	665
Teak	2	1087
Technetium	1	363
Teflon	2	967
Teflon, Duroid 5600	2	968
Tellurium	1	366
Tellurium + Arsenic + ΣX_i	1	1068
Tellurium + Selenium	1	805
Tellurium + Thallium	1	808
Terbium	1	372
Thallium	1	376
Thallium + Cadmium	1	811
Thallium + Indium	1	812
Thallium + Lead	1	815

Material Name	Vol.	Page	Material Name	Vol.	Page
Thallium - lead intermetallic compound			TiB_2	1	1358
$\quad Tl_2Pb$	1	1349	Tin	1	389
Thallium + Tellurium	1	818	Tin + Aluminum	1	823
Thallium + Tin	1	821	Tin + Antimony	1	824
Thallium bromide (TlBr)	2	570	Tin + Antimony + ΣX_i	1	1069
Thallium carbide (TlC)	2	625	Tin + Bismuth	1	827
Thiokel ST rubber	2	982	Tin + Cadmium	1	830
Thoria (see thorium dioxide)			Tin + Copper	1	833
Thorium	1	381	Tin + Copper + ΣX_i	1	1072
Thorium + Uranium	1	822	Tin + Indium	1	834
Thorium carbides			Tin + Lead	1	839
$\quad$ ThC	2	592	Tin + Mercury	1	842
$\quad ThC_2$	2	593	Tin - selenium intermetallic compound		
Thorium dioxide (ThO_2)	2	195	$\quad SnSe_2$	1	1352
Thorium dioxide + Graphite	2	557	Tin + Silver	1	845
Thorium dioxide + Uranium dioxide	2	413	Tin - tellurium intermetallic compound		
Thoron (see radon)			$\quad$ SnTe	1	1355
Thulium	1	385	Tin + Thallium	1	846
Thuringian glass	2	923, 924	Tin + Zinc	1	847
			Tin alloys (specific types)		
Ti-130 A	1	850	$\quad$ SAE bearing alloy 10	1	1070
Ti-140 A	1	1081	$\quad$ SAE bearing alloy 11	1	1070
Ti-150 A	1	1078, 1089	$\quad$ Soft solder	1	840
Ti-155 A	1	1074	$\quad$ White bearing metal	1	1070
Ti-2.5 Al-16V	1	1087	Tin anhydride [SnO_2] (see tin dioxide)		
Ti-3Al-11Cr-13V	1	1087	Tin ash [SnO_2] (see tin dioxide)		
Ti-4Al-4Mn (see titanium alloy C-130 AM, or titanium alloy RC-1308)			Tin dioxide (SnO_2)	2	199
			Tin dioxide + Magnesium oxide	2	416
Ti-4Al-3Mo-1V	1	1074, 1075	Tin dioxide + Magnesium oxide + ΣX_i	2	523
Ti-5Al-1.4Cr-1.5Fe-1.2Mo (see Ti-155 A)			Tin dioxide + Zinc oxide	2	419
Ti-5Al-2.5Sn (see titanium alloy A-110 AT)			Tin dioxide + Zinc oxide + ΣX_i	2	524
Ti-6Al-4V	1	1074	Tin peroxide [SnO_2] (see tin dioxide)		
Ti-2Cr-2Fe-2Mo (see Ti-140 A)			TiNi	1	1361
Ti-8Mn	1	850	TiNi + Cu	1	1433
Ti-13V-11Cr-3Al	1	1087	TiNi + Ni	1	1436

Material Name	Vol.	Page
Titania (see titanium oxide)		
Titanic acid anhydride [TiO$_2$] (see titanium dioxide)		
Titanic anhydride [TiO$_2$] (see titanium dioxide)		
Titanic oxide [TiO$_2$] (see titanium dioxide)		
Titanium	1	410
Titanium, iodide	1	411
Titanium + Aluminum	1	848
Titanium + Aluminum + ΣX_i	1	1073
Titanium - boron intermetallic compound		
TiB$_2$	1	1358
Titanium + Chromium + ΣX_i	1	1077
Titanium + Iron + ΣX_i	1	1080
Titanium + Manganese	1	849
Titanium + Manganese + ΣX_i	1	1083
Titanium - nickel intermetallic compound		
TiNi	1	1361
Titanium + Oxygen	1	852
Titanium + Vanadium + ΣX_i	1	1086
Titanium + ΣX_i	1	1089
Titanium alloys (specific types)		
120 VCA	1	1087
A-110 AT	1	1074
AMS 4908 (see titanium alloys Ti-8Mn)		
AMS 4925 A (see titanium alloys C-130 AM, or titanium alloys RC-1308)		
AMS 4926 (see titanium alloys A-110 AT)		
AMS 4928 (see titanium alloys Ti-6Al-4V)		
AMS 4929 (see titanium alloys Ti-155 A)		
AMS 4969 (see titanium alloys Ti-155 A)		
ASTM B 265-58 T, grade 6 (see titanium alloy A-110 AT)		
ASTM 265-58 T, grade 7 (see titanium alloy Ti-8Mn)		

Material Name	Vol.	Page
Titanium alloys (specific types) (continued)		
C-130 AM	1	1074
C-110 M (see Ti-8Mn)		
MSM-4Al-4Mn (see titanium alloy C-130 AM, or titanium alloy RC-1308)		
MSM-6Al-4V (see titanium alloy Ti-6Al-4V)		
MST-6Al-4V (see titanium alloy Ti-6Al-4V)		
MST-8Mn (see titanium alloy Ti-8Mn)		
RC-1308	1	1084
Ti-130 A	1	850
Ti-140 A	1	1081
Ti-150 A	1	1078, 1089
Ti-155 A	1	1074
Ti-2.5Al-16V	1	1087
Ti-3Al-11Cr-13V	1	1087
Ti-4Al-4Mn (see titanium alloy C-130 AM, or titanium alloy RC-1308)		
Ti-4Al-3Mo-1V	1	1074, 1075
Ti-5Al-1.4Cr-1.5Fe-1.2Mo (see titanium alloy Ti-155 A)		
Ti-5Al-2.5Sn (see titanium alloy A-110 AT)		
Ti-6Al-4V	1	1074
Ti-2Cr-2Fe-2Mo (see titanium alloy Ti-140 A)		
Ti-8Mn	1	850
Ti-13V-11Cr-3Al	1	1087
Titanium boride [TiB$_2$] (see titanium - boron intermetallic compound)		
Titanium carbide (TiC)	2	594
Titanium carbide - cobalt cermets	2	725
Titanium carbide - cobalt - niobium carbide cermets	2	726
Titanium carbide - nickel - molybdenum - niobium carbide cermets	2	727
Titanium carbide - nickel - niobium carbide cermets	2	730

Material Name	Vol.	Page
Titanium nitride (TiN)	2	668
Titanium dioxide (TiO_2)	2	202
Dense titania	2	204
Rutile	2	203
Tl_2Pb	1	1349
Toluene ($C_6H_5CH_3$)	3	242
Tool steel	1	1115, 1233
Tool steel, M1 high-speed	1	1195
Tool steel, M10 high-speed	1	1195
Topaz	2	251
Tourmaline	2	855
Tourmaline, Brazil	2	855
Transite	2	1107
Triangle beryllia	2	126
Trichlorofluoromethane [Cl_3CF] (see Freon 11)		
Trichloromethane [$CHCl_3$] (see chloroform)		
Trichlorotrifluoroethane [CCl_2FCClF_2] (see Freon 113)		
Trifluoroborane [BF_3] (see boron trifluoride)		
Trifluorotrichloroethane [CCl_2FCClF_2] (see Freon 113)		
Trinitrotoluene [$CH_3C_6H_2(NO_2)_3$]	2	1007
Tripolite brick	2	894
Tritium	3	87
Tuballoy (same as uranium)	1	429
Tuff	2	856
Tungsten	1	415
Tungsten - arsenic intermetallic compound		
W_3As_7	1	1364
Tungsten - boron intermetallic compound		
WB	1	1365
Tungsten + Iron + ΣX_i	1	1090
Tungsten + Nickel + ΣX_i	1	1091
Tungsten + Rhenium	1	855

Material Name	Vol.	Page
Tungsten - selenium intermetallic compound		
WSe_2	1	1368
Tungsten - silicon intermetallic compound		
WSi_2	1	1369
Tungsten - tellurium intermetallic compound		
WTe_2	1	1370
Tungsten + Thorium dioxide	1	1439
Tungsten alloy, ferrotungsten (Russian)	1	1090
Tungsten boride [WB] (see tungsten - boron intermetallic compound)		
Tungsten carbide (WC)	2	598
Tungsten trioxide (WO_3)	2	209
Tungsten trioxide + Zinc oxide	2	422
Tungsten diselenide [WSe_2] (see tungsten - selenium intermetallic compound)		
Tungsten disilicide [WSi_2] (see tungsten - silicon intermetallic compound)		
Tungsten ditelluride [WTe_2] (see tungsten - tellurium intermetallic compound)		
Tungstic acid anhydride [WO_3] (see tungsten trioxide)		
Tungstic anhydride [WO_3] (see tungsten trioxide)		
Tungstic oxide [WO_3] (see tungsten trioxide)		
UBe_{13} (see beryllium - uranium intermetallic compound)		
Uranic oxide [UO_2] (see uranium dioxide)		
Uranium	1	429
Uranium + Aluminum	1	858
Uranium + Chromium	1	859
Uranium + Iron	1	862
Uranium + Magnesium	1	863
Uranium + Molybdenum	1	864
Uranium + Molybdenum + ΣX_i	1	1094
Uranium + Niobium	1	867
Uranium + Silicon	1	868
Uranium + Uranium dioxide	1	1442
Uranium + Zirconium	1	871

Material Name	Vol.	Page	Material Name	Vol.	Page
Uranium + Zirconium + ΣX_i	1	1097	Vermiculite brick	2	894
Uranium carbides			Vermiculite mica, granulated	2	825
UC	2	601	Vitallium type alloy (see Haynes stellite alloy 21)		
UC_2	2	605			
Uranium carbide - uranium cermets	2	731	Viton rubber	2	983
Uranium - 3% fissium alloy	1	1095	Vitreous silica	2	184, 185, 187
Uranium - 5% fissium alloy	1	1095, 1097	Volcanic ash (see tuff)		
Uranium - 8% fissium alloy	1	1095	Vulcanized fiber	2	1088
Uranium - 10% fissium alloy	1	1095	Vycor-brand glass	2	926
Uranium nitride (UN)	2	672	W-2 chromalloy (see molybdenum - silicon intermetallic compound)		
Uranium oxides					
UO_2	2	210	Wallboard	2	1131
U_3O_8	2	237	Walnut	2	1089
Uranium dioxide (UO_2)	2	210	W_3As_7	1	1364
Uranium dioxide + Beryllium oxide	2	423	Water (H_2O)	3	120
Uranium dioxide + Calcium oxide	2	426	WB	1	1365
Uranium dioxide - chromium cermets	2	732	White bearing metal	1	1070
Uranium dioxide - molybdenum cermets	2	735	White cast iron	1	1130, 1135
Uranium dioxide - niobium cermets	2	738	White oak	2	1082
Uranium dioxide + (di)Niobium pentoxide	2	427	White pines	2	1083
Uranium dioxide - stainless steel cermets	2	741	White plate glass	2	923, 925
Uranium dioxide - uranium cermets	2	744	White temper cast iron	1	1137
Uranium dioxide + Yttrium oxide	2	428	White wood	2	1090
Uranium dioxide - zirconium cermets	2	746	Winchester crushed trap rock	2	829, 830
Uranium dioxide + Zirconium dioxide	2	429	Window glass	2	923, 924
(tri)Uranium octoxide (U_3O_8)	2	237	Wolfram (see tungsten)		
Uranous uranic oxide [U_3O_8] (see (tri)uranium octoxide)			Wolfamic acid, anhydrous [WO_3] (see tungsten trioxide)		
Vacromin F	1	1213	Wolframite [WO_3] (see tungsten trioxide)		
Valve bronze (see navy M)					
Vanadium	1	441	Wollastonite	2	859
Vanadium + Iron	1	874	Wood felt	2	1133
Vanadium + Yttrium	1	877	Wood fibers	2	1091
Vanadium alloy, ferrovanadium (Russian)	1	875	Wood's metal	1	939
Vanadium carbide (VC)	2	606	Wood products	2	1132
Vegetable fiberboards	2	1129			

Material Name	Vol.	Page	Material Name	Vol.	Page
Wool	2	1092	Zinc - silicon - arsenic intermetallic compound		
Angora	2	1092			
Sheep	2	1092	$ZnSiAs_2$	1	1374
Wrought iron	1	1185, 1219	Zinc alloys (specific types)		
			Zamak Nr 400	1	880
WSe_2	1	1368	Zamak Nr 410	1	1098
WSi_2	1	1369	Zamak Nr 430	1	1098
WTe_2	1	1370	Zinc dichloride ($ZnCl_2$)	2	626
X-metal (see uranium)			Zinc ferrate ($ZnFe_2O_4$)	2	314
X-ray protection glass	2	924	Zinc germanium phosphide ($ZnGeP_2$)	2	792
Xenon	3	88	Zinc oxide (ZnO)	2	243
Xenon - deuterium system	3	371	Zinc oxide + Magnesium oxide	2	435
Xenon - hydrogen system	3	374	Zinc oxide + Strontium oxide + ΣX_i	2	527
Xenon - nitrogen system	3	377	Zinc oxide + Tin dioxide	2	438
Xenon - oxygen system	3	379	Zinc oxide + Tin dioxide + ΣX_i	2	528
Yellow brass	1	981, 982	Zinc selenide [ZnSe] (see zinc - selenium intermetallic compound)		
Ytterbium	1	446	Zinc selenium arsenide [$ZnSiAs_2$] (see zinc - selenium - arsenic intermetallic compound)		
Yttria (see yttrium oxide)					
Yttrium	1	449	Zinc sulfate heptahydrate ($ZnSO_4 \cdot 7H_2O$)	2	694
Yttrium aluminate ($Y_3Al_5O_{12}$)	2	308	Zircaloy-2	1	888
Yttrium ferrate [$Y_3Fe_2(FeO_4)_3$]	2	311	Zircaloy-4	1	888
Yttrium iron garnet (see yttrium ferrate)			Zircon, Brazil	2	318
			Zircon 475	2	318
Yttrium oxide (Y_2O_3)	2	240	Zirconia (see zirconium dioxide)		
Yttrium oxide + Uranium dioxide	2	432	Zirconia, stabilized	2	522
"Z" nickel (see duranickel)					
Zamak Nr 400	1	880	Zirconia brick	2	535, 895, 905
Zamak Nr 410	1	1098			
Zamak Nr 430	1	1098	Zirconium	1	461
Zinc	1	453	Zirconium, iodide	1	462, 463
Zinc + Aluminum	1	880			
Zinc + Aluminum + ΣX_i	1	1098	Zirconium + Aluminum	1	882
Zinc + Cadmium	1	881	Zirconium + Aluminum + ΣX_i	1	1100
Zinc + Lead + ΣX_i	1	1099	Zirconium - boron intermetallic compound		
Zinc - selenium intermetallic compound			ZrB	1	1375
ZnSe	1	1371	Zirconium + Hafnium	1	883

Material Name	Vol.	Page	Material Name	Vol.	Page
Zirconium + Hafnium + ΣX_i	1	1101	Zirconium orthosilicate ($ZrSiO_4$) (continued)		
Zirconium + Molybdenum + ΣX_i	1	1104	Zircon	2	318
Zirconium + Niobium	1	886	Zircon tam	2	318
Zirconium + Tantalum + ΣX_i	1	1105	ZnSb + CdSb	1	1412
Zirconium + Tin	1	887	ZnSe	1	1371
Zirconium + Tin + ΣX_i	1	1108	$ZnSiAs_2$	1	1374
Zirconium + Titanium	1	890	ZrB	1	1375
Zirconium + Uranium	1	891			
Zirconium + Uranium + ΣX_i	1	1111			
Zirconium + Zirconium dioxide	1	1444			
Zirconium + ΣX_i	1	1112			
Zirconium alloys (specific types)					
Zircaloy-2	1	888			
Zircaloy-4	1	888			
Zirconium boride [ZrB] (see zirconium - boron intermetallic compound)					
Zirconium carbide (ZrC)	2	609			
Zirconium hydride (ZrH)	2	793			
Zirconium nitride (ZrN)	2	675			
Zirconium dioxide (ZrO_2)	2	246			
Zirconium dioxide + Aluminum oxide	2	441			
Zirconium dioxide + Calcium oxide	2	442			
Zirconium dioxide + Calcium oxide + ΣX_i	2	531			
Zirconium dioxide + Magnesium oxide	2	446			
Zirconium dioxide + Silicon dioxide + ΣX_i	2	534			
Zirconium dioxide - titanium cermets	2	749			
Zirconium dioxide + Yttrium oxide	2	449			
Zirconium dioxide + Yttrium oxide + ΣX_i	2	537			
Zirconium dioxide - yttrium oxide - zirconium cermets	2	753			
Zirconium dioxide - zirconium cermets	2	752			
Zirconium silicate [$ZrSiO_4$] (see zirconium orthosilicate)					
Zirconium silicate, natural (see zircon)					
Zirconium orthosilicate ($ZrSiO_4$)	2	317			
Brazil zircon	2	318			